MECHANISMS OF DEVELOPMENT

Mechanisms
of development

Richard G. Ham, Ph.D.

Department of Molecular, Cellular, and Developmental
Biology, University of Colorado, Boulder, Colorado

Marilyn J. Veomett, Ph.D.

Lincoln, Nebraska

With 1086 illustrations

THE C. V. MOSBY COMPANY

ST. LOUIS • TORONTO • LONDON   1980

This book is dedicated to the spirit
of mutual cooperation and respect that
made it possible.

Copyright © 1980 by The C. V. Mosby Company

All rights reserved. No part of this book may be reproduced
in any manner without written permission of the publisher.

Printed in the United States of America

The C. V. Mosby Company
11830 Westline Industrial Drive, St. Louis, Missouri 63141

Library of Congress Cataloging in Publication Data

Ham, Richard G   1932-
  Mechanisms of development.

  Bibliography: p.
  Includes index.
  1. Developmental biology. I. Veomett, Marilyn J.,
1946-   joint author. II. Title.
QH491.H33   574.3   79-9236
ISBN 0-8016-2022-8

C/VH/VH 9 8 7 6 5 4 3 2 1   01/C/087

PREFACE

☐ *Mechanisms of Development* was written specifically to serve as the textbook for a one-semester course on developmental mechanisms. Its emphasis is on molecular and cellular mechanisms, and we have included only the bare minimum of descriptive embryology that is needed to understand those mechanisms. There already exist a number of good textbooks of descriptive embryology, and we have tried to avoid unnecessary duplication of the material contained in them. Because of this complementary relationship, *Mechanisms of Development* can also be used in conjunction with a standard embryology text for a comprehensive two-semester course that emphasizes both developmental phenomena and the mechanisms that underlie them.

A living organism is a hierarchy of organization, beginning with the smallest subatomic particles and progressing through atoms, molecules, macromolecules, organelles, cells, tissues, organs, and finally to the intact organism as an integrated functional unit. Our approach in this book is to seek an explanation for each aspect of development at the most basic level of organization that directly affects it. Our attempts to explain developmental phenomena are aimed primarily at the levels of macromolecules, subcellular organelles, cells, and interactions among cells. We also recognize the importance of events at atomic and subatomic levels in determining the properties of the molecules that comprise living organisms, but in most cases we have chosen to view these events as part of the intrinsic nature of the molecules and thus more appropriately covered in texts on atomic physics or chemistry.

Mechanisms of Development is designed for use by upper division undergraduates in the biological sciences. We have assumed that the students using this book have completed a broadly based course in general biology and that they have a reasonable understanding of the fundamentals of chemistry. A number of concepts from molecular biology, biochemistry, and cell biology are essential for a detailed understanding of the mechanisms of development, and students who have completed those courses will find them helpful. However, to make our book comprehensible to students who may not have completed courses in some of these areas, we have also attempted to explain the basic concepts as they are needed, either in the text or in appendixes.

Our definition of development as "progress of an organism through its life cycle" (Chapter 1) is very broad. We are concerned with all phases of the life cycles of multicellular organisms, including gametogenesis, fertilization, embryogenesis, maturation, and senescence. To fit all these topics into a single textbook of manageable size, we have found it necessary to be selective in several ways. First, we have emphasized the biological mechanisms by which development occurs rather than the specific details of development. Second, our presentation is greatly biased toward the development of higher vertebrates, including humans. Other types of organisms are discussed when they are useful to illustrate particular points, but no attempt has been made to describe their development in detail. Third, we have given the greatest attention to those aspects of development for which research into mechanisms has been particularly active. This has resulted in strong emphasis on such areas as the molecular biology of gene expression, the behavior of differentiated cells in culture, and the generation of shape during embryonic development.

Developmental biology is a dynamic, rapidly changing field in which important new discoveries are constantly being made. We have attempted to capture the excitement of new discovery in this book. To do so, we must deal with areas of research in which the final answers are not yet available and, in some cases, in which the investigators are still looking for the right questions to ask. This may be disconcerting to students who are accustomed to textbooks that present their subject material as a compilation of facts and principles. However, we consider it far more important to bring our students to the frontiers of new research than to pretend that we already know everything worth knowing about the mechanisms of development.

Since we are dealing with current research, it is inevitable that some of the material in this book will be obsolete even before the book is off the press. We have tried to identify in the text those areas which we consider likely to change rapidly. In addition, we have included an appendix showing the interested student how to use the research literature to keep abreast with the latest progress in the field.

We consider it desirable for our readers to have some knowledge of overall patterns of development before attempting to understand the detailed mechanisms that are involved. The first section of this book therefore provides an overview of development and a brief description of the major phenomena that are involved. The first chapter presents our working definitions of "mechanism" and "development," together with a summary of the types of life cycles that occur in various organisms and the relationship between those life cycles and development. The second chapter provides a summary of the major events in early embryonic development, and the third chapter introduces the concept of cellular differentiation.

After this brief introduction attention is turned to basic control mechanisms that affect development at the molecular level of organization. Section Two describes molecular control of gene expression in considerable detail. Since many of the pioneering studies in that area have been done with prokaryotic organisms, the first chapter in the section provides a brief review of that work, including both the basic mechanisms of gene expression and the control systems that are responsible for switching specific genes on and off in prokaryotic cells. The next five chapters examine the molecular pathway of information flow in eukaryotic cells, starting with the organization of the genome and attempting to follow the information all the way to final differentiated products. The emphasis of this section is on the mechanisms believed to be responsible for regulating the expression of specific genes in differentiated cells.

Section Three deals with differentiation as a cellular phenomenon. The first two chapters are concerned with environmental influences that control the expression of differentiation in cultured cells and with possible mechanisms for such controls. Attention is then shifted to control mechanisms operating in the intact embryo, including intrinsic cellular determinants, embryonic induction, and hormonal regulation. Throughout this section the emphasis is on the nature of signals that a cell receives from its environment and on the intracellular changes triggered by those signals that ultimately lead to the expression of specific differentiated properties.

Section Four deals with the generation of specific form and shape during development. This section begins with a discussion of how cells change shape and how such changes can influence the shapes of tissues and organs. Attention is then given to mechanisms of cellular recognition, sorting, and aggregation. The last two chapters in the section deal with selective growth control and selective cell death as mechanisms that help to mold the final shape of the organism.

The separation of differentiation and morphogenesis in Sections Three and Four was introduced for simplicity of presentation and does not exist in real life, where the two are intimately intertwined. This is clearly illustrated in Section Five, where selected developmental systems are analyzed in terms of a variety of the mechanisms previously presented. The first chapter in this section deals with gametogenesis and fertilization and serves both to illustrate a high level of cellular adaptation for specialized function and to add material concerning parts of the life cycle that have received relatively little attention in other chapters. The second chapter is concerned with mammalian sex differentiation, a process that offers a fascinating example of hormonal control of development in many diverse tissues and at all levels of organization from molecules to whole-body morphology. The next chapter deals with vertebrate limb development, a prime example of a system in which multiple inductive interactions generate a complex structure from a rudimentary precursor with a minimum of external influence. This is followed by a chapter on insect development, which is of particular interest because of the extreme experimental manipulations to which insect embryos can be subjected and because it is one of the few developmental systems in which genetic techniques have been extensively used. Next is a chapter on the cellular slime mold *Dictyostelium discoideum,* an organism that provides an unusual example of conversion from individual free living cells to a differentiated multicellular organism and that has also been the subject of detailed studies on the molecular control of gene expression. The section ends with an analysis of aging, which many investigators are beginning to view as a part of an overall program for the entire life cycle and therefore inseparable in principle from embryonic development and maturation.

The final section attempts to bring together into a coherent picture all the developmental mechanisms that have been presented. In addition, it seeks to summarize some of the major unanswered questions about development that are now capable of being approached with modern analytical techniques of molecular and cellular biology. We expect that many of these questions will be answered as today's students become tomorrow's investigators. Hopefully this book will inspire some of its readers to take a closer look at what we do not yet know about development.

We thank Karen Brown for preparing the manuscript, Susan Jennings and Lisa Klaumann for doing library research and locating previously published figures, and Paul Ham and Cathy Verhulst for providing original illustrations. We are indebted to the many authors and publishers who have given us permission to reproduce their previously published illustrations. Thanks are also due the many people who have read and criticized portions of the manuscript, including Dr. Mary Bonneville, Dr. Dona Chikaraishi, Dr. Kathleen Danna, Dr. John David, Ms. Kaye Edwards, Dr. Larry Gold, Ms. Kathy Jones, Dr. Edwin McConkey, Dr. J. Richard McIntosh, Dr. Donna Peehl,

Dr. Keith Porter, Dr. Donald Riddle, Dr. Urs Rutishauser, Mr. Gary Shipley, Dr. James R. Smith, Dr. Larry Soll, Dr. Jonathan Van Blerkom, Dr. Ben Walthall, and Dr. Michael Yarus. We particularly thank Dr. George Veomett for many helpful suggestions and for the preparation of Appendix C. Special thanks are also due the 1976, 1977, and 1978 classes in developmental mechanisms at the University of Colorado for serving as experimental subjects and providing the feedback that helped this book to develop into its current form. Finally, we thank Mary, George, and our respective children for putting up with our preoccupation during the long gestation period of this book.

Richard G. Ham
Marilyn J. Veomett

CONTENTS

MECHANISMS OF DEVELOPMENT

SECTION ONE

Basic concepts

1 Life cycles and development
2 Embryology and morphogenesis
3 Cellular differentiation

☐ The detailed analysis of the mechanisms of development presented in subsequent sections of this book begins at the molecular level of organization and moves progressively to more and more complex developmental phenomena. There is an inherent danger in this approach: the details may completely obscure the overall phenomena that we are trying to explain. This introductory section seeks to minimize that risk by providing a brief overview of developmental biology, first at the level of the intact organism and then in terms of the organism's component parts.

One of the basic tenets of information theory is that the symbols used for communication have no inherent meaning in themselves. Any symbolic code, whether it is the number of lanterns in a church steeple, a series of dots and dashes, or the words on a printed page, fails to communicate until everyone involved has agreed to use a common set of definitions. If we, the authors, use a word such as "development" to mean one thing and our readers understand it to mean something different, communication will be seriously impaired. We have therefore chosen to use a substantial part of Chapter 1 to define as precisely as possible concepts such as "mechanism" and "development" and to be certain that we have established meaningful communication.

In addition to presenting basic concepts and definitions, the first chapter also contains a discussion of the continuity of life from generation to generation and of the role of developmental processes in maintaining that continuity. Chapter 2 focuses on the individual organism and on the generation of specific shapes (morphogenesis) as each new organism is formed from a fertilized egg cell. Chapter 3 shifts to the

1

cellular level and focuses on the diversity of shape and function that the various cells within each organism exhibit (cellular differentiation) and on the biochemical and ultrastructural bases for those differences.

Together these three chapters provide a road map to help the reader remain oriented during the detailed analysis of selective gene expression, cellular differentiation, and morphogenesis in later sections of this book.

CHAPTER 1

Life cycles and development

☐ The words "mechanism" and "development" are both used in many different contexts in the English language, and each has a long list of dictionary definitions. Even within the field of developmental biology, it sometimes seems as though there are nearly as many definitions of development as there are developmental biologists. In this chapter we present the rationale behind our own definitions of "mechanism" and "development," which are, respectively, "the fundamental details of how things work" and "progress of an organism through its life cycle." Together these two definitions state accurately the aim of this book, which is to summarize at molecular and cellular levels of organization what is known about the events responsible for the sequence of changes that living organisms go through during their life cycles. Later in this chapter we also delve into the various types of life cycles that exist and the major developmental changes that occur during those cycles.

Definitions from three different dictionaries provide a starting point for our attempts to convey exactly what we mean when we use the word "mechanism" in this book: (1) "the means or way in which something is done" (*World Book Encyclopedia Dictionary*); (2) "the agency or means by which an effect is produced or a purpose is accomplished" (*Random House College Dictionary*); (3) "the fundamental physical or chemical processes involved in or responsible for an action, reaction, or other natural phenomenon" (*Webster's New Collegiate Dictionary*). These definitions suggest that the overall concept of a mechanism includes both the means or machinery involved in a process and the way that the machinery functions to achieve the desired results. In simpler terms, what we are talking about is "how things work."

To understand the mechanisms involved in a particular process, it is generally necessary to examine events occurring at more fundamental levels of organization than that of the process itself. For exam-

MECHANISMS

ple, to describe the mechanisms responsible for changes in shape and function that are observed in developing embryos, we find it necessary to speak of events that take place at molecular and cellular levels of organization. Similarly, when a chemist explains the mechanisms responsible for reactions that occur at the molecular level, it is necessary to speak in terms of events occurring at atomic and subatomic levels. Thus the simple definition needs to be amended to read "the fundamental details of how things work."

The term "mechanism" also carries an implication of cause and effect. Each event in a developmental sequence tends to be influenced (and in some cases directly controlled) by those which precede it, and each in turn tends to influence those which follow. For example, the release of specific hormones is often triggered by neural factors or by other hormones, and, in turn, the hormones that are released exert a variety of specific effects on their target tissues.

Unfortunately, this type of mechanistic cause-and-effect relationship is often confused with purpose and motivation. A large part of this confusion results from the incorrect use of the word "why" in everyday patterns of speech and thought. When properly used, "why" refers to purpose, reason, or motivation. It can be used unambiguously to refer to voluntary actions or decisions and to refer to the results of such actions. The classical vaudeville question "Why did the chicken cross the road?" and its answer "To get to the other side" illustrate the first case, and the question "Why does this book contain so many definitions?" and its answer "Because the authors consider them important" illustrate the second.

Problems arise when "why" is used to ask a question about processes or things that lack the capability of decision making or voluntary action. One common response to such a question is to substitute immediate cause for motivation and to answer by describing a mechanism. In other words, a "why" question is answered in terms of "how." An example is the question "Why do men have beards and women not?" and the answer "Because their hormones are different." This answer actually describes the mechanism that is responsible for the difference in facial hair between men and women without dealing with cause or motivation in any fundamental sense.

Another common response to inappropriate "why" questions is to answer with teleological reasoning, which assumes (often incorrectly) that all actions are undertaken specifically to achieve goals. An example of this is the question "Why does messenger RNA (mRNA) bind to ribosomes?" and the answer "In order to be translated." In this case the answer implies that the mRNA molecule knows it should be translated and deliberately binds to ribosomes to accomplish that goal. In reality, there is built into the molecular structure of the mRNA an affinity for the ribosomal translation complex, which leads to the binding and subsequent translation of the mRNA (Chapter 9). That affinity is the product of prolonged evolutionary selection for ribosomes and mRNA molecules that work efficiently together and does not involve specific motivation on the part of the interacting molecules. In general, the teleological, or goal-directed, approach is not appropriate for analyzing the basic mechanisms of development, since the cells and molecules that are involved do not make deliberate choices or pursue specific goals.

We have deliberately avoided use of the word "why" in this book. True "why" questions that seek the fundamental reasons or causes for life and the universe being as they are go beyond the realm of experimental science and thus have no place here. Questions about mechanisms and cause-and-effect relationships are better asked in terms of "how" or "What is the mechanism responsible for?" The only time that we consider it legitimate to use "why" is when we are talking about conscious choice or motivation or about actions resulting from them.

DEVELOPMENT

The term "development" has undergone considerable evolution in the hands of developmental biologists, and, as we indicated in the introduction to this section, general agreement on the best definition is still lacking. Two dictionary definitions of the intransitive verb "to develop" provide a starting point for our own efforts: (1) "to grow into a more mature or advanced state" (*Random House*) and (2) "to grow from an embryonic or rudimentary stage into a more complex or adult stage" (*World Book*). These definitions fit well with everyday experience. Seeds develop into plants; buds develop into flowers; fertilized eggs develop into embryos. But are these definitions complete enough? Are there also other processes in living organisms that should be included?

In one sense all of life is development. Organisms are dynamic and constantly changing. If any living thing ceases developing or changing, it is dead or, at best, dormant. Perhaps the safest definition of development is "progress of an organism through its life cycle." Before we examine the implications of this definition, which is extremely broad, it will be useful to review some of the basic properties of organisms and their life cycles.

CONTINUITY OF LIFE

We are not concerned in this book with the origin of life or with evolution. Within the time scale that we are considering, life comes only from preexisting life of the same kind. What we are concerned with is the ongoing process of reproduction—the formation of new organisms that, except for minor genetic variations, are identical to their parents.

As human individuals, we tend to think of life as finite. We commonly speak of life "beginning" either at conception or at birth and "ending" at the death of the individual. This viewpoint is not adequate for a full understanding of developmental biology. Life is continuous and is passed from one generation to the next in an unbroken chain that lasts as long as the species survives. Each individual is only a temporary link in that chain. We obtain life from our parents and pass it on to our children. This goes on generation after generation. Life is a continuously repeating cycle which, barring extinction, proceeds indefinitely.

LIFE CYCLES   Nature has evolved an overwhelming diversity of life cycles. In every case, however, the overall process is a closed circle. Sometimes cycles have temporary diversions, such as spore formation during adverse environmental conditions, or even alternate cycles, such as a choice between sexual and asexual modes of reproduction. However, if we start with any organism at any stage of its life history (other than postreproductive senescence, which will be discussed later) and follow the progress of that organism and its progeny through subsequent developmental stages, we ultimately observe new individuals at identically the same stage with which we started.

The simplest life cycle consists of growth of an organism, including duplication of all its parts, followed by division into two new organisms, both of which are basically the same as the parent was prior to growth and duplication. Many unicellular organisms, ranging from bacteria to protozoa to cultured human cells, reproduce by this type of asexual cell growth cycle (Fig. 1-1), either exclusively or as one of their available options. In eukaryotic cells such a cycle is referred to as a mitotic cycle, and the division process itself is referred to as mitosis. (The mitotic cycle and methods of analyzing it are described in detail in Appendix C.) The terms "mitosis" and "mitotic cycle" are generally not used for prokaryotic cells, since the chromosomal events characteristic of mitosis (condensation of chromosomes, formation of a mitotic spindle, alignment of chromosomes in a metaphase plate, separation of chromosomes, etc.) do not occur in prokaryotic cells.

Unicellular organisms that reproduce by asexual cell growth cycles provide interesting model systems for the cellular multiplication that occurs during growth of more complex organisms. Also, changes in gene expression that occur during the adaptation of such "simple" organisms to altered environmental conditions are similar in many ways to the process of cellular differentiation that occurs in higher

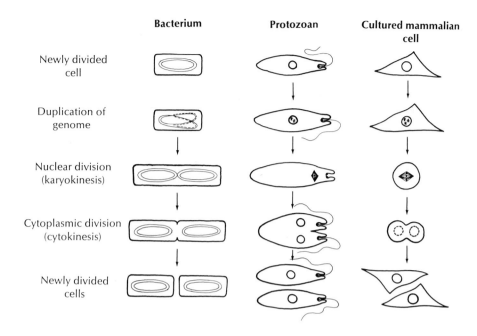

Bacterium   **Protozoan**   **Cultured mammalian cell**

Newly divided cell

Duplication of genome

Nuclear division (karyokinesis)

Cytoplasmic division (cytokinesis)

Newly divided cells

Fig. 1-1. Asexual growth cycles in unicellular organisms. Three very different types of organisms are compared — a procaryotic bacillus, a protozoan, and a cultured mammalian cell (whose behavior within its specialized culture milieu is that of a free-living microorganism). These schematic representations show duplication of the genome, division of the nucleus (or nucleoid in bacteria, which do not have a true nucleus), and, finally, division of the cytoplasm to yield two daughter cells, each of which is essentially the same as the starting cell. These representations are highly schematic and do not illustrate all the complexities that are actually involved. For example, the length of the DNA in the bacterial cell is more than 1000 times the cellular diameter, and a diploid human cell contains 1.74 m of DNA in a nucleus that is only a few microns (10^{-6} m) in diameter.

organisms. The controls over gene expression that operate in bacterial cells provide the starting point for the analysis of gene regulatory mechanisms presented in Section Two of this book and are reviewed in some detail in Chapter 4. Thus, although the primary emphasis in this book is on multicellular organisms, information regarding unicellular forms that reproduce primarily by asexual division is also important to our overall understanding of developmental mechanisms.

In most of the life cycles of more complex organisms the parental organism is partitioned unequally to generate immature progeny that undergo some kind of maturation (other than doubling of size and duplication of parts) to become equivalent to the parent(s). In some cases, such as the colonial alga, *Volvox,* the parent is destroyed during the reproductive process, but more commonly the parent retains its identity and can continue to produce progeny.

Most organisms in which a surviving parent can be distinguished from the progeny exhibit the phenomenon of aging. After a period of reproductive activity that can range from less than a day for some insects to thousands of years for some trees, the parental organisms lose their vigor and ultimately die without any obvious external cause. The postreproductive senescent period is not part of the closed cycle of growth and reproduction that maintains the continuity of life from generation to generation. However, the mechanisms involved in aging are not clearly distinguishable from those involved in development or even fully separated from them in time. Whether aging should be called a part of development is a matter of semantics. We have chosen to include a discussion of aging in this book (Chapter 24).

Another phenomenon that is closely related to development is regeneration. Most organisms have some ability to repair damage that they have suffered, and some are capable of extensive replacement of missing parts. Such repairs are not strictly part of the normal life cycle, but they involve mechanisms similar to those which occur during the normal process of forming a new individual. We have therefore chosen to include them in our broadly based definition of development.

Continuity of the germ cell line

The process of cellular differentiation results in the formation of various types of cells within a multicellular organism that differ from one another and perform specialized functions for the entire organism. One of the types of differentiation that occurs in organisms which reproduce sexually is a separation of "germ" cells from "somatic" cells. The germ cells are defined as those cells which are the direct precursors of progeny organisms. More precisely, the germ cells are those which provide the nuclear genome for the progeny, since somatic "nurse" cells may contribute significantly to the cytoplasm of large egg cells. Somatic cells are defined as all cells in the body of a multicellular organism that are not germ cells. In higher animals with obligate sexual reproduction, germ cells tend to be set aside early in embryonic development, whereas in plants and lower animals with alternative nonsexual reproductive pathways a clear distinction may not be obvious until the organism enters a sexual cycle.

In organisms such as ourselves, whose only means of reproduction is sexual, the physical basis for the continuity of life resides entirely in the germ cells. If we consider only the perpetuation of the human species, we find that our somatic cells serve primarily to nurture and protect our germ cells and to ensure that the male and female germ cells are brought together under circumstances optimal for their interaction. The somatic cells, which include everything that we commonly identify as ourselves, are not part of the continuing life cycle, and after reproduction has been accomplished, they (we) are expendable. As somatic individuals, we age and die while the progeny of our germ cells continue to reproduce themselves generation after generation. This relationship, as it occurs in sexually reproducing organisms, is diagrammed in Fig. 1-2.

Alternation of haploid and diploid states

The life cycles of most organisms include mechanisms for bringing together in a single individual new combinations of genetic information derived from two different parents. These range from obligate

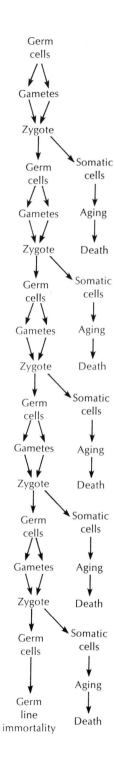

Fig. 1-2. Immortality of the germ line. The sequence on the left represents the continuity of the germ line, which consists of an endless repetition of three basic steps: (1) germ cells form gametes; (2) gametes fuse to form zygotes; and (3) zygotes develop into mature organisms containing both germ and somatic cells. Each generation of somatic cells has no further role after it has assisted the germ cells in gametogenesis and fertilization and has nurtured the zygote. The ultimate fate of the somatic cells is aging and death.

sexual reproduction in higher animals to optional cycles of conjugation and genetic exchange in various types of unicellular organisms.

In higher animals, including ourselves, nearly all the genetic material is present in two sets (except that in the sex chromosomes), and almost all the life cycle is spent in the diploid condition, that is, with two full sets of genetic information, one from each parent. Reduction divisions, known as meiosis, produce haploid gametes (sperm and egg cells), each containing only one set of genetic information. In humans and other multicellular animals the life span of the haploid stage is normally very short. The male and female gametes must be united quickly in the process called fertilization to yield the beginnings of a new diploid individual, the zygote, or else they perish.

As predominantly diploid organisms, we tend to look at life from a diploid point of view and regard the haploid state as a transitory event that occurs only briefly during reproduction. However, this is not necessarily true for other kinds of organisms. Many organisms spend nearly all their life cycle as haploids. They become diploid or partially diploid only briefly to exchange genetic information and then almost immediately revert to the haploid state. There are also some interesting groups of organisms, such as yeasts, mosses, and ferns, that spend a substantial portion of their life cycle in each of the two states. Examples of these various types of life cycles are diagrammed in Fig. 1-3.

All life cycles involving an alternation of haploid and diploid generations share in common five key features. In higher organisms these are referred to as fertilization, recombination, meiosis, mitosis, and growth. Some of these features have different names in lower organisms, but the principles involved are the same. Fertilization refers to the process of incorporating genes from two separate parents into a single cell. This can be accomplished either by fusion of a sperm cell with an egg cell or by direct transfer of genetic material from one cell to another in lower organisms. In either case the result is a cell with at least part of its genome present in two sets, one derived from each parent. When such a cell contains two complete haploid genomes, it is referred to as a zygote.

Recombination refers to the appearance in the offspring of new combinations of traits not found together in either of the parents. This

Fig. 1-3. Alternation of haploid and diploid phases in various types of organisms.
Six different life cycles with progressively greater emphasis on the diploid form are shown.

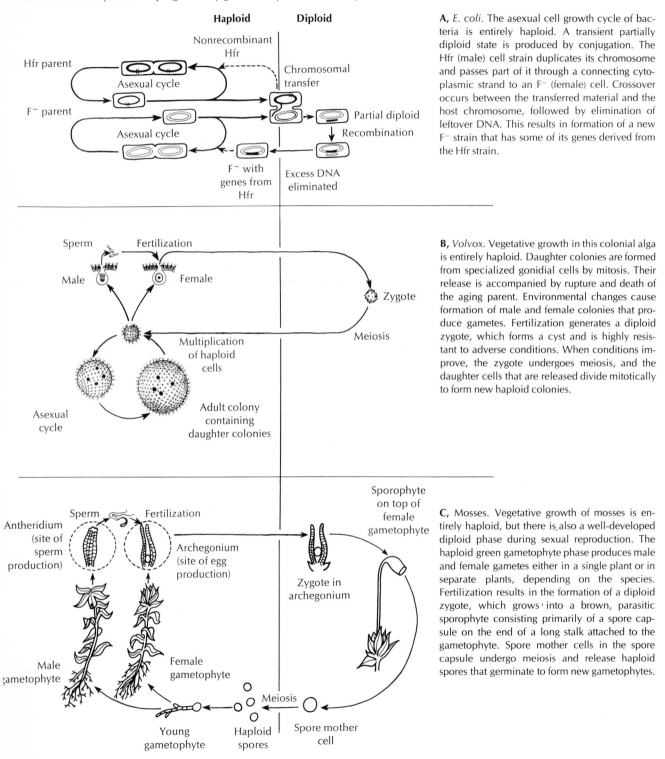

Haploid   **Diploid**

Nonrecombinant Hfr

Hfr parent

Asexual cycle

Chromosomal transfer

F⁻ parent

Asexual cycle

Partial diploid

Recombination

F⁻ with genes from Hfr

Excess DNA eliminated

A, *E. coli.* The asexual cell growth cycle of bacteria is entirely haploid. A transient partially diploid state is produced by conjugation. The Hfr (male) cell strain duplicates its chromosome and passes part of it through a connecting cytoplasmic strand to an F⁻ (female) cell. Crossover occurs between the transferred material and the host chromosome, followed by elimination of leftover DNA. This results in formation of a new F⁻ strain that has some of its genes derived from the Hfr strain.

Sperm   Fertilization

Male   Female

Zygote

Multiplication of haploid cells

Meiosis

Asexual cycle

Adult colony containing daughter colonies

B, *Volvox.* Vegetative growth in this colonial alga is entirely haploid. Daughter colonies are formed from specialized gonidial cells by mitosis. Their release is accompanied by rupture and death of the aging parent. Environmental changes cause formation of male and female colonies that produce gametes. Fertilization generates a diploid zygote, which forms a cyst and is highly resistant to adverse conditions. When conditions improve, the zygote undergoes meiosis, and the daughter cells that are released divide mitotically to form new haploid colonies.

Sporophyte on top of female gametophyte

Sperm   Fertilization

Antheridium (site of sperm production)

Archegonium (site of egg production)

Zygote in archegonium

Male gametophyte

Female gametophyte

Young gametophyte   Haploid spores   Spore mother cell   Meiosis

C, Mosses. Vegetative growth of mosses is entirely haploid, but there is also a well-developed diploid phase during sexual reproduction. The haploid green gametophyte phase produces male and female gametes either in a single plant or in separate plants, depending on the species. Fertilization results in the formation of a diploid zygote, which grows into a brown, parasitic sporophyte consisting primarily of a spore capsule on the end of a long stalk attached to the gametophyte. Spore mother cells in the spore capsule undergo meiosis and release haploid spores that germinate to form new gametophytes.

D, Yeasts. Asexual vegetative growth by budding can occur in either the haploid or the diploid phase of yeast, which is a single-celled eukaryote. Haploid cells of two different mating types can fuse to form a diploid zygote, which can either undergo immediate meiosis or begin vegetative growth as a diploid strain. Under unfavorable conditions diploid cells undergo meiosis to form haploid ascospores, which are later released from their ascus and germinate to form new haploid vegetative cultures.

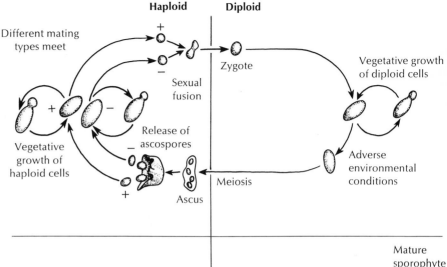

E, Ferns. The diploid sporophyte phase predominates in ferns. However, there is also a substantial haploid gametophyte phase. Haploid spores germinate and multiply to form a flat, multicellular gametophyte. Sperm and egg cells are produced in specialized structures. A diploid zygote is formed by fertilization and develops into the dominant leafy sporophyte. Sporogonia on the undersides of the leaves of the diploid sporophyte contain spore mother cells, which undergo meiosis to generate the haploid meiospores. New haploid gametophytes develop from the meiospores.

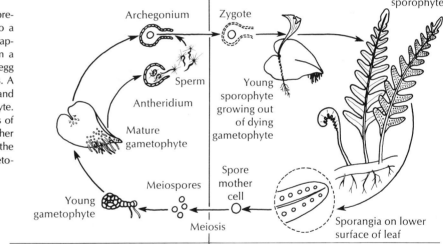

F, Higher animals, represented here by the frog. The diploid phase dominates the life cycle in higher animals (and also in higher plants). The haploid phase has only a brief existence in the gametes. Early stages of meiosis may be started in the germ cells (oogonia) of female mammals before birth, but the final meiotic division is generally not completed until the sperm nucleus has penetrated the egg cytoplasm. In the male, meiosis begins with sexual maturity and continues throughout the reproductive life span. However, individual sperm perish within a short time after their release if they do not participate in fertilization.

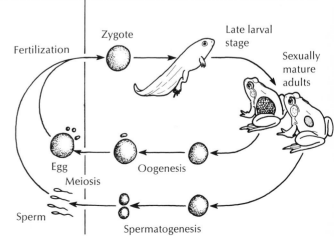

is accomplished by breaking and rejoining (crossover) of the genetic material within homologous chromosome pairs. In eukaryotic organisms with several chromosomes new combinations of traits also result from independent assortment of the chromosomes during meiosis.

Meiosis refers to the reduction division (or comparable processes in lower organisms) that results in the separation of duplicate genes and the generation of one or more exactly haploid sets of genetic information. The recombination events just described usually occur during the process of meiosis. Depending on the type of organism involved, the product of meiosis can either be a gamete with a transitory existence or a fully viable haploid organism.

As defined earlier, mitosis refers to division without net change in the amount of genetic material per cell. This can occur either in the haploid phase of the life cycle or in the diploid phase, or, in some cases, in both phases. The mitotic cycle and its equivalent in lower organisms are the primary mechanisms responsible for growth and also ultimately for the increase in number of organisms in any type of life cycle. Even in cases in which large numbers of progeny are produced sexually, mitotic division of germ cells is the mechanism responsible for generating the large number of gametes needed for such reproduction.

The overall concept of growth includes increase in number of cells, increase in mass, and increase in volume. It is difficult to define the term "growth" precisely, however, since its three components are frequently separated in time as an organism progresses through its life cycle. For example, an egg cell increases its mass and volume greatly during oogenesis and then after fertilization undergoes repeated rounds of division with little net change in mass or volume. Similarly, during bud formation in plants, extensive cell division occurs with little change in mass or volume, followed by a large increase in volume and mass with little cell division as the resulting shoot matures. Whether these disjointed segments of the overall process should be called growth is a matter of semantics. Ultimately, however, growth must include increases in all three parameters (mass, volume, and cell number).

APPROACHES TO THE STUDY OF DEVELOPMENT

The narrow dictionary definitions of development that we started with emphasized growth and the acquisition of increasing complexity. In higher organisms these processes are concentrated into the time period between the start of the diploid phase (fertilization) and the appearance of the mature adult. The study of embryology deals rather specifically with the formation from a fertilized egg of a miniature organism that possesses most, if not all, the essential structures of an adult. Developmental biology is generally considered to encompass a somewhat broader area. The question of how much broader constitutes the basis for most of the disagreement about definitions that we referred to earlier. Our own definition ("progress of an organism through its life cycle") is extremely broad. We are concerned in this book with all phases of the continuing life cycle—gametogenesis, fertilization, embryogenesis, and maturation—plus related phenomena such as aging and the repair of injuries that involve similar mechanisms.

Because of our mechanistic point of view we have chosen to examine development in terms of specific types of mechanisms and levels of organization, rather than in terms of phases of the life cycle. Thus, although we cite examples from all parts of the life cycle, our presentation is organized around topics such as molecular control of gene expression, environmental effects on cellular differentiation, and forces leading to change in shape. The relative emphasis that we are giving to each aspect of development is dictated largely by the amount of mechanism-oriented research that has been done on it.

Generally, investigators who are concerned with mechanisms have concentrated their efforts on a limited number of "model" systems. Our usual approach is to start with those model systems and then attempt to generalize. Frequently we also speculate about hypothetical mechanisms and the results of future investigations. We consider such speculation to be important. It would be a disservice to our readers not to make them aware of the ongoing nature of current research in developmental biology and the significance of answers that are likely to be found in the next few years. We have tried to label our speculation conspicuously so that our readers can keep fact and speculation clearly separated in their minds.

This chapter has sought to orient the reader to the general philosophy of this book and to provide a brief review of life cycles and the concept of development. The next two chapters describe the basic phenomena of embryonic development and cellular differentiation without attempting to delve deeply into their underlying mechanisms. At the end of Chapter 3 we will be ready to begin analyzing mechanisms, starting at the molecular level and working our way toward the level of the intact organism.

CHAPTER 2

Embryology
and morphogenesis

□ In all predominantly diploid multicellular organisms, including ourselves, numerous major developmental changes occur early in the diploid part of the life cycle. Collectively these processes transform a fertilized egg cell into an embryo that contains, in an immature form, virtually all the structures that are found in the adult (or in advanced larvae in those organisms which undergo metamorphosis late in development). This process is commonly referred to as embryogenesis, and the science associated with it is called embryology.

Embryonic development can conveniently be viewed as a composite of three overlapping processes—morphogenesis, cellular differentiation, and growth. Morphogenesis means the creation of shape or form and in embryology refers to the multitude of orderly changes of shape that occur as a mature organism is formed from an egg cell. Cellular differentiation refers to cells becoming different from one another in structure and function within a single organism. Growth, as used here, refers specifically to increase in mass and volume of the developing organism. However, in all except a few instances it also refers to increase in cell number, as discussed in Chapter 1. This chapter deals with embryonic development in terms of morphogenesis and controlled growth. Cellular differentiation is considered separately in Chapter 3.

MAJOR STEPS IN EMBRYONIC DEVELOPMENT

The details of the embryonic development of various types of organisms differ greatly from each other. However, certain generalized patterns can be seen. In particular, the following series of events occurs in similar fashion in the development of embryos of many different types of multicellular animals, both vertebrate and invertebrate:

1. A motile sperm cell from the male unites with an egg cell from the female in the process known as fertilization, giving rise to a diploid cell known as a zygote. (If the egg cell has completed

meiosis prior to fertilization, it is called an ovum; however, in most species, meiosis is not completed until after fertilization, and the unfertilized egg cell is called an oocyte.)

2. The zygote (fertilized egg cell) divides repeatedly, and often quite rapidly, in a process called cleavage, to form smaller cells, called blastomeres, without any overall increase in size of the embryo.

3. The cells generated by cleavage rearrange themselves into either a hollow ball, referred to as a blastula, or into a double-layered flat sheet, referred to as a discoblastula. (In mammals the future extraembryonic membranes form a hollow ball known as a blastocyst, and the embryo itself subsequently develops as a discoblastula from the "inner cell mass" of the blastocyst.)

4. A complex process of cellular migration, referred to as gastrulation, results in the formation of three primary germ layers (ectoderm, mesoderm, and endoderm) and the establishment of the body axis of the developing embryo.

5. The three germ layers undergo extensive cellular differentiation and morphogenesis to generate all the major body organ systems.

These processes are summarized in Fig. 2-1 for the sea urchin, frog, chicken, and human. Although the general principles are similar from species to species, the details, which are discussed later in this chapter, differ greatly from one type of animal to another. The pattern of development that is followed by eggs from a particular species is closely related to the size and architecture of the egg that is involved.

Egg cells are generally classified in terms of the amount of yolky material that they contain and its spatial distribution throughout the egg's cytoplasm (Fig. 2-2). The term "yolk" refers to food materials that are stored in the egg. The composition of yolk varies from animal to animal, but in general its main components are proteins and lipids.

EGG CHARACTERISTICS
Yolk distribution

Before we discuss the distribution of yolk in eggs, it will be useful to define the terms that are commonly used to describe locations in the egg. The location of the nucleus of the egg is indicated by the release of the second polar body (a miniature cell produced by the second meiotic division). The point where the second polar body appears is called the animal pole, and the point directly opposite it is called the vegetal pole. A line through the egg from one pole to the other is referred to as the animal-vegetal axis, and the region halfway between the poles is sometimes referred to as equatorial.

An isolecithal egg is one that has a uniform distribution of yolk. It

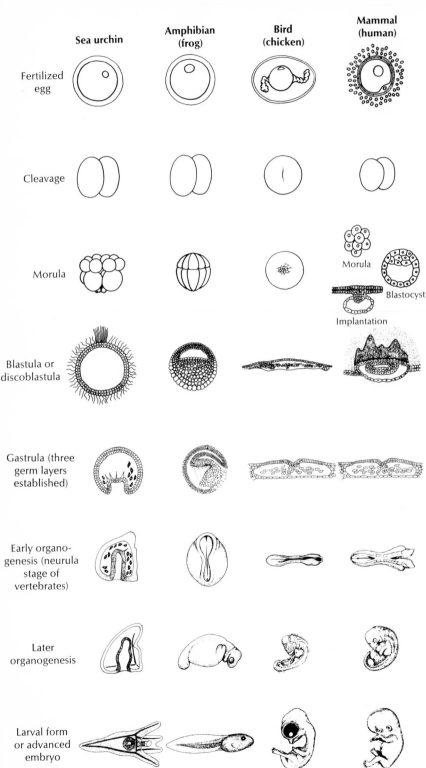

Sea urchin

Amphibian (frog)

Bird (chicken)

Mammal (human)

Fertilized egg

Cleavage

Morula

Morula

Blastocyst

Implantation

Blastula or discoblastula

Gastrula (three germ layers established)

Early organo-genesis (neurula stage of vertebrates)

Later organogenesis

Larval form or advanced embryo

Fig. 2-1. Comparison of embryonic development in sea urchins and three types of vertebrates. Drawings have been arranged to emphasize the common themes that occur in the developmental pathways of these four diverse organisms. The eggs and the adult organisms differ greatly from one another, but similar steps are involved in the development of each, as indicated on the left. From the cleavage stage onward, all external membranes have been removed from the embryos. Blastula and gastrula stages are shown in cross section; all others are external views. Additional stages associated with implantation of mammalian embryos are shown between the morula and discoblastula stages. The mammalian discoblastula is shown in relationship to the uterine wall just after implantation. In mammals an advanced embryo in which organogenesis is essentially completed (lower right) is commonly called a fetus.

also typically has a small amount of yolk. This type of egg is produced by animals whose embryos do not require large reserves of stored food for development. An example is the sea urchin (Fig. 2-1), whose embryos develop to a feeding stage in a short time (about 40 hours). The eggs of placental mammals (Fig. 2-2, *A*) also contain little yolk, since the developing embryos are able to obtain nourishment directly from the maternal organism via the placenta.

Eggs that contain larger amounts of yolk tend to have it distributed unevenly. Those whose yolk is concentrated primarily at one pole are referred to as telolecithal. An example is the amphibian egg, in which the yolk is located primarily in the vegetal hemisphere of the egg (Fig. 2-2, *B*). The eggs of birds are examples of extreme telolecithal eggs. In this case the bulk of the egg is yolk, and the actual egg

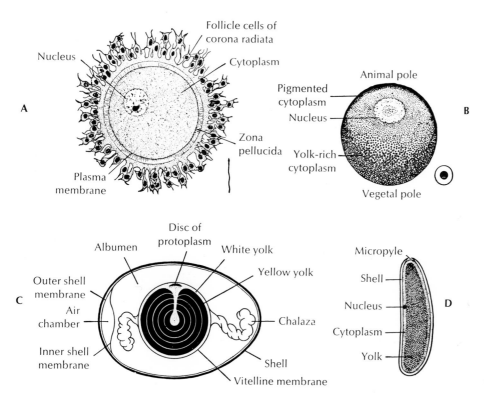

Fig. 2-2. Representative types of egg cells. **A,** Isolecithal human oocyte, freshly ovulated and still surrounded by noncellular zona pellucida and a halo of cells from the ovarian follicle, known as the corona radiata. A human sperm cell is shown at the lower right at the same scale. **B,** Moderately telolecithal egg of the frog. A frog egg with its surrounding jelly coat in place is shown 0.9 times actual size at the lower right. **C,** Highly telolecithal chicken egg. **D,** Centrolecithal egg of the fly. (**A,** ×180; **B,** ×13.5; **C,** ×0.9; **D,** ×31.5.) (From Arey, L. B., 1974. Developmental anatomy, 7th ed. W. B. Saunders Co., Philadelphia.)

cytoplasm, which contains the nucleus, is found as a thin cap lying on top of the yolk (Fig. 2-2, *C*). The albumen, or egg white, is not part of the egg cell itself, but is a noncellular secretion of the oviduct that is added after oogenesis is completed.

Another type of egg that contains a large amount of yolk is the centrolecithal egg, which is found mainly in the insects (Fig. 2-2, *D*). In this type of egg the yolk is located in a central position and is surrounded by a thin surface layer of cytoplasm. Typically the nucleus prior to fertilization is located in a small island of cytoplasm within the mass of yolk. Early development of centrolecithal eggs differs significantly from the general patterns outlined earlier and will be discussed separately later in this chapter and in greater detail in Chapter 22.

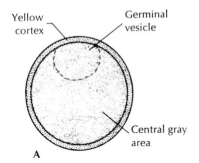

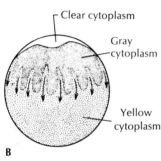

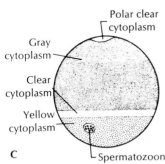

Fig. 2-3. Cytoplasmic streaming and formation of the organ-forming plasms in the egg of the ascidian *Styela* following fertilization. **A,** Unfertilized egg with large germinal vesicle (egg nucleus) at the animal pole. A thin cortical layer of yellow cytoplasm surrounds the entire egg. **B,** Immediately after fertilization the yellow cytoplasm and the clear cytoplasm released from the germinal vesicle flow toward the vegetal pole. **C,** Intermediate stage with clear cytoplasm stratified above yellow cytoplasm in the vegetal region. **D,** Side view of a slightly later stage, showing yellow and clear cytoplasms pushed into the crescent-shaped area in the posterior part of the future embryo. The arrow indicates the anterior-posterior axis of the future embryo (anterior at arrowhead). **E,** Distribution of organ-forming areas just before the first cleavage. The yolky gray cytoplasm has settled to the vegetal pole to become the future endoderm. The yellow pigmented cytoplasm is segregated in a posterior "yellow crescent" destined to become future mesodermal tissues, including muscle. The clear crescent material has diffused to the animal pole and is future epidermis. A light gray crescent has formed in the anterior ventral portion of the embryo directly opposite the yellow crescent. It will become both neural ectoderm and notochord. **F,** Distribution pattern after the first cleavage division. Bilateral symmetry is established, with the same distribution of organ-forming areas in both blastomeres. Further development and segregation of organ-forming areas are described in Chapter 12. (From Torrey, T. W. 1971. Morphogenesis of the vertebrates, 3rd ed. John Wiley & Sons, Inc., New York.)

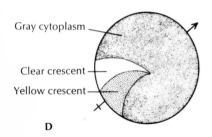

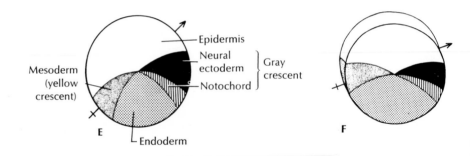

The cortex of a cell is the outer part of the cytoplasm, including the cell membrane. The cortical cytoplasm, located just under the plasma membrane, appears to play an important role in the development of many types of eggs, particularly in species in which development occurs outside the mother's body. Interesting changes in pigmentation during development are often associated with these same areas.

The eggs of the ascidian *Styela*, for example, contain yellow pigment granules in the cortical cytoplasm. On sperm entry, which occurs at the vegetal pole, this cortical cytoplasm starts to move, and all the yellow pigment streams toward the vegetal pole of the egg. Movement of the sperm pronucleus from the vegetal pole causes a further redistribution of pigment granules so that eventually three crescent-shaped zones with different pigmentation are seen in the vegetal half of the egg (Fig. 2-3). As will be described in Chapter 12, each of these zones is destined to develop into a particular part of the *Styela* embryo.

In the sea urchin *Paracentrotus lividus* the egg contains red pigment in the cortical cytoplasm. During egg maturation, this surface layer becomes concentrated as a band of pigment below the egg equator (Fig. 2-4). As in *Styela*, this pigmented region is later found in a specific part of the embryo.

One of the best-known examples of egg pigmentation occurs in the amphibian egg. Most amphibian eggs are dark brown to black, and again their pigment granules are located primarily in the egg cortex. Most of the pigment is concentrated at the animal pole, and the egg has two major zones of color—darker color at the animal pole, and light color at the vegetal pole with a rather diffuse boundary. On fertilization the cortical layer of cytoplasm contracts toward the animal pole and then rotates with respect to the more yolky internal cytoplasm. This results in the appearance of a gray crescent-shaped area, which is called the gray crescent, near the equatorial region on one side of the egg (Fig. 2-5). In normal fertilization the gray crescent always appears on the side of the egg opposite to the point of sperm entry. However, in parthenogenetically stimulated eggs its location is

Cortical cytoplasm and pigmentation

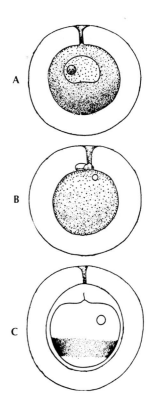

A

B

C

Fig. 2-4. Displacement of pigment in the egg of the sea urchin *Paracentrotus lividus*. All stages are shown with surrounding jelly coat in place. **A,** Oocyte. **B,** Immature unfertilized egg with relatively uniform distribution of pigment. **C,** Pigment redistribution that occurred during egg maturation is shown in a fertilized egg with raised fertilization membrane (middle circle). (From Balinsky, B. I. 1975. An introduction to embryology, 4th ed. W. B. Saunders Co., Philadelphia; originally published in Morgan, T. H. 1927. Experimental embryology. Columbia University Press, New York.)

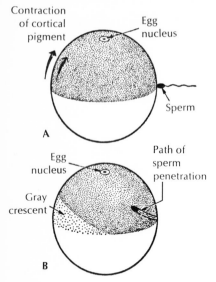

Contraction of cortical pigment

Egg nucleus

Sperm

A

Egg nucleus

Path of sperm penetration

Gray crescent

B

Fig. 2-5. Origin of the gray crescent in amphibian eggs. **A,** Fertilization. Contraction of cortical pigmented layer in the animal hemisphere on the side opposite the point of sperm entry is shown. **B,** Slightly later stage, showing gray crescent left behind as a result of the cortical shift. The gray crescent determines the location of the dorsal lip of the blastopore and therefore the orientation of the embryo relative to the egg. (From Deuchar, E. M. 1975. Cellular interactions in animal development. Chapman & Hall Ltd., London. © Elizabeth M. Deuchar.)

random with respect to the point of pricking. As will be discussed later in this chapter, and also in Chapter 12, the gray crescent region plays an important role in determining the site where gastrulation begins and in the location of the body axis of the developing embryo.

Although the pigmented regions of eggs often develop into specific structures or regions within the embryo, the pigment itself is not thought to have any determining role in such processes as cell differentiation or cell movement. For example, amphibian eggs which contain no pigment develop in the same manner as those of related species that are pigmented. However, pigmentation can be a visible reflection of some other cytoplasmic specialization that does have a determining role in the development of the embryo.

CLEAVAGE
Characteristics of the cleavage stage of development

Although the pattern of cleavage varies for different kinds of eggs, the general process and its end result are essentially the same for all. Cleavage is a period of cell division without growth. The embryo does not increase in mass but merely becomes divided into cells of smaller size; the same mass of cytoplasm that was initially contained in one cell is distributed among many cells. The general shape of the embryo changes little during this period, although toward the end of cleavage a fluid-filled cavity usually forms in the interior of the cell mass.

The number of cleavage divisions that occur depends on the size of the egg and on the size of an average cell in the mature form of that particular organism. Cleavage occurs until the resulting blastomeres are of a size that approximates the normal cell size of the organism. The rate of cleavage divisions is frequently rapid. The zebra fish egg cleaves about every 15 minutes, sea urchin eggs about every half hour, and frog eggs about every hour. Cleavage tends to be slower in mammalian embryos. For example, in the mouse embryo, cleavage occurs every 10 to 12 hours (except for the first cleavage, which takes about 24 hours). It has been demonstrated in marine invertebrates that the rate of cleavage is determined by cytoplasmic factors in the egg. If enucleated eggs of one species with a fast cleavage rate are fertilized with sperm of a second species with a low cleavage rate, the rate of cleavage in the resulting embryo is rapid.

Although there is no growth during cleavage, the cells are engaged in considerable synthetic activity. In particular, cleavage requires extensive synthesis of DNA, chromosomal proteins, and mitotic spindle proteins. The rapid DNA synthesis that occurs in cleavage-stage embryos apparently occurs by means of an altered pattern of DNA synthesis. In eukaryotic cells with large amounts of DNA the genome is organized into many separate replication units, known as replicons. Each replicon has a site for the initiation of DNA synthesis. Once synthesis is initiated, the DNA polymerase travels along the DNA at a constant rate until it reaches the end of the replicon. When replicating DNA is labeled with ^{3}H-thymidine and is stretched out on a microscope slide, which is then coated with a photographic emulsion and stored in the dark, the radioactive decay "exposes" the emulsion (autoradiography). When the emulsion is developed, the position of the radioactive DNA can be seen as linear arrays of silver grains visible

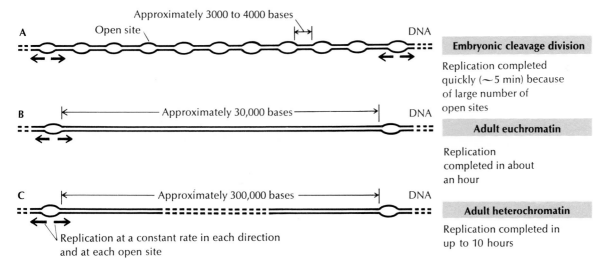

Fig. 2-6. Schematic representation of DNA replication in chromatin from embryonic and adult *Drosophila* tissue. **A,** Embryonic tissue. Replicons are located close together. DNA synthesis is completed in about 5 minutes. **B,** Euchromatin from cultured adult tissue. The replicons are much longer. DNA synthesis is completed in about an hour. **C,** Heterochromatin from cultured adult tissue. The replicons are extremely long. DNA synthesis may require up to 10 hours for completion. (Modified from Watson, J. D. 1976. Molecular biology of the gene, 3rd ed. The Benjamin/Cummings Publishing Co., Inc., Menlo Park. Calif.)

with a light microscope. When dividing cells are labeled for a short time with ^{3}H-thymidine and autoradiographed, one normally observes a large number of short segments of labeled DNA, which become longer as the labeling time is increased. When the DNA is labeled for a period longer than the time it takes for the DNA polymerase to travel the length of each replicon, very long continuously labeled DNA molecules are observed.

Recent comparisons of the length of replicons in cleavage-stage eggs of *Drosophila* with those in tissue culture cells derived from adult *Drosophila* have shown that during cleavage the replicons are much closer together and therefore much shorter (Fig. 2-6). The rate of travel of the DNA polymerase seems to be unchanged, but because there are more initiation sites and each DNA polymerase molecule has a shorter distance to travel, DNA synthesis is completed more rapidly. When somatic cell nuclei are transplanted into enucleated *Xenopus* eggs, their rate of DNA synthesis increases, presumably because of the increased number of replicons and active DNA polymerase molecules that are involved. Nuclei that have been "conditioned" by prior division in an egg cytoplasm support development of embryos from enucleated eggs more effectively than those transplanted directly from somatic tissues of *Xenopus*. This phenomenon is discussed in greater detail in Chapter 6.

Division of the egg cytoplasm into smaller cells during cleavage can result in the segregation of certain cytoplasmic components of the egg into a limited number of the daughter cells. When such components are essential for differentiation of specific types of cells, the segregation process determines which daughter cell will undergo those specific types of differentiation. An example is the localized germ plasm of the frog's egg, which is essential for the formation of germ cells (Chapter 19). Segregation of cytoplasmic components is discussed at greater length in Chapter 12.

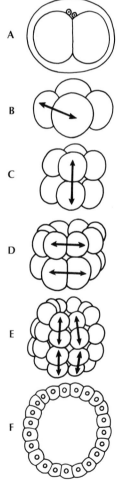

Fig. 2-7. Radial cleavage in the sea cucumber *Synapta digitata*. Fertilization membrane and polar bodies are shown only for the two-cell stage. Arrows connect pairs of new cells that have arisen from division of a single cell. Note the equal size of the blastomeres and their precise alignment above one another. **A** to **E,** The 2-, 4-, 8-, 16- and 32-cell stages resulting from synchronous cleavage divisions. **F,** Spherical blastula shown in cross section. (Modified from Korschelt, E. 1936. Vergleichende Entwicklungsgeschichte der Tiere. Verlag von Gustav Fischer, Jena.)

The cleavage pattern that an egg undergoes is rather directly related to the amount and distribution of yolk within the egg. Complete (holoblastic) cleavage, in which the entire egg is cleaved into blastomeres, occurs in isolecithal and many moderately telolecithal eggs. The cleavage of isolecithal eggs is both complete and equal. That is, the egg is completely divided into blastomeres of a single size (Fig. 2-7). In moderately telolecithal eggs, cleavage is complete but unequal. In the amphibian egg, for example, the yolky vegetal region cleaves more slowly than the animal region and produces fewer and larger cells. The first two cleavage divisions occur with furrows running from the animal to the vegetal pole and divide the cytoplasm equally. The third cleavage, however, occurs horizontally and is displaced toward the animal pole. The result is four large vegetal cells with four smaller animal cells atop them (Fig. 2-8).

In incomplete (or meroblastic) cleavage the entire egg does not divide. The cytoplasmic region cleaves into individual cells, but the yolky region remains undivided. This type of cleavage occurs in extreme telolecithal eggs and centrolecithal eggs. In extreme telolecithal eggs, such as the bird egg, cleavage is discoidal. That is, it occurs only in the small disc of cytoplasm at the animal pole of the egg. The blastomeres at the edge and bottom of the disc remain continuous with the yolk (Fig. 2-9). In centrolecithal eggs, cleavage is restricted to the outer surface. In the early stages of cleavage of insect eggs the nuclei divide mitotically (karyokinesis), but division of the cytoplasm (cytokinesis) does not occur, resulting in the formation of a syncytium (a mass of cytoplasm containing many nuclei). The nuclei move to the periphery of the egg to form what is called the syncytial blastoderm. Later, cleavage furrows develop between these nuclei at the beginning of the stage called the cellular blastoderm. These furrows form concurrently around all the nuclei and generate a large number of separate cells at once (Fig. 2-10).

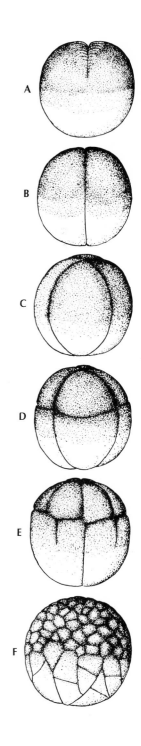

Fig. 2-8. Semidiagrammatic representation of cleavage in the frog egg. **A,** First cleavage, beginning at the animal pole and progressing into the yolk-laden vegetal hemisphere. **B,** Two-cell stage, showing equal division of the cytoplasm. **C,** Four-cell stage with the second division again from animal to vegetal poles. **D,** Eight-cell stage. The third division, which separates the animal and vegetal portions of the egg, occurs above the equator and forms animal pole cells that are smaller than the yolk-laden vegetal pole cells. **E,** Fourth division in progress, showing cleavage in the vegetal hemisphere lagging behind that near the animal pole. **F,** Morula stage, showing many small cells in the animal hemisphere and fewer larger cells in the vegetal hemisphere. (From Balinsky, B. I. 1975. An introduction to embryology, 4th ed. W. B. Saunders Co., Philadelphia.)

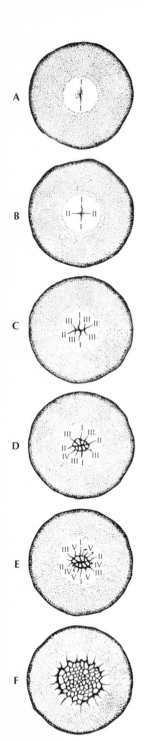

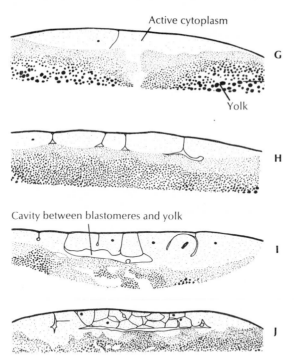

Active cytoplasm

Yolk

Cavity between blastomeres and yolk

Fig. 2-9. Cleavage in the chicken egg. **A** to **E,** First five cleavages shown in a surface view of the germinal disc on the top of the yolk. The cleavage furrows appearing at each division are indicated by Roman numerals. **F,** Early morula stage, showing cells incompletely cleaved from the surrounding yolk mass at the edges. **G** to **J,** Cross-sectional views of early cleavage stages. Formation of a cavity can be seen below those cells which have become completely separated by membranes from the surrounding yolk mass. (**A** to **F** from Patten, B. M., and B. M. Carlson. 1974. Foundations of embryology, 3rd ed. McGraw-Hill Book Co., New York. Copyright © 1974 by McGraw-Hill, Inc. Used with permission of McGraw-Hill Book Co.; **G** to **J** from Balinsky, B. I. 1975. An introduction to embryology, 4th ed. W. B. Saunders Co.)

The positioning of mitotic spindles during cleavage divisions has an important effect on the future morphology of the embryo. In most types of embryos the first two cleavage planes are along the animal-vegetal axis, whereas the third cuts across it (Fig. 2-11). It has been shown clearly in sea urchins that the position of the mitotic spindle is controlled by the amount of time that has elapsed since fertilization, independently of whether cleavage has actually occurred. Thus, if cleavage is delayed by treatments such as shaking or exposure to diluted seawater, the orientation of the cleavage spindle will shift at the normal time, and the delayed cleavage will occur in an abnormal orientation when it is allowed to proceed (Fig. 2-12). A normal pattern is therefore dependent on the synchronous occurrence of two independently controlled events, cleavage and spindle orientation.

At least three different patterns of holoblastic cleavage can be distinguished by the spatial relationship of the early blastomeres to one another. The first two, called radial and spiral cleavage, respectively, are characterized by synchronous division of all the blastomeres and by a precise geometric relationship among the blastomeres. The third, which occurs in mammals, does not have precisely synchronized division of the blastomeres or a definite spatial relationship among them.

Spiral and radial cleavage can be distinguished from each other by the geometric relationship of the cells in the eight-celled embryo. In both cases the first two cleavage planes are along the animal-vegetal axis, and the third is across it. In radial cleavage the third cleavage plane is perpendicular to those of the first two cleavage divisions (as illustrated in Fig. 2-11), and at the eight-cell stage, the four cells in the animal hemisphere are located directly above the corresponding cells in the vegetal hemisphere (Fig. 2-7). In spiral cleavage the layer of blastomeres in the animal hemisphere is shifted relative to those in the vegetal hemisphere so that the animal hemisphere cells lie over the boundaries between the vegetal hemisphere cells (Fig. 2-13, *A*). This arrangement results from the positioning of the mitotic spindles. At the third cleavage division in spirally cleaving embryos the four spindles are in an oblique position with respect to the animal-vegetal axis. In subsequent cleavages the spindles are also in an oblique position, but the direction changes with each division. Spiral cleavage occurs primarily in segmented worms and mollusks. In the freshwater snail, *Limnaea*, it has been shown that the direction of spiraling in the cleavage-stage embryo, which in turn determines the direction of coiling of the shell in the adult, is maternally inherited (Fig. 2-13, *B*). This means that the direction of coiling for any individual is controlled by factors in the egg cytoplasm and is therefore determined strictly by the genotype of its mother and is independent of its own genotype.

In mammals the cleavage divisions are somewhat asynchronous, resulting in the transient existence of embryos with cell numbers such as 3, 5, 6, 7, and 9, which are never found in the geometric progression (1, 2, 4, 8, 16, etc.) of cell numbers in synchronously dividing embryos. During and after early divisions in some mammalian species the blastomeres can be seen to move around within the membrane (zona pellucida) that surrounds the embryo. Experiments in which blastomeres have been added to or removed from embryos suggest that in mammalian development the fates of individual cells are not firmly determined during early cleavage. Thus the embryo can regulate its pat-

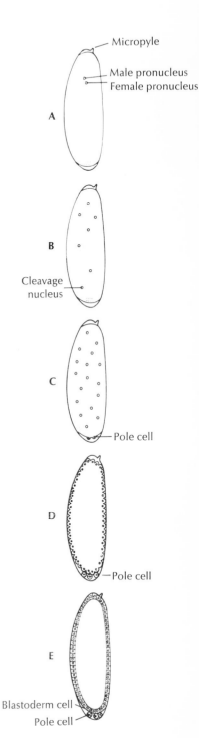

Fig. 2-10. Cleavage of the centrolecithal egg of *Drosophila*. All views are in cross section. **A,** Fertilized egg, prior to union of male and female pronuclei. The micropyle is a tube through the vitelline membrane that provides an entry route for the sperm cell. **B,** Early cleavage of nuclei in islands of cytoplasm in the yolk mass. **C,** Formation of pole cells, the future germ cells of the fly, after the seventh division. **D,** Migration of nuclei to the surface and formation of the syncytial blastoderm. **E,** Formation of cell walls around the nuclei, yielding the cellular blastoderm. (Modified from Graham, C. F., and P. F. Wareing. 1976. The developmental biology of plant and animals. W. B. Saunders Co., Philadelphia.)

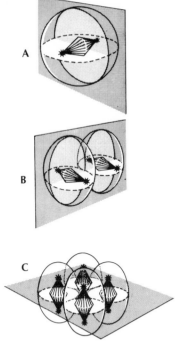

Fig. 2-11. Orientation of cleavage planes in eggs exhibiting radial cleavage. **A,** First division runs from the animal pole to the vegetal pole and divides the egg into equivalent halves. **B,** Second cleavage plane is also through the animal-vegetal axis and is perpendicular to the first, resulting in four equal blastomeres. **C,** Third cleavage is perpendicular to both of the first two and cuts across the animal-vegetal axis, dividing the egg into animal and vegetal halves. (From Torrey, T. W. 1971. Morphogenesis of the vertebrates, 3rd ed. John Wiley & Sons, Inc.)

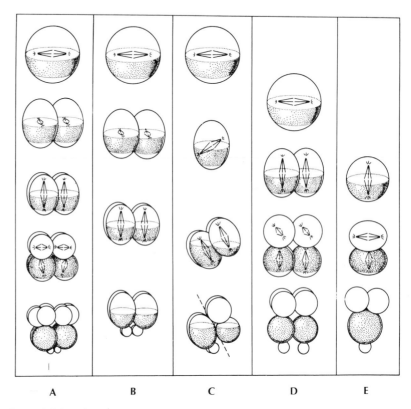

Fig. 2-12. Effects of delayed mitosis in the sea urchin. **A,** Normal cleavage pattern. The first two divisions are along the animal-vegetal axis and produce four equivalent blastomeres. The third division is across the axis and separates four animal blastomeres from four vegetal blastomeres. In the animal hemisphere the fourth division is parallel to the animal-vegetal axis and produces a ring of eight equal blastomeres. In the vegetal hemisphere it is perpendicular to the axis and unequal, producing four "macromeres" and four "micromeres." **B,** Third cleavage delayed, resulting in premature formation of micromeres. **C,** First cleavage delayed until spindle orientation is partially rotated, resulting in cleavages occurring at an oblique angle. **D,** Normal second cleavage is eliminated by delay, resulting in formation of an embryo with half the normal number of each type of cell. **E,** Normal first and second cleavages eliminated by delay, resulting in an embryo with a fourth the normal number of each type of cell. (Reprinted with permission of Macmillan Publishing Co., Inc. from Patterns and principles of animal development by Saunders, J. W., Jr. Copyright © 1970 by John W. Saunders, Jr.)

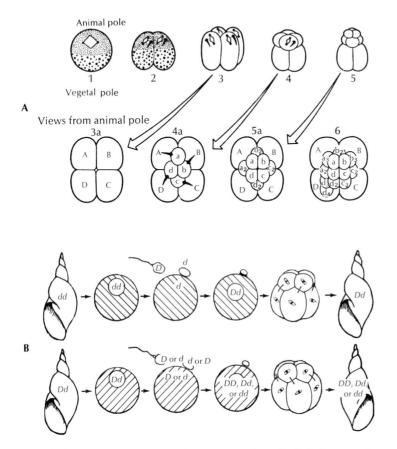

Fig. 2-13. Spiral cleavage. **A,** Diagrammatic representation of spiral cleavage. See text for details. **B,** Maternal inheritance of cleavage pattern and coiling of shell in the freshwater snail, *Limnaea.* The gene for right-handedness *(D)* is dominant over the gene for left-handedness *(d).* In the top row the ovum from a genetically homozygous left-handed female *(dd)* produces left-spiraling offspring, even when fertilized by a genetically dominant right-handed sperm. In the second row a phenotypically left handed female, genetically heterozygous for the dominant right-handed gene *(Dd),* produces ova that give rise only to right-spiraling offspring, no matter what kind of sperm they are fertilized with. Note that some of these right-spiraling individuals can be homozygous for the recessive left-handed gene *(dd).* In each case the direction of the spiraling of the shell is determined by the orientation of the cleavage planes at the third division when the spiral cleavage pattern first becomes evident. (**A** from Deuchar, E. M. 1975. Cellular interactions in animal development. Chapman & Hall Ltd., London. © 1975 Elizabeth M. Deuchar; **B** modified from Gilchrist, F. G. 1968. A survey of embryology. McGraw-Hill Book Co., New York. Copyright © 1968 by McGraw-Hill, Inc. Used with permission of McGraw-Hill Book Co.)

tern of development to compensate for major geometric changes such as loss of as much as three fourths of its original mass or even the fusion of two complete eight-celled mouse embryos, doubling the original mass (Chapter 12). However, late in the eight-cell stage in the mouse embryo, the cells become tightly compacted together and cellular position begins to be important. In particular, recent experiments suggest that as soon as one cell is completely surrounded by others, it begins to differentiate toward formation of the inner cell mass of the blastocyst.

FORMATION OF THE BLASTULA OR DISCOBLASTULA
Blastocyst formation and implantation of mammalian embryos

Comparison of the developmental stages that follow immediately after cleavage in the species that are summarized in Fig. 2-1 can become confusing as a result of the special stages that mammalian embryos go through as they become implanted in the wall of the uterus prior to extensive development of the embryo itself. As the holoblastic cleavage of mammalian embryos progresses, a cluster of rounded cells that looks somewhat like a berry is formed. This stage is called a morula from the Latin word for mulberry. The next step is the formation of the fluid-filled cavity within the mass of cells and rearrangement of cells so that only one or a few layers of cells surround the cavity. In mammals this structure is called a blastocyst, and it is functionally quite different from the similar-appearing structure known as a blastula, which is formed in a variety of other types of embryos ranging from sea urchins to frogs. We will discuss the developmental significance of the blastocyst first and then return to the blastula and to

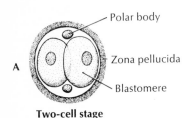

Two-cell stage

Polar body

Zona pellucida

Blastomere

Four-cell stage

Fig. 2-14. Cleavage stages of human development. **A** to **D,** Progressively more advanced cleavage stages. The cells are retained inside the membranous zona pellucida and exhibit no fixed geometric relationship to one another. Also, cleavage is not synchronous. **E** to **F,** Cross sections of the expanding blastocyst. After ''hatching'' from the zona pellucida, the blastocyst increases in size significantly as a result of accummulation of fluid in the blastocyst cavity. (From Moore, K. L. 1974. Before we are born—basic embryology and birth defects. W. B. Saunders Co., Philadelphia.)

Eight-cell stage

Morula

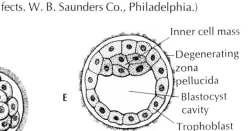

Inner cell mass

Degenerating zona pellucida

Blastocyst cavity

Trophoblast

Early blastocyst

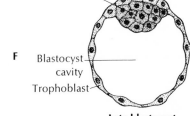

Inner cell mass

Blastocyst cavity

Trophoblast

Late blastocyst

stages in humans and other mammals that are comparable to the blastula but which do not occur until after implantation in the uterine wall has been achieved.

The mammalian embryo is dependent on maternal nutrition from the earliest stages of development onward, since it develops from an egg that does not contain significant nutrient reserves. In addition, since the female reproductive tract undergoes cyclic changes (e.g., the human menstrual cycle, the estrus cycle of rodents), the uterus is normally receptive to implantation of the embryo for only a limited period of time. Specifically in the human, the receptive state of the uterus is maintained for only about 12 days after ovulation, and the next menstrual bleeding will begin at 14 days if the embryo has not implanted and provided an alternate source of gonadotropins before the cyclic supply from the pituitary declines.

The mammalian blastocyst is a hollow sphere composed of a single layer of large, flattened cells known as the trophoblast (or trophectoderm) plus a cluster of smaller, rounded cells at one edge called the inner cell mass (Fig. 2-14). The trophoblast is a special evolutionary adaptation that functions in implantation and becomes part of the placenta. The actual embryo develops primarily from the inner cell mass, and at the time of blastocyst formation it is still in a primitive state comparable to an early morula in other organisms.

The blastocyst begins to form about the fifth day after fertilization in the human. At about the same time the membranous zona pellucida begins to break down and the blastocyst escapes or "hatches" from it. The blastocyst then undergoes a period of expansion caused by fluid accumulation. By the sixth day after fertilization in the human the trophoblast attaches to the wall of the uterus, and during the next few days it undergoes a period of rapid invasive growth so that by about 12 days it is completely embedded in the uterine lining. Blastocyst formation in the mammal appears to occur after a specific number of cell divisions, rather than in response to the number of cells present. If three of the four cells in a four-celled rabbit or mouse embryo are destroyed, the remaining cell continues to divide, and blastocyst formation occurs at the usual time, resulting in the formation of a blastocyst containing only one fourth the usual number of cells.

An efficient way for a cleaving embryo to monitor the number of cell divisions it has undergone would be to monitor the nuclear:cytoplasmic ratio in its blastomeres. As cleavage proceeds, this ratio becomes larger, and it is conceivable that a particular nuclear:cytoplasmic ratio is required for blastocyst formation. Evidence for such a mechanism has been derived from studies with haploid embryos produced by parthenogenic stimulation of unfertilized eggs. The haploid

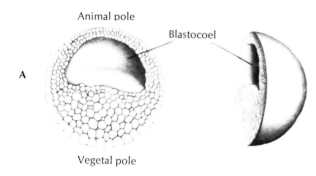

Animal pole

Blastocoel

A

Vegetal pole

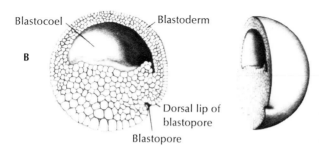

Blastocoel

Blastoderm

B

Dorsal lip of
blastopore

Blastopore

Fig. 2-15. Blastula and gastrula stages of frog development. Sections on the left are cut in the median plane and viewed from the embryo's left. The same sections are viewed obliquely from the dorsal posterior on the right. The blastopore is at the posterior end of the embryo. **A,** Late blastula stage, before the beginning of gastrulation. **B,** Early gastrulation. Cells are just beginning to turn under and move to the interior at the dorsal lip of the blastopore. **C,** Middle gastrulation. Extensive migration of cells over the dorsal lip of the blastopore has moved mesodermal tissue to the interior and has generated a new cavity, the archenteron or primitive gut, which is beginning to displace the blastocoel. The mesoderm on the midline will become the notochord. **D,** Late gastrulation. The spreading outer layer of ectoderm has engulfed most of the yolk, and invagination has also begun at the ventral and lateral lips of the blastopore, leaving only a small yolk plug exposed at the surface. See Fig. 2-20 for greater detail of this stage. (Modified from Balinsky, B. I. 1975. An introduction to embryology, 4th ed. W. B. Saunders Co., Philadelphia.)

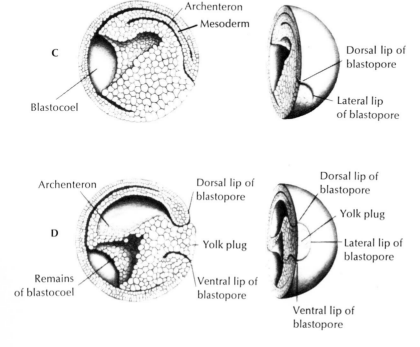

Archenteron

Mesoderm

C

Dorsal lip of
blastopore

Lateral lip
of blastopore

Blastocoel

Archenteron

Dorsal lip of
blastopore

Dorsal lip of
blastopore

Yolk plug

D

Yolk plug

Lateral lip of
blastopore

Remains
of blastocoel

Ventral lip of
blastopore

Ventral lip of
blastopore

nuclei contain only half as much chromatin as diploid nuclei, and blastocyst formation in the haploid embryos occurs about one cleavage division later than in normal diploid embryos. This extra cleavage would be required to produce the same nuclear:cytoplasmic ratio that the diploid embryos have when they initiate blastocyst formation.

In some mammals, embryos do not necessarily implant on reaching the uterus but may enter a resting stage or diapause, during which they remain unattached in crypts of the uterine lumen. This delay in implantation is obligatory in some mammals such as the roe deer and the black bear in which it is controlled by environmental conditions and assures that young will be born at a favorable seasonal time. In other mammals such as the mouse, rat, and kangaroo delayed implantation is facultative and occurs in response to the stimulus of suckling young. This stimulus inhibits the release of pituitary gonadotropins. Delayed implantation can also be induced artificially in laboratory mice and rats by ovariectomy or hypophysectomy early in pregnancy. When progesterone is supplied to such animals, the blastocysts are maintained but do not implant. They can be caused to implant by the administration of a small amount of estrogen.

Delayed implantation

Until implantation is completed, the actual embryonic tissue undergoes limited development and progresses only to the formation of a two-layered disc of cells comparable to the discoblastula formed early in the development of the chicken embryo. Before considering further development of human (and other mammalian) embryos, we will discuss blastula formation in other types of embryos.

The idealized blastula is composed of a single layer of cells called a blastoderm surrounding a fluid-filled cavity called a blastocoel. This pattern is followed most closely in some isolecithal eggs such as those of the sea urchin (Fig. 2-1) and sea cucumber (Fig. 2-7). In the frog embryo the blastocoel is displaced toward the animal pole. The roof of the blastocoel contains a few layers of small animal cells, whereas the floor of the blastocoel is composed of numerous layers of large, yolky cells (Fig. 2-15, *A*).

Hollow blastula

The centrolecithal eggs of insects do not form a fluid-filled cavity. However, the cellular blastoderm stage in which peripheral cells surround a yolky interior is analogous to the blastula of other organisms, except that the blastocoel is filled with yolk (Fig. 2-10).

Formation of a blastula by the avian embryo is complicated by the flattened disclike nature of the cellular part of the embryo. A developmental stage called a discoblastula, which is analogous to the blastula

Discoblastula

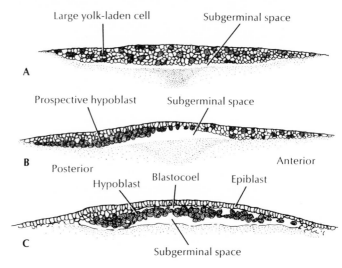

Fig. 2-16. Formation of the avian discoblastula. **A,** Section through the blastoderm, showing large, yolk-laden cells (future hypoblast) intermingled with smaller cells (future epiblast). **B,** Accummulation of larger cells in the subsurface position, particularly in the posterior part of the blastoderm. **C,** Organization of the larger cells into a distinct hypoblast layer concomitantly with the formation of a blastocoelic space between the hypoblast and epiblast and a thinning of the epiblast. There is also a subgerminal cavity between the yolk and the hypoblast that is distinct from the blastocoel. (From Torrey, T. W. 1971. Morphogenesis of the vertebrates, 3rd ed. John Wiley & Sons, New York.)

stage of other embryos, does occur, however, in which two distinct layers of cells are formed—the epiblast, which corresponds to the animal pole cells in other embryos, and the hypoblast, which is a thin layer closer to the yolk. A flattened fluid-filled blastocoel separates the two layers (Fig. 2-16). The inner cell mass in early postimplantation mammalian embryos becomes organized into a bilaminar structure that is much like the avian discoblastula. From this stage on, the development of avian and mammalian embryos is similar, even to the formation of a yolk sac, although it is empty in the case of the mammal. Such similarities presumably reflect the relatively recent evolutionary divergence of the two classes of organisms.

TERMINOLOGY USED TO DESCRIBE LOCATIONS IN EMBRYOS

Before we proceed to a description of gastrulation, in which the basic axial organization of embryos is established, it will be helpful to define the terms used to describe locations within vertebrate organisms and their embryos. Two sets of terminology are widely used, one based on location relative to specific anatomical features and the other based on the normal orientation of the organism. The latter becomes very confusing when comparing humans with most other vertebrates because of the upright posture of humans.

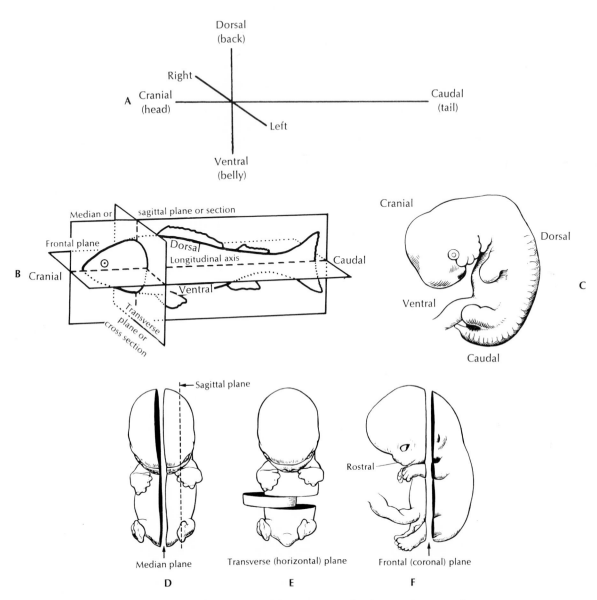

Fig. 2-17. Terminology used to describe anatomical locations. **A,** The three major axes of the vertebrate body, oriented with the vertebral column horizontal and the head at the left. **B,** Three axes and the sections that cut across them in a vertebrate species with a relatively straight cranial-caudal axis. **C,** Dorsal-ventral relationships in an embryo with a curved cranial-caudal axis. **D,** Sections cut on median and sagittal planes in an embryo. (Left is separated from right.) **E,** Sections cut on a transverse plane in an embryo. (Cranial is separated from caudal.) **F.** Sections cut on a frontal plane in an embryo. Dorsal is separated from ventral. (**B** modified from Storer, T. I. 1943. General zoology. McGraw-Hill Book Co., New York; **C** to **F** from Moore, K. F. 1974. Before we are born—basic embryology and birth defects. W. B. Saunders Co., Philadelphia.)

A. Ectoderm
  1. Epidermis, including
    a. Cutaneous glands
    b. Hair, nails, lens
  2. Epithelium of
    a. Sense organs
    b. Nasal cavity, sinuses
    c. Mouth, including oral
       glands and enamel
    d. Anal canal
  3. Nervous tissue, including
    a. Hypophysis
    b. Chromaffin tissue
B. Mesoderm (including
   mesenchyme)
  1. Muscle (all types)
  2. Connective tissue, cartilage,
     bone, notochord
  3. Blood, bone marrow
  4. Lymphoid tissue
  5. Epithelium of
    a. Blood vessels, lymphatics
    b. Body cavities
    c. Kidney, ureter
    d. Gonads, genital ducts
    e. Suprarenal cortex
    f. Joint cavities, etc.
C. Endoderm
  1. Epithelium of
    a. Pharynx, including
      (1) Root of tongue
      (2) Auditory tube, etc.
      (3) Tonsils, thyroid
      (4) Parathyroids, thymus
    b. Larynx, trachea, lungs
    c. Digestive tube and
       associated glands
    d. Bladder
    e. Vagina, vestibule
    f. Urethra and associated
       glands

*Modified from Arey, L. B. 1974. Developmental anatomy, 7th ed. W. B. Saunders Co., Philadelphia.

The terminology based on anatomical features is illustrated in Fig. 2-17. It is organized around a set of three orthogonal axes (Fig. 2-17, *A*):

1. Cranial-caudal. This is the longitudinal axis of the vertebrate body, running from head to tail along the vertebral column. In idealized cases it is quite straight (Fig. 2-17, *A* and *B*), but in many cases, particularly in embryos, it can be rather sharply curved (Fig. 2-17, *C*). *Cranial* refers to the head or the direction of the head, and *caudal* to the tail. Additional terms that are sometimes used are *cephalic*, which also refers to the head, and *rostral*, which refers to the snout or nasal part of the head (Fig. 2-17, *F*).

2. Dorsal-ventral. This axis runs from the back (vertebral) side of the animal to the belly side. *Dorsal* refers to locations near the back (e.g., in humans the scapulae, or shoulder blades, are dorsal). *Ventral* refers to the side away from the backbone (e.g., in humans the chest is a ventral surface).

3. Left-right. This axis runs across the animal from one side to the other. The term *median* refers to the centerline of the body, around which most vertebrates exhibit bilateral symmetry. *Medial* refers to locations toward the center and *lateral* to locations away from center and toward the sides.

The terms proximal and distal are used to refer to locations in appendages such as limbs. *Proximal* means close to the trunk (e.g., the shoulder), and *distal* refers to locations further from the trunk (e.g., the fingers).

The terminology used to describe sections cut through the body (e.g., for microscopic study) is also illustrated in Fig. 2-17.

1. A sagittal plane or section is parallel to the midline and separates right and left portions of the organism. If it is on the midline, it is called a median plane or section (Fig. 2-17, *B* and *D*).

2. A transverse section cuts across the long axis of the body and separates cranial and caudal portions (Fig. 2-17, *B* and *E*).

3. A frontal section separates dorsal and ventral areas from each other (Fig. 2-17, *B* and *F*).

The major positional terms based on orientation of the body are superior and inferior, referring to top and bottom, and anterior and posterior, which are widely used to refer to front and back. In vertebrate embryology, *anterior* is rather frequently used as a synonym for cranial and *posterior* for caudal. That is the only way in which those terms are used in this book. Anterior will never mean ventral (as it often does in adult human anatomy), and posterior will never mean dorsal.

The first stages of embryogenesis, cleavage and formation of the blastula, although important and dramatic in themselves, merely set the stage for the next major process, that of gastrulation. During gastrulation spectacular morphogenetic events occur in which cells undergo large-scale movements, change their shapes, and form new interactions with other cells. In vertebrate embryos the end result of these cellular rearrangements is twofold. First is the formation of the embryonic germ layers—the outermost layer, ectoderm, which will form the animal's neural tissue and epidermis; the middle layer, mesoderm, which will give rise to muscle, connective tissue, the vascular system, and various internal organs; and the inner layer, endoderm, which produces the lining of the entire gastrointestinal tube and its derivatives (see outline, p. 34). Second, as the germ layers assume their appropriate positions within the embryo, the basic axial organization of the embryo is also determined. The stage is set for formation of primary organ rudiments such as the notochord, neural tube, and archenteron and, subsequently, the complete course of organogenesis and cell differentiation. A similarly dramatic separation of germ layers and orientation for future development also occurs in invertebrates such as the sea urchin.

The rearrangement of cells that occurs during gastrulation is precisely ordered. This is apparent from the ability to construct fate maps of various embryos prior to gastrulation. From the location of a cell within the late blastula or early gastrula, one can accurately predict its movement, eventual location, and developmental fate within a normally developing embryo (Fig. 2-18). Although the presumptive fate of a cell can frequently be altered by experimental manipulation,

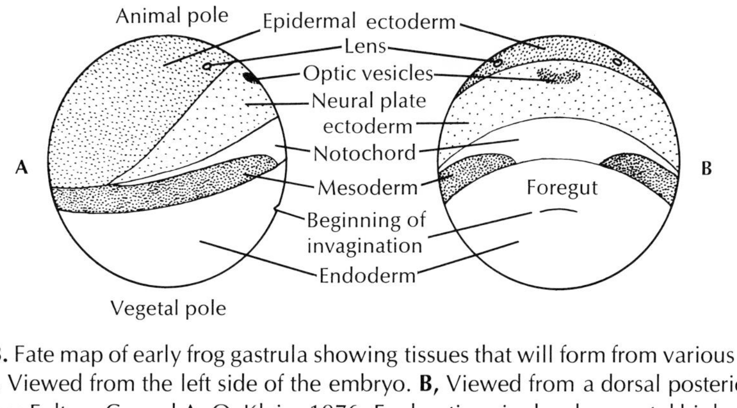

GASTRULATION
Developmental significance of gastrulation

Fig. 2-18. Fate map of early frog gastrula showing tissues that will form from various surface areas. **A,** Viewed from the left side of the embryo. **B,** Viewed from a dorsal posterior position. (From Fulton, C., and A. O. Klein. 1976. Explorations in developmental biology. Harvard University Press. Reprinted by permission.)

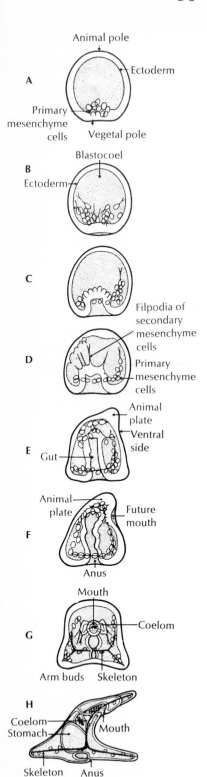

A

Animal pole

Ectoderm

Primary mesenchyme cells

Vegetal pole

B

Blastocoel

Ectoderm

C

D

Filpodia of secondary mesenchyme cells

Primary mesenchyme cells

E

Animal plate

Ventral side

Gut

F

Animal plate

Future mouth

Anus

G

Mouth

Coelom

Arm buds   Skeleton

H

Coelom

Stomach

Mouth

Skeleton   Anus

the cell's original fate is consistently achieved when development is allowed to proceed normally. Specific cellular mechanisms involved in gastrulation will be discussed in greater detail in Chapter 15. At this point we will concern ourselves only with a description of the basic process of gastrulation and the specific patterns of gastrulation that occur in selected embryonic systems.

Gastrulation in embryos with hollow blastulae

Gastrulation in the sea urchin embryo is a relatively simple process, somewhat analogous to punching in an underinflated basketball. Just prior to the actual beginning of gastrulation the vegetal pole of the blastula undergoes a flattening and thickening and its cells pulsate and appear to lose affinity for each other. Some of these cells move into the blastocoel and become the primary mesenchyme, which will give rise to the skeleton of the embryo. Cells at the vegetal end of the embryo, which is now termed a mesenchyme blastula, invaginate as a sheet, and the vegetal region indents into the blastocoel to a distance of about halfway to the animal pole (Fig. 2-19). At this point, cells at the animal end of the primitive gut formed by the invagination send out fine pseudopodia or filopodia that make contact with the roof of the blastocoel and pull the gut inward to its complete extent. These cells, which are termed secondary mesenchyme cells, detach and give rise to various mesodermal structures of the animal. Developmental stages in the sea urchin leading to formation of a feeding pluteus larva are summarized briefly in Figs. 2-1 and 2-19 and are discussed further in Chapter 15.

Gastrulation in the amphibian is more complex. Movement of the presumptive mesodermal and endodermal cells to the interior cannot

Fig. 2-19. Gastrulation and organogenesis in the sea urchin embryo. **A,** Blastula with primary mesenchyme beginning to separate from outer layer at the vegetal pole. For simplicity the apical ciliary tuft and the finer cilia over the entire embryo (Fig. 2-1) are not shown in this series. **B,** Late blastula with extensive separation of primary mesenchyme cells and slight inward curvature of the vegetal pole. **C,** Early gastrula with distinct invagination of the primitive gut. Primary mesenchyme cells are beginning to organize into skeletal elements. **D,** Midgastrula stage. Secondary mesenchyme cells can be seen stretching filopodia to the top of the blastocoel. **E,** Late gastrula with elongated gut. **F,** Early organogenesis with future mouth beginning to form. **G,** Slightly later stage with mouth and beginnings of arm buds. This view is from the side that has the mouth. This side is called ventral because it becomes the underside in the adult. **H,** Fully formed pluteus larva, with elongated arms and well-developed digestive system. Development beyond this stage requires an external source of nutrition. A complex process of metamorphosis transforms the pluteus into a radially symmetrical adult. (From Wolpert, L., and T. Gustafson. 1967. Endeavour **26**:86.)

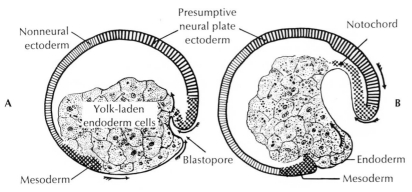

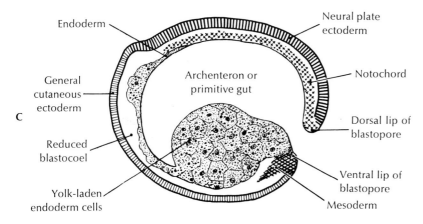

Fig. 2-20. Gastrulation in amphibian embryos. **A,** Early gastrulation, showing presumptive notochord cells just beginning to move past the dorsal lip of the blastopore to the interior of the embryo. **B,** Midgastrulation with the mesoderm just beginning to move to the interior at the ventral lip of the blastopore. **C,** Advanced stage of gastrulation, with ectoderm spread almost completely around the egg and nearly all mesoderm and endoderm on the interior. The size of the yolk mass has been reduced in this schematic representation for clarity. This stage corresponds to the yolk plug stages shown in Figs. 2-15, *D,* and 2-20, *D.* **D,** Median (sagittal) section of late gastrula, slightly rotated, showing spread of the endoderm and mesoderm to form a double-walled lining of the archenteron. Dorsal movement of the endoderm is shown by the solid arrow. Ventral movement of the mesoderm between the endoderm and ectoderm is shown by the dotted arrow. **E,** Neural fold stage (transverse section). The endoderm has formed a thin continuous layer over the dorsal side of the archenteron to give the primitive gut a complete endodermal lining. The mesoderm has spread completely around the endoderm and has differentiated into notochord, segmental plate (precursor to somites), and lateral mesoderm, which is beginning to separate into somatic and splanchnic layers, as described later in this chapter.

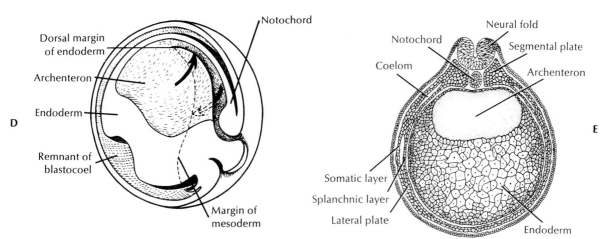

(**A to C** modified from Patten, B. M., and B. M. Carlson, 1974. Foundations of embryology, 3rd ed. McGraw-Hill Book Co., New York; **D** modified from Fulton, C., and A. O. Klein, 1976. Explorations in developmental biology. Harvard University Press. Reprinted by permission; **E** reprinted with permission of Macmillan Publishing Co., Inc. from Patterns and principles of animal development by Saunders, J. W., Jr. Copyright © 1970 by John W. Saunders, Jr.)

occur by a simple pushing-in type of invagination because of the presence of the thick yolky vegetal region. The more complicated cellular movements that occur during amphibian gastrulation can be readily followed, however, by marking cells with vital stains or carbon particles. One of the first events that is seen is a spreading of cells originating from the animal pole area over the surface of the egg. This process, which is called epiboly, reduces the thickness of the roof of the blastocoel from several layers to about two layers and results in a considerable increase in the area on the surface of the embryo that is covered by animal pole cells. The surface cells of the embryo, which are bound together in epithelial sheets, flow around the spherical egg and converge toward the newly formed dorsal lip of the blastopore. The dorsal

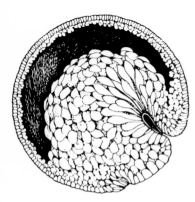

Fig. 2-21. Schematized section through an early amphibian gastrula showing elongated bottle-shaped cells. (From Trinkaus, J. P. 1969. Cells into organs—the forces that shape the embryo. Prentice-Hall, Inc., Englewood Cliffs, N.J. Reprinted by permission of Prentice-Hall, Inc., Englewood Cliffs, N.J.)

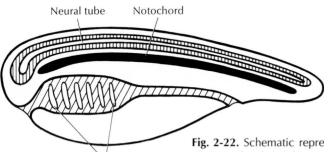

Neural tube   Notochord

Pharyngeal clefts

Fig. 2-22. Schematic representation of the diagnostic characteristics of a member of the phylum Chordata. Three properties, a notochord, a dorsal neural tube, and pharyngeal clefts, must be present in some part of the life cycle for an animal to be classified as a chordate. In humans and other mammals, only the dorsal neural tube remains prominent in the adult. The notochord serves an important inductive function in organizing the body axis (Chapter 13) and then regresses. The pharyngeal clefts and the blood vessels (aortic arches) that pass through the tissue between the clefts have significant roles during the course of development (Chapter 18), but the clefts are eliminated and the blood vessels greatly modified in the adult. (From Torrey, T. W. 1971. Morphogenesis of the vertebrates, 3rd ed. John Wiley & Sons, Inc., New York.)

lip of the blastopore appears as a cleft in the region originally occupied by the gray crescent, and it is here that cells first begin to involute from the surface to the interior (Figs. 2-15 and 2-20). Cells at the blastopore elongate and burrow interiorly into the presumptive endoderm. One end of each cell remains attached to the surface of the embryo for a while, and the cells assume a typical elongated shape from which they derive the name "bottle" cells (Fig. 2-21). As involution proceeds, the rim of the blastopore spreads toward the vegetal pole and eventually forms a complete circle that leaves just a few yolky cells, called the yolk plug, showing at the surface (Fig. 2-15, D). The yolk plug decreases in size as the rim of the blastopore contracts.

After their convergence to the blastopore and involution to the interior the cells continue their movements, sliding under the surface (ectodermal) layer to form a middle (mesodermal) layer of cells in characteristic locations. For example, presumptive notochord cells migrate from the blastopore, which is at the caudal end of the embryo, directly toward the cranial end. They come to rest along the dorsal midline, directly under the ectoderm that will later become the neural tube. In that location they become organized into the rod-shaped notochord, which is the distinguishing feature in the development of all members of the phylum Chordata from the most primitive protochordates to humans (Fig. 2-22).

The endoderm moves to the interior in a more passive manner as the ectoderm spreads around it, but once on the inside, it begins to organize itself into a tube that ultimately completely lines the primitive gut cavity (called the archenteron) that is formed by the inward movement of cells during gastrulation (Fig. 2-20, D). As the inward movements progress, the archenteron increases in size, and the blastocoel is gradually pushed aside and finally obliterated.

Presumptive mesoderm moving to the interior of the embryo on either side of the presumptive notochord, and also over the ventral lip of the blastopore at the yolk plug stage, slips into the space between the endoderm and the surface ectoderm (Fig. 2-20, D). As the presumptive mesoderm moves through the blastopore to the inside, the presumptive ectoderm, which originally occupied only a part of the surface of the embryo, thins and spreads until it covers the entire outer surface (Fig. 2-20, C).

Gastrulation in avian embryos is different from the process described for amphibian embryos, since it occurs entirely within the small disc of cells that lies atop the large yolk mass of the avian egg. In the chicken extensive cell division has already occurred before the egg is laid, and at the time of laying, the germinal disc is already composed

Primitive streak gastrulation

Fig. 2-23. Primitive streak gastrulation in the chicken embryo. **A,** Chicken blastoderm after 3 to 4 hours' incubation. The accumulation of cells in the posterior region that precedes primitive streak formation is already evident. **B,** Same blastoderm as in **A** after 5 to 6 hours of incubation. **C,** Same after 7 to 8 hours of incubation. **D,** Same after 10 to 12 hours of incubation. The accumulated cells have narrowed and elongated to yield a well-defined primitive streak. **E,** Median section of embryo after 17 hours of incubation. Mesodermal cells are now sinking below the surface in the primitive streak and spreading between the epiblast and the hypoblast. Cells moving in a cephalic direction from Hensen's node (which is functionally equivalent to the dorsal lip of the blastopore in amphibian embryos) condense to form a notochord (often called the "head process" in avian embryos). **F,** Transverse section at 17 hours, showing lateral movement of mesodermal cells. **G,** Schematic three-dimensional representation of the primitive streak embryo, cut in transverse section to show the spread of mesoderm between the epiblast and hypoblast. (**A** to **F** [**E** and **F** slightly modified] from Patten, B. M., and B. M. Carlson. 1974. Foundations of embryology, 3rd ed. McGraw-Hill Book Co., New York. Copyright © 1974 by McGraw-Hill, Inc. Used with permission of McGraw-Hill Book Co.; **G** from Balinsky, B. I. 1975. An introduction to embryology, 4th ed. W. B. Saunders Co., Philadelphia.)

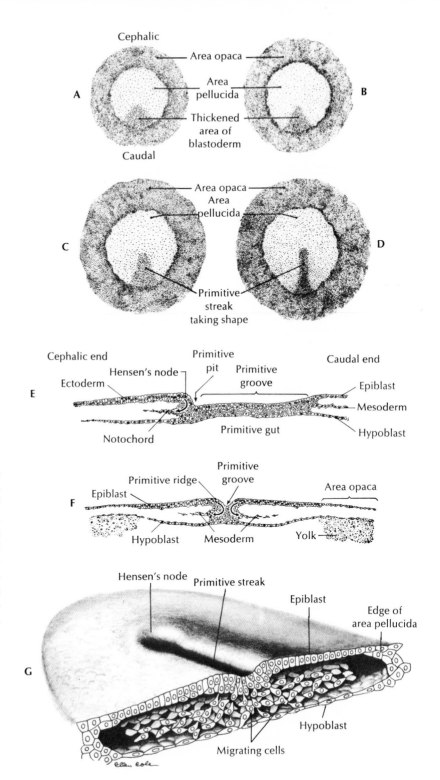

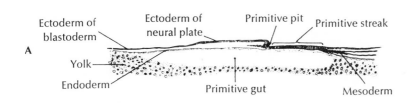

Ectoderm of
blastoderm

Ectoderm of
neural plate

Primitive pit

Primitive streak

A

Yolk

Endoderm

Primitive gut

Mesoderm

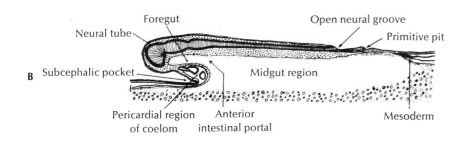

Foregut

Open neural groove

Neural tube

Primitive pit

B

Subcephalic pocket

Midgut region

Mesoderm

Pericardial region
of coelom

Anterior
intestinal portal

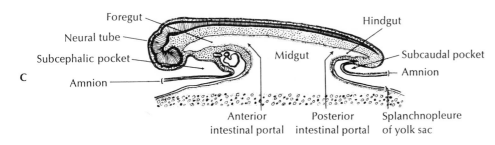

Foregut

Hindgut

Neural tube

Subcephalic pocket

Midgut

Subcaudal pocket

Amnion

C

Amnion

Anterior
intestinal portal

Posterior
intestinal portal

Splanchnopleure
of yolk sac

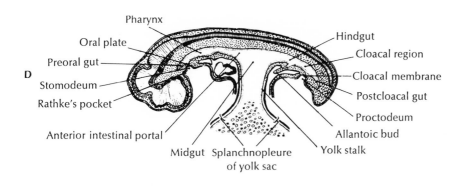

Pharynx

Oral plate

Hindgut

Preoral gut

Cloacal region

D

Stomodeum

Cloacal membrane

Rathke's pocket

Postcloacal gut

Proctodeum

Anterior intestinal portal

Allantoic bud

Yolk stalk

Midgut   Splanchnopleure
of yolk sac

Fig. 2-24. Formation of the gut in chicken embryos. **A,** Median section of the early primitive streak stage near the end of the first day of incubation. The embryo is still resting flat on the surface of the yolk with no regional differentiation of the primitive gut. **B,** Near the end of the second day the head fold has begun to lift the cephalic portion of the embryo off the surface of the yolk, establishing a foregut. **C,** At about 2½ days of incubation the tail fold has raised the caudal portion off the yolk, establishing a hindgut. The midgut, although still open to the yolk, has also begun to differentiate. **D,** At about 3½ days of incubation the head and tail folds are approaching one another and also beginning to close in laterally so that the embryo is now separated from the yolk at the end of a definite yolk stalk. The allantois, which later grows out to the inner surface of the egg shell and becomes an important respiratory organ (Fig. 2-30, C), is beginning to form as a bud from the hindgut. (Modified from Patten, B. M., and B. M. Carlson. 1974. Foundations of embryology, 3rd ed. McGraw-Hill Book Co., New York. Copyright © 1974 by McGraw-Hill, Inc. Used with permission of McGraw-Hill Book Co.)

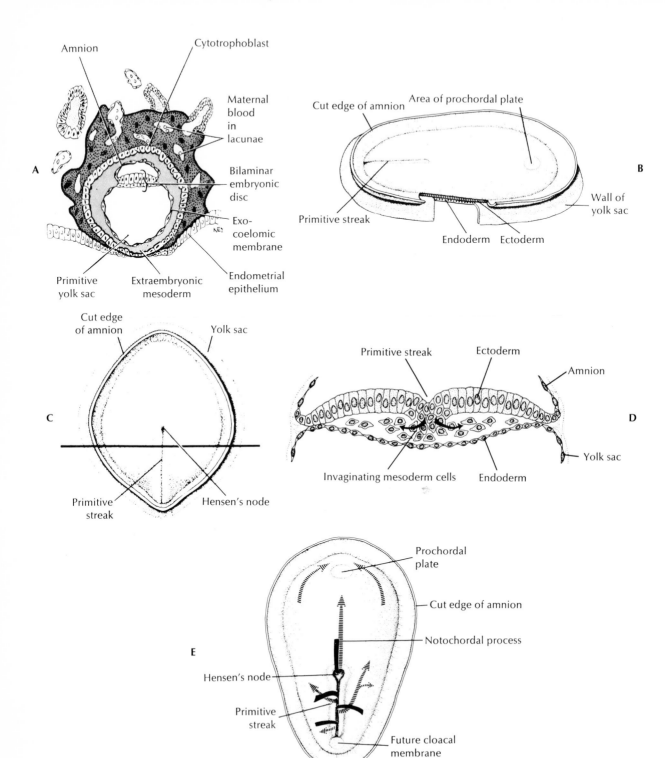

A

Amnion

Cytotrophoblast

Maternal blood in lacunae

Bilaminar embryonic disc

Exo-coelomic membrane

Endometrial epithelium

Extraembryonic mesoderm

Primitive yolk sac

B

Cut edge of amnion

Area of prochordal plate

Primitive streak

Endoderm   Ectoderm

Wall of yolk sac

C

Cut edge of amnion

Yolk sac

Primitive streak

Hensen's node

D

Primitive streak

Ectoderm

Amnion

Yolk sac

Invaginating mesoderm cells

Endoderm

E

Prochordal plate

Cut edge of amnion

Notochordal process

Hensen's node

Primitive streak

Future cloacal membrane

Fig. 2-25. Gastrulation in the human embryo. **A,** A human embryo, about 9 days after fertilization, implanted in the uterine wall. The amnion and yolk sac have already been formed, but the embryo proper is still at a bilaminar discoblastula stage, comparable to the blastoderm of the unincubated chicken egg. **B,** Schematic three-dimensional representation of the bilaminar embryonic disc at 14 days. Note the beginnings of the primitive streak. The prochordal plate is an area of tight adhesion between ectoderm and endoderm that ultimately becomes the mouth. **C,** Dorsal view of a 16-day embryo. **D,** Transverse section of a 16-day embryo, showing spread of mesoderm laterally from the primitive streak. Note the similarity to comparable stages in the chicken embryo (Fig. 2-23, *F*). **E,** Schematic representation of cell movement during primitive streak gastrulation in the human embryo, viewed from the dorsal surface. The solid lines represent cellular movement on the surface, and the dotted lines represent movement of mesodermal cells that have already passed through the primitive streak and are in the space between the ectoderm and endoderm. Note the movement of future notochord cells directly toward the cranial end of the embryo. (**A** from Moore, K. L. 1974. Before we are born—basic embryology and birth defects. W. B. Saunders Co., Philadelphia; **B** to **E** from Langman, J. 1969. Medical embryology, 2nd ed. The Williams & Wilkins Co., Baltimore.)

---

of two layers of cells, an upper epiblast and lower hypoblast. Cells on the surface of the epiblast move posteriorly and medially and form the primitive streak, a midline thickening with a central groove (Fig. 2-23). As migrating cells reach the primitive streak, they sink through it and migrate anteriorly and laterally between the epiblast and hypoblast and form a third layer called the mesoblast. Although groups of cells move coordinately, their movements are more individual than those of the tightly attached sheets of cells seen in the amphibian gastrula. Bottle-shaped cells are formed as cells move from the surface to the interior, as in the amphibian embryo. The presumptive notochord material enters the mesoblast layer at the anterior end of the primitive streak, which is called Hensen's node, and it moves in an anterior direction to form the notochordal plate and later a definitive notochord. Embryonic endoderm is formed from epiblast cells that invaginate through the primitive streak and assume a position over the hypoblast. The hypoblast forms the endodermal component of extraembryonic structures such as the yolk sac. As gastrulation proceeds in the bird embryo, cell invagination through the primitive streak ceases at the anterior end of the embryo while it is still continuing at the posterior end. This produces an apparent shrinkage and movement of the primitive streak toward the caudal end of the embryo, where it eventually disappears.

Although the primitive streak is analogous to the blastopore of other embryos in that it is the region at which surface cells move to the interior, it does not open into a gut cavity as do the blastopores of other embryos. Formation of the gut cavity in the avian embryo occurs in a different manner in which all three germ layers undergo a gradual folding-under, which occurs first in the head region and later in the tail and middle regions. These body folds contract in a purse-string fashion so that the embryo, which originally lies open on the mass of yolk, is eventually connected to it by only a thin stalk (Fig. 2-24).

Gastrulation in the mammalian embryo has not been studied as thoroughly as in lower vertebrates, but the basic process appears similar to that seen in its yolky relative, the avian embryo (Fig. 2-25). The inner cell mass becomes a flattened disc of columnar cells, called the epiblast. A layer of cells comparable to the hypoblast forms on the interior surface of the inner cell mass, but it is unclear whether it is derived from inner cell mass cells or trophoblast. An amnion is formed over the bilaminar embryonic disc at an early stage (Fig. 2-25, *A*). As in the bird, the hypoblast cells form a yolk sac, but in this case it contains no yolk. A primitive streak forms in the epiblast (Fig. 2-25, *B* and *C*), and gastrulation proceeds in a manner similar to that described for birds (Fig. 2-25, *D* and *E*).

BEGINNING OF ORGANOGENESIS

Gastrulation lays out the basic organizational framework of the organism. It sets the stage for cellular differentiation and morphogenesis and the orderly progression of developmental events that will produce a complete new individual. After gastrulation the tissues of the embryo are arranged in the overall pattern required for the entire sequence of organ formation (Fig. 2-26). Many of the mechanisms involved in producing the complex, fully formed individual from the relatively simple gastrula will be described in greater detail in Sections Three and Four. In order that they may be placed in perspective, we will briefly outline here some of the major events that occur in the three germ layers during vertebrate organogenesis.

Ectoderm

The first major event in vertebrate organogenesis is neurulation, or formation of the neural tube, the structure that gives rise to the brain and spinal cord (Fig. 2-27). In birds and mammals neurulation begins in the anterior part of the embryo while gastrulation is still proceeding in more posterior regions (Fig. 2-28). In amphibia the process is similar (Fig. 2-20, *E*) except that it occurs more synchronously. Ectoderm on the dorsal side of the embryo thickens to form the neural plate. The edges of the plate become raised and form the neural folds, which fuse in the midline, detach from the overlying ectoderm, and form the neural tube (Fig. 2-29). Neural crest cells, which are left between the neural tube and the overlying skin ectoderm after tube formation, migrate to various parts of the body and are the source of pigment cells, cartilage of the face, and various ganglia (Fig. 2-29). The noto-

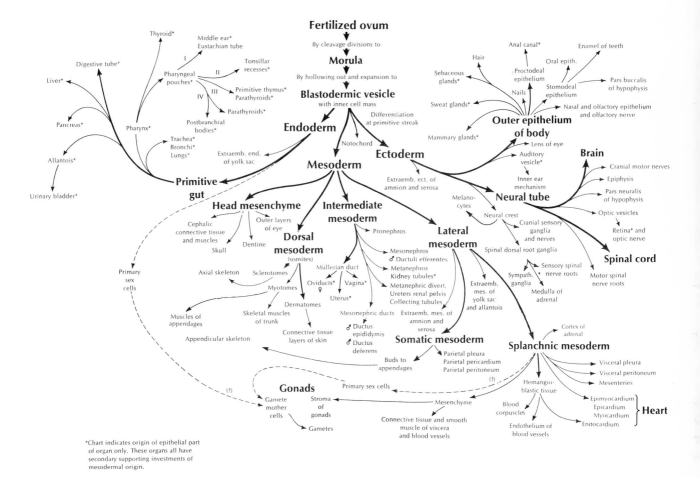

Fig. 2-26. Developmental origins of various parts of the body. Note that the origin of all organs and systems in the adult body can be traced back to the three primary germ layers. (Modified from Patten, B. M., and B. M. Carlson. 1974. Foundations of embryology, 3rd ed. McGraw-Hill Book Co., New York. Copyright © 1974 by McGraw-Hill, Inc. Used with permission of McGraw-Hill Book Co.)

chord plays an important role in inducing neural differentiation and organization of the body axis, as will be discussed in Chapter 13.

The anterior part of the neural tube develops into the various regions of the brain. Aside from forming the definitive brain, these regions also interact with ectoderm of the head to form the primordia of the optic, auditory, and olfactory organs. Cranial ganglia are produced by cells that detach from the brain tube along with neural crest cells. The more posterior part of the neural tube gives rise to the spinal cord. The ectoderm that encloses the embryo is the source of the outer layer of the skin and its derivatives such as hair and feathers.

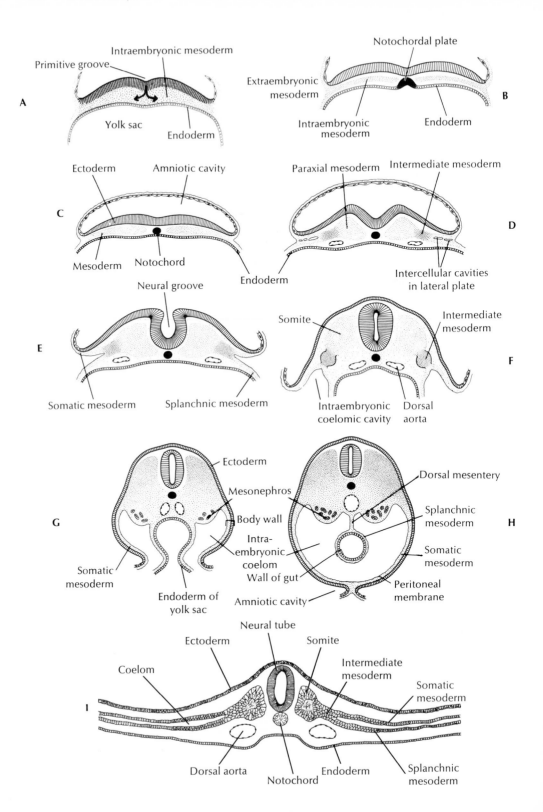

Fig. 2-27. Formation of the neural tube and the gut in human and avian embryos. Note that primitive streak gastrulation proceeds in a cephalic to caudal direction (Fig. 2-28) and that several of the stages shown here can represent either progression through time or movement from caudal to more cephalic positions in the same embryo. All illustrations represent transverse sections. **A,** Primitive streak gastrulation in the human embryo (comparable to Fig. 2-25, *D*). **B,** Early stages of organization of the notochord as its cells move in a cephalic direction from Hensen's node. **C,** Later stage with well-formed cylindrical notochord. **D,** Neural folding in progress. Early stages of separation of somatic and splanchnic mesodermal layers can also be seen. **E,** Nearly closed neural tube. The layers of the lateral mesoderm are clearly separated. **F,** Neural tube completely closed. Somites have formed as condensations of the mesoderm closest to the neural tube, and the intermediate mesoderm has become distinct from the lateral (somatic plus splanchnic) mesoderm. **G,** Partial closure of the gut and ventral side of the body. The intermediate mesoderm has given rise to a primitive kidney structure, the mesonephros (Chapter 18). **H,** Complete closure of the gut and body wall. The basic arrangement of the trunk section of the human body has now been achieved with both the gut and the peritoneal cavity fully closed. This represents about 2 weeks of development beyond the early primitive streak stage or a total of about 4 weeks from fertilization. **I,** Chicken embryo soon after closure of the neural tube (about 28 hours of incubation). Note the overall similarity to the comparable stage of human development shown in **F.** (**A** to **H** from Langman, J. 1975. Human embryology, 3rd ed. The Williams & Wilkins Co., Baltimore; **I** reprinted with permission of Macmillan Publishing Co., Inc. from Animal morphogenesis by Saunders, J. W., Jr. Copyright © 1968 by John W. Saunders, Jr.)

Mesoderm

The mesoderm occupies the middle position between ectoderm and endoderm after gastrulation, and it provides the greatest mass of body tissues. The mesoderm that underlies the presumptive neural tube forms the cylindrical notochord that extends along the anterior-posterior axis of the embryo. It is primarily an embryonic structure and makes only a minor contribution to the axial skeleton of the adult vertebrate.

The somites are segmented blocks of mesoderm that condense along each side of the notochord. These form the axial skeleton and the skeletal muscles, except those of the head, and also contribute to the connective tissue layer of the skin. The myotome, or muscle-producing portion of the somite, originates in a dorsal position but then spreads ventrally just under the surface until the entire body cavity is surrounded by muscle tissue. The myotome also contributes to limb musculature in at least some species.

The intermediate mesoderm is further from the midline than the somites (Fig. 2-27). It gives rise to the urinary tract, gonads, reproductive ducts, and adrenal cortex. The mesoderm in the most lateral position is called the hypomere or simply the lateral mesoderm and is split into two layers. The outer layer, which is referred to as somatic or parietal, is the source of the peritoneum that lines the body cavity and is

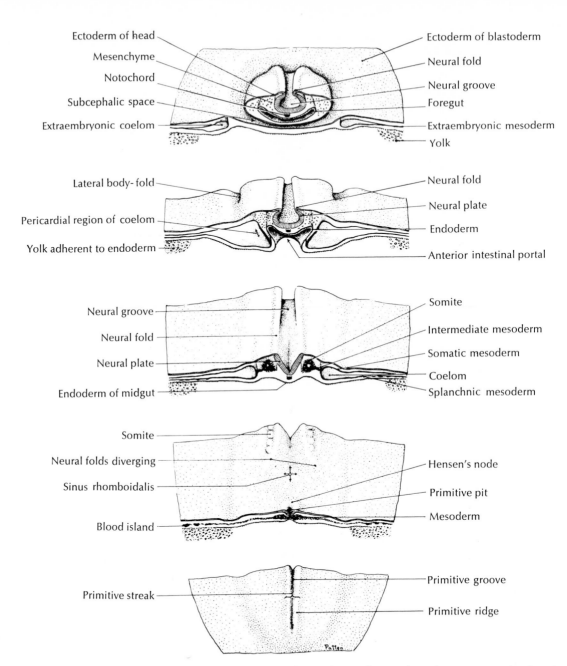

Ectoderm of head
Mesenchyme
Notochord
Subcephalic space
Extraembryonic coelom

Ectoderm of blastoderm
Neural fold
Neural groove
Foregut
Extraembryonic mesoderm
Yolk

Lateral body-fold
Pericardial region of coelom
Yolk adherent to endoderm

Neural fold
Neural plate
Endoderm
Anterior intestinal portal

Neural groove
Neural fold
Neural plate
Endoderm of midgut

Somite
Intermediate mesoderm
Somatic mesoderm
Coelom
Splanchnic mesoderm

Somite
Neural folds diverging
Sinus rhomboidalis
Blood island

Hensen's node
Primitive pit
Mesoderm

Primitive streak

Primitive groove
Primitive ridge

Patten

Fig. 2-28. Stereogram of chicken embryo after 24 hours of incubation. A nearly closed neural tube and a lifting of the head from the yolk by the head fold (Fig. 2-24) are evident at the cephalic end of the embryo while primitive steak gastrulation is still in progress at the caudal end. Somite formation and early mesodermal differentiation are clearly evident in the middle sections. (From Patten, B. M., and B. M. Carlson. 1974. Foundations of embryology, 3rd ed. McGraw-Hill Book Co., New York. Copyright © 1974 by McGraw-Hill, Inc. Used with permission of McGraw-Hill Book Co.)

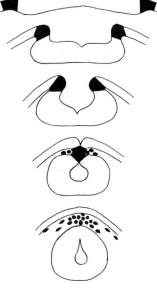

Fig. 2-29. Origin of neural crest cells. Successive stages in the closure of the neural tube in amphibian embryos are shown. The cells that become the neural crest are shown in black. As the closed neural tube separates from the overlying ectoderm, the neural crest cells become separated from both. They are highly migratory and ultimately differentiate into a variety of forms scattered through the body, as indicated in the text and in Fig. 2-26. (From Balinsky, B. I. 1975. An introduction to embryology, 4th ed. W. B. Saunders Co., Philadelphia.)

also the source of mesoderm for the early outgrowth of the limb bud. The inner layer, which is referred to as splanchnic or visceral, produces the heart and the outer part of the digestive tube and its derivatives and mesenteries. The mesoderm is also the source of the body's connective tissue and blood vessels.

All parts of the limbs except their epidermal covering are of mesodermal origin. The limb bud originates as a slight thickening of the dorsal part of the somatic (parietal) layer of lateral mesoderm, followed by separation of loose mesenchymal cells from it. These mesenchymal cells aggregate and push the ectoderm outward, resulting in the formation of a limb bud. Complex inductive interactions between the limb ectoderm and mesoderm lead to rapid outgrowth of the developing limb and sequential determination of structures along its proximal-distal axis. Limb development is described in detail in Chapter 21.

Endoderm

The innermost germ layer of the gastrula, the endoderm (or entoderm), provides the functional inner layer of the digestive tube and all its derivatives such as the liver and pancreas. Endodermal evaginations from more anterior parts of the digestive tube give rise to the lung buds as well as various glands, including the thyroid, thymus, and anterior pituitary. Finally, in many species the endoderm is the source of the primordial germ cells, which, as described in Chapter 19, are first seen in a location remote from the mesodermal gonad and colonize it only after extensive migration.

Extraembryonic membranes

The development of both avian and mammalian embryos includes extensive development of extraembryonic structures that function in nutritional, excretory, and respiratory roles and are discarded when the animal hatches or is born. Four distinct structures, the amnion, the chorion, the yolk sac, and the allantois, are involved. These structures are illustrated schematically for mammals in Fig. 2-30, *A*. The

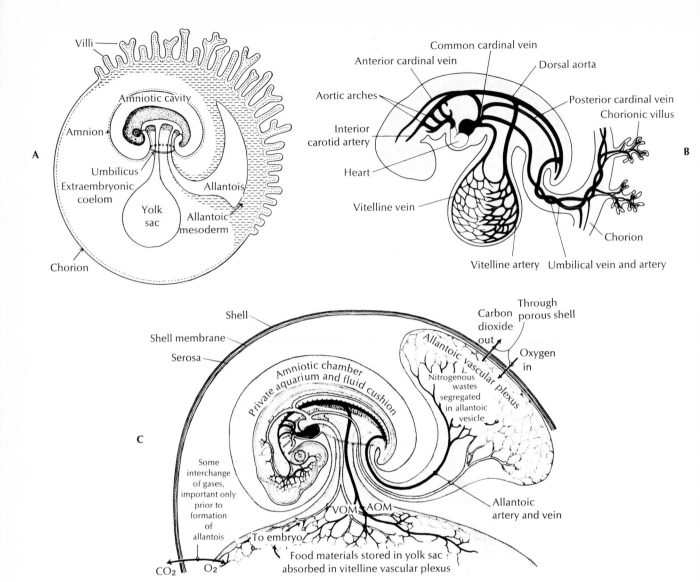

Fig. 2-30. Extraembryonic membranes of avian and mammalian embryos. **A,** Generalized diagram of the extraembryonic membranes of a placental mammal. **B,** Circulatory patterns in extraembryonic membranes of the human embryo. The allantoic circulation spreads through the villi of the placenta, which are shown schematically in **A,** and exchanges respiratory gases, nutrients, and waste products with the maternal circulation. **C,** Circulatory patterns in the extraembryonic membranes of the chicken embryo. Note the similarity with the mammalian patterns in **A** and **B.** The allantois serves both as a respiratory organ and a repository for nitrogenous wastes. Except for the fact that nutrients are derived from the yolk via the yolk sac in the avian embryo and from the mother via the allantoic circulation in the mammal, the functions of the extraembryonic membranes are similar in both. *AOM* = Omphalomesenteric artery; *VOM* = omphalomesenteric vein. (**A** from Balinsky, B. I. 1975. An introduction to embryology, 4th ed. W. B. Saunders Co., Philadelphia; **B** from Langman, J. 1969. Medical embryology, 2nd ed. The Williams & Wilkins Co., Baltimore; **C** modified from Patten, B. M., and B. M. Carlson. 1974. Foundations of embryology, 3rd ed. McGraw-Hill Book Co. Copyright © 1974 by McGraw-Hill, Inc. Used with permission of McGraw-Hill Book Co.)

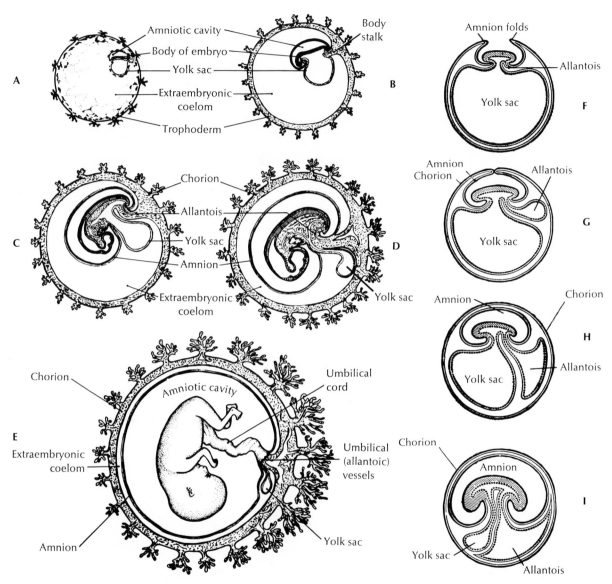

Fig. 2-31. Progressive development of extraembryonic membranes in human and chicken embryos. **A** to **E**, Human. **F** to **I** Chicken. Note that the yok sac is initially an extension of the midgut in both and that the allantois begins as a bud from the hindgut. As development proceeds, the midgut becomes closed, leaving the allantois and the yolk sac connected to the embryo by a stalk (called the umbilical cord in the case of mammals). Also note the similarity of overall pattern as development proceeds (e.g., compare **D** and **I**). (**A** to **E** from Patten, B. M., and B. M. Carlson. 1974. Foundations of embryology, 3rd ed. McGraw-Hill Book Co., New York. Copyright © 1974 by McGraw-Hill, Inc. Used with permission of McGraw-Hill Book Co.; **F** to **I** from Arey, L. B. 1974. Developmental anatomy, 7th ed. W. B. Saunders Co., Philadelphia.)

striking similarity of circulatory patterns in the yolk sac and allantois of mammals and birds can be seen by comparing Fig. 2-30, *B* (human embryo) and *C* (chicken embryo). The sequential development of human and chicken extraembryonic membranes is shown in Fig. 2-31.

The amnion is a protective fluid-filled sac that completely surrounds the developing embryo. It is formed either by a folding process, as shown in Fig. 2-31, *F* for birds, or by separation of the inner cell mass from the overlying trophoblast in some mammals (Fig. 2-25, *A*).

The chorion is the outermost membrane surrounding the embryo. In mammals it is derived initially from the trophoblast and becomes a major portion of the placenta, which is the embryonic organ that exchanges nutrients, oxygen, and wastes between the maternal and fetal circulations (Fig. 2-31, *A* to *E*). In birds it is derived from the same folding process that gives rise to the formation of the amnion (Fig. 2-31, *F* to *I*).

The yolk sac in birds grows over the surface of the yolk and becomes filled with blood vessels, which transport stored nutrients from the yolk to the embryo. Early blood cell differentiation also occurs in the yolk sac. At one stage in their development mammalian embryos also have a highly vascularized yolk sac. A portion of the mammalian yolk sac becomes the gut (Fig. 2-27, *G*, and *H*) and the yolk sac is also involved in blood formation in mammalian embryos.

The allantois arises as an outgrowth of the caudal portion of the gut. In birds it becomes highly vascularized (Fig. 2-30, *C*) and forms a layer just under the eggshell membrane, where it functions as a respiratory organ. Its membranes fuse with those of the chorion to generate the highly vascularized chorioallantoic membrane, which is often used experimentally as a site for short-term growth of transplanted tissues. The hollow allantoic vesicle also serves as a storage area for nitrogenous wastes. In mammals a similar outgrowth follows the connecting stalk between the chorion and the embryo proper. Vasculature similar to that in the allantoic stalk of birds connects the chorionic placenta to the embryo and becomes the vasculature of the umbilical cord as the embryo develops (Fig. 2-30, *B* and Fig. 2-31, *E*).

OVERVIEW   This chapter has summarized the brief segment of the life cycle of typical multicellular animals in which the most intense developmental activity takes place. The following events all occur within a very short time (relative to the total life span of the diploid individual) after initiation of the diploid phase of the life cycle by the union of the male and female gametes:

1. Repeated *cleavage* of the egg cell reduces the average cell size to approximately that found in the adult organism and provides a

sufficient number of cells for differentiation and morphogenesis to form miniature replicas of adult structures.

2. Complicated rearrangement of the cells, accompanied by the beginnings of differentiation, results in the formation of a hollow or flattened *blastula* or *discoblastula*. This is followed by movement of cells to the interior of that structure to generate a *gastrula* containing the three primary germ layers, *ectoderm*, *mesoderm*, and *endoderm*, from which all adult structures are subsequently derived.

3. The establishment of the *main axis* of the body and the carrying out of an intense program of *organogenesis* lead to formation of all major adult structures, at least in rudimentary form.

The sequential nature of these developmental processes is summarized in Fig. 2-26, which traces the developmental histories of the major systems of the vertebrate body. Fig. 2-1 compares the patterns of development of four different species, three vertebrate and one invertebrate, emphasizing the basic similarities that can be seen, despite major differences in details and final results.

BIBLIOGRAPHY
Books and reviews

Arey, L. B. 1974. Developmental anatomy, 7th ed. W. B. Saunders Co., Philadelphia.

Balinsky, B. I. 1975. An introduction to embryology, 4th ed. W. B. Saunders Co., Philadelphia.

Berrill, N. J., and G. Karp. 1976. Development. McGraw-Hill Book Co. New York.

Deuchar, E. M. 1975. Cellular interactions in animal development. Chapman & Hall Ltd., London.

Gilchrist, F. G. 1968. A survey of embryology. McGraw-Hill Book Co., New York.

Graham, C. F., and P. F. Wareing, eds. 1976. The developmental biology of plants and animals. W. B. Saunders Co., Philadelphia.

Hamilton, W. J., and H. W. Mossman. 1972. Hamilton, Boyd and Mossman's human embryology—prenatal development of form and function, 4th ed. The Williams & Wilkins Co., Baltimore.

Johnson, K. E. 1974. Gastrulation and cell interactions. *In* J. Lash and J. R. Whittaker, eds. Concepts of development. Sinauer Associates, Inc., Sunderland, Mass.

Johnson, M. H., ed. 1977. Development in mammals. Vols. 1 and 2. North-Holland Publishing Co., Amsterdam.

Langebartel, D. A. 1977. The anatomical primer—an embryological explanation of human gross morphology. University Park Press, Baltimore.

Langman, J. 1975. Medical embryology: human development—normal and abnormal, 3rd ed. The Williams & Wilkins Co., Baltimore.

McLaren, A. 1972. The embryo. *In* C. R. Austin and R. V. Short, eds. Reproduction in mammals. Book 2. Embryonic and fetal development. Cambridge University Press, Cambridge.

Moore, K. L. 1974. Before we are born: basic embryology and birth defects. W. B. Saunders Co., Philadelphia.

Patten, B. M., and B. M. Carlson. 1974. Foundations of embryology, 3rd ed. McGraw-Hill Book Co., New York.

Saunders, J. W., Jr. 1970. Patterns and principles of animal development. The Macmillan Co., New York.

Stearns, L. W. 1974. Sea urchin development: cellular and molecular aspects. Dowden, Hutchinson & Ross, Inc., Stroudsburg, Pa.

Van Blerkom, J. and P. Motta. 1979. The cellular basis of mammalian reproduction. Urban and Schwarzenberg, Munich.

CHAPTER 3

Cellular differentiation

☐ The previous chapter described the development of a complex organism from a single fertilized egg cell, with emphasis on the progressive generation of shape (morphogenesis) and the formation of specialized functional structures (organogenesis). These developmental processes occur in a stepwise sequential manner, beginning with cleavage and followed by gastrulation and separation of the three primary germ layers and then by generation of many different types of tissues from these germ layers, often as a result of cooperative interactions. The developmental history of many of the organs that are formed in the mammalian body is summarized in Fig. 2-26.

A closer examination of these organs and the individual tissues that comprise them reveals the presence of many different types of cells (Fig. 3-1). Each organ is composed of particular types of cells best suited for its specialized functions. Estimates of the total number of different types of cells that can be identified in the mammalian body vary considerably with the criteria used for identification. The distinctions between different types of cells range from obvious morphological differences such as those between skeletal muscle cells and secretory glandular cells to subtle variations such as those between two

Fig. 3-1. Cellular differentiation. The diversity of size and morphology that occurs among differentiated cells in the mammalian body is illustrated. All cells are drawn at the same magnification. The scale bar is 20 μm (0.02 mm). **A,** Portion of multinucleate striated muscle cell. **B,** Schwann cell with its membranes wrapped around a nerve axon to form a myelin sheath. **C,** Large neuron of reticular formation. **D,** Purkinje cell of cerebellar cortex. **E,** Collagen-forming cell from fibrous connective tissue. **F,** Smooth muscle cells. **G,** Rods and cones from retina of eye with cell bodies. **H,** Ciliated epithelial cell. **I,** Polymorphonuclear neutrophil (a type of white blood cell). **J,** Antibody-secreting cell from lymph node. **K,** Cross section of capillary (composed of a single endothelial cell) with a red blood cell in the lumen. **L,** Red blood cells. **M,** Chondrocyte. **N,** Hepatocyte. **O,** Mucous cell of stomach. **P,** Fat cell from loose connective tissue with large lipid inclusion. **Q,** Oocyte and surrounding follicle cells prior to ovulation. **R,** Spermatozoon.

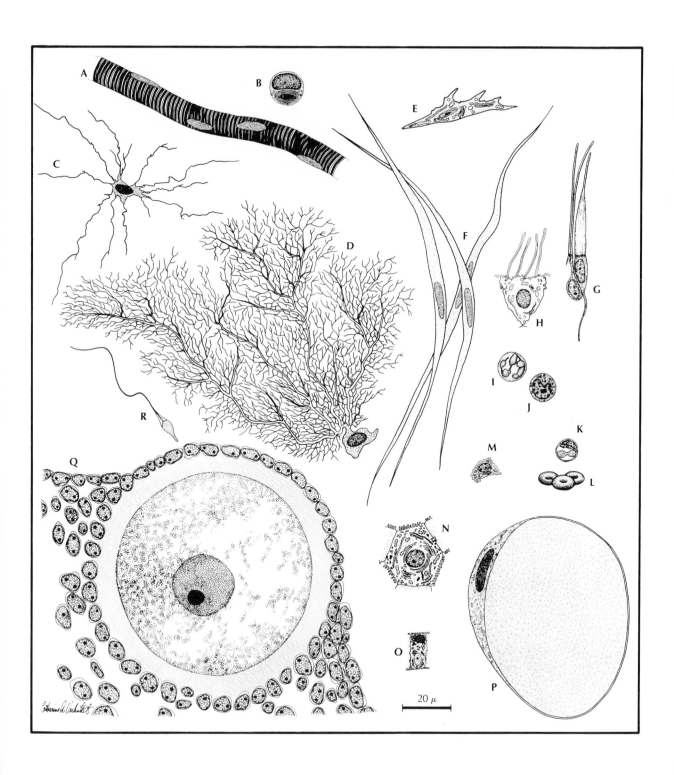

20 μ

types of glandular cells that are morphologically similar but secrete different products. The most conservative count, based on morphological criteria alone, is somewhat in excess of a hundred different types of cells in the mammalian body, and that number becomes considerably larger if subtle biochemical and functional criteria are also taken into consideration.

The process whereby many different types of cells are formed from a single fertilized egg cell during development is called cellular differentiation. The basic concept is relatively simple — generation by a single parental cell of daughter cells that differ from one another. However, as we shall see later in this chapter, a precise definition of cellular differentiation is much more difficult. The major problems that are involved in defining differentiation are that (1) differentiation is a stepwise process with many intermediate stages that are neither undifferentiated nor fully differentiated; (2) environmental conditions affect expression of differentiated properties, particularly in cultured cells; (3) commitment to a particular pathway of differentiation is not always accompanied by expression of the particular set of differentiated properties associated with that pathway; and (4) the simple definition just given does not exclude a number of phenomena that are not normally considered to be examples of differentiation (e.g., mutation in unicellular organisms).

CRITERIA OF DIFFERENTIATION

Before we attempt to define cellular differentiation more precisely, it will be useful to examine the criteria commonly employed to distinguish one type of cell from another. These criteria fall into three broad classes: structure, chemistry, and function. Although we will examine them one at a time, it is important to remember that they are interrelated and that any particular differentiated state is clearly reflected in all three types of properties. In fact, one definition that has been offered for differentiation is "the collective specializations of structure and chemistry that equip the nucleus and cytoplasm for a particular array of functional activities."*

The criteria of differentiation that are discussed in the following paragraphs apply primarily to cells that fully express their particular set of specialized properties. Cases in which a cell is determined or committed to a particular kind of differentiation but is not fully expressing it involve more subtle intracellular changes and must be evaluated in terms of different criteria, which will be discussed later in this chapter and in more detail in Chapters 10 to 12.

---

*Hay, E. 1966. Embryologic origin of tissues. *In* R. O. Greep, ed. Histology, 2nd ed. McGraw-Hill Book Co., New York.

The primary tools for examining cellular structure are the light and electron microscopes. A detailed discussion of the theory and practice of microscopy is beyond the scope of this book. However, some of the major techniques used in light and electron miscroscopy have been summarized in the following outline, together with information on procedures that extend the range of usefulness of microscopy by allowing data on the chemistry and function of individual cells and their organelles to be collected through microscopy.

Cellular structure

A. Light microscopy (LM). The following techniques are based on use of light wave optics to achieve magnification and, in some cases, image enhancement. Resolution is limited by the wavelength of light.
   1. Bright field. This technique is most effective with sections cut from preserved material. The technique depends on staining for contrast. It is generally not effective for unstained living specimens.
   2. Ultraviolet. This technique is the same as bright field, except ultraviolet illumination is used. No stain is needed for structures that contain nucleic acids. Theoretical resolution is greater; television camera or photographic film is used to detect image.
   3. Fluorescent. The object is illuminated with ultraviolet light and viewed by fluorescent emission of visible light. The technique is used with natural fluorescence and with fluorescent stains, including antibodies with attached fluorescent molecules.
   4. Dark field. Intense illumination shines on the object but does not enter the objective lens of the microscope. The object is seen illuminated on a dark background. This is useful for objects too small to resolve in bright-field microscopy.
   5. Phase contrast. Phase differences caused by slower passage of light through the object are used to enhance contrast by constructive or destructive interference. This is particularly useful for observing living specimens without staining.
   6. Interference contrast. The principle is similar to phase contrast, but more complex optics and a split light beam are used to obtain image contrast. It can also provide measurement of the mass of objects viewed. Nomarski differential interference contrast produces particularly vivid images with a distinctive three-dimensional appearance.
   7. Polarization. The object is illuminated with polarized light and viewed through a polarizing filter at right angles. Ordered arrays of molecules in crystalline and paracrystalline cellular inclusions appear illuminated on a dark background.
B. Electron microscopy (EM). The following techniques are based on use of the optical properties associated with accelerated electrons to achieve the desired magnification. Much greater resolution is possible than with the light microscope.
   1. Transmission. This technique is analogous to bright field light microscopy. Very thin sections are used, and staining with electron opaque materials (e.g., heavy metals) enhances contrast.
   2. High voltage. A powerful electron beam penetrates thicker specimens. Thick sections or even whole cultured cells can be observed. Stereoscopic images can be obtained by taking two pictures with the specimen tilted

at slightly different angles. Three-dimensional reconstruction of intracellular structures at a high level of resolution is possible.

3. Scanning. Secondary electrons emitted by surfaces as a sharply focused electron beam scans across them are collected and converted to an image on a television tube. Features of the cellular surface can be visualized precisely with a three-dimensional appearance. Stereoscopic images can be obtained by taking two pictures at different angles of specimen orientation.

C. Special microscope techniques. Special techniques of specimen preparation extend the usefulness of microscopy. Most of the following techniques can be used in both LM and EM, although freeze fracture requires the high resolution of EM.

1. Cytochemistry. Judicious choice of stains together with special biochemical techniques make possible selective staining of components of cells that have particular biochemical properties. It is useful in both LM and EM, although a wider range of techniques are available for LM.

2. Autoradiography. A thin coating of photographic emulsion detects the localization of radioactive isotopes within thin sections of biological material. Selective incorporation of labeled biochemicals provides information concerning biosynthesis and differentiated biochemical functions in cells and their organelles.

3. Labeled antibodies. Antibodies prepared against cellular antigens bind selectively to those antigens. Bound antibodies can be visualized with fluorescence microscopy by chemical attachment of fluorescent molecules or in EM by attachment of large electron-dense molecules such as ferritin.

4. Partial isolation. Structures of interest are sometimes visualized by destroying or disrupting other parts of the cell. In LM hypotonic swelling and controlled rupture of cells is used to obtain well-separated mitotic chromosomes. In EM, DNA and RNA are often partly or completely separated from proteins so that their extended structures can be seen.

5. Freeze fracture. When frozen cells are fractured, breakage occurs preferentially in the hydrophobic centers of membranes. When this is followed by subliming of the ice, structures that are not easily seen by other means are revealed.

D. X-ray diffraction. This technique is particularly useful for crystals and also can sometimes be used to provide direct information concerning cellular structure at levels below the resolution of the electron microscope. The double helical structure of DNA was originally resolved by x-ray diffraction.

Prior to the introduction of the electron microscope, bright-field microscopy of sectioned and stained material was the primary research tool of histology (the study of tissues and the cells from which they are formed). The resolving power of the light microscope is limited by the wavelength of light, and unless special techniques such as dark-field or ultraviolet microscopy are employed, structures smaller than about one fifth of a micron (μ), or micrometer (μm), in diameter cannot be seen (1 μm $= 10^{-6}$ m or 10^{-3} mm). The electron microscope has a much greater resolving power, and, under favorable conditions, individual molecules of proteins or nucleic acids can be visualized, although this is usually not possible within an organized structure.

It is important to remember during the discussion of cellular structure that cells are composed of molecules and that each structure that is discussed is in fact an aggregate of molecules. The discussions themselves will generally be organized in terms of the particles, fila-

Fig. 3-2. Cellular ultrastructure. The drawing is a three-dimensional representation of a hypothetical cell prominently displaying a wide variety of organelles and ultrastructural features seen in many different types of cells with the electron microscope. (From Anthony, C. P., and N. M. Koltoff. 1975. Textbook of anatomy and physiology, 9th ed. The C. V. Mosby Co., St. Louis.)

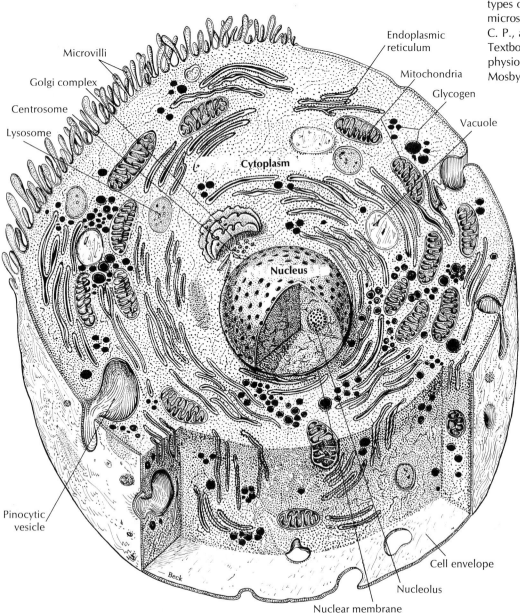

Microvilli

Golgi complex

Centrosome

Lysosome

Endoplasmic reticulum

Mitochondria

Glycogen

Vacuole

Cytoplasm

Nucleus

Pinocytic vesicle

Beck

Cell envelope

Nucleolus

Nuclear membrane

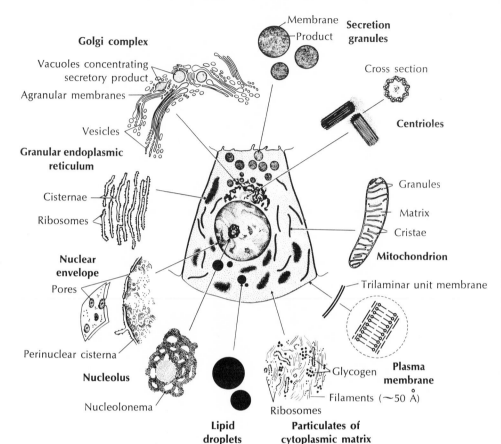

Golgi complex
Vacuoles concentrating secretory product
Agranular membranes
Vesicles

Granular endoplasmic reticulum
Cisternae
Ribosomes

Nuclear envelope
Pores

Perinuclear cisterna

Nucleolus

Nucleolonema

Lipid droplets

Membrane
Product
Secretion granules

Cross section

Centrioles

Granules
Matrix
Cristae
Mitochondrion

Trilaminar unit membrane

Glycogen
Plasma membrane
Filaments (~50 Å)

Ribosomes
Particulates of cytoplasmic matrix

Fig. 3-3. Features of a pancreatic acinar cell as seen with light and electron microscopy. In the center are illustrated the major features observed in a pancreatic acinar cell with light microscopy. Around the periphery are representations of the same components as they are seen with the electron microscope. Many of the individual features are described in the text. (From Bloom, W., and D. W. Fawcett. 1975. A textbook of histology, 10th ed. W. B. Saunders Co., Philadelphia.)

ments, membranes, and more complex organelles that can actually be seen with the electron microscope, rather than directly in terms of molecules, whose role is either inferred or demonstrated only by special microscopic techniques.

The major structures that can be seen in mammalian cells are illustrated in Figs. 3-2 to 3-5. Other animal cells tend to be similar, and many of the same organelles are also found in plant cells. Fig. 3-2 is a composite three-dimensional view of the arrangement of structures found in mammalian cells in general. Fig. 3-3 depicts the major structural features that can be seen with the light microscope, together with drawings of their appearance at higher magnification in the electron microscope. Fig. 3-3 shows a rather typical secretory cell, the pancreatic acinar cell, which synthesizes and secretes zymogen precursors for many of the digestive enzymes that function in the small intestine. Fig. 3-4 depicts a hepatocyte, or liver cell, which has many

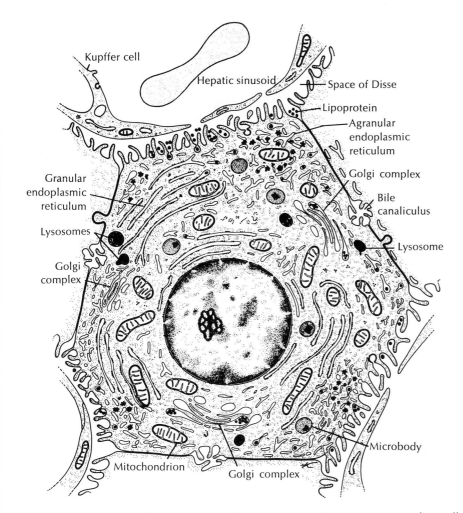

Labels on figure:

Kupffer cell

Hepatic sinusoid

Space of Disse

Lipoprotein

Agranular endoplasmic reticulum

Golgi complex

Bile canaliculus

Lysosome

Granular endoplasmic reticulum

Lysosomes

Golgi complex

Microbody

Mitochondrion

Golgi complex

Fig. 3-4. Diagram of a liver parenchymal cell and its relationship to surrounding cells in the liver. The complexity of metabolic functions performed by liver cells is indicated by the prominence of both rough and smooth endoplasmic reticulum, as well as prominent mitochondria. Complex surface specializations are also evident. Where two parenchymal cells come together, a bile canaliculus connected to the bile duct system is formed, with tight junctions on either side (see Fig. 3-13). The hepatic sinusoids are blood-filled cavities lined with endothelial cells and phagocytic Kupffer cells. The surfaces of the liver cell adjacent to the hepatic sinusoids are covered with irregularly oriented microvillae whose tips rest lightly against the endothelial cells, with an intervening space through which blood plasma can flow freely as a result of openings or pores in the endothelial lining of the sinusoid. (From Bloom, W., and D. W. Fawcett. 1975. A textbook of histology, 10th ed. W. B. Saunders Co., Philadelphia.)

Fig. 3-5. Diagram of the ultrastructural features of a myelocyte. This cell is a precursor of the most common type of white blood cell, the neutrophilic leukocyte, whose major function appears to be to remove the foreign particulate material from the bloodstream by phagocytosis. Membrane-bound lysosomes, which are filled with intracellular digestive enzymes, are prominently visible in the cytoplasm, giving it a distinct granular appearance in the light microscope. The nucleus is prominent and bilobed. The amount of cytoplasm is small, and it contains relatively few ribosomes and only a modest amount of granular endoplasmic reticulum, in sharp contrast to the cells shown in Figs. 3-3 and 3-4. (From Greep, R. O., ed. 1965. Histology, 2nd ed. McGraw-Hill Book Co. Copyright © 1965 by McGraw-Hill, Inc. Used with permission of McGraw-Hill Book Co.)

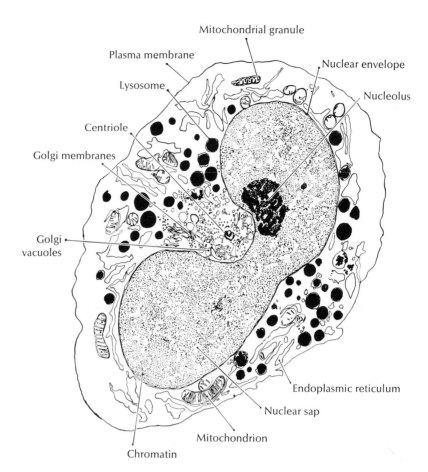

diverse biochemical functions, including the synthesis of blood proteins. Fig. 3-5 shows a myelocyte, which is a precursor of the neutrophilic leukocyte, the most common type of white blood cell. The magnifications of Figs. 3-4 and 3-5 have been adjusted so that the cells are depicted at about the same size, although in actuality the myelocyte is significantly smaller.

As we summarize the major structural features shared by mammalian and other animal cells, it will be evident that differentiation has qualitative or quantitative effects on virtually every cellular organelle. The three types of cells in Figs. 3-3, 3-4, and 3-5 contain few unique structural elements and do not exhibit extreme morphological specialization of the type that is evident in cells from tissues such as bone, muscle, nerve, and the lining of the intestine. However, each of these three cell types, the pancreatic acinar cell, the hepatocyte, and

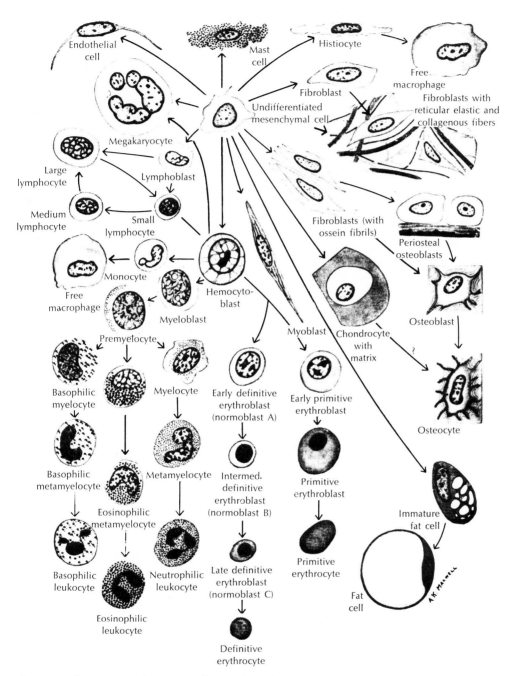

Fig. 3-6. Differentiation of various cell types from the undifferentiated mesenchymal cell. With the exception of the myoblast (muscle cell precursor), all cells depicted here are classified as connective tissue cells. Note the variability not only of cytoplasmic morphology but also of nuclear morphology. (From Hamilton, W. J., and H. W. Mossman. 1972. Hamilton, Boyd and Mossman's human embryology, 4th ed. The Williams & Wilkins Co., Baltimore. By permission of Macmillan, London and Basingstoke.)

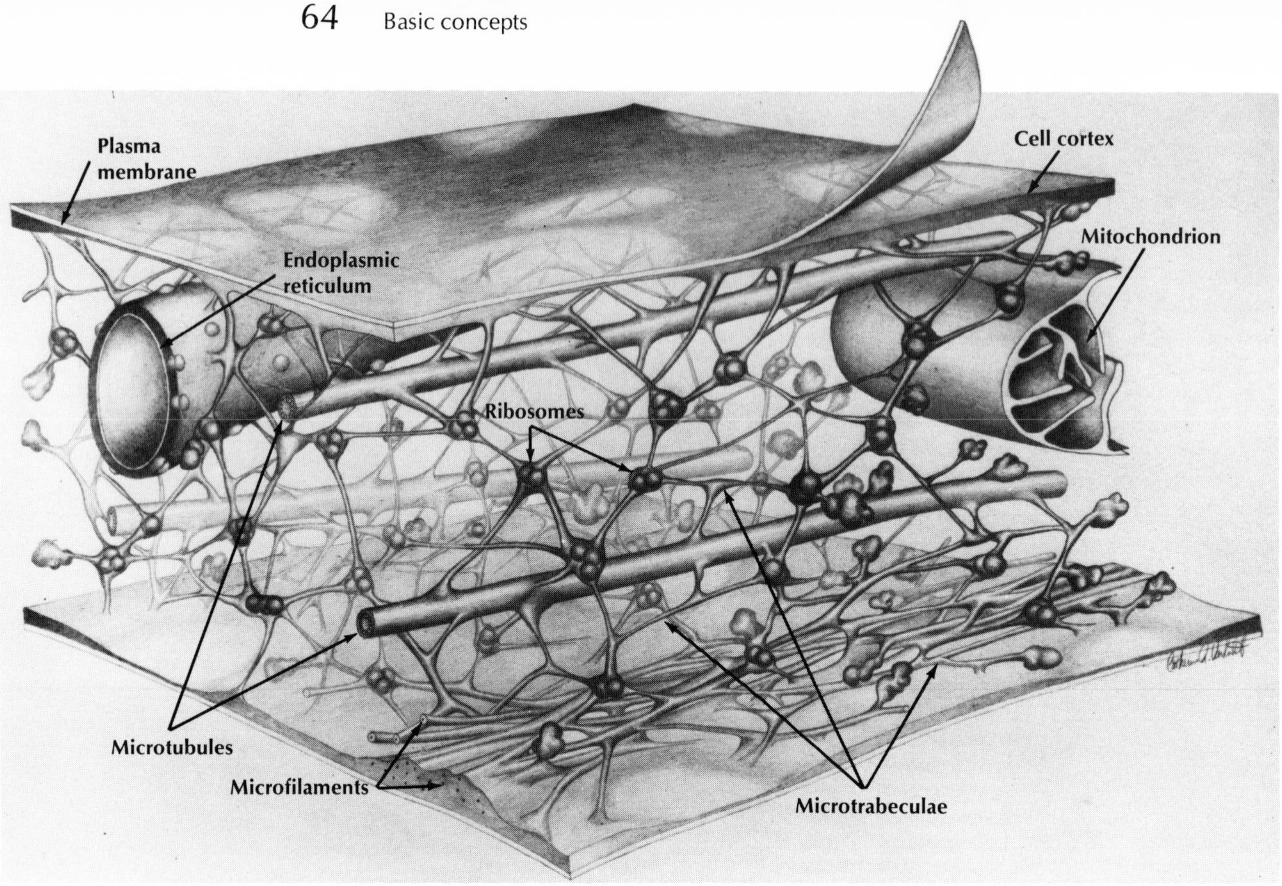

Fig. 3-7. Three-dimensional diagram of the microtrabecular lattice in a relatively thin portion of the cytoplasm of a cultured human diploid fibroblast. Diagram is based on stereoscopic photographs taken with the high-voltage electron microscope. (From Wolosewick, J. J., and K. R. Porter. 1979. J. Cell Biol. **82:**114.)

the myelocyte, has a characteristic appearance, which permits it to be identified by an experienced histologist and distinguished not only from the other two types of cells that are pictured but also from any of the more than a hundred morphologically distinct types of cells that comprise the mammalian body.

We will not attempt a detailed analysis of cellular structure and function in this book. However, a brief summary of major cellular structures and their functions will be helpful in understanding current concepts of cellular differentiation, including the extent to which overall structural organization is affected by differentiation.

Nucleus. The distinguishing feature of eukaryotic cells is a membrane-bound nucleus, with the genome on the inside and the machinery for synthesis of proteins on the outside (Chapter 5). The nucleolus

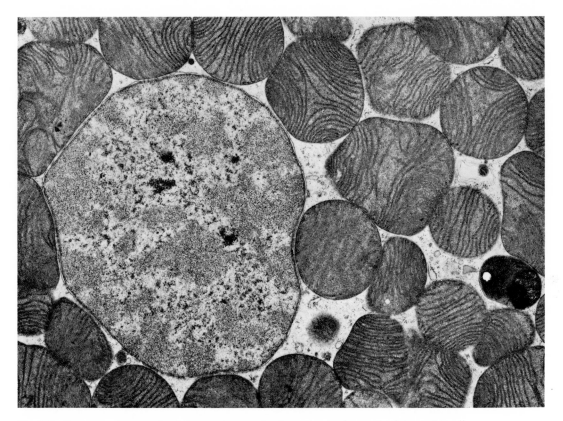

Fig. 3-8. Electron micrograph of the nucleus and adjacent cytoplasm of a brown fat cell from a bat recently aroused from hibernation. The cytoplasm is crowded with large spherical mitochondria containing unusually long and prominent membranous cristae. (From Bloom, W., and D. W. Fawcett. 1975. A textbook of histology, 10th ed. W. B. Saunders Co., Philadelphia.)

is a dense granular structure within the nucleus that is involved in the synthesis of ribosomal RNA. The double membrane or envelope that surrounds the nucleus contains many complex "pores," which are believed to be involved in the passage of materials between the nucleus and the cytoplasm. Differences in the size and shape of the nucleus and in the degree of condensation of the chromatin (DNA plus protein) within it are commonly observed among differentiated cells. Such differences are clearly evident in Fig. 3-6, which depicts differentiation of a variety of types of cells from primitive embryonic mesenchymal cells, as seen with the light microscope. The neutrophilic leukocyte is frequently referred to as "polymorphonuclear" because of the extreme convolutions of its nucleus. A distinct bilobed nuclear morphology can frequently be seen in its precursor, the myelocyte (Fig. 3-

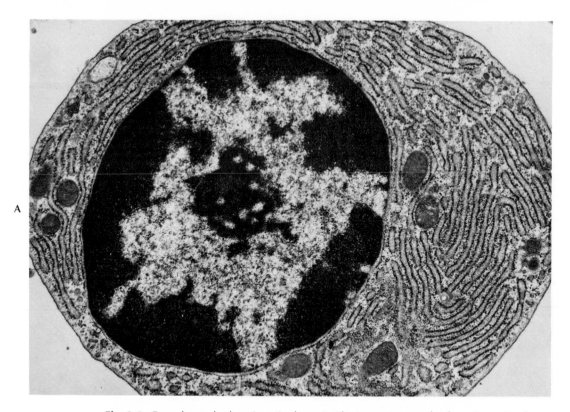

Fig. 3-9. Granular endoplasmic reticulum. **A,** Electron micrograph of a guinea pig plasma cell. The cytoplasm is almost completely filled with granular endoplasmic reticulum. Note also the heavy condensation of chromatin in the nucleus adjacent to the nuclear membrane.

5). Megakaryocytes, which are the precursors from whose cytoplasm nonnucleated blood platelets are derived, also have polymorphous nuclei. Major changes in the nucleus also occur as cells undergo mitosis, including condensation of chromosomes and complete breakdown of the nuclear envelope.

Cytoplasm. This is a general term for everything in a cell between the outer cellular membrane and the nuclear envelope. The cytoplasm is composed of a fluid phase, the cytosol, and a variety of filamentous, membranous, and particulate inclusions. Recent studies with the high-voltage electron microscope have shown a three-dimensional network of interlinking filamentous structures, known as the microtrabecular lattice, running through the cytoplasm and making its structure far more organized than was previously realized (Fig. 3-7). The cytoplasm also contains contractile microfilaments and rigid microtubules, whose role in cellular shape and motility is discussed in

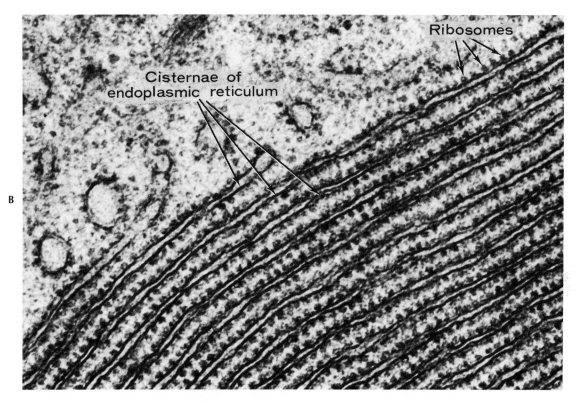

Fig. 3-9, cont'd. B, Electron micrograph of a basophilic area of cytoplasm from a pancreatic acinar cell. The layers of granular endoplasmic reticulum are closely packed in parallel arrays. (From Bloom, W., and D. W. Fawcett. 1975. A textbook of histology, 10th ed. W. B. Saunders Co., Philadelphia.)

Chapter 15. Numerous organelles of varying complexity are found in the cytoplasm. Examples of particular interest are discussed separately in the following paragraphs. The ratio of cytoplasmic volume to nuclear volume varies greatly from one type of differentiated cell to another, as is clearly evident when one compares the myelocyte in Fig. 3-5 with the hepatocyte in Fig. 3-4 and the pancreatic acinar cell in Fig. 3-3.

Mitochondria. These complex membranous organelles are the site of oxidative metabolism and a major source of biochemical energy in the form of adenosine triphosphate (ATP), which is essential for many different cellular functions. Both their shape and their number vary significantly from one type of cell to another. Cells that are active in oxidative metabolism or biosynthesis, such as liver, muscle, and gland cells, tend to contain relatively large numbers of mitochondria. Extremely high concentrations of mitochondria are found in the brown fat cells of animals that hibernate (Fig. 3-8). One of the major functions of such cells is to generate heat to raise the body temperature when the animals emerge from hibernation.

Ribosomes. The cellular organelles responsible for protein synthesis are dense ribonucleoprotein particles known as ribosomes, which are described in greater detail in Chapter 9. Proteins synthesized for use within the cell are generally synthesized by ribosomes that are considered to be "free" in the cytoplasm (although recent evidence suggests that they may actually be associated with the microtrabecular lattice, as shown in Fig. 3-7). Proteins synthesized by glandular cells for export, on the other hand, are synthesized by ribosomes attached to a membranous intracellular structure known as the endoplasmic reticulum (Fig. 3-9). The concentration of ribosomes in a cell tends to parallel its level of protein synthesis, either for intracellular use or for export.

Granular endoplasmic reticulum. The combination of endoplasmic reticulum and attached ribosomes is known variously as granular endoplasmic reticulum (GER), rough endoplasmic reticulum (RER), and ergastoplasm, which is a term from light microscopy used to describe an area of the cytoplasm that stains intensely with basic dyes that bind preferentially to acidic substances such as nucleic acids. Large amounts of GER are characteristic of cells that synthesize large amounts of protein for export. This is clearly evident, for example, in plasma cells, which synthesize antibodies (Fig. 3-9, *A*), and in the basal area of pancreatic acinar cells, where great quantities of zymogens are being synthesized (Fig. 3-9, *B*). Proteins that are to be exported from the cell pass into the internal spaces (cisternae) of the endoplasmic reticulum as they are synthesized and are kept segregated from the rest of the cytoplasm.

Agranular endoplasmic reticulum. Clusters of membranous tubules or vesicles rather similar to the granular endoplasmic reticulum except for their lack of ribosomes occur in many types of cells (Fig. 3-10). They are believed to function in lipid and cholesterol metabolism and in detoxification of lipid-soluble substances. The ratio of granular to agranular endoplasmic reticulum in different cells varies greatly with the type of differentiation. Pancreatic acinar cells contain almost exclusively the granular form, whereas muscle cells contain almost entirely the agranular form. Liver cells are interesting in that they contain substantial amounts of both.

Golgi complex. Proteins produced for export pass from the GER to the Golgi complex (Fig. 3-11), which consists of a characteristically oriented cluster of membranous vesicles, generally located near the nucleus. The Golgi complex is involved in the packaging of proteins

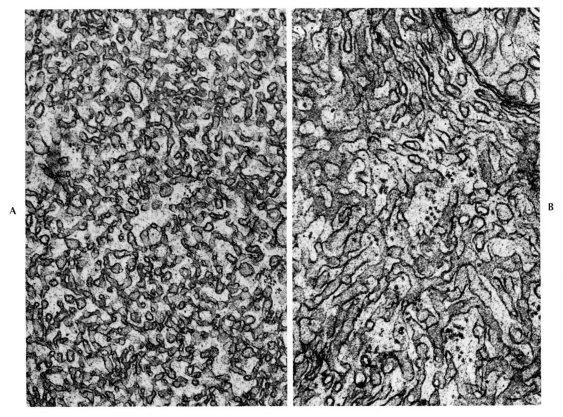

Fig. 3-10. Electron micrographs of agranular endoplasmic reticulum from, **A,** hamster liver and, **B,** human adrenal cortex. (**A,** ×34,000; **B,** ×50,000.) (Courtesy J. Long, from Bloom, W., and D. W. Fawcett. 1975. A textbook of histology, 10th ed. W. B. Saunders Co., Philadelphia.)

into secretory granules and is also believed to be responsible for the addition of carbohydrates to glycoproteins used in the assembly of the cell surface. The Golgi complex tends to be prominent in cells that export large amounts of protein.

Cytoplasmic granules. The zymogen storage granules depicted in Figs. 3-3 and 3-11 are the final storage site of zymogens (enzyme precursors), which will be released from pancreatic acinar cells. Such granules are characteristic of glandular cells that store their products and release them suddenly on demand. A variety of other types of granules also occur in the cytoplasm of various types of cells, including lipid droplets and glycogen granules, which serve as energy reserves. In adipocytes (fat cells), the lipid droplets become the dominant feature of the cell, with everything else condensed into a thin layer around the outside. Pigmented cells contain pigment granules or

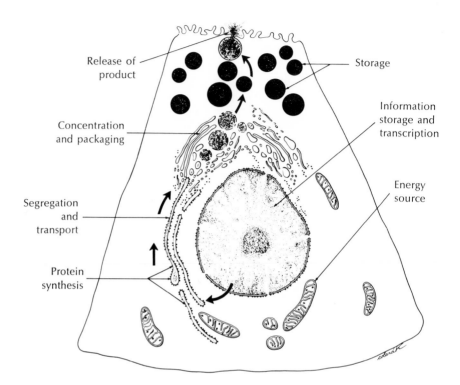

Fig. 3-11. Schematic representation of the role of various intracellular organelles in the synthesis, packaging, storage, and release of a secretory protein. (From Bloom, W., and D. W. Fawcett. 1975. A textbook of histology, 10th ed. W. B. Saunders Co., Philadelphia.)

melanosomes. A fluorescent pigment, lipofuscin, accumulates in granules during aging in a variety of tissues (Chapter 24).

Lysosomes. Many cells are able to engulf materials from their environment, either as solids (by phagocytosis) or as liquids (by pinocytosis). In these processes part of the cell membrane surrounds a solid particle and/or a portion of the aqueous environment enclosing it in an internal vesicle (Figs. 3-2 and 3-12). In either case the vacuole containing the ingested material merges with membrane-bound intracellular vesicles known as lysosomes, which contain a variety of hydrolytic enzymes and appear to function as intracellular digestive organelles. Lysosomes are also involved in the digestion of discarded cellular organelles (Fig. 3-12). The entire intracellular digestive system is particularly well developed in macrophages, whose function in the body is to remove foreign material and the remnants of degenerating cells.

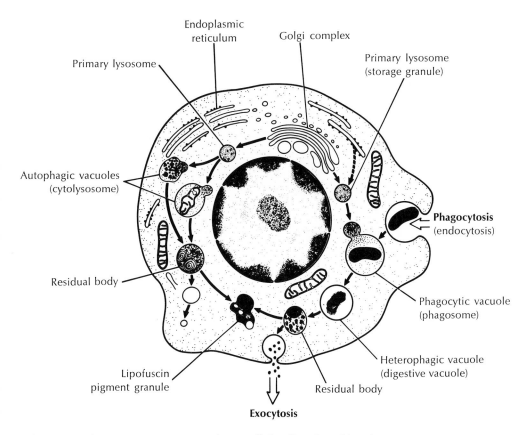

Fig. 3-12. Schematic representation of intracellular digestion. Digestive enzymes are synthesized by the granular endoplasmic reticulum and packaged by the Golgi complex into membrane-bound storage granules known as primary lysosomes. Extracellular materials that are to be digested are taken into the cells either by phagocytosis or pinocytosis. Intracellular materials or organelles that are to be digested become enclosed in autophagic vacuoles. In each case primary lysosomes merge with the vacuoles, and their enzymes perform the digestion. Indigestible material is either retained in the cell in residual bodies or expelled by exocytosis. Lipofuscin granules, which accumulate in certain tissues during aging, are believed to be derived from residual bodies. (From Bloom, W., and D. W. Fawcett. 1975. A textbook of histology, 10th ed. W. B. Saunders Co., Philadelphia.)

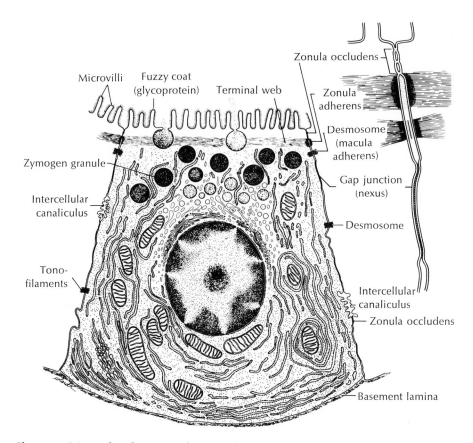

Microvilli   Fuzzy coat
(glycoprotein)   Terminal web

Zonula occludens

Zonula
adherens

Desmosome
(macula
adherens)

Zymogen granule

Gap junction
(nexus)

Intercellular
canaliculus

Desmosome

Tono-
filaments

Intercellular
canaliculus

Zonula occludens

Basement lamina

Fig. 3-13. Principal surface specializations that may be found on a simple columnar epithelial cell. The junctional complex that forms around the edges of the cell to prevent leakage of fluid between the epithelial cells is shown at higher magnification at the right. The zonula occludens and the zonula adherens are continuous around the cell. Desmosomes are spot junctions that occur in a row just below the zonula adherens and are also randomly distributed elsewhere around the cell. The gap junction, or nexus, allows for free flow of small molecules and electric currents between neighboring cells. The intercellular canaliculus is an extension of the free surface and communicates with it. The edges of the intercellular canaliculus are sealed with zonula occludens. (From Hay, E. D. 1973. *In* R. O. Greep and L. Weiss, eds. Histology, 3rd ed. McGraw-Hill Book Co., New York. Copyright © 1973 by McGraw-Hill, Inc. Used with permission of McGraw-Hill Book Co.)

Junctions. A variety of specialized modifications of the cell membrane and immediately adjacent cytoplasm occur when cells come into contact with one another (Fig. 3-13). The "junctional complex," which occurs in tissues such as the lining of the intestine where a tight seal against leakage between cells is important, is particularly interesting in that it contains three distinctly different types of junctions (Fig. 3-14). The first is the tight junction, or zonula occludens, in which the membranes of the two cells are in close contact. Zonula oc-

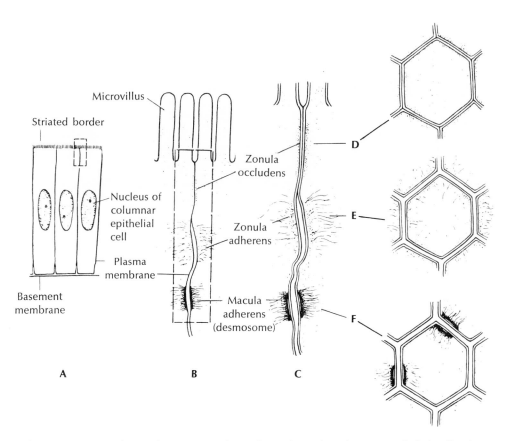

Fig. 3-14. Junctional complex. **A,** Interrelationships of simple columnar epithelial cells of the small intestine as determined by light microscopy. **B,** Structure of the junctional complex in the region outlined in **A** as seen with the electron microscope. **C,** Details of the junctional complex as seen in high-resolution electron microscopy. **D,** Schematic representation of a cross section of a columnar epithelial cell at the level of the zonula occludens. Note that the outer leaflet of the plasma membranes of the adjacent cells are in close contact around the entire circumference of the cell, forming a tight seal. **E,** Cross section of a columnar epithelial cell at the level of the zonula adherens showing separation of the cellular membranes by extracellular matrix material and fibers radiating into the surrounding cytoplasm. The zonula adherens, like the zonula occludens extends completely around the cell. **F,** Cross section of a columnar epithelial cell at the level of the macula adherens. Note that, in contrast to the junctions depicted in **D** and **E,** the macula adherens are spot junctions and do not extend completely around the cell. Note also that there is extracellular material between the cells and a dense cytoplasmic condensation at each macula adherens or desmosome. The plasma membranes have been enlarged out of proportion to the cell area in **D** to **F** to illustrate details more clearly. (From Copenhaver, W. M., D. E. Kelly, and R. L. Wood. 1978. Bailey's textbook of histology. 17th ed. The Williams & Wilkins Co., Baltimore.)

cludens are also found in other locations where a tight seal is needed. The second type of junction in the junctional complex is the intermediate junction, or zonula adherens, in which the membranes of the two cells are separated by extracellular matrix material and which has a fibrous structure in the adjacent cytoplasm. These two types of junctions are zonular—that is, they extend all the way around the cells to form a complete seal (Fig. 3-14, *D* and *E*). The third type of junction, the desmosome, or macula adherens (Fig. 3-14, *F*), is restricted to localized spots on the cell membranes (macula is Latin for spot). The desmosome is characterized by a dense plaque in the cytoplasm just under the plasma membrane, with radiating tonofilaments. Like the intermediate junction, it contains extracellular matrix between the

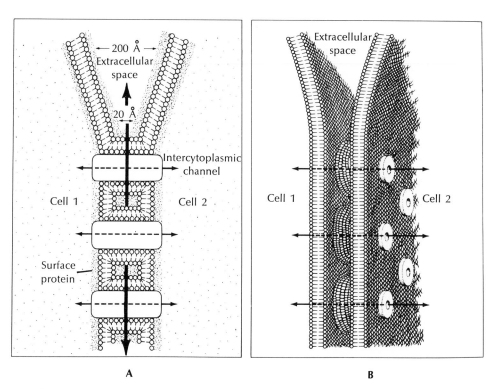

A   B

Fig. 3-15. Schematic representations of the structure of the gap junction. **A,** Molecular architecture in cross section, as inferred from freeze-fracture and physiological studies. Note the intercellular channels that allow for free movement of small molecules and electric currents from one cell to the next with no connection to the extracellular space. **B,** Three-dimensional representation of the communicating channels shown in **A.** Proteins coating the membrane surfaces (shown as fine stippling in **A**) have been omitted. (Drawing by B. Tagawa from Pappas, G. D. 1975. Hosp. Practice **8:**90 and Pappas, G. D. 1975. Junctions between cells. In G. Weissmann and R. Claiborne, eds. Biochemistry, cell biology and pathology. HP Publishing Co., Inc., New York.)

cell membranes. Desmosomes are also found in many tightly adherent tissues in the absence of the other elements of the junctional complex.

A fourth type of junction of particular interest in the study of development is the gap junction, or nexus (Fig. 3-13). This special type of tight junction has communicating channels that allow the passage of small molecules and electrical currents between the adjacent cells with no connection to the external environment (Fig. 3-15). Gap junctions form in many embryonic tissues and are generally believed to play important roles in communication among the cells of the embryo during development.

Surface modifications. Free cell surfaces that are not in contact with other cells often develop small cytoplasmic projections known as microvilli. These are particularly prominent in the epithelium of the small intestine, where they are so dense they give the surface a brush-like appearance and are known as the "brush border" or striated border. Epithelial cells of mucous membrane often have cilia on their surfaces.

Extracellular matrix. Except in areas where cellular membranes are in close contact, as in tight and gap junctions, cell surfaces in general are coated with varying amounts of glycoprotein, known as glycocalyx. In certain types of tissues such as cartilage, tendon, and bone the amount of extracellular matrix surrounding the cells is greatly increased and sometimes also mineralized. Production of large amounts of matrix is one of the characteristic differentiated properties of the cells in such tissues.

Summary. The foregoing discussion is far from complete. Its purpose has been to provide a brief description of major cellular structures for students not already familiar with them and to demonstrate the multiplicity of structural modifications, both qualitative and quantitative, that accompany the process of cellular differentiation. Specialized structures that are unique to certain kinds of differentiated cells have not been included. They will be discussed in relation to the cells that contain them later in this chapter. Students wishing a more detailed understanding of cell structure and its relationship both to organismal structure and to cellular function are referred to the textbooks on histology and cell biology cited at the end of this chapter.

Cellular chemistry

Together with the structural specialization just discussed, chemical specialization is one of the characteristic features of cellular differentiation. Because of difficulties in precise quantification of all except the most extreme structural features of differentiated cells, chemical properties are frequently used to measure the extent of differentiation.

It is convenient, even if sometimes misleading, to measure differentiation in terms of unique products that are synthesized only by the particular kind of cell under consideration or in terms of products that are made in vastly greater amounts by the differentiated cell than by other types of cells. There is a distinct tendency among investigators studying the process of cellular differentiation to select extreme cases, in which a differentiated cell devotes a major portion of its total synthetic activity to the formation of one or a few specialized products. Prominent examples that can be cited include the production of hemoglobin by red blood cells, the production of contractile proteins by muscle cells, the synthesis of ovalbumin (the major protein of egg white) by the chicken oviduct, the production of zymogens by pancreatic acinar cells, the synthesis of an extracellular matrix consisting of collagen and chondromucoproteins rich in chondroitin sulfate by cartilage cells, the production of crystallins by cells in the lens of the eye, and synthesis of keratin (the major fibrous protein of skin) by epidermal keratinocytes. Although synthesis of such specialized products by the appropriate types of cells is a valid criterion of differentiation, exclusive study of such extreme cases may lead to false conclusions about the mechanisms responsible for cellular differentiation.

Differentiation does not consist simply in activating the gene responsible for a particular differentiated product. Rather, it involves multiple changes in cellular biochemistry. Each type of cell in the mammalian body, including those with a heavy commitment to one particular biosynthetic pathway, tends to have a characteristic pattern of metabolic activities. This pattern includes not only qualitative characteristics (those which involve turning on or off the synthesis of particular products) but also a variety of quantitative characteristics which reflect the relative proportions of various products that are synthesized by the cell. Thus the regulatory mechanisms responsible for cellular differentiation must include a complex "program" for each type of differentiated cell, with quantitative as well as qualitative controls.

In some cases the overall picture is complicated even further by the shift from one form of a particular product to another during differentiation. For example, all cells contain a certain amount of the proteins actin and myosin as components of the contractile microfilament system (Chapter 15). Muscle cells contain much larger amounts of actin and myosin as components of their contractile myofibrils. On closer examination it is found that the actin and myosin of muscle cells are different in their electrophoretic mobility (movement in an electric field) from the actin and myosin of nonmuscle cells. In the case of actin it has been shown specifically that the proteins from muscle and nonmuscle sources have different amino acid sequences and are therefore coded for by different genes. To complicate the story even further, not all muscle actins are alike. The actins from skeletal and heart muscle differ in their amino acid sequences and appear to be the products of two different genes. Two kinds of actin with different electrophoretic mobility have also been found in nonmuscle cells, but differences in amino acid sequence have not been reported, and it is possible that the differences could be caused by biochemical modification of the proteins after they are synthesized. However, it is clear that there are at least three different genes for actin in rats and that at least two are specifically associated with different types of muscle differentiation.

Fine changes in gene expression of this sort are not unusual. Another example can be found in the different kinds of collagen that are produced at different locations in the body and at different developmental ages (Table 3-1). The common collagen of skin, tendon, and bone (Type I) is synthesized in vastly different amounts in different tissues. It is a trimer, composed of two identical peptide chains designated $\alpha_1(I)$ plus a different chain designated α_2. Cartilage tissue produces a totally different type of collagen referred to as Type II collagen. It is also a trimer but is composed of three identical chains designated $\alpha_1(II)$ that are different from either of the chains in Type I collagen. Human skin also contains Type III collagen, which consists of a trimer of three identical chains designated $\alpha_1(III)$. These chains differ from all the other collagen monomer chains in that they contain the amino acid cysteine. Type III collagen predominates in fetal skin, whereas in adult skin there is a 3:1 ratio of Type I collagen to Type III collagen. The basement membrane that underlies most types of epithelial cells contains yet another type of collagen referred to as Type IV, which is composed of three identical chains designated $\alpha_1(IV)$. In

Table 3-1. Types of collagen found in mammalian tissue*

Type	Monomer(s)	Trimer	Tissues
I	$\alpha_1(I), \alpha_2$	$[\alpha_1(I)]_2 [\alpha_2]$	Bone, tendon, dermis, dentin
II	$\alpha_1(II)$	$[\alpha_1(II)]_3$	Cartilage
III	$\alpha_1(III)$	$[\alpha_1(III)]_3$	Fetal dermis (major component), adult dermis (minor component), cardiovascular system
IV	$\alpha_1(IV)$	$[\alpha_1(IV)]_3$	Basement membrane, cultured endothelial cells

*Modified from Miller, E. J., and V. J. Matukas. 1974. Fed. Proc. **33**:1197.

the case of collagen formation, changes in the particular genes expressed occur both from one type of cell to another and from one developmental age to another.

Still another example of complex switching of genes for closely related proteins occurs in the synthesis of hemoglobin. The functional hemoglobin molecule is a tetramer, which is nearly always composed of two molecules each of two different kinds of protein chains. Greek letters are used to identify the chains. The major hemoglobin of adult blood (hemoglobin A) contains two alpha (α) chains and two beta (β) chains. At all developmental ages, most of the hemoglobin molecules contain α-chains, although early in embryonic development an alternate form designated zeta (ζ) is found in place of α-chains in a small fraction of the hemoglobin. The second pair of chains (mostly β in the adult) is much more variable. A total of four alternate types of chains, designated beta (β), gamma (γ), delta (δ), and epsilon (ϵ) have been identified. These fall into three major time classes, associated with early embryonic development (ϵ), later fetal development (γ), and postnatal development (β and δ), respectively (Fig. 3-16).

The shifting pattern of synthesis of hemoglobin chains involves quantitative as well as qualitative controls. As will be seen in Chapter 9, the regulatory mechanisms involved in maintaining this balance operate at several different levels between the genes and the final expression of their products and include "translational" controls that act specifically at the level of protein synthesis.

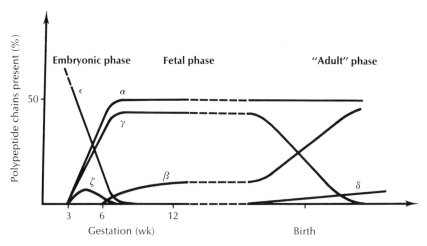

Fig. 3-16. Changes in relative frequency of human hemoglobin chains during embryonic and postnatal development. The amount of each chain is shown as a percentage of total polypeptide chains present. (From Nigon, V., and J. Godet. 1976. Int. Rev. Cytol. **46**:79.)

The chemical differences characteristic of differentiated cells can be divided into two broad categories. Many of the products that we have been discussing are specific proteins whose amino acid sequences are encoded directly in the genome. These include enzymes, structural proteins, and a variety of special purpose proteins such as actin, myosin, hemoglobin, serum albumin, and antibodies. The remainder of the differentiated chemical properties are products of the action of specific enzymes. For example, the complex polysaccharide chondroitin sulfate, which is found in large amounts in cartilage matrix, is synthesized by a set of seven specific enzymes. Since enzymes are also proteins, we can propose that the chemical aspects of cellular differentiation are all either direct or indirect consequences of the synthesis of specific types of proteins in appropriate amounts by the differentiated cells.

In the next chapter we will review the evidence showing that the arrangement of amino acids into specific sequences in the polypeptide chains of proteins is encoded in the DNA of the genome. From that conclusion it will follow that selective synthesis of specific proteins consists of selective expression of certain genes that are contained within the total genome. Thus, as we study the regulation of gene expression in eukaryotic cells in Chapters 5 to 9, we will, in fact, be studying the basic mechanisms responsible for cellular differentiation.

One of the end results of cellular differentiation is the production of cells that are capable of performing particular specialized functions. Although functional capacity is inherent in the structure and chemistry of these cells, it is sometimes useful to examine the combination of structure, chemistry, and function as a single, integrated unit. For example, the intimate mingling of structure and chemistry with function is clearly shown in Fig. 3-17, which summarizes the synthesis and packaging of glycoprotein secretory products by a secretory cell, and in Fig. 3-11, which summarizes zymogen synthesis and secretion by a pancreatic acinar cell. One can see within such cells an organization almost like an assembly line for the production, packaging, storage, and release of the glandular products. Each area within the cytoplasm possesses both the chemical specialization and the structural organization necessary to achieve its particular task.

Cellular function

Functional organization of this kind is evident to some degree in every type of specialized cell. Three examples will serve to illustrate the importance of functional organization:

1. Skeletal muscle consists of long, multinucleate cells known as myotubes, which are arranged in parallel bundles (Fig. 3-18, *A* and *B*).

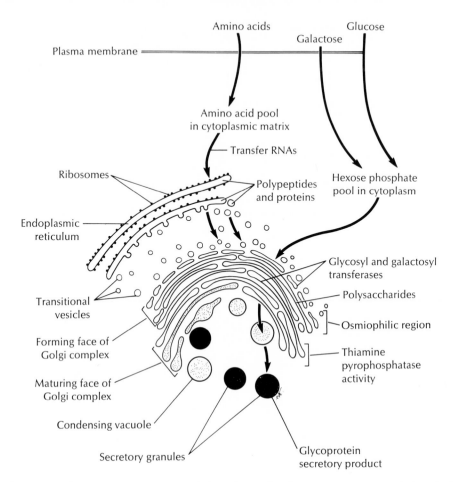

Fig. 3-17. Schematic representation of the relationship between structure and chemical function during the synthesis of glycoproteins for export from the cell. Proteins are synthesized in the endoplasmic reticulum and transferred to the Golgi complex where the carbohydrates are added. The resultant glycoproteins are then packaged into storage vacuoles until their release from the cell is called for. (From Bloom, W., and D. W. Fawcett. 1975. A textbook of histology, 10th ed. W. B. Saunders Co., Philadelphia.)

Each cell (Fig. 3-18, *C*) contains many myofibrils (Fig. 3-18, *D*), which are composed of the contractile proteins actin and myosin together with other molecular species needed for contraction. Intimately mingled among the contractile filaments are mitochondria to generate metabolic energy plus a system of transverse (T) tubules and a membranous sarcoplasmic reticulum (Fig. 3-19). The T tubules carry the electrical excitation needed for contraction throughout the cell to ensure that myofibrils at all depths in the fiber contract at the same time, and the sarcoplasmic reticulum reversibly releases cal-

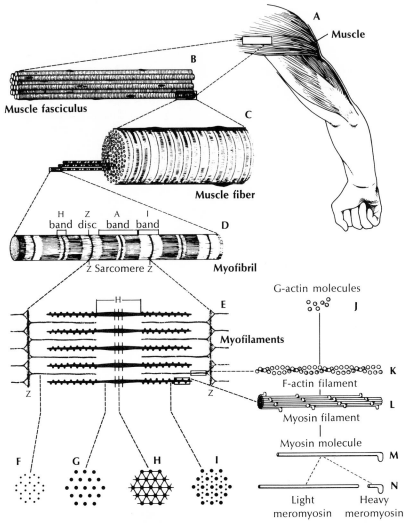

G-actin molecules

F-actin filament

Myosin filament

Myosin molecule

Light
meromyosin

Heavy
meromyosin

Fig. 3-18. Schematic representation of functional specialization in muscle cells. **A,** Intact muscle. **B,** Bundle of multinucleate muscle cells (myotubes). **C,** Enlargement of a single myotube showing individual myofibrils. **D,** Enlargement of a myofibril showing the pattern of banded sarcomeres that can be seen with the electron microscope. **E,** Enlargement of a single sarcomere showing the overlapping thick and thin filaments of which it is composed. The thick filaments are myosin and the thin filaments actin. **F,** Cross section through the thin actin filaments. **G,** Cross section through the thick myosin filaments. **H,** Section through the midpoint of the A band, showing cross connections among the thick myosin filaments. **I,** Cross section of the overlapping region of thick and thin filaments, showing hexagonal array of thin filaments around each thick filament. **J,** Globular subunits of actin. **K,** Actin monomers polymerized into an actin filament. **L,** Thick myosin filament with globular projections. **M,** Individual myosin molecule, rod-shaped, with globular projection at one end. **N,** Myosin molecule cleaved by brief exposure to trypsin, forming rod-shaped light meromyosin and globular heavy meromyosin, which is the part of the molecule that interacts with actin. (From Bloom, W., and D. W. Fawcett. 1975. A textbook of histology, 10th ed. W. B. Saunders Co., Philadelphia.)

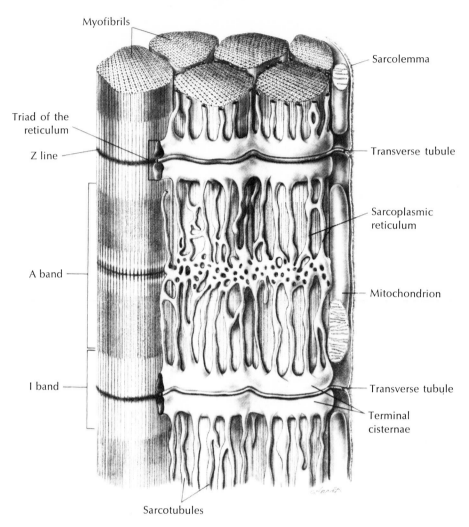

Myofibrils

Sarcolemma

Triad of the reticulum

Z line

Transverse tubule

Sarcoplasmic reticulum

A band

Mitochondrion

I band

Transverse tubule

Terminal cisternae

Sarcotubules

Fig. 3-19. Schematic representation of myofibrils with associated sarcoplasmic reticulum and transverse tubules in amphibian skeletal muscle. Note also the juxtaposition of mitochondria and myofibrils. (From Bloom, W., and D. W. Fawcett. 1975. A textbook of histology, 10th ed. W. B. Saunders Co., Philadelphia.)

cium ion, which triggers the actual contraction. The filaments of actin and myosin (Fig. 3-18) are precisely arranged next to each other in an overlapping pattern. Contraction is believed to be achieved by these two types of protein filaments sliding past each other (Fig. 3-20). Additional details of this fascinating system can be found in *A Textbook of Histology* by Bloom and Fawcett cited at the end of this chapter.

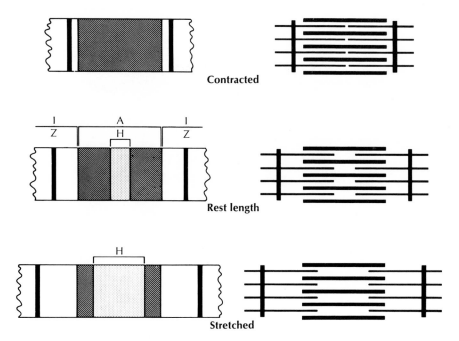

Fig. 3-20. Schematic representation of the current interpretation of the changing appearance of cross striations of skeletal muscle in different phases of contraction. Contraction is believed to be achieved by a sliding motion of the thin filaments deeper into the space between the thick filaments. The A band (composed of the thick filaments) remains a constant length, whereas the I band (portion of thin filaments not overlapping thick filaments) becomes shortened during contraction. The central H band (portion of the thick filaments not overlapped by thin filaments) also becomes smaller or disappears during contraction. (From Bloom, W., and D. W. Fawcett. 1975. A textbook of histology, 10th ed. W. B. Saunders Co., Philadelphia.)

2. Capillaries are formed from specialized cells known as endothelial cells that have extremely spread and flattened cytoplasms with relatively few organelles (Fig. 3-21). Where two endothelial cells come together, they form tight junctions with considerable overlap, and when cell boundaries are visualized in whole capillaries, they are often irregular and somewhat interdigitated. In addition, many capillary cells contain pores or fenestrations covered by only an extremely thin membrane to permit free exchange of fluids without loss of red blood cells (Fig. 3-22).

3. One of the most extreme cases of specialization occurs in the mammalian erythrocyte, or red blood cell. It has become so completely adapted to its specialized function that it has lost even its nucleus and consists of little more than a flattened biconcave membrane-bound bag of hemoglobin (Fig. 3-23). The extreme contrast of functional spe-

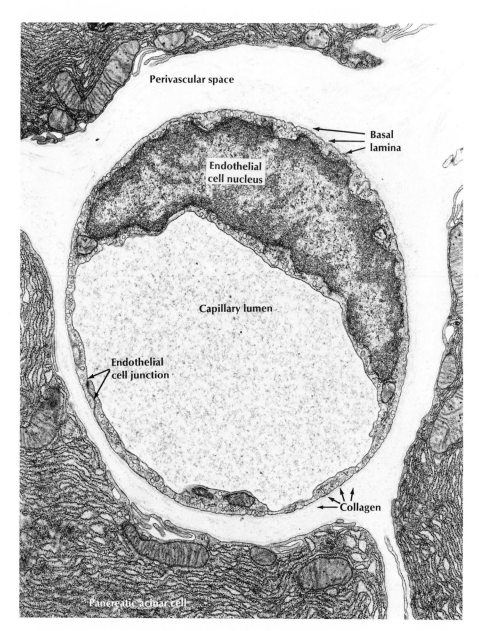

Fig. 3-21. Electron micrograph of a typical capillary from guinea pig pancreas. In this case the entire circumference is made up of a single endothelial cell. Note the flattened nature of the endothelial cell cytoplasm and the relative absence of organelles. (From Bloom, W., and D. W. Fawcett. 1975. A textbook of histology, 10th ed. W. B. Saunders Co., Philadelphia; modified from Bolender, R. 1974. J. Cell Biol. **61:**269.)

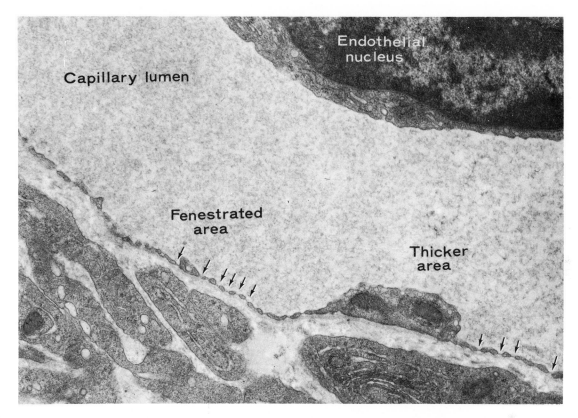

Fig. 3-22. Electron micrograph of a typical fenestrated capillary. Note the extremely thin membrane covering the pores of the fenestrated area and the alternation of fenestrated areas with thicker areas. (From Bloom, W., and D. W. Fawcett. 1975. A textbook of histology, 10th ed. W. B. Saunders Co., Philadelphia.)

cialization among muscle, endothelial, and red blood cells is particularly evident in Fig. 3-24, which shows a portion of a capillary containing a red blood cell and located next to a heart muscle cell.

Many other cells are less obviously specialized than these three extreme cases, but for virtually any type of cell, careful examination of structure and chemistry in terms of specialized function will reveal a clear-cut relationship among these three features.

We will frequently refer to differentiation in this book in terms of specific structural, chemical, or functional specializations of the cell. Although these may serve as convenient "markers" for the measurement of differentiation, it is important to keep in mind the fact that they actually represent only small facets of the complex and highly integrated process of cellular differentiation, which affects virtually every aspect of the cell.

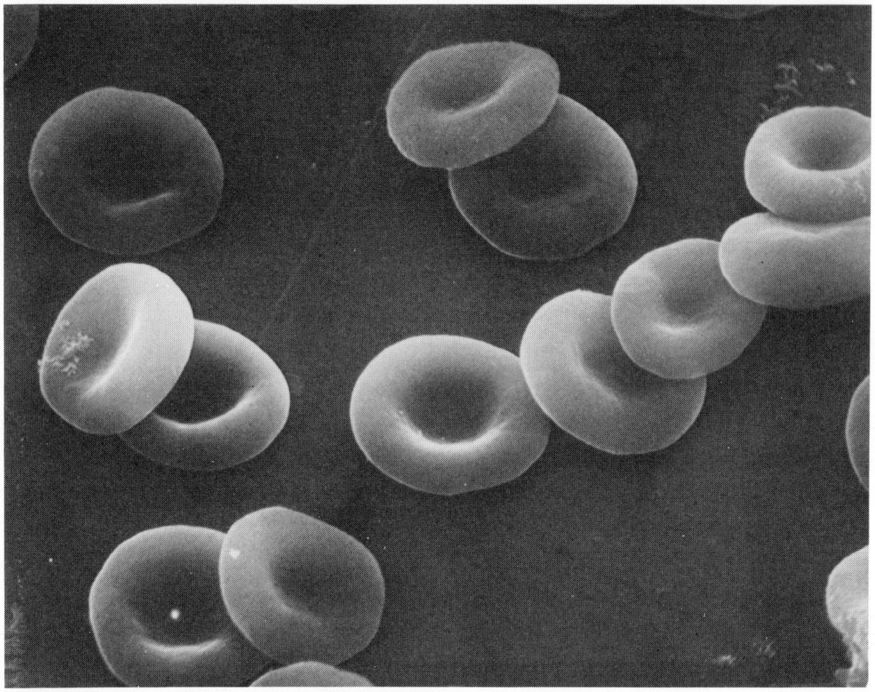

Fig. 3-23. Intact human erythrocytes (red blood cells) as seen with the scanning electron microscope. Note their flattened nature and the surface concavity of each. (×3800.) (From Morel, F. M. M., R. F. Baker, and H. Wayland. 1971. J. Cell Biol. **48**:91.)

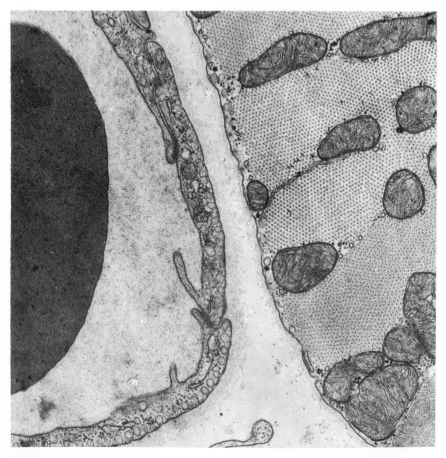

Fig. 3-24. Three diverse types of functional specialization in differentiated cells. This ultrathin section of heart muscle shows, left to right, part of an erythrocyte, part of a capillary endothelial cell, and part of a heart muscle cell. The heart muscle cell has been sectioned perpendicular to the myofilaments and shows the intimate relationship between energy-providing mitochondria and the contractile elements. The endothelial cell displays its typical very thin cytoplasm with interdigitating junctions. The complete absence of organelles in the erythrocyte is clearly evident. (From Bloom, W., and D. W. Fawcett. 1975. A textbook of histology, 10th ed. W. B. Saunders Co., Philadelphia.)

Now that we have dealt with the structural, chemical, and functional criteria of differentiation, we can attempt to define the phenomenon of cellular differentiation more precisely. The two definitions that were introduced early in the chapter describe the overall process reasonably well: (1) the generation from a single parental cell of daughter cells that differ from one another and (2) the collective specializations of structure and function that equip the nucleus and cytoplasm for a particular array of functional activities. As long as we restrict ourselves to differences among cells within a single multicellular organism, these definitions both work reasonably well, although questions can be raised about short-term physiological responses (e.g., caused by hormones). It is when we attempt to develop a definition that is applicable to all types of cells that major problems begin.

One of the most difficult distinctions regarding cellular differentiation is whether changes that occur in single-celled organisms should be referred to as differentiation. One type of change in single-celled organisms that clearly should not be called differentiation is that due to simple mutation. Many aspects of the set of phenomena widely accepted as true differentiation are not consistent with a simple mutational mechanism. These range from the nonrandom nature of the process of differentiation to the fact that no true genetic changes can be found, at least for some types of differentiated cells. The most convincing evidence for the latter point comes from the finding that certain types of differentiated cells, or their nuclei, can be used to generate complete organisms that are normal in every respect and fully fertile. The evidence that the genome remains intact and unmutated during differentiation will be discussed in Chapter 6.

If we accept the premise that the genotype remains unchanged during differentiation, it becomes possible to propose the following definition of cellular differentiation: *display of different phenotypes by cells of the same genotype*. (The terms "genotype" and "phenotype" are used by geneticists to refer to an organism's genetic makeup and the inherited properties that it actually displays, respectively.) This definition is rather broad and can be construed to include any nonmutational change in the properties exhibited by a cell.

The next question that arises is whether differentiation-like changes that occur in cellular phenotype during the life cycles of various unicellular organisms should or should not be called cellular differentiation. Three examples will serve to illustrate the kinds of changes that can occur:

1. Certain bacteria such as *Bacillus subtilis* undergo a process of sporulation in response to adverse environmental conditions. The spore cell that is formed is inert metabolically and differs

DEFINITIONS OF DIFFERENTIATION

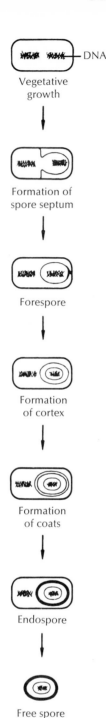

Vegetative
growth

Formation of
spore septum

Forespore

Formation
of cortex

Formation
of coats

Endospore

Free spore

Fig. 3-25. Diagram showing the stepwise formation of a spore of a typical bacillus. In the mature spore the cell membrane is surrounded by a thick cortex and by heavy layers of an outer coat material. The spore is extremely resistant to adverse environmental conditions. (From Davis, B. D., et al. 1973. Microbiology, 2nd ed. Harper & Row, Publishers, Hagerstown, Md.)

greatly both in structure and chemistry from active vegetative cells (Fig. 3-25).

2. Certain protozoa undergo complex morphological and functional changes as a part of their normal life cycles. For example, members of a genus of ciliates, *Tokophyra*, go through most of their life cycles as attached sessile organisms (Fig. 3-26). However, their reproduction involves the formation of free-swimming larval forms, which swim to new locations and then undergo metamorphosis into attached sessile forms similar to the parents (Fig. 3-27).

3. The life cycles of parasitic protozoa, such as the organism responsible for malaria, tend to be extremely complex, with several different morphological forms occurring at different times in the cycle and in different hosts (Fig. 3-28).

Although many pages have been written on whether these types of cellular changes should be called cellular differentiation, no firm conclusions have been reached. Whether these changes are called differentiation or merely interesting model systems similar to differentiation is primarily a matter of semantics. The changes are clearly differentiation-like and are part of normal life cycles. However, some investigators do not like to call them differentiation because they are not occurring in a multicellular organism and are not causing sister cells to become different from one another.

In this book our attention is focused primarily on multicellular organisms. Therefore we will generally refer to differentiation in terms of processes occurring in such organisms. Thus, operationally, we will define differentiation as the *display of different phenotypes by cells of the same genotype within a single multicellular organism.*

The stepwise nature of cellular differentiation provides another problem for one tying to distinguish differentiation from other cellular changes, that is, that of determining precisely when a cell is or is not differentiated. Embryologists have long distinguished between "determination" and "cytodifferentiation." Determination refers to the processes whereby a cell becomes committed to a particular pathway of differentiation, whereas cytodifferentiation (or sometimes just differ-

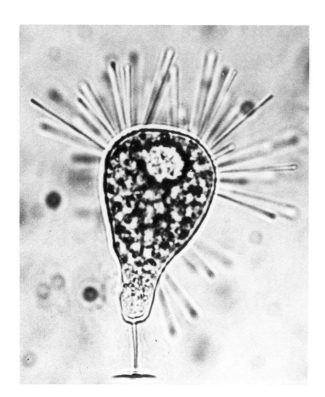

Fig. 3-26. Light micrograph of the complex unicellular protozoan *Tokophyra infusionum,* seen from the side, showing attachment disc, stalk, and tentacles. The cell is about 60 μm in length. (From Curtis, H. 1968. The marvelous animals. The Natural History Press, Garden City, N.Y.)

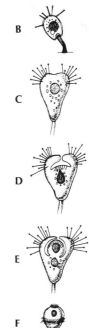

Fig. 3-27. Life cycle of a typical member of the genus *Tokophyra.* **A,** Free-swimming ciliated larva. This stage of the life cycle lasts at most for only a few hours and is followed by metamorphosis into the attached adult form. **B,** Young attached form. All cilia have been lost, and an attachment disc, stalk, and tentacles have formed. The tentacles are not cilia but, rather, are complex tubular structures that serve first to capture prey and then to withdraw nutrients from the prey. **C,** Adult form prior to appearance of brood pouch. **D,** Formation of brood pouch and beginnings of nuclear division. **E,** Development of larva inside brood pouch. Cilia are formed prior to "birth" of the larva, and after the larva severs cytoplasmic connection with the adult, it spins freely in the brood pouch for a few minutes before it is "born." **F,** Free-swimming larva ready to begin another generation. After release of the larva, the adult reorganizes itself and soon begins to produce another larva. Unlike the case in most protozoans, parents and offspring are clearly distinguishable in *Tokophyra.* During the height of its reproductive activity *Tokophyra* may produce as many as 12 embryos each 24 hours, but after a few days the cell begins to age. Its ability to feed and reproduce declines, and ultimately the parental organism dies. (**A** and **C** to **F** from Curtis, H. 1968. The marvelous animals. The Natural History Press, Garden City, N.Y.; **B** from Kudo, R. R. 1966. Protozoology. Charles C Thomas, Publisher, Springfield, Ill.)

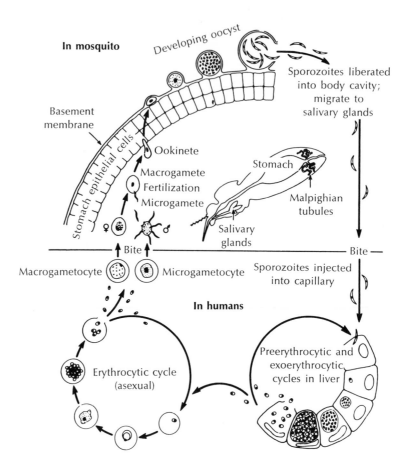

Fig. 3-28. The life cycle of *Plasmodium vivax.* This unicellular parasite, which is responsible for malaria in humans, undergoes a complex life cycle that involves several different morphological forms in human liver and red blood cells and in the stomach epithelium and salivary glands of the *Anopheles* mosquito. (From Adam, K., J. Paul, and V. Zaman. 1971. Medical and veterinary zoology. An illustrated guide. Churchill Livingstone, Edinburgh.)

entiation) is used to refer to the overt expression of differentiated properties by a cell. In the time course of embryonic development, determination often occurs significantly before cytodifferentiation is evident, although in certain cases such as early distinctions between cells destined to become inner cell mass or trophectoderm, the reverse can also be true. Biochemical differences can be detected among the cells of very early mouse embryos (8 to 16 cells) before the cells are permanently committed to either pathway.

Some investigators have resolved problems of this kind by adopting the view that there is no such thing as an undifferentiated cell—rather, there exist only different states of differentiation. From this point of view, the sperm and egg cells are differentiated, the zygote formed by their union is differentiated, the early cleavage blastomeres are differentiated, the embryonic ectoderm is differentiated, etc. Each type of cell has its own set of structural and functional properties, and each is different from all the others. Each step along the pathway to the terminally differentiated adult state can thus be viewed as a differentiated state itself, differing from the terminally differentiated state primarily by its transient nature and its ease of alteration into other differentiated states. This is particularly evident in some of the pathways of blood cell differentiation depicted in Fig. 3-6.

Other problems regarding the distinction between determination and cytodifferentiation arise when differentiated cells are studied in artificial culture systems. In many cases the expression of differentiated properties is dependent on a particular set of environmental conditions and is freely reversible. Determination, or commitment to a particular type of differentiation, on the other hand, is stable and is inherited from one cell generation to the next, even when the environmental conditions do not permit its expression. These relationships will be examined in greater detail in Chapters 10 and 11.

We are left with the conclusion that there is no universally accepted definition for the broad class of phenomena that are usually referred to as cellular differentiation. However, this will not create serious problems for us, since we will normally not be dealing with the borderline cases that make the definition so difficult. In most cases the term "cellular differentiation" as it is used in this book will be compatible with both of the definitions that we have proposed: (1) display of different cellular phenotypes by cells of the same genotype within a single multicellular organism and (2) the collective specializations of structure and function that equip the nucleus and cytoplasm for a particular array of functional activities.*

Developmental biologists generally think of tissues and organs in terms of their origins from the three primary germ layers, as diagrammed in Fig. 2-26. Histologists, on the other hand, usually employ a classification scheme based on the properties of the tissues themselves, rather than on their embryological origins. Following are the four major classes of tissues (and their component cells) that are commonly recognized by histologists:

HISTOLOGICAL CLASSIFICATION OF CELLS AND TISSUES

1. *Epithelium.* This term is used to refer to all surface coverings and to glandular tissue derived from such coverings. The cells tend to be closely opposed laterally with little cementing substances between them, and they are arranged either as sheets covering or lining a sur-

---

*Students who wish to pursue the definition of differentiation further are referred to the review articles by Paul Gross and Chandler Fulton cited at the end of the chapter. The article by Fulton has threaded through it a fictionalized conversation between a physicist and a developmental biologist that makes particularly clear the problems encountered in trying to define differentiation precisely.

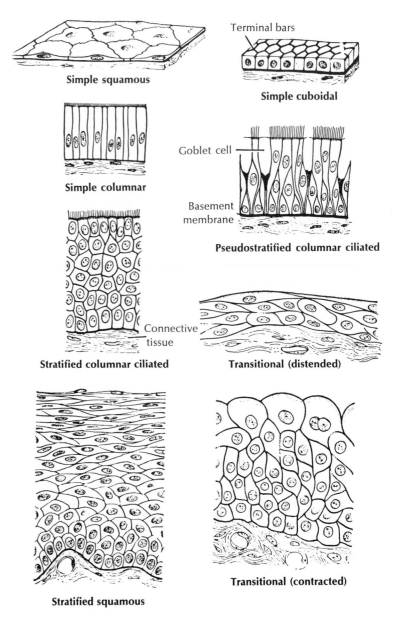

Fig. 3-29. Schematic representation of common types of epithelium. In each case the epithelium is shown with some of the underlying connective tissue. (From Copenhaver, W. M., D. E. Kelly, and R. L. Wood. 1978. Bailey's textbook of histology, 17th ed. The Williams & Wilkins Co., Baltimore.)

face or else as masses of cells in glands (Fig. 3-29). Epithelia are classified as squamous (composed of flattened cells), cuboidal (composed of cubelike cells), or columnar (composed of tall cells). In addition, they are classified as simple (composed of one layer of cells) or stratified (composed of multiple layers). Glands are also considered to be epithelial, although they are often in more complex masses, as in the liver. They are classified as exocrine or endocrine, depending on whether their secretory products are released into an external duct system or directly into the bloodstream.

2. *Muscle*. This term is used to refer to all tissues whose primary function is to generate movement. There are three basic types—voluntary skeletal muscle, with a characteristic cross-striated appearance; cardiac muscle, which has the special property of spontaneous rhythmic contraction; and smooth muscle, which is responsible for the slower involuntary movement of many organs.

3. *Nervous tissue*. This term refers to specialized tissues whose function is to conduct electrical impulses and coordinate bodily functions. It consists of neurons with elongated cytoplasmic processes capable of conducting electrical impulses, plus various types of related supporting and protective cells.

4. *Connective tissue*. This term refers to a broad category of cells that tend to be widely separated from one another by extracellular materials. It encompasses all types of supporting, connecting, and padding tissues, including bone, cartilage, tendons, ligaments, fat, and loose, fibrous tissue such as that immediately below the skin. Blood is generally also classified as a specialized form of connective tissue, consisting of cells surrounded by an extracellular substance, the blood plasma, that is fluid rather than solid. A few histologists prefer to classify blood as a fifth type of tissue, but they are in the minority. Most (but not all) connective tissues have a common origin from loose mesenchymal cells, as depicted in Fig. 3-6. Others, such as the cornea of the eye, originate from epithelial tissues in the early embryo.

The differences between the histological and embryological classification systems are important. Classes based on the three germ layers refer only to developmental origin and are not precisely correlated with adult functions. Some functional types of cells can be derived from more than one germ layer. For example, most muscle is mesodermal in origin, but certain muscles associated with the eye, as well as the smooth muscles associated with sweat glands, are of ectodermal origin. Similarly, epithelial tissues that cover surfaces can be derived from any one of the three primary germ layers. Hence histologists prefer to ignore developmental origins and to classify tissues and cells on a purely functional basis.

A detailed analysis of the many different types of cells that occur in the mammalian body and their organization into functional tissues and organs is beyond the scope of this text. Students who wish to pursue that topic further are referred to the textbooks in histology and cell biology listed at the end of this chapter. Our attention in the remainder of this chapter will be restricted to a limited number of specialized cells and to prominent differentiated properties that are frequently employed as models for studying the overall process of cellular differentiation.

MODEL SYSTEMS   As mentioned earlier, most studies of differentiation have been done with a limited number of types of cells that possess "markers" of differentiation which are particularly amenable to study. Despite the problems that are inherent in using such systems as models for the overall process of differentiation, the bulk of currently available data is derived from them. It is therefore appropriate to review briefly some of the most widely studied cell types and the particular differentiated properties that they exhibit.

Keratinocytes. Keratinocytes are the principal cell type in the epithelial layer of the skin. They form a multilayered, or stratified, epithelium with continual proliferation of cuboidal stem cells near the basement membrane and progressive flattening in the upper layers to form a squamous epithelium as the cells undergo terminal differentiation near the free surface. The terminal differentiation is characterized by the formation of large amounts of a tough, fibrous protein, keratin, and finally by drying and loss of cellular viability. This process, known as keratinization, generates the tough, fibrous, nonliving outer layer of the skin. The affinity of keratin for acidic dyes is sometimes used to follow the progress of keratinization, particularly in cultured cells.

Melanocytes. Melanocytes, which are characterized by synthesis of the dark pigment melanin, originate from the neural crest and migrate throughout the body. They are particularly concentrated in the pigmented retina and choroid of the eye, in hair follicles, and in the epidermis. They are characterized structurally by the presence of pigment granules known as melanosomes, which have a characteristic appearance in the electron microscope, although it may be obscured by the pigment that they contain (Fig. 3-30). Chemically, both the presence of melanin pigment and of the enzyme required for its synthesis from tyrosine, tyrosinase, are used as markers. Melanocytes can frequently be identified visually with low-power microscopy by the presence of pigment. However, pigment synthesis is regulated by a variety of physiological conditions, and cells structurally identifiable as melanocytes may fail to synthesize pigment. This also occurs in

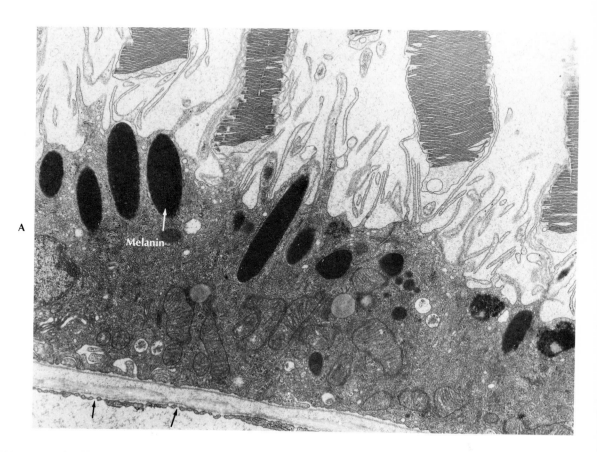

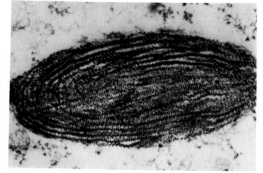

Fig. 3-30. Melanosomes. **A,** Electron micrograph showing melanosomes in the pigmented retina of the eye. The pigmented epithelium contains dense melanosomes whose internal detail is completely obliterated by the melanin pigment they contain. At top, photoreceptor outer segments with stacked lamellar discs can be seen interdigitating with microvilli of the pigmented epithelium. At bottom, fenestrated capillary endothelium is visible (arrows). Major contrasts in functional specialization are clearly evident in these three cell types. **B,** Electron micrograph of a developing human melanosome from the retina, showing the periodicity of its structural framework. When the melanosome is fully developed, its interior structure will be obscured by the accumulated melanin. (×63,000.) (**A** modified from Cogan, D. G., and T. Kuwabara. 1973. The eye. *In* R. O. Greep and L. Weiss. Histology, 3rd ed. McGraw-Hill Book Co., New York. Copyright © 1973 by McGraw-Hill, Inc. Used with permission of McGraw-Hill Book Co.; **B** from Bloom, W., and D. W. Fawcett. 1975. A textbook of histology, 10th ed. W. B. Saunders Co., Philadelphia.)

certain types of inherited albinism in which the defect is in pigment synthesis rather than in melanocyte differentiation.

Liver. The liver arises embryologically as an outgrowth of the endoderm of the gut, and histologically it is classified as glandular epithelial tissue. The liver has both an exocrine function (production of bile) and endocrine functions (production of blood proteins and hormones such as somatomedin). In addition, it performs many special metabolic functions for the entire body. Its structural organization is rather like that of an endocrine gland, although it also has an external duct system, the bile canaliculi, running between its cells (Fig. 3-4) and emptying into the bile duct. The chemical markers for differentiation of liver cells that are most frequently studied are production of serum albumin and the presence of various enzymes related to its specialized metabolic functions. In some cases it is possible to distinguish a liver-specific form of an enzyme from other forms of the same enzyme that are synthesized elsewhere in the body and are apparently coded for by different genes. An example is the B form of fructose 1,6-diphosphate aldolase, more frequently referred to simply as liver aldolase or aldolase B. Tissue-specific forms of enzymes, such as aldolase B, are frequently detected by differences in their electrophoretic mobility.

Mammary gland. The secretory epithelium of the mammary gland is formed by an inward budding of the ectoderm during embryonic development. With proper hormonal stimulus (Chapter 14) the mammary epithelial cells synthesize milk. The usual markers employed to measure functional differentiation of such cells are casein (a phosphoprotein containing unusually high levels of phosphate), α-lactalbumin, β-lactoglobin, and the enzymes responsible for synthesis of the milk sugar, lactose.

Chicken oviduct. The oviduct in the chicken is the site of synthesis of the proteins found in egg white. When stimulated with estrogen, cells of the oviduct differentiate to form the tubular glands that are responsible for synthesis of these proteins. Over 60% of the protein synthesized by fully stimulated tubular gland cells is a single species, ovalbumin. The protein avidin, which can be identified by its strong binding to biotin (one of the B vitamins) is also sometimes used as a specific marker of oviduct differentiation.

Pancreas. The pancreas arises as an outgrowth of the gut endoderm, and its functional differentiation has been studied extensively, particularly in terms of inductive relationships between the endodermal epithelium and the underlying mesodermal connective tissue of the gland. Three different types of pancreatic cells have been studied in parallel. The exocrine cells are responsible for synthesis of a number of zymogens, including those for intestinal amylase, chymotrypsin,

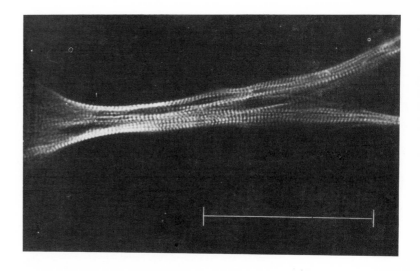

Fig. 3-31. A single cultured myotube seen with the polarizing microscope. The cross striations due to molecular orientation in the myofibrils are plainly evident. The scale marker represents 0.1 mm. (From Konigsberg, I. R. 1963. Science **140:**1273. Copyright 1963 by the American Association for the Advancement of Science.)

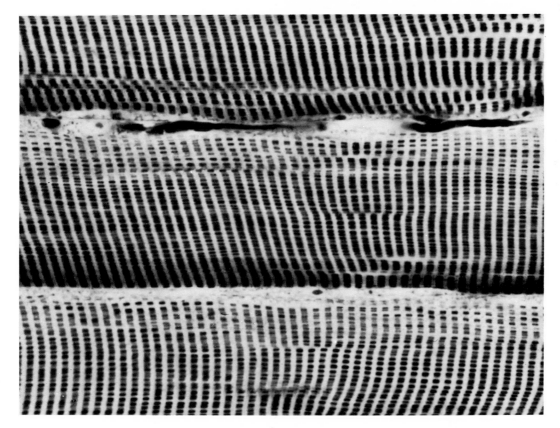

Fig. 3-32. Stained muscle fibers seen with the bright field light microscope. (Iron-hematoxylin; ×1200). (From Bloom, W., and D. W. Fawcett. 1975. A textbook of histology, 10th ed. W. B. Saunders Co., Philadelphia.)

trypsin, and lipase. The α-endocrine cells synthesize the hormone glucagon, which increases circulating levels of glucose in the bloodstream, and the β-endocrine cells synthesize insulin, which functions to decrease circulating glucose and also to stimulate various aspects of metabolism.

Skeletal muscle. Skeletal muscle cells, which are of mesodermal origin, are highly specialized, both morphologically and chemically, for their contractile function, as described earlier in this chapter (Figs. 3-18 to 3-20). Because of their abundance of visual and chemical markers they are frequently studied as model systems for differentiation. Mature skeletal muscle cells consist of complex multinucleate myotubes, which are formed by end-to-end fusion of a large number of mononucleate precursor cells known as myoblasts. The cytoplasm of the myotubes is almost completely filled with contractile myofibrils, which have a characteristic cross-striated appearance when viewed live with phase contrast or with the polarizing microscope (Fig. 3-31) or in stained preparations with the ordinary light microscope (Fig. 3-32). The myofibrils also have a characteristic appearance in the electron microscope (Fig. 3-33). The most frequently studied chemical markers of muscle differentiation are muscle actin and muscle myosin. However, a number of other biochemical markers, including tropomyosin, troponin, three forms of actinin, creatine phosphokinase, glycogen phosphorylase, myokinase, acetyl choline receptors, and acetyl cholinesterase, have also been employed by some investigators. In addition, skeletal muscle can be subdivided into red and white, depending on whether the respiratory protein myoglobin is present in large amounts. Many of these same biochemical markers are also present in cardiac and smooth muscles. In the case of actin, as described earlier (p. 77), minor differences in amino acid sequence have been reported between the form found in skeletal muscle and that from cardiac muscle, suggesting that different genes for actin synthesis are activated during differentiation of these two classes of muscle cells.

Fig. 3-33. Electron micrograph of glycerin-extracted skeletal muscle. This treatment improves the contrast of the cross-striated myofibrils, but damages the mitochondria and the sarcoplasmic reticulum. Note the uniform diameter of the myofibrils. The corresponding bands of adjacent myofibrils are usually in register across a muscle fiber. When they are out of register, as at upper left, this is usually an artifact of specimen preparation. (From Bloom, W., and D. W. Fawcett. 1975. A textbook of histology, 10th ed. W. B. Saunders Co., Philadelphia.)

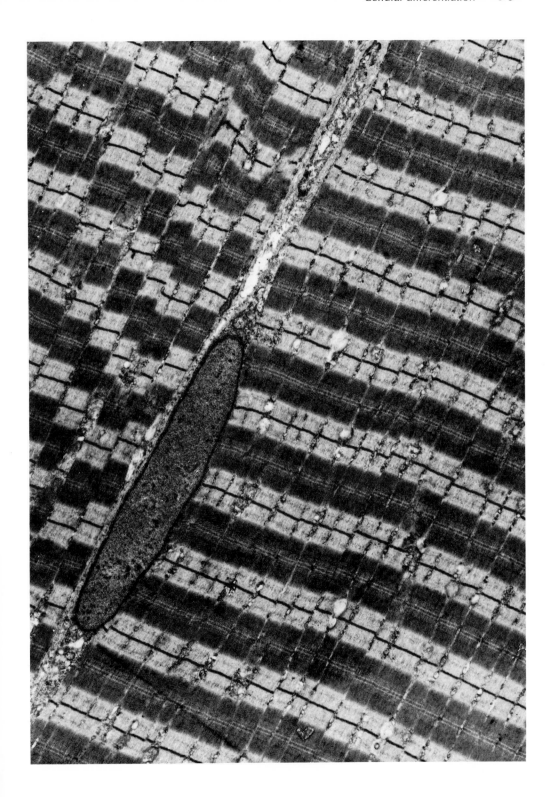

Neurons. Embryologically, the central nervous system arises by infolding of the ectoderm to form the neural tube. Many of the associated ganglia arise from neural crest tissue. Neurons are highly differentiated cells that function by responding to stimuli and conducting electrochemical impulses over long distances via specialized cytoplasmic extensions known as axons (Fig. 3-34). Mature neurons arise from precursors known as neuroblasts. Outgrowth of the axon, which is often used as a morphological marker for nerve differentiation, involves both an active movement of the tip away from the cell body, mediated by cytoplasmic microfilaments, and maintenance of rigidity of the growing axon, mediated by microtubules (Chapter 15). The axons of neurons are often surrounded by a multilayered membranous

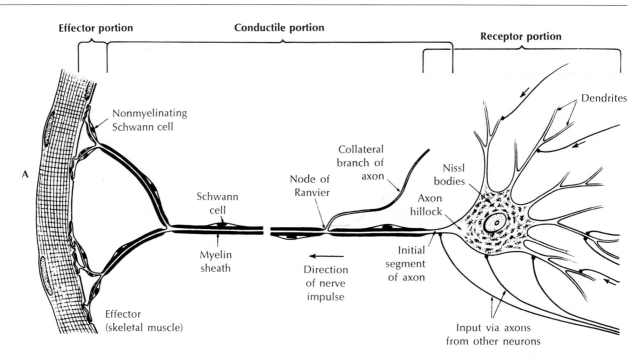

Fig. 3-34. A, Diagrammatic representation of a large motor neuron. The dendrites of the receptor portion of the cell receive input from axons of other neurons. The axon of a motor neuron to a limb muscle in humans may be as much as 2 to 3 feet long. A myelin sheath, which is formed by an adjacent Schwann cell, acts as "insulation" around the axon and serves to increase its conduction velocity. **B,** Electron micrograph showing multiple layers of myelin wrapped around an axon. The inset shows the myelin layers at higher magnification. (**A** from Bloom, W., and D. W. Fawcett. 1975. A textbook of histology, 10th ed. W. B. Saunders Co., Philadelphia; modified from Copenhaver, W. M., R. P. Bunge, and M. B. Bunge. 1971. Bailey's textbook of histology, 16th ed. The Williams & Wilkins Co., Baltimore; **B** from Webster, H. 1974. *In* J. Hubbard, ed. The vertebrate peripheral nervous system. Plenum Press, New York.)

sheath of myelin, which does not originate from the neuron but rather from accessory cells of the nervous system known as Schwann cells, whose membranous extensions are wrapped repeatedly around the axons. Many types of morphological specialization occur both in the cells of the central nervous system, which can assume complex shapes (Figs. 3-1 and 3-35), and in receptor cells that feed information to the central nervous system. Chemical markers that are used to identify neuronal differentiation include synthesis and hydrolysis of the neurotransmitter substance acetylcholine, and sometimes a second set of neurotransmitters, the catecholamines. Functional measurements are sometimes made of the ability of neuronlike cells to respond to specific stimuli and to conduct electrochemical impulses. Neuro-

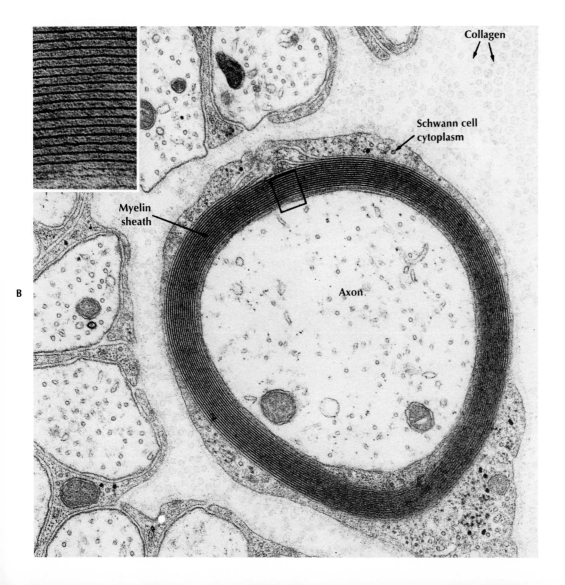

B

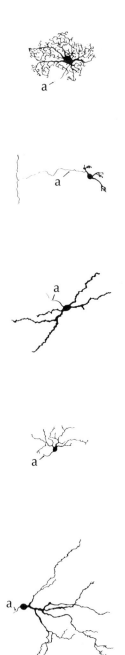

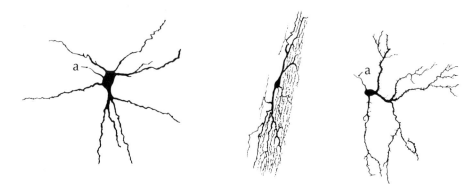

Fig. 3-35. Some characteristic types of neurons found in the central nervous system. The *a* indicates the axon. Other processes are dendrites. A remarkable diversity of cell form is evident. (Courtesy Dr. Clement Fox, from Truex, R. C., and M. B. Carpenter. 1969. Human neuroanatomy, 6th ed. The Williams & Wilkins Co., Baltimore.)

blastoma tumor cells, which have retained the ability to differentiate into neuronlike cells under certain conditions, are also rather widely studied.

Chondrocytes. Cartilage cells, or chondrocytes, are identified primarily by their copious secretion of extracellular matrix material consisting of a mixture of Type II collagen and chondromucoproteins rich in the acidic polysaccharide, chondroitin sulfate. In many cases the matrix completely surrounds each cell and totally isolates it from contact with other cells (Fig. 3-36). The highly acidic properties of chondroitin sulfate make possible the use of special stains such as acidic solutions of Alcian blue or Alcian green to assay for cartilage differentiation. Incorporation of radioactive sulfate into the matrix is also used as a marker of differentiation. A set of seven enzymes is needed for synthesis of chondroitin sulfate. All seven of these enzymes are present in large amounts in differentiated cartilage cells. Synthesis of Type II collagen is also a highly specific marker for cartilage differentiation.

Tendon cells. Fibroblastic cells from tendons are characterized by a commitment to the synthesis of Type I collagen (up to 30% of their total protein synthesis under some conditions). This particular marker is quantitative, rather than qualitative, since Type I collagen is synthesized in smaller amounts by many other types of connective tissue, including dermis and bone.

Erythroblasts. Erythroblasts are nucleated precursors of erythrocytes, or red blood cells. During the later stages of their development, hemoglobin is virtually the only protein that they synthesize. As noted earlier in this chapter, sequential changes occur in the type of hemoglo-

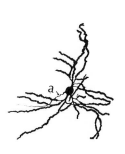   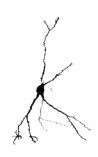

bin synthesized in humans as the individual passes from early embryonic stages to later fetal development and then to postnatal development.

Plasma cells. Differentiation of type B lymphocytes into antibody-synthesizing plasma cells is sometimes studied as a model system, although the gene activation that is involved is extremely complex and may use regulatory mechanisms other than those involved in more typical cases of differentiation.

CULTURED CELLS

The multiplicity of different events occurring at the same time in a developing embryo makes it difficult to study any one in detail. One of the methods that is widely used to get around this problem is to explant cells from living organisms into culture dishes containing nutrient media. This process is referred to as cell culture, tissue culture, or organ culture, depending on the level of organization that is maintained in the explanted cultures. Since the original work was generally done in glass culture vessels, such cultures are often referred to as "in vitro,"* although now the bulk of the work is done in specially treated plastic containers, rather than in glass. The corresponding term for studies done in intact organisms is "in vivo," or sometimes "in situ," to indicate that the natural location of the tissue has not been disturbed.

The use of cell and tissue culture to study differentiation will be examined in Chapters 10 and 11, after the detailed analysis of molecular mechanisms of differentiation presented in the following section of

---

*The term "in vitro," which means "in glass," is a potential source of confusion, since it can have two rather different meanings. As used here, it refers to intact cells or to tissue or organ fragments that contain intact cells growing in a culture vessel that is fully separated from the body of the donor organism. Biochemists and molecular biologists frequently use the same term to refer to experimental conditions in which cell membranes have been ruptured and enzymes or other macromolecules are in free solution (e.g., protein synthesis by cell-free extracts is referred to as occurring in vitro, as opposed to occurring in intact cells).

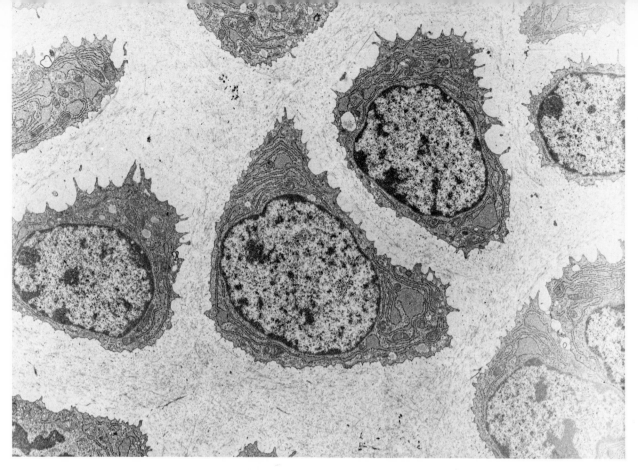

Fig. 3-36. Electron micrograph of the cartilage of mouse trachea. The cells are completely separated from one another by the extracellular matrix. Mature chondrocytes typically have an irregular outline and well-developed granular endoplasmic reticulum. Some types of cartilage cells also contain large fat droplets, although only small lipid inclusions are evident in this picture. (From Seegmiller, R., C. Ferguson, and H. Sheldon. 1972. J. Ultrastruct. Res. **38**:288.)

this book. However, since experiments involving cultured cells are frequently mentioned in the chapters on molecular mechanisms, it will be helpful at this time to summarize briefly some of the major points.

In general terms, cultured cells can be divided into two broad classes: (1) established cell lines that have undergone a major process of adaptation, which generally includes chromosomal changes, and have become capable of indefinite growth in culture and (2) normal diploid cells, which have retained many of the properties of their tissue of origin and are generally capable of only a limited number of divisions in culture. The restricted reproductive potential of normal cells is often studied as a model of aging and will be examined in Chapter 24.

Most established cell lines are either of malignant origin, or else have acquired malignant properties during their adaptation to long-term culture growth. Because of their relative ease of handling, established lines have been studied much more extensively than normal cells, although much of the work with differentiation in culture is done with normal diploid cells. Many established lines do not exhibit differentiated properties, and those which do often exhibit them constitutively in a manner that suggests that normal regulatory mechanisms may not be fully operational.

During the period when cell culture methods were being developed, the media and techniques that were available were sufficiently crude that extensive evolution of the cells in vitro was needed to obtain a cell line that would grow well. Among the changes that typically occur during the development of cell lines are chromosomal changes, in particular an increase in the total amount of genetic material. The cells also tend to acquire rapid growth properties similar to those found in cancerous cells if the starting tissue was not already malignant. Several of the cell lines that were established during this early period were shipped to many different laboratories and became widely used as research tools. Two lines in particular, mouse L and HeLa, are still widely used.

Established cell lines

Mouse L cell line. This line was derived from subcutaneous connective tissue removed from a C3H mouse in 1940. Its adaptation to culture included long-term exposure to a chemical carcinogen, 20-methylcholanthrene, which resulted in acquisition of malignant properties and the ability to grow much more rapidly in culture. The letter "L" refers to experiment L in an alphabetically designated series of different times of exposure to and concentrations of the carcinogen. Descendants of the original mouse L line are still widely used.

HeLa cell line. This line was derived from a malignant carcinoma of the cervix, removed surgically from a patient and designated by the first two letters of her first and last names. HeLa was the first established line of human origin to become widely used. The patient from whom the tumor was removed was of African descent, and one of the biochemical markers that the HeLa cell carries is a variant of glucose-6-phosphate dehydrogenase designated Type A, which is found primarily in black populations. HeLa is a vigorously growing cell line, which has, on a number of occasions, completely displaced other cell lines when it was accidently introduced into cultures as a contaminant. Despite its malignant properties and abnormal karyotype (chromosomal constitution), the HeLa cell line has become the standard experimental tool in a number of laboratories studying the molecular biology of gene expression in mammalian cells, and results of experiments done with HeLa cells will be cited frequently in Chapters 5 to 9.

Differentiated tumor lines. There exist a variety of tumors, both in humans and in experimental animals, that continue to exhibit differentiated properties while growing malignantly. Cell lines that exhibit differentiated properties in culture have been established from such tumors and have been studied as possible model systems for the normal process of differentiation. In some cases these cells appear to exhibit relatively normal control over the expression of differentiated properties, whereas in others the differentiation is always expressed, and normal controls do not appear to be functional. Differentiated tumor lines that are frequently studied include adrenal and pituitary tumors that continue to produce their respective hormones, hepatomas with liverlike properties, neuroblastoma cells that retain the ability to form axons and undergo neuronlike differentiation in suitable environments, melanomas that produce pigment (and others with melanosomes but no pigment), myelomas that synthesize large amounts of specific antibodies, and erythroleukemia lines that synthesize hemoglobin when properly stimulated.

Teratocarcinoma. Another group of tumor lines of particular interest are those derived from teratocarcinomas, which are embryonal tumors, presumably derived from germ cells or yolk sac. Teratocarcinoma cells have the unusual property of being able to undergo many different types of differentiation, both in culture and when injected into test animals. They thus appear to retain many of the developmental potentialities of early embryonic cells, although they lack the regulatory mechanisms needed to form fully organized embryos.

Normal cells in culture

Modern culture techniques and media (discussed in Chapter 10) have in recent years made it possible to grow in culture many different types of normal differentiated cells. The complexity of the undefined supplements that have generally been necessary in the culture media for growth of such cells has thus far hampered major studies of environmental signals controlling expression of differentiated properties by such cells. However, progress is beginning to be made toward better culture systems for normal cells, and normal cells in culture are being used with increasing frequency as model systems for studying the control of differentiation.

Most of the studies that have thus far been done with normal differentiated cells in culture have concentrated on control of the expression of differentiated properties, rather than on the initial commitment to a particular pathway of differentiation during embryonic development. These studies have shown clearly that expression is under reversible environmental control and that cells tend to retain their commitment to a particular type of differentiation even when they are not actually expressing specific differentiated properties. Such studies will be described in Chapter 10.

Without agreeing on an exact definition, we have seen that differentiation is basically the process by which cells within an organism become different from one another. The differences that are observed involve virtually every aspect of cellular structure and chemistry and can also be described in terms of functional specialization. Differentiation does not consist of a single change in one particular cellular property. Rather, it appears to involve all aspects of the cell, both structural and chemical, suggesting that there may exist encoded somewhere within the genome a set of detailed "programs" for each particular type of cellular differentiation. The mechanisms responsible for activating such programs remain almost completely unknown, and the existence of the programs themselves is still highly speculative. One of the major challenges that still exists in developmental biology is to find out how cells that start with a common genome and presumably use the same types of mechanisms for control of gene expression are able to exhibit such radically different properties within a single developing embryo.

MAJOR CONCEPTS OF DIFFERENTIATION

• • •

We have now completed our preliminary summarization of life cycles, embryonic development, and the major concepts of cellular differentiation. The remainder of this book is dedicated to an examination in greater depth of the mechanisms involved in these processes. In the following section (Chapters 4 to 9) we will examine differentiation in terms of selective gene expression. In so doing, we will be focused primarily on chemical aspects of cellular differentiation, since at the present time, it is far easier to relate chemical changes to genetic events than it is to try to explain morphological changes in terms of known mechanisms for the control of gene expression. After we have completed our analysis at the molecular level, we will return to the cell and begin examining phenomena and regulatory mechanisms that have been observed at the cellular level of organization but have not yet been explained adequately at the molecular level.

It is particularly important to keep in mind during our large excursion into molecular mechanisms that we are still a long way from being able to explain in molecular terms many types of developmental phenomena observed at the cellular and the organismic levels. Thus, although it is one of our basic articles of faith that such explanations will ultimately be forthcoming, we must not lose our perspective while becoming immersed in the rapid progress that is currently being made in certain areas of the molecular biology of gene expression.

BIBLIOGRAPHY
Books and reviews

Banerjee, M. R. 1976. Responses of mammary cells to hormones. Int. Rev. Cytol. **47**:1.

Bevelander, G., and J. A. Ramaley. 1974. Essentials of histology. The C. V. Mosby Co., St. Louis.

Bloom, W., and D. W. Fawcett. 1975. A textbook of histology, 10th ed. W. B. Saunders Co., Philadelphia.

Copenhaver, W. M., D. E. Kelley, and R. L. Wood. 1978. Bailey's textbook of histology, 17th ed. The Williams & Wilkins Co., Baltimore.

Cox, R. P. and J. C. King. 1975. Gene expression in cultured mammalian cells. Int. Rev. Cytol. **43**:281.

Fulton, C. 1977. Cell differentiation in *Naegleria gruberi*. Annu. Rev. Microbiol. **31**:597.

Greep, R. O. and L. Weiss, eds. 1973. Histology, 3rd ed. McGraw-Hill Book Co., New York.

Gross, P. R. 1968. Biochemistry of differentiation. Annu. Rev. Biochem. **37**:631.

Ham, A. W. 1969. Histology, 6th ed. J. B. Lippincott Co., Philadelphia.

Leeson, T. S., and C. R. Leeson. 1970. Histology, 2nd ed. W. B. Saunders Co., Philadelphia.

Levitt, D., and A. Dorfman. 1974. Concepts and mechanisms of cartilage differentiation. Curr. Top. Dev. Biol. **8**:103.

Martin, G. R. 1975. Teratocarcinoma as a model system for the study of embryogenesis and neoplasia. Cell **5**:229.

Merlie, J. P., M. E. Buckingham, and R. G. Whalen. 1977. Molecular aspects of myogenesis. Curr. Top Dev. Biol. **11**:61.

Miller, E. J., and V. J. Matukas. 1974. Biosynthesis of collagen. The biochemist's view. Fed. Proc. **33**:1197.

Nigon, V., and J. Godet. 1976. Genetic and morphogenetic factors in hemoglobin synthesis during higher vertebrate development: an approach to cell differentiation mechanisms. Int. Rev. Cytol. **46**:79.

Porter, K. R., and M. A. Bonneville. 1968. Fine structure of cells and tissues, 3rd ed. Lea & Febiger, Philadelphia.

Prasad, K. N. 1975. Differentiation of neuroblastoma cells in culture. Biol. Rev. **50**:129.

Rutter, W. J., R. L. Pictet, and P. W. Morris. 1973. Toward molecular mechanisms of developmental processes. Annu. Rev. Biochem. **42**:601.

Sherman, M. I., and D. Solter, eds. 1975. Teratomas and differentiation. Academic Press, Inc., New York.

Staehelin, L. A., and B. Hull. 1978. Junctions between living cells. Sci. Am. **238**: 141 (May).

Thorp, F. K., and A. Dorfman. 1967. Differentiation of connective tissues. Curr. Top. Dev. Biol. **2**:151.

Weatherall, D. J., and J. B. Clegg. 1976. Molecular genetics of human hemoglobin. Annu. Rev. Genet. **10**:157.

Weatherall, D. J., and J. B. Clegg. 1979. Recent developments in the molecular genetics of human hemoglobin. Cell **16**:467.

Wigley, C. B. 1975. Differentiated cells in vitro. Differentiation **4**:25.

Selected original research articles

Lu, R., and M. Elzinga. 1977. Comparison of amino acid sequences of actins from bovine brain and muscles. *In* R. Goldman, R. Pollard, and J. Rosenbaum, eds. Cell motility. Cold Spring Harbor Conferences on Cell Proliferation. Vol. 3B. Cold Spring Harbor Laboratory, Cold Spring Harbor, N. Y.

Martin, G. R., L. M. Wiley, and I. Damjanov. 1977. The development of cystic embryoid bodies in vitro from clonal teratocarcinoma stem cells. Dev. Biol. **61**:230.

Storti, R. V., and A. Rich. 1976. Chick cytoplasmic actin and muscle actin have different structural genes. Proc. Natl. Acad. Sci. U.S.A. **73**:2346.

Molecular mechanisms that control gene expression

☐ In the preceding chapter we have seen that one of the major features of cellular differentiation is differential synthesis of specialized proteins. For example, red blood cells synthesize hemoglobin, muscle cells synthesize actin and myosin, cartilage cells produce collagen and the enzymes for synthesis of chondroitin sulfate, and glandular cells produce their particular products for export. The information that guides the synthesis of each of these differentiated products is coded in the genome. This has been shown particularly well in the case of human hemoglobin, for which many genetic variants are known, reflecting both simple mutations that result in single amino acid substitutions and a variety of complex mutations resulting in large-scale changes in amino acid sequence (some are described in Chapter 8).

Differential expression of selected parts of the vast library of information that is carried in the genome is thus one of the major aspects of cellular differentiation. The next six chapters are dedicated to an exploration of the molecular mechanisms that are involved in selective gene expression. These mechanisms are not simple, and there is no way to make them sound simple without deceiving our readers. Even the process of differentiation itself is not as simple as we made it sound in the previous paragraph. Although gross biochemical markers of differentiation, such as hemoglobin synthesis, are the easiest and most convenient to study, innumerable subtle quantitative changes also occur to such an extent that the entire metabolic pattern of the cell becomes differentiated. In addition, each type of differentiated cell acquires its own characteristic morphological organization, which, at present, is difficult to explain fully in terms of differential synthesis of specific proteins.

In the interest of clarity we will look one at a time at the many different molecular mechanisms that are involved in cellular differentiation. However, we ask our readers to keep in mind the fact that in reality all these mechanisms (and still others that we have not mentioned) operate concurrently within differentiated cells.

Living cells are classified into two fundamental types—prokaryotic and eukaryotic. The primary distinguishing feature between them is the presence of a nuclear envelope in eukaryotic cells and its absence in prokaryotic cells. The cells of all higher plants and animals are eukaryotic, as are also many unicellular organisms such as protozoa, yeast, and most kinds of algae. Prokaryotic organisms are typically single celled, such as bacteria and blue-green algae, although some, such as the moldlike actinomycetes, form relatively complex multicellular aggregates. Prokaryotic cells are much smaller than eukaryotic cells and have a relatively simple internal structure without endoplasmic reticulum, mitochondria, or chloroplasts. In addition, their genomes are much smaller than those of typical eukaryotic cells.

Our major concern in this book is with eukaryotic organisms. However, we have elected to begin this section with a chapter describing the molecular mechanisms of gene expression and its control as they occur in prokaryotic organisms. Our reason is that much of the pioneering research in molecular genetics was done with these "simpler" systems. At the present time they are better understood than corresponding eukaryotic systems, and much of the current research that is being done on eukaryotic gene expression is modeled on the foundation that has been acquired by studying prokaryotic organisms. An understanding of the fundamentals of gene expression and gene regulation in prokaryotic cells is therefore essential as a starting point for understanding current research that is being done on eukaryotic cells.

Our readers should not infer from the previous paragraph that all the major questions regarding gene expression in prokaryotes have already been answered. This is simply not the case. Although we think we understand most of the general principles, the exact details of the regulatory interactions among the different macromolecules that participate in the control of gene expression have barely begun to be explored. In the case of eukaryotic cells we probably do not even know about all the major principles that will ultimately prove to be important in the regulation of gene expression. Significant new discoveries continue to be reported frequently for both types of cells, and the material that is presented in this book must be regarded only as the current "state of the art."

The major compensation for the lack of firmly established information in this area of study is the excitement of discovery that currently permeates the entire field of gene regulatory mechanisms. Significant findings are constantly being published in the current research literature, and we expect that trend to continue for many years after this book has been published. It is therefore important for the reader who wishes to keep up with new discoveries in the field to read both the original research literature and review journals regularly. A guide to the use of those materials is provided in Appendix E.

The first chapter in this section (Chapter 4) summarizes the basic mechanisms that are involved in the flow of information from DNA to RNA to protein (often referred to as the "central dogma" of molecular biology) and surveys specific mechanisms that are responsible for the selective control of gene expression in prokaryotic organisms. Emphasis is given to basic mechanisms that are well established, but we have also tried to give our readers an honest view of the many unanswered questions that still exist even for these so-called simple organisms.

The next five chapters are dedicated to a systematic analysis of genetic information in eukaryotic cells and of known or potential sites where the regulatory processes responsible for selective expression of that information might occur. A typical eukaryotic genome is far larger than seems "reasonable" from our limited point of view. We can draw an analogy between the eukaryotic genome and a vast library of potential information, most of which is never used by the average reader and some of which may be read only rarely, if at all.

In Chapter 5 we explore the physical organization of the library, which is contained inside a membrane-bound nucleus and is thus fully separated from the protein-synthesizing machinery of the cytoplasm, which must ultimately translate its nucleotide sequence information into amino acid sequences in protein molecules. The library itself is composed of coded DNA sequences and is intimately associated with a variety of histone and nonhistone proteins. These proteins are thought to play both structural and regulatory roles.

In Chapter 6 we ask whether each cell in a multicellular organism contains a complete copy of the entire library and whether all of it is in an accessible form. Specialized cases in which parts of the library are selectively amplified, eliminated, or inactivated permanently are examined, but the generalized conclusion is that most differentiated cells within a single organism contain the same basic genetic information, even though they express it differently.

In Chapter 7 we begin asking how those portions of the library which are to be translated into protein sequences are selected. Evidence is presented that some of the selectivity occurs in the transcription of information from DNA sequences in the genome into RNA sequences and that transcriptional regulation is important to the overall process of cellular differentiation. Specific mechanisms involved in the regulation of transcription are also examined.

In Chapter 8 we focus on the fact that the original RNA copied from the eukaryotic genome must undergo extensive processing in the nucleus before it is transported to the cytoplasm and made available to the protein-synthesizing machinery. Only a small fraction of the information that is transcribed into RNA sequences in the nucleus ever reaches the cytoplasm as functional messenger RNA capable of supporting protein synthesis. Recent evidence suggests that controls acting at this level may play a major role in differentiation.

In Chapter 9 we examine protein synthesis itself. The translation of RNA sequences into amino acid sequences is one of the most complicated processes that occurs in any biological system, and it offers many additional opportunities for control of gene expression. Also, even after synthesis of the protein chains is completed, major modification is often needed to transform them into functional enzymes and structural proteins. Regulatory mechanisms acting at the translational and posttranslational levels appear to be used rather extensively in the "fine tuning" of controlled gene expression.

The final discussion in Chapter 9 presents a brief overview of the complete process of gene expression in eukaryotic cells and the many points along the pathway of information flow where known or potential regulation of that gene expression can occur. Readers may wish to look ahead to Fig. 9-14 from time to time to maintain an overall perspective as the individual steps that are involved in the flow of information from gene to functional protein and the mechanisms believed responsible for regulating that flow are examined in detail.

Finally, to the reader who asks, "Is it really necessary to have all this molecular material in a textbook on developmental biology?" we reply, "Yes, we think it is." As we emphasized in the preface to this book, all living organisms are composed of molecules. We will never have a true understanding of how organisms operate until we know what happens at the molecular level. The control of gene expression is an area where particularly rapid progress is being made toward that goal. Thus it is important for any student who is preparing to understand the progress of future research on the mechanisms of development to gain a thorough understanding of the molecular biology of gene expression and its regulation.

CHAPTER 4

Gene expression and its regulation in prokaryotic cells

☐ This chapter is concerned with what genes are, how they work, and how their function is controlled in prokaryotic cells. The major themes that are developed in this chapter are the following:

1. The primary function of most genes is to provide sequence information for the synthesis of biologically active protein molecules (e.g., enzymes and structural proteins).

2. Genes are composed of deoxyribonucleic acid (DNA) or, in some viruses, of ribonucleic acid (RNA).

3. There is a direct linear relationship between the nucleotide sequence of each gene and the amino acid sequence of the protein that it specifies. Each amino acid in the protein is coded for by three adjacent nucleotides, commonly referred to as a codon.

4. DNA does not serve directly as a template for protein synthesis. The sequence information from DNA is first "transcribed" into "messenger" RNA (mRNA) sequences, which are then "translated" into amino acid sequences. This flow of information from DNA to RNA to protein is often referred to as the "central dogma" of molecular biology.

5. Not all proteins are synthesized at the same time or in the same amounts. Complex regulatory mechanisms permit particular genes to be expressed or repressed selectively, with relatively little effect on the remainder of the genome.

Because first-year courses in biology vary so greatly in their treatment of molecular genetics, the material in this chapter will be almost entirely review for some of our readers and almost entirely new for others. In either case we urge readers to be certain that they understand these essential background concepts thoroughly before proceeding to the following chapters on gene regulatory mechanisms in eukaryotic cells. Students with no previous exposure to these topics may wish to study them in greater detail, using the references cited at the end of this chapter. In particular, we recommend Chapters 6 to 15 of the third edition of *Molecular Biology of the Gene* by Watson.

BASIC CONCEPTS OF MOLECULAR GENETICS
Brief history

The entire field of molecular genetics is relatively new. The chemical nature of the genetic material was not known until 1944 when Oswald T. Avery, Colin M. MacLeod, and Maclyn McCarty demonstrated that DNA extracted from dead bacteria was capable of transferring genetic information to a different strain of living bacteria. The double helical structure of DNA, which provided a basis for understanding how genes duplicate themselves precisely, was worked out 9 years later in 1953 by James Watson and Francis Crick. Although these and other pioneering discoveries set the stage, it was not until 1961, within the lifetimes of the first students who will use this book, that gene expression and its regulation in prokaryotic cells began to be understood in the form we know today.

Three related events in 1961 provided the basis for our current understanding. The first was demonstration that sequence information for protein synthesis is carried by mRNA molecules that are entirely separate from the ribosomal RNA (rRNA) and transfer RNA (tRNA) molecules that are involved in the mechanical aspects of assembling protein molecules. The second was cracking of the genetic code, whose polynucleotide triplet nature had previously been only a matter of speculation. The third was development of the operon model for regulation of gene expression, which introduced the concept of "operator" sites in the genome that respond to external signals and control the expression of adjacent "structural" genes. The operon model also brought the entire pattern of information flow from gene to protein together into a single coherent picture that was much more precise than that previously available.

Our knowledge of gene function has escalated rapidly in the years since 1961. Many details have been added to the original operon model, and additional control pathways involving different mechanisms have also been described. At the time this book is being written, exciting progress is occurring in the development and application of techniques for sequence analysis of proteins and nucleic acids. These techniques in combination with fine-structure genetics are making it possible to construct models of regulatory mechanisms with a whole new order of precision.

What are genes?

In general terms, a gene can be defined as a fundamental unit of heredity that is responsible for transmission of a particular heritable trait to future generations. Thus genes are carriers of information. Superficially, genes appear to carry many different types of information, as evidenced by the great diversity of structural, physiological, and metabolic traits that are known to be inherited. However, modern research is making it appear increasingly likely that only two types of information are actually carried in the genome. The first consists of coded information for the synthesis of specific proteins (and functional RNA that is not translated into protein sequences, such as tRNA and rRNA). The second class consists of regulatory information that in various ways controls gene function. In addition, there is a substantial amount of DNA, particularly in the very large genomes of some eukaryotic organisms, that currently appears not to have a genetic function. Whether this is "junk" DNA (e.g., evolutionary leftovers or side products that are not functional) or whether it has a yet-to-be-discovered function remains to be determined by future research.

Relationship of genes to protein synthesis. It has long been evident to careful observers that genes and enzymes are closely related. Edmund B. Wilson wrote in 1896, "Inheritance is the recurrence in successive generations of like forms of metabolism," and by 1909 Archibald Garrod was able to make an accurate association between a recessively inherited metabolic defect in humans, alkaptonuria, and loss of the ability to metabolize homogentisic acid, a breakdown product of the amino acid tyrosine. After the isolation of mutant strains of the mold *Neurospora crassa* with specific biochemical defects was reported by George Beadle and Edward Tatum in 1941, many writers began to refer to a "one gene–one enzyme" relationship.

Subsequent discoveries that genes are made of DNA and that DNA carries coded information for protein synthesis made it possible to refer to a gene specifically as the coded information for the synthesis of one enzyme. However, enzymes are sometimes composed of several different polypeptide subunits, each of which can be under the control of a separate gene. In addition, structural and regulatory proteins that do not possess enzymatic or catalytic activity are also coded for by genes. Hence it is now more accurate to describe a gene as the coded information for the synthesis of a single polypeptide chain.

However, even this definition is not fully satisfactory, since there are also specific genes that do not code for polypeptides. These are the genes that code for nonmessenger RNA molecules, such as rRNA and tRNA, which are never translated into proteins. In addition, it is now known that in certain cases genes can overlap (discussed later in this chapter), and they can sometimes contain noncoding sequences inserted into the middle of coding sequences (Chapters 5 and 8).

The term "cistron" is frequently used to describe the coding unit for a single peptide chain, although strictly speaking it refers to the small-

A

B

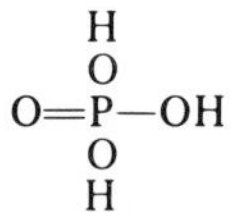

H
|
O
‖
O=P—OH
|
O
|
H

Phosphoric acid

HOCH$_2$ O   OH
5 | 4   3   2   1 |
H   H   H   H
| 3   2 |
OH   H

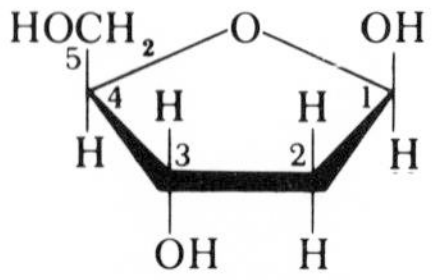

D-2-Deoxyribose

(β-D-2-deoxyribofuranose)

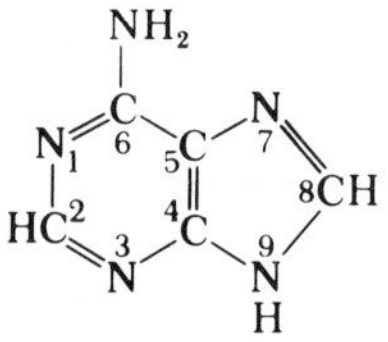

NH$_2$
|
C
N$_1$ 6   5   C   7
‖   C   N
HC$_2$ 3   4   C   8CH
N   9
N
H

Adenine

(6-aminopurine)

O
‖
C
HN$_1$ 6   5   C   7
| N
H$_2$N—C$_2$ 3   4   C   8CH
N   9
N
H

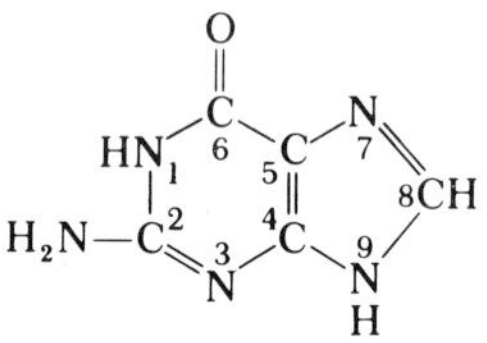

Guanine

(2-amino-6-oxypurine)

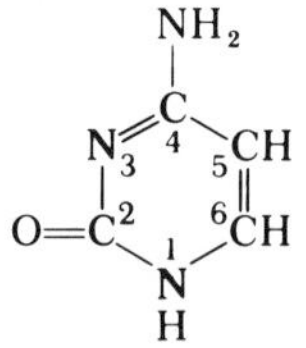

NH$_2$
|
C
N$_3$ 4   5   CH
‖   6
O=C$_2$ 1   CH
N
H

Cytosine

(2-oxy-4-aminopyrimidine)

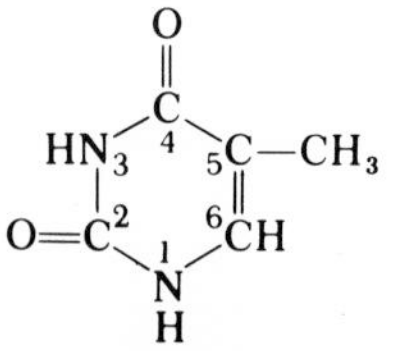

O
‖
C
HN$_3$ 4   5   C—CH$_3$
| 6
O=C$_2$ 1   CH
N
H

Thymine

(5-methyl-2,4-dioxypyrimidine)

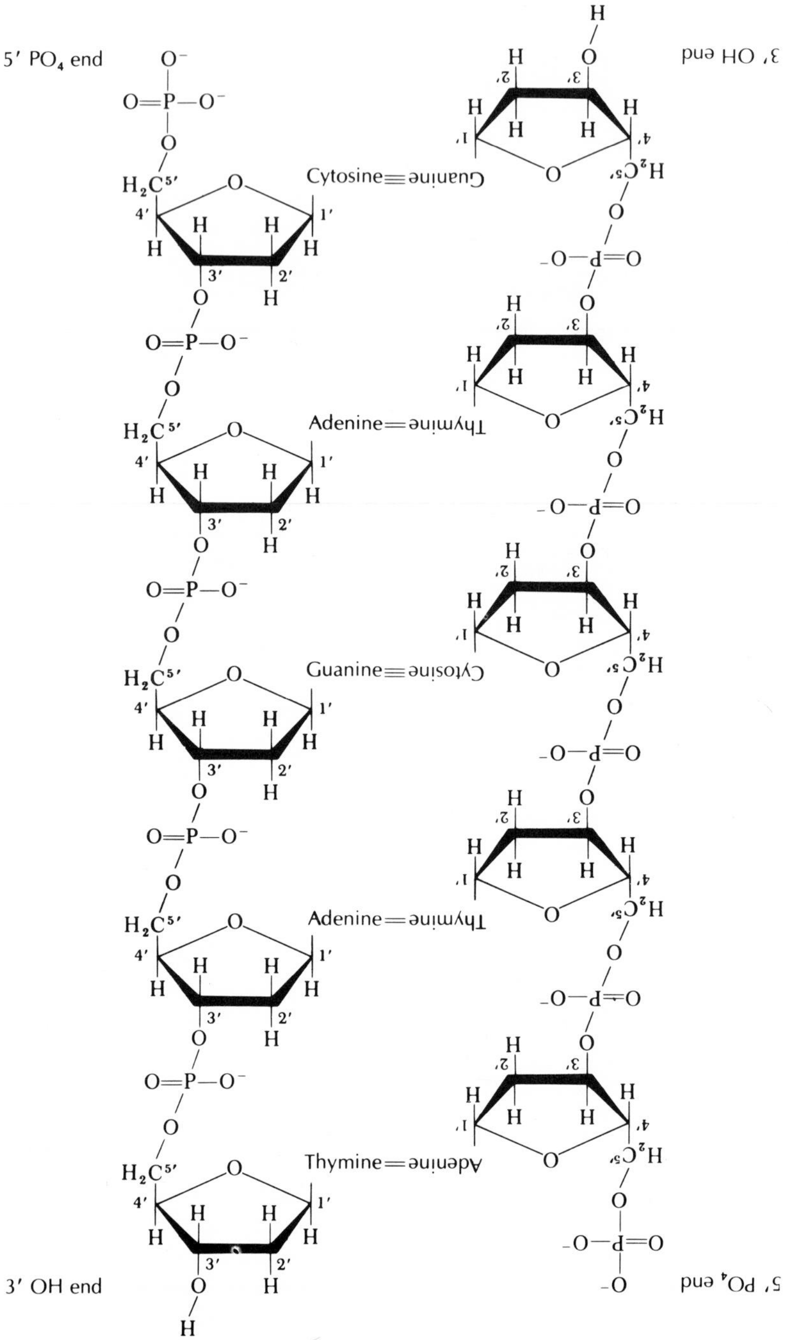

5' PO$_4$ end

O$^-$
|
O=P—O$^-$
|
O
|
H$_2$C 5'   O
4'   H   H   1'   Cytosine≡Guanine
H   H
3'   2'
O   H
|
O=P—O$^-$
|
O
|
H$_2$C 5'   O
4'   H   H   1'   Adenine=Thymine
H   H
3'   2'
O   H
|
O=P—O$^-$
|
O
|
H$_2$C 5'   O
4'   H   H   1'   Guanine≡Cytosine
H   H
3'   2'
O   H
|
O=P—O$^-$
|
O
|
H$_2$C 5'   O
4'   H   H   1'   Adenine=Thymine
H   H
3'   2'
O   H
|
O=P—O$^-$
|
O
|
H$_2$C 5'   O
4'   H   H   1'   Thymine=Adenine
H   H
3'   2'
O   H
|
H

3' OH end

C   **D**

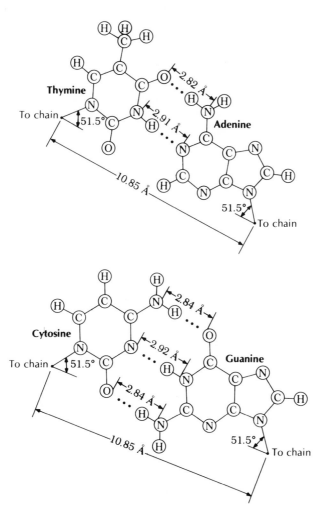

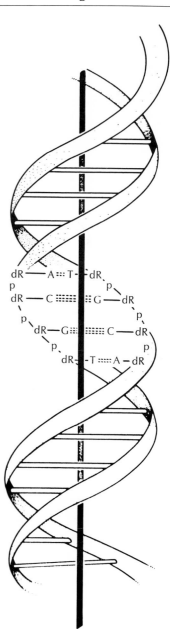

Fig. 4-1. Structure of DNA. **A,** Structural components of DNA. **B,** Backbone structure of Watson-Crick double helix, showing opposite polarity of two chains. The dimensions of the component parts have been distorted in this two-dimensional view. **C,** Hydrogen-bonded base pairs found in the DNA double helix. The resonant double bonds in the bases, which are responsible for their planar shape, are not shown in this diagram. Note the similarity of dimensions and overall shape of the adenine-thymine and guanine-cytosine base pairs. **D,** Schematic representation of the double helical structure showing the spiral twisting of the two backbone chains and the central position of the stacked base pairs. (**A** [modified] and **C** from White, A., P. Handler, and E. L. Smith. 1973. Principles of Biochemistry, 5th ed. McGraw-Hill Book Co., New York, used with permission of McGraw-Hill Book Co.; **B** modified from Suzuki, D. T., and A. J. F. Griffiths. 1976. An introduction to genetic analysis. W. H. Freedman & Co., Publishers, San Francisco; **D** reprinted with permission of Macmillan Publishing Co., Inc. from Animal morphogenesis by Saunders, J. W., Jr. Copyright © 1968 by John W. Saunders, Jr.)

est functional genetic unit that can be detected by the "cis-trans" test.* The mRNA transcribed from an operon (a cluster of physically adjacent and functionally related genes whose transcription is turned on and off coordinately) is often referred to as "polycistronic," meaning that it carries the coded information for several protein molecules within a single large mRNA molecule.

Chemical structure of genes. It has been known for a long time that eukaryotic chromosomes, which are rich in DNA and protein, carry genetic information. However, as described earlier in this chapter, the first direct experimental demonstration that DNA carried genetic information was not reported until 1944. Prior to that time many investigators believed that proteins, with their seemingly greater diversity, were more likely to be the carriers of genetic information. However, within a few years after the original demonstration that bacteria could be genetically "transformed" by purified DNA, evidence from a variety of sources removed all doubt concerning the genetic role of DNA. Today it is possible to isolate specific genes, to make direct comparisons of the nucleotide sequences of those genes with the amino acid sequences of the proteins for which they code, and even to carry out in a test tube the entire series of biochemical steps leading from the gene to a functional protein.

The chemical structure of DNA has been studied in detail (Fig. 4-1). DNA is a polymer whose backbone contains phosphate and the five-carbon sugar 2-deoxyribose in equal amounts. Each phosphate group joins the 3' hydroxyl group of one deoxyribose to the 5' hydroxyl group of the next in a phosphodiester linkage (Fig. 4-1, *B*). Either a purine base (adenine or guanine) or a pyrimidine base (cytosine or thymine) is attached to the 1' position of each deoxyribose. (Note that in nucleic acids, ordinary numbers are used to describe positions on the purine and pyrimidine rings and numbers with prime ['] are used for positions on the sugars.)

Most naturally occurring DNA (except in certain viruses) is in the form of the double helix, whose structure was determined by Watson

---

*The term "cistron" is derived from the "cis-trans" test, which is used to determine whether two mutations are both located within the same functional genetic unit. When applied to structural genes, it determines whether the mutations are within the same peptide-coding unit. If two mutations in different coding units are introduced into the same partially diploid cell, normal function is observed, since one nonmutant copy of the coded information for each peptide chain is still present. However, if both mutations are in the same coding unit, neither is able to code for synthesis of a functional peptide chain as long as the two mutations remain in separate genomes, or "trans" to each other. It is only when genetic "crossover" brings the nonmutant parts of the two coding units together into a single genome, or "cis" to each other, that a functional protein can be made.

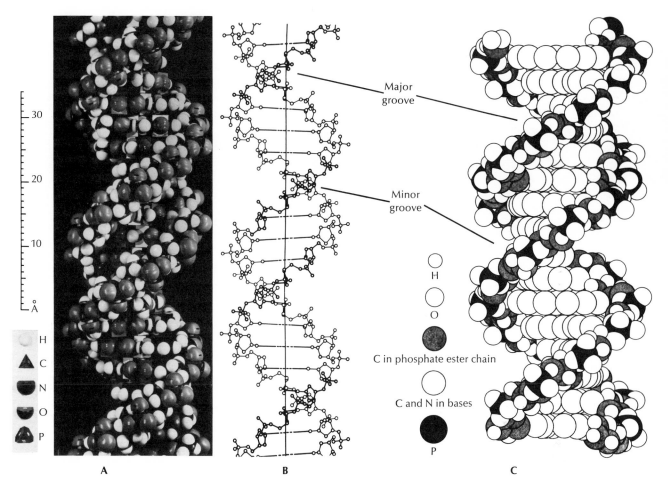

Fig. 4-2. Molecular models of double helical DNA. **A,** Space-filling model with deoxyribose-phosphate spiraling around the outside and base pairs stacked in the center. **B,** Skeleton model of deoxyribose-phosphate backbones, showing major and minor grooves between the backbones. The slight tilt of the base pairs is also evident. **C,** Interpretative drawing of space-filling model. The stacking of the flat base pairs is particularly evident. (**A** courtesy Professor M. H. F. Wilkins, Biophysics Department, King's College, London; **B** reprinted with permission of Macmillan Publishing Co., Inc. from Principles of genetics by Herskowitz, I. H. Copyright © 1973 by Irwin H. Herskowitz [courtesy M. H. F. Wilkins]; **C** from Genetics by Ursula Goodenough and Robert Paul Levine. Copyright © 1974 by Holt, Rinehart & Winston, Inc. Reprinted by permission of Holt, Rinehart & Winston and M. H. F. Wilkins.)

and Crick in 1953. The double helix contains two deoxyribose phosphate backbones, with their 3′ to 5′ polarity in opposite directions, as shown in Fig. 4-1, *B*. A key feature of its structure is the formation of hydrogen-bonded base pairs between purine and pyrimidine bases from opposite chains. Specifically, adenine pairs with thymine, and guanine pairs with cytosine (Fig. 4-1, *C*). The overall dimensions of these two types of base pairs, which are commonly referred to as AT and GC, are virtually identical. They can be freely interchanged, both backward and forward, in the double helical structure without distorting it. The three-dimensional structure of the double helix is depicted in Figs. 4-1, *D*, and 4-2. The two deoxyribose phosphate chains with opposite polarities form a spiral around the outside of the flat AT and GC base pairs, which are stacked like a pile of coins in the center.

The double helical structure explains many of the properties of DNA and of genes. The complementary base pairing allows each strand to serve as a template for the synthesis of the other. Thus genes can be duplicated precisely by separating the strands of the double helix and synthesizing a new complementary strand adjacent to each. In addition, since base pairs can be freely interchanged within the structure, any one of the base pairs can occur next to any other, and its complementary base will fit correctly into the other strand. Thus there are no mechanical constraints on the base sequences that can be used to carry coded information in genes.

In some viruses RNA replaces DNA as the genetic material. RNA is similar chemically to DNA. It has the five-carbon sugar ribose (which has a hydroxyl group in the 2' position) in place of deoxyribose and the pyrimidine base uracil (which lacks a methyl group in the 5 position) in place of thymine (Fig. 4-3). A strand of RNA can form a double helix with a strand of DNA that has a complementary base sequence. In addition, two complementary strands of RNA can form a double helix, as occurs in certain viruses whose genomes consist of double-stranded RNA. RNA, like DNA, has a definite polarity, based on the 3' to 5' orientation of its ribose phosphate backbone.

In addition to its primary genetic role in certain viruses, RNA also serves as an intermediate carrier of genetic information during protein synthesis in cells with DNA genomes. DNA-directed RNA synthesis occurs by a process similar to DNA replication, with one of the strands of the DNA serving as a template for the synthesis of the RNA strand by complementary base pairing.

Chemical structure of proteins

The diversity of types of proteins that are produced by living cells is extensive. Some are hydrophilic (highly water soluble), whereas others are relatively hydrophobic and tend to form insoluble aggregates or become associated with lipid-rich structures such as membranes. Some have catalytic (enzymatic) activity, and others serve in regulatory or structural roles. Some are acidic and others are basic; still others are neutral. Despite this seeming heterogeneity, however, all proteins have two properties in common: they are all coded for by

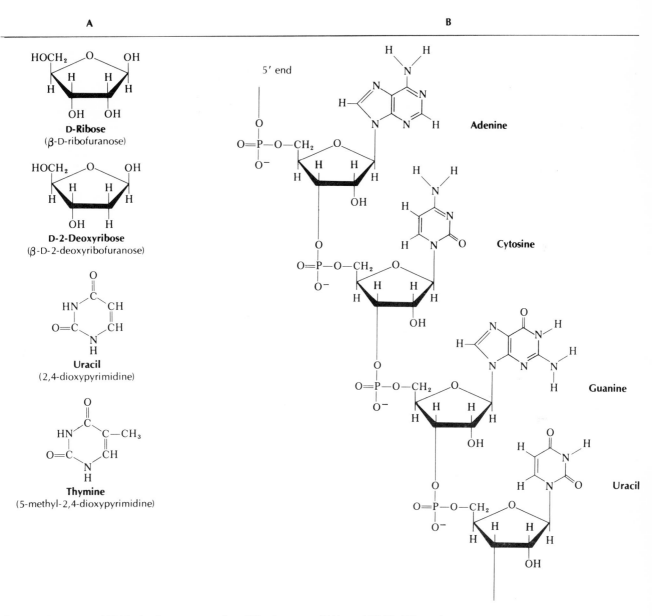

Fig. 4-3. Structure of RNA. **A,** Components that differ between RNA and DNA. Ribose in RNA has a hydroxyl group in the 2′ position, which is missing from the deoxyribose of DNA. Uracil in RNA lacks the methyl group in the 5 position, which characterizes thymine in DNA. **B,** Ribose-phosphate backbone of RNA with attached bases. (**A** from White, A., P. Handler, and E. L. Smith. 1973. Principles of biochemistry, 5th ed. McGraw-Hill Book Co., New York, used with permission of McGraw-Hill Book Co.; **B** modified from James D. Watson, Molecular biology of the gene, 3rd ed., copyright © 1976, 1970, 1965 by James D. Watson and The Benjamin/Cummings Publishing Co., Inc., Menlo Park, Calif.)

Relatively hydrophilic, neutral
(No net charge at pH 6 to 7)

Asparagine
(Asn)
N (2)

$$H_2N-\overset{\displaystyle }{\underset{\displaystyle O}{C}}-CH_2-\overset{\displaystyle NH_3^+}{\underset{\displaystyle COO^-}{C}}-H$$

Cysteine
(Cys)
C (2)

$$HS-CH_2-\overset{\displaystyle NH_3^+}{\underset{\displaystyle COO^-}{C}}-H$$

Glutamine
(Gln)
Q (2)

$$H_2N-\overset{\displaystyle }{\underset{\displaystyle O}{C}}-CH_2-CH_2-\overset{\displaystyle NH_3^+}{\underset{\displaystyle COO^-}{C}}-H$$

Glycine
(Gly)
G (4)

$$H-\overset{\displaystyle NH_3^+}{\underset{\displaystyle COO^-}{C}}-H$$

Serine
(Ser)
S (6)

$$HO-CH_2-\overset{\displaystyle NH_3^+}{\underset{\displaystyle COO^-}{C}}-H$$

Threonine
(Thr)
T (4)

$$H_3C-\overset{\displaystyle OH}{\underset{\displaystyle H}{C}}-\overset{\displaystyle NH_3^+}{\underset{\displaystyle COO^-}{C}}-H$$

Tyrosine
(Tyr)
Y (2)

$$HO-\bigcirc-CH_2-\overset{\displaystyle NH_3^+}{\underset{\displaystyle COO^-}{C}}-H$$

Relatively hydrophilic, basic
(Net positive charge at pH 6 to 7)

Arginine
(Arg)
R (6)

$$\overset{\displaystyle H_2N}{\underset{\displaystyle H_2^+N}{C}}-NH-CH_2-CH_2-CH_2-\overset{\displaystyle NH_3^+}{\underset{\displaystyle COO^-}{C}}-H$$

Histidine
(His)
H (2)

$$H^+N\cdots\cdots-CH_2-\overset{\displaystyle NH_3^+}{\underset{\displaystyle COO^-}{C}}-H$$

Lysine
(Lys)
K (2)

$$H_3N^+-CH_2-CH_2-CH_2-CH_2-\overset{\displaystyle NH_3^+}{\underset{\displaystyle COO^-}{C}}-H$$

Relatively hydrophilic, acidic
(Net negative charge at pH 6 to 7)

Aspartic acid
(Asp)
D (2)

$$\overset{\displaystyle ^-O}{\underset{\displaystyle O}{C}}-CH_2-\overset{\displaystyle NH_3^+}{\underset{\displaystyle COO^-}{C}}-H$$

Glutamic acid
(Glu)
E (2)

$$\overset{\displaystyle ^-O}{\underset{\displaystyle O}{C}}-CH_2-CH_2-\overset{\displaystyle NH_3^+}{\underset{\displaystyle COO^-}{C}}-H$$

Relatively hydrophobic, neutral
(No net charge at pH 6 to 7)

Alanine
(Ala)
A (4)

$$H_3C-\overset{\displaystyle NH_3^+}{\underset{\displaystyle COO^-}{C}}-H$$

Isoleucine
(Ile)
I (3)

$$H_3C-CH_2-\overset{\displaystyle H}{\underset{\displaystyle CH_3}{C}}-\overset{\displaystyle NH_3^+}{\underset{\displaystyle COO^-}{C}}-H$$

Leucine
(Leu)
L (6)

$$H_3C-\overset{\displaystyle }{\underset{\displaystyle CH_3}{CH}}-CH_2-\overset{\displaystyle NH_3^+}{\underset{\displaystyle COO^-}{C}}-H$$

Methionine
(Met)
M (1)

$$H_3C-S-CH_2-CH_2-\overset{\displaystyle NH_3^+}{\underset{\displaystyle COO^-}{C}}-H$$

Phenylalanine
(Phe)
F (2)

$$\bigcirc-CH_2-\overset{\displaystyle NH_3^+}{\underset{\displaystyle COO^-}{C}}-H$$

Proline
(Pro)
P (4)

$$\overset{\displaystyle CH_2}{H_2C}\quad\overset{\displaystyle NH_2^+}{\underset{\displaystyle COO^-}{C}}-H$$

Tryptophan
(Trp)
W (1)

$$\bigcirc\bigcirc-CH_2-\overset{\displaystyle NH_3^+}{\underset{\displaystyle COO^-}{C}}-H$$

Valine
(Val)
V (4)

$$H_3C-\overset{\displaystyle }{\underset{\displaystyle CH_3}{CH}}-\overset{\displaystyle NH_3^+}{\underset{\displaystyle COO^-}{C}}-H$$

genes, and they are all formed from the same 20 amino acids. It is true that some proteins undergo chemical modification of their amino acids and/or the addition of other substances such as phosphate, carbohydrate, or lipid after they are synthesized from amino acids. Nevertheless, most of their diversity is due to the relative amounts of the different kinds of amino acids and the genetically determined sequence arrangement of those amino acids within the protein molecules.

The 20 different amino acids that proteins are made from are shown in Fig. 4-4. Each has an amino group, a carboxyl group, and a side chain, all attached to the "α" carbon atom. Proteins are linear polymers of amino acids, with the carboxyl group of each amino acid joined to the amino group of the next amino acid by an amide linkage, which is commonly referred to as a peptide bond.

The side chains that protrude from the linear polymer give each protein its characteristic properties. Some side chains are nonpolar and hydrophobic, making the region of the protein molecule that is rich in them poorly soluble in water and inclined to interact with lipid-rich membranes or with other hydrophobic proteins. Other side chains contain acidic, basic, or hydroxyl groups, which make protein chains that contain them more soluble in water. The relative amounts of acidic and basic side chains determine whether the protein has acidic, basic, or neutral properties. The —SH groups in the side chain of cysteine oxidize readily to form —SS— cross-links that join adjacent polypeptides or hold different parts of the same protein molecule in particular folded configurations. Hydrophobic interactions among side chains in various parts of a protein molecule also help the protein to assume the proper three-dimensional configuration to achieve its enzymatic or structural function. Thus the sequence information coded for in genes confers a wide variety of specific properties on proteins, and small changes in amino acid sequence caused by mutation can sometimes drastically alter the functional properties of a protein.

Fig. 4-4. Structural formulas of the 20 amino acids that are involved in protein synthesis. The ionic forms that predominate at pH 6.0 to 7.0 are shown. For all asymmetrically bonded (optically active) carbon atoms the vertical lines represent bonds projecting into the page, and the horizontal lines represent bonds projecting out from the page. The amino acids are depicted with the portion of the molecule that forms the polypeptide chain oriented vertically at the right and the side chain oriented horizontally at the left. The amino acids are grouped according to the chemical properties of their side chains. Under the names of each amino acid are given the three-letter abbreviations commonly used, the one-letter abbreviations that are used to describe long protein sequences, and the number of RNA codons (Table 4-7) specifying that amino acid. (Modified from Lehninger, A. L. 1975. Biochemistry, 3rd ed. Worth Publishers, Inc., New York, and Herskowitz, I. H. 1977. Principles of genetics, 2nd ed. Macmillan Publishing Co., Inc., New York.)

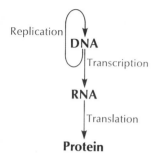

Fig. 4-5. Basic concepts of the "central dogma" of molecular biology. DNA is the primary carrier of genetic information. New DNA is replicated from DNA templates. During gene expression, sequence information from DNA is first transcribed into RNA sequences, which are then translated into amino acid sequences in protein molecules. Nucleotide sequences in DNA and RNA are colinear with amino acid sequences in proteins. The flow of information is always from nucleic acids to proteins and never from proteins to nucleic acids.

Central dogma of molecular biology

The study of gene expression and its regulation, in both prokaryotic and eukaryotic cells, is centered around the role of DNA as a template for the synthesis of RNA and the role of mRNA in determining the amino acid sequence of proteins. These two steps, which are referred to as transcription and translation, respectively, have come to be known as the "central dogma" of molecular biology (Fig. 4-5). The key feature of the central dogma is the fact that DNA does not serve directly as a template for protein synthesis. Instead, the nucleotide sequence of the DNA is transcribed into a similar sequence in mRNA, and that RNA sequence is, in turn, translated into the very different language of amino acid sequence in a protein molecule.

Continued use of the term "dogma," which was introduced in the mid-1950s when many of the relationships were still speculative, is somewhat misleading. Every step in the pathway of information flow from gene to peptide chain in prokaryotic cells has been studied carefully (as will be seen later in this chapter) and there no longer exists any doubt that the pathway of information flow is as depicted in Fig. 4-5, at least in cells that are not infected with viruses.

The diversity of viral genomes, which can be composed of single- or double-stranded DNA or single- or double-stranded RNA, and the diverse patterns of gene expression and replication by viruses have made it necessary to expand the original central dogma, as shown in Fig. 4-6. However, the original premise that information always flows from nucleic acids to proteins, and never from proteins to nucleic acids, remains valid.

The additions to the central dogma in Fig. 4-6 are all concerned with transfer of information among various forms of single- and double-stranded DNA and RNA. Viruses that have genomes consisting of

Fig. 4-6. Added complexities of the central dogma in virally infected cells. These pathways do not all occur with the same kinds of viruses. See text for details.

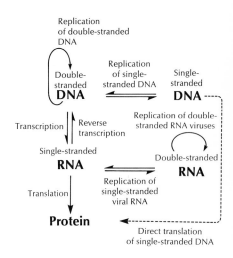

single-stranded DNA, such as φX174, must first synthesize a complementary negative strand of DNA, which then serves as a template for mRNA synthesis and for the synthesis of more single-stranded viral DNA. Single-stranded DNA can also serve as a direct template for protein synthesis under certain laboratory conditions. Certain viruses that have genomes consisting of single-stranded RNA, such as the RNA tumor viruses of eukaryotic cells, possess an enzyme called reverse transcriptase, which allows them to use RNA as a template for the synthesis of single-stranded DNA, which in turn is converted to double-stranded DNA and serves as a template for synthesis of more viral RNA. In a third group of viruses, known as reoviruses, the genome consists of complementary double-stranded RNA, which serves both as a template for its own replication and as a template for mRNA synthesis, with no involvement of DNA at any time. A fourth group, the RNA bacteriophages, whose genome consists of single-stranded RNA, replicate by synthesis of a complementary (negative) strand of RNA, which then serves as the template for synthesis of positive-strand RNA molecules that function both as the genome and as mRNA. Finally, many RNA viruses of animal cells contain negative-strand RNA in the virion (virus particle), together with an enzyme that permits positive strands that function as mRNA and replicative intermediates to be transcribed directly from the negative strand.

Although any of these mechanisms could conceivably be involved in some aspect of differential gene expression in noninfected cells, they are generally regarded as "exceptions" rather than the rule. Nearly all the regulatory mechanisms that we will be concerned with in this book, both in prokaryotic and eukaryotic cells, will involve the standard central dogma, with the primary flow of information from double-stranded DNA to single-stranded mRNA and then to proteins, as depicted in Fig. 4-5.

As we now proceed to examine the mechanisms behind the central dogma in prokaryotic organisms, we will examine the following questions in greater detail:

1. What is the nature of the information carried in the genome?
2. How does the genome make exact replicas of itself?
3. How is information transferred from DNA to RNA?
4. How is information transferred from RNA to protein?
5. What are the mechanisms that permit preferential use of selected portions of the information that is available in the genome?
6. What are the major unanswered questions about this entire system?

The latter is perhaps the most important of all for the student contemplating a career in experimental biology. If the time ever comes when there are no significant questions left to be answered, this area of study will lose much of the challenge and the excitement of discovery that currently characterize it.

Nature of coded information in the genome

At least two major classes of information are coded into the DNA of the genome. The first consists of nucleotide sequences that specify amino acid sequences in polypeptide chains and also the sequences that code for functional RNA products; the second, which is far more complex and less well defined, provides the various types of regulatory signals that are needed for use of the sequence information.

Nucleotide triplet code. One of the important corollaries of the central dogma is that the sequence information of protein-coding DNA is colinear with the sequence of amino acids in the protein specified by that DNA. This has now been verified in a variety of ways in prokaryotes, including direct sequence analysis of genes, mRNA, and proteins. As first demonstrated experimentally by Marshall Nirenberg in 1961, each amino acid is coded for by a three-base "triplet" in the mRNA. Since mRNA is synthesized by base pairing with the "negative" strand of the DNA, it has the same triplet sequence as the "positive" strand of the DNA of the gene. (In comparing DNA and RNA sequences, thymine and uracil are considered to be equivalent, since both form base pairs equally well with adenine.)

Four different bases taken three at a time provide a potential dictionary of 64 triplet code words. Since only 20 amino acids are involved in protein synthesis, there are considerably more code words than there are amino acids. This permits a redundancy in the genetic code, with as many as six different triplet code words, or codons, coding for some of the more frequently used amino acids (Fig. 4-7). Sixty-

A

Second base

		U	C	A	G	
First base	**U**	UUU ⎤Phe UUC ⎦ UUA ⎤Leu UUG ⎦	UCU ⎤ UCC ⎥Ser UCA ⎥ UCG ⎦	UAU ⎤Tyr UAC ⎦ UAA Term UAG Term	UGU ⎤Cys UGC ⎦ UGA Term UGG Trp	U C A G
	C	CUU ⎤ CUC ⎥Leu CUA ⎥ CUG ⎦	CCU ⎤ CCC ⎥Pro CCA ⎥ CCG ⎦	CAU ⎤His CAC ⎦ CAA ⎤Gln CAG ⎦	CGU ⎤ CGC ⎥Arg CGA ⎥ CGG ⎦	U C A G
	A	AUU ⎤ AUC ⎥Ile AUA ⎦ AUG Met	ACU ⎤ ACC ⎥Thr ACA ⎥ ACG ⎦	AAU ⎤Asn AAC ⎦ AAA ⎤Lys AAG ⎦	AGU ⎤Ser AGC ⎦ AGA ⎤Arg AGG ⎦	U C A G
	G	GUU ⎤ GUC ⎥Val GUA ⎥ GUG ⎦	GCU ⎤ GCG ⎥Ala GCA ⎥ GCG ⎦	GAU ⎤Asp GAC ⎦ GAA ⎤Glu GAG ⎦	GGU ⎤ GGC ⎥Gly GGA ⎥ GGG ⎦	U C A G

Third Base

B

Amino acid	Codons					
Alanine	GCU	GCC	GCA	GCG		
Arginine	CGU	CGC	CGA	CGG	AGA	AGG
Asparagine	AAU	AAC				
Aspartic acid	GAU	GAC				
Cysteine	UGU	UGC				
Glutamic acid	GAA	GAG				
Glutamine	CAA	CAG				
Glycine	GGU	GGC	GGA	GGG		
Histidine	CAU	CAC				
Isoleucine	AUU	AUC	AUA			
Leucine	UUA	UUG	CUU	CUC	CUA	CUG
Lysine	AAA	AAG				
Methionine	AUG					
Phenylalanine	UUU	UUC				
Proline	CCU	CCC	CCA	CCG		
Serine	UCU	UCC	UCA	UCG	AGU	AGC
Threonine	ACU	ACC	ACA	ACG		
Tryptophan	UGG					
Tyrosine	UAU	UAC				
Valine	GUU	GUC	GUA	GUG		
Initiation	AUG	GUG				
Termination	UAA	UAG	UGA			

Fig. 4-7. Genetic code. **A,** Amino acids specified by RNA triplet codons. Dotted lines connect codons with different first and/or second bases that code for the same amino acid. "Term" indicates a translation terminating codon. **B,** Codons listed by the amino acid that they specify. The number of codons per amino acid ranges from one to six. Note that GUG specifies *N*-formylmethionine when it serves as an initiation codon and valine when it is in an internal position in mRNA. (**A** reprinted [modified] with permission of Macmillan Publishing Co., Inc. from Principles of genetics, 2nd ed. by Herskowitz, I. Copyright © 1977 Irwin H. Herskowitz; **B** reprinted [modified] with permission of Macmillan Publishing Co., Inc. from Animal morphogenesis by Saunders, J. W., Jr. Copyright © 1968 John W. Saunders, Jr.)

one different codons are used to specify amino acids. The other three (UAA, UAG, UGA) are specific chain termination signals that stop the translation process when synthesis of a particular peptide chain is completed.

There is no punctuation between codons in an mRNA molecule. In prokaryotic cells, synthesis of a protein chain always begins with N-formylmethionine at an AUG (or sometimes GUG) codon. The next three bases specify the second amino acid, the following three the third, and so on until a termination codon is reached. For example, the mRNA sequence -AUGACCAUGAUUACGGAU- specifies the amino acid sequence (N-formylmethionine)-threonine-methionine-isoleucine-threonine-aspartic acid- at the N-terminal end of the enzyme β-galactosidase in *Escherichia coli*. In most cases the N-formylmethionine is removed from the peptide chain soon after it is synthesized.

An interesting class of mutations known as "frameshift" mutants occurs when one nucleotide is added or subtracted in a coding sequence. From the point of the mutation on, nucleotide triplets are read one nucleotide out of phase, resulting in a completely different set of amino acids being inserted into the protein. For example, addition of uracil between nucleotides 7 and 8 in the sequence for β-galactosidase would result in the sequence -AUGACCAUUGAUUACGGAU-, which would be read as (N-formylmethionine)-threonine-isoleucine-aspartic acid-tyrosine-glycine-. Study of frameshift mutations played an important role in establishing the nonoverlapping triplet nature of the genetic code.

An interesting phenomenon closely related to frameshifting—the presence of overlapping genes that are read in different frames—occurs in the genomes of certain small viruses. The genomes of ϕX174, a bacteriophage with 5375 nucleotides of single-stranded DNA, and of simian virus 40 (SV40), an animal cell virus with 5224 nucleotide pairs of double-stranded DNA, have both been completely sequenced. In ϕX174 there are two separate cases of overlapping, genes A and B and genes D and E. In each case translation in two different reading frames yields two totally different amino acid sequences from a single nucleotide sequence, as shown for genes D and E in Fig. 4-8. In SV40, recent data appear to show overlapping both in the same and in different reading frames. The amino acid sequence of viral protein VP-3 is the same as the middle and C-terminal regions of a larger protein VP-2. The initiation codon for translation of VP-3 lies inside the coding sequence for VP-2 in the same reading frame, and the same termination codon is used for both proteins. In addition, the N-terminal coding sequence for protein VP-1, which is translated in a different reading frame, overlaps the last 110 nucleotide pairs of the C-terminal coding

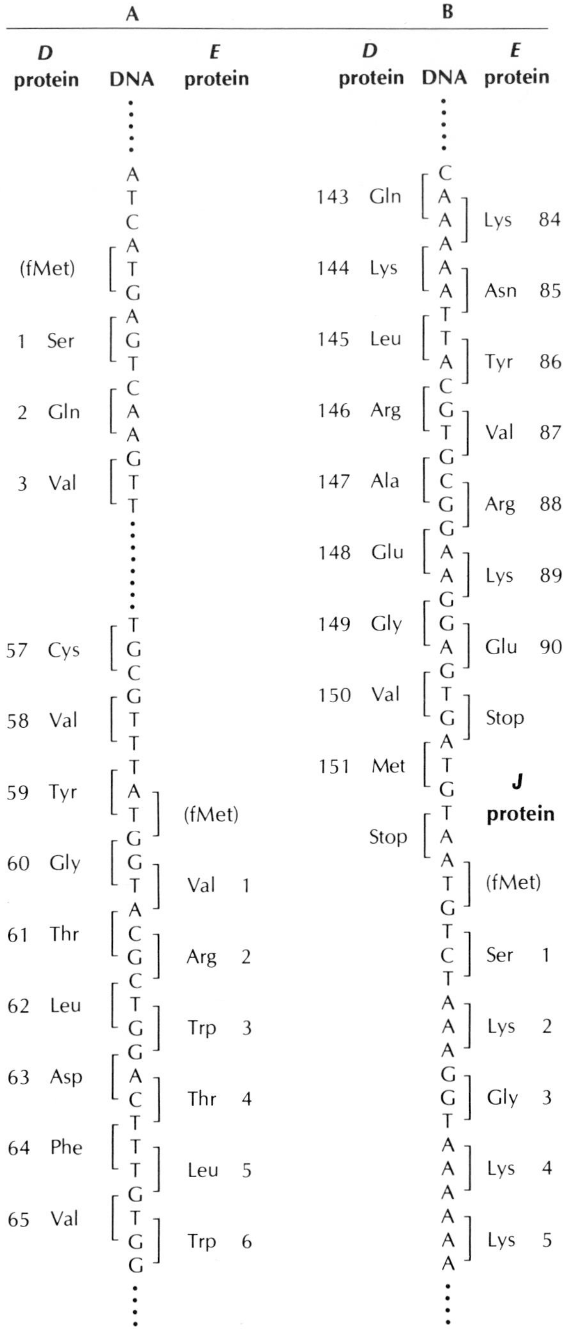

A

D protein	DNA	E protein
	⋮	
	A	
	T	
	C	
(fMet)	A	
	T	
	G	
1 Ser	A	
	G	
	T	
2 Gln	C	
	A	
	A	
3 Val	G	
	T	
	T	
	⋮	
57 Cys	T	
	G	
	C	
58 Val	G	
	T	
	T	
59 Tyr	T	
	A	
	T	(fMet)
60 Gly	G	
	G	
	T	Val 1
61 Thr	A	
	C	
	G	Arg 2
62 Leu	C	
	T	
	G	Trp 3
63 Asp	G	
	A	
	C	Thr 4
64 Phe	T	
	T	
	T	Leu 5
65 Val	G	
	T	
	G	Trp 6
	G	
	⋮	

B

D protein	DNA	E protein / J protein
	⋮	
	C	
143 Gln	A	
	A	Lys 84
144 Lys	A	
	A	
	A	Asn 85
145 Leu	T	
	T	
	A	Tyr 86
146 Arg	C	
	G	
	T	Val 87
147 Ala	G	
	C	
	G	Arg 88
148 Glu	G	
	A	
	A	Lys 89
149 Gly	G	
	G	
	A	Glu 90
150 Val	G	
	T	
	G	Stop
151 Met	A	
	T	
	G	**J protein**
Stop	T	
	A	
	A	(fMet)
	T	
	G	
	T	
	C	Ser 1
	T	
	A	
	A	Lys 2
	A	
	G	
	G	Gly 3
	T	
	A	
	A	Lys 4
	A	
	A	
	A	Lys 5
	A	
	⋮	

Fig 4-8. Overlapping coding sequences in bacteriophage φX174. **A,** Initiation of translation of genes D and E in different reading frames. The base sequences are for DNA and therefore have T rather than U in the codons for the individual amino acids. The numbers indicate the number of amino acids from the N-terminal position in each protein. **B,** Termination of translation of genes D and E and initiation of gene J. The initiation codon for gene J overlaps the termination codon for gene D. The reading frame for gene J is different from that of either gene D or gene E. Note that all three possible reading frames are utilized within a distance of six nucleotides. (Based on data from Barrell, B. G., G. M. Air, and C. A. Hutchison, III. 1976. Nature **264:**34.)

sequence shared by VP-2 and VP-3. The evolutionary restraints imposed by overlapping must be severe, since single-base mutations would often affect the amino acid sequences of more than one protein. It seems probable that an evolutionary adaptation of this sort would occur only in a very small genome where maximum use of coding capacity is of particular importance.

Regulatory signals. The "regulatory" information coded into the genome is not yet fully understood, but as a minimum it includes signals for the following:

1. Initiation of RNA synthesis at precisely determined points
2. Termination of RNA synthesis at precisely determined points
3. Ribosomal attachment and initiation of protein synthesis at precisely determined points in mRNA sequences (which in turn have been transcribed from the genome)
4. Termination of protein synthesis at precisely determined points
5. Attachment sites for regulatory molecules that control transcription
6. Attachment sites (coded into mRNA sequences) for regulatory molecules that control translation
7. Initiation of DNA synthesis at precisely determined locations

In eukaryotic cells there are also a variety of signals related to posttranscriptional processing of the initial precursor of mRNA, as will be described in Chapter 8. Similar events also occur to a lesser extent in prokaryotes.

With the possible exception of termination of protein synthesis, which appears to occur whenever a termination codon (UAA, UAG, or UGA) is encountered, all these signals seem to involve more than simple triplet code recognition. Initiation of protein synthesis, for example, always occurs at AUG or GUG codons, but those same codons also occur at internal positions in peptide sequences where initiation does not occur. The GUG sequence is particularly interesting, since as an initiation codon it specifies N-formylmethionine, whereas at internal sites it specifies valine. The reason why some AUG and GUG codons initiate protein synthesis while others do not is not certain. Currently popular theories suggest that (1) there may be sequences near the initiation codon that form base pairs with the 3' end of the 16S ribosomal RNA and (2) secondary structure due to folding and base pairing within the mRNA chain may place certain potential initiation codons in particularly favorable configurations to react with other components of the initiation complex, which is described later in this chapter.

Complex sequences of base pairs appear to be involved in the initiation and regulation of transcription, as well as in its termination. These are discussed in later sections of this chapter. Initiation of DNA synthesis occurs at a well-defined point in the circular chromosome of *E. coli* and spreads around it in both directions.

These various regulatory sites which do not code for amino acid sequences occupy a significant portion of the total genome in prokaryotic organisms. In eukaryotic cells, where the total amount of DNA per genome is much larger, only a small percentage of the DNA sequences appear to code for amino acid sequences, but it is not certain how much of the remainder (more than 80% of the total genome) is actually involved in regulatory functions (see Chapter 5).

Synthesis of new DNA

At first glance, replication of DNA might seem to be a simple matter consisting of separation of the chains of the double helix, followed by base pairing of free deoxyribonucleoside triphosphates to the existing chains and then the linking together of the new complementary polynucleotide chains to yield two complete double helical DNA molecules. However, it is actually much more complicated. The two complementary chains in DNA have opposite 3' to 5' polarities, as illustrated in Fig. 4-1, *B*. When the strands separate to yield a branch point during replication, one of the new complementary strands that is formed needs to grow in a 5' to 3' direction, and the other needs to grow in a 3' to 5' direction. All known DNA-synthesizing enzymes add 5' nucleoside triphosphates to hydroxyl groups at the 3' ends of existing chains and thus support chain growth only in a 5' to 3' direction. Direct chain growth in a 3' to 5' direction does not appear to occur. Current evidence suggests that short fragments of DNA, about 1000 nucleotides in length, known as Okazaki fragments, are synthesized in a 5' to 3' direction along both strands of a separated DNA double helix and that these short fragments are then joined to previously synthesized complementary chain fragments by specific "ligase" enzymes (Fig. 4-9).

Recent studies with highly purified DNA polymerase enzymes have added still further complications to the question of how DNA is replicated. These enzymes appear to require the synthesis of short segments of complementary RNA, which then serve as primers from which the new DNA strands grow. Evidence for the actual existence of such primers in growing DNA chains is still weak, but it is widely

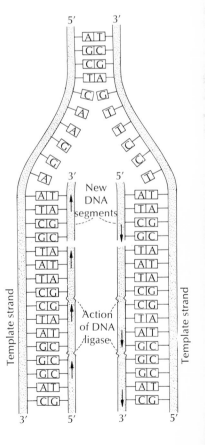

Fig. 4-9. Schematic representation of the discontinuous nature of DNA synthesis. After separation of the parental strands at the replication fork, complementary daughter strands are synthesized by DNA polymerase in short segments, called Okazaki fragments, which are subsequently joined into continuous DNA strands by a second enzyme called DNA ligase. The average length of Okazaki fragments is about 1000 nucleotides. (Reprinted [modified] with permission of Macmillan Publishing Co., Inc. from The science of genetics, 3rd ed. by Burns, G. W. Coypright © 1976 by George W. Burns.)

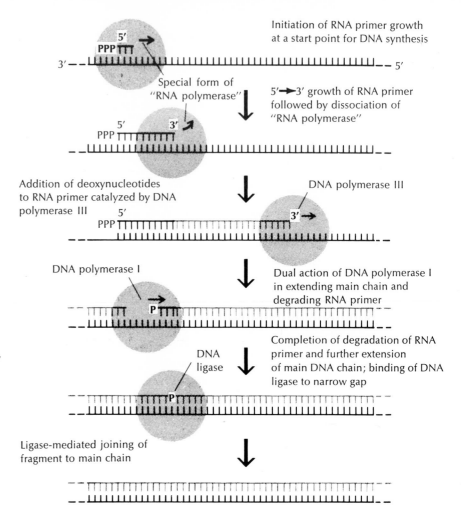

Fig. 4-10. Postulated role of RNA-containing intermediates in DNA replication. (From James D. Watson, Molecular biology of the gene, 3rd ed., copyright © 1976, 1970, 1965 by James D. Watson and The Benjamin/Cummings Publishing Co., Inc., Menlo Park, Calif.)

Initiation of RNA primer growth at a start point for DNA synthesis

Special form of "RNA polymerase"

$5' \rightarrow 3'$ growth of RNA primer followed by dissociation of "RNA polymerase"

Addition of deoxynucleotides to RNA primer catalyzed by DNA polymerase III

DNA polymerase III

DNA polymerase I

Dual action of DNA polymerase I in extending main chain and degrading RNA primer

DNA ligase

Completion of degradation of RNA primer and further extension of main DNA chain; binding of DNA ligase to narrow gap

Ligase-mediated joining of fragment to main chain

believed that such primers are inserted and then quickly excised and replaced by DNA. A model that has been proposed for the sequence of events involved in the initiation of new DNA synthesis is shown in Fig. 4-10.

In addition to these biochemical problems, the mechanical problems that appear to be involved in the duplication of a bacterial genome are staggering to the imagination. For example, the genome of *E. coli* consists of a DNA double helix containing approximately 4.2×10^6 base pairs and forming a single closed circle. Its total length (circumference) is 1.4 mm, well over a thousand times the diameter of the cell. A highly complex arrangement of coiling and supercoiling is required to fit that much DNA into the cell. The remnants of that organization can be seen in Fig. 4-11 in DNA released from a disrupted *E. coli* cell.

Fig. 4-11. The membrane-attached, coiled, and folded chromosome from a ruptured *E. coli* cell (×11,000). (With permission from Delius, H., and A. Worcel. 1974. J. Mol. Biol. **82:**107. Copyright by Academic Press Inc. [London] Ltd.)

DNA replication is believed to occur at specific sites attached to the cell membrane in *E. coli*. If this view is correct, the DNA must actually move past those sites in a manner analogous to the movement of magnetic tape past a recording head. However, it is also known that DNA replication proceeds around the circular chromosome in both directions from the point of origin. Thus the pattern of movement of the DNA past the polymerase sites must be complex. Another major problem is the separation of the two strands of the double helix, which are twisted around each other once for each 10 base pairs, or about 420,000 times in the *E. coli* genome. This process appears to involve special "swivelase" enzymes, but it is not yet fully understood.

Despite all these problems the genome of *E. coli* is able to duplicate itself once every 20 minutes in rapidly growing cultures without becoming tangled or forming knots. Similar mechanical problems are probably also faced by eukaryotic cells. A diploid human cell, for example, contains a total of nearly 2 meters of DNA in a nucleus that is only a few microns in diameter.

Synthesis of RNA
The first step in the actual expression of genetic information is transcription of the nucleotide sequence of the gene into RNA. This is accomplished by complementary base pairing of ribonucleoside triphosphate molecules along the "negative" strand of the DNA and the formation of 5' to 3' phosphodiester linkages by splitting out pyrophos-

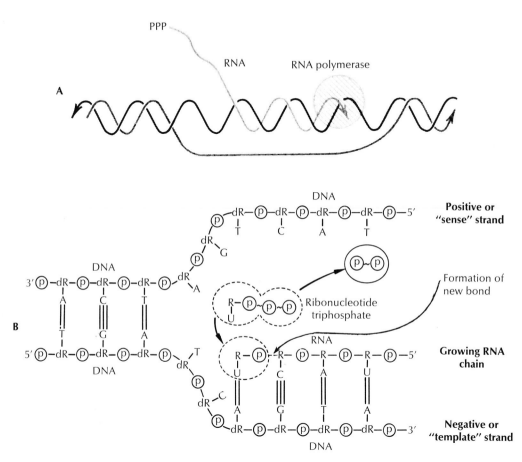

Fig. 4-12. DNA-dependent RNA synthesis. **A,** RNA polymerase temporarily separates the strands of the DNA double helix. A single-stranded RNA molecule with a nucleotide sequence identical to that of the positive strand of the DNA (except that uracil replaces thymine) is synthesized from 5'-ribonucleoside triphosphates by complementary base pairing with the negative strand of the DNA. **B,** Molecular details of RNA chain elongation. Growth is from 5' to 3' through addition of nucleoside triphosphate groups to hydroxyl groups at the 3' end of the growing chain with loss of pyrophosphate. An unreacted triphosphate group remains at the 5' end of newly synthesized RNA molecules, as shown in **A.** (**A** from James D. Watson, Molecular biology of the gene, 3rd ed., copyright © 1976, 1970, 1965 by James D. Watson and The Benjamin/Cummings Publishing Co., Inc., Menlo Park, Calif.; **B** reprinted [modified] with permission of Macmillan Publishing Co., Inc. from Animal morphogenesis by Saunders, J. W., Jr. Copyright © 1968 John W. Saunders, Jr.)

phate (Fig. 4-12). The DNA-dependent RNA polymerase that carries out this reaction in *E. coli* is a complex enzyme consisting of a "core polymerase" and a dissociable factor called the "sigma" (σ) factor that increases its specificity. The core polymerase possesses the ability to attach to a template DNA molecule at random locations and to synthesize RNA complementary to one of the DNA chains. The core polymerase is composed of four subunits, two α, one β, and one β'. Addition of the σ factor reduces nonspecific binding of the core polymerase to DNA and greatly enhances binding and initiation of RNA synthesis at the specific promoter sites where transcription is normally initiated.

The recognition sites for RNA polymerase binding for a number of different bacterial and viral genes are shown in Fig. 4-13. These show certain common properties but no absolute sequence specificity. Transcription always begins with a purine six or seven bases downstream (in the direction of transcription) from the binding site. After promoter recognition and the beginning of transcription the σ factor is released from the polymerase complex and is then available to function in another specific initiation with another core polymerase (Fig. 4-14). During bacterial sporulation and bacteriophage infection, changes often

RNA polymerase binding site sequences

fd	T G C T T C T G A C	T A T A A T A	G A C A G G G T A A A G A C C T G A T T T T T G A
T7 A3	A A G T A A A C A C G G	T A C G A T G	T A C C A C A T G A A A C G A C A G T G A G T C A
T7 A2	A G T A A C A T G C A G	T A A G A T A	C A A A T C G C T A G G T A A C A C T A G C A G
Lac-UV-5	G C T T C C G G C T C G	T A T A A T G	T G T G G A A T T G T G A G C G G A T A A C A A
Lambda P_R	A C C T C T G G C G G T	G A T A A T G	G T T G C A T G T A C T A A G G A G G T T G
SV40	T T T A T T G C A G C T	T A T A A T G	G T T A C A A A T A A A G C A A T A G C A T C
Lambda P_L	A C C A C T G G C G G T	G A T A C T G	A G C A C A T C A G C A G G A C G C A C T G A C
E. coli Tyr tRNA	C G T C A T T T G A	T A T G A T G	C G C C C C G C T T C C C G A T A A G G G A G C A G
Lac wildtype	G C T T C C G G C T C G	T A T G T T G	T G T G G A A T T G T G A G C G G A T A A C A A

Favored sequence   **T A T PuA T G**

Fig. 4-13. Nucleotide sequences of firm binding sites for RNA polymerase in various promoters. Only the positive strand sequence is shown, and the portion that is transcribed is underlined. The sequences immediately adjacent to the initiation of transcription show little similarity. However, a seven-base sequence with striking similarity is found in all cases beginning with the sixth or seventh nucleotide "upstream" from the initiation point. (A second region about 35 nucleotides upstream that is not shown here also appears to be involved). *Pu* = purine, with no preference between adenine and guanine. (Modified from James D. Watson, Molecular biology of the gene, 3rd ed., copyright © 1976, 1970, 1965 by James D. Watson and The Benjamin/Cummings Publishing Co., Inc., Menlo Park. The "favored" sequence was proposed by David Pribnow on the basis of an analysis of firm binding site sequences of various promoters (1975. Proc. Natl. Acad. Sci. U.S.A. **72:**784.)

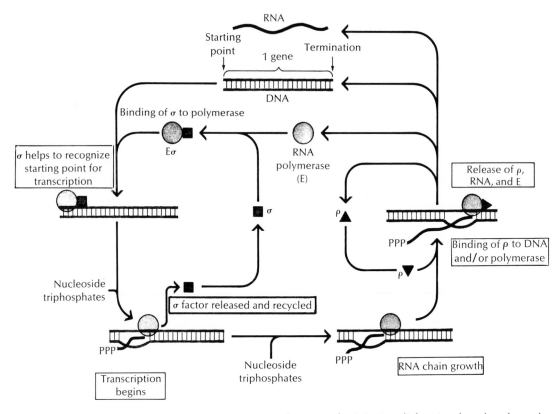

Fig. 4-14. Schematic representation of RNA synthesis in *E. coli* showing the roles of σ and ρ factors. The σ and ρ factors are needed only for initiation and termination, respectively, and after each has served its function, it is released from the core polymerase and "recycled." ⬤ = RNA polymerase; ▶ = ρ factor; ■ = σ factor. (From James D. Watson, Molecular biology of the gene, 3rd ed., copyright © 1976, 1970, 1965 by James D. Watson and The Benjamin/Cummings Publishing Co., Inc., Menlo Park, Calif.)

occur in the specificity of promoter recognition that result in the transcription of different genes. This can involve either the modification of the bacterial RNA polymerase or the synthesis of a new virus-specific RNA polymerase. These changes are discussed in the section on transcriptional control later in this chapter.

It is obvious that any given RNA transcript is copied from only one strand of the DNA double helix. From this, it might seem reasonable to think that one of the strands in a bacterial or viral genome is the positive or "sense" strand and the other is a negative copy used only for template purposes. However, convincing evidence has been accumulated, both for bacteria and for bacteriophages, that different genes are transcribed from different strands of the genome. In circular chromosomes such as that of *E. coli* some genes are transcribed from one strand in a clockwise direction from their promoters, whereas others are transcribed from the other strand in a counterclockwise direction. Cases are also known in which transcription proceeds in both directions on opposite strands from a common control site.

Termination of transcription occurs at specific sites that are coded into the genome. The exact DNA sequence that signals termination is not certain, but it generally appears to include a sequence of T's in the positive strand of the DNA, so that the transcript ends with a series of U's. There is also some evidence suggesting that a GC-rich area capable of forming secondary structures (loops) may occur just before the series of U's. Termination occurs spontaneously at some sites, but reliable termination of most transcripts at the proper location requires the association of a specific protein factor, termed rho (ρ) with the RNA polymerase molecule. When the completed RNA strand and the core polymerase are separated from the DNA and from each other, the ρ factor is also released and recycled, as shown in Fig. 4-14.

Frequently the structural and regulatory genes for a set of related functions are linked together on the bacterial chromosome in a cluster known as an operon. Operons usually have only one promoter and are transcribed into polycistronic mRNA molecules that code for the synthesis of all the enzymes involved in the operon. Examples include the lactose operon, with three structural genes transcribed as a single mRNA, and the histidine operon, where the coding information for 10 different enzymes is transcribed as a single very large mRNA. Details of the structural organization and regulation of transcription of operons are discussed later in this chapter.

The synthesis of protein is probably the most complex metabolic process that occurs in any living cell. Four major steps are involved in the formation of a peptide chain from individual amino acids:

Synthesis of protein

1. Activation of the 20 different amino acids and their attachment to small RNA molecules known as tRNA, which contain "anticodons" capable of complementary base pairing with the codons in the mRNA

2. Formation of an initiation complex consisting of mRNA, large and small ribosomal subunits, and a special initiator tRNA carrying the first amino acid (N-formylmethionine) of the peptide chain that is to be made.

3. Elongation of the peptide chain by systematic addition of amino acids (carried on tRNA molecules), as specified by the triplet code in the mRNA

4. Termination of protein synthesis when a termination codon in the mRNA is reached

Formation of aminoacyl-tRNA complexes. As the word "translation" implies, it involves conversion of information from the language of nucleotide sequences in RNA to the totally different language of amino acid sequences in a polypeptide chain. We have already examined the dictionary that is used for this translation in our discussion of the genetic code (Fig. 4-7). The "interpreters" who do the actual translation are the 20 aminoacyl-tRNA synthetase enzymes. These enzymes perform the following steps: (1) recognize each of the 20 different amino acids, (2) activate them with adenosine triphosphate (ATP) to form enzyme-bound adenosine monophosphate (AMP)–amino acid complexes, and (3) attach the activated amino acids to tRNA molecules that contain anticodons which will interact with the codons in mRNA that correspond to the correct amino acids. Each activating enzyme must possess at least three distinct recognition sites: (1) for the amino acid that is activated, (2) for ATP, and (3) for the set of tRNA molecules (which in turn may recognize as many as six different codons) that correspond to that particular amino acid.

Together, the aminoacyl-tRNA synthetase enzyme and the tRNA bridge the gap between the nucleotide triplet sequence in the mRNA and the amino acid sequence in the protein. The aminoacyl-tRNA synthetase is the only molecule known to recognize both a specific amino acid and an RNA molecule that has nucleotide triplet-coding specificity. Once the amino acid has been attached to the tRNA, all further specificity is based on codon-anticodon recognitions. If the amino acid becomes attached to the wrong kind of tRNA, it will be inserted in the wrong place in the protein.

Transfer RNA molecules are typically about 75 to 80 nucleotides in length and contain a number of modified purine and pyrimidine bases that are not generally found in other types of RNA. Complementary base pairing within the tRNA molecule results in a characteristic pattern of loop formation. The loops can conveniently be represented in two dimensions by the cloverleaf pattern in Fig. 4-15, *A*. This pattern has been highly conserved in evolution, and each part of the structure is thought to have a definite function.

Amino acid attachment occurs at the 3' end of the tRNA, which always has the sequence CCA–3'OH, and is located at the end of the "stem" of the cloverleaf. Binding of the amino acid is by an ester linkage between its carboxyl group and the free 3'OH group of the tRNA. The amino group of the amino acid remains free for peptide bond formation (except for *N*-formylmethionine, which becomes the *N*-terminal amino acid in prokaryotic proteins and has no need for a reactive free amino group).

The "anticodon," which recognizes the codon for the amino acid in

Fig. 4-15. Structure of tRNA molecules. **A,** Cloverleaf pattern. Transfer RNA molecules in general are characterized by the presence of substantial numbers of modified bases and by internal base pairing, which permits them to be depicted in a "cloverleaf" pattern consisting of three major loops and a stem. The major loops, in sequence from the 5' to the 3' end of the RNA chain, are called the D loop, which consistently contains dihydrouracil; the anticodon loop, which contains the anticodon that pairs with the mRNA codon; and the TψC loop, which consistently contains the sequence thymine, pseudouridine, cytosine. The amino acid binding site at the 3' terminus of the RNA chain is located at the end of the stem. The sequence depicted here is for alanine tRNA from yeast. (From Goodenough, U., and R. P. Levine. 1974. Genetics. Holt, Rinehart, & Winston, Inc., New York [modified from Watson, J. D. 1970. Molecular biology of the gene, 2nd ed. The Benjamin/Cummings Publishing Co., Inc., Menlo Park, Calif.]).

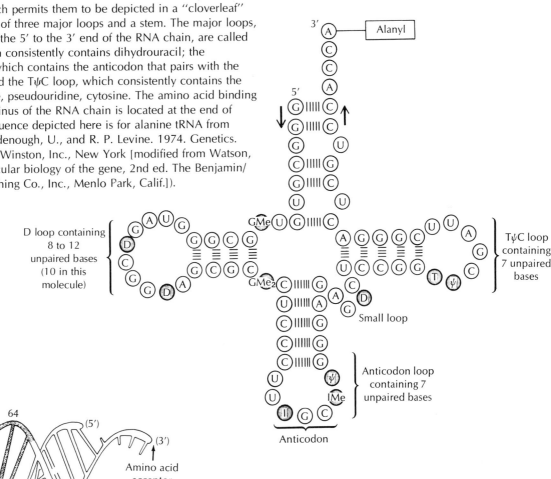

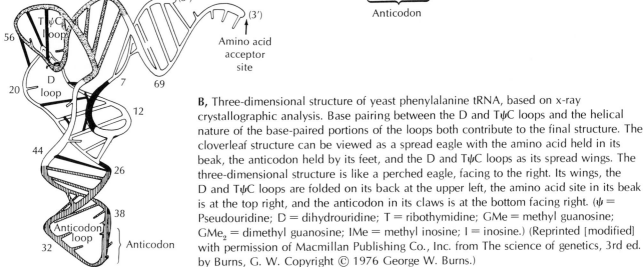

B, Three-dimensional structure of yeast phenylalanine tRNA, based on x-ray crystallographic analysis. Base pairing between the D and TψC loops and the helical nature of the base-paired portions of the loops both contribute to the final structure. The cloverleaf structure can be viewed as a spread eagle with the amino acid held in its beak, the anticodon held by its feet, and the D and TψC loops as its spread wings. The three-dimensional structure is like a perched eagle, facing to the right. Its wings, the D and TψC loops are folded on its back at the upper left, the amino acid site in its beak is at the top right, and the anticodon in its claws is at the bottom facing right. (ψ = Pseudouridine; D = dihydrouridine; T = ribothymidine; GMe = methyl guanosine; GMe₂ = dimethyl guanosine; IMe = methyl inosine; I = inosine.) (Reprinted [modified] with permission of Macmillan Publishing Co., Inc. from The science of genetics, 3rd ed. by Burns, G. W. Copyright © 1976 George W. Burns.)

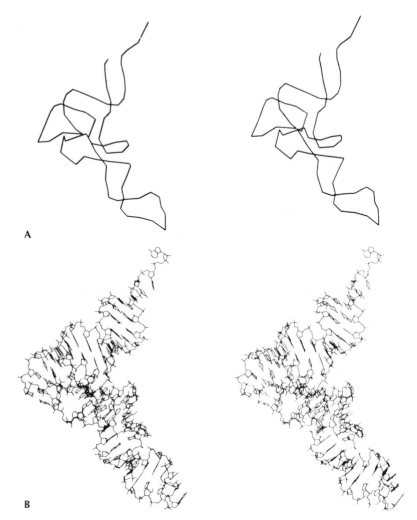

A

B

Fig. 4-16. Stereo pairs showing three-dimensional structure of tRNA. **A,** Ribose-phosphate backbone of the yeast phenylalanine tRNA molecule shown diagrammatically in Fig. 4-15, *B.* **B,** Detailed molecular structure of the same tRNA. These figures are best viewed with a stereo viewer with lenses to facilitate fusion of the two images into a single three-dimensional picture. Such viewers are available in many biology departments. However, most individuals can see a three-dimensional image without special equipment by following simple instructions: Begin with **A,** which is simpler. Hold the page at normal reading distance. Cross your eyes just enough so that the image of the right-hand drawing seen with your left eye overlaps the image of the left-hand drawing seen with your right eye. With a little practice and twisting of the book to line the images up exactly, most people can see a clear three-dimensional image. It is important to concentrate on the drawings and ignore peripheral visual clues that are trying to tell you to uncross your eyes. Purists will realize that they are seeing a mirror image of the structure depicted in Fig. 4-15, since the right-hand picture is viewed with the left eye and the left-hand picture with the right eye. However, the three-dimensional twisting of the helixes and folding of the molecule remains just as impressive. (From Kim, S. H. 1976. Prog. Nucleic Acid Res. Mol. Biol. **17:**181.)

the mRNA, is located in the loop of the cloverleaf farthest from the stem, at the bottom in Fig. 4-15, *A*. The loops to the left and right in Fig. 4-15, *A* contribute to the overall geometry required for recognition of the tRNA by aminoacyl-tRNA synthetase and for interaction of the tRNA with the ribosome. Additional base-pairing interactions between the left and right loops cause them to fold together into a complex three-dimensional structure. This plus twisting of the base-paired double helical loops gives the whole molecule an L-shaped configuration, as shown in Fig. 4-15, *B*. The amino acid attachment site is located at the end of one arm of the L, and the anticodon is at the end of the other arm, some 70 Å distant (1 Å is 10^{-8} cm, or approximately the diameter of a hydrogen atom). The three-dimensional structure can be seen more clearly in the stereo pairs in Fig. 4-16. The amino acid binding site is at the extreme upper right, and the three bases of the anticodon protrude from the right side of the loop at the extreme lower right.

Initiation of translation. For protein synthesis to occur, the codons of the mRNA and the anticodons of the aminoacyl tRNAs must be brought into juxtaposition. In addition, the amino acids bound to the tRNAs must be brought into a physical relationship that will permit formation of peptide bonds. These functions are carried out by complex ribonucleoprotein particles known as ribosomes. Each ribosome is composed of a small subunit (referred to as 30S on the basis of its rate of sedimentation in Svedberg units) and a larger subunit (50S). The large subunit of prokaryotic ribosomes is composed of two RNA molecules (23S and 5S) and about 33 different proteins. The smaller has a single species of RNA (16S) and about 21 proteins. The precise architecture of these complicated particles has not been worked out fully. Under laboratory conditions each of the ribosomal subunits can be made to reassemble spontaneously from its constituent RNA and protein molecules.

In addition to tRNA, mRNA, and large and small ribosomal subunits, the actual translational process in prokaryotic cells also requires a number of accessory protein factors (three initiation factors, two elongation factors, and two release factors) plus energy in the form of guanosine triphosphate (GTP). Still other factors "recycle" the catalytic factors and enhance reaction rates.

The initiation stage of translation involves the formation of an initiation complex consisting of the 30S and 50S ribosomal subunits, mRNA, and the initiator tRNA with bound N-formylmethionine. Three initiation factors, IF-1, IF-2, and IF-3, function catalytically in the formation of the complex, and hydrolysis of GTP is required. The sequence of events thought to be involved is shown in Fig. 4-17, although there is still some uncertainty about the order of addition of

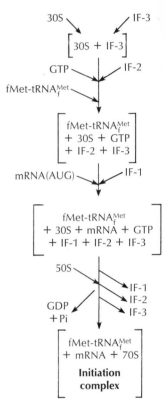

Fig. 4-17. Prosposed sequence for formation of the 70S initiation complex in *E. coli* (see Fig. 4-18, *A*). The exact sequence of addition of mRNA, fMet-tRNA$_f^{Met}$, and IF-3 to the 30S ribosomal subunit is still controversial.

mRNA and fMet-tRNA$_f^{Met}$ to the 30S subunit. Eukaryotic initiation is similar in principle, although more initiation factors are involved and the initiator tRNA carries ordinary methionine rather than *N*-formylmethionine (see Fig. 9-1).

Elongation. Growth of the peptide chain, which is commonly referred to as elongation, is diagrammed in Fig. 4-18. Each 70S ribosome has two sites for aminoacyl-tRNA attachment, designated A and P, respectively, for amino acid and peptide. At the beginning of the elongation process (Fig. 4-18, *A*) the tRNA carrying *N*-formylmethionine is locat-

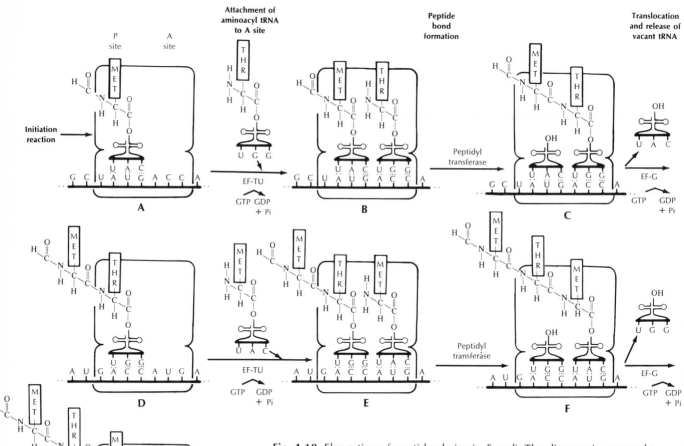

Fig. 4-18. Elongation of peptide chains in *E. coli*. The diagram is arranged to emphasize the repetitive nature of the elongation process. Note that the second line repeats the steps in the first line with the next amino acid, and that these would again repeat for another amino acid in the third line if it were shown in its entirety. The mRNA and protein sequences depicted are those for the *N*-terminal portion of β-galactosidase from *E. coli*. The initiation reaction was depicted by Fig. 4-17. MET and THR represent the side chains of methionine and threonine, respectively. Note that the first methionine has an *N*-formyl substitution.

ed in the P site, where it attaches during the initiation process just described (Fig. 4-17). The initial attachment of all other tRNA molecules carrying single amino acids occurs at the A site (Fig. 4-18, *B* and *E*). Elongation factor EF-Tu and the hydrolysis of a molecule of GTP are required for each such attachment. (A second factor, EF-Ts, is needed to remove GDP from EF-Tu so that the EF-Tu can be reused, but EF-Ts does not function directly in elongation.)

Elongation begins with attachment at the A site of the tRNA carrying the second amino acid of the protein (Fig. 4-18, *B*). Peptide bond formation is catalyzed by a peptidyl transferase enzyme, which is one of the proteins in the large ribosomal subunit. The peptide bond is formed by transferring the carboxyl group of the *N*-formylmethionine from its tRNA to the free amino group of the amino acid in the A site (Fig. 4-18, *C*). The new dipeptide remains attached at its carboxyl end to the tRNA that brought the second amino acid to the ribosome.

The tRNA carrying the dipeptide is then translocated to the P site, displacing the vacant tRNA that had previously carried the *N*-formylmethionine (Fig. 4-19, *D*). The translocation step requires another elongation factor, EF-G, and the hydrolysis of another molecule of GTP. A tRNA carrying the next amino acid coded for by the mRNA then enters the A site (Fig. 4-18, *E*), and the carboxyl end of the dipeptide is transferred to the free amino group of that amino acid (Fig. 4-18, *F*). The tRNA carrying the tripeptide is then translocated to the P site (Fig. 4-18, *G*), freeing the A site to receive the next aminoacyl tRNA.

This cyclic process of peptide chain growth, catalyzed by two elongation factors plus EF-Ts and peptidyl transferase and requiring the hydrolysis of two GTPs per amino acid added to the growing chain, continues until the chain reaches its full length. The growing chain always remains attached to the tRNA molecule that carried the most recently added amino acid.

Termination. When a termination codon enters the A site, the completed peptide chain is released from the ribosome and from the final tRNA that it was attached to. Termination requires the presence of specific release factors and the hydrolysis of GTP.

"Nonsense" mutations, in which the codon for an amino acid has been converted by base substitution into a termination codon, have been used widely to study gene-protein relationships. Such mutations are particularly valuable because their effects can be suppressed by other mutations (suppressors) that alter particular tRNA molecules so that their anticodons pair with the termination codons and insert amino acids into the growing chain in place of termination. Suppressor mutations also interfere with normal termination to some extent, but their impact on cellular survival tends to be minimal, both because they are "leaky" (they do not reverse termination every time) and because prokaryotic messages frequently end with two termination codons in tandem (so that reversal of one of them does not prevent termination from occurring at approximately the right peptide chain length). The use of nonsense mutations and their suppressors makes it possible to develop and maintain stocks of mutations that would otherwise be lethal and therefore impossible to work with. The term "conditional lethal" is used to describe such mutations.

Mutation can also convert a termination codon into an amino acid–specifying codon. This results in the formation of an abnormally long protein whose growth continues until the next termination codon is reached. Such mutations are observed more frequently in eukaryotic cells, which tend to have only single termination codons. Examples are discussed in Chapter 8.

CONTROL OF GENE EXPRESSION

The following is a partial list of the types of changes in gene expression that occur in prokaryotic organisms and their viruses:

1. *Enzyme induction.* Bacteria such as *E. coli* possess the ability to use a variety of energy substrates (e.g., glucose, lactose, arabinose). In many cases the genes coding for the enzymes needed for use of a particular substrate are linked in an operon that is transcribed only when specifically "induced" by the presence of the substrate.

2. *Repression.* Many bacteria possess the ability to synthesize everything they need for growth in a simple medium containing only a source of energy and carbon (such as glucose) and a mixture of inorganic salts. However, when other nutrients such as amino acids are available from an exogenous (external) source, they are used preferentially, and operons coding for enzymes needed for endogenous (internal) synthesis of these nutrients are "repressed" (turned off). This type of repression is sometimes referred to as "end product" repression, since it is under control of the end product of the sequence of enzymatic reactions that is affected.

3. *Catabolite repression.* When glucose and other substrates such as lactose are both available to bacteria such as *E. coli,* the glucose is used preferentially, and induction of the enzymes that would permit use of the second substrate is repressed. This phenomenon is termed "catabolite repression," since the effect is actually due to catabolites (products of metabolism) derived from glucose, rather than to glucose itself. The mechanism responsible for this effect is distinctly different from that of end product repression, and it actually involves a positive, rather than a negative, control process, as will be explained in the discussion of transcriptional control.

4. *Stringent response.* When the supply of one or more amino acids in a bacterial cell falls to such a level that the tRNA for that amino acid is not kept fully charged, a series of major metabolic changes is brought about by the so-called stringent control mechanism. These changes, which are mediated by guanosine-5′-diphosphate-3′-diphosphate (ppGpp), include shutting off ribosomal and tRNA synthesis and enhancement of amino acid synthesis.

5. *Metabolic regulation.* As they respond to varying environments, bacterial cells must make quantitative adjustments in the rates of many different metabolic pathways (and therefore in rates of enzyme synthesis) to maintain suitable internal equilibrium relations and metabolic balance.

6. *Bacterial sporulation.* Certain types of bacteria have the ability to form spores that are highly resistant to adverse environmental conditions. Complex changes in gene expression occur during this process.

7. *Bacteriophage infection.* Bacterial viruses partially or completely use transcriptional and translational machinery from the host bacterial cell. A variety of regulatory mechanisms are involved in the inactivation of host protein synthesis and in the sequential expression of different bacteriophage genes during the course of the infective cycles of various types of bacteriophages.

Questions of whether these changes in gene expression in prokaryotic organisms should be classified as differentiation, or, in fact, of whether differentiation can even occur in single-celled organisms, can be argued indefinitely on a semantic level. We will bypass such arguments by simply stating that these processes are of interest as potential models for the changes in gene expression that occur during cellular differentiation in multicellular organisms.

The mechanisms involved in the regulation of gene expression in prokaryotic organisms act not only at the level of transcription but also at translational and posttranslational levels. Examples of controls occurring at each of these three levels will be considered in the following discussions.

Control mechanisms acting at the level of RNA synthesis play a major role in the regulation of gene expression in prokaryotic organisms. The types of transcriptional control that are known to occur are summarized on p. 146. All transcriptional controls must function in one of two basic modes: (1) negative control, in which transcription naturally occurs in the absence of the control elements and is turned off or reduced in amount when the control operates or (2) positive control, in which transcription does not occur or occurs only at a low level

Transcriptional control

SUMMARY OF TRANSCRIPTIONAL CONTROLS
IN PROKARYOTIC ORGANISMS

A. Negative control. Unaltered RNA polymerase recognizes the unaltered promoter, and continuous transcription occurs except when turned off by specific negatively acting regulatory mechanisms.

   1. Transcription is turned off by a regulatory molecule that binds to an operator site and prevents attachment or movement of RNA polymerase.

      a. Induction. The regulatory protein blocks transcription except when an effector molecule (coinducer) renders it incapable of binding to the operator and thus "induces" transcription (e.g., induction of the lactose operon by lactose).

      b. Repression. The regulatory protein must be modified by an effector molecule (corepressor) to bind to the operator and stop transcription (e.g., repression of the operon for tryptophan synthesis by tryptophan).

   2. RNA polymerase is altered or inactivated so that it no longer recognizes the original promoters (e.g., inactivation of host RNA polymerase during bacteriophage T7 infection).

B. Positive control. The unaltered promoter is not recognized by unaltered RNA polymerase, and transcription occurs only when it is specifically activated.

   1. Specific regulatory molecules bind at a site near the promoter and alter its interaction with RNA polymerase so that transcription occurs.

      a. Induction. The regulatory molecule must be modified by an effector molecule (coinducer) to activate transcription (e.g., induction of the arabinose-utilization operon by arabinose; positive control of the lactose operon by cyclic AMP (cAMP) and cAMP receptor protein (CRP); positive effect of ppGpp on the operon for histidine synthesis).

      b. Repression. The regulatory molecule is rendered incapable of activating transcription when an effector molecule (corepressor) is bound to it (e.g., inhibition of the operon for histidine synthesis by histidyl tRNA).

   2. Preexisting RNA polymerase is modified to recognize new promoters (e.g., modification of E. coli RNA polymerase by addition of proteins coded by bacteriophage T4 genes 33 and 55, which enables the polymerase to transcribe T4 genes expressed late in the infective cycle).

   3. A new RNA polymerase with different promoter specificity is synthesized (e.g., synthesis of a new RNA polymerase that recognizes promoters for late genes in bacteriophage T7 infection).

C. Mixed controls. Expression of a single operon is controlled by several different mechanisms (e.g., expression of the lactose operon requires both a positively controlled induction by cAMP and a negatively controlled induction by lactose).

until it is specifically activated or stimulated by a regulatory mechanism.

In both cases changes in transcription can be achieved either by binding a regulatory molecule to the genome in the vicinity of the promoter or by changing the recognition specificity of the RNA polymerase. Generally speaking, metabolic adjustments such as induction, end product repression, catabolite repression, and the stringent response are achieved by the action of regulatory molecules, whereas more drastic changes such as those in bacterial sporulation and bacteriophage infection often involve changes in the RNA polymerase.

As is evident from the outline (p. 146), induction and repression both can be achieved by either negative or positive control mechanisms. In practice the same operon is often influenced by several different effector molecules, and it is not unusual for both negative and positive mechanisms to be involved in its overall control. For simplicity we will consider the four possibilities (negatively controlled induction, negatively controlled repression, positively controlled induction, and positively controlled repression) one at a time.

Induction with negative control. The lactose operon, which consists of three enzymes whose synthesis is induced by lactose, is one of the best studied examples of negatively controlled induction. This system, which was used by François Jacob and Jacques Monod in 1961 in their original description of the operon concept, is diagrammed in Fig. 4-19, *A*. The three structural genes are closely linked together and are transcribed as a single polycistronic mRNA.

A regulatory molecule, called the lactose repressor, is coded for by a separate gene located close to the lactose operon but not transcriptionally linked to it. The lactose repressor protein, which is a tetramer, is always made by the cells in small amounts. In the absence of lactose it binds tightly to an operator site on the genome just downstream (in terms of the direction of transcription) from the promoter (Fig. 4-19, *B*). The presence of a bound repressor protein alters the structure of the promoter region so that initiation of transcription does not occur. An intermediate in the breakdown of lactose (allolactose) binds to sites on the repressor protein molecule and alters its conformation so that it is no longer capable of binding tightly to the operator site on the genome (Fig. 4-19, *C*). Allolactose thus turns on transcription of the lactose operon by reversing the negative effect of the lactose repressor protein (Fig. 4-19, *D*).

Changes in the conformation of protein molecules that cause them to undergo changes in biological activity are referred to as allosteric transformation. The small molecule that causes an allosteric transformation is referred to as an allosteric effector. In the case

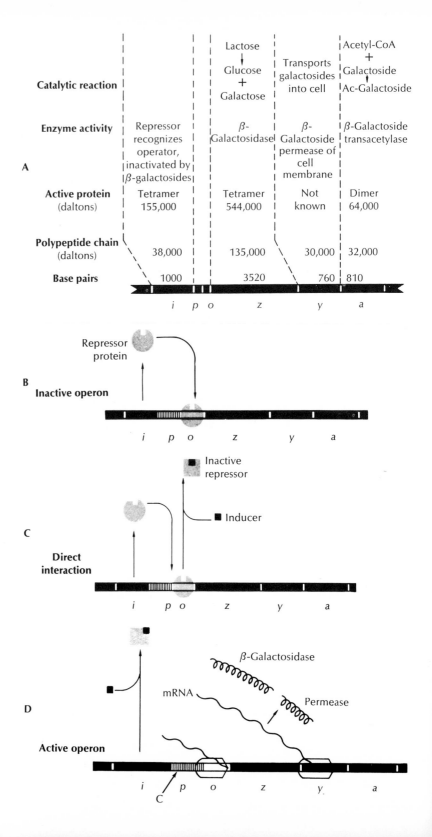

Fig. 4-19. The lactose operon of *E. coli*. **A,** Diagram of the lactose operon and the closely linked gene for its repressor protein. *Z, y,* and *a* are structural genes for the indicated enzymes; *i* is the structural gene for the repressor protein; *p* is the promoter for transcription of *z, y,* and *a;* and *o* is the operator that controls their transcription. **B,** Diagram of inactive operon in which the repressor protein binds to the operator site and prevents transcription of the operon. **C,** Binding of allolactose (or synthetic analogs of lactose) to the repressor protein, causing it to dissociate from the operator. **D,** Diagram of active operon. The inducer keeps the repressor protein molecules dissociated from the operator. This allows RNA polymerase to initiate transcription at the promoter site. Translation begins before transcription is completed. Efficient transcription also requires site C to be occupied by a complex of cAMP and a binding protein, as is described in the discussion of induction with positive control. (Reproduced with permission, from Gene Expression-I: Bacterial Genomes, by B. Lewin. Copyright © 1974, John Wiley & Sons Ltd.)

of the lactose operon the end result of the allosteric transformation of the repressor protein by allolactose is its dissociation from the operator site and "induction" of synthesis of mRNA for the enzymes of the lactose operon (Fig. 4-19, *C* and *D*). This operon is also subject to positive control by cAMP in association with its receptor protein (CRP), as will be explained in the discussion of induction with positive control.

One of the enzymes induced by lactose is a lactose permease, which increases the efficiency of lactose transport into the induced cells. This enzyme can be used to construct a crude but interesting model of the type of permanent change in gene expression that occurs in differentiation (Fig. 4-20). *E. coli* cells are grown with succinate as their energy source (so there is no catabolite repression). In the presence of high concentrations of thiomethyl-β-D-galactoside

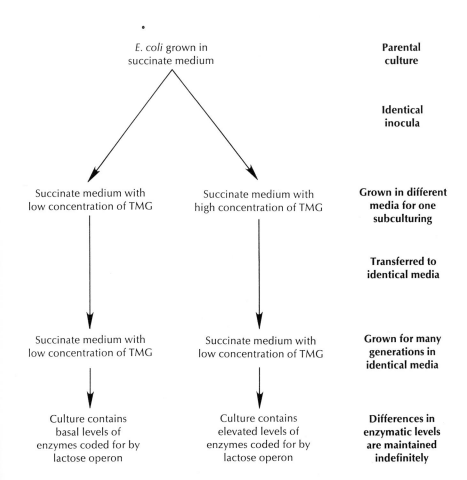

Fig. 4-20. Stable differences in expression of lactose operon of *E. coli* in identical media. Exposure to a low concentration of thiomethyl-β-galactoside (TMG) fails to induce the lactose operon even when cells are maintained in it indefinitely, as illustrated on the left. However, a brief exposure of the cells to a higher concentration of TMG induces the enzymes of the lactose operon, including a permease. When such cells with active permease are transferred to media containing a low concentration of TMG, the permease maintains an intracellular concentration of TMG that keeps the enzymes of the lactose operon, including the permease, induced indefinitely, as shown on the right. Thus, as a result of their past culture history, genetically identical cells grown in identical media can continue to exhibit different states of induction of the lactose operon indefinitely. (Based on data from Novick, A., and M. Weiner. 1957. Proc. Natl. Acad. Sci. U.S.A. **43**:533.)

(TMG), a synthetic analog of lactose that is not metabolized, the enzymes of the lactose operon are fully induced. If the fully induced cells are then transferred to a low concentration of TMG, they remain fully induced indefinitely. However, if cells that have not been previously induced are placed in the same medium, they do not become induced.

Thus two genetically identical cultures grown in identical media can exhibit different properties, depending on their past history. In this case the difference is due to lactose permease. The induced cells have a high level of permease, which permits them to maintain a sufficiently high intracellular level of TMG to maintain the induction when grown with a low level of TMG in their medium. The uninduced cells, on the other hand, lack the permease and without it never achieve a high enough intracellular level of TMG to become induced.

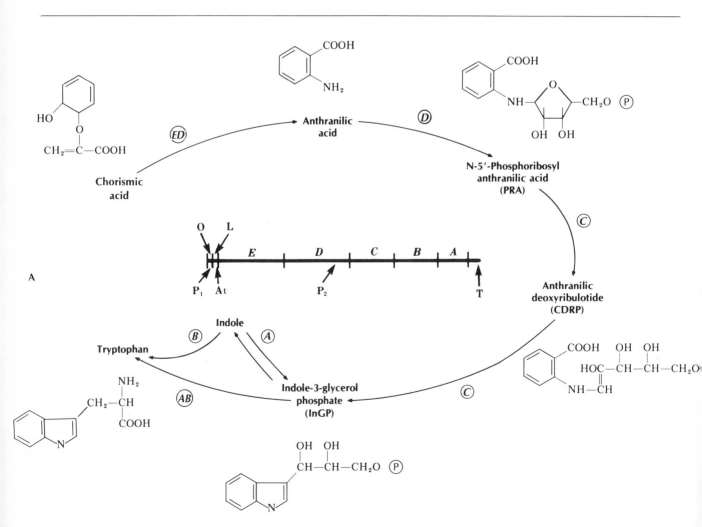

Repression with negative control. The tryptophan synthesis operon consists of five separate genes that are coordinately repressed in the presence of tryptophan. This system has a well-characterized repressor protein (called an aporepressor), which is inactive alone but becomes an active repressor when combined with tryptophan. The activated repressor binds to an operator site and reduces initiation of transcription of the operon approximately 70-fold in the presence of excess tryptophan (Fig. 4-21). By itself, this system fits the definition of repression with negative control precisely. However, for this particular operon, nature did not stop there. There also exists a secondary control called the tryptophan attenuator system, which, in the presence of high concentrations of tryptophan, causes premature termination of transcription prior to the beginning of the coding region for the first gene

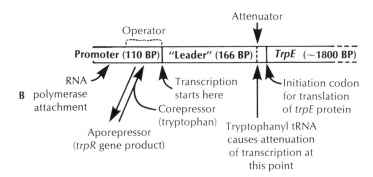

Fig. 4-21. The tryptophan operon of *E. coli.* **A,** Diagram of the genes and metabolic steps involved in tryptophan synthesis. *A, B, C, D,* and *E* are structural genes that code for five separate proteins whose catalytic roles are indicated by the circled letters in the metabolic pathways. (Note that two proteins functioning together are needed for certain steps.) P_1 is the major promoter for transcription of the operon. P_2 is a secondary promoter of lower efficiency whose effect is negligible except when P_1 is repressed. O is the operator locus for repression at P_1. L is a transcribed leader sequence between P_1 and gene *E.* T is the normal termination site. *At* is the attenuation site at which premature termination of transcription occurs in the presence of excess tryptophanyl tRNA. The relative sizes of the regulatory and coding regions are drawn to scale. **B,** Details of the transcriptional control region. The promoter-operator sequence is about 110 base pairs (BP) in length and contains both a site for RNA polymerase attachment and an operator locus that interacts with regulatory proteins. An aporepressor, coded for by the *trpR* gene, combines with tryptophan to form an active repressor that reduces initiation 70-fold when sufficient tryptophan is present. The "leader" sequence of 166 base pairs includes an attenuator site, at which transcription is reduced another eightfold to tenfold by being terminated before reaching the structural genes when tryptophanyl tRNA is available. (**A** modified from Calhoun, G. H., and G. W. Hatfield. 1975. Annu. Rev. Microbiol. **29:**275, and Lewin, B. 1974. Gene expression-I. Bacterial genomes. John Wiley & Sons Ltd., London; **B** based on data from Yanofsky, C. 1976. *In* D. P. Nierlich, W. J. Rutter, and C. F. Fox, eds. Molecular mechanisms in the control of gene expression. Academic Press, Inc., New York.)

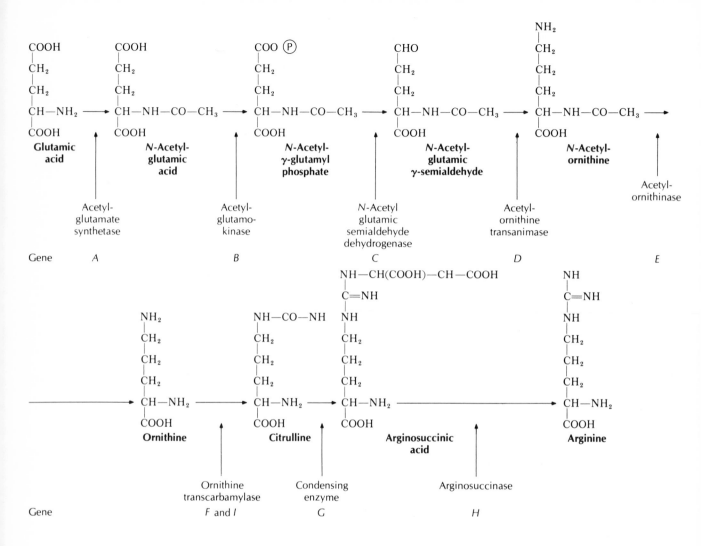

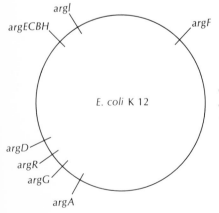

Fig. 4-22. The arginine regulon of *E. coli* K12. Nine structure genes, *argA*, *argB*, *argC*, *argD*, *argE*, *argF*, *argG*, *argH*, and *argI*, located in six different regions of the genetic map, code for the enzymes involved in arginine biosynthesis. *ArgF* and *argI* probably arose by duplication, since they code for similar enzymes and many bacterial strains have only one of them. Gene *argR* codes for a regulatory protein that combines with arginine to repress transcription of the entire set of structural genes. Each structural gene or gene cluster appears to have its own operator site that is responsive to the arginine-repressor complex. See Fig. 4-23 for details of the *argECBH* gene cluster. (Reproduced with permission, from Gene Expression-I: Bacterial Genomes, by B. Lewin. Copyright © 1974, John Wiley & Sons Ltd.)

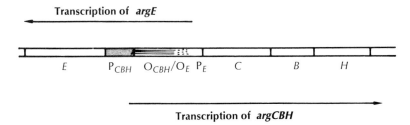

Fig. 4-23. The *argECBH* gene cluster. *ArgE* is transcribed to the left from one strand of the DNA while *argCBH* is transcribed to the right from the other strand. The operators and promoters for both are located in the region between *argE* and *argC* and are thought to overlap. The exact molecular details of this system have not yet been worked out. (Reproduced with permission, from Gene Expression-I: Bacterial Genomes, by B. Lewin. Copyright © 1974, John Wiley & Sons, Ltd.)

in the operon. This attenuation, which does not occur under conditions of tryptophan deficiency, results in another 8- to 10-fold reduction in message transcription so that together the two systems can cause a 500- to 700-fold decrease in transcription of the structural genes for tryptophan synthesis when excess tryptophan is present.

The genes for arginine synthesis are also repressed coordinately, apparently by a negatively acting repressor molecule that functions only in the presence of arginine. However, in this case the nine genes that are involved are located in six different parts of the bacterial chromosome (Fig. 4-22). Each gene or gene cluster has its own operator region, and all appear to be controlled by the same repressor protein in combination with arginine. Such a set of operons all under the control of a common regulatory mechanism is sometimes referred to as a "regulon."

One cluster of four genes designated *argE*, *argC*, *argB*, and *argH* is particularly interesting. The operator site for this cluster is between genes *argE* and *argC*. Three genes, *argC*, *argB*, and *argH*, are transcribed to the right from one strand of the DNA, whereas *argE* is transcribed to the left from the other strand. The operator and promoter sites for these genes seem to involve both strands of the DNA in the region between *argE* and *argC* (Fig. 4-23).

Induction with positive control. Arabinose is a sugar that induces transcription of an operon containing three enzymes involved in its use. In contrast to the lactose system, in which a maximum rate of transcription occurs in mutants lacking a functional lactose repressor protein, no induction of the arabinose operon occurs in the absence of the regulatory protein. Thus the regulatory protein with arabinose bound to it serves as a positive activator of transcription. However, detailed

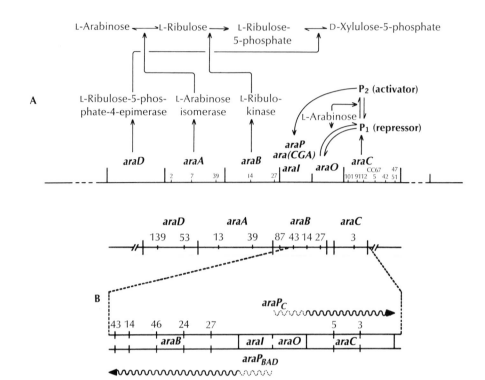

Fig. 4-24. Regulation of transcription of the arabinose *BAD* gene cluster. **A,** Genes *araA*, *araB*, and *araD* code for enzymes involved in arabinose utilization, as shown. Closely linked gene *araC* codes for a regulatory protein with a dual function. In the absence of arabinose it binds to an operator, *araO*, and blocks transcription of *araBAD*. Arabinose converts the regulatory protein from a repressor (P_1) to an activator (P_2) that binds to an initiator site *(araI)* and enhances transcription. In the absence of the regulatory protein, transcription cannot be induced above a basal level. *Ara(CGA)* refers to the binding site for a cAMP-protein complex (p. 155) that is also needed for *araBAD* transcription. *AraP* is the promoter for *araBAD*. Neither has been located precisely relative to the ends of the structural genes for *araB* and *araC*. Transcription of two other genes, *araE* and *araF*, which are involved in arabinose transport and are located in other parts of the genetic map, is also controlled by arabinose and the *araC* protein. **B,** Transcription of *araBAD* and *araC*. The *araBAD* operon is transcribed to the left while the gene for the regulatory protein, *araC*, is transcribed to the right. The distance between their initiation points, measured by EM, is about 150 base pairs. The total distance between genes *B* and *C*, including the "leader" sequences of their mRNA molecules, is estimated to be about 300 base pairs. Within that limited amount of DNA are located promoters for *araBAD* and *araC*, the cAMP receptor binding site *(ara* [*CGA*]), the operator site for negative control by the *araC* protein, and the initiator site for positive control by the *araC* protein. The numbers refer to map locations of specific mutations within the various genes. (**A** from Englesberg, E., and G. Wilcox. 1974. Annu. Rev. Genet. **8:**219. Reproduced, with permission, from the Annual Review of Genetics, Volume 8. © 1974 by Annual Reviews, Inc.; **B** from Wilcox, G., et al. 1974. Proc. Natl. Acad. Sci. U.S.A. **71:**3635.)

studies have shown that in the absence of arabinose the regulatory protein also functions secondarily as a negative control repressor much like the repressor of the lactose operon. Thus the arabinose operon is under a dual control system in which the same regulatory protein serves both in a negative and a positive control role (Fig. 4-24).

Recent data suggest that the regulon for maltose metabolism, which consists of three separate operons, is regulated in a strictly positive manner. A gene designated *malT*, which is located adjacent to one of the operons, produces a regulatory protein that is inactive alone but becomes a positive-acting inducer of all three operons when combined with maltose. Unlike the arabinose system, the maltose regulatory protein does not appear to have a negative effect on transcription in the absence of maltose.

Another example of positively controlled induction is the so-called catabolite repression system. The low molecular weight effector for this system is cAMP, whose concentration in *E. coli* drops when glucose is being metabolized and adequate amounts of glucose catabolites are present within the cell. When the supply of glucose is exhausted and the level of catabolites falls, the intracellular concentration of cAMP increases by mechanisms that are still not well understood. The cAMP binds to a protein referred to by various investigators as CRP (cAMP receptor protein), CAP (catabolite activator protein), or CGA (catabolite gene activator). The promoter regions of sensitive operons, including the lactose and arabinose operons that have been described, contain receptor sites for the cAMP-CRP complex that must be occupied for high levels of initiation of mRNA synthesis to occur in response to inducing substrates such as lactose and arabinose (Fig. 4-19, *D*).

Binding of the cAMP-CRP complex to the promoter appears to change its conformation so that it interacts more favorably with the σ subunit of RNA polymerase. Certain mutations in the promoter region render the initiation of transcription independent of the cAMP-CRP complex, apparently by making the conformation of the promoter region more like that of other promoters whose initiation does not require the cAMP-CRP complex. Recently it has also been found that mutational changes in the σ subunit can lead to efficient initiation of transcription of the lactose operon in the total absence of the cAMP-CRP complex. Thus, improved compatibility between the promoter and σ-containing RNA polymerase can be achieved by modification either of the promoter or of the σ subunit. The role of the cAMP-CRP complex appears to be to facilitate such interaction without permanent modification of either component.

A fourth example of induction with positive control is the effect of ppGpp on the operon for histidine synthesis. Bacteria that possess the so-called stringent growth control system, generate ppGpp in an ATP-consuming reaction whenever uncharged tRNA molecules enter the A sites of ribosomes engaged in peptide chain elongation. One of the many effects of ppGpp on cellular metabolism is to enhance transcription of the histidine synthesis operon by positive control mechanisms that are not yet fully understood.

Repression with positive control. In a positive control system, repression is achieved when the effector molecule binds to the regulatory molecule and renders it incapable of activating transcription. This type of mechanism appears to operate in a number of systems, but none have been fully defined. One example is the inhibition by histidyl tRNA of the operon for histidine synthesis by mechanisms that appear to be separate from the effects of ppGpp just described. Another possible case occurs in the biosynthesis of isoleucine and valine, in which the enzyme threonine deaminase, which is required for the first step in isoleucine synthesis, appears to serve as a positive regulator, causing the operon to be transcribed at an enhanced level at all times except when the enhancement is blocked by the end products of the biosynthetic sequence. The enzymes that are involved in isoleucine formation and the proposed positive control mechanism for repression of their synthesis are diagrammed in Fig. 4-25.

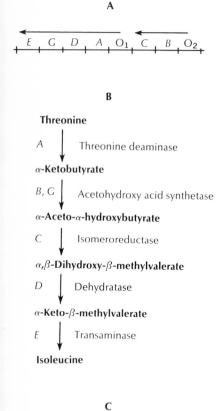

Fig. 4-25. Proposed positive control mechanism for end product repression of transcription of mRNA for enzymes involved in isoleucine biosynthesis. **A,** Genetic map of isoleucine-valine gene cluster *(ilv)* of *E. coli* K12. O_1 is the operator site for structural genes *A, D, G,* and *E;* and O_2 is the operator site for structural genes *B* and *C.* **B,** Enzymatic steps involved in the biosynthesis of isoleucine. The enzymes that are involved are identified on the right and the genes that code for them on the left. Synthesis of valine proceeds by a parallel pathway involving all these enzymes except threonine deaminase. Most of the same enzymes are also involved in leucine synthesis. **C,** Positive control mechanism proposed by Levinthal, et al. (1976. J. Mol. Biol. **102:**453.) The product of gene *A* (which continues to be synthesized at a reduced rate when the system is repressed) enhances transcription of the *ilvADGE* operon. Isoleucine (and also valine and leucine) interact with the activator protein and prevent it from enhancing transcription. Note that an allosteric interaction between isoleucine and threonine deaminase also results in direct inhibition of the enzymatic activity of threonine deaminase (Fig. 4-26, *B*).

Altered promoter specificity. Although induction and repression appear to be accomplished primarily by interaction of regulatory molecules with the genome, more drastic changes in gene expression, such as those which occur in bacterial sporulation and in bacteriophage infection, frequently involve initiation of transcription at different promoter sites. This can be accomplished either by modification of preexisting RNA polymerase or by the synthesis of an entirely new RNA polymerase.

The genome of bacteriophage T4 appears to contain genes with at least three different promoter specificities. These genes are termed "early," "middle," and "late" on the basis of when in the course of the infective cycle they are first transcribed. The early genes are transcribed by the normal RNA polymerase of the *E. coli* host cell. Proteins coded for by those genes modify the host polymerase so that it becomes able to initiate transcription at the middle and late promoters. The changes involved in acquiring late promoter specificity are the best understood. Genes 33 and 55 of phage T4 code for two proteins of molecular weight 12,000 and 22,000, which become incorporated into the host RNA polymerase and alter its specificity. The σ factor appears not to be displaced or inactivated during this process.

Bacterial sporulation also involves changes in RNA polymerase specificity, but there have been conflicting reports on the mechanism that is involved.

Some bacteriophages switch from early to late genes by inactivating the host RNA polymerase and replacing it with a totally new virally coded RNA polymerase. In bacteriophage T7, for example, one of the early genes transcribed by the host RNA polymerase codes for a viral RNA polymerase that consists of a single polypeptide chain with a molecular weight of 110,000 daltons. Two other early genes code for proteins that inactivate the host polymerase. The T7 polymerase recognizes promoter sites that are found only in the T7 genome and transcribes genes that are expressed only during the later parts of the infective cycle of T7.

Translational controls

Gene expression is controlled at the level of translation in a number of cases. The genome of the single-stranded RNA bacteriophage R17 includes one gene each for RNA replicase and for coat protein. These two proteins are needed in different amounts.

One replicase molecule can sythesize many copies of the viral RNA (which serves both as genome and as mRNA), whereas some 180 molecules of coat protein are needed per RNA in each new virus particle. Preferential synthesis of coat protein is achieved in two ways. First, the configuration of the RNA is such that translation of coat protein is initiated more frequently than that of the replicase. In addition, when enough bacteriophage RNA has been synthesized so that coat protein begins to accumulate in the infected cell, the coat protein attaches to the RNA coding for the replicase and specifically inhibits its translation. Coat protein translation continues to occur, unaffected by the accumulation of its own end product.

It can be argued that this is a special case, since there is no opportunity for transcriptional control in RNA phages. However, translational control also occurs in *E. coli* cells infected with bacteriophage T4. In this case the phage genome is double-stranded DNA, and opportunities for transcriptional control do exist. Gene 32 of bacteriophage T4 codes for a protein that binds to newly synthesized bacteriophage DNA and is essential for normal bacteriophage replication. When the DNA becomes saturated with gene 32 protein and an excess begins to accumulate in the infected cells, it binds specifically to its own mRNA and inhibits further translation of itself in an autoregulatory feedback loop. This control mechanism allows the infected cells to synthesize the amount of gene 32 protein that is needed and then to shut off further synthesis at the translational level.

Differences in message half-life are important in determining the amounts of particular proteins that are translated in eukaryotic cells. Similar processes are also thought to operate in prokaryotic cells, although the whole time scale is much faster. The average half-life of mRNA in *E. coli* at 37° C is only about 2 minutes. There are also differences in the efficiency of initiation of translation, even for different messages within a single polycistronic mRNA molecule. For example, the three enzymes coded for by the lactose operon (β-galactosidase, galactoside permease, and galactoside acetylase) are synthesized in a ratio of approximately 10:5:2.

Over the years there have been many attempts to link the availability of tRNA molecules that recognize different codons for the same amino acid to gene regulatory mechanisms. In theory, control could be achieved by regulating the supply of tRNA for a particular codon that

Fig. 4-26. Feedback inhibition of enzymatic activity. **A,** Schematic representation of the inhibition of an enzymatic reaction due to allosteric transformation of the enzyme by a metabolic end product with no direct structural relationship to the substrates of the enzyme. **B,** End product inhibition of the first enzymatic step in the biosynthesis of isoleucine from threonine. Note that in this synthetic pathway a closely related allosteric transformation is also involved in repression of enzyme synthesis at the transcriptional level (Fig. 4-25). **C,** End product inhibition of the first enzymatic step in pyrimidine biosynthesis. (Modified from James D. Watson, Molecular biology of the gene, 3rd ed., copyright © 1976, 1970, 1965 by James D. Watson and The Benjamin/Cummings Publishing Co., Inc., Menlo Park, Calif.)

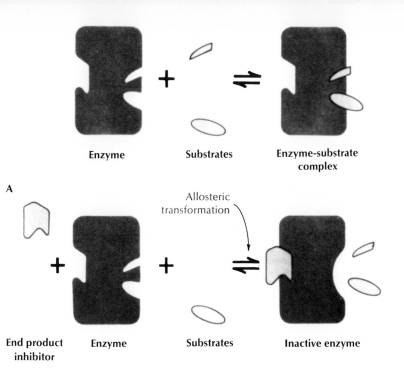

A

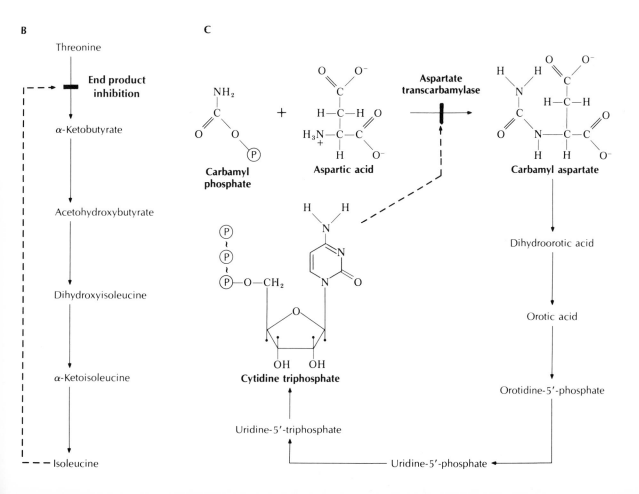

B

Threonine

End product inhibition

α-Ketobutyrate

Acetohydroxybutyrate

Dihydroxyisoleucine

α-Ketoisoleucine

Isoleucine

C

Carbamyl phosphate

Aspartic acid

Aspartate transcarbamylase

Carbamyl aspartate

Dihydroorotic acid

Orotic acid

Orotidine-5′-phosphate

Cytidine triphosphate

Uridine-5′-triphosphate

Uridine-5′-phosphate

is only used to code for certain classes of proteins. It is known that the genomes of certain bacteriophages, such as T4, code for phage-specific tRNA molecules and also that host cell tRNA molecules are modified during bacteriophage infection and during bacterial sporulation. However, the precise role of these changes is uncertain. They may simply improve the efficiency of the translational process by providing a spectrum of tRNA molecules whose relative amounts of various codons more closely match those of the messages being translated. Ribosomal modifications of unknown significance also occur during bacteriophage infection.

Posttranslational controls   Even after completion of protein synthesis, functional expression of particular genes can be further regulated. One of the most common forms, which has already been mentioned, is allosteric control, in which a low molecular weight allosteric effector binds to a specific regulatory site on a protein molecule and causes it to undergo a change in physical conformation so that its functional properties are altered.

One particular type of allosteric control that is important in the overall regulation of gene expression is feedback inhibition, in which the end product of a series of metabolic reactions allosterically inhibits the first enzyme in the series, thus functionally turning off the entire sequence of reactions (Fig. 4-26). Such mechanisms permit a rapid reduction in the rate of synthesis whenever sufficient amounts of the end product are accumulated. This phenomenon is similar in principle to the end product repression of transcription discussed earlier, but it acts directly at the enzymatic level with no involvement of RNA or protein synthesis. It can thus be more rapidly responsive to the immediate needs of the cell, although it is less efficient in terms of cellular expenditure of energy, since mRNA and enzymes continue to be made even when they are not needed.

Another effect that occurs entirely at the enzymatic level is the stabilization of certain enzymes by their substrates. One widely cited example of this is the enzyme tryptophan pyrrolase. At one time, synthesis of this enzyme was thought to be induced by its substrate, tryptophan. However, detailed studies have shown that the effect is entirely due to a reduced rate of degradation of the enzyme when it is bound to tryptophan.

Finally, there is the unusual case in which certain viruses use host proteins for a totally different function than they had in the uninfected cell. Small RNA bacteriophages such as $Q\beta$ form an RNA replicase enzyme complex from one newly synthesized protein plus three preexisting host proteins. The new protein has a molecular weight of 65,000 and contains the actual catalytic activity. The three host proteins are S1, which is a normal component of the small ribosomal subunit, and elongation factors EF-Ts and EF-Tu, which are normally involved in growth of peptide chains. In the viral RNA replicase complex these three proteins appear to be involved in specific interactions with the viral RNA.

In this chapter we have emphasized the "central dogma" of molecular biology, which states that information flows from nucleic acids to proteins but *never* from proteins to nucleic acids. In its original form the information flow was described as from double-stranded DNA to single-stranded RNA to protein. This is still valid for prokaryotic and eukaryotic cells, but various complications occur during virus replication, including a reverse flow from RNA to DNA, the existence of single-stranded DNA and double-stranded RNA, and some systems in which DNA is not involved at all.

All living organisms need to synthesize different amounts of different proteins. A wide range of mechanisms has been evolved for this purpose, both in prokaryotic cells, which were discussed in this chapter, and in eukaryotic cells, which will be discussed in the following five chapters. From a teleological point of view the most economical point for control is to regulate transcription of DNA sequences into RNA sequences so that only as much mRNA is synthesized as is needed. Control over gene expression is frequently, but by no means always, accomplished at this level. Control can be negative, with transcription specifically blocked by negative-acting regulatory proteins, or positive, with transcription occurring only when specifically activated by special regulatory factors or by RNA polymerase enzymes with altered promoter recognition properties.

Control over gene expression can also occur at the translational level, at the level of enzyme function through allosteric feedback inhibition, and by altering enzyme stability. Thus, even in supposedly "simple" prokaryotic organisms, controls can be exerted at several different levels in the flow of genetic information. As has been emphasized throughout this chapter, many of the controls themselves are far from simple. The same set of genes is often influenced by several different types of effectors, including the overall metabolic state of the cell. In addition, although it might not seem so to the reader, we have deliberately avoided discussion of mechanisms that are particularly controversial, confusing, or complicated. Relatively recent advances in sequencing techniques are currently adding a flood of new information, which in most cases is raising new questions more rapidly than it is providing new answers.

RELATIONSHIP OF PROKARYOTIC GENE REGULATORY MECHANISMS TO CELLULAR DIFFERENTIATION

We emphasized at the beginning of this chapter that molecular genetics is a new field of science. With the progress that has been made, it can no longer be regarded as in its infancy; however, it still has many of the properties of adolescence: it has grown rapidly, it is at times confused, awkward, and inconsistent, and it sometimes fails to recognize how much it still has to learn. This last point is a particular hazard both to us, the authors, and to our readers. In our attempts to understand the molecular genetics of prokaryotic organisms, we find it easy to be awed by the fantastic accomplishments that have been achieved in a few short years and to assume that science must already know almost everything that is worth knowing in the field. The reality, however, appears to be that molecular biologists have barely reached the threshold of understanding the vast maze of interactions among highly complex molecules that constitute these so-called simple organisms.

The importance to eukaryotic cellular differentiation of the mechanisms of gene expression and its regulation in prokaryotic cells that we have studied in this chapter is that of a foundation to a skyscraper. The foundation does not look much like the visible parts of the skyscraper, but it is absolutely necessary as a starting point. Similarly, the details of gene regulatory mechanisms in prokaryotic cells appear to differ significantly from those of eukaryotic cells insofar as the latter have been studied, but they provide an essential conceptual foundation for research with eukaryotic cells.

Virtually every phase of gene expression that is currently beginning to be analyzed in eukaryotic cells has already been studied extensively in prokaryotic cells. There is no indication that this trend will change in the foreseeable future. We have emphasized repeatedly in this chapter the many unanswered questions that still remain for prokaryotic cells. Given the head start that investigators working with prokaryotic cells already have and the many inherent advantages of working with prokaryotic cells, it is likely that prokaryotic cells will continue to provide many of the conceptual advances that will guide research on eukaryotic cells for years to come.

In addition, there is a growing trend toward the use of prokaryotic cells to study eukaryotic problems. Advances in recombinant DNA techniques are making it increasingly easy to transfer bits of eukaryotic genomes to prokaryotic cells. Thus many of the inherent advantages of prokaryotic cells, ranging from short generation times to small genomes and well-defined genetic systems, may be able to be applied to the study of eukaryotic genes and their expression.

In the following five chapters the eukaryotic genome and mechanisms currently believed to be related to control of its expression are analyzed. The basic concepts of gene expression and its regulation in prokaryotic cells are referred to repeatedly as we build toward the more complex eukaryotic systems from the foundation we have gained in this chapter on prokaryotic systems.

BIBLIOGRAPHY
Significant older works cited in historical discussions

Avery, O. T., C. M. MacLeod, and M. McCarty. 1944. Studies on the chemical nature of the substance inducing transformation of pneumococcal types. J. Exp. Med. **79**:137. (First demonstration that DNA carries genetic information.)

Beadle, G. W. 1945. Biochemical genetics. Chem. Rev. **37**:15. (Early review of one gene–one enzyme concept.)

Beadle, G. W., and E. L. Tatum. 1941. Genetic control of biochemical reactions in *Neurospora*. Proc. Natl. Acad. Sci. U.S.A. **27**:499. (First experimental evidence for one gene–one enzyme relationship.)

Garrod, A. E. 1909. Inborn errors of metabolism. Oxford University Press, London. (Reprinted with a supplement by H. Harris, Oxford University Press, 1963.) (Evidence [Chapter 3] that human alkaptonuria, which is inherited as a Mendelian recessive trait, is caused by absence of ability to metabolize homogentisic acid.)

Jacob, F., and J. Monod. 1961. Genetic regulatory mechanisms in the synthesis of proteins. J. Mol. Biol. **3**:318. (An important review article unifying the then new concepts of mRNA and regulation of gene expression and developing the operon model essentially as it is known today.)

Nirenberg, M. W., and J. W. Mathaei. 1961. The dependence of cell-free protein synthesis in *E. coli* upon naturally occurring or synthetic polyribonucleotides. Proc. Natl. Acad. Sci. U.S.A. **47**:1588. (Evidence that poly [U] codes for polyphenylalanine–the first step in breaking the genetic code.)

Watson, J. D., and F. H. C. Crick. 1953. Molecular structure of nucleic acids. A structure for deoxyribose nucleic acid. Nature **171**:737. (First description of the double helical structure of DNA.)

Watson, J. D., and F. H. C. Crick. 1953. Genetical implications of the structure of deoxyribonucleic acid. Nature **171**:964. (A discussion of the complementary nature of the strands in the DNA double helix and the fact that each can serve as a template for the other.)

Wilson, E. B. 1896. The cell in development and inheritance. The Macmillan Co., Publishers, New York. (Reprinted with a new introduction by H. J. Muller, Johnson Reprint Corp., New York, 1966.) (A discussion of the relationship between inheritance and metabolism and of the critical role of the nucleus in development as Wilson visualized them in 1896 [pp. 326-327].)

Books and reviews

Bloomfield, V. A., D. M. Crothers, and I. Tinoco, Jr. 1974. Physical chemistry of nucleic acids. Harper & Row, Publishers, New York.

Chamberlin, M. J. 1974. The selectivity of transcription. Annu. Rev. Biochem. **43**:721.

Crick, F. H. C. 1966. The genetic code–yesterday, today, and tomorrow. Cold Spring Harbor Symp. Quant. Biol. **31**:3.

Dressler, D. 1975. The recent excitement in the DNA growing point problem. Annu. Rev. Microbiol. **29**:525.

Gefter, M. 1975. DNA replication. Annu. Rev. Biochem. **44**:45.

Gilbert, W. 1976. Starting and stopping sequences for RNA polymerase. *In* R. Losick and M. Chamberlain, eds. RNA polymerase. Cold Spring Harbor Laboratory, Cold Spring, N.Y.

Goldberger, R. F., R. G. Deeley, K. P. Mullinix. 1976. Regulation of gene expression in prokaryotic organisms. Adv. Genet. **18**:1.

Herskowitz, I. H. 1977. Principles of genetics, 2nd ed. The Macmillan Co., Publishers, New York.

Kim, S. H. 1976. Three-dimensional structure of transfer RNA. Prog. Nucleic Acid Res. Mol. Biol. **17**:181.

Kozak, M., and D. Nathans. 1972. Translation of the genome of a ribonucleic acid bacteriophage. Bacteriol. Rev. **36**:109.

Kurland, C. G. 1977. Structure and function of the bacterial ribosome. Annu. Rev. Biochem. **46**:173.

Lewin, B. 1974. Gene expression. Vol. 1. Bacterial genomes. Wiley-Interscience Div., New York.

Lewin, B. 1977. Gene expression. Vol. 3. Plasmids and viruses. Wiley-Interscience Div., New York.

Losick, R., and M. Chamberlain, eds. 1976. RNA polymerase. Cold Spring Harbor Laboratory, Cold Spring Harbor, N. Y.

Marmur, J., and G. Zubay, eds. 1973. Papers in biochemical genetics. Holt, Rinehart & Winston, Inc., New York.

Rich, A., and S. H. Kim. 1978. The three-dimensional structure of transfer RNA. Sci. Am. **238**:52 (Jan.).

Stent, G. S. 1971. Molecular genetics. W. H. Freeman & Co., Publishers, San Francisco.

Watson, J. D. 1976. Molecular biology of the gene, 3rd ed. The Benjamin Co., Inc., Menlo Park, Calif.

Weissbach, H., and N. Brot. 1974. The role of protein factors in the biosynthesis of proteins. Cell **2**:137.

Selected original research articles

Artz, S. W., and J. R. Broach. 1975. Histidine regulation in *Salmonella typhimurium*: an activator-attenuator model of gene regulation. Proc. Natl. Acad. Sci. U.S.A. **72**:3453.

Barrell, B. G., G. M. Air, and C. A. Hutchison III. 1976. Overlapping genes in bacteriophage ϕX174. Nature **264**:34.

Bertrand, K., and C. Yanofsky. 1976. Regulation of transcription termination in the leader region of the tryptophan operon of *Escherichia coli* involves tryptophan or its metabolic product. J. Mol. Biol. **103**:339.

Contreras, R., et al. 1977. Overlapping of the VP_2-VP_3 gene and the VP_1 gene in the SV40 genome. Cell **12**:529.

Debarbouille, M., et al. 1978. Dominant constitutive mutations in *malT*, the positive regulator gene of the maltose regulon in *Escherichia coli*. J. Mol. Biol. **124**:359.

Fiers, W., et al. 1978. Complete nucleotide sequence of SV40 DNA. Nature **273**:113.

Gold, L., P. Z. O'Farrell, and M. Russel. 1976. Regulation of gene 32 expression during bacteriophage T4 infection of *Escherichia coli*. J. Biol. Chem. **251**: 7251.

Hirsh, J., and R. Schleif. 1977. The *ara C* promoter: transcription, mapping and interaction with the *ara BAD* promoter. Cell **11**:545.

Levinthal, M., M. Levinthal, and L. S. Williams. 1976. The regulation of the *ilvADGE* operon: evidence for positive control by threonine deaminase. J. Mol. Biol. **102**:453.

Novick, A., and M. Weiner. 1957. Enzyme induction as an all-or-none phenomenon. Proc. Natl. Acad. Sci. U.S.A. **43**:553.

Pribnow, D. 1975. Nucleotide sequence of an RNA polymerase binding site at an early T7 promoter. Proc. Natl. Acad. Sci. U.S.A. **72**:784.

Russel, M., et al. 1976. Translational, autogenous regulation of gene 32 expression during bacteriophage T4 infection. J. Biol. Chem. **251**:7263.

Sanger, F., et al. 1977. Nucleotide sequence of bacteriophage ϕX174 DNA. Nature **265**:687.

Smith, M., et al. 1977. DNA sequence at the C termini of the overlapping genes A and B in bacteriophage ϕX174. Nature **265**:702.

Stephens, J. C., S. W. Artz, and B. N. Ames. 1975. Guanosine 5'-diphosphate-3'-diphosphate (ppGpp): positive effector for histidine operon transcription and general signal for amino acid deficiency. Proc. Natl. Acad. Sci. U.S.A. **72**:4389.

Travers, A. A., et al. 1978. A mutation affecting the σ subunit of RNA polymerase changes transcriptional specificity. Nature **273**:354.

CHAPTER 5

Organization of the eukaryotic genome

☐ The genomes of eukaryotic cells have a number of characteristic properties that distinguish them from the genomes of prokaryotes. An understanding of these characteristics is essential to any analysis of the mechanisms responsible for control of gene expression in the cells of developing eukaryotic organisms. Precise details of genomic organization vary somewhat from one species to another, but the following properties are general characteristics of eukaryotic cells:

1. The genome is enclosed within a nuclear envelope, which physically separates the site of RNA synthesis in the nucleus from the site of protein synthesis in the cytoplasm.
2. The DNA is in the form of chromatin—a complex of DNA, protein, and small amounts of RNA.
3. The DNA is divided into a number of discrete packaging units, called chromosomes.
4. In diploid organisms, where most of the development we will be analyzing occurs, each cell normally contains two of each kind of chromosome (except the sex chromosomes).
5. The assembly of DNA and proteins into chromatin fibers and the organization of the chromatin fibers into chromosomes are both highly ordered.
6. Eukaryotic DNA contains repeated nucleotide sequences that are present in amounts ranging from a few copies to a million or more per haploid genome.
7. The repeated sequences are in most cases interspersed between unique sequences (sequences that appear not to be repeated within the entire haploid genome).
8. The amount of DNA per haploid genome in many types of eukaryotic cells is large. Lower eukaryotes such as yeast may have genomes as small as three times that of *E. coli*. However, cells of higher eukaryotes contain far more. The human haploid genome, for example, contains about 600 times the DNA of the *E. coli* genome. Even if repetitive sequences are disregarded, the

amount of unique-sequence DNA is much greater than would seem to be necessary to code for all needed proteins, and it also appears to be too large to be compatible with currently accepted rates of gene mutation.

These characteristic features of eukaryotic cells suggest many possibilities for controlling gene expression. The nuclear envelope, the presence of firmly bound proteins on the DNA, and the tight structural organization of the chromosomes all provide potential means for blocking access of RNA polymerase to the DNA. The nuclear membrane could also restrict the access of newly synthesized mRNA to the protein-synthesizing machinery in the cytoplasm. The large excess of DNA and the interspersed repetitive sequences could easily be used in specific control mechanisms. It is generally believed that the selective expression of specific genes that occurs in differentiated cells is controlled by mechanisms related to the structural organization of the eukaryotic genome.

NUCLEAR ENVELOPE   Compartmentalization of eukaryotic DNA within the nucleus complicates the overall process of transcription and translation compared to that found in prokaryotes, where ribosomes bind to the 5′ end of an mRNA while the rest of the molecule is still being transcribed and immediately begin protein synthesis (Fig. 5-1). In eukaryotes, transcription must be completed and the mRNA must cross the nuclear envelope before it is available for translation. The initial product of transcription in eukaryotic cells is modified extensively before it leaves the nucleus. These modifications generally include removal of excess nucleotides and additions to both the 5′ and the 3′ end of the message (Fig. 5-2). In higher eukaryotes with large genomes, only a small fraction of the RNA that is synthesized in the nucleus ever reaches the cytoplasm. The presence of a nuclear envelope thus provides many potential points for posttranscriptional control of gene expression. These relationships are discussed in detail in Chapter 8.

CHROMATIN   The genes of eukaryotic cells are not as accessible to enzymes or to regulatory molecules as are those of prokaryotic cells. When eukaryotic DNA is isolated at neutral pH, a tightly bound complex of DNA, RNA, and protein is obtained. Dissociating conditions such as high salt, acid, base, or urea are required to disrupt the complex. The protein components can be separated into two major classes: the histones, which are a family of basic proteins that can be extracted from chromatin with acid, and the nonhistone chromosomal (NHC) proteins, which are a much more diverse group of proteins left after the histones are removed. The RNA consists of a mixture of incomplete

Fig. 5-1. Electron micrograph of concurrent transcription and translation in *E. coli*. A segment of the bacterial chromosome can be seen as a thin vertical line. The arrow points to what is believed to be an RNA polymerase molecule attached near the point of initiation of transcription. Below it are progressively longer mRNA molecules with attached ribosomes. Translation begins at the 5′ ends of the mRNA (away from the chromosome) and proceeds toward the chromosome at the same time that the mRNA is being elongated. The number of attached ribosomes on the mRNA at the bottom of the picture suggests that its 5′ end has already been translated more than 20 times, although synthesis of its 3′ end is not yet completed. (From Miller, O. L., Jr., et al. 1970. Science **169**:392. Copyright 1970 by the American Association for the Advancement of Science.)

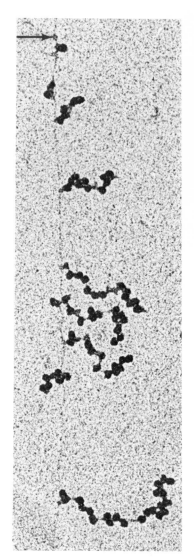

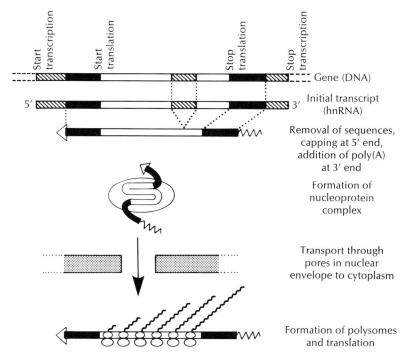

Fig. 5-2. Schematic representation of the formation of mRNA in most types of eukaryotic cells. The initial product of transcription, called heterogeneous nuclear RNA (hnRNA), is significantly longer than the final mRNA. Sequences may be removed from the 5′ end, from interior regions, or from both, and possibly also from the 3′ end. A special "cap" of 7-methyl-guanosine is added to the 5′ end of the mRNA by a novel 5′ to 5′ triphosphate link, and polyadenylic acid is added to the 3′ end. After these steps occur in the nucleus, the mRNA is complexed to proteins and transported through the nuclear membrane to the cytoplasm, where it is translated. These processes are described in detail in Chapter 8. In lower eukaryotes such as yeasts, slime molds, and water molds, the initial product of transcription is smaller and there is far less removal of sequences. ■ = Nontranslated regions of mRNA; □ = message sequence; ▧ = possible locations of nonconserved segments of hnRNA.

transcripts plus a family of reasonably well-defined small chromosomal RNA molecules whose function is not yet known. The relative amounts of these components vary with the method of preparation. The weight ratio of DNA to histone is usually about 1:1. The weight ratio of DNA to NHC protein is also usually on the order of 1:1, but the amount of NHC protein shows more variability than that of histone. The amount of RNA that is found is highly variable.

Histone

The histones have been well characterized in recent years. They are all relatively small proteins with molecular weights ranging between 11,000 and 21,000. They typically contain about 25% basic amino acids and only about 13% acidic amino acids, giving them a strong net positive charge under physiological conditions. In addition, the positive charges tend to be clustered near the ends of their peptide chains, with the greatest concentration near the N-terminal end, whereas the centers of the molecules are relatively rich in hydrophobic groups. The positively charged ends are believed to interact with phosphate groups in DNA, and the hydrophobic regions contribute to intramolecular folding of the histones and protein-protein interactions.

Chromatin from most tissues in higher organisms contains five distinct types of histone (Table 5-1). The literature contains a variety of systems for naming the histones, based on the use of various fractionation techniques and on the chemical composition of the histones. We will use the designations H1, H2a, H2b, H3, and H4, which have been used widely by recent reviewers.

Histone H1 actually refers to a heterogeneous group of histones that are all rich in lysine. The amount of H1 relative to DNA frequently varies from tissue to tissue within a single species, but the other

Table 5-1. Characterization of the five major histones

	H1	H2a	H2b	H3	H4
Older names	I, f1, KAP	IIb1, f2a2, ALG	IIb2, f2b, KSA	III, f3, ARE	IV, f2a1, GRK
Total amino acids	About 215	129	125	132	102
Molecular weight	About 21,000	14,004	13,774	15,324	11,282
Lysine:arginine ratio	22.0	1.17	2.50	0.72	0.79
Class	Very lysine rich	Lysine rich	Lysine rich	Arginine rich	Arginine rich
Other characteristics	Heterogeneous; some tissue specificity	Also rich in arginine and glycine		Highly conserved in evolution	Also rich in glycine; highly conserved in evolution

four (H2a, H2b, H3, and H4) exhibit much less variability. Microheterogeneity of H2a, H2b, H3, and H4 can be observed in some preparations. Until recently this was thought to be entirely due to modification of specific amino acids within the peptide chains. Modifications that are frequently found in the histones include acetylation of the N-terminal amino acid, acetylation or methylation of ϵ-amino groups of lysine, and phosphorylation of serine. Such modifications are often incomplete — that is, only some of the molecules in a given population are modified. These modifications reduce the net positive charge of the histone molecules and therefore alter the rates at which they move in an electric field during electrophoresis. Recent sequence analysis, however, has revealed that in addition to these modifications, the primary amino acid sequences of certain variants of H2a, H2b, and H3 differ from one another by one to three amino acids in the central hydrophobic regions of the molecules.

The overall amino acid sequences of H3 and H4 have been highly conserved in evolution. H2a and H2b show relatively more evolutionary divergence but have also been moderately conserved. The estimated mutation rate for H4 is 0.06 substitutions per 100 amino acids per 100 million years. This is the lowest mutation rate yet observed in a protein. The amino acid sequence of H4 from calf thymus is shown in Fig. 5-3, together with the differences between it and H4 from pea seedlings. The only differences in amino acid sequence are two highly conservative substitutions, isoleucine for valine at residue 60 and arginine for lysine at residue 77. The only other differences observed are absence of the dimethyl modification of lysine in pea seedling H4 at residue 20 and a decreased incidence of acetylation of lysine at residue 16 in pea seedling H4.

Sequence analysis of purified histone H4 mRNA from two species of sea urchins indicates that evolutionary selection for unaltered protein structure rather than for unaltered nucleotide sequences is directly responsible for the extreme conservation of amino acid sequence. Although the amino acid sequences of H4 from the two species are thought to be the same, 11.5% of the bases in their mRNA's are different. Base changes that do not affect amino acid coding are limited primarily to the third, or "wobble," base of codons (Fig. 4-7). Thus approximately a third of the bases that can undergo substitution without altering amino acid sequence have done so during the estimated 6×10^7 years since the species shared a common ancestor. The contrast between the rate of nucleotide substitutions (about 3×10^{-9} per codon per year) and the rate of amino acid substitution (about 6×10^{-12} per codon per year) suggests that there is strong selective pressure to keep all parts of the H4 protein unchanged.

A

B

Fig. 5-3. Conservation of amino acid sequence of histone H4 during evolution. **A,** Amino acid sequence of H4 from calf thymus, together with posttranslational modifications of amino acids 1, 16, and 20. This schematic representation is not intended to depict the three-dimensional configuration of the histone molecule. **B,** Differences in amino acid sequences between H4 of calf thymus and that of pea seedlings. Pea seedling H4 also lacks the dimethyl modification of lysine at position 20 and has a lower incidence of acetylation of the lysine at position 16. (Based on data from DeLange, R. J., et al. 1969. J. Biol. Chem. **244:**5669.)

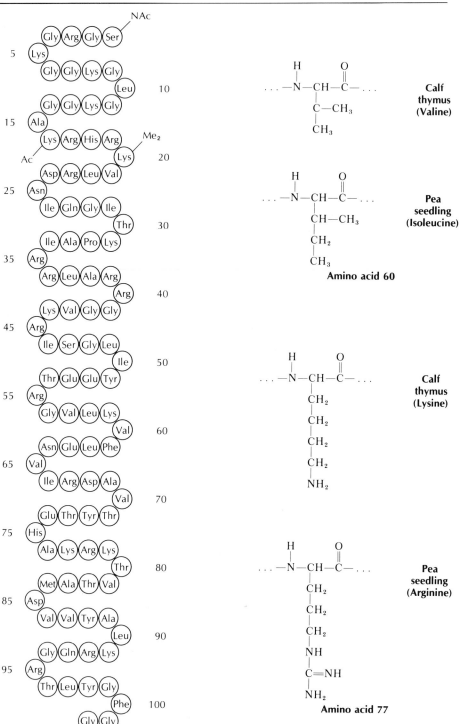

The limited polymorphism that occurs among the individual histones probably serves a specific function, since it appears to be conserved from species to species. The sequence variants of H2a, H2b, and H3 are found in different relative amounts in various tissues, but the same kinds and relative amounts of variants tend to be present in each species examined. In sea urchin embryos specific changes are reported to occur at the time of gastrulation in the types of H1, H2a, and H2b that are synthesized.

Aside from the limited polymorphism of H2a, H2b, and H3 and the variability of H1, only two clear examples of major changes of histones or histonelike proteins in differentiation are known. The first occurs in the nucleated erythrocytes of birds, where a basic protein designated histone 5 appears to be involved in the repression of RNA synthesis as the cells mature and their nuclei become inactive. Histone 5 is phosphorylated as it is synthesized. Cessation of RNA synthesis in the erythrocyte precursor cell is correlated in time with removal of the bound phosphate. The second example occurs in the highly condensed chromatin in sperm of some animals, where the normal histones are replaced with very basic proteins known as protamines, which in extreme cases can be up to 75% arginine (in fish) or 50% lysine (in mollusks).

Except for a few specialized cases such as those described, a general role for the histones in regulation of gene expression seems unlikely. Their minimal diversity, together with their extreme evolutionary conservation, suggests that their biological function is probably structural. A regulatory role would presumably require far more diversity from one tissue to another and from one type of organism to another to achieve specific control over the large number of genes that must be individually regulated. Recent data on chromatin structure, which will be discussed later in this chapter, support this view. H2a, H2b, H3, and H4 have been strongly implicated as integral structural components of chromatin beads (nucleosomes), and the H1 group appears to be involved in the condensation of chromatin. In addition, according to one popular theory, which is discussed in Chapter 7, histones may play an indirect role in the regulation of gene expression by nonspecifically inhibiting transcription from all genes except those which are specifically activated by certain of the NHC proteins.

The NHC proteins consist of all the proteins of chromatin that are not removed by acid extraction. Because of their complexity these proteins have not been characterized as well as the histones. A number of enzyme activities expected to be associated with DNA and histones are found in this fraction, including DNA and RNA polymerases, vari-

Nonhistone chromosomal (NHC) protein

Table 5-2. Examples of enzyme activities found in chromatin*

Type of enzyme activity	Source of chromatin
Enzymes involved in DNA metabolism	
DNA polymerase	Rat liver, rat ascites hepatoma, sea urchin, calf thymus
DNA endonuclease	HeLa cells
DNA ligase	Rabbit bone marrow
DNAase	Rat liver
Terminal-DNA nucleotidyltransferase	Calf thymus
Enzymes involved in RNA metabolism	
RNA polymerase	Rat liver, mouse melanoma, hen oviduct, coconut
Poly(A) polymerase	Wheat
Enzymes involved in poly (adenosine diphosphate) ribose metabolism	
Poly ADP-ribose polymerase	Rat liver
Poly ADP-ribose glycohydrolase	Rat liver
Enzymes involved in histone metabolism	
Histone acetyltransferase	Rat thymus
Histone methylase	Calf thymus
Histone kinase (acid-labile phosphate)	Rat carcinosarcoma
Histone protease	Rat liver, calf thymus
Enzymes involved in NHC protein metabolism	
NHC protein kinases	Rat liver

*Modified from Elgin, S. C. R., and H. Weintraub. 1975. Annu. Rev. Biochem. **44**:725.

ous nucleases, and protein phosphokinases. Table 5-2 lists some of the enzyme activities that have been found associated with chromatin and the cell types in which they have been detected. Complete characterization of NHC proteins will be a huge task and will probably require many years. Fractionation and analysis of the NHC proteins suggest that about 15 to 20 major proteins comprise some 50% to 70% of the starting material on a weight basis. However, sensitive two-dimensional separation techniques have resolved nearly 500 spots, each representing at least one protein species or protein modification (Fig. 5-4).

The types of NHC protein that are found show more variation from tissue to tissue than do the histones. The tissue-specific differences in NHC protein that have thus far been demonstrated are generally not dramatic, but in all except the two-dimensional separation mentioned previously, the analytical techniques that have been used are not sensitive enough to detect regulatory proteins that are present in very small numbers per genome. Nevertheless, a regulatory role for compo-

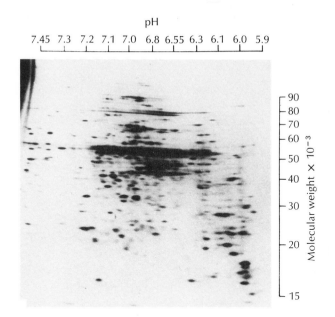

Fig. 5-4. Two-dimensional separation of nonhistone chromosomal proteins from HeLa cells in polyacrylamide gels. The horizontal separation is by isoelectric point (isoelectric focusing), and the vertical separation is by molecular weight (SDS gel electrophoresis). Autoradiography is used to detect very small amounts of radioactively labeled proteins in each "spot." (From Peterson, J., and E. H. McConkey. 1976. J. Biol. Chem. **251:**548.)

nents of the NHC protein fraction has been strongly inferred from experiments in which altered patterns of transcription are obtained when chromatin is dissociated and the DNA and histone are reconstituted with NHC protein from different tissues (see Chapter 7). Phosphorylation and other modifications of NHC protein may also play a significant role in the regulation of gene expression.

Nuclear RNA

In addition to its primary constituents of DNA, histone, and NHC protein, eukaryotic chromatin also contains RNA. Some of this undoubtedly consists of nascent chains of RNA in the process of transcription. However, there are also a number of well-characterized small RNA molecules (90 to 100 nucleotides in length) in the nucleus, including some that are presumed to have structural roles. A regulatory role for chromosomal RNA in the control of transcription has been suggested by some investigators, but this view has been strongly challenged and remains highly controversial.

CHROMOSOMES

The term "chromosome" is sometimes reserved for the condensed structures seen in cells at mitosis or meiosis. However, the DNA with-

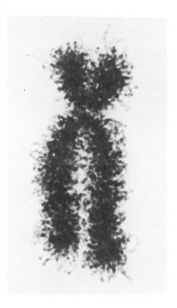

Fig. 5-5. Whole-mount electron micrograph of human chromosome 12 at mitotic metaphase. The chromosome is composed of two sister chromatids held together at the centromere. Chromatin fibers can be seen looping out from both chromatids. (From DuPraw, E. J. 1970. DNA and chromosomes. Holt, Rinehart & Winston, Inc., New York.)

in the nucleus seems to be organized in the same units (although in a less condensed form) throughout the cell cycle. The phenomenon of linkage, in which all the genes on one chromosome tend to remain together and segregate in a linked fashion during meiosis, suggests that all the DNA of a chromosome must in some way remain structurally joined together throughout the cell cycle. It has recently been demonstrated that, in at least some organisms, each chromosome contains a single, long, continuous DNA molecule, rather than a number of smaller pieces of DNA joined together by some other component such as protein.

The structural organization of DNA into the chromosomes of eukaryotic cells is no minor engineering feat. If all the DNA in a single diploid human cell were lined up end to end, it would form a double helix approximately 1.74 m in length, or roughly the height of an average adult human. Each human cell contains this amount of DNA packed into a nucleus only a few micrometers in diameter. If all the DNA from the approximately 10^{14} cells in an average adult human body were lined up end to end, it would extend about 1.7×10^{11} km—more than a thousand times the distance from the earth to the sun.

Electron micrographs of mitotic chromosomes show only a tangled mass of fibers, as seen in Fig. 5-5. However, there are a number of reasons for believing that the arrangement of DNA and protein in chromosomes must be highly ordered. First, genetic crossing over of DNA between homologous chromosome pairs requires that the position of comparable DNA sequences in the two chromosomes be essentially the same. Second, when metaphase chromosomes are stained under certain conditions, they show characteristic patterns of banding (Fig. 5-6). This pattern is the same for both chromosomes of a homologous pair as well as for the comparable chromosomes from all other cells in the same individual.

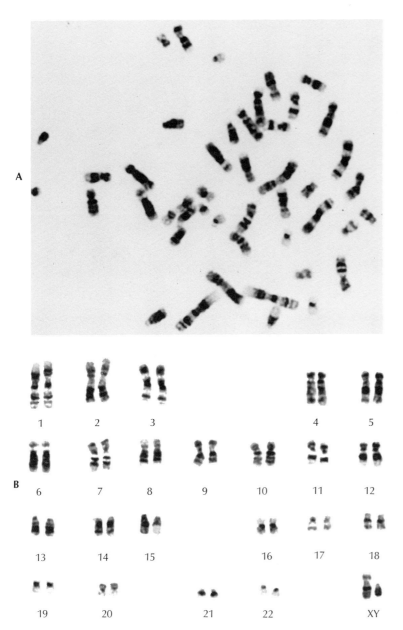

Fig. 5-6. Human chromosomes stained to show G banding. **A,** Chromosomes from human cell at metaphase. Cells have been fixed with acetic acid and methanol, subjected to hypotonic swelling, and stained with Giemsa stain. **B,** The same chromosomes arranged in pairs according to the standardized numbering system for human chromosomes. The similarity of banding of homologous chromosome pairs is clearly evident. The chromosomes from a particular cell arranged in this manner are referred to as the karyotype of the cell. The presence of one X and one Y chromosome indicates that this cell was from a male. (From Nagle, J. J. 1979. Heredity and human affairs, 2nd ed. The C. V. Mosby Co.)

Fig. 5-7. Nucleosomes and condensation of chromatin. **A,** Electron micrograph of partially "opened" chromatin fibers from chicken erythrocyte nuclei, showing "beads on a string" nucleosome structure. **B,** Diagram of nucleosomes, showing the composition of the particles and the connecting strings. **C,** Schematic representation of packing ratios. A fully extended DNA double helix is represented by a line at the bottom and sides of the illustration. Packing ratios of 7:1 (nucleosomes), 50:1 (interphase chromatin fibers), and 100:1 (mitotic chromatin fibers) are shown drawn at the same scale. It is not possible to depict the 10,000:1 packing ratio of a metaphase chromosome directly, since the chromosome would be only 0.03 mm long at the same scale. If the DNA of the metaphase chromosome in Fig. 5-5 were shown fully extended at the same magnification as in Fig. 5-5, its length would be 600 m, or about one third of a mile. Its diameter at that magnification would be about 0.05 mm, or roughly that of a human hair. (**A** from Olins, A. L., et al. 1977. *In* P. Ts'o, ed. The molecular biology of the mammalian genetic apparatus. Elsevier/North-Holland Biomedical Press, Amsterdam.)

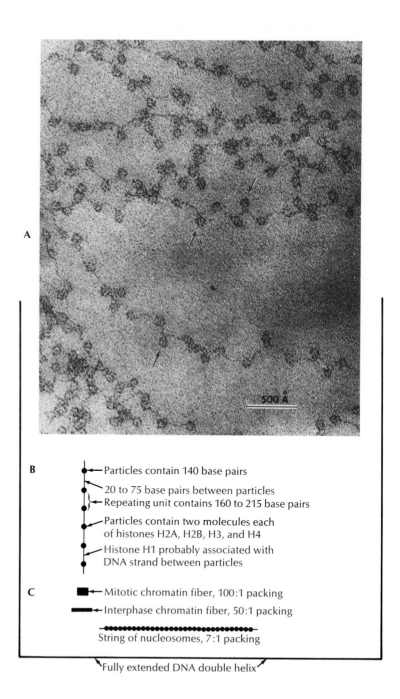

A

B

Particles contain 140 base pairs

20 to 75 base pairs between particles

Repeating unit contains 160 to 215 base pairs

Particles contain two molecules each of histones H2A, H2B, H3, and H4

Histone H1 probably associated with DNA strand between particles

C

Mitotic chromatin fiber, 100:1 packing

Interphase chromatin fiber, 50:1 packing

String of nucleosomes, 7:1 packing

Fully extended DNA double helix

Fig. 5-8. Electron micrograph of a metaphase chromosome from a HeLa cell treated to remove histone and release DNA from nucleosomes. The vast amount of DNA originally contained in the chromosome can be seen spilling out from the residual core of nonhistone proteins, which is approximately the size of the original chromosome. (From Paulson, J. R., and U. K. Laemmli. 1977. Cell **12**:817. Copyright © MIT; published by the MIT Press.)

Chromatin fibers   The continuous DNA molecule of each chromosome is packaged into a chromatin fiber of the type visible in Fig. 5-5. The length of these chromatin fibers is much shorter than that of the DNA they contain. The ratio of DNA length to chromatin fiber length is referred to as the packing ratio. Chromatin fibers in interphase nuclei generally have a packing ratio of about 50:1, whereas the somewhat more dense fibers in metaphase chromosomes exhibit ratios near 100:1 (Fig. 5-7). The overall packing ratio of DNA in a metaphase chromosome (condensation of the DNA into a chromatin fiber plus condensation of that fiber into a chromosome) is often greater than 10,000:1. This is illustrated dramatically in Fig. 5-8, which shows the DNA spilling out of a metaphase chromosome treated to remove histones.

Rapid progress has been made in the last few years toward understanding the organization of DNA within chromatin fibers. Data from a variety of sources collectively indicate that chromatin has a regular repeating substructure. Early x-ray diffraction studies of chromatin had suggested a repeating structure. When the relationship of histone to this repeating pattern was examined, it was found that all histones except group H1 had to be present to obtain the characteristic repeating pattern.

Electron microscope studies also suggested a repeating structure for chromatin. Linear arrays of spherical particles were seen in samples prepared by swelling nuclei with water and centrifuging their chromatin onto carbon films. This pattern, which has been popularly labeled "beads on a string," is shown in Fig. 5-7. The beads, which are referred to as nucleosomes, have a diameter of about 100 Å and are separated by short stretches of fully extended DNA fiber.

Recent biochemical studies have confirmed the substructure suggested by the x-ray and EM studies. Each bead contains a nuclease-resistant core composed of two molecules each of H2a, H2b, H3, and H4 with about 140 nucleotide pairs of DNA wrapped tightly around the outside. The exact structure of these core particles has not been worked out, but most models suggest that the hydrophobic central portions of the histone molecules interact with one another while the positively charged terminal regions interact with the phosphate groups of the DNA. The fully extended "spacer" seen in Fig. 5-7 is probably an artifact of the preparation technique. Essentially all the 160 to 215 nucleotide pairs contained in each repeating unit are normally closely associated with nucleosomes in chromatin fibers of approximately 100 Å diameter. Lower eukaryotes such as yeast tend to have smaller repeat units. This is because of shorter connecting strands, since the amount of DNA in the nuclease-resistant core particles has been found to be about the same for all species and tissues that have been examined.

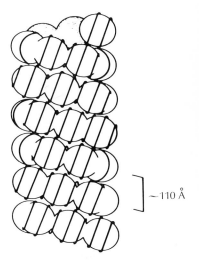

Fig. 5-9. Schematic drawing depicting possible folding of nucleosomes into 300 Å diameter chromatin fibers. The nucleosomes are stacked to form a solenoid or helical structure with a pitch of 110 Å. The thin line wound as a helix around the nucleosomes is intended to represent the folding of the DNA double helix on the outside of the histone core of the nucleosome particles. This representation is schematic, since the exact pattern of folding is not known. (From Finch, J. T., and A. Klug. 1976. Proc. Natl. Acad. Sci. U.S.A. **73:**1897.)

~110 Å

The packing ratio of the DNA in the beads of the fully condensed 100 Å fiber is about 7:1. These fibers in turn undergo further orders of packing to achieve the 50:1 ratio observed in 300 Å diameter fibers from interphase chromatin that has not been stretched. Histone H1 is thought to be involved in the latter packing, and some investigators have suggested that phosphorylation of H1 may be involved in changes in fiber length. The structural details of chromatin fibers are not yet certain, but they probably involve some kind of helical "supercoil." One model based on electron microscope studies is shown in Fig. 5-9.

The nucleosome structure of chromatin is self-assembling. It can be reconstituted by combining DNA and histones H2a, H2b, H3, and H4. Even DNA that normally lacks histone such as that of bacteriophage lambda will form nucleosomes when combined with these histones. This suggests that in the formation of nucleosomes the histone molecules must interact with the fundamental structure of DNA, rather than with any specific base sequences.

REPEATED DNA SEQUENCES

A substantial fraction of the DNA in a typical eukaryotic organism consists of nucleotide sequences that are present in multiple copies per haploid genome. In some cases the degree of repetition, or reiteration, as it is also called, is as high as 10^6 copies per haploid genome.

Before we examine repeated sequences and their developmental implications in detail, it will be helpful to outline how the degree of reiteration is measured. This is done by renaturation reactions in which double-stranded DNA is fragmented into pieces of a standard size (usually about 400 to 500 nucleotide pairs long), separated into single strands (denatured) by high temperature and/or denaturing agents, and then slowly cooled or freed from denaturing agents, allowing complementary strands to renature into double helixes (Fig. 5-10).

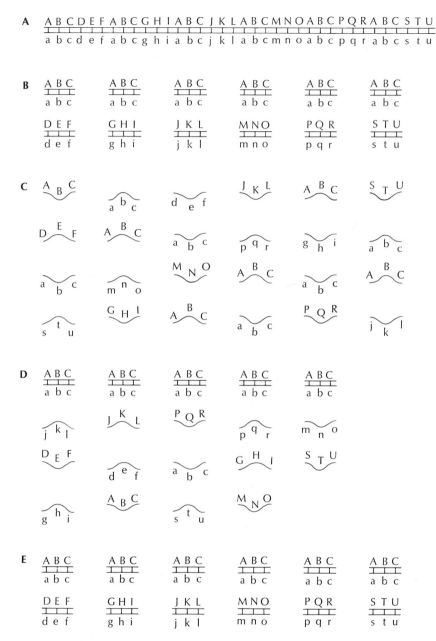

Fig. 5-10. Schematic representation of renaturation of DNA containing a mixture of unique and repeated sequences. Native DNA (**A**) is sheared into fragments about 500 nucleotides long (**B**). These are then separated into single strands by high temperature or denaturing agents (**C**). When conditions that permit base pair formation are restored, the repetitious sequences, which are present at the highest effective concentrations, renature the most rapidly (**D**). If renaturation is allowed to proceed long enough, all the DNA fragments eventually renature, including those of unique sequence (**E**).

Renaturation of DNA that has been broken into relatively small pieces and denatured is achieved by random collisions of single-stranded fragments containing complementary sequences. The rate of renaturation is therefore dependent on the effective concentration of each of the reacting species. When a genome contains repeated sequences, the effective concentrations of those repeated sequences are higher than would be predicted from the total amount of DNA in the genome. Renaturation of the repeated sequences therefore proceeds more rapidly than that of the unique sequences (Fig. 5-10, D).

During a renaturation reaction the extent of renaturation that has occurred depends both on the initial concentration of denatured DNA fragments and on the amount of time that the reaction has progressed. The product of initial DNA concentration (C_0) and the amount of time that the reaction has been in progress (t) has proved to be particularly useful in relating renaturation rates to genomic complexity. By convention, C_0 is expressed as total moles per liter of nucleotides contained in the denatured DNA. This convention allows molar concentrations to be used without concern about the actual size of the fragmented DNA molecules or concentrations of individual components. The effective concentrations of the individual reacting molecular species are a function both of total DNA concentration and of the sequence complexity of the DNA. Sequence complexity is usually expressed in terms of the total length of DNA that could be constructed from all the different sequences without repeating any of them. For a fixed total amount of DNA (C_0) a higher total sequence complexity results in lower effective concentrations of the individual sequences and therefore causes renaturation to be slower. Thus the sequence complexity is reflected in the rate constant (k) for the renaturation of each individual sample. The units of the product C_0t (which is often written "cot") are moles times seconds per liter.

At half reaction, that is, when half the original amount of denatured DNA has undergone renaturation, a simple relationship exists between the cot value and the reaction rate constant (k). This can be expressed as $cot_{1/2} = \frac{1}{k}$. As the sequence complexity of the DNA (genome size) becomes larger, k becomes smaller, and the product $cot_{1/2}$ becomes larger. Thus there is a direct linear relationship between the sequence complexity of a DNA preparation and its $cot_{1/2}$ value. It is possible both to calculate theoretical $cot_{1/2}$ values for DNA preparations of known complexity and to determine complexity (genome size) from measured $cot_{1/2}$ values. Fig. 5-11 shows renaturation curves and $cot_{1/2}$ values for a variety of DNA preparations of

Cot values

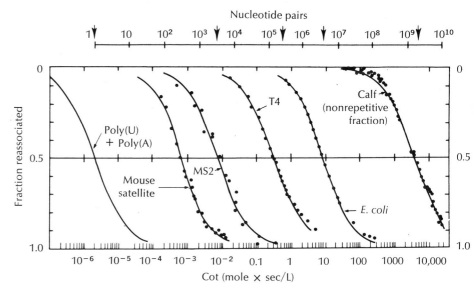

Fig. 5-11. Cot curves for reassociation of DNAs from different organisms. Larger cot values are required for renaturation of DNA of greater sequence complexity. Total sequence complexity in nucleotide pairs can be estimated from the cot values for one-half renaturation ($cot_{1/2}$) by use of the top scale. Arrows indicate the unique-sequence complexity of the DNAs. MS2 and T4 are small and large bacteriophages, respectively. Poly(U) + Poly(A) refers to a mixture of polyuridylic and polyadenylic acids, which has an effective sequence complexity of 1 base pair. (From Britten, R. J., and D. E. Kohne. 1968. Science **161**:529. Copyright 1968 by the American Association for the Advancement of Science.)

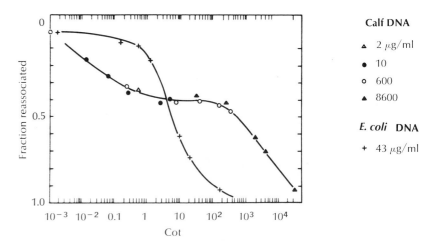

Fig. 5-12. Comparison of cot curves for reassociation of DNA from calf thymus and *E. coli*. The bacterial DNA renatures as a uniform population of molecules with a single cot value. The calf DNA contains a rapidly renaturing component and a second component that renatures much more slowly. These correspond to repeated-sequence DNA, which has a high effective concentration in the renaturation reaction, and unique-sequence DNA, whose individual species are present in very low concentrations. The similarity of cot curves for widely differing concentrations of calf DNA emphasizes the constancy of the $C_o \times t$ relationship, irrespective of the absolute values of either. (From Britten, R. J., and D. E. Kohne. 1968. Science **161**:529. Copyright 1968 by the American Association for the Advancement of Science.)

widely differing complexities. A more detailed explanation of renaturation kinetics and a full mathematical derivation of the relationship between $cot_{1/2}$ values and k are presented in Appendix A.

The presence of repeated sequences causes the DNA of higher organisms to exhibit smaller $cot_{1/2}$ values than would be predicted from the amount of DNA per haploid genome. In addition, the mixture of unique and repeated sequences causes the overall renaturation reaction to be spread over a much wider range of cot values than would be expected for a more homogeneous preparation. This is illustrated in Fig. 5-12, which compares the renaturation of calf thymus DNA, which contains both unique and repeated sequences, with that of *E. coli* DNA, which is essentially all unique. The calf DNA renatures over a wide range of cot values. The highly repeated DNA sequences renature much faster than the *E. coli* DNA, whereas the unique sequences renature much more slowly.

Fractionation of the annealing mixture after the renaturation reaction has progressed to any desired cot value permits separation of components that renature at different rates. The effective sequence

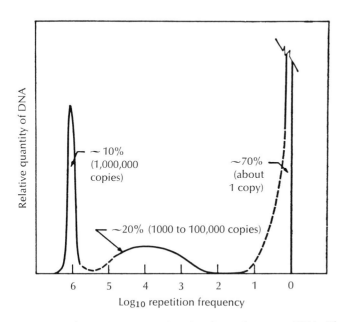

Fig. 5-13. Frequency of sequence repetition in sheared mouse DNA. Three distinct classes of DNA can be detected. About 70% of the genome is unique-sequence DNA. Roughly 20% is repeated between 10^3 and 10^5 times per genome. About 10% of the genome consists of a highly repeated "satellite" fraction, which is repeated about 10^6 times per genome. (From Britten, R. J., and D. E. Kohne. 1968. Science **161**:529. Copyright 1968 by the American Association for the Advancement of Science.)

length, or complexity, in nucleotide pairs of each isolated component can be determined by denaturing it and measuring its $cot_{1/2}$ value. This information, together with data on the total number of nucleotide pairs in the original genome and the percentage of the genome represented by the component, allows calculation of how many times the repeating unit is present per genome.

Separation of the DNA of higher organisms on the basis of renaturation kinetics usually results in three distinct fractions—"unique" sequences that are present in only one (or possibly a few) copies per haploid genome, moderately repeated sequences that are present in about 10^2 to 10^5 copies per genome, and highly repeated sequences that are present at about 10^6 copies or more per haploid genome. The distribution of DNA sequences among unique and repeated fractions in the mouse genome is shown in Fig. 5-13.

Highly repeated DNA sequences

The highly repetitious DNA (repeated about 10^6 times or more per haploid genome) is frequently given the label "satellite" DNA. This designation was originally based on results obtained with cesium chloride (CsCl) density gradient centrifugation. When DNA is centrifuged for a prolonged time in such a gradient, it accumulates in a thin layer, or "band," corresponding to its density, which is a function of its overall base composition (GC base pairs are more dense than AT base pairs, and the overall buoyant density of DNA fragments is normally proportional to their ratio of $[G + C]:[A + T]$). When the DNA of higher organisms is broken into moderate-sized pieces, as normally happens during preparation for centrifugation, one often finds that a portion of the DNA forms one or more bands in the gradient with densities distinct from that of the major band. These were called satellites because they formed additional peaks at the edge of the main DNA peak when DNA concentration was plotted against buoyant density (Fig. 5-14). Most satellites are composed of very short nucleotide sequences repeated many times, although somewhat larger sequences that are repeated fewer times, such as the GC-rich ribosomal RNA genes, can also form satellites under certain conditions.

Not all highly repeated DNA can be detected as satellites in CsCl buoyant density centrifugation. In the human genome, for example, nearly all the DNA that renatures at low cot values forms bands with densities that are so close to that of the main band that they cannot be detected by CsCl centrifugation of unfractionated human DNA. This has lead to a confusion of nomenclature. The term "satellite" is frequently used to refer to any highly repeated short nucleotide sequence, whether or not it is actually detectable as a physical satellite in CsCl gradient centrifugation.

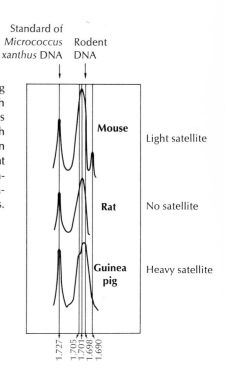

Fig. 5-14. Differences in buoyant density of satellites detected in mouse, rat, and guinea pig DNA by cesium chloride buoyant density centrifugation. A CsCl solution centrifuged at high speed forms a density gradient due to a balance between centrifugal force, which pushes the heavy cesium ions toward the bottom of the tube, and the forces of diffusion, which tend to move the ions away from the higher concentration at the bottom of the tube. When a DNA preparation is centrifuged in a CsCl solution, it accumulates in narrow "bands" that correspond to the bouyant density of each of its components. The graphs show DNA concentration plotted against density of the CsCl solution in fractions collected after centrifugation to equilibrium. (From Lewin, B. 1974. Gene expression-2. Eucaryotic chromosomes. John Wiley & Sons, Inc., New York.)

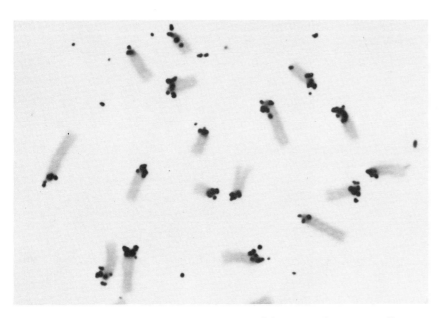

Fig. 5-15. Autoradiograph showing chromosomal location of mouse satellite DNA. Radioactive cRNA copied in vitro from mouse satellite DNA was hybridized to mouse metaphase chromosomes whose DNA had been denatured by exposure to alkaline pH. The silver grains in the photographic emulsion show that nearly all the labeled RNA hybridized in the centromeric regions of the chromosomes. (From Pardue, M. L. 1971. Chromosomes Today **3**:47.)

One of the most studied satellites is mouse satellite DNA, which comprises 10% of the mouse genome and consists of a sequence of about 110 to 140 nucleotides repeated more than 10^6 times per haploid genome. Recent sequence analysis of this satellite suggests that it evolved from an even shorter repeating unit (less than 20 nucleotides in length) that underwent sequence divergence. The degree of random mutation in this satellite suggests considerable evolutionary constraint, although the exact nucleotide sequence seems not to be critical.

These highly repeated short nucleotide sequences presumably provide some selective value to the organism, since they have not been eliminated during the evolution of individual species. However, there is great variability from species to species, and no generalized physiological role of satellites has been found. They appear not to be transcribed, since base sequences corresponding to them have not been found in cellular RNA. It has been suggested that they may play a role in chromosome structure. Mouse satellite DNA is specifically localized in the region of the centromere (Fig. 5-15). This centromeric location of mouse satellite sequences does not necessarily indicate that they are all arranged tandemly in one small portion of the chromosomal DNA molecule, since folding of the chromatin fiber could bring widely separated sequences together. However, other evidence, such as the fact that relatively large DNA fragments can be isolated that contain only satellite sequences, indicates that at least some of them are indeed arranged in tandem.

The human Y chromosome has recently been reported to contain a highly repeated simple nucleotide sequence that is not found in other human chromosomes. This male-specific repeated DNA has properties similar to those of other satellite DNAs, and according to one estimate may comprise as much as 50% of the total amount of DNA in the human Y chromosome.

Palindromes and restriction endonucleases

Renaturation with a very low $cot_{1/2}$ value does not always indicate the presence of highly repeated DNA sequences. Extremely rapid renaturation also occurs in another type of DNA known as "snapback" DNA, or a palindrome. Such DNA contains a self-complementary sequence within each strand. For example, the sequence -CTCGA-ATTCGAG- can fold back on itself to form $\frac{\text{-CTCGAA-}}{\text{-GAGCTT-}}$]. Since the complementary sequences are joined to each other by covalent bonds within a single DNA fragment, they renature into double helical configurations even more rapidly than highly repeated satellite sequences. Current data suggest that snapback sequences may account for 5% or more of the total genomes of many eukaryotes. Snapback sequences were not distinguished from repeated sequences in most older reports on the amounts of repeated sequences in genomes.

The presence of palindromes or snapback sequences makes possible the formation of side loops on DNA molecules, as illustrated in Fig. 5-16. Such loops can easily be formed in solutions of DNA, but their occurrence in intact cells has been questioned. RNA transcribed from foldback DNA also has the ability to fold back on itself and form double-stranded "hairpin" loops. RNA with that property has been reported to occur in the nuclei of eukaryotic cells. Transfer RNA molecules also have areas of self-complementarity that permit them to take on their characteristic "cloverleaf" configurations (Fig. 4-15).

Another suggested role for palindromes is to serve as recognition sites for regulatory proteins. The symmetrical distribution of sequences around the center of the palindrome may match precisely the distribution of interacting groups on a symmetrically arrayed dimeric or tetrameric protein complex. For example, the lactose operator sequence, described in Chapter 4, is an imperfect palindrome, and the lactose repressor that interacts with it is a tetramer composed of four identical protein subunits.

Enzymes from bacterial cells that selectively hydrolyze double-stranded DNA at certain palindromic sequences have become important research tools in molecular genetics. These enzymes, which are called restriction endonucleases, break both strands of native double helical DNA at special recognition sites that have a twofold symmetry and a particular base sequence. For example, an enzyme from *E. coli* (Eco R1) breaks double-stranded DNA at the sequence $\frac{-GAATTC-}{-CTTAAG-}$, which is at the center of the palindrome used in the examples just cited. An enzyme from *Hemophilus* influenzae (Hind III) recognizes $\frac{-AAGCTT-}{-TTCGAA-}$. Enzymes such as these permit large genomes to be cut into a limited number of pieces of more manageable size for detailed analysis and sequencing.

Another valuable property of the restriction endonucleases is that many of them do not cleave both strands of the DNA at exactly the same point. For example, Eco R1 cleaves the sequence $\frac{-GAATTC-}{-CTTAAG-}$ between the G and the adjacent A in each strand. The resulting ends, $\frac{-G}{-CTTAA}$ and $\frac{AATTC-}{G-}$, each have an unpaired 5′ terminal sequence AATT-. These short complementary sequences cause the ends to be "sticky" because of base pairing, and together with the enzyme polynucleotide ligase they permit pieces from two dissimilar DNAs that

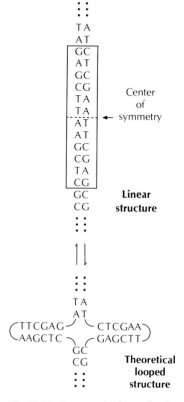

Fig. 5-16. Symmetrical snapback sequence in DNA. In theory, such a sequence can fold back on itself to form double-stranded hairpin loops, as illustrated. RNA transcribed from such a sequence also possesses the ability to fold back on itself.

have both been cleaved with the same restriction endonuclease to be joined together (DNA recombination). The technique of DNA recombination has been used to insert specific DNA sequences from eukaryotic cells into prokaryotic genomes where large numbers of identical copies of the sequence (clones) can be produced. The insertion is usually made into the DNA of small autonomous genetic units called plasmids, rather than into the much larger bacterial chromosome, to facilitate purification of the cloned DNA.

As an alternative to the direct isolation of a particular gene (which is often difficult to accomplish), it is possible to begin with purified mRNA. Certain RNA viruses (Fig. 4-6) produce an enzyme known as reverse transcriptase, which is able to use RNA as a template for the synthesis of a single-stranded DNA molecule whose sequence is complementary to the RNA, commonly referred to as cDNA. Another enzyme can then be used to convert the single-stranded cDNA into double-stranded DNA, which can then be incorporated into a plasmid and cloned.

DNA recombination and cloning is already contributing to major advances in our understanding of eukaryotic gene regulatory mechanisms and will undoubtedly play a major role in the next few years. Despite the great potential of DNA recombination as a research tool, the technique is the subject of major concern because of the potential danger of creating organisms with new combinations of genes that might pose serious health hazards. Strict guidelines have been established for such recombinant DNA research to minimize the danger.

Intervening sequences in genes

One particularly noteworthy and exciting finding that has come to light as a result of the ability to clone recombinant DNA molecules in prokaryotes is the discovery of intervening noncoding sequences* in the middle of the protein-coding sequences of genes such as those for chicken ovalbumin and mouse and rabbit β-globin. Comparison of

---

*The discovery of intervening sequences is so recent that the nomenclature used to describe them is still evolving. They are sometimes referred to as "gene insertions," "insertion sequences," or simply "inserts." However, those terms are somewhat misleading, since they have previously been used to refer to a rather different phenomenon in prokaryotic organisms and since the sequences are not inserted but instead are part of the genome and are removed during processing of the mRNA after transcription. Another recently proposed system uses the term "intron" to refer to "intragenic regions" that are not expressed as mRNA and the term "exon" for those sequences which are actually expressed as mRNA. However, the fact that introns are excluded from the mRNA but exons are included in it tends to be confusing, and it remains to be seen whether this nomenclature will be generally accepted. For the present we will continue to use the somewhat cumbersome terms "intervening sequence" or "noncoding sequence" to refer to sequences that are eliminated from the mRNA, and "coding sequence" or "mRNA sequence" to describe those which are retained in the mRNA.

cloned genes with cDNAs prepared from mRNAs has revealed the presence in the genes of additional sequences that are absent from their mRNAs and the corresponding cDNAs. The additional sequences can be detected in several ways, including differences in sizes of fragments generated by restriction endonucleases, the presence in the genes of additional restriction endonuclease sensitive sites that are missing in the cDNA, formation of loops visible with the electron microscope when mRNA or cDNA is hybridized with DNA from intact genes, and direct comparison of nucleotide sequences of the genes and the cDNA or mRNA.

At this time new data are appearing in the literature so rapidly that it is difficult to present a coherent picture of intervening sequences. One of the earliest reports showed the presence of an intervening sequence of about 600 base pairs near the 3′ end of the structural gene for rabbit β-globin. The mouse β-globin gene contains a similar sequence, variously reported as 550 to 780 base pairs in length, and apparently also a smaller second intervening sequence. Chicken ovalbumin appears to contain six or seven intervening sequences within the coding sequence, including two large segments of noncoding DNA 1.0 to 1.5 kb* in length. The combined length of those two sequences (2.5 kb) is about twice the length of the coding sequence (1.2 kb) and significantly longer than the entire ovalbumin mRNA (1.9 kb). One recent report suggests that the ovalbumin mRNA sequence is broken into eight different parts by noncoding sequences and is spread over a distance of approximately 6.0 kb in the intact gene. Comparison of DNA from tissues where the ovalbumin gene is and is not expressed appear to show no differences in intervening sequences.

Split genes are also known to occur in viral genomes. In adenovirus and in SV40 the 5′ untranslated leader sequences of several messages are not contiguous with the coding sequences in the genome. In the adenovirus system, and probably also in the SV40 system, the intervening sequence is transcribed together with the leader and coding sequences, and mRNA is subsequently formed by excising the intervening sequence and "splicing" the message pieces together. A similar pattern of transcription of the intervening sequence, followed by excision and splicing, also occurs in the formation of other mRNAs, including that for mouse β-globin (Chapter 8). Genes for some of the transfer RNAs of yeast also contain intervening sequences that are removed after transcription. Some of the multiple copies of the gene coding for 28S rRNA in *Drosophila* contain an intervening sequence,

---

*The abbreviation "kb" (kilobase) is used for 1000 nucleotide pairs of double-stranded DNA (or RNA) and also for 1000 nucleotides of single-stranded RNA (or DNA).

but the significance of this is unclear, since other copies of the 28S gene do not contain such sequences.

It is too early to tell how widespread the phenomenon of intervening sequences will prove to be. However, it appears likely that not all genes contain them. Sequence analysis of sea urchin histone genes, for example, has not detected intervening sequences in any of the genes for the five histones H1, H2a, H2b, H3, and H4.

Moderately repeated DNA sequences

Sequences that are moderately repeated in the genome (10^2 to 10^5 times) have received considerable attention recently as potential sites for regulation of gene expression. Their distribution in the genome has been examined in a variety of organisms, and in most cases they have been found to be interspersed with unique-sequence DNA. Over half the human genome, for example, consists of a rather regular distribution of unique DNA sequences with an average length of about 2 kb alternating with moderately repeated sequences with an average length of about 0.4 kb. Similar patterns also occur in organisms as diverse as sea urchins and *Xenopus*. A related pattern is also seen in *Drosophila*, but the average lengths are longer, both for the unique-sequence segments (>13 kb) and for the repeated sequences (about 5 to 6 kb).

It has been suggested that this arrangement may reflect a regulatory role for the moderately repeated sequences. They could, for example, serve as binding sites for molecules that regulate transcription of adjacent unique-sequence genes. Possible regulatory roles of moderately repeated sequences are discussed further in Chapter 7. In addition, the intermixing of unique and moderately repeated sequences in

Table 5-3. Numbers of genes coding for ribosomal and transfer RNAs*

Organism	Number of genes per haploid genome			Size of genome (daltons)
	18S and 28S rRNA	5S rRNA	tRNA	
Escherichia coli (bacteria)	6	6	50	2.8×10^9
Saccharomyces cerevisiae (yeast)	140	160 to 230	320 to 400	1.2×10^{10}
Drosophila melanogaster (fly)	130 to 250	160 to 200	750 to 860	1.2×10^{11}
Xenopus laevis (toad)	400 to 600	About 25,000	7800	1.8×10^{12}
HeLa (altered human)†	300‡	2000	1300	3.1×10^{12}

*Based primarily on data from Lewin, B. 1974. Gene expression. Vol. 2. Wiley-Interscience, New York, and Tartof, K. D. 1975. Annu. Rev. Genet. **9**:355.
†The HeLa cell is an established line derived originally from a human cancer. It has an altered genome with substantially more DNA than normal diploid human cells. Complete data for normal cells from humans or other mammals do not appear to be available. If all types of DNA in the HeLa cell have been increased proportionately, values for normal human cells should be approximately 60% of those shown for HeLa cells.
‡Most values reported for 18S and 28S rRNA in normal human cells are about 180 to 220 copies per haploid genome. However, a recent report suggests that the actual value might be as low as 50 copies per genome when systematic errors common to the earlier measurements are eliminated.

the DNA of cellular slime molds and its possible regulatory implications are discussed in Chapter 23.

The moderately repeated fraction of DNA also contains coding sequences. The genes for histones are present in approximately 500 copies per genome in the sea urchin and in tens to hundreds of copies per genome in mammals and in *Drosophila*. This gene repetition probably reflects the need to synthesize large amounts of histones quickly during the period of rapid cell division that occurs early in embryogenesis. The genes for transfer and ribosomal RNAs are also repeated in eukaryotic genomes (Table 5-3).

The distribution of genes for 5S rRNA is particularly interesting in some species. In humans there are about 2,000 copies, most of which are located in one region on chromosome number 1. However, in *Xenopus laevis* there are about 25,000 copies, and they are found at the ends (telomeres) of many different chromosomes (Fig. 5-17). In *Xenopus* the 5S genes and nontranscribed "spacers" are arranged tandemly within each gene cluster. The situation in *Xenopus* is further complicated by the presence of two different classes of 5S rRNA genes with slightly different nucleotide sequences. One of the classes is expressed only in oocytes, whereas the other appears to be expressed both in oocytes and in somatic cells.

The 5S sequences tend to be conserved during evolution, but the spacers show major divergence, even among closely related species of *Xenopus* (Fig. 5-18). Within each species, however, all copies of the spacers tend to remain similar. The relative absence of divergence within a single organism (or even a species) of multiple copies of genes coding for the same function is a common occurrence. It has

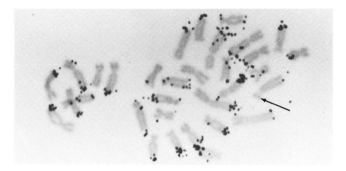

Fig. 5-17. Autoradiograph of metaphase chromosomes from *Xenopus laevis* showing hybridization of radioactive 5S cRNA to the telomeres of the long arms of most of the chromosomes. The arrow indicates the location of the nucleolar organizer, which contains the genes for 5.8S, 18S, and 28S rRNA. (From Pardue, M. L., et al. 1973. Chromosoma **42**:191.)

Fig. 5-18. Comparison of genes and spacers for somatic cell 5S rRNA of two species of *Xenopus*. The 5S rRNAs from both species are 120 nucleotides long, and the *X. laevis* somatic form differs by only one nucleotide from the major somatic form in *X. borealis*. The spacers, on the other hand, are different, both in size and composition, for the two species. Lengths of spacers and repeat units are approximate. *BP* = Base pairs; *% CG* = percentage of cytosine plus guanine base pairs. (*X. borealis* was mistakenly identified in the original papers as *X. mulleri* [Brown, D. D., et al. 1977. Dev. Biol. 59:266].) (Based on data from Brown, D. D., and K. Sugimoto. 1973. J. Mol. Biol. **78**:379, and Ford, P. J., and R. D. Brown. 1976. Cell **8**:485.)

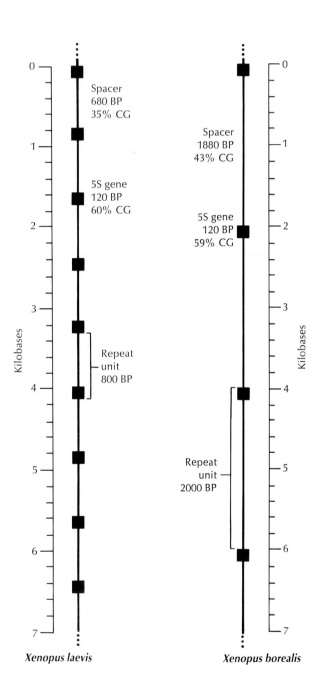

been observed in genes for 5S, 18S and 28S rRNA, tRNA, and histones. Such genes tend to be clustered together in tandemly repeated sequences as just described. Closely related species often exhibit significant differences in nucleotide sequences, particularly in noncoding spacers and in "wobble" bases of codons, but within a single species, and particularly within a single individual, all the copies of tandemly repeated genes and their spacers tend to exhibit only minimal differences.

A variety of models have been proposed to explain how an organism might eliminate the diversity that must arise by mutation in such genes, but little solid evidence is available for any of them. The simplest model conceptually is the master-slave theory, in which all the genes except one master copy are periodically destroyed and multiple new copies are made from the master copy. However, experimental evidence in support of the model is lacking, and differences that do arise within the tandem repeat pattern seem to be inherited, at least on a short-term basis. The model that many investigators currently favor is unequal crossover, which is claimed to alternately increase and decrease the number of copies of the repeated unit until a mutation either becomes established throughout the entire set of repeated sequences or is eliminated (Fig. 5-19). Despite computer modeling and complicated mathematical analysis, this model must be considered as

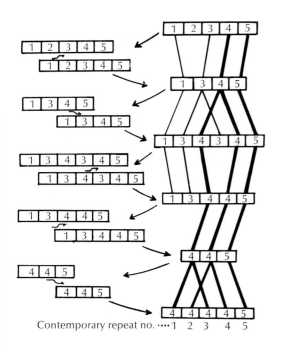

Contemporary repeat no. ····1 2 3 4 5

Fig. 5-19. Schematic representation of unequal crossover. Evolution of a segment of a chromosome that initially contains five tandem repeats of a gene is shown. Crossover events are indicated by small curved arrows. Mismatching of position due to the repeated nature of the genes results in loss of gene 2 during the first crossover. In subsequent crossovers, 3 and 4 are doubled, one 3 is eliminated, 1 and 3 are eliminated, and the double 4 is doubled again. The net result is total elimination of the descendents of the original genes 1, 2, and 3 and the generation of tandem repeats of the original gene 4. (From Smith, G. P. 1973. Cold Spring Harbor Symp. Quant. Biol. 38:507.)

only speculative. Until strong data are available to support a specific model, the sequence uniformity of tandemly repeated genes will remain a fascinating problem to follow in the research literature.

Families of closely related but not identical genes that all code for the same or similar gene products are also known. Switching from one of these genes to another often occurs either as development proceeds or as one type of cell becomes differentiated from another. Examples include the multiple forms of hemoglobin that appear during mammalian development (Fig. 3-16), multiple forms of collagen that are related both to type of tissue and to developmental stage, multiple forms of actin in various tissues, variant forms of histones described earlier in this chapter, and multiple forms of antibody molecules.

The DNA coding for feather keratin in chickens is organized into a somewhat different class of moderately repeated sequences. Hybridization of cDNA prepared from keratin mRNA with reverse transcriptase indicates that there are between 100 and 240 genes coding for keratin in the haploid genome of the chicken. However, there are 25 to 35 different species of keratin mRNA with minor differences in sequence represented among those genes so that on the average there are only a few copies of each different keratin gene per genome. There is also preliminary evidence suggesting that the transcribed "spacers" attached to the coding portions of the keratin genes are unique-sequence DNA. This is what would be expected if no correction mechanisms were operating to prevent independent evolution of each spacer.

Finally, it should be noted that much of the DNA that is commonly regarded as moderately repeated may not consist of exact repeats. With more precise hybridization techniques the amount of moderately repetitive DNA that is detected declines, whereas the fraction of "unique" sequences increases. Thus there appear to be families of DNA sequences that are only slightly different from one another.

Unique DNA sequences

Unique DNA sequences are the most prevalent type in most eukaryotic genomes. In mammals they are usually in the range of 50% to 70% of the total genome. Molecular hybridization techniques have demonstrated that most, but not all, of the mRNA in eukaryotic cells is transcribed from unique-sequence DNA. This includes messages for proteins that are made in large amounts in differentiated cells. Certain of these messages (e.g., those for hemoglobin, ovalbumin, silk fibroin) can be isolated in relatively pure form. Such messages, or cDNA prepared from them with reverse transcriptase, have been hybridized to DNA isolated from the corresponding differentiated cells. In every case the hybridization kinetics indicate that these messages are transcribed from genes that are present in only one or a few copies per haploid genome. In addition, for several types of cells the entire population of cytoplasmic mRNA molecules has been hybridized to corresponding DNA preparations and found to react primarily with the unique-sequence fraction of DNA.

Detailed genetic analysis of loci coding for the various polypeptide chains found in human hemoglobin suggests that there are two copies of the gene for the α-chain per haploid genome, whereas β- and δ-chain genes seem to be present only in a single copy per genome. There are at least two genes that vary slightly from one another for γ-chains, and there may be two copies of each per haploid genome. On the basis of this limited sampling, it appears that low levels of repetition may occur among those genes which appear to be "unique" in molecular hybridization studies.

A comparison of the amounts of DNA present in prokaryotic and eukaryotic cells shows a huge difference in coding potential (Table 5-4). The molecular weight of the *E. coli* genome is about 2.8×10^9,

DNA CONTENT OF EUKARYOTIC CELLS

Table 5-4. Comparison of *E. coli* and human haploid genomes*

	E. coli	Human haploid
Mass of DNA (g)	4.7×10^{-15}	2.8×10^{-12}
Mass of DNA (daltons)	2.8×10^9	1.7×10^{12}
Length of DNA (Å)	1.4×10^7	8.7×10^9
Length of DNA (mm)	1.4	870
Number of nucleotide pairs	4.2×10^6	2.6×10^9
Number of gene equivalents†	4.2×10^3	2.6×10^6
Unique-sequence DNA (%)	99.7	70
Unique-sequence gene equivalents	4.2×10^3	1.8×10^6
Number of separate chromosomes	1	23

*For each genome a single, widely cited value has been selected from the literature (*E. coli* genome = 2.8×10^9 daltons; human haploid genome = 2.8 pg). All other figures in the table have been calculated to ensure consistency within the table. The following conversion factors have been used: 1 g = 6.023×10^{23} daltons (Avogodro's number); 1 average nucleotide pair = 660 daltons = 3.4 Å of double helical DNA. (A recent report [Griffith, J. D. 1978. Science **201**:525] suggests that a value of 2.9 Å per base pair may be more accurate for DNA, both in dilute solution and in EM. If this is confirmed by further study, all previous interconversions of DNA length and number of base pairs, which have been based on 3.4 Å, will have to be recalculated.) There is significant variation among source references, and all figures in this table must be viewed as approximations. Because of this uncertainty, values are frequently rounded to one significant figure in the text (e.g., the human genome is referred to as 3 pg, or 3×10^6 gene equivalents).

†We have followed the widely used practice of equating a "gene equivalent" to 1000 nucleotide pairs (enough to code for 333 amino acids or a protein whose molecular weight is about 35,000). This correspohds to 6.6×10^5 daltons of DNA, with a length of 3400 Å. Some authors prefer the slightly larger value of 1.0×10^6 daltons, which corresponds to 1500 base pairs or a protein containing 500 amino acids. Recent data on sizes of mRNA molecules and their precursors in eukaryotic cells (Chapter 7) suggest that both of these equivalency values, which are based only on theoretical coding capability, are substantially smaller than the total amount of DNA actually associated with individual functional genes in eukaryotic cells.

which is enough DNA to code for about 4×10^3 proteins. The molecular weight of the human haploid genome is about 1.7×10^{12}, which is enough DNA to code for almost 3×10^6 proteins. Pride in our human capabilities might make us think we are hundreds of times more complex than a bacterium, but further examination of the sizes of genomes in various other organisms will quickly deflate such delusions. For example, the molecular weight of the genome of *Euglena* is 1.8×10^{12}, quite close to that of the human. The size of the genome of *Amphiuma* (a salamander) is 30 times that of the human; the size of the genome of the African lung fish is 15 times that of the human.

These large amounts of DNA in eukaryotic genomes cannot be explained in terms of repetitive sequences. The human genome contains 70% unique-sequence DNA or enough to code for about 1.5 to 1.8 million proteins. In the salamander *Amphiuma* only 20% of the nucleotide sequences are unique, but because of the huge size of the genome, they correspond to some 16 million gene equivalents.

Some investigators have attempted to correlate the minimum size of genome for particular types of organisms with their evolutionary complexity (Fig. 5-20). However, such data must be interpreted cau-

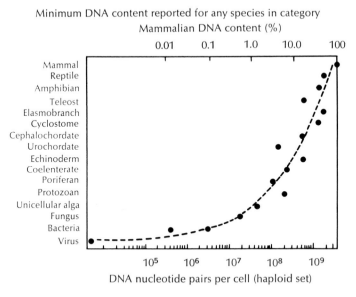

Fig. 5-20. Possible evolution of minimum DNA content of eukaryotic species. The values reported here are the smallest known for haploid genomes within each taxonomic group or species. Most of the groups have relatively wide ranges of DNA contents, as shown in Fig. 5-21. The minimum DNA content for each group tends to increase with the complexity of that group of organisms. (Modified from Britten, R. J. and E. H. Davidson 1969. Science **165**:349. Copyright 1969 by the American Association for the Advancement of Science.)

tiously. Within any one group there is often great diversity in the amounts of DNA present in different species (Fig. 5-21). In addition, birds, which were omitted from Fig. 5-20, appear in Fig. 5-21 to have genomes that are somewhat too small for the evolutionary position that they are normally assigned.

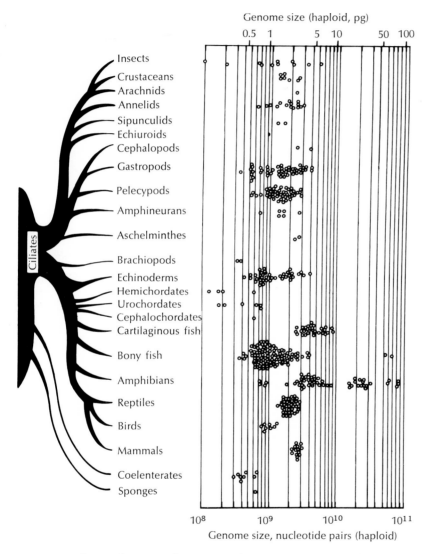

Fig. 5-21. Range of sizes of genomes for various eukaryotic organisms. Note the wide range of sizes for some types of organisms. By contrast, mammals, including the human species, all have genome sizes that fall within a narrow range. Note also that birds have too little DNA and cartilaginous fish too much to fit smoothly into the evolutionary sequence suggested in Fig. 5-20. (From Lewin, B. 1974. Gene expression-2. Eucaryotic chromosomes. John Wiley & Sons, Inc., New York.)

Mammals, unlike many of the other taxonomic groups depicted in Fig. 5-21, exhibit a striking constancy in sizes of their genomes from species to species, with a narrow range of DNA content, usually between 2 and 3 pg (roughly 2 to 3×10^9 base pairs). Thus the data given in Table 5-3 for the human genome are also approximately correct for all other mammals whose genomes have been studied carefully.

Attempts have been made to determine how much of the genome is transcribed by measuring the sequence complexity of cellular RNA (i.e., the number of different RNA sequences that can be detected by hybridization to DNA). For example, when total cellular RNA (nuclear plus cytoplasmic) from the sea urchin gastrula is examined, about 20% to 30% of the total genome appears to be transcribed. Similar values have also been reported for mouse embryos. However, as we

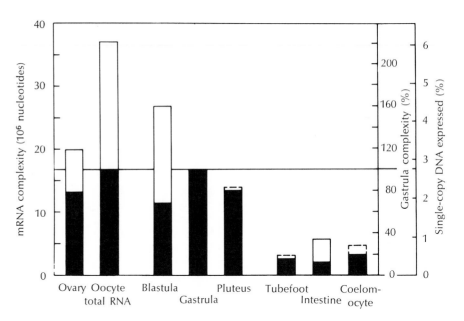

Fig. 5-22. Gene expression in sea urchin embryos and adult tissues. The amount of gene expression is shown in three different types of units. On the left, it is given in terms of number of nucleotides of unique-sequence genome. On the right it is given as amount relative to that expressed during the gastrula stage and as percentage of the total unique-sequence DNA of the genome that is expressed. The height of each bar indicates the total complexity (number of different sequences expressed) of cytoplasmic mRNA from the stage or tissue indicated. (Complexity of total RNA is shown for the oocyte.) The solid portion of the bar indicates those sequences shared in common with the gastrula stage. The white portion indicates mRNA sequences found in the stage or tissue that cannot be detected in gastrulas. Dashed lines indicate the maximum amount of sequences different from gastrula that could have escaped detection in those cases in which no such sequences were found. (From Galau, G. A., et al. 1976. Cell **7**:487. Copyright © MIT; published by the MIT Press.)

shall see in Chapter 8, much of the RNA that is transcribed apparently never leaves the nucleus. If only the cytoplasmic mRNA sequences from sea urchin embryos at the gastrula stage are examined, they are found to represent about 3% of the genome, which is enough to code for 10,000 to 15,000 proteins. The amount of the sea urchin genome expressed as cytoplasmic mRNA varies with the developmental stage. The largest amount of expression occurs in the oocyte, where 6% of the genome (enough to code for 20,000 to 30,000 proteins) is expressed. Fig. 5-22 illustrates the complexity of mRNA sequences found at different developmental stages and in different tissues in the sea urchin.

Comparable studies in the mouse indicate that the most extensive gene expression occurs in the brain, where a cytoplasmic population of apparent mRNA molecules (identified by the presence of polyadenylic acid, which will be discussed in Chapter 8) has been detected that is large enough to code for 50,000 to 100,000 average-sized proteins. This value corresponds to about 5.3% of the mouse genome or 7.6% of the nonrepetitive sequences.

The genome of *Drosophila*, which is composed of about 10^8 base pairs and could code for about 100,000 proteins, appears to contain only about 5000 genes that are normally expressed. This estimate is based on the number of bands seen in the giant chromosomes of the salivary gland (Chapter 6) and the apparent 1:1 correlation between those bands and functional genes (Chapter 7).

Theoretical considerations also suggest that much of the eukaryotic genome probably does not code for proteins. Starting with currently accepted values for mutation rates, population geneticists have calculated that if the entire human genome served an essential coding function and was therefore subject to mutation, the number of deleterious mutations occurring per generation would be too large to be eliminated by natural selection and the species would not survive. From such calculations it has been postulated that the human genome cannot contain more than about 40,000 functional genes and that the remaining nearly 99% of the genome must have some noncoding function for which nucleotide sequence is less critical.

Opinions vary about how many genes are necessary to code for all the enzymes and structural proteins that occur in a typical eukaryotic organism. The usual estimates range from 5000 to 20,000, although one popular biochemistry textbook suggests that the total number of different proteins in a complex organism could be as large as 100,000. Even if this last figure is valid, most eukaryotic genomes still contain far more DNA than is needed to code for all the amino acid sequences in that many proteins.

Much of the unique fraction of DNA probably does not code for protein. In the human genome, for example, unique sequences account for enough DNA to code for 1.8 million average-sized proteins. As just discussed, it is unlikely, both on theoretical and experimental grounds, that more than a small fraction of this potential genetic information is ever expressed as proteins. A certain amount of the "excess" DNA might serve in regulatory roles. As will be described in Chapter 8, substantial numbers of both unique and moderately repeated DNA sequences appear to be transcribed in the nucleus, followed by rapid degradation of the transcripts, which never reach the cytoplasm as functional mRNA. However, the amount of the genome that could serve in a regulatory role is limited by the same mutational load considerations that limit the number of possible coding sequences. In addition, we are still left with an amount of unique-sequence DNA equivalent to more than a million average-sized genes that is apparently never transcribed and whose function and value to the organism remain completely unknown.

IMPLICATIONS FOR CONTROL OF GENE EXPRESSION

As we stated at the beginning of this chapter, the structural organization of the eukaryotic genome provides many opportunities for the regulation of gene expression in cellular differentiation. The fact that the DNA is associated with tightly bound proteins and densely packed into highly organized structures, even in interphase cells, suggests that the availability of specific sequences for transcription must be greatly restricted. It is, in fact, feasible to propose that transcription of the eukaryotic genome may occur only where it is specifically activated, perhaps by specific alterations of the generalized chromatin structure (see Chapter 7).

The physical isolation of the genome from the rest of the cell also has important implications. The role of the nuclear envelope in regulatory processes is not at all clear, but it represents a barrier that must be traversed by any signal that carries information either to the genome or from the genome to the cytoplasm. The extensive processing of RNA transcripts that occurs in the nucleus before they are released to the cytoplasm in the form of functional mRNA will be discussed in Chapter 8.

The significance of the vast amount of seemingly excess DNA in the eukaryotic genome is not understood. Some of it may function in structural or regulatory roles, but the total amount involved seems to be far too large for such functions alone. One speculation that has been advanced is that organisms whose genomes contain a large excess of DNA that is not essential for their survival have more opportunities for evolutionary exploration, since most chance mutations would not be deleterious and might occasionally confer a selective advantage. However, since most of the excess DNA appears never to be transcribed into RNA or expressed as proteins, that explanation remains questionable. It would be necessary to postulate first an advantageous mutation and then chance activation of its transcription before it would confer any selective advantage to the organism. At this time all we can do is speculate that the "excess" DNA must have some direct or indirect survival value, since its synthesis reflects a considerable expenditure of metabolic energy, which, in the absence of specific advantages, would seem likely to be of negative value under highly competitive conditions.

Now that we have described the overall structure of the eukaryotic genome, we will ask in the next chapter whether any significant change occurs in the DNA that it contains during the process of cellular differentiation. Subsequent chapters will also examine the role of chromosomal proteins in the control of gene expression and the nature of the processing that occurs before newly synthesized mRNA leaves the nucleus and becomes available for translation by the ribosomes in the cytoplasm.

BIBLIOGRAPHY
Books and reviews

Bostock, C. 1971. Repetitious DNA. *In* D. Prescott, L. Goldstein, and E. McConkey, eds. Advances in cell biology. Vol. 2. Plenum Press, Inc., New York.

Britten, R. J., and D. E. Kohne. 1970. Repeated segments of DNA. Sci. Am. **222:**24 (April).

Brown, D. D. 1973. The isolation of genes. Sci. Am. **229:**21 (Aug.).

Comings, D. 1972. The structure and function of chromatin. Adv. Hum. Genet. **3:**237.

Dutrillaux, B., and J. Lejeune. 1975. New techniques in the study of human chromosomes: methods and applications. Adv. Hum. Genet. **5:**119.

Elgin, S. C. R., and H. Weintraub. 1975. Chromosomal proteins and chromatin structure. Annu. Rev. Biochem. **44:**725.

Kedes, L. H. 1976. Histone messengers and histone genes — review. Cell **8:**321.

Kornberg, R. 1974. Chromatin structure: a repeating unit of histones and DNA. Science **184:**868.

Kornberg, R. D. 1977. Structure of chromatin. Annu. Rev. Biochem. **46:**931.

Lewin, B. 1974. Gene expression. Vol. 2. Wiley-Interscience, New York (Chapters 1, 3, and 4).

Lewin, B. 1974. Sequence organization of eucaryotic DNA: defining the unit of gene expression. Cell **1:**107.

Tartof, K. D. 1975. Redundant genes. Annu. Rev. Genet. **9:**355.

Weatherall, D. J. and J. B. Klegg. 1976. Molecular genetics of human hemoglobin. Annu. Rev. Genet. **10:**157.

Williamson, B. 1977. DNA insertions and gene structure. Nature **270:**295.

Selected original research articles

Bantle, J. A., and W. E. Hahn. 1976. Complexity and characterization of polyadenylated RNA in the mouse brain. Cell **8:**139.

Berget, S. M., C. Moore, and P. A. Sharp. 1977. Spliced segments at the 5′ terminus of adenovirus 2 late mRNA. Proc. Natl. Acad. Sci. U.S.A. **74:**3171.

Britten, R. J., and D. E. Kohne. 1968. Repeated sequences in DNA. Science **161:**529.

Cohen, L. H., K. M. Newrock, and A. Zweidler. 1975. Stage specific switches in histone synthesis during embryogenesis of the sea urchin. Science **190:**994.

Cold Spring Harbor Symp. Quant. Biol. 1973. Vol. 38. Chromosome structure and function. (Entire volume contains numerous pertinent research articles.)

Cooke, H. 1976. Repeated sequence specific to human males. Nature 262:182.

Finch, J. T., et al. 1977. Structure of nucleosome core particles of chromatin. Nature 269:29.

Finch, J. T., and A. Klug. 1976. Solenoidal model for superstructure in chromatin. Proc. Natl. Acad. Sci. U.S.A. 73:1897.

Ford, P. J., and R. D. Brown. 1976. Sequences of 5S ribosomal RNA from Xenopus mulleri and the evolution of 5S gene-coding sequences. Cell 8:485.

Franklin, S. G., and A. Zweidler. 1977. Non-allelic variants of histones 2a, 2b and 3 in mammals. Nature 266:273.

Grunstein, M., P. Schedl, and L. Kedes. 1976. Sequence analysis and evolution of sea urchin (Lytechinus pictus and Strongylocentrotus purpuratus) histone H4 messenger RNAs. J. Mol. Biol. 104:351.

Kemp, D. J. 1975. Unique and repetitive sequences in multiple genes for feather keratin. Nature 254:573.

Kinniburgh, A. J., J. E. Mertz, and J. Ross. 1978. The precursor of mouse β-globin messenger RNA contains two intervening RNA sequences. Cell 14:681.

Knapp, G., et al. 1978. Transcription and processing of intervening sequences in yeast tRNA genes. Cell 14:221.

Lai, E. C., et al. 1978. The ovalbumin gene: structural sequences in native chicken DNA are not contiguous. Proc. Natl. Acad. Sci. U.S.A. 75:2205.

Lohr, D., et al. 1977. Comparative subunit structure of HeLa, yeast, and chicken erythrocyte chromatin. Proc. Natl. Acad. Sci. U.S.A. 74:79.

Mandel, J. L., et al. 1978. Organization of coding and intervening sequences in the chicken ovalbumin split gene. Cell 14:641.

Manning, J. E., C. W. Schmid, and N. Davidson. 1975. Interspersion of repetitive and nonrepetitive DNA sequences in the Drosophila melanogaster genome. Cell 4:141.

Morris, N. R. 1976. A comparison of the structure of chicken erythrocyte and chicken liver chromatin. Cell 9:627.

Oudet, P., M. Gross-Bellard, and P. Chambon. 1975. Electron microscopic and biochemical evidence that chromatin structure is a repeating unit. Cell 4:281.

Perlman, S., C. Phillips, and J. O. Bishop. 1976. A study of foldback DNA. Cell 8:33.

Perelson, A. S., and G. I. Bell. 1977. Mathematical models for the evolution of multigene families by unequal crossing over. Nature 265:304.

Schaffner, W., et al. 1978. Genes and spacers of cloned sea urchin histone DNA analyzed by sequencing. Cell 14:655.

Schmid, C. W., and P. L. Deininger. 1975. Sequence organization of the human genome. Cell 6:345.

Tilghman, S. M., et al. 1978. Intervening sequence of DNA identified in the structural portion of a mouse β-globin gene. Proc. Natl. Acad. Sci. U.S.A. 75:725.

CHAPTER 6

Does the genome change during differentiation?

☐ Cellular differentiation is a process in which cells derived mitotically from a common ancestor become different from one another both in their function and in their morphology. Differentiation is frequently characterized by the synthesis of large amounts of a limited number of specialized cellular products. For example, skeletal muscle cells produce large amounts of actin and myosin, cartilage cells produce and excrete chondroitin sulfate and collagen, red blood cells produce hemoglobin, and cells of the exocrine pancreas produce precursors for digestive enzymes. This type of cellular specialization is the result of differential gene expression. That is, products coded for by certain genes are produced in disproportionately large amounts in the specialized cells.

Theoretically one of the ways that this could be accomplished is by direct alteration of the genomes of the differentiated cells. An animal's germ cells would, of course, have to retain the full complement of genes to be transmitted to the next generation, but no such constraints exist for specialized somatic cells. Three possible mechanisms can be visualized whereby genetic alteration could play a role in cellular differentiation. These are diagrammed in Fig. 6-1.

First, those genes needed for specialized functions could be increased in number or "amplified" in the differentiated cells. Thus, for example, amplification of the number of genes coding for hemoglobin in a red blood cell precursor might explain how large amounts of hemoglobin are synthesized in developing red blood cells but not in other types of cells. Such a mechanism alone could not fully account for cell specialization, however, since it would not explain the apparent total absence of expression of some genes in specialized cells.

A second possibility is that as development proceeds, those genes which are not necessary for specialized functions or for survival of the differentiated cells could be eliminated. For example, a nerve cell with

203

Fig. 6-1. Scheme showing three possible modifications of the genome that in theory could favor selective expression of genes *A* or *B*, which are initially present in one active copy each. See text for details.

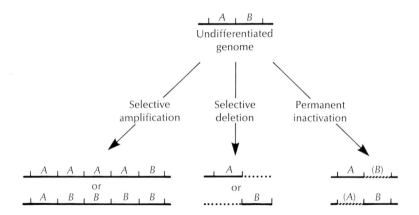

no apparent need for hemoglobin or insulin might eliminate the genes for these proteins to concentrate more fully on synthesis of its own specialized products.

Finally, differentiated cells might permanently inactivate unneeded genes, rather than totally eliminating them. Such genes would be retained in the genomes of the cells but in a completely inactivated form so that they are never transcribed.

There is convincing evidence that all three of these processes operate in certain cases during development. However, on the basis of the best information currently available, gene amplification, gene elimination, and large scale inactivation of major blocks of genes seem to occur only in specialized cases and do not appear to be generally used as mechanisms for cellular differentiation. In this chapter we will examine the evidence for occurrence of each of these phenomena and the evidence that they happen only in specialized cases.

GENE AMPLIFICATION

Before we examine the evidence for gene amplification, it is essential to distinguish clearly between amplified genes and the repeated DNA sequences discussed in Chapter 5. Repeated sequences are a fundamental part of the organism's basic genome and occur with the same degree of repetition per haploid genome in all its cells, including the haploid gametes. Such sequences may or may not be transcribed. Amplified genes, on the other hand, are specific sequences whose number is increased relative to the rest of the genome to fulfill a particular need in certain cells. Before amplification, they may be present in the genome either in a single copy or in multiple copies. The distinguishing features of amplification are (1) that it is a selective process involving only a portion of the total genome and (2) that it occurs only during a portion of the total life cycle and only in certain types of cells.

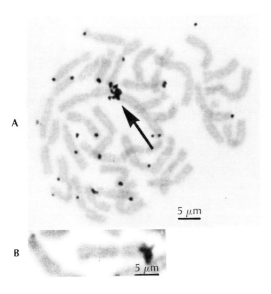

Fig. 6-2. Autoradiographs showing localization of 18S and 28S rRNA genes in *Xenopus mulleri*. **A,** The ends of the two chromosomes carrying rDNA tend to stick together in most chromosome spreads. **B,** Single chromosome showing location of rRNA genes. (From Pardue, M. L. 1973. Cold Spring Harbor Symp. Quant. Biol. **38:**475.)

rRNA genes

One of the best-documented cases in which differential gene expression is accomplished by alteration of the genetic material is amplification of the genes for 18S and 28S rRNA in oocytes of *Xenopus laevis*, a South African swimming toad. No new rRNA is synthesized in *Xenopus* embryos during the period of rapid cell division immediately after fertilization, although considerable protein synthesis occurs during that period. Instead, large amounts of rRNA, sufficient to carry the embryo through its early development, are made and stored in the egg cell during oogenesis.

Prior to amplification the haploid genome for *Xenopus* already contains about 500 copies of the genes for the 18S and 28S rRNAs. These genes are located in the nucleolar organizer, a specific chromosomal region present once in each haploid set of chromosomes. The site can be visualized as shown in Fig. 6-2 by in situ hybridization of metaphase chromosomes with radioactively labeled rRNA prepared from purified ribosomes. The DNA coding for the 18S and 28S rRNAs (rDNA) can also be separated from the rest of the DNA of the cell by centrifugation in a cesium chloride density gradient. Its high content of guanine and cytosine makes it more dense than the other DNA.

The genes for 18S and 28S rRNAs are arranged tandemly along the DNA molecule in transcription units, as diagrammed in Fig. 6-3, *A*. If one reads from the 5′ end of the RNA product, each transcription unit contains (1) a "spacer" region that is transcribed into RNA and later degraded, (2) the 18S rRNA gene, (3) a second spacer that is transcribed and later partially degraded (a portion is retained as the so-called 5.8S

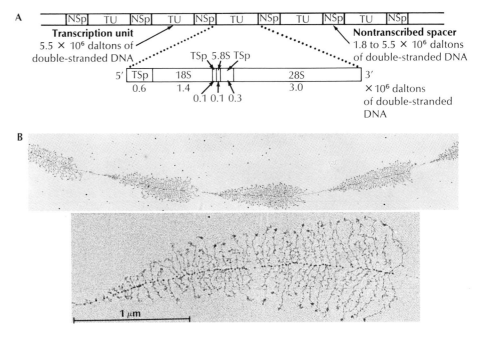

Fig. 6-3. Organization of genes for 18S, 5.8S, and 28S rRNA in amphibia. **A,** Organization in *Xenopus laevis. TSp* = Transcribed spacer; *NSp* = nontranscribed spacer; *TU* = transcriptional unit. **B,** Electron micrograph of tandemly arranged RNA genes from *Triturus viridescens* undergoing transcription (arranged to match 5'-to-3' orientation in **A**) and a micrograph of a single rRNA gene at high magnification. (Low-magnification photograph from Watson, J. D. 1976. Molecular biology of the gene, 3rd ed. Benjamin/Cummings Publishing Co., Inc., Menlo Park, Calif. [courtesy O. L. Miller, Jr.]; high-magnification photograph from Miller, O. L., Jr., and B. R. Beatty. 1969. J. Cell Physiol. **74** [supp. 1]:225.)

rRNA, which attaches to the 28S rRNA by complementary base pairing, and is usually considered to be part of the 28S rRNA), and (4) the 28S rRNA gene. Between each transcription unit there is a longer spacer sequence that is not transcribed. Approximately 500 rRNA transcription units and their nontranscribed spacers are clustered together in each nucleolar organizer region. A segment of rDNA with tandemly arranged transcription units and spacers is shown in Fig. 6-3, *B*.

The extent of amplification of the 18S and 28S rRNA genes in the amphibian oocyte is approximately 1000 times. Recent data on an analysis of variations in the nontranscribed spacer regions suggest that only a small subset of the roughly 500 copies of the 18S and 28S genes that are present per haploid genome are actually amplified. Thus the extent of amplification of those genes would be substantially

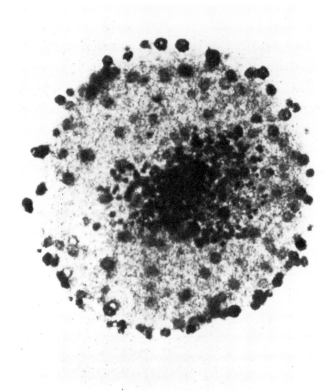

Fig. 6-4. Photomicrograph of an isolated nucleus from an oocyte of *Xenopus laevis*. Its diameter is about 400 μm. The deeply stained spots are some of the hundreds of nucleoli. (From Brown, D. D., and I. B. Dawid. 1968. Science **160**:272. Copyright 1968 by the American Association for the Advancement of Science.)

greater than the overall 1000-fold increase that occurs in the amount of rDNA.

Amplification occurs early in oogenesis, soon after the female toad undergoes metamorphosis into an adult form. Since the oocytes are in early stages of meiosis, and therefore contain four copies of each chromosome, they begin with four nucleolar organizer regions containing approximately 2000 genes for 18S and 28S rRNA and undergo amplification to approximately 2 million such genes. Depending on the size of nontranscribed spacer (Fig. 6-3, *A*), the total mass of the amplified DNA would be between 1.3×10^{13} and 2.2×10^{13} daltons, which is somewhat larger than the total mass of the four copies of the genome that are contained in the oocyte. The amplified genes are present as circular molecules and are found in about 1000 to 1500 extrachromosomal nucleoli (Fig. 6-4). These extra nucleoli are present throughout oocyte maturation until the time of the first meiotic division, when they lose their structural organization. The amplified DNA is slowly lost during early cleavage.

The number of 18S and 28S rRNA genes produced by amplification does not seem to depend on the number present prior to amplification. The entire nucleolar organizer region is deleted in the anucleolate mutant of *Xenopus* (0 *nu*), which is a recessive lethal gene. Heterozygous females (1 *nu*) have only one nucleolar organizer and therefore only one nucleolus per somatic cell. Their premeiotic oocytes contain only two nucleolar organizer regions and therefore have only half the number of genes for 18S and 28S rRNA that are present in normal oocytes. However, after amplification has occurred, the oocytes of the heterozygotes contain the same amount of rDNA as normal oocytes. Thus there appears to be a specific regulatory mechanism that increases the number of 18S and 28S rRNA genes to a level adequate for the massive rRNA synthesis that occurs during oogenesis.

The mating of two heterozygous (1 *nu*) animals can result in an embryo (0 *nu*) that has no genes for 18S and 28S rRNA and thus is incapable of rRNA synthesis. Such embryos proceed normally through those phases of development during which no rRNA synthesis is required and continue to develop at a reduced rate, using stored maternal rRNA for some time after rRNA synthesis would normally be initiated. They will usually survive long enough to hatch and become somewhat abnormal swimming tadpoles before ultimately dying from lack of 18S and 28S rRNA synthesis.

Amplification of genes does not occur during oogenesis for all the RNA components of ribosomes. In addition to the 18S and 28S rRNA molecules that are amplified together in a single transcriptional unit during oogenesis, ribosomes also contain 5S rRNA molecules that are coded at separate genetic sites in most eukaryotic genomes. (In bacteria and in some primitive eukaryotes such as the cellular slime mold *Dictyostelium discoideum*, the 5S gene is directly linked to the genes for the larger rRNAs.) The 5S genes are present in the same number of copies per haploid genome in oocytes as they are in somatic cells.

The genome of *Xenopus laevis* contains about 25,000 genes for 5S rRNA located at the ends of numerous chromosomes (Fig. 5-17). Complex regulatory mechanisms balance the rates of transcription so that roughly equal numbers of 5S, 18S, and 28S rRNA molecules are always found, both in diploid somatic cells that contain about 50,000 5S genes and 1000 18S/28S genes per cell, and in oocytes that contain about 100,000 5S genes and 2,000,000 18S/28S genes per cell. Part of the regulation in *Xenopus* appears to involve the transcription in oocytes of 5S sequences that are not expressed in somatic cells. These oocyte-specific 5S genes show a sequence divergence from somatic cell 5S sequences that is greater than the difference between the somatic cell 5S sequences of closely related species of *Xenopus*.

Transcription of oocyte-specific sequences does not completely explain the balanced rate of 5S and 18S/28S rRNA synthesis in oocytes, however, since even with all available 5S sequences activated, the oocyte still contains a 20-fold excess of amplified 18S/28S sequences.

The mechanism by which rDNA is amplified in oocytes has not been completely determined. One interesting theory that has been suggested is that templates might be produced by a reverse transcrip-

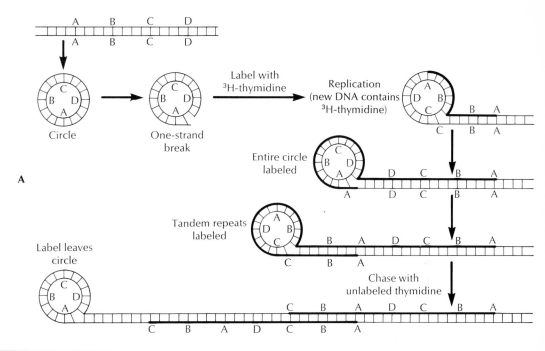

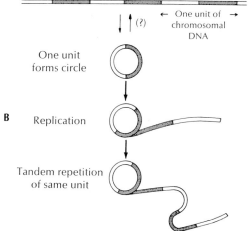

Fig. 6-5. Rolling circle form of DNA replication. **A,** Basic mechanism. The DNA forms a circle; then one strand breaks and forms a replication fork. One of the newly synthesized strands remains in the circle, and the other becomes part of a growing linear chain. **B,** Rolling circle replication of a ribosomal gene transcription unit showing generation of tandem repeats from a single unit. (**B** from Hourcade, D., et al. 1974. Cold Spring Harbor Symp. Quant. Biol. **38:**537.)

tase which copies special RNA transcripts of the entire rDNA region (including the spacer regions that are normally not transcribed). However, this is a highly controversial view, and many investigators believe that a more conventional DNA-dependent replication process is involved. There is also evidence that after the initial formation of extrachromosomal templates, the rDNA circles may be replicated by a "rolling circle" type of replication, similar to that which occurs in some viruses. A possible sequence of events involved in the amplification of rDNA is shown in Fig. 6-5. Such a mechanism would allow a large number of copies to be made easily from each extrachromosomal template and would also explain the selective amplification of a small subset of the nontranscribed spacers described on p. 206.

Although rDNA amplification has been studied most thoroughly in amphibian oocytes, it also occurs frequently during oogenesis in other species, including fish, mollusks, worms, and insects. Another amplification-like phenomenon is observed in *Drosophila*. Under certain circumstances the "bobbed" mutant, which has a partial loss of rRNA genes, can undergo a process called "magnification" in which the number of genes for rRNA is restored to normal. Details of this phenomenon, which occurs in the germ line and is inherited, are poorly understood.

Polytene chromosomes   A different type of gene amplification occurs during the development of very large cells in the salivary gland and certain other tissues of some insect larvae. In these cells homologous chromosomes come together and line up in pairs in a manner similar to the pairing that normally occurs during meiosis. They then undergo multiple rounds of DNA replication until each chromosome contains about a thousand parallel DNA molecules. These giant chromosomes, which are called polytene chromosomes, exhibit characteristic banded staining patterns when examined with the light microscope (Fig. 6-6). Initially it might seem that this process is simply a duplication of the total genome and that it should therefore not be referred to as gene amplification. However, there is little or no duplication of some of the highly repeated DNA sequences, and the rRNA genes are only duplicated about one sixth as much as the rest of the chromosome during the formation of a polytene chromosome. Thus the process of polytene chromosome formation can be considered to be selective amplification of unique-sequence DNA. However, since all the functional genes except those for rRNA appear to be amplified to the same extent, polytene chromosome formation should not be viewed as a mechanism for differential gene expression. Polytene chromosomes have been widely studied because their large size makes minor aberrations readily visi-

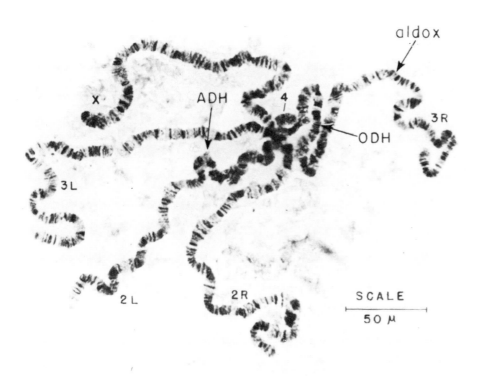

aldox

X   ADH   4   3R

3L   ODH

2L   2R

SCALE

50 μ

Fig. 6-6. Giant polytene chromosomes from the salivary gland of *Drosophila melanogaster*. The four chromosomes are closely connected at their centromere regions in a compact chromocenter, from which the opposite arms protrude limply. (From Ursprung, H., et al. 1968. Science **160**:107 5. Copyright 1968 by the American Association for the Advancement of Science.)

ble and also because sites of active transcription can be seen as "puffs" on them, even at light microscope magnifications (Chapter 7).

The question of whether gene amplification is a significant factor in cell differentiation has been approached recently by the technique of molecular hybridization. It is possible to isolate fairly pure mRNA species from highly specialized cells that produce predominantly one product. Examples include silk fibroin mRNA from the posterior silk gland of the silk moth, globin mRNA from erythrocyte precursors of a variety of animals, and ovalbumin mRNA from the chicken oviduct. Radioactively labeled mRNA, or more commonly, labeled single-stranded DNA copies prepared with reverse transcriptase (cDNA), can be hybridized with cellular DNA both from the differentiated cells and from undifferentiated controls, and the kinetics of hybridization can be used to calculate the approximate number of genes coding for the specific mRNA that are present per genome in each type of cell. The measurements that are involved are similar to those used to determine cot values from DNA:DNA hybridization (Chapter 5 and Appendix A). In all cases tested so far, highly differentiated cells have been

Are genes amplified during differentiation?

found to contain only one or a few copies per haploid genome of the genes coding for their specialized products (p. 194). Furthermore, this number is no greater than the number found in other cells from the same animal that do not make the specialized product. For example, fetal mouse liver, which is an erythropoietic tissue and synthesizes hemoglobin, contains no more copies of the globin genes per haploid genome than does mouse sperm.

The amplification of the genes for 18S and 28S rRNA thus seems to be a specialized case. This could be related to the fact that the end product of these genes is RNA rather than protein and is therefore made in a one-step process with no opportunity for further amplification at the translational (protein synthesis) level. The final products coded for by most other genes are proteins, which are made in a two-step process, transcription followed by translation. As we shall see in later chapters, considerable "amplification" of the genetic message for differentiated products is often achieved by stabilization and repeated reuse of the mRNAs for such products. Several investigators have made theoretical calculations, based on known rates of transcription and translation, which verify that gene amplification is not necessary for the synthesis of large amounts of specialized proteins, provided that mRNA is stabilized and reused.

However, despite strong evidence that structural genes for differentiated products are not amplified, there are a number of reports in the literature suggesting that a small degree of amplification of portions of the genome may occur during differentiation. Some of these suggest that moderately repeated sequences, which are involved as control elements in many models for control of gene expression (Chapter 7), may be amplified. Such reports are highly controversial and not generally accepted at the present time. However, they should be followed carefully during the next few years, since if they prove to be valid, they will force a reevaluation of our conclusion that gene amplification is not involved in cellular differentiation.

Gene amplification does appear to be involved in the acquisition of resistance to the drug methotrexate in cultured cells, however. The drug is an inhibitor of the enzyme dihydrofolate reductase, and the rate of synthesis of that enzyme is greatly increased in the resistant lines. The increased enzyme synthesis has been shown to be due to a corresponding increase in the number of copies of the gene coding for the enzyme.

The techniques that were employed to demonstrate the increase are a good example of methods that can be used to study the gene for a protein that is never made in large amounts. The enzyme was first purified, and then rabbit antibodies against it were prepared and purified. Polysomes were obtained from cells that were synthesizing the enzyme. The rabbit antibody recognized and bound to incomplete enzyme molecules still attached to the polysomes. Goat antibody to rabbit γ-globulin was then used to precipitate the polysome-antibody complex, and mRNA was extracted from the precipitated material. Reverse transcriptase was used to prepare radioactive cDNA, which was in turn used as a hybridization probe to determine the number of copies of the dihydrofolate reductase gene by analysis of cot values. A strain with 200- to 250-fold increase in enzyme synthesis showed an increase of about 200-fold in the number of gene copies. Lines that had reverted to a less resistant state in the absence of the drug showed a corresponding decrease in the number of gene copies that they contained.

Portions of the genome are eliminated during the development of a number of invertebrates and also in rare cases of vertebrates. A classically studied case occurs in the nematode *Ascaris*, where all the somatic cells lose large amounts of chromatin early in development and only the germ cells retain the complete genome. During the first cleavage division the fertilized egg appears to have two very long chromosomes that undergo a normal mitotic division. At the second cleavage the chromosomes in one of the two cells break up into numerous small chromosome fragments (Fig. 6-7, *A*). Those fragments from the central region of the original chromosome each have their own centromere and move to the poles of the spindle. However, the fragments from the ends of the original chromosome lack centromeres and are lost in the cytoplasm. The other cell, which is the precursor of the germ line, undergoes a normal mitotic division with no loss of chromatin. At the third cleavage division one of the daughters of the presumptive germ cell undergoes chromatin diminution, but the other does not. The same thing occurs at the fourth cleavage division, but not at the fifth, so that after five divisions are completed, the embryo has two presumptive germ cells that contain the entire chromosome complement and 30 presumptive somatic cells which have undergone loss of chromatin. The course of chromatin diminution during cleavage in *Ascaris* is diagrammed in Fig. 6-7, *B*. The two cells that retain the entire chromosome complement ultimately become the germ cells of the animal.

The amount of DNA lost from the somatic cells of *Ascaris* is substantial. In one species that has been studied in detail about 27% of the DNA is eliminated. The eliminated DNA is rich in repeated sequences but also contains a significant number of unique sequences. The DNA in the germ line contains about 23% repeated sequences,

GENE ELIMINATION
Chromatin diminution in Ascaris

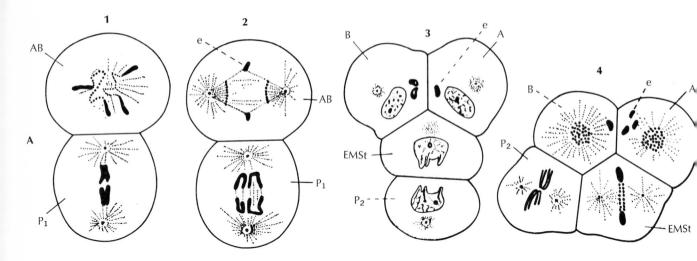

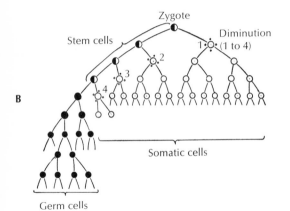

Fig. 6-7. A, Partial chromosome loss in somatic cells of *Ascaris*. *1* = Beginning of the second cleavage; *2* = later stage of the same cleavage, with the chromosomes becoming fragmented, and their ends *(e)* lost in the cytoplasm; *3* = four-cell stage, viewed from the animal pole; *4* = later stage of the four-cell stage; chromosome fragmentation and loss of ends is occurring in cell EMSt. **B,** Diagrammatic representation of chromosome diminution in *Ascaris* cells during cleavage. (**A** from Waddington, C. H. 1956. Principles of embryology. George Allen & Unwin Ltd., London; **B** reprinted with permission of Macmillan Publishing Co., Inc. from The cell in development and heredity, 3rd ed. by Wilson, E. B. Copyright 1925 by Macmillan Publishing Co., Inc., renewed 1953 by M. K. Wilson.)

whereas the DNA remaining in the somatic cells contains only about 10% repeated sequences. Thus nearly half the DNA eliminated from the somatic cells consists of unique sequences.

Elimination of chromosomes in insects and higher organisms

Selective loss of DNA from somatic cells also occurs in various other animal species. Whole chromosomes are lost from the somatic cells of certain dipteran insects in the family Cecidomyiidae (gall midges, fungus gnats). During the early development of insect embryos no cytoplasmic division occurs, but the nuclei divide and distribute themselves throughout the egg cell to produce a syncytium with many nuclei in a common cytoplasm. In the gall midge, *Mayetiola destructor*, the first four cleavage divisions are normal, resulting in 16 nuclei. Two of these nuclei enter a cytoplasmic region called the "germ plasm" or "pole plasm," which is located at the extreme posterior end of the egg. These two nuclei, which are destined to form the

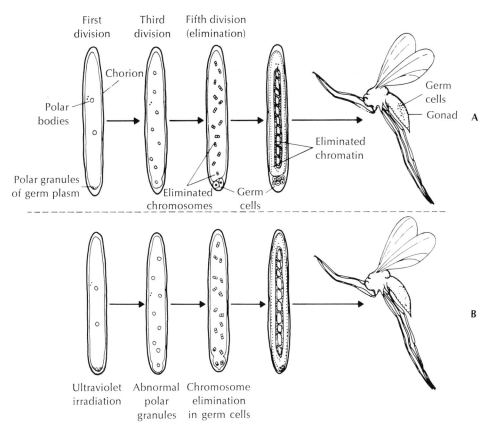

Fig. 6-8. Chromosome elimination in the gall midge *Mayetiola destructor*. **A,** Normal development. The nucleus of the fertilized egg undergoes four normal mitotic divisions. During the third and fourth divisions two nuclei become located in the posterior pole of the egg and the cytoplasm contracts to separate the nuclei from the rest of the egg. At the fifth division the remaining 14 nuclei lose 32 out of 40 chromosomes. The germ cells develop normally from the nuclei in the germ plasm, and the adult insect is fertile. **B,** Destruction of germ plasm. If the germ plasm is irradiated with ultraviolet light, all nuclei undergo chromosome elimination at the fifth division. The germ cells do not develop normally, and the adult animal is sterile. (With permission from Bullough, W. S. 1967. The evolution of differentiation. Copyright by Academic Press Inc. [Ltd.] London.)

animal's germ cells, do not divide at the fifth division, which is synchronous for the rest of the nuclei. The other 14 nuclei undergo an abnormal division during which they lose all but 8 of their original 40 chromosomes. The eliminated chromosomes do not travel to the spindle poles during mitosis but remain at the equator, coalesce, and then disintegrate. This series of events is depicted in Fig. 6-8, *A*. If nuclei are kept from entering the germ plasm region by ligating the egg (tying a fine thread or hair tightly around it), or if the germ plasm is irradiated as shown in Fig. 6-8, *B*, all 16 nuclei undergo the fifth division, and all lose chromosomes. This results in an adult animal that is sterile because of the lack of germ cells. Experiments of this kind suggest that the eliminated chromosomes contain information that is necessary for germ cell formation and perpetuation of the species, but that is not necessary for normal functioning of somatic cells.

Protection of the germ cell nuclei from chromosome elimination seems to be conferred by "polar granules" that are present in the cytoplasm of the germ plasm region. If these granules are displaced from the polar region by centrifugation, the nuclei that they become associated with in their new location do not undergo chromosome elimination. The polar granules contain RNA that is probably essential to their function, since ultraviolet irradiation of the pole plasm (a technique known to alter nucleic acids) destroys its ability to prevent chromosome elimination.

Some types of insects undergo complex patterns of chromosomal elimination in conjunction with sex determination. For example, the zygote of the dipteran fly *Sciara* contains two sets of autosomes and three X chromosomes. One of the X chromosomes is of maternal origin. Two are of paternal origin and result from the lack of separation of chromatids at the second meiotic division during spermatogenesis. During embryogenesis in *Sciara*, females lose one of the paternal X chromosomes, both from the germ cells and the somatic cells. Males undergo a similar loss from the germ cells but lose both paternal X chromosomes from the somatic cells. Thus germ cells of both sexes have identical chromosomal complements, whereas the somatic cells in the male contain one less X chromosome than those of the female. The decision of whether one or two X chromosomes are to be eliminated from the somatic cells is made in the egg and apparently is not the result of prior chromosome imprinting of the paternal Xs during spermatogenesis. This conclusion is drawn from the finding that there are two types of *Sciara* females. One type produces primarily daughters and the other primarily sons.

There are also rare examples of chromosome elimination occurring in vertebrates. In the bandicoot, a marsupial, female somatic cells eliminate one X chromosome and male somatic cells eliminate the Y chromosome, giving both sexes an XO karyotype in their somatic cells.

In the cases mentioned above, chromosome elimination could be responsible, at least in part, for the differentiation of somatic cells from germ cells, or for sex determination. However, it could not be the mechanism by which somatic cells become differentiated from one another, since all somatic cells in each case appear to have the same chromosomal complement.

The large size of the eukaryotic genome, particularly in higher plants and animals, makes it difficult to determine by direct analytical means whether some genes have been eliminated from differentiated cells. It therefore becomes necessary to ask in functional terms whether the genome of a differentiated cell retains all the genetic information necessary to form a complete fertile organism. Obviously, in those animals whose somatic cells lose major amounts of chromosomal material, this is not the case. However, such examples of chromatin loss appear to be the exception rather than the rule.

Experiments designed to test the totipotency of single somatic cells or nuclei have provided evidence that genes are not routinely lost as cells differentiate during development. An exciting experiment performed by Frederick C. Steward with cultured cells from carrots demonstrated that a single differentiated carrot cell contains all the information necessary to form a complete fertile adult carrot plant. Tiny

Evidence that the entire genome is retained in differentiated cells

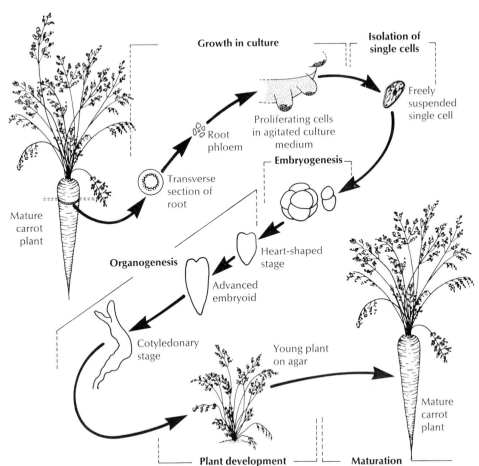

Fig. 6-9. Experiment designed to demonstrate the totipotency of carrot root cells. This diagram shows the stages by which an isolated root cell can be grown into a complete plant.

cubes of phloem from carrot root were incubated in a shaking culture system, and individual cells were released from the tissue. Single isolated cells were then cultured under a series of hormonal conditions for a number of weeks and eventually multiplied and formed embryolike structures, which later developed into complete carrot plants with normal flowers and seeds (Fig. 6-9). Such experiments have also been done with pith tissue from tobacco plants with similar results.

Differentiated cells in amphibia have also been shown to contain all the genetic information necessary for the complete development of a mature, fertile animal. However, an experimental approach different from that used with plant cells is required. The differentiated state in animal cells is stable, and it is only rarely that cells can be induced to change from one type into another. Therefore, in experiments designed to determine if differentiated animal cells contain a complete genome, nuclei are removed from the influence of their differentiated cytoplasm and transferred to a new cytoplasmic environment—that of an egg cell.

Because of the size of an amphibian egg and the fact that its nucleus is close to the surface, the host nucleus can be physically removed with a needle or functionally removed by irradiation with ultraviolet light without serious damage to the rest of the egg. In the latter case, where the irradiated host nucleus is still present, host and donor nuclei are distinguished by a visible genetic marker (number of nucleoli) so the experimenter can be certain that any development that occurs is in fact supported by the transplanted nucleus. There is no necessity for fertilization, since the donor nucleus is already diploid and the amphibian egg can be stimulated to begin development parthenogenetically.

In 1962 John Gurdon did an experiment in which nuclei from cells of the intestinal epithelium of a feeding tadpole of *Xenopus laevis* were transplanted into enucleated eggs. The steps involved in this experiment are shown in Fig. 6-10. The cells from which the donor nuclei were taken were obviously differentiated, with all the characteristics of intestinal epithelium, including a well-developed brush border on the surface facing the intestinal lumen. These nuclei are believed to have an advantage in that they come from cells that divide rapidly in the intact organism and therefore require less adjustment to accomplish the rapid divisions characteristic of the amphibian egg during early cleavage. A small percentage of these nuclei were able to support the development of normal-appearing tadpoles from functionally enucleated eggs. Some of these tadpoles were maintained in the laboratory until they formed fully fertile and apparently normal male and female adult frogs.

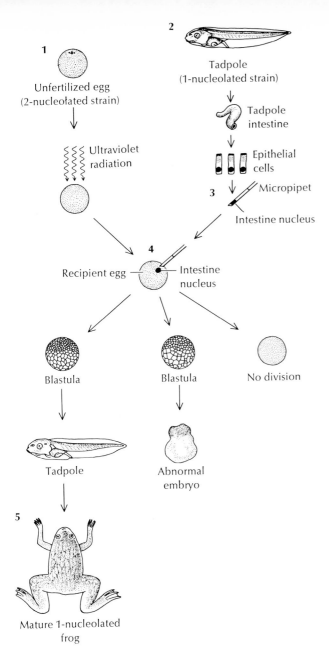

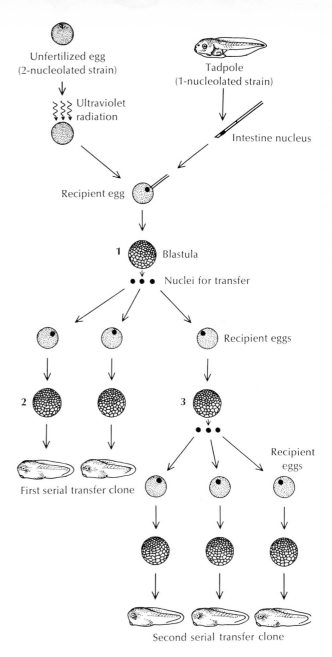

Fig. 6-10. Transplantation of somatic nucleus into frog egg. Transplant procedure starts with preparation of a frog's unfertilized egg (1) for receipt of a cell nucleus by destroying its own nucleus through exposure to ultraviolet radiation. Next, intestine is taken from a tadpole that has begun to feed (2), and cells are taken from its epithelial layer. A single epithelial cell is then drawn into a micropipet; the cell walls break (3), leaving the nucleus free. The intestine cell nucleus is transplanted into the prepared egg (4), which is allowed to develop. In about 1% of transplants the egg develops into a frog that has one nucleolus in its nucleus (5). (From Gurdon, J. 1968. Sci. Am. **219**(6):24. Copyright © 1968 by Scientific American, Inc. All rights reserved.)

Fig. 6-11. Serial transplantation of nuclei. Serial transplants involve the same first steps as the usual transplantation procedure. At the blastula stage (1) the cells of a transplant embryo are dissociated. The genetically identical nuclei from these cells are then transplanted into enucleated eggs, giving rise to a clone: a population of genetically identical individuals (2). The procedure can be continued indefinitely (3). (Modified from Gurdon, J. 1968. Sci. Am. **219**(6):24. Copyright © 1968 by Scientific American, Inc. All rights reserved.)

Nuclei from cells of adult frog tissue also have been shown to support development when transplanted into enucleated eggs. A major problem in using nuclei from adult tissues, however, is that they normally proceed through the cell cycle of growth and division too slowly to keep up with the cleavage divisions of the egg cytoplasm. Ronald Laskey and John Gurdon resolved this problem by placing differentiated cells into culture, a procedure that results in more rapid proliferation, and then making the nuclear transplants from cultured cells. When most differentiated cells are placed in culture, however, they tend to lose their differentiated appearance. It is therefore necessary to demonstrate by other means that the donor cells actually are differentiated. In this case adult skin cells were used and were shown to be differentiated by a number of criteria, including their ability to bind fluorescent antibody to keratin.

In all these nuclear transplantations the frequency of successful development is greater when a serial transfer procedure is used. In serial transfers the egg that receives the donor nucleus is allowed to cleave and form a blastula, which is then dissociated and used as a source of donor nuclei for a second transplantation. As shown in Fig. 6-11, this procedure can be continued indefinitely by taking nuclei from blastulae of each transfer generation and introducing them into enucleated eggs. It is thought that the greater success obtained with blastula nuclei is due to the fact that they are more capable of the rapid DNA synthesis required for early embryonic development than are nuclei taken directly from adult cells.

The fact that nuclei transplanted from differentiated cells are able to support development indicates that none of the genes that are needed for any type of specialized function in the entire animal are lost or permanently inactivated during differentiation. A similar conclusion is also implied from a very different type of experimental procedure, cell fusion. Cell fusion can be brought about by exposure of cells to inactivated Sendai virus, which causes an alteration in the cell membrane so that the membranes of two or more cells can fuse. The cytoplasmic components of the cells mix, and at mitosis the nuclear components also mix, generating a "hybrid" cell. (This process is described in greater detail in Chapter 11.) After fusion of mouse fibroblast (3T3) cells with rat hepatoma cells, some clones of hybrid cells were found to produce mouse serum albumin. Obviously the mouse fibroblast cells had retained the gene for albumin even though albumin is normally synthesized only in the liver. Under the right cytoplasmic influences in the hybrid cells the long-silent genes for mouse albumin can be expressed once again.

Finally, as has already been discussed on pp. 194 and 211, hybridization with mRNA or cDNA has been used to measure the number of copies of specific genes in various types of cells. In addition to showing that the genes that are expressed during a particular type of differentiation are not amplified, studies of this type have also shown that genes which are not being expressed in differentiated cells do not decrease in number in those cells.

Thus the available evidence indicates that unused genes remain present in differentiated cells and that these genes are still capable of future function. We must therefore conclude that cellular differentiation is not caused primarily by either an increase or a decrease in the number of copies of specific genes per cell.

GENE INACTIVATION

The evidence that has been presented, which argues against gene elimination as a routine occurrence during differentiation, also argues against completely stable and permanent gene inactivation as a routine mechanism of differentiation. There are, nevertheless, a number of interesting cases of apparently permanent gene inactivation that are known to occur in developing systems.

X chromosome inactivation

The inactivation of one X chromosome in each somatic cell of female mammals is a particularly interesting example of gene inactivation that has been studied in detail. It is widely believed that the purpose of the inactivation is to accomplish gene dosage compensation so that both males and females end up with the same ratio of functional X chromosomal genes to autosomal genes. However, it also appears that portions of both X chromosomes remain functional, since the complete absence of one X chromosome in a human female results in Turner's syndrome (described in Chapter 20), which involves a number of developmental abnormalities, including ovarian dysgenesis and failure of secondary sexual development.

In 1949 Murray Barr and co-workers described a distinct dark-staining region of chromatin that can be seen in the nuclei of female cells during interphase (Fig. 6-12). The number of these regions (termed Barr bodies) is always one less than the number of X chromosomes. Thus no Barr body is found in cells of XO (Turner's syndrome) females or of XY (normal) males; one Barr body is found in cells of XX (normal) females and XXY (Kleinfelter's syndrome) males; and two Barr bodies are found in XXX (superfemale) females.

In cells of female mammals the two X chromosomes, in contrast to other homologous chromosome pairs, do not replicate at the same time in the cell cycle. The timing of replication of individual chromosomes is determined by the following procedure. Cells synchronized with respect to their position in the cell cycle (see Appendix C) are labeled

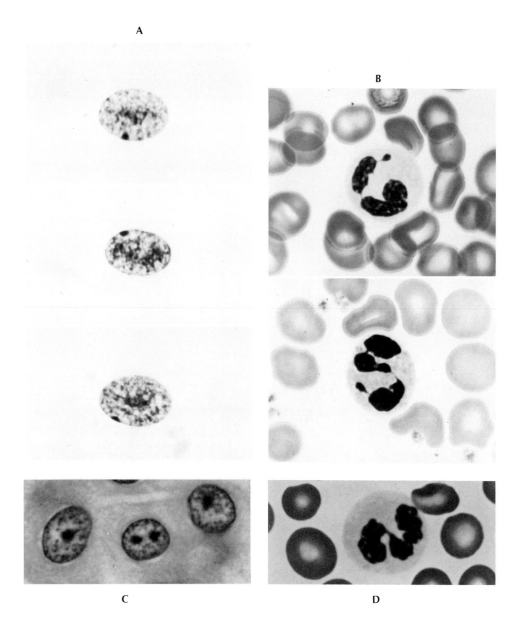

Fig. 6-12. Heterochromatic X chromosomes. **A,** Barr bodies adjacent to the nuclear membrane in the nuclei of cells from an XX female. **B,** Human neutrophils (type of white blood cell) from an XX female, showing the heterochromatic X chromosome as a "drumstick" attached to one of the lobes of the highly convoluted nucleus. **C** and **D,** Comparable nuclei from cells of XY males, showing absence of Barr body and drumstick. (**A** and **B** from Greenhill, J. P. 1965. Obstetrics, 3rd ed. W. B. Saunders Co., Philadelphia; **C** courtesy M. L. Barr from Dyson, R. D. 1978. Cell biology, 2nd ed. Allyn & Bacon, Inc., Boston; **D** courtesy M. L. Barr.)

with ³H-thymidine at different times during the S phase (the period in the cell cycle during which DNA synthesis occurs), collected at metaphase, treated with hypotonic saline solution, and flattened on microscope slides so that individual chromosomes can be distinguished. Autoradiography then permits identification of the chromosomes or parts of chromosomes that were synthesizing DNA at the time of exposure to the isotope. Specific regions of the various chromosomes replicate in a definite time sequence during the S phase. One entire X chromosome does not replicate until late in the S period, far out of phase with the other X chromosome. If cells labeled late in the S phase are examined by autoradiography during the following interphase, the Barr body is consistently found to be labeled. The X chromosome that forms the Barr body is therefore the one that replicates late.

In 1961 Mary Lyon developed a theory regarding X chromosome inactivation that has since been named the Lyon hypothesis. She observed that female mice heterozygous for coat color genes that are carried on the X chromosome have a variegated coat pattern (Fig. 6-13, *A*). She suggested that one of the X chromosomes is inactivated randomly in each cell early in development, resulting in the formation of clones (groups of cells descended from a common ancestor), each expressing only one of the X-linked coat color genes, as shown in Fig. 6-13, *B*.

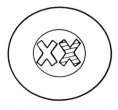

Female zygote with two X chromosomes

Early cleavage—both X chromosomes active in all cells

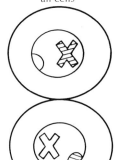

Random inactivation of one X chromosome in all cells of embryo

Fig. 6-13. A, Female mouse heterozygous for X-linked coat color gene, dappled. **B,** Schematic diagram depicting random X chromosome inactivation in XX female mammals. The zygote contains one maternal X chromosome (clear) and one paternal X chromosome (shaded). The paternal X chromosome is reactivated along with the rest of the chromosomes from the sperm nucleus soon after fertilization. Both X chromosomes remain active during cleavage stages. At about the time of implantation there is a random inactivation of one X chromosome in each cell of the developing embryo. The exact timing appears to vary from one species to another. (**A** from Lyon, M. 1966. Adv. Teratol. **1:**25.)

A

B

Conclusive proof of X chromosome inactivation was obtained by examination of genes for X-linked biochemical markers. The gene for glucose-6-phosphate dehydrogenase (G-6-PD) is carried on the X chromosome in humans and a number of other mammals. Mutant forms of the enzyme (isozymes) are known that can be distinguished from one another by electrophoresis. When skin cells of females heterozygous for different forms of G-6-PD are placed in culture and single cells are isolated and allowed to multiply to form clones, each clone of cells produces only one of the two possible forms of the enzyme (Fig. 6-14). Thus, although the individual that the cells were taken from produces both isozymes of G-6-PD, the descendants of any one cell from that individual make only one form of the enzyme, implying that the gene for the other isozyme has been stably inactivated.

A definitive demonstration that the late-replicating X is also the inactive X was obtained by examining the G-6-PD isozymes in cells cultured from the offspring of horse-donkey crosses. Donkey and horse X chromosomes are cytologically distinguishable (the donkey X is acrocentric; the horse X is submetacentric), and their G-6-PD enzymes are electrophoretically distinct. When cells of the female offspring of a horse-donkey cross are cultured and cloned, it is found that the horse X

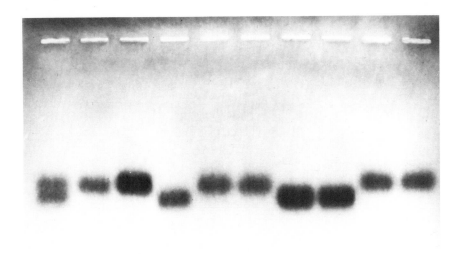

Fig. 6-14. Electrophoretic pattern of G-6-PD from sonicates of cultured cells. Samples were run singly, starting from the origin at the top of the figure. From left to right are the AB phenotype of the cell culture prior to cloning and the single bands of nine clones derived from the original cell lines. Variation in intensity of staining is due to inequality of enzyme concentration applied to the starch gel. (From Davidson, R. G., et al. 1963. Proc. Natl. Acad. Sci. U.S.A. **50**:484.)

chromosome is late replicating in clones that produce the donkey enzyme and vice versa.

In most mammals, X chromosome inactivation occurs at random; either the paternal or maternal chromosome may be inactivated. However, in marsupials, the paternal X chromosome is always inactivated. In mules and hinnies (the offspring of horse-donkey crosses) the donkey chromosome appears to be inactivated more frequently than the horse chromosome, although some investigators believe that this conclusion may be an artifact due to cells with active horse X chromosomes growing more readily in cell culture than cells with active donkey X chromosomes.

Once X chromosome inactivation has occurred, it is normally permanent. However, one exception to this rule appears to occur in female germ cells. Experiments in which germ cells of females heterozygous for variants of the X-linked enzyme G-6-PD were tested for the presence of hybrid enzyme suggest that mitotic germ cells contain only one active X chromosome. However, both X chromosomes appear to be functional in developing oocytes that have ceased mitotic division and entered meiosis. Furthermore, activities of X-linked enzymes have been reported to be twice as high in XX mouse oocytes as they are in XO mouse oocytes (unlike XO humans, XO mice are fully fertile). Such a difference in activities is not seen in the somatic cells of these animals.

As mentioned earlier, there appears to be a small region of the human X chromosome that is not inactivated in either homolog. Two types of evidence suggest this. First, red blood cells from females heterozygous for the X-linked blood group antigen Xg[a] cannot be separated into two types by means of agglutinating antibodies. All the cells appear to display the same antigens.

Second, individuals with abnormal numbers of X chromosomes (caused by nondisjunction or failure of chromosomes to separate during meiosis in the germ cells of their parents) exhibit certain abnormalities that would not be expected if all but one of the X chromosomes were completely inactivated in any individual possessing more than one X chromosome. Females with one X and no Y chromosome (Turner's syndrome) exhibit a number of malformations (described in Chapter 20), at least some of which appear to be related to a deficiency of genes carried on the short arm of the X chromosome. If the second X chromosome were completely inactivated in normal XX females, its total absence would be expected to have no effect on XO individuals. Some of the defects in Turner's syndrome may be caused by the apparent need for two active X chromosomes during oogenesis, but there are also other effects that appear to be totally unrelated to oogenesis or normal ovarian development. Individuals who possess one complete X chromosome plus the short arm of a second X chromosome do not express these abnormalities.

Similarly, males who possess two X plus one Y chromosome (Klinefelter's syndrome) also exhibit abnormalities (described in Chapter 20). Some of these defects can be traced to abnormal germ cell development, but many of the defects would not be expected if the supernumerary X chromosome were totally inactivated.

The time in development at which X inactivation occurs has been a point of considerable experimental interest. Inactivation appears to occur at a rather precisely fixed time in development for each species that has been studied. In the mouse, Barr bodies are not apparent early in cleavage but appear around the time of blastocyst formation. However, absence of a Barr body is not always definitive evidence that both X chromosomes are active.

Attempts have been made to calculate the time of X inactivation by examination of the phenotypes of females heterozygous for X-linked characteristics. Using simple probability theory, one can calculate the number of cells present at the time of inactivation from the frequency of heterozygotes that appear phenotypically homozygous. If n cells are present at the time of inactivation, the chances are $2 \times (1/2)^n$ that the same X chromosome will be inactivated in all cells and that a phenotypically homozygous individual will be obtained.*

Calculations of this kind were done with data on phenotypes of red blood cells of women known to be heterozygous for G-6-PD. The fact that they were genetically heterozygous was established from the G-6-PD phenotypes of their male offspring. The frequency of genetic heterozygotes whose red blood cells appear to be homozygous was found to be about 1%. This indicates that approximately eight red blood cell precursors were present at the time of X chromosome inactivation ($2 \times (1/2)^8 = 2 \times 1/256 =$ about 1%). Since some of these women exhibited heterozygous phenotypes in other tissues, the calculated values can be applied only to the number of red blood cell precursors and not to the total number of cells present in the embryo at the time of inactivation.

A second approach to determining the timing of X inactivation has been taken by Richard Gardner in collaboration with Mary Lyon. A single embryonic mouse cell heterozygous for an X-linked coat color

---

*In each cell the probability that either X^A or X^B will be inactivated is the same and equals one half. For two cells the probability that X^A will be inactivated in both is $1/2 \times 1/2$ or $(1/2)^2$. For three cells the probability is $1/2 \times 1/2 \times 1/2$ or $(1/2)^3$. For n cells the probability is $(1/2)^n$. The probability that a heterozygous individual will appear homozygous is $2 \times (1/2)^n$, since there are two possible homozygous phenotypes — all cells expressing A, and all cells expressing B.

was injected into the blastocoel of a second embryo that would normally develop into an animal with a single coat color distinct from either of the colors produced by the genes in the donor cell. If the donor cell had already undergone inactivation before injection, all its offspring would express the same coat color, and the resulting mouse would exhibit two colors, one from the host and one from the remaining active X chromosome of the donor cell. If the donor cell had not undergone inactivation, its offspring would undergo random inactivation after injection, and the resulting mouse would exhibit three coat colors — one from the host and two others from the two genes carried by the donor cell. Cells from mouse embryos of different ages were used as donors, and the results obtained indicated that inactivation had not occurred by 3½ days of development (early blastocyst stage) but that it occurred shortly thereafter.

It has been shown experimentally that X inactivation is permanent, at least in somatic cells. Most cells have the option of either synthesizing their own purines de novo or using preformed purines from their environment through use of the X-linked enzyme hypoxanthine-guanine phosphoribosyltransferase (HGPRT). When cells from an individual heterozygous for a defect in HGPRT are grown in a medium containing 8-azaguanine, those cells which express the $HGPRT^+$ gene incorporate the analog and are killed, and the cells expressing the $HGPRT^-$ gene survive. If these $HGPRT^-$ cells are then transferred to a medium in which de novo purine synthesis is blocked, they are unable to use an external source of purine (hypoxanthine) for growth because they lack functional HGPRT. This system can be used to generate a strong selective advantage for any $HGPRT^-$ cells that might be able to reverse the X chromosome inactivation and express the $HGPRT^+$ gene. However, even under severe selective pressure the X chromosome containing the $HGPRT^+$ gene apparently never becomes reactivated.

Susumu Ohno and his co-workers have found a mutation in the mouse that seems to be concerned with X inactivation. The O^{hv} mutation is located on the X chromosome, and in heterozygotes the X chromosome containing that mutation preferentially remains active. This can be shown by the expression of other genetic markers linked to O^{hv} on the same X chromosome. For example, individuals with the genes *Tfm*, O^{hv}, and *Blo* on one X chromosome and wild-type genes on the other X chromosome mostly express the *Blo* coat color and the *Tfm* phenotype. (*Tfm* stands for testicular feminization, which is discussed in Chapter 20.)

In a speculative model for X inactivation, Ohno has proposed that the O^{hv} locus is a receptor site recognized by molecules responsible for the X chromosome remaining active. The key concept of his theory is that all X chromosomes except those specifically protected from inactivation are inactivated. The protection is said to be due to a small number of "activating" molecules that exhibit high cooperativity in their binding to the receptor site so that only one X is protected by them. Chromosomes carrying the O^{hv} mutation are claimed to have receptors with higher than normal affinity for the activating molecules and thus are preferentially activated. When both X chromosomes have receptor sites with equal affinity for the activating molecules, the chances of either becoming protected are equal, and random X inactivation is predicted by the model.

Chromosome inactivation in male mealy bugs

Another system in which permanent chromosome inactivation occurs is the male mealy bug. At the sixth cleavage division of the mealy bug embryo the entire paternal set of chromosomes is inactivated in some of the embryos. These embryos develop into males, and those which do not undergo chromosomal inactivation become females.

The inactivated paternal chromosomes are retained in the genome of the males but are eliminated during spermatogenesis. Thus the

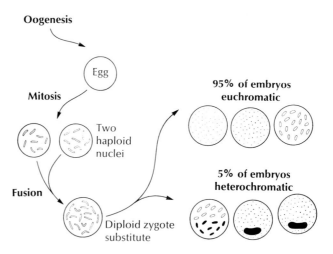

Fig. 6-15. Development of embryos in parthenogenetic species of mealy bug. Following meiosis the haploid egg nucleus divides to yield two daughter nuclei, which unite and produce a diploid zygote substitute. About 5% of the embryos produced undergo heterochromatization of one chromosome set. (Modified from Brown, S. W., and V. Nur. 1964. Science **145:**130. Copyright 1964 by the American Association for the Advancement of Science.)

sperm cells that are produced contain only chromosomes of maternal origin.

The process of sex determination by chromosomal inactivation appears to be controlled entirely by the egg rather than the sperm. Evidence for this comes from certain parthenogenetic species in which unfertilized eggs divide to produce two haploid daughter cells, which then fuse and undergo development as completely homozygous diploid individuals (Fig. 6-15). Some of the embryos produced in this manner undergo chromosome inactivation and become males, whereas others retain two active sets of chromosomes and become females.

It is clear that the cases of gene inactivation discussed are not directly responsible for cellular differentiation. They involve large blocks of genes, ranging in size from most of the X chromosome in female mammals to a complete haploid genome in the male mealy bug. However, they do represent a stable inactivation of genes that is "inherited" from one cell generation to another. In that sense they are similar to the stable inactivation of unused genes that appears to occur on a smaller scale in cellular differentiation. Thus these large-scale inactivations may be suitable for study as models of the more limited inactivation observed in differentiation.

Relationship between inactivation of genes and differentiation

From the evidence presented regarding gene amplification, gene elimination, and gene inactivation, it appears that physical modification of the genome is not the primary mechanism responsible for cellular differentiation. Changes of the types that have been discussed occur only in a limited number of specialized cases such as production of large reserves of rRNA during oogenesis, differentiation of somatic cells from germ cells, compensation of gene dosage between sex chromosomes and autosomes, and sex determination. Also, many of them occur only in a limited number of species.

One case in which an altered genome is associated with gene expression occurs in the immune system, where partial antibody-coding sequences that are widely separated from one another in germ line DNA are joined together in the DNA of anti-body-producing cells. This gene-splicing phenomenon, which has been observed only in a few specialized systems, has led to some interesting speculative models. However, substantial evidence suggests that much of the control of gene expression in differentiation occurs along the pathway of information flow from an unaltered genome to functional protein molecules. In the next three chapters we will examine the potential points of control along that pathway.

DIFFERENTIAL EXPRESSION OF AN UNALTERED GENOME

BIBLIOGRAPHY
Books and reviews

Beermann, W., ed. 1972. Developmental studies on giant chromosomes. Results and problems in cell differentiation. Vol. 4. Springer-Verlag, Inc., New York.

Brown, D. D. 1973. The isolation of genes. Sci. Am. **229**:20 (Aug.).

Brown, D. D., and I. Dawid. 1968. Specific gene amplification in oocytes. Science **160**:272

Brown, S. W., and U. Nur. 1964. Heterochromatic chromosomes in the coccids. Science **145**:130.

Cattanach, B. M. 1975. Control of chromosome inactivation. Annu. Rev. Genet. **9**:1.

Chandra, H. S., and S. W. Brown. 1975. Chromosome imprinting and the mammalian X-chromosome. Nature **253**:165.

Chapman, V. M., J. D. West, and D. A. Adler. 1977. Genetics of early mammalian embryogenesis. *In* M. I. Sherman, ed. Concepts in mammalian embryogenesis. M.I.T. Press, Cambridge, Mass.

Gall, J. G. 1969. The genes for rRNA during oogenesis. Genetics **61**(supp.):121.

Gartler, S. M., and R. J. Andina. 1976. Mammalian X-chromosome inactivation. Adv. Hum. Genet. **7**:99.

Gurdon, J. B. 1974. The control of gene expression in animal development. Harvard University Press, Cambridge, Mass.

Hood, L., H. V. Huang, and W. J. Dreyer. 1977. The area code hypothesis: the immune system provides clues to understanding the genetic and molecular basis of cell recognition during development. J. Supramol. Struct. **7**:531.

Lewin, B. 1974. Gene expression. Vol. 2. Wiley-Interscience, New York.

Lyon, M. F. 1972. X-chromosome inactivation and developmental patterns in mammals. Biol. Rev. **47**:1.

Miller, O. L., Jr. 1973. The visualization of genes in action. Sci. Am. **228**:34 (March).

Steward, F. C. 1970. From cultured cells to whole plants: the induction and control of their growth and morphogenesis. Proc. R. Soc. Lond. (Biol.) **175**:1.

Tartof, K. D. 1975. Redundant genes. Annu. Rev. Genet. **9**:355.

Weatherall, D. J., and J. B. Clegg. 1976. Molecular genetics of human hemoglobin. Annu. Rev. Genet. **10**:157.

Selected original research articles

Alt, F. W., et al. 1978. Selective multiplication of dihydrofolate reductase genes in methotrexate-resistant variants of cultured murine cells. J. Biol. Chem. **253**:1357.

Bantock, C. R. 1970. Experiments on chromosome elimination in the gall midge, *Mayetiola destructor.* J. Embryol. Exp. Morph. **24**:257.

Boveri, T. 1899. Die Entwicklung von *Ascaris megalocephala* mit besonderer Rucksicht auf die Kern verhaltnisse. Festschrift F. C. von Kupffer, Jena.

Dawid, I. B., D. D. Brown, and R. H. Reeder. 1970. Composition and structure of chromosomal and amplified ribosomal DNAs of *Xenopus laevis.* J. Mol. Biol. **51**:341.

Drews, U., et al. 1974. Genetically directed preferential X-activation seen in mice. Cell **1**:3.

Gardner, R. L., and M. F. Lyon. 1971. X-chromosome inactivation studied by injection of a single cell into the mouse blastocyst. Nature **231**:385.

Geyer-Duszynska, I. 1959. Experimental research on chromosome elimination in Cecidomyidae (Diptera). J. Exp. Zool. **141**:391.

Gurdon, J. B. 1962. Adult frogs derived from the nuclei of single somatic cells. Dev. Biol. **4**:256.

Kedes, L. H., and M. L. Birnsteil. 1971. Reiteration and clustering of DNA sequences complementary to histone mRNA. Nature New Biol. **230**:165.

Lyon, M. F. 1961. Gene action in the X-chromosome of the mouse. Nature **190**:372.

Paul, J., et al. 1973. The globin gene: structure and expression. Cold Spring Harbor Symp. Quant. Biol. **38**:885.

Suzuki, Y., L. P. Gage, and D. D. Brown. 1972. The genes for silk fibroin in *Bombyx mori.* J. Mol. Biol. **70**:637.

Tobler, H., K. D. Smith, and H. Ursprung. 1972. Molecular aspects of chromatin elimination in *Ascaris lumbricoides.* Dev. Biol. **27**:190.

Wellauer, P. K., et al. 1976. The arrangement of length heterogeneity in repeating units of amplified and chromosomal ribosomal DNA from *Xenopus laevis.* J. Mol. Biol. **105**:487.

CHAPTER 7

Control of gene expression at the transcriptional level

☐ In the previous chapter, evidence was presented that most cases of cellular differentiation do not involve significant changes in the genome. It is therefore likely that preferential synthesis of specific gene products in differentiated cells is due primarily to differential expression of an unvarying set of genes. This chapter deals with controls of gene expression that operate at the level of transcription. Differential control of DNA-dependent RNA synthesis is known to be the primary mechanism responsible for differential gene expression in bacteria. Transcriptional control also has been shown to be of major importance in eukaryotic cells. However, many other possible means for controlling gene expression occur in eukaryotic cells, including selective processing of nuclear RNA, unequal transport of various RNA molecules to the cytoplasm, differential stabilization of specific mRNAs, and preferential translation of certain mRNA molecules. The relationship among these potential points of control and the relative importance of each in the overall regulation of gene expression have not yet been fully worked out for any type of cellular differentiation.

This chapter begins with evidence that transcriptional control is an important aspect of the overall process of cellular differentiation in various types of eukaryotic organisms. Specific molecular mechanisms thought to be involved in transcriptional control are then analyzed, and the Britten and Davidson model of cellular differentiation, which is based on controls operating at the transcriptional level, is discussed.

Although some of the evidence that is presented in this chapter for the role of transcriptional control in cellular differentiation is highly specific, many of the older data are derived from experiments that show differences in cytoplasmic mRNA populations among differentiated cells. Such data do not reliably distinguish between controls operating at the level of transcription and those which involve events occurring after transcription has been completed. The posttranscriptional steps that are involved in mRNA formation are analyzed in Chapter 8. Gene expression can also be controlled to some extent by mechanisms that operate at the level of protein synthesis, or it can even be controlled during the formation of functional catalytic or structural protein molecules from the newly synthesized polypeptide chains. These potential levels of control are considered in Chapter 9.

Clarity of presentation demands that these various processes and potential control mechanisms be presented one at a time in this and the next two chapters. However, it is important for the reader to keep

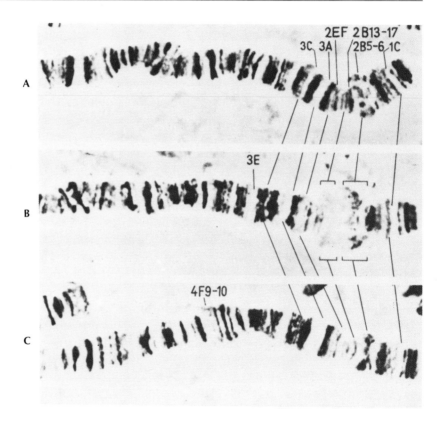

Fig. 7-1. Puffing sequence of the distal end of a *Drosophila melanogaster* X chromosome. **A,** Puff stage 1 (110-hour larva, female), **B,** Puff stage 8 (120-hour larva, female), **C,** Puff stage 10 (0-hour prepupa, female), **D,** Puff stage 15 (4-hour prepupa, female), **E,** Puff stage 18 (8-hour prepupa, female), **F,** Puff stage 20 (10-hour prepupa, male), **G,** Puff stage 21 (12-hour prepupa, female). (From Ashburner, M. 1969, Chromosoma **27**:47.)

in mind the fact that these events all occur more or less simultaneously in real life as components of a highly complex and precisely integrated biological system.

   Strong evidence for control of gene expression at the transcriptional level comes from studying the polytene chromosomes that occur in certain larval tissues of some insects (Chapter 6). When examined cytologically, these chromosomes appear to be composed of a series of differentially staining band and interband regions (Fig. 6-6). During normal larval development, or in response to certain experimental

EVIDENCE FOR DIFFERENTIAL TRANSCRIPTION
Polytene chromosome puffs in insect larvae

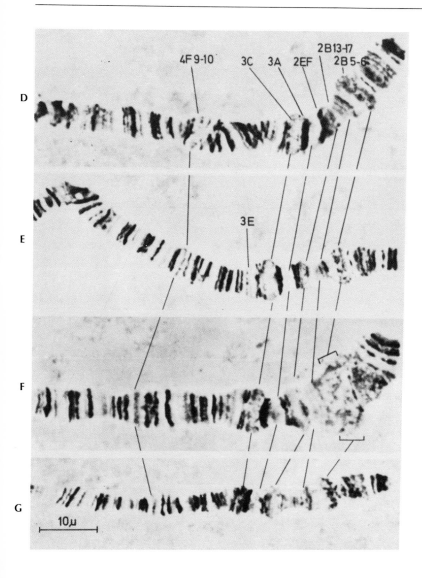

conditions such as hormone treatment or exposure to high tempera- ture, some of the bands assume a "puffed" appearance in which dif- fuse material extends away from the axis of the chromosome. It has been demonstrated by autoradiography of chromosomes labeled with radioactive RNA precursors that puffs are sites of active RNA syn- thesis.

The pattern of chromosome puffs seen varies both with the type of larval tissue and its stage of development. The sequential appearance of puffs suggests that sequential transcription of RNA from specific genes is occurring during development. The puffing sequence found on a portion of the X chromosome of *Drosophila* salivary gland tissue during prepupal development is shown in Fig. 7-1. Major changes in patterns of puffing also occur when larval tissues are exposed to specific hormones that control moulting and metamorphosis (Chap- ter 22).

The RNA product from Balbiani ring 2 (BR2), one of the large puffs in the salivary gland of *Chironomus tentans*, has been isolated after microdissection of the Balbiani ring. The RNA extracted from this puff is a single molecular species of about 15 to 35×10^6 daltons and hybridizes only with the BR2 locus when hybridized in situ with *C. tentans* chromosomes (Fig. 7-2). It has been shown recently that in- duction of specific puffs by heat treatment in *Drosophila* salivary glands is followed by the synthesis of approximately the same number

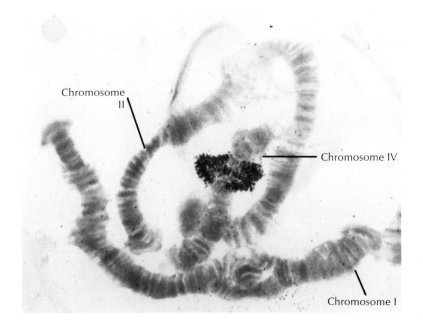

Fig. 7-2. Autoradiograph of cytological hybridization between polytene chromosomes of *Chironomus tentans* salivary glands and RNA transcribed from the BR2 locus of chromosome IV. Heavy label is located in the BR2 region. Only chromosomes I, II, and IV are shown. (With permission from Lambert, B. 1972. J. Mol. Biol. **72**:65. Copyright by Academic Press Inc. [London] Ltd.)

of new proteins as there are induced puffs. Also, the appearance of salivary secretory proteins during *Drosophila* development has been correlated with puffing activity in regions of the chromosomes known from genetic studies to carry the genes for these products. Thus there appears to be a 1:1 correlation between puffs and active expression of specific genes in polytene chromosomes.

Molecular hybridization techniques have been used to examine the specificity of transcription in animal tissues. RNA is isolated from differentiated tissues, or else chromatin preparations from differentiated tissues are transcribed by RNA polymerase in vitro. The tissue specificity of the RNA is then determined by the technique of competition hybridization, which is diagrammed in Fig. 7-3. Basically the pro-

Evidence from DNA-RNA hybridization studies

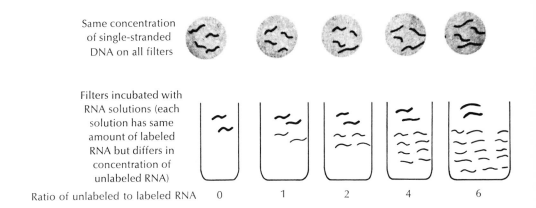

Same concentration of single-stranded DNA on all filters

Filters incubated with RNA solutions (each solution has same amount of labeled RNA but differs in concentration of unlabeled RNA)

Ratio of unlabeled to labeled RNA   0   1   2   4   6

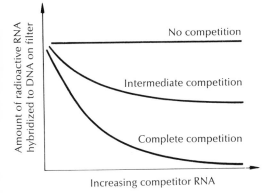

Amount of radioactive RNA hybridized to DNA on filter

No competition

Intermediate competition

Complete competition

Increasing competitor RNA

Fig. 7-3. Competition assay for RNA sequences. Filters with single-stranded DNA are incubated with a series of preparations of RNA. Each preparation contains the same amount of radioactively labeled RNA but a different amount of unlabeled competitor RNA. The graph shows three possibilities. If the labeled and unlabeled RNA preparations share no sequences in common, they do not compete for sites on the DNA, and the same amount of labeled RNA is retained on all filters, irrespective of the concentration of unlabeled RNA. At the other extreme, if the labeled and unlabeled RNA preparations are identical, the unlabeled RNA competes for all the sites on the DNA that are complementary to the labeled RNA. In an intermediate case, in which the unlabeled RNA contains some, but not all, of the sequences in the labeled RNA, there is partial competition. The extent of displacement of the labeled RNA is proportional to the fraction of its sequences that are shared in common with the unlabeled RNA. (Modified from Lewin, B. 1974. Gene expression-II. Eucaryotic chromosomes. John Wiley & Sons, Inc., New York, and Markert, C. L., and H. Ursprung. 1971. Developmental genetics. Prentice-Hall, Inc., Englewood Cliffs, N.J. Reprinted by permission.)

cedure involves determining how effectively an unlabeled RNA preparation competes with a fixed amount of radioisotopically labeled RNA for specific sites on a fixed amount of denatured DNA. If the unlabeled RNA has sequences similar to those in the labeled RNA, increasing the concentration of the unlabeled RNA increases the competition for sites complementary to those sequences and reduces the amount of labeled RNA that is hybridized. This can be seen clearly in the lowest curve in Fig. 7-4, where unlabeled RNA from prism-stage sea urchin embryos competes effectively with labeled RNA from the same source for sites on sea urchin DNA. If none of the sequences in the labeled RNA are represented in the unlabeled RNA, there is no competition for DNA sites and the amount of label hybridized to the DNA is independent of the concentration of the unlabeled RNA. This is evident in

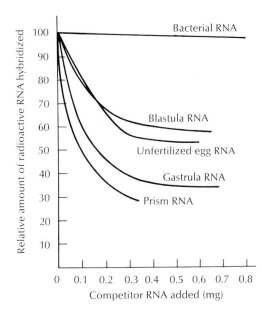

Fig. 7-4. Example of competition hybridization. Labeled RNA from sea urchin embryos at the prism stage (between the gastrula and pluteus) is shown in competition with unlabeled RNA from various sources. *Top line,* Unlabeled bacterial RNA—no competition. *Middle lines,* Unlabeled RNA from unfertilized sea urchin eggs or from blastula or gastrula stages—intermediate competition. *Bottom line,* Unlabeled RNA from prism larva—competition is greater than with other RNAs. Lack of complete displacement of the labeled RNA is due to at least two factors: (1) the labeled RNA was collected after a 1-hour pulse and is probably enriched in short-lived nuclear sequences that are poorly represented in the unlabeled RNA, and (2) there is some nonspecific retention of labeled RNA to the DNA-agar mixture used in these experiments. (From Markert, C. L., and H. Ursprung. 1971. Developmental genetics. Prentice-Hall, Inc., Englewood Cliffs, N.J. Reprinted by permission.)

the top line in Fig. 7-4, which shows the lack of competition between sea urchin prism-stage RNA and a control preparation of unlabeled bacterial RNA. Since most assays involve RNA preparations that contain some competing sequences, the usual concern is with how effective the competition is. Intermediate levels of competition with labeled prism-stage RNA can be seen in Fig. 7-4 for unlabeled RNA from unfertilized eggs and from blastula and gastrula stage embryos.

Competition hybridization experiments have also been used to demonstrate that tissue specificity is at least partially preserved when isolated chromatin is transcribed in vitro by bacterial RNA polymerase. For example, one early experiment demonstrated that labeled RNA transcribed in vitro from bone marrow chromatin was competed with better by RNA extracted from bone marrow cells than by RNA extracted from thymus cells (Fig. 7-5). When thymus chromatin was transcribed in vitro, the reverse relationship was found.

It was discovered later that early hybridization experiments such as those diagrammed in Figs. 7-4 and 7-5 were done under conditions that allowed hybridization of RNA only to repeated DNA sequences (described in Chapter 5). Thus the tissue and stage specificity demonstrated in those figures is in transcription from repeated sequences. This finding is of interest in itself in light of the postulated regulatory

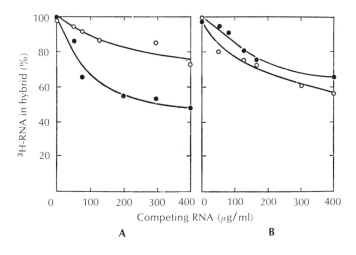

Fig. 7-5. Competition hybridization between labeled RNA transcribed in vitro from bone marrow chromatin (**A**) or thymus chromatin (**B**) with unlabeled RNA extracted from thymus (O——O) or bone marrow (●——●). Note that these experiments were done under conditions in which only repeated sequences formed hybrids. Also note that in vitro transcription generated many sequences that were not effectively competed with by the preparations of RNA extracted from tissues. (With permission from Paul, J., and R. S. Gilmour. 1968. J. Mol. Biol. **34**:305. Copyright by Academic Press Inc. [London] Ltd.)

role for repeated sequences, but it tells us little about the specificity of transcription of protein-coding sequences, which occur primarily as unique-sequence DNA.

However, more recent work has demonstrated convincingly that there is indeed tissue-specific transcription of particular protein-coding sequences. John Paul and his co-workers have examined the transcription of the globin genes (which code for the protein portion of hemoglobin) from chromatin of different tissues. Chromatin preparations from fetal mouse liver (an erythropoietic or red blood cell-producing tissue) and mouse brain tissue were transcribed in vitro with *E. coli* RNA polymerase. The transcription products of these reactions were tested for ability to hybridize to DNA copies of the globin message (globin cDNA) that had been made from purified globin mRNA with reverse transcriptase. As shown in Fig. 7-6, the RNA that had been transcribed from the fetal liver chromatin hybridized to globin cDNA, whereas that transcribed from brain chromatin did not. Numerous control experiments were done to show that the globin message was not present in the original chromatin preparation from fetal liver and that it had indeed been transcribed in vitro. Thus this experiment demonstrated conclusively not only tissue-specific transcription of the gene for globin but also that the transcriptional specificity was retained in isolated chromatin.

Transcription of the globin message has also been examined in Friend leukemia cells, a line of cultured erythroid cells that can be experimentally induced to produce hemoglobin. The amount of RNA in the nuclei of these cells that is capable of hybridizing with globin cDNA was found to increase five to six times following induction, thus

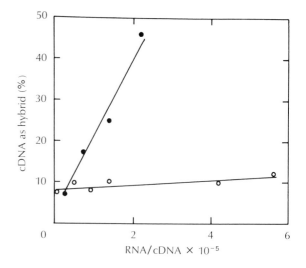

Fig. 7-6. Hybridization of RNA transcribed from mouse fetal liver chromatin (●———●) and mouse brain chromatin (○———○) with globin cDNA. A fixed amount of cDNA was hybridized with variable amounts of RNA. Hybridization was assayed by measuring the amount of cDNA that became resistant to a nuclease that specifically degrades single-stranded DNA. RNA transcribed from fetal liver chromatin contains many sequences that form hybrids with globin cDNA and protect it from the nuclease, whereas RNA transcribed from brain contains few such sequences. (From Paul, J., et al. 1973. Cold Spring Harbor Symp. Quant. Biol. **38**:885.)

providing evidence for specific message transcription as a result of induced expression of differentiated properties. Similar data have also been obtained for transcription of the ovalbumin gene in cells of the chicken oviduct in response to stimulation by estrogen.

The data in Fig. 7-4 suggest a pattern of changing RNA synthesis in developing sea urchin embryos. Recent studies have verified that different mRNA sequences are expressed when such changes occur. These experiments, which were designed to detect both quantitative and qualitative changes in mRNA composition have already been discussed in a different context in Chapter 5. The extent of gene expression in sea urchins at different developmental stages was determined by hybridizing mRNA isolated from purified polysomes with unique-sequence DNA that had been separated from the rest of the sea urchin genome. As shown in Fig. 5-22, the largest degree of message sequence complexity was found in oocyte mRNA, which represents about 6% of the single-copy sequences in the sea urchin genome. The portion of the genome represented in mRNA decreased as the embryos progressed to the blastula, gastrula, and pluteus stages and was smallest in adult tissues. Also, new mRNA species not present in gastrula mRNA appeared as development progressed.

Unfortunately, data based on mRNA changes do not distinguish between selective transcription and selective processing of transcripts. Recent studies of nuclear RNA sequences suggest that selective posttranscriptional processing may be quite important in eukaryotic cells. In the sea urchin, for example, blastula-stage mRNA contains many sequences that are not expressed in adult tissues as cytoplasmic mRNA. However, small amounts of essentially all of the blastula-stage mRNA sequences have been found in the nuclear RNA of cells in adult tissues. Current evidence suggests that genes coding for mRNAs in the "complex" (only a few copies per cell) and "moderately prevalent" (hundreds of copies per cell) classes are transcribed nonspecifically in all types of cells and that cytoplasmic mRNA formation is controlled primarily by selective processing of the transcripts (Chapter 8).

Transcriptional control is well documented, however, for the "superprevalent" class of mRNA sequences (more than 10^4 copies per cell) and also for certain repeated DNA sequences whose transcripts never become cytoplasmic mRNA. These transcripts, which are capable of forming RNA:RNA duplexes, have been suggested as possible control elements in selective mRNA processing. Transcriptional control is thus clearly important, but it is possible that the controls described in this chapter may prove to be valid only for RNA sequences that code for proteins made in unusually large amounts and for special classes of regulatory RNA molecules. This possibility is analyzed in detail in a review by Davidson and Britten (1979) cited at the end of this chapter.

MECHANISMS FOR CONTROL OF TRANSCRIPTION

If differential gene expression is due at least in part to differential synthesis of RNA, how is this transcriptional control achieved? One can envision two main mechanisms by which differential transcription might occur. The first mechanism involves alterations in the RNA polymerase enzyme or its associated factors that would result in specific transcription at certain gene loci only. For example, different σ factors (Chapter 4) might be required for the transcription of different sets of genes. The second type of mechanism by which specificity of transcription might be accomplished is control of template activity at specific gene loci. That is, specific alterations in chromatin could occur that would render only certain parts of the genome available for transcription by RNA polymerase. A combination of these two mechanisms could also be involved.

Role of RNA polymerases in specificity of transcription

We will first examine the evidence for transcriptional control by means of RNA polymerase. The enzymes found thus far in eukaryotic cells are large and complex, as is the case with bacterial RNA polymerases. They have molecular weights of about 500,000 and consist of a number of subunits. Three distinct RNA polymerase activities are found in eukaryotic nuclei, but they seem to distinguish only among genes coding for different classes of RNA (18S/28S rRNA, mRNA, 5S rRNA and tRNA). In addition to their transcriptional specificity, these enzymes can also be distinguished by their localization within the nucleus, their drug sensitivities, and their ionic requirements for optimal activity. Major properties of these three enzymes are summarized in Table 7-1.

RNA polymerase I is located in the nucleolar fraction of cells. It is insensitive to the drug α-amanitin and is inhibited by low concentra-

Table 7-1. Properties of eukaryotic RNA polymerases*

	I	II	III
Nuclear localization	Nucleolus	Nucleoplasm	Nucleoplasm
Sensitivity to α-amanitin	None	Low concentrations ($10^{-9}-10^{-8}$M)	High concentrations ($10^{-5}-10^{-4}$M)
Mn^{2+} : Mg^{2+} activity ratio	1:1	5:1 to 50:1	2:1 to 3:1
Ionic strength (optimum)	0.05	0.10 to 0.15	0 to 0.2
Type of RNA synthesized	rRNA (18S and 28S)	hnRNA (message precursor)	tRNA, 5S RNA

*Modified from Chambon, P. 1975. Annu. Rev. Biochem. **44**:613, and Rutter, W. J., et al. 1974, *In* J. Paul, ed. Biochemistry of cell differentiation. International review of biochemistry. Vol. 9. University Park Press, Baltimore.

tions of actinomycin D. Its activity is favored by Mg^{2+} and low ionic strength. When cells are incubated under these conditions and the other two RNA polymerases are blocked with α-amanitin, the RNA that is transcribed has the characteristics of 18S and 28S rRNA. It has the high G-C content characteristic of rRNA, and it is efficiently displaced from DNA by purified rRNA in competition hybridization experiments. In addition, autoradiographic studies show that such RNA is made in the nucleolus, which is known to be the site of rRNA synthesis.

RNA polymerase II is located in the nucleoplasmic fraction of cells. It is sensitive to α-amanitin and is inhibited by actinomycin D, but only at high concentrations. This enzyme functions most efficiently in the presence of Mn^{2+} and at high ionic strength. When cells are incubated in conditions favorable to the activity of RNA polymerase II, the RNA that is transcribed has a base composition like that of total cellular DNA and it is competed with efficiently by total nuclear RNA but not by rRNA in competition hybridization experiments. The labeled RNA made by these cells is seen over the entire nucleus after autoradiography and is the precursor of cytoplasmic mRNA.

RNA polymerase III is a minor activity that has been found in the nucleoplasm of many, but not all, tissues. It is insensitive to low concentrations of α-amanitin but is inhibited by larger amounts. Its activity is favored by Mn^{2+} and is relatively independent of ionic strength. This enzyme appears to be responsible for the synthesis of small RNA molecules (e.g., 5S rRNA, tRNA). Synthesis of these RNAs is inhibited by α-amanitin only at high concentrations, and the inhibition curves for their synthesis are identical to the curves for inhibition of RNA polymerase III activity.

Thus different RNA polymerase molecules seem to provide a general control with regard to the class of RNA that is synthesized, but there is little evidence that they exert any fine control over RNA transcription. No σ-like factors that might alter the specificity of polymerases have been isolated from eukaryotic cells, although there are some reports of protein fractions that stimulate RNA polymerase activity. One indirect argument against altered RNA polymerase specificity as a mechanism for cellular differentiation in eukaryotes is the fact that differences in transcription are still obtained when chromatin from different tissues is transcribed in vitro with RNA polymerase from *E. coli*.

If RNA polymerase is not directly responsible for transcriptional specificity, this role must reside in the template itself. It has been shown experimentally that chromatin is a much less efficient tem-

Role of template availability in specificity of transcription

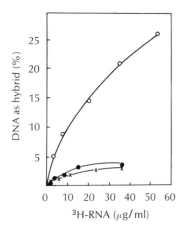

Fig. 7-7. Kinetics of hybridization to rabbit embryo DNA of labeled RNA transcribed in vitro from deproteinized rabbit embryo DNA (O——O), rabbit bone marrow chromatin (●——●), or rabbit thymus chromatin (x——x). The filters were loaded with 1 μg of denatured DNA, and labeled RNA preparations were incubated with the filters at the concentrations shown. Unlike competition experiments, this experiment asks how many of the sites in a restricted amount of DNA can be hybridized by the RNA under saturation conditions. Saturation values were as follows: thymus chromatin, 4.7%; bone marrow chromatin, 6.8%; and rabbit DNA, 47%. The RNA transcribed from naked DNA contains sequences capable of hybridizing with far more of the DNA sequences than does the RNA transcribed from either of the chromatins. (With permission from Paul, J., and R. S. Gilmour. 1968. J. Mol. Biol. **34:**305. Copyright by Academic Press Inc. [London] Ltd.)

plate for RNA synthesis than is naked DNA. In addition, the RNA transcribed in vitro from chromatin hybridizes to a much smaller percent of the genome than that transcribed from naked DNA (Fig. 7-7). This and similar observations have demonstrated conclusively that the molecules associated with DNA in chromatin profoundly influence its transcription.

Numerous experiments have been done to determine just which components of chromatin are responsible for specificity of transcription. When histones are selectively removed from chromatin, the diversity and rate of transcription are both increased significantly, although the diversity of the RNA sequences that are made is still somewhat less than that seen with naked DNA. Further evidence for suppression of RNA synthesis by histones has been obtained from chromatin reconstitution experiments. In such studies chromatin is separated into DNA, histones, and nonhistone chromosomal proteins (NHC proteins) in conditions of high salt plus urea. The individual components are remixed in high concentration of salt and urea and then dialyzed against successively lower concentrations until a normal physiological salt concentration is reached. The proteins associated with the DNA in chromatin are generally thought to return to their original positions, since hybridization competition experiments indicate that the RNA transcribed from reconstituted chromatin is similar to that transcribed from native chromatin. However, if only histones are added back to the DNA, transcription is almost totally suppressed, and the small amount of polyribonucleotide that is made shows essentially no hybridization with the DNA (Fig. 7-8). If the DNA is reconstituted only with NHC protein, the RNA sequences that are transcribed from it are more diverse than those transcribed from native chromatin but less diverse than those transcribed from naked DNA (Fig. 7-8). These results suggest that histones nonspecifically repress template

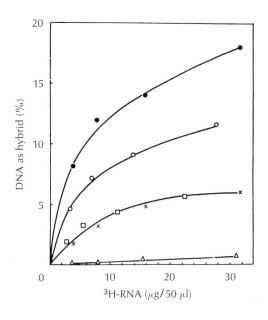

Fig. 7-8. Hybridization of a limited amount of calf thymus DNA to labeled RNAs transcribed in vitro from calf thymus DNA (●——●), calf thymus chromatin (x——x), DNA plus NHC proteins (○——○), DNA plus histone (△——△), and DNA plus histone plus NHC proteins (□——□). Note that chromatin reconstituted with both NHC proteins and histone is similar in template activity to native chromatin. (With permission from Gilmour, R. S., and J. Paul. 1969. J. Mol. Biol. **40:**137. Copyright by Academic Press Inc. [London] Ltd.)

activity and that NHC protein decreases this repression, possibly in a specific manner.

Possible control by NHC proteins

    The possible role of NHC proteins as specific regulators of transcription has been examined for globin, histone, and ovalbumin genes. Control of globin gene transcription was studied by reconstituting DNA and histone that had been prepared from brain chromatin with NHC protein samples prepared either from brain chromatin or from fetal liver chromatin. Globin cDNA, prepared from globin mRNA with reverse transcriptase, was used to test for the presence of globin-specific sequences in the transcripts obtained with bacterial RNA polymerase. As shown in Fig. 7-9, RNA transcribed from reconstituted chromatin containing NHC protein from fetal liver hybridized with the globin cDNA and thus contained globin-specific sequences, whereas the RNA transcribed from chromatin reconstituted with brain NHC protein did not. Control over transcription of the globin message thus appears to reside in the NHC protein fraction from fetal liver cells.

    Histones are synthesized only during the S phase of the cell cycle (phases of the cell cycle and their analysis are described in Appendix C). Analysis of polysomal mRNA by hybridization with histone cDNA has been used to demonstrate that the polysomes contain histone mRNA only during the S phase of the cell cycle. There are conflicting reports in the literature concerning the relative importance of transcriptional and posttranscriptional events in the regulation of histone mRNA levels in the cytoplasm. Chromatin from cells in the G_1 phase

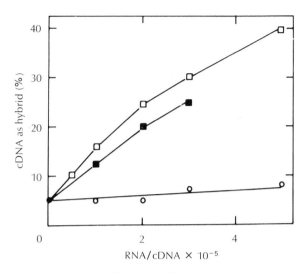

Fig. 7-9. Hybridization of globin cDNA with RNA transcribed from reconstituted brain chromatin (O——O), reconstituted liver chromatin (□——□), and brain chromatin reconstituted in the presence of NHC proteins from fetal liver (■——■). Hybridization was assayed by measuring the amount of cDNA resistant to a nuclease that degrades single-stranded DNA. Note that the presence of NHC proteins from fetal liver permits transcription of sequences that hybridize with globin cDNA. (From Paul, et al. 1973. Cold Spring Harbor Symp. Quant. Biol. **38**:885.)

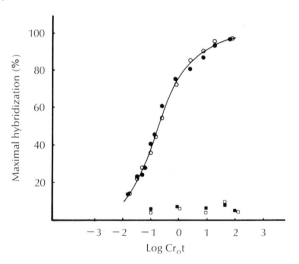

Fig. 7-10. Kinetics of hybridization of histone cDNA to RNA transcribed from the following types of chromatin: ● = native S phase chromatin; O = chromatin reconstituted with S phase NHC proteins; □ = native G_1 phase chromatin, and ■ = chromatin reconstituted with G_1 phase NHC proteins. Note that S phase NHC proteins permit transcription (or accumulation) of sequences that hybridize with histone cDNA. Cr_ot = Initial concentration of RNA × time. (From Stein, G., et al. 1975. *In* G. S. Stein and L. J. Kleinsmith, eds. Chromosomal proteins and their role in the regulation of gene expression. Academic Press, Inc., New York.)

of the cell cycle (prior to DNA synthesis) is claimed not to serve as a template for transcription in vitro of histone-specific RNA sequences, whereas S phase chromatin is active. Chromatin reconstitution experiments by the same investigators appear to show that histone-specific RNA is transcribed only when the reconstituted chromatin contains NHC protein isolated from S phase cells (Fig. 7-10). Another laboratory has shown that histone-specific transcripts are still made in the G_2 phase of the cell cycle (after DNA synthesis) and also when DNA synthesis is blocked, but are degraded rapidly under these conditions. Cytoplasmic mRNA for histones is also degraded rapidly when DNA synthesis stops. Additional study is needed to reconcile these conflicting views and to gain a more complete understanding of the mechanisms that coordinate histone synthesis with DNA synthesis.

A fundamental question in the area of transcriptional control that is just beginning to be dealt with is whether the mechanisms that regulate transcription act primarily in a negative fashion, as seems to be the favored mode in prokaryotic systems (Chapter 4), or whether they are positive in their action. It is tempting on the basis of experiments similar to those just described to draw a rather simple model of transcriptional control in eukaryotic cells, in which histones indiscriminately turn off all transcription while NHC protein molecules selectively reverse that inhibition and activate specific genes. Recent findings from the histone system appear to support this model. When a mixture of NHC protein from G_1 and S phase cells is incorporated into reconstituted chromatin, the histone genes are readily transcribed. This "dominance" of the S phase NHC protein suggests that a positive signal activates transcription of the histone genes. NHC protein from G_1 cells does not block transcription of the histone genes in the presence of NHC protein from S phase cells. Similar results have also been obtained for estrogen-stimulated synthesis of ovalbumin mRNA in the chicken oviduct. NHC protein from stimulated cells functions as a positive activator of specific transcription, and NHC protein from nonstimulated cells has no negative effect on activated transcription of the ovalbumin gene.

Caution should be exercised, however, in drawing generalized conclusions from these few examples. Various bits of evidence suggest that NHC protein molecules may also be able to act as negative regulators that specifically suppress certain genes. As described, the addition of NHC protein to naked DNA reduces the diversity of transcription that can be obtained from it, although not all individual genes are affected. There is also suggestive evidence from the histone system, where protein synthesis is needed to block the accumulation of histone-specific RNA when DNA synthesis ceases. However, this could be a simple negative feedback due to accumulation of free histone. Also, there is currently uncertainty, as mentioned, that direct transcriptional controls are responsible for stopping the accumulation of histone-specific RNA sequences after DNA synthesis is inhibited.

Results from cellular hybridization experiments (Chapter 11) also suggest that there are negative controls of gene expression, although they do not make it clear whether such controls operate at the transcriptional level. In these experiments a cell that exhibits one or more specific differentiated properties in culture is fused with a nondifferentiated cell. Typically, the expression of differentiated properties is inhibited in the initial hybrids but later begins to reappear as chromosomes derived from the nondifferentiated parent are lost from the hybrid cells. Those chromosomes appear to carry negative genetic elements that prevent expression of specific differentiated properties while they are present in the genomes of the hybrid cells.

Experiments of this kind do not disprove the positive activator role proposed for NHC proteins, but they do remind us that we do not yet fully understand the mechanisms that are involved in cellular differentiation. It seems likely that we will ultimately find that those mechanisms are far more complex than we currently realize and that they will involve both positive and negative controls operating in a precisely balanced system. Until we gain that type of understanding, we must keep in mind the fact that there is still plenty of room for alternate explanations of much of the current experimental data. If NHC protein does in fact function as a positive regulator of transcription at specific sites within the genome, we are still left with a major unanswered question: What regulates the formation and activation of specific NHC protein molecules? Ultimately, NHC protein must be translated from messages that have themselves been transcribed, and NHC protein activation (if it is needed) must be accomplished by other molecules such as kinases or phosphorylases. Thus, if we attribute control of transcription to NHC protein molecules, we are only pushing the question of control back one step further. Ultimately we must examine complex feedback loops and balance relationships and perhaps also some totally new mechanisms to bridge the gap from coded DNA sequences and cytoplasmic information to selective expression of specific genes in specific parts of the embryo at specific times during development.

RELATIONSHIP OF TRANSCRIPTIONAL CONTROL TO CHROMATIN STRUCTURE

It is clear from the data that have been presented that the control of transcription in eukaryotic cells is a complex phenomenon. There is evidence that both negative and positive control elements are involved. Mixing experiments suggest that the positive controls are dominant and that they specifically activate transcription of selected genes, whereas the negative controls appear to be more generalized (e.g., indiscriminate inhibition of transcription by histones).

Recent physicochemical studies of chromatin suggest that the relationship among histones, NHC proteins, and the genome may be far more complex than a simple reversal of the negative effects of histones by specific activator molecules in the NHC protein fraction. Interactions between the histones and the NHC proteins appear to have a significant effect on the way that both interact with DNA. In the presence of histones the binding of most classes of NHC protein molecules to DNA is actually increased by an order of magnitude in comparison to their binding to naked DNA. At the same time the NHC protein fraction significantly weakens the interaction between histones and DNA. These cooperative interactions suggest that highly specialized complexes that directly involve all three types of molecules must occur in chromatin and during chromatin reconstitution.

The relationship between the beaded nucleosome structure of chromatin and activation of transcription has not yet been determined. However, preliminary data suggest that the nucleosome structure may be modified in transcriptionally active areas. The beaded structure is still present in such areas, but DNA that is capable of transcription is preferentially digested by pancreatic deoxyribonuclease I. Thus, for example, a digestion procedure that digests only 10% of total nuclear DNA will digest 70% of the sequences coding for ovalbumin in the oviduct nuclei of laying hens. No such preferential digestion is observed in liver nuclei of hens, where the ovalbumin sequences are not being transcribed. Since NHC protein has been shown to activate ovalbumin transcription in chromatin reconstitution experiments, we may speculate that the process rendering a gene available for transcription is linked to a modification of the nucleosome complex that normally forms between the ovalbumin structural gene DNA and histones.

At the present time specific components of the NHC protein fraction are the most likely candidates for the role of specific gene activators in eukaryotic cells. When considering the role of NHC proteins in the control of transcription, we must keep in mind the complexity of the NHC protein fraction, which was emphasized in Chapter 5. Many of the components of NHC protein have well-defined enzymatic roles, and there may also be structural components. Individual activators need only to interact with a limited number of structural or regulatory genes, and they are likely to be present in NHC protein only in small amounts that will be difficult to detect among the other proteins.

None of the details of the regulation of transcription in eukaryotic cells are yet understood at the level of specific molecular interactions. Such information is just beginning to become available for prokaryotic systems, where the genome is much smaller and has fewer types of associated proteins. There is a natural tendency among those who propose models to begin with the prokaryotic system and embellish it as necessary to account for the special properties of eukaryotic systems. There is some risk in this approach due to the major differences in organization of prokaryotic and eukaryotic genomes, which were discussed at the beginning of Chapter 5. However, in most cases there is not sufficient information available to construct totally new models for eukaryotic cells. In addition, observations such as the fact that bacterial RNA polymerases recognize at least some of the tissue specificity of eukaryotic chromatin suggest that the two systems must share some features in common.

Most current models suggest that the immediate control of transcription involves interactions among RNA polymerase, DNA, and specific regulatory molecules, which occur at chromosomal sites closely adjacent to the actual coding sequences. Major differences from the prokaryotic system that must be accounted for in eukaryotic models include the general absence of clustered groups of functionally related genes (operons), the initial production of transcripts that are larger than the final mRNA, and the presence of repetitious DNA interspersed among the actual coding sequences.

In the following discussion, we will consider in detail a model for the control of transcription in eukaryotic cells that has been proposed by Roy Britten and Eric Davidson. As we do so, it is important to remember that models are the product of scientific imagination and that they should be regarded only as proposals until they have been verified by carefully designed experiments. The special value of models lies in the new insight into experimental observations that they may offer and the conceptual framework that they can provide for the design of future experiments. When a model appears to fit the known facts well and becomes widely quoted, as is the case with the Britten-Davidson model, it is easy to forget that it is just a model and that many aspects of it have not yet been proved to be valid. The reader is cautioned not to make this mistake.

BRITTEN-DAVIDSON MODEL OF TRANSCRIPTIONAL CONTROL

The Britten-Davidson model of gene regulation was originally proposed in 1969. It has undergone a certain amount of modification since then but still retains essentially the same overall form that it had when it was first presented. Its longevity as a model is due in part to its accurate prediction of certain aspects of sequence organization in eukaryotic genomes. The difficulty of testing some if its predictions with present experimental techniques has probably also contributed to its long life.

The fundamental features of the Britten-Davidson model, which is based on data and concepts from a number of investigators, are diagrammed in Fig. 7-11. The model proposes that there are four basic kinds of DNA regions in the genome. The first is the producer gene, which is directly analogous to the structural gene of prokaryotic operons. Producer genes code for all gene products except the regulatory molecules that control transcription.

The second type of region in the DNA is the receptor site. The receptor site is physically linked to the producer gene. The receptor sequence is recognized by specific regulatory molecules that stimulate transcription of its associated producer gene. The receptor site is somewhat analogous to the operator locus in prokaryotic operons.

The third type of region, which is visualized as physically separated from the region carrying the producer gene and receptor site, is the integrator gene. This sequence codes for the regulatory molecules that interact with receptor sites. In the original version of the Britten-Davidson model these regulatory molecules were considered likely to be RNAs, since interaction of a regulatory RNA molecule with a receptor site would require only nucleotide sequence complementarity. However, little evidence for the existence of such RNAs has been found and components of the NHC protein fraction have come into prominence as possible transcriptional regulators. In more recent reviews of their model, Britten and Davidson have placed increased emphasis on the idea that the integrator genes code for activator proteins, which then interact with the receptor sites. This concept was also included as an alternative possibility in the original model.

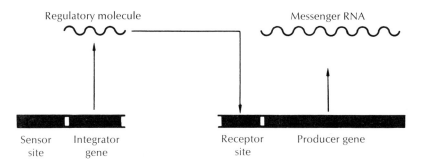

Fig. 7-11. Britten-Davidson model of gene regulation. Stimulation of a sensor site results in transcription of the adjacent integrator gene, which codes for a regulatory molecule (RNA or protein). The regulatory molecule interacts with a receptor site and causes the transcription of the adjacent producer gene. (Modified from Lewin, B. 1974. Gene expression-II. Eucaryotic chromosomes. John Wiley & Sons, Inc., New York.)

The final DNA region proposed by the model is the sensor site. The sensor site, which is proposed to be closely linked to the integrator gene, provides a binding site for inducing agents from the intracellular or extracellular environment. The best example to date of an inducing agent that is compatible with the model is the steroid hormone-receptor protein complex, which enters the nucleus and appears to interact directly with the genome (Chapter 14).

Thus the overall picture of gene regulation in the Britten-Davidson model involves binding of an inducing agent to the sensor sequence, resulting in activation of the integrator gene linked to it. The integrator gene is transcribed to produce an mRNA coding for an activator protein (or to produce an activator RNA directly). The resulting activator molecule then binds to the receptor site, resulting in transcription of the producer gene.

One of the interesting features of the Britten-Davidson model is the suggestion that the integrator and/or receptor sites may be composed of repeated-sequence DNA. This postulation of multiple integrator genes and/or multiple receptor sites allows for considerable elaboration of the basic model to generate schemes that permit many different genes to be controlled by one stimulus and to permit many different stimuli to influence the same gene. Some of these possibilities are diagrammed in Fig. 7-12.

The presence of multiple receptor sites associated with each producer gene (PG) would allow the producer to be transcribed in response to different environmental agents and would allow coordinate activation of two or more PGs by the same environmental agent. For example, in Fig. 7-12, *A*, PG-*A* can be transcribed in response to induction of either sensor site 1 or sensor site 2. In Fig. 7-12, *B*, PG-*A* and PG-*B* are transcribed coordinately in response to induction of sensor site 1. The existence of multiple integrator genes associated with a single sensor site could cause some of the same results. For example in Fig. 7-12, *C*, PG-*A* and PG-*B* are transcribed coordinately in response to induction of sensor site 1. Multiple integrators also allow a PG to be activated, together with different sets of other PGs, depending on which sensor site is activated. For example, in Fig. 7-12, *C*, PG-*A* is activated coordinately with PG-*B* in response to induction of sensor site 1, and it is activated coordinately with PG-*C* in response to induction of sensor site 2. Finally, as diagrammed in Fig. 7-12, *D*, one can combine both possibilities, multiple integrator genes and multiple receptor sites. This allows an even greater degree of complex coordination of gene activation.

The Britten-Davidson model of gene regulation can explain a number of features known to characterize the eukaryotic genome and the

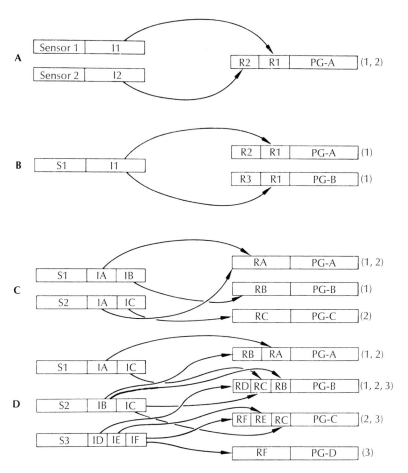

Fig. 7-12. Britten-Davidson model showing interactions possible with multiple receptor sites (**A** and **B**), multiple integrator genes (**C**), and multiple receptor sites plus multiple integrator genes (**D**). S = Sensor site; I = integrator gene; R = receptor site; PG = producer gene. Numbers in parentheses to the right of each producer gene indicate which sensor sites are able to activate that producer gene. (Modified from Truman, D. E. S. 1975. Biochemistry of cytodifferentiation. John Wiley & Sons, Inc., New York.)

regulation of its expression. For example, simple external signals such as hormones or embryonic inducers can cause the activation of a broad range of different genes. Many of the genes that tend to be coordinately activated in eukaryotes are not located next to one another in the genome, as is often the case in prokaryotes. The existence of multiple integrator and/or receptor genes would allow these genes to be functionally linked.

The Britten-Davidson model also offers at least a partial explanation for the huge amounts of DNA present in eukaryotic cells and for the presence of repeated sequences in DNA. Large amounts of DNA would be required to code for integrator genes as well as producer genes; in addition, the noncoding sensor sites and receptor sites could comprise sizeable amounts of DNA. The noncoding receptor sites, if present in multiple copies, could account for a portion of the moderately repeated fraction of the eukaryotic genome. (Integrator genes could also comprise part of the moderately repeated fraction of DNA if identical activator molecules are coded for by integrator genes adjacent to many different sensor sites, as illustrated in Figs. 7-12, C and D.)

Since this model was originally proposed, studies on the organization of unique and repeated sequences in eukaryotic genomes have generated conclusions that are in general agreement with the predictions of the model. For example, it has been shown that most of the unique-sequence DNA in a variety of eukaryotic organisms is interspersed with short sequences of repetitive DNA (Chapter 5). Furthermore, it has recently been demonstrated in several species that the majority of the nonrepeated DNA sequences that code for proteins are located immediately adjacent to repeated sequences in the genome. This has been demonstrated by comparing the hybridization of polysomal mRNA with the total population of unique DNA sequences in the genome and with those unique sequences which are contiguous with repeated sequences. (If the recent finding of intervening noncoding sequences in genes [Chapter 5] should prove to be a general phenomenon, such sequences may account at least partially for the interspersion of repeated sequences and protein-coding unique-sequence DNA.) One possible exception to this general agreement with the predictions of the theory is the genome of *Drosophila,* in which the repeated-sequence DNA is in longer segments and each segment is separated by a substantially longer stretch of unique-sequence DNA than is the case for a variety of other species, including slime molds, sea urchins, amphibia, and mammals.

The Britten-Davidson model predicts that it should be possible to isolate two distinct classes of mutations that influence the rate or amount of synthesis of an enzyme without altering the enzyme protein itself. Mutations at the regulator site would be closely linked to the structural (producer) gene and would be *cis* active—that is, they would affect only the gene that they were directly linked to and would not affect allelic genes on homologous chromosomes. On the other hand, mutations in the sensor-integrator complex would generally not be linked to the structural gene. Mutations of the first type have been detected in the mouse. Mutations in closely linked regulatory genes that do not alter the enzyme molecules are known for the enzymes δ-aminolevulinate dehydratase and β-glucuronidase. In addition, the regulatory gene for β-glucuronidase has been shown to be *cis* active.

These observations are consistent with the Britten-Davidson model but cannot be used as proof for it, since they are also compatible with any other model that has a regulatory element directly linked to the structural gene, including an arrangement similar to the bacterial operon.

The unique feature of the Britten-Davidson model that remains to be verified is the genetically separate sensor-integrator combination. If this part of the model is valid, there must be synthesis of an activator substance (either RNA or protein) to activate transcription. Fragmentary data suggest that this may not always be the case. For example, in the activation of histone mRNA synthesis, described earlier in this chapter, there are data suggesting that the immediate triggering event may be increased phosphorylation of specific components of NHC protein. Similarly, short-term stimulation of transcription by steroid hormone-receptor complexes appears not to require protein synthesis. Thus the most vulnerable part of the Britten-Davidson model is the sensor-integrator concept, which has not yet been verified experimentally and which may be in conflict with experimental observations for at least some types of regulatory systems.

In this chapter, evidence has been presented suggesting that differences in patterns of transcription play a significant role in the regulation of expression of specific genes in cellular differentiation. Until recently many investigators have tended to attribute substantially all the regulation to transcriptional control and to write off as relatively unimportant the other steps in the flow of information from genes to functional proteins. However, there is substantial evidence that these other steps do in fact play significant roles in the overall regulation of gene expression, although their specific mechanisms generally have not been analyzed in as much detail as those of transcriptional control. The next two chapters are dedicated to an analysis of events at posttranscriptional, translational, and posttranslational levels and to an investigation of known and potential controls that may operate at those levels.

Many gaps still exist in our knowledge of the mechanisms of transcriptional control. Current data suggest that the mechanisms responsible for changes in transcriptional activity may turn out to be posttranscriptional and perhaps even posttranslational. In particular, preliminary indications that changes in phosphorylation of NHC protein components may play an important role in triggering transcriptional changes are interesting and should be examined carefully.

There is a critical need for more detailed molecular studies of transcriptional control in eukaryotic cells. A full understanding of the

TRANSCRIPTIONAL CONTROLS AND CELLULAR DIFFERENTIATION

mechanisms that are involved in transcriptional control probably will not be achieved until genes can be isolated with their promoter and regulatory regions intact. Once such genes are available, detailed studies of their interactions at the molecular level with regulatory proteins and RNA polymerase can be undertaken. The recent surprising demonstration that genes for some eukaryotic proteins have noncoding sequences interspersed with coding sequences demonstrates dramatically that fuller knowledge of the details of gene structure in eukaryotes is essential to an understanding of the mechanisms that control transcription of eukaryotic genes. Rapid progress can be anticipated during the next few years toward a detailed understanding of the relationship among genes, their initial products of transcription, and cytoplasmic mRNA, as well as of the regulatory mechanisms that control transcription and processing of the transcript.

BIBLIOGRAPHY
Books and reviews

Ashburner M. 1972. Puffing patterns in *Drosophila melanogaster* and related species. *In* W. Beerman, ed. Results and problems in cell differentiation. Vol. 4. Developmental studies on giant chromosomes. Springer-Verlag, Inc., New York.

Chambon, P. 1975. Eukaryotic nuclear RNA polymerases. Annu. Rev. Biochem. **44**:613.

Davidson, E. H., and R. J. Britten. 1973. Organization, transcription and regulation in the animal genome. Q. Rev. Biol. **48**:565.

Davidson, E. H., and R. J. Britten. 1979. Regulation of gene expression: possible role of repetitive sequences. Science **204**:1052.

Davidson, E. H., et al. 1975. Comparative aspects of DNA organization in Metazoa. Chromosoma **51**:253.

Lewin, B. 1974. Gene Expression. Vol. 2. Wiley-Interscience, New York.

Paigen, K., et al. 1975. The molecular genetics of mammalian glucuronidase. J. Cell Physiol. **85**:379.

Paul, J., et al. 1973. The globin gene: structure and expression. Cold Spring Harbor Symp. Quant. Biol. **38**:885.

Rutter, W. J., M. I. Goldberg, and J. C. Perriard. 1974. RNA polymerase and transcriptional regulation in physiological transitions. *In* J. Paul ed. Biochemistry of cell differentiation. International review of biochemistry. Vol. 9. University Park Press, Baltimore.

Stein, G., et al. 1975. Regulation of histone gene transcription during the cell cycle by non-histone chromosomal proteins. *In* G. S. Stein, and L. J. Kleinsmith, eds. Chromosomal proteins and their role in the regulation of gene expression. Academic Press, Inc., New York.

Stein, G. S., T. C. Spelsberg, and L. J. Kleinsmith. 1974. Non-histone chromosomal proteins and gene regulation. Science **183**:817.

Original research articles

Britten, R. J., and E. H. Davidson. 1969. Gene regulation for higher cells: a theory. Science **165**:349.

Davidson, E. H., et al. 1975. Structural genes adjacent to interspersed repetitive DNA sequences. Cell **4**:217.

Galau, G. A., et al. 1976. Structural gene sets active in embryos and adult tissues of the sea urchin. Cell **7**:487.

Garel, A., and R. Axel. 1976. Selective digestion of transcriptionally active ovalbumin genes from oviduct nuclei. Proc. Natl. Acad. Sci. U.S.A. **73**:3966.

Georgiev, G. P. 1969. On the structural organization of operon and the regulation of RNA synthesis in animal cells. J. Theor. Biol. **25**:473.

Jansing, R. L., J. L. Stein, and G. S. Stein. 1977. Activation of histone gene transcription by nonhistone chromosomal proteins in WI-38 human diploid fibroblasts. Proc. Natl. Acad. Sci. U.S.A. **74**:173.

Lapeyre, J.-N., and I. Bekhor. 1976. Chromosomal protein interactions in chromatin and with DNA. J. Mol. Biol. **104:** 25.

Lewis, M., P. J. Helmsing, and M. Ashburner. 1975. Parallel changes in puffing activity and patterns of protein synthesis in salivary glands of *Drosophila*. Proc. Natl. Acad. Sci. U.S.A. **72:**3604.

Manning, J. E., C. W. Schmid, and N. Davidson. 1975. Interspersion of repetitive and non-repetitive DNA sequences in the *Drosophila melanogaster* genome. Cell **4:**141.

Melli, M., G. Spinelli, and E. Arnold. 1977. Synthesis of histone messenger RNA of HeLa cells during the cell cycle. Cell **12:** 167.

Schmid, C. W., and P. L. Deininger. 1975. Sequence organization of the human genome. Cell **6:**345.

Tsai, S. Y., et al. 1976. Effect of estrogen on gene expression in the chick oviduct: control of ovalbumin gene expression by non-histone proteins. J. Biol. Chem. **251:** 6475.

Wold, B. J., et al. 1978. Sea urchin embryo mRNA sequences expressed in the nuclear RNA of adult tissues. Cell **14:**941.

CHAPTER 8

Origin of cytoplasmic messenger RNA

☐ In the previous chapter we examined transcription of the eukaryotic genome and the mechanisms thought to be responsible for the selective nature of transcription in differentiated cells. In this chapter we will examine the conversion of the initial product of transcription into functional mRNA and the transport of that mRNA from the nucleus to the cytoplasm.

The organization of the eukaryotic genome was described in Chapter 5, together with a brief discussion of the implications of the physical separation of the site of transcription from the site of translation. In prokaryotic cells, where the sites are not separated, the product of transcription appears to function immediately as mRNA with little or no modification. Translation begins at the 5' end of newly synthesized RNA sequences long before transcription of their 3' ends is completed (Fig. 5-1). In eukaryotic cells, however, the biogenesis of mRNA is far more complex, as was summarized briefly in Fig. 5-2.

Although the precise details are still being worked out, there is ample evidence that the initial product of transcription is quite different from the final cytoplasmic message. The operations required to convert the newly synthesized RNA sequences into functional mRNA have been collectively termed "processing." They include removal of excess nucleotides, modification of both ends of the remaining message-containing sequence, and its transport across the nuclear membrane to the cytoplasm, probably as a ribonucleoprotein complex.

The recent finding that the genes for certain eukaryotic proteins contain noncoding sequences interspersed with coding sequences has some potentially important implications for processing of transcribed RNA into mRNA. There are three possible ways whereby mRNA transcripts that lack the intervening sequences could be produced from the insertion-containing genes. The first mechanism would involve looping out of the DNA during transcription so that the noncoding sequences would not be transcribed, and the mRNA would be made as a single continuous molecule. This mechanism would not have implications for processing, but might involve specific transcriptional controls. The other two possible mechanisms that do involve processing are (1) that the gene is transcribed in pieces that are later joined together or (2) that the entire gene, including noncoding sequences, is transcribed and the noncoding sequences are later removed with a splicing together of the coding sequences. Recent data strongly favor the latter mechanism, which has been shown to occur in such diverse systems as adenovirus RNA, certain tRNAs, and β-globin mRNA. Although it is not yet known how generally intervening noncoding sequences occur in genes, they have already been found often enough to make them of extreme interest to those concerned with the regulatory events of mRNA processing.

Extensive research is still being done in all areas involved in RNA processing, and new concepts are appearing in the literature frequently. This chapter deals with the details of processing as they are currently envisioned and with how they may influence gene expression in differentiating cells.

It is characteristic of most, if not all, classes of RNA in eukaryotic cells that the initial product of transcription contains extra nucleotides that are later removed. Base modifications and nontranscriptive additions of nucleotides are also frequently involved in generation of the final functional molecules. The first type of RNA processing to be studied in detail in eukaryotic cells and the best understood at the present time, is the series of reactions leading to formation of the 18S, 28S, and 5.8S rRNAs from a single large transcript. That processing is summarized in the remainder of this discussion, together with brief comments on current views of the processing of tRNA and 5S rRNA. These cases provide an introduction to the more extensive discussion of mRNA processing that follows.

The organization of the genes coding for 5.8S, 18S, and 28S rRNA, which was discussed briefly in Chapter 6, is diagrammed in Fig. 8-1, A. All three sequences are closely linked within a single transcription unit, which also includes three transcribed spacer sequences that are subsequently degraded (Fig. 8-1, B). The transcription units alternate with nontranscribed spacers in a tandemly repeated pattern at certain locations in the genome known as nucleolar organizers.

Under most conditions the precursor of the 5.8S, 18S, and 28S rRNAs is the most plentiful class of newly synthesized nuclear RNA in eukaryotic cells. It consists of a single large molecule whose size is about 40S in amphibia and 45S in mammals. Its processing consists of

PROCESSING OF rRNAs AND tRNAs

selective methylation in those sections of the molecule destined to become rRNA, followed by specific nuclease splitting to separate the 5.8S, 18S, and 28S sequences from the remaining sections of the precursor, which are rapidly degraded in the nucleus. The cleavages and degradations involved in this process are depicted in Fig. 8-1. Methylation appears to protect the sequences that are to be preserved.

The other RNA molecules involved in the mechanical aspects of protein synthesis (tRNA and 5S rRNA) are also generated by processing larger initial transcripts. In the case of the tRNAs, 15 to 35 nucleotides are removed from the precursor RNAs and extensive modification of specific bases occurs, probably mainly in the cytoplasm. Recent studies on tRNA precursors from yeast have shown that the extra nucleotides are present as an insert in the middle of the tRNA sequence. Processing of 5S rRNA has been difficult to demonstrate, but in heat-shocked cultured *Drosophila* cells an apparent precursor containing

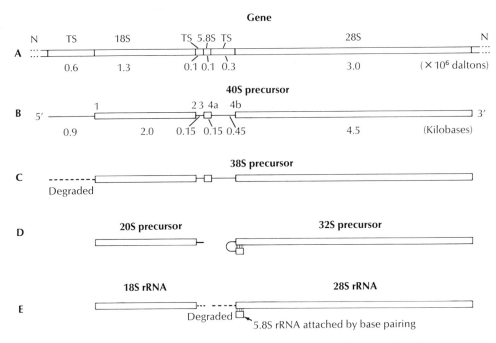

Fig. 8-1. Gene organization, transcription, and processing of 5.8S, 18S, and 28S rRNA in *Xenopus laevis*. The organization in most other eukaryotes is similar except for differences in the sizes of transcribed spacers. **A,** Organization of gene. *N* = Nontranscribed spacer; *TS* = transcribed spacer region. Molecular weights for each segment are given in daltons × 10^6 of double-stranded DNA. **B,** Initial product of transcription (40S precursor). The size of each segment is indicated in kilobases. *1, 2, 3, 4a,* and *4b* = Cleavage sites for processing. **C** to **E,** Successive steps in processing. In some types of cells, cleavage occurs at site 2 before site 3, and no 20S intermediate is detected.

an additional 15 nucleotides has been observed. The precursors for 5S rRNA and the tRNAs comprise only a small portion of the newly synthesized radioactive RNA that can be detected in the nucleus after a short exposure of cells to labeled nucleosides. Students wishing to understand the processing of rRNA and tRNA molecules in greater detail are referred to the review by Robert Perry cited at the end of this chapter.

Before we examine the processing reactions that lead to the formation of mRNA, it will be helpful to examine mRNA itself. One can easily be misled by the name "messenger" into thinking that mRNA is a simple molecule that contains little other than the base sequence needed to specify a particular protein. However, as shown in the examples in Table 8-1, every mRNA for a specific protein that has been isolated and studied has proved to contain more nucleotides than would be needed just for coding. In addition, the 5′ ends of nearly all mRNA species have specialized "cap" structures, and the 3′ ends of most have an added polyadenylic acid (poly[A]) sequence.

The overall structure of eukaryotic mRNA suggested by recent research reports is shown in Fig. 8-2. Characteristic portions of the molecule, identified by the letters a to g in Fig. 8-2, are the following:

a – 5′ Cap. It has recently been found that essentially all eukaryotic

STRUCTURE OF mRNA

Table 8-1. Comparison of lengths of mRNA molecules and their coding sequences*†

Cell	Protein	Coding length	mRNA length
Rabbit red blood cell	α-globin	423	650
	β-globin	438	589
Mouse myeloma	Light immunoglobin	660	1200, 1250, 1300
	Heavy immunoglobin	1350	1800
Chicken oviduct	Ovalbumin	1158	1859
Calf lens	α-A2-Crystallin	520	1460
	δ-Crystallin	1260	2000
Bombyx mori silk gland	Fibroin	14,000	16,000
Lytechinus pictus (sea urchin)	Histone H4	310	370 to 400

*Modified from Lewin, B. 1975. Cell **4**:77.
†Coding lengths are the number of nucleotides required to specify each protein, estimated from its number of amino acids or molecular weight. The lengths of the mRNAs are those determined experimentally; where more than one value is shown, each represents an independent determination. Values for α- and β-globin and ovalbumin are based on sequence analysis.

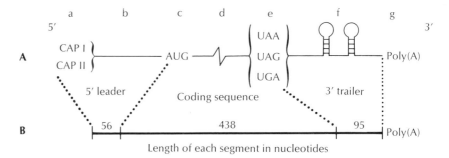

Fig. 8-2. Schematic representation of eukaryotic mRNA. **A,** Expanded view of noncoding sections of a typical mRNA. a = Cap I (m^{7}G(5′)pppX^{m}pYpZp . . .) or cap II (m^{7}G(5′) pppX^{m}pY^{m}pZp . . .); b = untranslated region at 5′ end; c = initiation codon; d = translated sequence; e = termination codon; f = untranslated sequence at 3′ end, including possible self-complementary loop forming sequences; g = poly(A) (not present in all mRNAs). **B,** Scale drawing of rabbit β-globin mRNA with loops straightened out.

mRNA molecules (except small viral RNAs that can serve as messages) have special modifications at their 5′ ends. These modifications have been termed "caps," and their structure is shown in Fig. 8-3. Cap I consists of 7-methylguanosine connected in an unusual 5′-to-5′ triphosphate linkage to the 5′ end of the mRNA molecule, plus a 2′-O-methyl substitution on the first nucleoside in the regular portion of the RNA. Cap II is the same except that the second nucleoside is also methylated. In abbreviated nomenclature,* these are represented as m^{7}G(5′)pppX^{m}pYpZp. . . and m^{7}G(5′)pppX^{m}pY^{m}pZp . . . , where X, Y, and Z are the first three nucleosides in the regular 5′-to 3′-linked portion of the mRNA molecule.

$b-5′$ Untranslated segment. After the cap there is an untranslated "leader" segment before the actual message begins. This is 56 nucleosides long, including the AUG initiation sequence, in the β-globin message, which is one of the types of mRNA that has been analyzed in detail. Similar "leader" sequences are also found in prokaryotic organisms, where, for example, in *E. coli* the mRNA for the lactose operon contains an untranslated segment of 38 nucleotides at its 5′ end, and the mRNA for the tryptophan operon contains an even longer untranslated 5′ sequence of 166 nucleotides.

---

*The backbone of RNA is a regular repeating structure with the 3′ position of one ribose connected to the 5′ position of the next by a phosphodiester linkage. The bases are attached at the 1′ position. By convention, the 5′ side of the nucleoside is written to the left. Thus pppG means guanosine-5′-triphosphate, and ApCpGpU refers to a short piece of RNA with a free 5′-OH on adenosine at the left and a free 3′-OH on the uridine at the right.

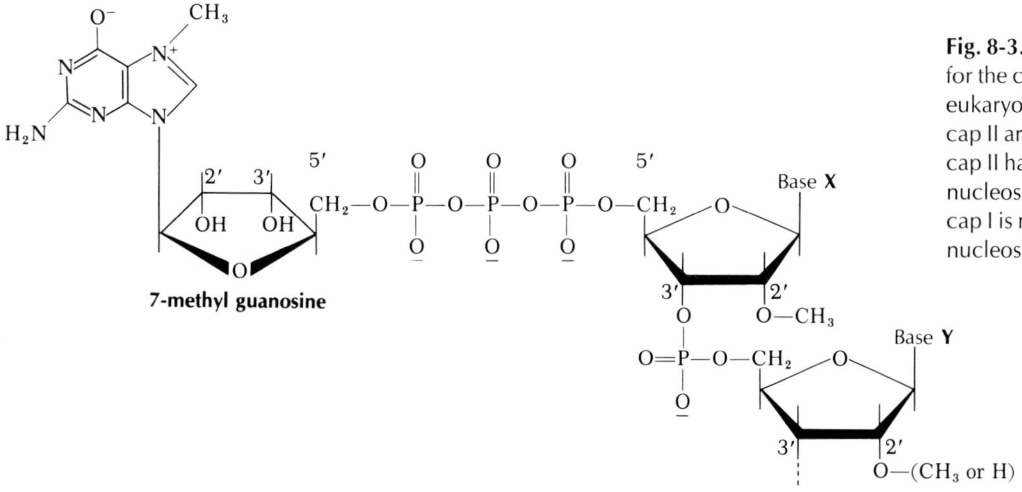

Fig. 8-3. Generalized structure for the cap at the 5′ end of eukaryotic mRNA. Cap I and cap II are identical, except that cap II has 2′-0-methyl groups on nucleosides X and Y, whereas cap I is methylated only on nucleoside X.

c – Initiation codon. The start of translation is signaled by the codon AUG, which inserts a methionine residue at the *N*-terminal end of the protein. In the case of β-globin, that residue is later removed, leaving the penultimate amino acid, valine, at the *N*-terminal position in the mature peptide chain. Removal of *N*-terminal methionine also occurs in most other eukaryotic proteins, and it is not unusual for a substantial number of amino acids to be removed from the *N*-terminal ends of newly synthesized peptide chains (see Chapter 9).

d – Translated sequence. The actual message is located in the center of the mRNA, well away from either end. In the case of the rabbit β-globin message, the total length of the complete mRNA without poly(A) is 589 nucleotides, whereas the coding sequence that it contains is only 438 nucleotides long, excluding the initiation and termination codons (Fig. 8-4). Similarly, human α-globin mRNA has a total length of about 650 nucleotides, of which only 423 normally code for amino acids.

e – Termination codon. The signals UAA, UAG, and UGA function specifically to stop translation when a polypeptide chain of the correct length has been made. There is some evidence that termination codons may occur in tandem pairs at the ends of prokaryotic genes. However, this is not necessarily the case in eukaryotic cells. A human hemoglobin variant, hemoglobin Constant Spring, has been described in which the α-chain contains 31 additional amino acids beyond the normal *C*-terminal amino acid. This variant appears to have arisen by mutation in a single termination codon. Its first amino acid beyond the normal *C*-terminus is glutamine, which could have arisen by substitution of C for U in a UAA or UAG termination codon.

f−3′ Untranslated sequence. In all eukaryotic mRNAs that have been examined, there is a relatively long untranslated "trailer" sequence at the 3′ end of the actual message. The Constant Spring variant (just described) of the human α-globin chain, with its 31 extra amino acids, provides direct evidence for a normally untranslated

mGppp pAC ACU UGC UUU UGA CAC AAC UGU GUU UAC UUG CAA UCC CCC AAA ACA GAC AGA AUG

GUG CAU CUG UCC AGU GAG GAG AAG UCU GCG GUC ACU GCC CUG UGG GGC AAG GUG AAU GUG
Val His Leu Ser Ser Glu Glu Lys Ser Ala Val Thr Ala Leu Trp Gly Lys Val Asn Val

GAA GAA GUU GGU GGU GAG GCC CUG GGC AGG CUG CUG GUU GUC UAC CCA UGG ACC CAG AGG
Glu Glu Val Gly Gly Glu Ala Leu Gly Arg Leu Leu Val Val Tyr Pro Trp Thr Gln Arg

UUC UUC GAG UCC UUU GGG GAC CUG UCC UCU GCA AAU GCU GUU AUG AAC AAU CCU AAG GUG
Phe Phe Glu Ser Phe Gly Asp Leu Ser Ser Ala Asn Ala Val Met Asn Asn Pro Lys Val

AAG GCU CAU GGC AAG AAG GUG CUG GCU GCC UUC AGU GAG GGU CUG AGU CAC CUG GAC AAC
Lys Ala His Gly Lys Lys Val Leu Ala Ala Phe Ser Glu Gly Leu Ser His Leu Asp Asn

CUC AAA GGC ACC UUU GCU AAG CUG AGU GAA CUG CAC UGU GAC AAG CUG CAC GUG GAU CCU
Leu Lys Gly Thr Phe Ala Lys Leu Ser Glu Leu His Cys Asp Lys Leu His Val Asp Pro

GAG AAC UUC AGG CUC CUG GGC AAC GUG CUG GUU AUU GUG CUG UCU CAU CAU UUU GGC AAA
Glu Asn Phe Arg Leu Leu Gly Asn Val Leu Val Ile Val Leu Ser His His Phe Gly Lys

GAA UUC ACU CCU CAG GUG CAG GCU GCC UAU CAG AAG GUG GUG GCU GGU GUG GCC AAU GCC
Glu Phe Thr Pro Gln Val Gln Ala Ala Tyr Gln Lys Val Val Ala Gly Val Ala Asn Ala

CUG GCU CAC AAA UAC CAC UGA GAU CUU UUU CCC UCU GCC AAA AAU UAU GGG GAC AUC AUG
Leu Ala His Lys Tyr His

AAG CCC CUU GAG CAU CUG ACU UCU GGC UAA UAA AGG AAA UUU AUU UUC AUU GC- polyA

Fig. 8-4. Nucleotide sequence of coding and noncoding portions of rabbit β-globin mRNA. Amino acids corresponding to the codons in the coding sequence are also shown. (From cover of Cell **10.** 1977. [April].)

sequence of at least 96 nucleotides (including a second termination codon) beyond the 3′ end of the normal α-globin coding sequence. Recent sequence analysis of normal human α-globin mRNA has shown that there are 16 nucleotides between the Constant Spring termination codon (UAA) and the beginning of the 3′ poly(A) sequence. Thus the total length of the normally untranslated 3′ sequence in the human α-globin mRNA molecule is 112 nucleotides. Preliminary data for a variety of other types of mRNA indicate that the 3′ untranslated segment can be either larger or smaller. In rabbit β-globin mRNA, for example, the length is 95 nucleotides, whereas in chicken ovalbumin mRNA the length is 637 nucleotides.

Small amounts of 6-N-methyladenine are found in some, but not all, eukaryotic mRNAs, presumably in the 3′ or 5′ untranslated sequences. Sequence analysis of several different vertebrate mRNA molecules has revealed that each has the same sequence, 5′ AAUAAA3′, a short distance from the 3′ poly(A). Comparison of 3′ sequences of rabbit α-globin mRNA and mouse immunoglobulin light chain mRNA has revealed two self-complementary sequences that in theory could form double hairpin loop structures with similar shapes (Fig. 8-5). Comparison of rabbit and human α-globin sequences suggests that less evolutionary divergence has occurred between their 3′ untranslated se-

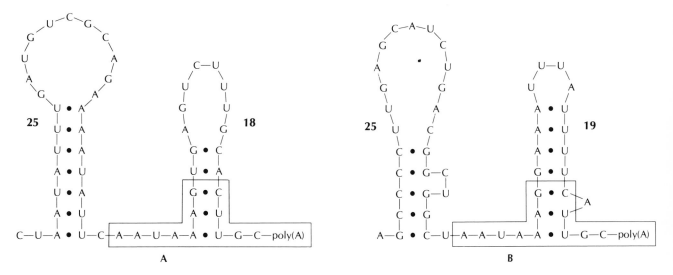

Fig. 8-5. Comparison of the 3′ end of **(A)** mouse immunoglobulin light chain mRNA with that of **(B)** rabbit β-globin mRNA. The sequence in the boxes denotes homology between the two mRNAs. The number adjacent to each hairpin loop denotes the total number of nucleotides in that loop. (From Proudfoot, N. J., and G. G. Brownlee. 1974. Nature **252**:359.)

quences than among third bases of codons that can mutate without causing amino acid substitutions. This suggests that there is some degree of evolutionary selection maintaining the untranslated sequence. It is reasonable to propose that both the 3' and 5' untranslated sequences of eukaryotic mRNAs may serve specific functions, but there is not yet enough data to speculate meaningfully on the exact nature of those functions.

g — Poly (A). Many, but not all, types of eukaryotic mRNA molecules contain poly(A) sequences at their 3' ends. The poly(A) sequence is added posttranscriptionally, and its length can vary from about 30 to 200 nucleotides. About 70% of all cytoplasmic mRNA in HeLa cells (a commonly used human cell line) contains poly(A), but the remainder, which appears to be transcribed from different sites on the genome, does not. In most species, histone mRNA does not contain poly(A), although there has been a recent report that histone mRNA from *Xenopus* oocytes is polyadenylated. Poly(A) does not appear to be essential for mRNA function, although it may improve message stability.

HETEROGENEOUS NUCLEAR RNA AND ITS RELATIONSHIP TO mRNA

Since transcription of the eukaryotic genome occurs in the nucleus, it is obvious that some type of nuclear RNA must be the precursor for cytoplasmic mRNA. The likely candidate for this role appears to be the class of RNAs called heterogeneous nuclear RNA (hnRNA). It receives its name by virtue of the fact that its size distribution is heterogeneous and its location is strictly nuclear. Among the major classes of RNA in the nucleus, only hnRNA has a base composition similar to the overall composition of the DNA, as is the case with mRNA. Also, the kinetics of labeling hnRNA and mRNA with radioactive isotopes are compatible with a precursor role for hnRNA, which becomes labeled very rapidly, whereas the appearance of label in cytoplasmic mRNA is appreciably delayed. The exact relationship of hnRNA to mRNA has been difficult to demonstrate, however, because much of the hnRNA is degraded rapidly without ever leaving the nucleus. The average half-life of hnRNA molecules is only about 25 minutes, and in higher eukaryotes, at least 80% of hnRNA sequences never reach the cytoplasm.

If typical hnRNA molecules serve as precursors of mRNA, substantial removal of excess nucleotides must occur during processing. The exact size distribution of hnRNA is controversial, but most investigators consider it to be significantly larger, on the average, than mRNA. Older reports suggested that hnRNA was extremely large, ranging from 5000 to 50,000 nucleotides (5.0 to 50 kilobases [kb]) in length. More recent studies, performed under denaturing conditions, seem to indicate that the bulk of it is actually much smaller. There is still not a clear consensus on its exact size, but most recent reports agree that the average molecule is less than 5 kb in length. Even under strict dissociating conditions, however, there does seem to be a small fraction that is quite large, probably in excess of 15 kb.

About half the molecules in a typical mRNA population are less than 1200 nucleotides in length, and few exceed 4 kb. There is an appreciable, and possibly substantial, overlap in size ranges between hnRNA and mRNA, but it is still necessary to postulate either that only the smaller hnRNA molecules serve as precursors of mRNA or else that some type of posttranscriptional removal of nucleotides from typical hnRNA molecules occurs to generate mRNA.

The genes for histones offer an interesting, but unusual, case. The genes for H1, H2a, H2b, H3, and H4 are all located close together in a gene cluster that is tandemly repeated. The repeat distance varies from 6 kb in sea urchins to 14.5 kb in HeLa cells. Recent data appear to indicate that in HeLa cells the entire histone gene cluster is transcribed into a single 14.5 kb RNA molecule, which is subsequently processed to yield the individual histone messages.

This pattern of transcription of the histone genes, which is reminiscent of polycistronic operons in prokaryotic cells, is clearly not the usual case in eukaryotic cells, where most genes are believed to be transcribed individually. However, it provides an interesting opportunity to compare the size of a transcript with the size of the coding sequences that it contains. The five histones together contain about 700 amino acids (Table 5-1). Their combined coding sequences therefore require about 2100 nucleotides plus another 30 for five initiation codons and five termination codons. In species in which they have been studied, the total lengths of the histone mRNAs are somewhat less than twice the lengths of their corresponding coding sequences. Thus in a 14.5 kb transcript roughly 2.1 kb, or about 15%, of the molecule actually contains coding sequences, and probably no more than 4 kb, or about 28%, is preserved as mRNA.

Synthesis of both hnRNA and mRNA shows the same drug sensitivities. For example, a low concentration of actinomycin D, which inhibits the synthesis of 18S and 28S rRNA, does not affect production of either hnRNA or mRNA. The range of higher actinomycin D concentrations that inhibits synthesis of hnRNA is the same as that which inhibits synthesis of mRNA. Similarly, the synthesis of both hnRNA and mRNA is inhibited by low levels of α-amanitin that are insufficient to inhibit synthesis of rRNA and tRNA precursors. On the basis of these drug sensitivities, hnRNA and mRNA appear to be transcribed by RNA polymerase II (Chapter 7).

A comparison of base sequences in the two types of molecules provides some of the most convincing evidence for a precursor-product relationship between hnRNA and mRNA. Hybridization to cDNA prepared from specific mRNA with reverse transcriptase has shown that hnRNA molecules contain specific message sequences. For example, cDNA prepared from globin mRNA hybridizes with hnRNA from erythroblasts. Likewise, ovalbumin cDNA hybridizes with chicken oviduct hnRNA. Thus the hnRNAs of these two cell types contain the mRNA sequences that code for their respective differentiated products. Also, in more generalized studies no base sequences have been found in cytoplasmic mRNA that are not also represented in hnRNA from the same type of cell.

A direct demonstration that mRNA sequences are present in hnRNA has been achieved by injecting hnRNA from fetal mouse liver (a hemopoietic tissue) into oocytes of *Xenopus*. The oocytes form functional mRNA from the hnRNA and synthesize detectable amounts of mouse globin. No mouse globin can be detected when mouse brain hnRNA is injected into the oocytes.

Selective conversion of hnRNA to mRNA   Although the bulk of evidence seems to suggest that hnRNA serves as a precursor of mRNA, this relationship is probably not direct and straightforward. One must still deal with the discrepancy in size ranges of the two classes of molecules and the high rate of turnover of hnRNA in the nucleus. A variety of experimental approaches have indicated that many of the sequences found in hnRNA are not found in the cytoplasm. Some of these sequences are repetitive, and some are nonrepetitive. In the sea urchin embryo the sequence complexity of hnRNA is ten times greater than that of mRNA contained in polyribosomes in the cytoplasm. In various mammalian cells the sequence complexity of hnRNA tends to be 3.5 to 6.0 times that of mRNA from the same cells.

Just how sequences are selected for entry into the cytoplasm is a question of considerable importance to the control of gene expression. The sequences that are retained in the nucleus may be noncoding and have only a structural or regulatory function. On the other hand, it is possible that some coding sequences are selectively transported to the cytoplasm while others are retained in the nucleus and degraded.

Evidence suggesting that some degree of control over gene expression may operate at the message-processing or transport level has been found in experiments using globin cDNA as a probe to detect minute amounts of globin message sequences in the nuclear and cytoplasmic RNAs of a variety of cells. As shown in Table 8-2, globin mRNA sequences were detected in the nuclear RNA of both erythroid and nonerythroid cells, although the amounts found were far less in

Table 8-2. Distribution of RNA molecules complementary to mouse globin cDNA in nuclei and cytoplasms of various types of mouse cells*

Type of cell	Molecules of 9S globin RNA per cell			Total globin RNA of cell (%)	
	Total	In nucleus	In cytoplasm	In nucleus	In cytoplasm
Fetal liver (14 day)	70,600	4600	66,000	7	93
Adult liver	145	75	70	52	48
Adult brain	380	80	300	24	76
Cultured LS fibroblast	~3	~2	~1	~66	~33
Cultured 3T3 cells	74	70	4	95	5

*Based on data from Humphries, S., J. Windass, and R. Williamson. 1976. Cell 7:267.

the nonerythroid cells. When cytoplasmic RNA was examined in the same cells, the great majority (93%) of the globin sequences in the erythroid cells were found in the cytoplasm, whereas in the nonerythroid cells the proportion in the cytoplasm was much less. If no artifacts resulted from working with the very low concentration of message-specific sequences involved in this study, these data suggest that in addition to the primary control operating at the transcriptional level, a secondary control mechanism operates at the level of message processing or transport to reduce the probability of unwanted transcripts becoming cytoplasmic mRNA. Caution must be exercised with such an interpretation, however, since similar results would also be obtained if globin mRNA were degraded more rapidly in the cytoplasm of nonerythroid cells than in the cytoplasm of erythroid cells. This is possible, since selective stabilization of mRNA for differentiated products is known to occur in differentiated cells (see Chapters 9 and 14).

Further evidence that message processing or transport may have an important influence on gene expression comes from the results of recent work done with sea urchin RNA. In these experiments a radioactive DNA tracer that was greatly enriched for sequences complementary to blastula mRNA was hybridized to nuclear RNA and to cytoplasmic mRNA prepared from adult and embryonic sea urchin tissues. The DNA tracer hybridized much less to cytoplasmic mRNA from adult tissue than to blastula mRNA, indicating that only a small fraction of the blastula mRNA sequences were present in the adult mRNA. However, it hybridized as well with nuclear RNA of adult tissues as it did with cytoplasmic mRNA of blastula cells. This finding implies that many blastula mRNA sequences that are not found in the mRNA of adult tissues are nevertheless represented in the nuclear RNA of adult cells. These observations argue strongly in favor of selective processing or transport of mRNA sequences in the adult tissues.

On the basis of these and other observations Eric Davidson and Roy Britten have proposed a new model for the regulation of gene expression at the level of posttranscriptional processing. Since their model was published very recently, we cannot judge how widely it will be accepted. However, the experimental data on which it is based clearly indicate that a new model is needed, and even if some of the details of this model prove to be incorrect, the data that led to its development will continue to have a major impact on thinking about the mechanisms of cellular differentiation in the next few years.

Specifically the new model seeks to accommodate the following observations:

1. Kinetic analysis of transcription and turnover of nuclear RNA suggests that most structural genes may be transcribed continuously at a rate that is constant and characteristic for the species and the type of cell. The only exceptions are genes that code for mRNAs that are produced in extremely large amounts ("superprevalent" mRNAs), such as the genes for ovalbumin in the oviduct of the laying hen.

2. The efficiency of conversion of structural gene transcripts to cytoplasmic mRNA varies greatly from one gene to another. Particular gene transcripts may be expressed as "moderately prevalent" mRNA (hundreds of copies per cell) or as "complex" mRNA (only a few copies per cell) or not expressed at all as mRNA (down to current limits of detection).

3. Different sets of gene transcripts are processed at different levels of efficiency in the various types of differentiated cells. Thus each type of differentiated cell has a unique qualitative and quantitative distribution of cytoplasmic mRNAs derived from a relatively constant set of structural gene transcripts in the nucleus.

4. Major tissue-specific differences are observed in the nuclear concentration of transcripts of moderately repeated DNA sequences that are never expressed as cytoplasmic mRNA. These nuclear transcripts are complementary to both strands of the moderately repeated DNA and thus capable of forming RNA:RNA duplexes.

5. A substantial part of the repeated-sequence DNA is interspersed throughout the genome in such a way that repeated sequences are thought to be located adjacent to the structural gene sequences that become mRNA.

6. Most of the rapidly turning over nuclear RNA molecules contain repetitive sequences attached to unique sequences in an arrangement similar to that in the genomic DNA.

The new model proposes that, except for the superprevalent mRNA class whose control is distinctly different, all structural genes are transcribed at a constant rate and that gene expression is controlled at the level of mRNA processing by regulatory RNA molecules that are transcribed from repeated sequences in the genome in a tissue-specific manner. These regulatory RNAs are thought to form RNA: RNA duplexes with complementary sequences adjacent to the mRNA sequences in the nuclear transcripts of the structural genes. Formation of the duplexes is postulated to protect specific transcripts from rapid degradation and/or to allow them to be processed into mRNA and transported to the cytoplasm with a high level of efficiency.

The degree to which a particular structural gene transcript is expressed as cytoplasmic mRNA is claimed to be dependent on the rate at which the transcript forms complexes with specific regulatory RNAs. That rate, in turn, depends on the concentration of the regulatory RNA in the nucleus, which is postulated to result partially from the size of the repeated gene family to which the regulatory RNA belongs and partially from tissue-specific control over the rate of transcription of that repeat family.

The regulatory RNAs in this new model are assumed to be transcribed from sites in the genome that are analogous to the integrator genes in the Britten and Davidson model of transcriptional control described in Chapter 7. Transcription of the regulatory RNA is assumed to be controlled by adjacent sensor sites similar to those which were proposed in the transcriptional model. Also, the new model assumes that the expression of large batteries of genes can be regulated by a single family of sensor-integrator sites. Thus, except that control over expression of structural genes is exerted at a posttranscriptional level in the new model, there are many similarities between it and the earlier Britten and Davidson model for transcriptional control described in Chapter 7.

There are also recent reports suggesting that processing changes may play a significant role in changes of histone mRNA synthesis that accompany changes in DNA synthesis. Histone hnRNA appears to continue to be made when DNA synthesis stops and the level of cytoplasmic histone mRNA drops sharply. However, the relationship between transcriptional control, processing, and stability of mRNA for this complex system is not yet fully worked out.

Some hnRNA molecules contain poly(A) sequences at their 3′ ends similar to those found on many cytoplasmic mRNAs. Although many hnRNA molecules contain transcribed sequences of repeated adenylic

Poly(A)

acid up to 25 nucleotides in length (oligo [A]), the long 3′ terminal poly(A) segments, which are up to 200 residues in length, are not transcribed from the genome. Instead, they are added after transcription by nuclear enzymes one nucleotide at a time without a template. Under most conditions the poly (A) sequences of cytoplasmic mRNA appear to be synthesized in the nucleus, although some poly(A) synthesis can also occur in the cytoplasm. When cells are incubated for short periods of time with ³H-adenosine, more than 95% of the labeled poly(A) is found in the nucleus. After longer periods of labeling or after a "chase" with unlabeled adenosine, labeled poly(A) is also found in substantial amounts in cytoplasmic mRNA.

Poly(A) synthesis can occur in the cytoplasm under special circumstances. Mitochondrial mRNA and the mRNA of viruses that replicate entirely in the cytoplasm both contain poly(A). Also, in sea urchin eggs some cytoplasmic mRNA is polyadenylated after fertilization. However, these all appear to be special cases, and the bulk of the evidence indicates that the poly(A) found in cytoplasmic mRNA normally comes from the nucleus.

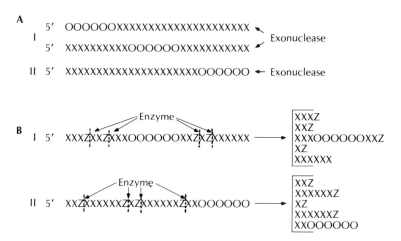

Fig. 8-6. Experimental approaches indicating the 3′ location of poly(A) in hnRNA and mRNA molecules. XXXXXX = Nucleotides of mixed base composition; OOOOOO = poly(A). **A,** RNA is treated with exonuclease that digests in a 3′-to-5′ direction. *I,* If poly(A) sequence is internal or at the 5′ end, digestion will release mixed-sequence nucleotides before poly(A). *II,* If the poly(A) sequence is at the 3′ end, digestion will release poly(A) prior to mixed-sequence nucleotides. **B,** RNA is treated with enzymes that cleave at the 3′ sides of guanine, cytosine, or uridine nucleotides. Z = Nucleotide recognized by enzyme. *I,* If the poly(A) sequence is internal or at the 5′ end, digestion will release the poly(A) segment attached to nucleotide recognized by enzyme. *II,* If the poly(A) sequence is at the 3′ end, digestion will release the poly(A) segment devoid of nucleotide recognized by enzyme.

Poly(A) is found in all size classes of hnRNA molecules. There is substantial evidence that it only occurs at the 3' ends of such molecules (as would be expected from the mode of synthesis just described). For example, an exonuclease* that digests RNA one nucleotide at a time in the 3' to 5' direction, as diagrammed in Fig. 8-6, A, removes poly(A) from hnRNA before significant amounts of nucleotides of mixed base composition are released. Enzymes that cleave RNA molecules on the 3' side of guanine, cytosine, or uridine nucleotides, as diagrammed in Fig. 8-6, B, release poly(A) segments devoid of these nucleotides. This demonstrates that there are no other nucleotides beyond the 3' ends of the poly(A) sequences.

The discovery of 7-methylguanosine caps at the 5' ends of mRNA molecules is still recent, and details of their synthesis are not yet completely worked out. The available evidence indicates that the addition of 7-methylguanosine and methylation of the adjacent ribose normally occur in the nucleus prior to processing. Caps have been found on the 5' ends of hnRNA molecules of all sizes in HeLa and Mouse L cells, including some very large molecules (>10 kb). Radioactively labeled caps move rapidly from the nucleus to the cytoplasm. The second methylation to yield a "cap II" structure has been reported to occur in the cytoplasm.

5' Capping

The recent and unexpected discovery of intervening noncoding sequences within coding sequences for proteins (Chapter 5) has made it necessary to reevaluate all previous models for the conversion of hnRNA into mRNA. For example, arguments about whether nucleotides are removed from the 5' or 3' ends of hnRNA may be irrelevant in view of the third possibility that they may actually be removed from the middle. It has recently been shown, for example, that a major aspect of the processing of mouse 15S β-globin hnRNA to form mature 10S β-globin mRNA involves the removal of a 550-nucleotide intervening sequence from an internal position in the coding sequence.

Formation of mRNA from hnRNA

Many previously paradoxical experimental observations now appear to make sense when interpreted in terms of excision of internal sequences. In the past there have been many reports that appear to indicate (1) that poly(A) is added to the 3' end of hnRNA and remains closely linked to the coding sequence during processing and (2) that 5'

---

*An exonuclease is an enzyme that starts at one end of a polynucleotide molecule and digests it by removing one nucleotide at a time. It is distinguished from an endonuclease, which is able to cleave internal bonds and thus split a large polynucleotide into smaller polynucleotides.

caps are also added before processing and remain closely linked to the coding sequences. When viewed in terms of removal of intervening sequences, such observations are easy to reconcile, since portions of the coding sequence can be located close to both the 5′ end and the 3′ end of the precursor molecule before excision of the intervening noncoding sequence. However, at the same time, intervening sequences give rise to equally puzzling new questions, including the mechanism of their removal and their biological significance.

It is not safe to assume, however, that all removal of nucleotides during processing occurs from central regions of the hnRNA molecule. For example, the presence of caps attached to pyrimidine nucleotides at the 5′ ends of certain mRNA molecules suggests that cleavage of the initial transcript must sometimes occur prior to capping. It is generally accepted that transcription always begins with a purine nucleotide. If this is universally true, then a pyrimidine next to the 5′ cap indicates modification (presumably removal of bases) of the 5′ ends of certain transcripts before they are capped. Similarly, there is no strong evidence to preclude removal of nucleotides from the 3′ end after transcription. Synthesis of poly(A) requires an oligo(A) primer at the 3′ end of the RNA, and one speculative model has suggested that cleavage adjacent to transcribed oligo(A) sequences after transcription is completed could be the source of that primer. Thus the possibility must be kept open that processing may remove nucleotides from any part of the hnRNA molecule.

The question of the size of the initial product of transcription for various genes is still controversial. So far, all the messages for differentiated products that have been examined have proved to come from relatively small hnRNA precursors. However, there is direct evidence that at least some cytoplasmic mRNA is actually derived from large nuclear precursors. Large nuclear RNA (15 kb or more in length) was isolated from HeLa cells under stringent dissociating conditions and was cleaved to smaller pieces. Fragments containing poly(A) and the base sequences adjacent to it were collected on poly(U) Sepharose, and reverse transcriptase was used to prepare cDNA from those sequences. About 40% of the cDNA hybridized with cytoplasmic mRNA. Thus at least some cytoplasmic mRNA molecules appear to contain the same sequences as the 3′ ends of very large poly(A)-containing hnRNA molecules. However, probably not all mRNA arises in that fashion. Until initial products of transcription can be identified with certainty, the question of how many nucleotides are removed after transcription will of necessity remain speculative.

Processing of mRNA precursors proceeds rapidly, and it is possible that it may begin even before transcription is completed. If this is the

case, it may ultimately prove necessary to block processing or to analyze transcription in vitro with purified enzyme systems that do not
allow processing to occur to find out with certainty the size of the precursor. Rapid progress is being made in the isolation of intact eukaryotic genes and their introduction into bacterial plasmids, from which
they can be purified in large amounts. Studies of such "cloned" genes
have already provided much of our current information on intervening
sequences and will undoubtedly provide important insights into the nature of the initial product of transcription and its subsequent processing in the near future. At present, however, we are unable to present
firm answers to many of the more interesting questions about processing or even to be sure that we know which questions we should be
asking.

Despite the rather recent discovery of the caps at the 5′ ends of
eukaryotic mRNAs, there is already evidence suggesting that they
play a role in message translation. Removal of 7-methylguanosine
from the 5′ end of poly(A)-containing mRNA from rabbit reticulocytes
reduces its ability to direct in vitro synthesis of globin. When reovirus
mRNA is transcribed in vitro in a system deficient in methyl group
donors, transcripts are obtained that lack the 5′ 7-methylguanosine or
that contain an unmethylated guanosine. Neither of these forms will
initiate translation efficiently. However, when the inactive messages
are incubated with a methylating system, they are modified to contain
the sequence m^7GpppG and become fully active in initiating translation. These results indicate that the presence of 7-methylguanosine is
essential for efficient initiation of translation.

Further support for the need of a 7-methylguanosine cap for initiation of translation has been inferred from the fact that 7-methylgua-
nosine-5′-monophosphate inhibits initiation, apparently by competition with the cap structure. However, these findings must all be
viewed cautiously, since certain viral RNAs that lack caps initiate
translation readily, and one research group has claimed that if caps
are removed carefully without damage to other parts of the mRNA
molecule, it is possible to obtain translation without a cap. Preliminary
studies also indicate that caps may contribute to mRNA stability. One
recent report of particular interest suggests that mRNA molecules
with "cap II" structures ($m^7GpppX^mpY^mpZp$ . . .) have longer half-
lives than those with "cap I" structures (m^7GpppX^mpYpZp . . .). If verified in other laboratories, this observation could be highly significant
in the study of cellular differentiation, since the mRNAs that code for
differentiated products that are synthesized in large amounts tend to
have long half-lives.

**POSSIBLE FUNCTIONS OF
MODIFIED ENDS OF mRNA
5′ Caps**

Poly(A)   Typical eukaryotic cells appear to contain three distinct classes of mRNA: (1) poly(A)-containing mRNA, which comprises about 70% of the total mRNA in HeLa cells; (2) histone mRNA, which lacks poly(A) and is normally a minor fraction, except in very rapidly cleaving embryos, where its amount can be substantial (e.g., 60% of the total mRNA in late cleavage stage sea urchin embryos); and (3) nonhistone, nonpoly(A)-containing mRNA, which comprises about 30% of the mRNA in HeLa cells.

In HeLa cells, histone mRNA is characterized by very rapid entry into the cytoplasm after transcription, whereas the other two classes appear in the cytoplasm only after a lag of about 20 minutes. The possibility that poly(A) may be involved in the transport of messages from the nucleus to the cytoplasm has been proposed. The drug cordycepin (3' deoxyadenosine), which selectively blocks the synthesis of poly(A) but not of hnRNA, drastically inhibits the entry of new mRNA into the cytoplasm. The poly(A)$^+$ class is almost totally absent, as expected from inhibition of poly(A) synthesis, and the poly(A)$^-$ class is also significantly reduced. If poly(A) were not important for processing or transport, the amount of the poly(A)$^-$ class should have been increased by addition to it of normally poly(A)$^+$ species whose polyadenylation had been blocked by cordycepin. The fact that cordycepin also reduces the appearance of poly(A)$^-$ mRNA in the cytoplasm below control levels is not easy to explain in terms of inhibiting poly(A) synthesis, however. Thus cordycepin may also affect other processes essential to message transport. Alternately, the poly(A)$^-$ mRNA might actually contain a poly(A) sequence that is too short or too transitory to be detected by the usual technique of binding to poly(U) Sepharose.

When cells that have been maintained in culture in a nongrowing state are stimulated to begin growth, there is a substantial increase in the amount of cytoplasmic poly(A)$^+$ mRNA per cell, which cannot be accounted for by changes in the total amount of nuclear poly(A)$^+$ hnRNA, changes in its rate of synthesis, or changes in the half-life of cytoplasmic poly(A)$^+$ mRNA. The major change appears to be increased efficiency of conversion of nuclear poly(A)$^+$ hnRNA into cytoplasmic mRNA. Thus a higher order of control that goes beyond the simple presence of poly(A) may be involved.

Poly(A) is probably involved in processes other than movement of mRNA out of the nucleus, since it is also added to mitochondrial and viral messages that are transcribed in the cytoplasm. A number of investigators have suggested that poly(A) may affect either the stability of messages or their frequency of translation. Poly(A) is not an absolute requirement for translation, since messages that naturally lack poly(A) are readily translated, and removal of poly(A) does not alter the efficiency of translation of messages that normally contain it. However, when globin mRNA lacking poly(A) is incubated in a cell-free translation system in the absence of amino acids, its activity decays more rapidly than that of globin mRNA with poly(A). This and similar results suggest that poly(A) may increase message stability. It has also been observed that the poly(A) attached to a message becomes progressively shorter as that message is repeatedly translated, and speculative models have suggested that this could be a means of regulating how many times mRNA molecules are translated before they are degraded.

Both hnRNA and mRNA are found complexed with protein molecules when they are extracted from cells. These RNA-protein complexes, which are termed ribonucleoprotein (RNP) particles, appear not to be artifacts of isolation. It is considered likely that the proteins that bind to RNA play significant roles in processing, transport, or translation of the RNAs to which they attach. Characterization of the proteins found in RNP particles is currently an area of active research, and determination of specific functions for some of these proteins may soon be possible.

RIBONUCLEOPROTEIN COMPLEXES CONTAINING hnRNA AND mRNA

Association of hnRNA with protein has been shown both biochemically and by cytological observation. Considerable evidence indicates that hnRNP complexes actually occur in nuclei of living cells and that they are not artificially produced by the procedures used for their isolation. For example, as seen in Fig. 8-7, the particles can be visualized in electron micrographs of HeLa cell nuclei lysed with mild detergent procedures that are known not to disrupt polynucleotide structures in other cell types. Also, the sedimentation characteristics and protein composition of hnRNPs are found to be the same when they are isolated by a variety of different procedures. Finally, the proteins found in the hnRNP particles are different from those found in other cell fractions, including the proteins of chromatin, and hnRNP particles cannot be produced artificially by adding hnRNA molecules to intact or disrupted nuclei or to nuclear proteins.

hnRNP

The protein composition of hnRNP particles has been a point of disagreement among various investigators. G. P. Georgiev and co-workers in the Soviet Union, who have been studying ribonucleoprotein particles of rat liver for about 10 years, claim to find RNP particles that release a single protein species of 40,000 daltons on disruption. This protein, which they have named informatin, is associated with RNA in particles of a discrete size, which they call informofers. They envision each hnRNA molecule as being associated with a large

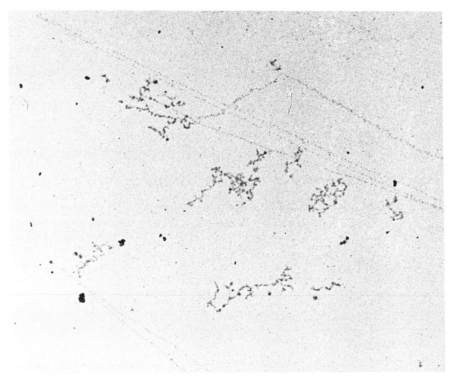

Fig. 8-7. Electron micrograph showing RNP fibrils in HeLa nuclei dispersed by mild detergent procedures. Some fibrils remain attached to the chromatin, and others are free. (From Miller, O. L., Jr., and A. H. Bakken. 1972. *In* E. Diczfalusy, ed. Gene transcription in reproductive tissue. Fifth Karolinska Symposium on Research Methods in Reproductive Endocrinology. Karolinska Institutet, Stockholm.)

number of these informofers in a manner similar to that diagrammed in Fig. 8-8.

These findings, however, have been recently challenged by other workers in the field who find a more heterogeneous population of proteins in the hnRNP particles of different cell types, including rat liver cells. For example, in HeLa cells a large number of protein species varying in molecular weight from 39,000 to 180,000 have been recovered from hnRNP particles. Complex distributions of proteins are also claimed to be found in hnRNP particles from three other lines of cultured cells and from rat liver cells. However, when the fraction of hnRNP that is devoid of poly(A) is analyzed separately, a single species of protein with a molecular weight the same as that of informofer protein is found. Reports that different cell types contain different RNA-bound protein populations have led to speculation that there

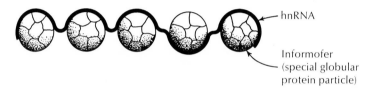

hnRNA

Informofer
(special globular
protein particle)

Fig. 8-8. Proposed association of hnRNA with informofer particles. (With permission from Samarina, O. P., et al. 1968. J. Mol. Biol. **33:**251. Copyright by Academic Press Inc. [London] Ltd.)

may be specific interactions between certain proteins and certain specific message sequences.

Thoru Pederson and co-workers have recently compared the proteins found in hnRNP and mRNP particles of HeLa cells. They isolated ribosome-associated mRNP particles by binding them to oligo(dT) cellulose. Thus they were examining only one fraction of cytoplasmic mRNP particles—those bound to ribosomes and containing poly(A) in a configuration free for binding. Three major protein species were found in the mRNP particles isolated in this fashion, including a protein that specifically binds to poly(A) sequences. None of the three were major components of hnRNP. Although the techniques used in examining the proteins of the RNP particles in these cells revealed only the major protein species, these experiments suggest that mRNP particles contain proteins that are distinct from those in the hnRNP particles.

Relatively little is known about the exact functions of the proteins that form complexes with hnRNA and mRNA. However, it has been found that the binding of globin mRNP particles to ribosomes is more efficient than that of free globin mRNA. Also, there has been extensive speculation concerning possible roles of such proteins in the "masking" of messages and in translational controls, as will be discussed in the following chapter. It is possible that some proteins associated with mRNA in mRNP particles may have a stimulatory effect on their translation while others have an inhibitory effect.

In this chapter we have examined the initial product of transcription in eukaryotic cells, its relationship to functional cytoplasmic mRNA, and the complex series of processing events that are involved in generating mRNA from the transcript and in transporting that mRNA to the cytoplasm. The size of each eukaryotic gene transcription unit is definitely larger than the actual protein-coding sequence that it contains, although the question of how much larger remains

mRNP

RELATIONSHIP OF PROCESSING TO REGULATION OF GENE EXPRESSION

controversial. Nucleotide sequences, which some investigators feel may be several times longer than the final mRNA, are removed from the initial transcript during mRNA formation. In addition, the mRNA itself contains substantial nontranslated sequences located at both ends of the actual message sequence.

The exact contribution of processing and transport of mRNA to the specificity of gene expression is still not fully understood. Although selective control of transcription is clearly of importance in cellular differentiation, it is also well established that much of the RNA that is transcribed in the nucleus never becomes cytoplasmic mRNA. Recent evidence suggests that processing may serve as the primary point of control for all structural genes except those expressed in unusually large amounts, and that a substantial number of potential protein-coding sequences are actually destroyed during processing. Addition of 7-methylguanosine cap structures at the 5' ends of mRNA molecules and polyadenylation of the 3' ends both appear to contribute to message stability. Messages for differentiated products that are made in large amounts are known to have long half-lives, but the exact contribution of caps and poly(A) to that stability is not clear. Preliminary observations suggesting that stable messages may have a special kind of 5' cap emphasize the need for further study in this area. The observation that poly(A) becomes progressively shorter as mRNA ages also needs to be investigated further.

The repeated DNA sequences that are interspersed among unique coding sequences in the eukaryotic genome appear to be represented to a much greater extent in hnRNA than in mRNA. Either these repeated sequences are removed during processing or they occur only in hnRNA molecules that are not destined to become mRNA. If the former is true, as seems likely, the repeated sequences may be related to regulatory mechanisms, as suggested by the Britten-Davidson model in Chapter 7, and the new Davidson-Britten model described in this chapter.

The rapid progress that is being made in sequencing techniques both for DNA and for RNA is beginning to make it possible to compare the limited sequences of cytoplasmic mRNA molecules with more extended sequences of the genes from which they are transcribed. Already this approach has revealed the presence of noncoding sequences within the coding sequences of several genes. In the future such comparisons should also be helpful in identifying the regulatory sequences that are associated with specific structural genes. These include the signals for beginning and ending transcription, recognition sites for regulatory molecules that determine whether transcription occurs, sequences that determine where the hnRNA is cleaved to generate mRNA, recognition sequences for capping and the addition of poly(A), recognition sites for ribosomal attachment to mRNA, and, finally, the initiation and termination codons for translation of the actual coding sequences. Improved fractionation techniques and the use of cDNA probes to identify message sequences in larger nuclear RNA molecules have already made it possible to isolate precursors for specific kinds of mRNA from hnRNA and to analyze them. Also, it should soon be possible to determine whether there exists any substantial fraction of hnRNA molecules that do not contain sequences found in cytoplasmic mRNA (and that are not degradation products of larger transcripts that originally contained message sequences).

This chapter has described an area of research in which particularly rapid progress has been made during the last few years. During the time required to plan and write this book, there were reported in the literature two major unexpected findings that challenged previously accepted concepts and made the mRNA processing step far more important than had previously been recognized. The first was the discovery that most structural genes contain intervening noncoding sequences that are transcribed and subsequently removed during processing. The second was the observation that most structural genes appear to be transcribed continuously in differentiated cells, with the apparent elimination during processing of all coding sequences that are not expressed as cytoplasmic mRNA in the particular type of cell. These findings have generated far more questions than they have answered, and it currently appears likely that additional significant discoveries will be made in the near future concerning the processing of hnRNA to form mRNA and the transport of that mRNA to the cytoplasm.

Even after mRNA reaches the cytoplasm, there are still many opportunities for regulation of gene expression. These include "masking" of messages as protein complexes, selective stabilization of certain messages, differential translation of particular messages, a variety of modifications of newly synthesized peptide chains, and stabilization or selective degradation of particular amino acid sequences. These mechanisms and their possible relationship to control of gene expression are discussed in the following chapter, together with a brief overview of the entire scheme of molecular control of gene expression in eukaryotic cells.

BIBLIOGRAPHY
Books and reviews

Abelson, J. 1979. RNA processing and the intervening sequence problem. Annu. Rev. Biochem. 48:1035.

Cohn, W. E., and E. Volkin, eds. 1976. Progress in nucleic acid research and molecular biology. Vol. 19. mRNA: the relation of structure to function. Academic Press, Inc., New York. (This entire volume deals with topics related to the synthesis and function of mRNA. Several articles from it that are of particular interest are cited separately in this list.)

Crick, F. 1979. Split genes and RNA splicing. Science 204:264.

Darnell, J. E. 1976. mRNA structure and function. Prog. Nucleic Acid Res. Mol. Biol. 19:493.

Darnell, J. E., W. R. Jelinek, and G. R. Molloy. 1973. Biogenesis of mRNA: genetic regulation in mammalian cells. Science 181:1215.

Davidson, E. H., and R. J. Britten. 1979. Regulation of gene expression: possible role of repetitive sequences. Science 204:1052.

Edmonds, M., et al. 1976. Transcribed oligonucleotide sequences in HeLa cell hnRNA and mRNA. Prog. Nucleic Acid Res. Mol. Biol. 19:97.

Georgiev, G. P., and O. P. Samarina. 1971. D-RNA containing ribonucleoprotein particles. Adv. Cell Biol. 2:47.

Kedes, L. H. 1976. Histone messengers and histone genes. Cell 8:321.

Lewin, B. 1974. Gene expression. Vol. 2. Wiley-Interscience, New York.

Lewin, B. 1975. Units of transcription and translation: the relationship between hnRNA and mRNA. Cell 4:11.

Lewin, B. 1975. Units of transcription and translation; sequence components of hnRNA and mRNA. Cell 4:77.

Perry, R. P. 1976. Processing of RNA. Annu. Rev. Biochem. 45:605.

Perry, R. P., et al. 1976. The relationship between hnRNA and mRNA. Prog. Nucleic Acid Res. Mol. Biol. 19:275.

Proudfoot, N. J., C. C. Cheng, and G. G. Brownlee. 1976. Sequence analysis of eukaryotic mRNA. Prog. Nucleic Acid Res. Mol. Biol. 19:123.

Rottman, F., A. J. Shatkin, and R. P. Perry. 1974. Sequences containing methylated nucleosides at 5′ termini of mRNA's: possible implications for processing. Cell 3:197.

Shatkin, A. J. 1976. Capping of eucaryotic mRNAs. Cell 9:645.

Weatherall, D. J., and J. B. Clegg. 1976. Molecular genetics of human hemoglobin. Annu. Rev. Genet. 10:157.

Selected original research articles

Both, G. W., A. K. Banerjee, and A. J. Shatkin. 1975. Methylation-dependent translation of viral mRNA's in vitro. Proc. Natl. Acad. Sci. U.S.A. 72:1189.

Dawid, I. B., and P. K. Wellauer. 1976. A reinvestigation of 5′-3′ polarity in 40S ribosomal RNA precursor of *Xenopus laevis*. Cell 8:443.

Efstratiadis, A., F. C. Kafatos, and T. Maniatis. 1977. The primary structure of rabbit β-globin mRNA as determined from cloned DNA. Cell 10:571.

Griffin, B. 1976. Eukaryotic mRNA: trouble at the 5′-end. Nature 263:188.

Hickey, E. D., L. A. Weber, and C. Baglioni. 1976. Inhibition of initiation of protein synthesis by 7-methylguanosine-5′-monophosphate. Proc. Natl. Acad. Sci. U.S.A. 73:19.

Humphries, S., J. Windass, and R. Williamson. 1976. Mouse globin gene expression in erythroid and non-erythroid tissues. Cell 7:267.

Kinniburgh, A. J., and T. E. Martin. 1976. Oligo(A) and oligo(A)-adjacent sequences present in nuclear ribonucleoprotein complexes and mRNA. Biochem. Biophys. Res. Commun. 73:718.

Kinniburgh, A. J., J. E. Mertz, and J. Ross. 1978. The precursor of mouse β-globin messenger RNA contains two intervening RNA sequences. Cell 14:681.

Knapp, G., et al. 1978. Transcription and processing of intervening sequences in yeast tRNA genes. Cell 14:221.

Kumar, A., and T. Pederson. 1975. Comparison of proteins bound to hnRNA and mRNA in HeLa cells. J. Mol. Biol. 96:353.

Lai, E. C., et al. 1978. The ovalbumin gene: structural sequences in native chicken DNA are not contiguous. Proc. Natl. Acad. Sci. U.S.A. 75:2205.

Lizardi, P. M. 1976. The size of pulse-labeled fibroin messenger RNA. Cell 7:239.

Mandel, J. L., et al. 1978. Organization of coding and intervening sequences in the chicken ovalbumin split gene. Cell 14:641.

McReynolds, L., et al. 1978. Sequence of chicken ovalbumin mRNA. Nature 273:723.

Melli, M., et al. 1977. Presence of histone mRNA sequences in high molecular weight RNA of HeLa cells. Cell 11:651.

Melli, M., G. Spinelli, and E. Arnold. 1977. Synthesis of histone messenger RNA of HeLa cells during the cell cycle. Cell 12:167.

Milcarek, C., R. Price, and S. Penman. 1974. The metabolism of a poly(A) minus mRNA fraction in HeLa cells. Cell 3:1.

Molloy, G. R., et al. 1974. Arrangement of specific oligonucleotides within poly(A) terminated hnRNA molecules. Cell 1:43.

Nemer, M. 1975. Developmental changes in the synthesis of sea urchin embryo messenger RNA containing and lacking polyadenylic acid. Cell 6:559.

Pederson, T. 1974. Proteins associated with hnRNA in eukaryotic cells. J. Mol. Biol. 83:163.

Perry, R. P., and D. E. Kelley. 1976. Kinetics of formation of 5′ terminal caps in mRNA. Cell 8:433.

Proudfoot, N. J., and G. G. Brownlee. 1974. Sequence at the 3′ end of globin mRNA shows homology with immunoglobin light chain mRNA. Nature 252:359.

Proudfoot, N. J., and J. I. Longley. 1976. The 3′ terminal sequences of human α and β globin messenger RNAs: comparison with rabbit globin messenger RNA. Cell 9:733.

Schwartz, H., and J. E. Darnell. 1976. The association of protein with the polyadenylic acid of HeLa cell messenger RNA: evidence for a "transport" role of a 75,000 molecular weight polypeptide. J. Mol. Biol. 104:833.

Sippel, A. E., et al. 1974. Translational properties of rabbit globin mRNA after specific removal of poly(A) with ribonuclease H. Proc. Natl. Acad. Sci. U.S.A. 71:4635.

Spohr, G., G. Dettori, and V. Manzari. 1976. Globin mRNA sequences in polyadenylated and nonpolyadenylated nuclear precursor-messenger RNA from avian erythroblasts. Cell 8:505.

Stein, G., et al. 1977. Evidence that the coupling of histone gene expression and DNA synthesis in HeLa S_3 cells is not mediated at the transcriptional level. Biochem. Biophys. Res. Commun. 77:245.

Tilghman, S. M., et al. 1978. The intervening sequence of a mouse β-globin gene is transcribed within the 15S β-globin mRNA precursor. Proc. Natl. Acad. Sci. U.S.A. 75:1309.

Wold, B. J., et al. 1978. Sea urchin embryo mRNA sequences expressed in the nuclear RNA of adult tissues. Cell 14:941.

CHAPTER 9

Controls acting at the cytoplasmic level

☐ In Chapters 7 and 8, we have seen that only a small portion of the genetic information contained in the genome is normally expressed as cytoplasmic mRNA. Control mechanisms acting at the level of transcription and at the level of posttranscriptional processing to form mRNA both appear to be important in determining the amounts and kinds of mRNA that reach the cytoplasm. In this chapter we will examine additional control mechanisms that act after the mRNA has reached the cytoplasm.

DO CYTOPLASMIC CONTROLS MAKE SENSE?

Questions are sometimes raised concerning the need for cytoplasmic controls over gene expression and the seeming lack of efficiency of such controls. Why should an organism that is capable of transcriptional control waste the energy involved in synthesis, processing, and transport of mRNA and then not fully use that mRNA? Questions of this sort can be misleading for two reasons:

1. They are phrased in teleological terms (see Chapter 1). The organism does not go through a reasoning process to decide which is the most effective means of regulating gene expression. The controls function as they do as the result of prolonged evolutionary selection for maximum survival value. When an organism exhibits a particular property, we have no real basis for challenging whether that property "makes sense." It obviously has made sense from an evolutionary point of view, or it would not be there.

2. A process may appear to us to be inefficient or wasteful simply because we do not fully understand its details or the subtle advantages that have led to its evolutionary establishment. Evolution is based on stochastic (random chance) events coupled with selection for survival advantage and not on deliberate engineering criteria. Organisms sometimes become locked into ancestral patterns that in themselves have no contemporary survival value but are preserved because they

provide a necessary starting point for developmental programs that have evolved more recently and do have survival value. An example is the appearance and disappearance during mammalian development of two types of primitive kidneys, the pronephros and the mesonephros. Although not functional in postembryonic life, these kidneys and their duct systems have important roles in the development of the metanephros, which is the functional kidney of mammals, and also in development of the male and female reproductive systems (Chapters 18 and 20).

We are concerned in this chapter with message-specific controls that affect translational or posttranslational steps in the production of functional gene products. Controls that act at the cytoplasmic level appear in general to have these major types of effect. First, they operate in cases in which a rapid response to some stimulus is necessary. Second, they operate to produce a precise balance in the production of two or more interacting molecules. Finally, they act in some systems to reinforce or amplify the effects of primary controls acting at the level of RNA synthesis. There are also a variety of modifications in the translational system that act to increase or decrease the overall rate of translation, with little or no selectivity for individual types of proteins. We are concerned with such mechanisms in this chapter only in cases in which they constitute part of a differentiated response. An example of such a change is the overall enhancement of translational activity that occurs in the chicken oviduct after stimulation of ovalbumin synthesis by administration of estrogen (Chapter 14).

The presentation of cytoplasmic control systems in this chapter is organized around the specific translational or posttranslational mechanisms that are involved. It begins with a general overview of the mechanisms of translation and then considers, in the following order, mechanisms affecting the actual translational process, mechanisms influencing availability and stability of mRNA, and mechanisms involving posttranslational modification of peptide chains. The chapter ends with a brief summary of the overall pathway of information flow from genes to differentiated products and of the many points along that pathway where the flow of information can be regulated either in practice or in theory.

Before we examine the specific controls over gene expression that act at the cytoplasmic level in eukaryotic cells, it will be helpful to review briefly the general scheme of protein synthesis as it is currently understood for these cells. This description is, of necessity, fairly complex. However, we believe that the student will benefit by exposure to this complexity even though a detailed understanding of all the steps

MECHANISMS OF TRANSLATION

Table 9-1. Molecules involved in translation in eukaryotic cells

Molecular species	Number
mRNA	1
40S ribosomal subunit (1 RNA, 25 to 30 proteins*)	26 to 31
60S ribosomal subunit (3 RNAs, 35 to 45 proteins*)	38 to 48
Initiation factors†	18 to 20
Elongation factors*	2
Release factor	1
tRNAs*	About 43
Amino acids	20
Aminoacyl-tRNA synthetases	20
ATP	1
GTP	1
Mg^{++}	1
*Total different molecules required for translation**	172 to 189

*Some values reported are estimates based on best current data.
†eIF-3 is a complex of 9 to 11 peptides (Table 9-2).

involved in protein synthesis is not critical to understanding the control mechanisms discussed later in the chapter.

Translation in eukaryotic cells is a highly complex process. Altogether, about 175 different molecular species are involved in some aspect of translation (Table 9-1). The overall process of translation in eukaryotes is similar to that in prokaryotes (Chapter 4) and can conveniently be divided into three phases—initiation, elongation, and termination.

Initiation

The initiation process in eukaryotes will be discussed rather thoroughly, since it differs, at least in detail, from prokaryotic initiation and also because it appears to be the phase that is most frequently involved in translational controls. Unfortunately, initiation is also the least completely understood phase of translation in eukaryotes. Studies on initiation fall into two general groups—those which use natural mRNA and those which use artificial substitutes for mRNA such as the homopolymer polyuridylic acid (poly[U]) and the trinucleotide ApUpG. Results obtained with these artificial model systems differ in several respects from those obtained with natural mRNA, but the reasons for the differences are not fully understood. Since we are concerned with control of translation, most of our attention in this chapter will be given to studies using natural messages.

In eukaryotic cells, as in prokaryotic cells, synthesis of protein chains begins with the amino acid methionine at the N-terminal end and proceeds toward the C-terminal end. However, the N-terminal amino group is not blocked in eukaryotes as it is in prokaryotes, where translation always begins with *N*-formylmethionine. The *N*-terminal methionine in eukaryotic protein synthesis is carried by a special initiator tRNA (Met-tRNA$_i$). New peptide chains are started at AUG initiation codons, and translation proceeds in a 5'-to-3' direction along the mRNA molecules. As described in the previous chapter, the initiation codon is usually located a substantial distance from the 5' end of the mRNA.

Numerous initiation factors are needed for assembly of the initiation complex. The names, sizes, and probable functions of these factors are summarized in Table 9-2. Unfortunately, several different sets of names have been used for these factors until recently. In this book we use the standardized nomenclature adopted at the International

Table 9-2. Summary of eukaryotic initiation factors

Name*	Previous nomenclature†			Molecular properties‡	Proposed function§
	Basel (Staehelin)	N.I.H. (Anderson)	Weissbach and Ochoa		
eIF-1	IF-E1	—	—	15,000 daltons	Binding of met-tRNA$_i$ to 40S subunit; binding of mRNA to 40S complex
eIF-2	IF-E2	IF-MP	eIF-2	3 proteins — 32,000, 47,000, and 50,000 daltons	Forms ternary complex with met-tRNA$_i$ and GTP; needed for binding of met-tRNA$_i$ to 40S subunit
eIF-3	IF-E3	IF-M5	—	Complex of 9 to 11 peptides totaling 500,000 daltons	Blocks premature joining of 40S and 60S subunits; stabilizes complex of 40S subunit with met-tRNA$_i$, GTP and eIF-2; essential for natural mRNA binding
eIF-4a	IF-E4	IF-M4	—	50,000 daltons	All members of eIF-4 group are thought to be involved in mRNA recognition and binding to 40S complex; eIF-4b hydrolyzes ATP during mRNA binding
eIF-4b	IF-E6	IF-M3	eIF-3	80,000 daltons	
eIF-4c	IF-E7	IF-M2B$_\beta$	eIF-2a$_3$	19,500 daltons	
eIF-4d	—	IF-M2B$_\alpha$	eIF-2a$_2$	16,500 daltons	
eIF-5	IF-E5	IF-M2A	eIF-2a$_1$	170,000 daltons	Joining of 40S complex to 60S subunit; hydrolysis of GTP
eIF-M1	—	IF-M1	eIF-2'	65,000 daltons	Formation of initiation complex with artificial messages such as poly (U) or ApUpG; not needed with natural mRNA; may be an evolutionary remnant

*The new standardized nomenclature for eukaryotic initiation factors adopted by the International Symposium on Protein Synthesis, National Institutes of Health, Oct. 18, 1976, Bethesda, Md., is used in this book. (From Anderson, W. F., et al. 1977. F.E.B.S. Lett. **76:**1.)

†These three systems of nomenclature were all in use prior to the symposium cited above. All three are used in references cited at the end of this chapter.

‡The molecular properties listed here are tentative in most cases. Some of the initiation factors for translation in eukaryotic cells have been discovered recently, and studies on them are incomplete. Most numerical values are from Staehelin, T. et al. 1975. Proc. Tenth F.E.B.S. Meeting. **39:**309 or from Weissbach, H., and S. Ochoa. 1976. Annu. Rev. Biochem. **45:**191. Values from Staehelin et al. have been used in cases of conflicts. Weissbach and Ochoa list a mass of 120,000 daltons for eIF-5 and approximately 400,000 daltons for eIF-4b.

§The proposed functions for eIF-1 and the eIF-4 group are tentative.

Symposium on Protein Synthesis held in October, 1976. However, most of the source references cited at the end of this chapter use one or another of the three older systems also listed in Table 9-2.

The major steps in eukaryotic initiation are depicted in Fig. 9-1. One molecule each of Met-tRNA$_i$, GTP, and eIF-2 form a ternary complex, which then attaches to the 40S ribosomal subunit. Attachment of mRNA to the 40S complex is promoted by several additional initiation factors and requires the hydrolysis of ATP, catalyzed by eIF-4b. The eIF-3 factor, which is essential for mRNA attachment, consists of about 9 to 11 different protein molecules aggregated together rather tightly. The composition of the eIF-3 complex is variable, and there are suggestions in the literature that different eIF-3–like complexes may be required for translation of different mRNAs, as will be discussed later in this chapter. Formation of the completed 80S initiation complex is achieved by linking a 60S ribosomal subunit to the previously formed 40S initiation complex. This requires a specific linking factor, eIF-5, which hydrolyzes the GTP that was bound to the 40S complex when it first was formed. All the initiation factors are displaced from the 40S complex when the 80S complex is formed.

In model systems in which initiation complexes are formed in the absence of natural mRNA through use of ApUpG, an initiation factor designated eIF-M1, which appears to be different from any of the factors previously discussed, is required. This factor is found in most eukaryotic cells, but its exact role (if any) in normal initiation of mRNA translation is not known, and some investigators view it as a nonfunctional evolutionary remnant.

Recent data suggest that the 7-methylguanosine caps at the 5′ ends of mRNA molecules play important roles in initiation, since removal of the caps greatly reduces initiation and since 7-methylguanosine-5′-monophosphate will competitively inhibit initiation. Studies of capped

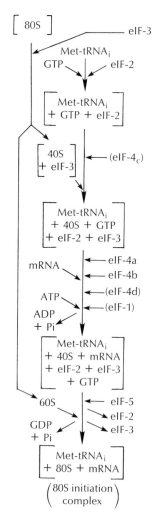

Fig. 9-1. Formation of 80S initiation complex in eukaryotic cells. Reactants are shown on the left and catalytic initiation factors on the right. Brackets surround stable intermediate complexes. Arrows angled downward indicate entry of components into complexes or release of reaction products from complexes. Note that initiation factors eIF-2 and eIF-3 are incorporated into complexes and later released during the final step leading to the formation of the 80S initiation complex. Horizontal arrows indicate catalytic action without formation of stable complexes. Factors that are stimulatory but have not been shown to be absolutely required are enclosed in parenthesis. One mole each of ATP and GTP are hydrolyzed during formation of the 80S complex. Although the principles are similar, many details of this series of reactions differ from those involved in the formation of the 70S initiation complex in prokaryotic cells (Fig. 4-17). (Modified from Benne, R., and J. W. B. Hershey. 1978. J. Biol. Chem. **253**:3078 [with modifications by Walthall, B., personal communication].)

synthetic polynucleotides suggest that efficient formation of initiation complexes requires both the presence of a cap and an adjacent AU-rich sequence. The substantial separation between the cap and the AUG initiation codon in mRNAs that have been studied carefully is puzzling. It is not obvious how the cap and the AUG codon can both participate in initiation. Perhaps the secondary structures of the mRNA molecules are such that the AUG codon is brought close to the cap by loop formation or by specific binding sites on ribosomal sub-units or initiation factors. Sequence analysis of 5′ leaders may provide a definite answer in the near future.

The elongation process, which is responsible for the actual assem-   **Elongation**
bly of protein chains, is similar in eukaryotes and prokaryotes (see Chapter 4 and Fig. 4-18 for the details of the process). In eukaryotes,

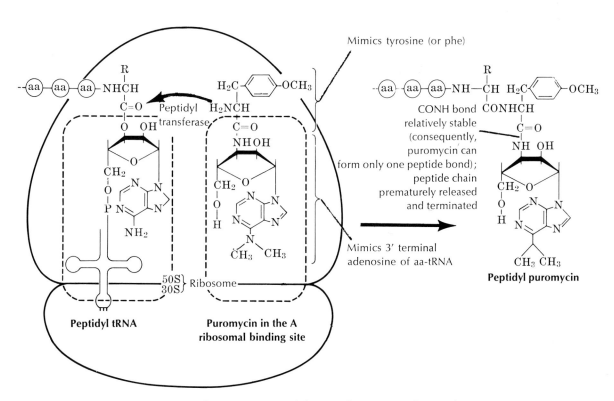

Fig. 9-2. Diagram showing action of puromycin in inhibition of protein synthesis. The structure of puromycin is similar to the 3′ end of an aminoacyl tRNA, and it is able to bind to the A site of the ribosome. The growing chain is then transferred to puromycin by pepti-dyl transferase and prematurely terminated. aa = Aminoacyl. (From James D. Watson, Molecular biology of the gene, 3rd ed., copyright © 1976, 1970, 1965 by James D. Watson and The Benjamin/Cummings Publishing Co., Inc., Menlo Park, California.)

elongation factor EF-1 is needed to bind the aminoacyl tRNA to the A site of the ribosome, and elongation factor EF-2 is involved in translocation of the peptidyl tRNA from the A site to the P site.

Many inhibitors of protein synthesis block some phase of elongation. Cycloheximide, which is active only in eukaryotic systems, binds to the 60S ribosomal subunit and blocks the process of chain elongation. Puromycin, which is active in both prokaryotic and eukaryotic systems, has a structure sufficiently similar to the 3' end of aminoacyl tRNA that it binds to the A site of the large ribosomal subunit. The growing peptide chain from the P site is transferred to a free amino group on the puromycin molecule. This results in premature termination of growth of the peptide chain, since the puromycin does not attach to the P site and does not have a carboxyl group for the addition of further amino acids (Fig. 9-2). Diphtheria toxin inhibits elongation in eukaryotes, but not in prokaryotes, by catalyzing the transfer of part of the coenzyme nicotinamide-adenine dinucleotide (NAD) to EF-2, rendering it nonfunctional.

Termination

Termination of protein synthesis in eukaryotes is also similar to termination in prokaryotes, although only a single release factor (RF) has been identified in eukaryotes.

The complex mechanisms involved in the three phases of translation (initiation, elongation, and release), together with mechanisms involving the availability and stability of mRNA, offer many opportunities for modifying the flow of information from genes to final differentiated products. A variety of examples are discussed in the following sections.

CONTROLS ACTING AT THE TRANSLATIONAL LEVEL

There are reports in the literature suggesting that control over gene expression can occur at several different points in the overall process of translation. Unfortunately, the data are frequently preliminary and incomplete, and in some of the more recently reported cases it is not yet possible to say which specific aspect of protein synthesis is affected.

Discrimination among messages at the level of initiation is well documented, but the nature of the process that is involved remains controversial. Under conditions of mRNA excess, in which the rate of translation is limited by components of the translational machinery, some types of message are definitely used more effectively than others. Some investigators, including Harvey Lodish, whose reviews are cited at the end of this chapter, believe that this is the primary, if not the only, type of specific regulation that occurs at the translational level. However, others claim to have found factors influencing translation that are specific for particular types of mRNA.

Controversy also surrounds the role of tRNA in determining translational specificity. It is well established that the relative amounts of the tRNAs for the various amino acids are adjusted to accommodate for the amino acid compositions of proteins that are made in particularly large amounts. There also are reports that translation of certain messages is controlled by the availability of specific tRNAs. However, these claims have been challenged repeatedly, and the final answers are not yet available.

The "masking" of messages by macromolecules that bind tightly to them and render them unavailable for translation is well established, particularly in early embryos. However, the question of whether message-specific selectivity is involved is still being debated. There are reports that specific types of messages that are not being translated have been detected in the cytoplasm of certain types of cells, but the data are not yet sufficiently complete to distinguish between masking and more direct effects on the translational process itself.

A variety of experimental approaches have been used to study competition between human α-globin and β-globin mRNAs for initiation of translation. The α- and β-globin molecules are needed in equal amounts for assembly of hemoglobin A ($\alpha_2\beta_2$), which comprises 98% of normal adult hemoglobin; and only a small additional amount of α-globin is needed for the minor components, hemoglobin A_2 ($\alpha_2\delta_2$), which accounts for most of the remaining 2%, and hemoglobin F ($\alpha_2\gamma_2$), which is a fetal form that is normally present only in trace amounts in adult hemoglobin. In reticulocytes (immature red blood cells that are actively engaged in hemoglobin synthesis), α- and β-chains are normally synthesized in a ratio of about 1.04:1. However, polysomes engaged in the synthesis of β-chains have been found to contain an average of five ribosomes, whereas polysomes making α-chains contain an average of three ribosomes. Since the mRNAs are similar in size, and since the numbers of α- and β-chains made per unit of time are essentially the same, this observation must mean either that ribosomes move more rapidly along the α-message or that there are more α-messages than β-messages and that translation of the α-messages is initiated less frequently. These two possibilities are depicted in Fig. 9-3.

The latter possibility, that there are more α-globin messages, with each message being initiated less frequently, has been shown to be correct by experiments using inhibitors of elongation. These inhibitors slow the rate of elongation to such a degree that initiation is no longer the rate-limiting factor. Both types of messages become saturated with ribosomes, and the relative amounts of the two types of globin chains that are synthesized depend only on the relative amounts of the

Competition among messages at initiation of translation

A

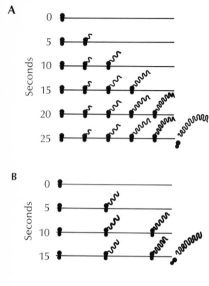

B

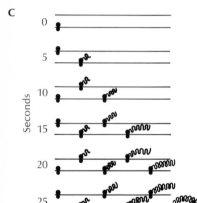

Fig. 9-3. Two theoretical models to account for differences in the sizes of polysomes synthesizing α- and β-globin chains. In both cases the sizes of the α- and β-globin mRNAs are assumed to be equal, and equal numbers of both proteins are assumed to be synthesized. **A,** Synthesis of β-chains for both models. An elongation rate of six amino acids per second and initiation once every 5 seconds lead to a steady state in which most polysomes synthesizing the 146−amino acid protein chain contain five ribosomes and one chain is completed every 5 seconds. **B,** Synthesis of α-chains in the first model. This model assumes that the frequency of initiation is the same for both chains and that the rate of elongation of α is 2 × that of β. Steady state conditions are two to three ribosomes per polysome, and one chain is completed every 5 seconds. **C,** Synthesis of α-chains in the second model. This model assumes (1) that there are more α-mRNAs than β, (2) that α-translation is initiated less often than β, and (3) that elongation rates are equal. For simplicity, this model depicts twice as much α-mRNA as β and initiation of α-translation only half as often as β. Steady state conditions are two to three ribosomes per polysome, and one chain is completed every 10 seconds for each mRNA. Current experimental data favor a model similar to this, except that the actual ratio of α- to β-mRNA is about 3:2, with initiation of translation on each α-mRNA occurring on the average only two thirds as often as initiation on each β-mRNA.

C

two types of mRNA that are present. The results of this experiment, which have also been confirmed by other experimental approaches, show that the ratio of α-globin mRNA to β-globin mRNA in reticulocytes is about 1.5:1 (Table 9-3).

It has also been demonstrated that the β-globin message has a greater affinity for initiation factors than the α-globin message. This finding has come from experiments done with a partially purified in vitro protein-synthesizing system in which the relative amounts of mRNA and initiation factors can be varied. At very low concentrations of mRNA, the ratio of α- to β-globin chains synthesized is 1.5:1, the same as the ratio of messages. However, as the concentration of mRNA is increased and the availability of initiation factors becomes rate limiting, the β-message competes more effectively for them than the α-message, and the relative amount of α-chain that is synthesized declines. At extremely high levels of mRNA the ratio of α- to β-chains synthesized is only 0.03:1.

Each globin chain in hemoglobin has covalently linked to it an iron-containing heme group that is responsible for the oxygen transport function of hemoglobin. Synthesis of globin chains in intact reticulocytes and in cell-free reticulocyte lysates is controlled at the initiation level by the availability of heme groups in the form of hemin. Deficiency in hemin results in the rapid appearance (1 to 2 minutes) of an ATP-dependent kinase activity that inhibits the action of eIF-2 by

Table 9-3. Ratio of α- to β-globin synthesized in the presence of inhibitors of chain elongation*

	Ratio of α- to β-globin	
Drug	Digest of [^{35}S]fMet-labeled protein†	Carboxymethylcellulose chromatography of [^{14}C]leucine-labeled protein
None	0.98; 1.04	1.03; 1.01
Sparsomycin	1.29	1.18
Blasticidin	1.60	1.48
Anisomycin	1.47	1.37
Fusidic acid	1.34	1.23
Streptovitacin	1.46	1.23
Cycloheximide	1.44	1.39
Emetine	1.51	1.40

*Modified from Lodish, H. F. 1971. J. Biol. Chem. **246:**7131.
†E. Coli enzymes were used to charge the initiator tRNA in this experiment. Thus the chains synthesized in vitro were initiated with N-formylmethionine.

phosphorylating the smallest of its three subunits. A second ATP-dependent kinase activity that phosphorylates specific proteins in the 40S ribosomal subunit has also been reported but is still uncertain. Phosphorylation of eIF-2 significantly inhibits initiation of globin mRNA translation. The exact mechanism responsible for the very rapid activation of the inhibitory kinase when the concentration of hemin is reduced is not known, but the speed of the reaction suggests that the kinases must already be present in the cytoplasm and maintained in an inactive form by direct or indirect action of hemin. In cases in which translation is only partially blocked by hemin deficiency, synthesis of α-chains is inhibited more severely than that of β-chains, a finding that is consistent with the preferential initiation of β-chains under competitive conditions, as just discussed. The addition of eIF-2 to hemin-deficient systems restores the synthesis of globin and also returns the relative amounts of the two chains synthesized to a normal ratio (Fig. 9-4).

An interesting case of translational discrimination between α- and β-globin messages, whose mechanism remains unknown, occurs in the human hereditary disease β_0-thalassemia Ferrara. The β_0-thalassemia syndrome is characterized by the absence of hemoglobin A ($\alpha_2\beta_2$), due to failure of synthesis of β-globin chains, and a compensatory increase in hemoglobins A$_2$ ($\alpha_2\delta_2$) and F ($\alpha_2\gamma_2$). Usually the condition is caused by either a deletion of the β-globin gene or a failure to pro-

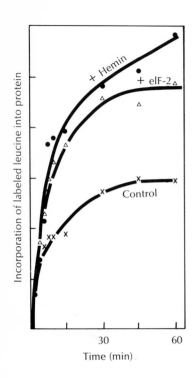

Fig. 9-4. Effect of hemin and elF-2 on protein synthesis by a reticulocyte lysate. The incorporation of radioactive leucine into protein was used to measure the rate of synthesis of protein (mostly hemoglobin) by a preparation of lysed reticulocytes (red blood cell precursors). The control curve shows the rate of protein synthesis without added hemin. Addition of either hemin or elF-2 markedly stimulates protein synthesis. (Modified from Clemens, M. J., et al. 1974. Proc. Natl. Acad. Sci. U.S.A. **71:**2946.)

duce functional β-globin mRNA. However, in a form of β_o-thalassemia discovered in Ferrara, Italy, no synthesis of β-globin occurs despite the presence of seemingly normal β-globin mRNA. Translation of that mRNA occurs in lysates of reticulocytes from the patients after the addition of supernatant fractions from normal reticulocytes. Synthesis of β-globin can also be caused to occur in vivo in the patients by transfusion with normal blood. Synthesis of α-, γ-, and δ-chains by the patients appears to be normal. The exact reason for failure of β-chain synthesis has not been worked out, but it appears to involve a distinction between β-globin mRNA and other globin mRNAs that is mediated by a soluble component of the translational machinery. It is interesting to note that the β-globin mRNA, which usually competes the most effectively for initiation, is the one that is not translated in this case.

When a population of mRNAs with differing abilities to compete for initiation is subjected to changes in the rate of initiation, complex changes in the pattern of protein synthesis can be obtained. An example of this occurs in the cellular slime mold *Dictyostelium discoideum* during transition from the growth phase, in which the cells behave as unicellular amoebas, to the aggregation phase, in which the cells converge to form a multicellular organism that progresses through a number of developmental stages that culminate in spore formation (Chapter 23). While the cells are well fed and growing actively, 90% of their ribosomes are found in polysomes. However, only 5 minutes after transfer to starvation medium, which stimulates aggregation, only 40% of the ribosomes are in the form of polysomes, and the average number of ribosomes per polysome is greatly reduced, suggesting a major reduction in the frequency of initiation. If the starved cells are resuspended in growth medium, new polysomes are formed, even

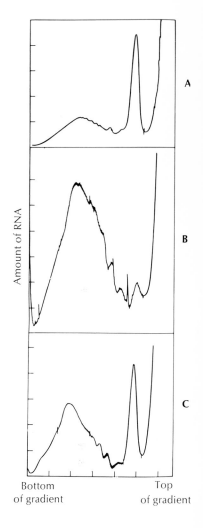

Fig. 9-5. Reformation of polysomes in *D. discoideum* after cells are switched from starvation conditions to growth medium. The graphs show the distribution of ribosomes in a sucrose density gradient. The larger polysomes sediment more rapidly and form a broad peak in the left-hand portion of the graph, whereas single ribosomes form a distinct peak in the right-hand portion of the graph. **A,** Distribution of ribosomes in cells after 75 minutes of starvation. **B,** Distribution of ribosomes after 75 minutes of starvation followed by 15 minutes in growth medium. **C,** Distribution of ribosomes in cells after 75 minutes of starvation followed by 15 minutes in growth medium containing inhibitors of RNA synthesis. (From McMahon and Fox, Volume II, Developmental biology, Copyright 1975, Benjamin/Cummings, Menlo Park, Calif., p. 380, Fig. 7, Reprinted with permission from Section title "RNA and Protein Synthesis During Differentiation of the Slime Mold *Dictyostelium and Discoideum*" by Harvey F. Lodish et al.)

when RNA synthesis is blocked (Fig. 9-5). This indicates that mRNA remains available in the cytoplasm in the starved cells despite the fact that it is not translated efficiently.

When the rate of synthesis of the protein actin is examined in starved cells of *Dictyostelium*, no decline is detected, and because the synthesis of other proteins is declining, the percentage of actin in newly synthesized protein actually increases approximately three-fold during the first hour of starvation. The basis for this change appears to be the ability of actin mRNA to compete more effectively than other kinds of mRNA for the limited amount of initiation that occurs under starvation conditions.

Major shifts in translational activity have been found in a variety of developmental phenomena. For example, fertilization in sea urchins is accompanied by a 30-fold increase in use of messages already present in the cytoplasm. Dormant embryos of the brine shrimp *Artemia salina* have an extremely low endogenous translation activity. Initiation factors eIF-2, eIF-3, eIF-4b, eIF-4c, eIF-4d, and eIF-5 are lacking in the dormant embryo, and when added to the lysate from dormant embryos, they greatly stimulate translation. However, there is no evidence for message-specific changes in translation in either of these systems, other than the type of competition described earlier that results from the inherent initiation efficiency of different messages.

The question of whether any true message-specific translational controls exist remains highly controversial. The message discrimination just described is based only on ability of messages to compete for initiation and, at best, can only rank mRNAs into general classes.

The search for message-specific initiation factors

There have been a number of reports of preferential translation of specific messages, but most of these can be explained on the basis of nonspecific competition for initiation, and the few remaining examples have generally not been confirmed.

Stuart Heywood and co-workers have published a series of papers presenting evidence for the existence of specific initiation factors. They have examined the translation of myosin and globin mRNAs in heterologous cell-free systems (i.e., globin mRNA in a system prepared from embryonic muscle and myosin mRNA in a system prepared from chick erythroblasts). Translation of each type of mRNA in the heterologous system requires a partially purified initiation factor (designated IF-3 and probably similar to eIF-3) from the same type of cell as the mRNA. A similar factor is also claimed to be required for translation of myoglobin mRNA from red muscle cells in a translational system prepared from white muscle cells, which do not synthesize appreciable amounts of myoglobin. The initiation factor fraction from red muscle cells that is responsible for the stimulation of myoglobin synthesis is distinct from that which stimulates myosin synthesis. If verified by other workers, these findings will be significant, but until then they must be viewed cautiously.

There are also interesting, but unconfirmed, reports from several other laboratories suggesting the possible existence of message-specific factors. For example, a new set of initiation factors is claimed to be necessary for synthesis of stage-specific proteins when the larvae of the insect *Tenebrio molitor* enter the pupal stage. Also, in chicken embryos, there are claims that translation of mRNA for hemoglobins E and P is favored in lysates of erythrocytes from 4-day embryos, whereas in lysates of 17-day embryos, translation of hemoglobin species A and D is favored.

Another type of message-specific translational control that has been reported is the so-called autogenous control. The effect of iron on the synthesis of ferritin appears to be an example of such a control, although the data are still somewhat incomplete. Ferritin is an iron-binding protein synthesized in the liver. Under conditions of iron deficiency, ferritin mRNA is not translated, but when iron is added to deficient rat liver cells, there is a dramatic increase in ferritin synthesis that does not require new mRNA synthesis. The proposed mechanism for the activation suggests that ferritin monomers that lack iron specifically bind to the ferritin mRNA and prevent its translation. The role of iron, according to this model, is to aggregate the ferritin monomers, thereby decreasing the binding of ferritin mRNA and permitting its translation. A somewhat similar mechanism has also been proposed for inhibition of synthesis of immunoglobulin heavy chains by intracellular accumulation of tetramers containing two heavy and two light chains, but neither of these proposed mechanisms has yet been verified experimentally. (Compare these eukaryotic "autogenous" controls with the T4–gene 32 control discussed in Chapter 4.)

A different approach to the question of specific translational factors has been taken by John Gurdon and colleagues. These workers have examined the translation of heterologous messages after their injection into *Xenopus* oocytes. Oocytes are injected with a specific mRNA population, incubated in the presence of labeled amino acids, homogenized, and assayed for the presence of the protein coded for by that mRNA. The rationale for these experiments is that if a message can be translated as efficiently in an oocyte as it can in its cell of origin, it must not require any cell-specific factors for its translation. Messages from a variety of different species and cell types, including messages for rabbit, mouse, and duck hemoglobin, trout testis protamine, and honey bee promelittin (a component of bee venom) were found to be translated accurately into the appropriate protein molecules when injected into oocytes.

Evidence against message-specific initiation

It thus appears that accurate translation of messages for specific differentiated products can occur in the absence of any cell-specific or even species-specific translation factors. However, another question of obvious importance in these experiments is the efficiency of translation of the foreign messages in the oocytes. The rate of β-globin synthesis in *Xenopus* oocytes appears to be more than 25% of that in reticulocytes, although this calculation is rough at best because of the number of assumptions required in making it.

Further support for the idea that message-specific translation factors may not be of general importance comes from the finding that foreign messages are efficiently translated in several eukaryotic cell-free protein-synthesizing systems. For example, the cell-free system from reticulocytes can efficiently translate more than a dozen different viral and cellular mRNA species. Systems derived from wheat germ are also capable of translating diverse types of mRNA. Thus, although the suggestions of message-specific initiation that have appeared in the literature are interesting, our position at the moment must be one of cautious skepticism, since none of the claims are well verified or generally accepted and since there is evidence that no specific factors are needed for initiation of synthesis of differentiated products in at least some systems. However, at the same time, there is no overwhelming evidence against message-specific initiation and we need to remain receptive to the possibility that such factors might in fact exist.

tRNA and translational control

Another possible mechanism by which cells could specifically control message translation is by changes in the tRNA population that would differentially influence the translation of certain messages. Since many amino acids are coded for by more than one codon, changes in the relative amounts of the different isoaccepting tRNA species (those which carry the same amino acid but recognize different codons in the mRNA) could differentially alter the efficiency of translation of different messages. For example, if the codons used in the mRNAs for proteins A and B are such that lysine is inserted into protein A primarily by $tRNA_1^{lys}$ and into protein B by $tRNA_2^{lys}$, changes in the relative amounts of these isoaccepting tRNAs will alter the relative efficiency of translation of the messages for proteins A and B.

Considerable evidence exists for changes in tRNA populations in different tissues during development. However, in most of these cases the changes appear only to improve the efficiency of translation, rather than to exert specific control over message translation. For example, in the silk gland of the silk worm the tRNA molecules that carry the major amino acids of silk fibroin (i.e., alanine and glycine) increase about 30-fold as fibroin production begins. However, this occurrence simply reflects a kind of supply-and-demand response in which the tRNA population is altered to match the amino acid (and codon) composition of the protein that is being synthesized.

An interesting example of translational control by changes in the tRNA population has been reported to occur in the development of the yellow mealworm, *Tenebrio molitor*. In common with many other insects, *Tenebrio* undergoes metamorphosis from larva to pupa to adult in a series of molts. The stage in development at which translational control has been proposed in this insect is the pupal stage, which normally lasts about a week. Near the end of the pupal stage, the mealworm synthesizes large amounts of adult cuticular proteins, whose amino acid composition differs significantly from that of pupal cuticle. Although day 1–pupae do not normally synthesize adult cuticular protein, a cell-free ribosomal system from these pupae is claimed to produce adult cuticle if supplied with tRNA and an aminoacyl-tRNA-synthetase fraction from day 7–pupae. These results have been interpreted to indicate that the day 1–pupae contain the message for adult cuticular protein but are unable to translate it because of the lack of necessary tRNA species.

It has also been reported that the change in the tRNA population that allows synthesis of adult cuticle late in the pupal stage is under the control of an insect hormone, juvenile hormone. Juvenile hormone influences the type of molt that an insect undergoes, and when pupae are treated with this hormone, they undergo a second pupal molt instead of the usual adult molt. (This phenomenon is discussed in Chapter 22.) If tRNA and enzyme from day 7-pupae that have been treated with juvenile hormone are added to the protein-synthesizing system from normal day 1-pupae, they do not stimulate the synthesis of adult cuticular protein. This suggests that production of the tRNA species required for the synthesis of adult cuticular protein occurs only at the low level of juvenile hormone normally found in pupae and is inhibited by higher levels of that hormone.

These results, which are all from a single laboratory, have been seriously questioned recently. Other workers report that the aminoacylating conditions used in these experiments were suboptimal and thereby resulted in differential charging of different leucyl tRNA species. If this is the case, the apparent translational control could have been caused by poor aminoacylation of certain specific tRNA molecules rather than by their absence. A further challenge to the earlier results obtained in the *Tenebrio* system comes from the reported finding that the assay used to determine the level of adult cuticular protein was inaccurate. Thus, although there are interesting suggestions in the literature, there is at the present time no well-substantiated evidence that alterations in tRNA populations do anything more than improve the efficiency of differentiated processes that are actually controlled at other places in the flow of genetic information.

Translation control RNA

Recent evidence from the laboratory of Stuart Heywood has implicated a special class of cytoplasmic RNA molecules in the control of message translation. This RNA has been named translation control RNA (tcRNA), and two types have been reported. The first type of tcRNA, which is found in a particular initiation factor fraction, is claimed to inhibit the translation of heterologous messages. For example, tcRNA from erythroblasts inhibits the translation of myosin message, and tcRNA from muscle inhibits the translation of globin message. Since these heterologous messages are not normally expected to be present in the cells that contain the tcRNA, the significance of such a control molecule is unclear. However, muscle cell tcRNA is also reported to have a slight stimulatory effect on the translation of homologous muscle message. The second class of tcRNA molecules appears to affect the availability of messages for translation and will be discussed in the next section. It remains to be seen whether other laboratories will verify the existence of tcRNA.

Message availability

The most concrete example of translational control in eukaryotic cells involves the "masked" messages found in unfertilized eggs of different animals. These messages are located in the cytoplasm but are inactive and separated from the cell's translational machinery. Masked messages have been studied most thoroughly in the sea urchin but have also been found in numerous other animals, including insects, amphibians, and mammals. Unfertilized sea urchin eggs contain a stable mRNA population that is not translated. These messages become available for translation at the time of fertilization, when a rapid rise in protein synthesis occurs, even in the presence of inhibitors of RNA synthesis (Fig. 9-6) and in enucleate eggs. Embryonic development can proceed normally to the gastrula stage in the absence of new RNA synthesis. The maternal messages seem to contain all the information required for early stages of embryonic development.

The failure of these messages to be translated in the unfertilized egg is due to their existence in the form of cytoplasmic ribonucleoprotein (RNP) particles. In particular, the messages for histone proteins, whose synthesis increases markedly following fertilization, have been isolated from RNP particles in the unfertilized egg. When the messages are separated from the proteins in these particles, they are able to direct the synthesis of histones in cell-free systems. Thus the low level of protein synthesis in the unfertilized egg can be explained by a sequestration of template molecules into RNP particles. This is in contrast to the low level of translation in dormant brine shrimp embryos, mentioned earlier, which appears to be caused by absence of initiation factors.

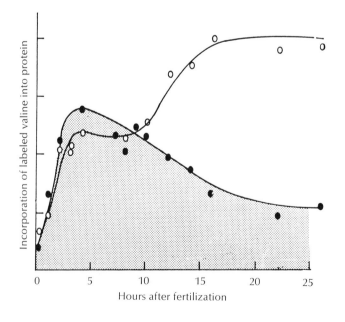

Fig. 9-6. Protein synthesis in fertilized eggs of the sea urchin *Arbacia punctulata* in the presence and absence of an inhibitor of RNA synthesis. Protein synthesis was assayed by the incorporation of labeled valine into protein. ○ = Incorporation in eggs incubated in normal artificial sea water; ● = incorporation in eggs incubated in the presence of actinomycin D from 3 hours prior to fertilization to time of assay. (Modified from Gross, P. R., L. J. Malkin, and W. A. Moyer. 1964. Proc. Natl. Acad. Sci. U.S.A. **51**:407.)

The occurrence of masked messages has also been demonstrated recently in cultured fibroblast cells. A large amount of polyadenylated mRNA is found in ribonucleoprotein particles in resting cells, but after cells are stimulated to grow by the addition of serum to the culture medium, these mRNAs are found in the polysome fraction. Similarly, as discussed, mobilization into polysomes of the cytoplasmic message for the iron-containing protein, ferritin, has been reported to occur in rat liver after administration of iron to rats. There is also recent evidence suggesting that myosin mRNA accumulates in myoblasts (muscle cell precursors) prior to their final differentiation and the initiation of high levels of myosin synthesis. Current data do not make it clear whether the activation of myosin synthesis is due to a change in masking molecules or to a change in components of the translational system itself.

As previously mentioned, it has recently been proposed that special RNA molecules called tcRNA may have a role in controlling message translation in eukaryotic cells. The second class of tcRNA that was studied was isolated from free cytoplasmic RNP particles. It is reported to inhibit translation of the messages contained in these particles, but it does not affect the translation of ribosome-bound messages. Since such a molecule could obviously be of importance in controlling message translation by sequestering specific messages from the ribosomes, it is expected that this line of evidence will be actively pursued in the near future.

Another mechanism by which gene expression can be regulated in cells is specific control of mRNA stability. Changing the half-life of a particular message can alter the number of times that message is translated before it is degraded. Such changes can greatly influence the relative amounts of different proteins made by a cell. Evidence from a variety of systems indicates that eukaryotic cells contain mRNA species with widely differing stabilities and that, in addition, the stability of particular types of mRNA can be changed when the state of the cell changes.

In general, it appears that mRNAs coding for proteins that are produced in large amounts by highly specialized cells have unusually long half-lives. For example, in the glandular cells of the silkworm that are specialized to produce large amounts of the zymogen for cocoonase (the enzyme used by the silkworm moth to digest its cocoon on completion of metamorphosis), the half-life of the message for the zymogen is nearly 100 hours, whereas the average half-life of the other cellular messages is only about 3 hours. Similarly, in the estrogen-stimulated chicken oviduct the message for ovalbumin has a half-life

MESSAGE STABILITY

of about 24 hours, whereas other cellular messages have an average half-life of about 5 hours.

There is also evidence in a number of systems for a change in the stability of specific messages under different cellular conditions. For example, following the withdrawal of estrogen stimulation in the hen oviduct the half-life of ovalbumin message decreases from about 24 hours to 2 to 3 hours. Similarly, histone messages undergo a radical change in stability after the cessation of DNA synthesis. Histones are normally synthesized only during DNA replication (the S phase of the cell cycle). The half-life of the histone message in a line of actively growing cultured mouse cells is about 11 hours, which is approximately the length of the S phase in those cells. However, when DNA synthesis is inhibited, histone message disappears rapidly, with a half-life of only about 13 minutes. It is possible that some factor necessary for stabilizing these messages or permitting ribosome binding is present only during DNA synthesis and its loss leaves the histone messages open to degradative enzymes. A third example of a specific change in message stability may occur during the differentiation of myoblasts in culture. When myoblasts fuse to form myotubes, the synthesis of muscle-specific proteins increases dramatically. When message stability is examined in this system, it is found that at myoblast fusion the half-lives of a number of mRNA species, including the putative message for the large subunit of myosin, increase from about 10 to 50 hours, whereas many other cell messages retain a shorter half-life. However, this finding has been clouded somewhat by the discovery of myosin-like molecules in nonmuscle cells and uncertainty as to which myosin the short half-life mRNA codes for.

In all these cases no precise means for the stabilization of specific messages has been demonstrated. One system in which the mechanism for alteration of mRNA stability has been studied fairly extensively is the synthesis of tyrosine aminotransferase (TAT) in hepatoma tissue culture (HTC) cells that have been stimulated by treatment with glucocorticoids (a class of steroid hormones that promotes breakdown of amino acids and synthesis of glucose). Treatment of HTC cells with glucocorticoids results in an increase both in TAT activity and in the concentration of TAT molecules that can be detected by immunoassay. This increase is due in part to transcriptional control (i.e., an increase in the synthesis of TAT message), but certain characteristics of the TAT induction and deinduction processes argue for translational control of TAT production as well.

When HTC cells are deinduced by the removal of the hormone, the level of TAT normally decreases (dotted line, Fig. 9-7). However, when deinduction is done in the presence of actinomycin D, which inhibits

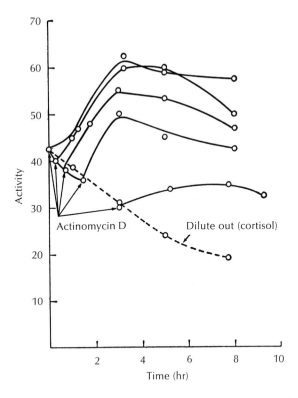

Fig. 9-7. Activity of TAT in HTC cells. Samples of cells that had been induced by incubation in glucocorticoid were deinduced by transfer to medium free of steroid. Samples were removed at various times (arrows) and incubated in the presence of actinomycin D. (From Tomkins, G. M., et al. 1969. Science **166:** 1474. Copyright 1969 by the American Association for the Advancement of Science.)

RNA synthesis, the level of TAT synthesis is maintained, and if actinomycin D is added after the deinduction is in progress, there is actually a partial restoration of TAT synthesis (solid lines, Fig. 9-7). These effects have been postulated by Gordon Tomkins and co-workers to result from the existence of a labile repressor molecule that inhibits the translation of TAT mRNA and also destabilizes it. In the presence of actinomycin D, synthesis of this short-lived repressor would cease, and translation of preexisting TAT message would be unimpeded. During normal induction, glucocorticoid would stimulate TAT production by preventing the action of the postulated repressor, and during normal deinduction, removal of the hormone would allow resumption of repressor function and consequent loss of TAT message translation. The message "rescue" that results in partial restoration of TAT synthesis when actinomycin D is added while deinduction is in progress is proposed to be caused by more rapid decay of the repressor than of the TAT mRNA bound to it. This model for the control of TAT synthesis is diagrammed in Fig. 9-8. Although the model has been criticized by a number of workers, there is still considerable circumstantial evidence in its favor. Conclusive evidence for its accuracy awaits actual isola-

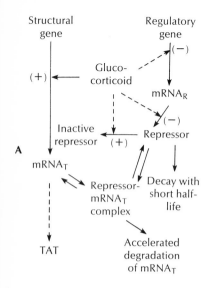

Fig. 9-8. Model showing hypothetical mechanism for control of TAT induction by gluco-corticoids. **A,** Basic model. Structural gene *(SG)* is transcribed to messenger RNA *(M_T)*, which is translated to yield TAT. Regulatory gene *(RG)* codes for repressor protein *(R)*, which combines with M_T to form complex MR, making M_T unavailable for translation and also hastening its degradation. Glucocorticoid *(GC)* enhances transcription of M_T and also blocks the effect of R, either by preventing its synthesis or converting it to an inactive form *(IR)*. R is assumed to have a short half-life and to decay much more rapidly than free M_T (which is relatively stable in the absence of R but undergoes accelerated degradation when complexed with R). Formation of the complex MR is reversible, so that decay of free R can lead to release of M_T that has not yet been degraded. **B,** Uninduced state. Both M_T and R are synthesized freely. Most of the M_T is complexed with R and degraded, leaving only a basal level of synthesis of TAT. **C,** Induced state. GC enhances transcription of M_T and also inactivates the R system. In the absence of functional R, M_T is stable. The combination of increased synthesis and increased stability of M_T results in a large increase in TAT synthesis. **D,** Message rescue. Removal of GC reactivates the R system and initiates a return to the uninduced state shown in **B.** If transcription is blocked by actinomycin D before deinduction is completed, further synthesis of R is blocked, and the R that has already formed decays more rapidly than M_T. This leads to "rescue" from the MR complex of M_T that has not yet been degraded and thus results in the return to a higher level of synthesis of TAT shown in Fig. 9-7. (Modified from Tomkins, G. M., et al. 1969. Science **166:**1474, and Baxter, J. D., et al. 1973. *In* J. K. Pollak, ed. The biochemistry of gene expression in higher organisms, D. Reidel Publishing Co., Inc., Hingham, Mass.)

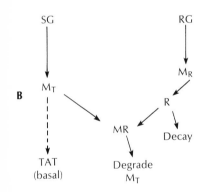

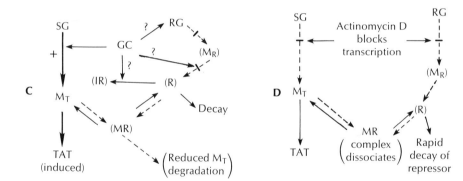

tion of the TAT message and a study of its behavior during induction and deinduction.

A labile repressor of message translation has also been postulated to operate in the control of interferon synthesis. Interferons are glycoproteins that are made and secreted by cells in response to viral infection or treatment with certain other agents (e.g., double-stranded RNA). Interferons interact with uninfected cells and render them resistant to viral infection. The ability of interferons to inhibit viral growth is the basis for their assay and their name (i.e., particle of interference). Because interferons are not produced in detectable amounts until cells have been infected, they are inducible proteins.

However, certain features of their inductive process differ from other inducible systems. First, there is no basal, constitutive level of interferon synthesis. More important, interferons are synthesized for only a relatively short time after induction; that is, the cells do not synthesize interferon continuously while in the presence of the inducing agent. This property is different, for example, from the induction of TAT in hepatoma cells, where TAT is produced in high levels as long as glucocorticoid is present.

Actinomycin D has been found to either inhibit or prolong the production of interferon in induced cells, depending on whether the drug is added early (the first 1 to 2 hours) or late (3 hours or later) in the induction period. The inhibition of interferon synthesis that occurs if actinomycin D is added early in induction is explained by the fact that synthesis of the mRNA for interferon is inhibited. The prolonged synthesis of interferon that occurs when actinomycin D is added late in induction is thought to occur by a mechanism similar to that postulated for the action of actinomycin D during TAT deinduction. That is, a labile repressor of the translation of the interferon message is normally produced and is responsible for the normal shut-off of interferon synthesis. In the presence of actinomycin D, however, the synthesis of the labile repressor is inhibited and the preexisting repressor decays, allowing a prolonged synthesis of interferon.

Following transcription and processing of mRNA and its translation into protein, many gene products undergo further modifications that influence their biological function. In some cases the initial product of translation must undergo some type of alteration such as cleavage, modification of amino acids, or the addition of prosthetic groups to become functional. For example, many digestive enzymes are synthesized as inactive precursors that are converted to their active form as needed by enzymatic cleavage. Current nomenclature favors referring to these as "pro-" enzymes, although some that have been known for a long time are referred to as zymogens (e.g., trypsinogen, chymotrypsinogen). This type of activation ensures that these highly potent substances do not begin to function before they are needed and also allows their function to be turned on rapidly.

Formation of functional molecules may occur in a single step or in a cascade reaction requiring a number of consecutive steps. For example, a two-stage cascade reaction is involved in the activation of pancreatic digestive enzymes in the intestine (Fig. 9-9). The zymogen trypsinogen is converted to trypsin by the enzyme enterokinase. Then trypsin in turn performs the cleavages necessary for converting chymotrypsinogen, proelastase, procarboxypeptidase, and prophospholi-

POSTTRANSLATIONAL MODIFICATION OF PROTEINS

Fig. 9-9. Activation of zymogens in the small intestine occurs by a two-step cascade reaction. Pancreatic trypsinogen is converted to trypsin by enterokinase and in turn converts the other zymogens to active enzymes. (From Neurath, H., and K. A. Walsh. 1976. Proc. Natl. Acad. Sci. U.S.A. **73**:3825.)

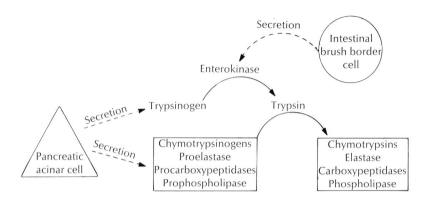

pase into their active forms. This type of cascade response permits amplification of the signal to activate the system. For example, one molecule of enterokinase can activate a large number of trypsinogen molecules, and each trypsin molecule that is produced can in turn activate a large number of its target zymogens. Complicated proteolytic cascade processes involving many steps occur during activation of blood coagulation and during activation of the complement system, which is involved, together with the immune system, in defending the body against "foreign" cells.

Many peptide hormones have been shown to be synthesized as larger precursors, including insulin, corticotropin, parathyroid hormone, placental lactogen, growth hormone, prolactin, and glucagon. A precursor of serum albumin, termed proalbumin, can be detected in the cisternae of the rough endoplasmic reticulum of liver cells that are actively synthesizing albumin in vivo.

Recently it has been found that many proteins, including prohormones, proenzymes (zymogens), immunoglobulin chains, and proalbumin, are initially translated as even larger precursors that are normally present only transiently. These transient forms can usually be detected only through the use of in vitro translation systems. The transient precursors are given the prefix "pre-," and when there also exists a "pro-" form of the protein, they are referred to as "prepro-," as in "preproinsulin" (Fig. 9-10) or "preproalbumin."

A variety of proteins that are produced for export from their cells of origin have similar hydrophobic sequences at the N-terminal end of their transient "pre-" form. A "signal hypothesis" has been proposed by Gunter Blobel and Bernhard Dobberstein, in which they suggest that the N-terminal hydrophobic region serves as a specific signal that causes the ribosome carrying the nascent peptide chain to attach to the membrane of the rough endoplasmic reticulum and the end of the

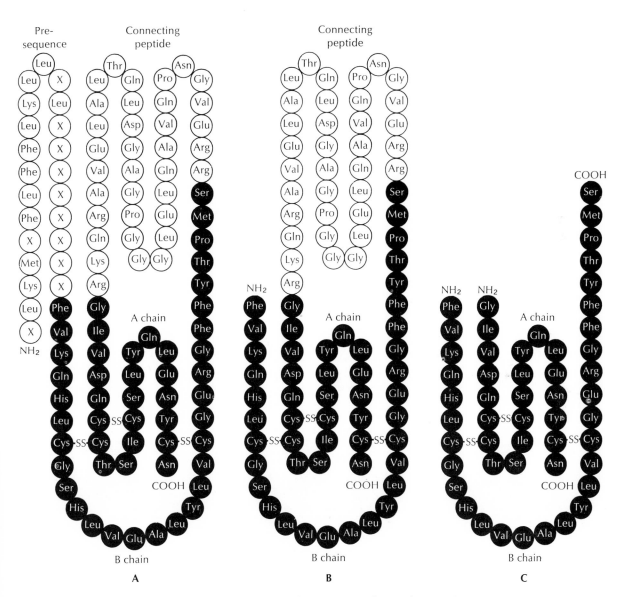

Fig. 9-10. Amino acid sequence of, **A,** rat preproinsulin, **B,** proinsulin, and **C,** insulin. There are two forms of rat preproinsulin. Only form II is shown here. X = Unidentified amino acids. (Based on data from Chan, S. J., et al. 1976. Proc. Natl. Acad. Sci. U.S.A. **73:**1964, and De Haën, C., et al. 1976. J. Mol. Biol. **106:**639.)

growing peptide chain to pass through the membrane and into the cisternal space inside the endoplasmic reticulum. The proposed sequence of events is illustrated in Fig. 9-11. The "signal" sequence is removed by an endopeptidase soon after the *N*-terminal portion of the polypeptide chain enters the cisternae. The rest of the growing peptide chain is proposed to thread its way into the intracisternal space as it is synthesized. Since the signal peptide is removed even before translation has been completed, the "pre-" form of the protein can be detected only through the use of in vitro translation systems in which processing of the protein has been disrupted. Additional processing of the "pro-" protein is thought to occur more slowly in the cisternae of the endoplasmic reticulum or in the Golgi apparatus, making it possible to detect "pro-" forms that are synthesized in intact cells. Although the signal hypothesis has not yet been verified directly, its predictions are consistent with current experimental data, and it is generally accepted as valid for a variety of exported proteins.

Recent studies of the synthesis of ovalbumin, which is produced for export in massive amounts by tubular gland cells of the chicken oviduct (Chapter 14), appear to show, however, that a signal sequence is not an essential prerequisite for protein export. Ovalbumin translation in reticulocyte lysates begins with methionine derived from Met-

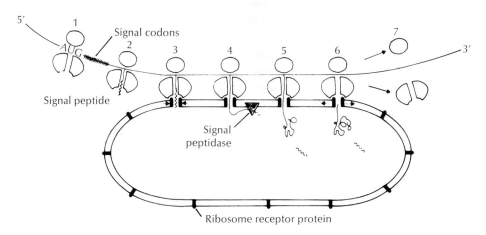

Fig. 9-11. Diagram showing scheme of "signal hypothesis" for transfer of protein precursor molecules into cisternae of the endoplasmic reticulum *(ER)*. After translation of the signal codons into the signal peptide, the ribosome becomes attached to the ER membrane. The signal sequence causes ribosome receptor proteins on the membrane to form a tunnel through which the nascent polypeptide enters the ER cisternae. The signal sequence, soon after its entry into the lumen of the ER, is removed from the remainder of the chain by a signal peptidase. (From Blobel, G. 1977. *In* B. R. Brinkley and K. R. Porter, eds. International Cell Biology. Rockefeller University Press, New York.)

tRNA$_i$. The next amino acid added is the N-terminal amino acid of ovalbumin, glycine, and it is followed by the rest of ovalbumin in normal sequence. The methionine is removed when the nascent peptide chain is about 20 amino acids long, and the new N-terminal glycine is acetylated when the polypeptide is about 44 amino acids long. The mechanisms responsible for the segregation of ovalbumin for export remain to be worked out.

There are several well-documented cases in which posttranslational cleavage generates more than one functional protein molecule from a newly translated polypeptide chain. A particularly interesting case occurs during the synthesis of corticotropin (adrenal corticotropic hormone [ACTH]). The precursor of ACTH, like most other precursors of peptide hormones, has little hormonal activity but can be isolated by immunoprecipitation with antibodies prepared against the active hormone. In the case of corticotropin, several large precursors were found after translation of mRNA from cultured mouse pituitary cells in a reticulocyte lysate. The largest, which appears to be the initial product of translation, has a molecular weight of 28,500 and contains about 260 amino acids, whereas there are only 39 amino acids in the active hormone. Careful examination of the precursor has revealed that in addition to corticotropin it also contains the sequences of three other pituitary products (Fig. 9-12). The largest of these is β-lipotropin (β-LPH), a pituitary protein that was initially thought to stimulate lipid breakdown but whose role has remained poorly defined. It is now evident that β-LPH is a precursor both for β-melanotropin (β-MSH), a pituitary hormone that is stimulatory to pigment cells, and β-endorphin, a peptide with potent opiate-like action that is normally produced by the pituitary in small amounts.

Another example of a common precursor for more than one protein is vitellogenin (also called serum lipophosphoprotein or SLPP), which is cleaved to form two distinctly different proteins found in yolk platelets in *Xenopus* eggs, phosvitin and lipovitellin. The synthesis of vitellogenin occurs in the livers of *Xenopus* females and can also be in-

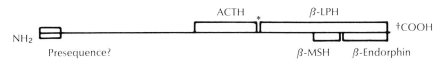

Fig. 9-12. Structure of the common precursor to corticotropin, β-LPH, β-MSH, and β-endorphin produced by translation of mRNA from a mouse pituitary cell line (AtT-20/D-16v) in a reticulocyte cell free system. * = Spacer peptide less than 20 amino acids in length; † = COOH-terminal peptide of about five amino acids. (From Roberts, J. L., and E. Herbert 1977. Proc. Natl. Acad. Sci. U.S.A. **74**:5300.)

Fig. 9-13. Translation of polycistronic mRNA of polio virus type I and posttranslational cleavage of the resulting peptide chain. The large initial translation product undergoes cleavage to form seven distinct protein products. (This type of polycistronic mRNA differs from that found in bacteria in that it has single initiation and termination sites for translation.) (From James D. Watson, Molecular biology of the gene, 3rd ed., copyright © 1976, 1970, 1965 by James D. Watson and The Benjamin/Cummings Publishing Co., Inc., Menlo Park, Calif.)

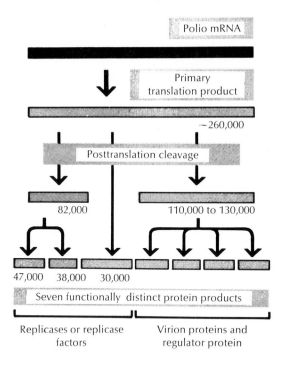

duced in the livers of males by administration of estrogen. Similar proteins also occur in chickens. In *Xenopus*, vitellogenin is synthesized as a single long peptide chain with a molecular weight of about 200,000. Extensive addition of phosphate, lipid, and carbohydrate occurs within the liver. The protein is then secreted into the bloodstream, where it appears to exist as a dimer. It is taken up by oocytes in the ovary and cleaved to form phosvitin, which has a molecular weight of 32,000 and is heavily phosphorylated, plus lipovitellin, which is composed of two subunits with molecular weights of 120,000 and 31,000, and contains essentially all the lipid from the vitellogenin. Still another example of synthesizing a single large peptide chain and then breaking it into more than one functional protein molecule is the synthesis of polio virus protein, in which seven or more different functional proteins are thought to be translated as one giant polyprotein, as the polycistronic precursors are often called (Fig. 9-13).

The possibility that posttranslational modifications may be specifically controlled has been examined through use of the technique of message injection into living oocytes. Messages for a variety of proteins that require posttranslational alteration have been translated in *Xenopus* oocytes, and the oocytes have been found to perform properly a number of different types of peptide chain modification (Table 9-4). For example, specific amino acids in calf lens crystallin and in trout

Table 9-4. Protein modifications obtained after injection of mRNA into frog oocytes*

Protein	Cell type from which mRNA is prepared and in which protein is synthesized	Normal protein modification	Protein modification in message-injected oocytes
αA2 crystallin	Calf lens epithelium	Acetylation of N-terminal methionine	N-acetyl-methionine in N-terminal peptide
Stable proteins of EMC virus	EMC virions from infected mouse ascites cells	Cleavage of primary polypeptide into mature proteins	At least six mature EMC proteins derived from different regions of primary polypeptide
Protamine	Trout testis	Phosphorylation of certain serine residues	Phosphorylated peptides
Light chain immunoglobulin	Mouse myeloma cells	About 15 amino acids removed from end of primary polypeptide	Mature light chains
Collagen	Cultured mammalian fibroblasts	Hydroxylation of proline in protocollagen	Hydroxylation of ^{3}H-protocollagen (injected directly or synthesized from injected mRNA).†

*Modified from Gurdon, J. B. 1974. The control of gene expression in animal development. Harvard University Press, Cambridge.
†Since uninjected oocytes synthesize collagen, this test was carried out in unfertilized eggs, which do not do so to a detectable extent.

testis protamine were acetylated and phosphorylated, respectively. A sequence of amino acids was removed from the end of mouse light chain immunoglobin. A polypeptide product of encephalomyocarditis (EMC) virus was cleaved into six final protein products. In each of these cases the possibility that additional messages coding for specific modifying enzymes might have been injected as contaminants of the mRNA preparation was shown to be unlikely.

These results seem to argue against the existence of highly specific protein-modifying enzymes in particular types of cells. Rather, they suggest that the modifications are carried out by enzymes that are more generally distributed and that the amino acid sequences of the peptide chains contain specific recognition sites which signal where the modifications are to be made. It should be noted, however, that none of the messages that were studied coded for the type of highly active protein, such as a hormone, whose function would be expected to be tightly controlled. Indeed, highly specific modifying enzymes are known to be involved in some activation processes such as the cascade of proteolytic cleavages that leads ultimately to the conversion of fibrinogen to fibrin during blood clotting.

A number of developmentally important molecules have been shown to be synthesized in an inactive form. Examples include collagenase, which is involved in tail resorption during amphibian metamorphosis; cocoonase, which is used by the silk moth to escape from its cocoon; and enzymes found in the acrosome of mammalian sperm, which have important roles in fertilization. Many of the peptide hormones discussed earlier in this chapter also have important roles in development (Chapter 14).

A variety of other types of reactions besides removal of peptide sequences are involved in posttranslational modification. We have already seen examples of a number of these, including the addition of heme groups to form hemoglobin, phosphorylation (which can either activate or inactivate an enzyme), acetylation and methylation (both of which were discussed in relation to histones in Chapter 5), lipidation, and glucosylation. In collagen, covalent cross-links are formed between adjacent polypeptide chains by oxidative modification of lysine residues. In elastin, groups of four lysine residues condense to form cyclic linking molecules called desmosine and isodesmosine.

Interaction among several subunits is sometimes necessary for the formation of a functional protein such as lactate dehydrogenase or hemoglobin. Also, a variety of cofactors are often needed for enzymatic activity, and enzyme activity may be further controlled by feedback or allosteric controls. Finally, the degree of stability of a functional protein plays an important role in determining the amount of that protein that is accumulated under equilibrium conditions.

TYPES OF CONTROLS THAT ACT AT THE CYTOPLASMIC LEVEL

The types of cytoplasmic controls that were discussed in this chapter can be divided into four general categories:

1. *Controls that enhance the effects of regulation at the levels of transcription and processing.* This type of control reinforces primary control mechanisms acting at the levels of transcription and processing and further enhances the synthesis of specific differentiated products. Examples include stabilization of specific mRNAs and provision of optimum amounts of specific aminoacyl-tRNAs for their translation.
2. *Controls that balance rates of synthesis.* In this type of control, synthesis of specific protein molecules is linked to the availability of other molecules with which they must interact. Examples include regulation of globin synthesis by the availability of heme for the formation of hemoglobin and synthesis of equal numbers of α- and β-globin chains, even though their mRNAs are present in unequal amounts.
3. *Controls that ensure precise timing of changes in gene expression.* This type of control operates where accumulation of mRNA begins well before synthesis of the new protein is initiated, as occurs in masked messages. Also, the cessation of synthe-

sis of particular proteins is sometimes accompanied by an abrupt reduction in the half-lives of their mRNAs, as in estrogen withdrawal.

4. *Controls that act to activate physiological responses.* This type of control acts in rapidly responding physiological processes that are held in reserve until they are needed (e.g., release of digestive enzymes, blood clotting). Activation of these processes often involves posttranslational events such as removal of peptides or phosphorylation of specific proteins. Certain hormonal responses also appear to involve cytoplasmic regulation, probably at the translational level.

We have examined in this chapter the last of the long series of steps extending from the vast library of information carried in the genome to the molecular expression of specific differentiated properties. Although major decisions concerning the pathway of differentiation appear to occur primarily at stages prior to the formation of cytoplasmic mRNA, extensive "fine tuning," which tends to amplify the effects of the controls acting at earlier steps, occurs at the translational stage and beyond. Although there is little evidence for highly selective controls, phenomena such as competition among messages for initiation can cause major shifts in patterns of protein synthesis. Messenger RNA can be stored in an inactive, masked form until it is needed, and it appears that precise timing of certain developmental changes may be achieved at the translational level, although the exact mechanisms remain obscure. Differences in message stability play a major role in determining the extent of synthesis of differentiated products, and in cases in which transcription ceases during terminal differentiation, they can lead to major shifts in the patterns of protein synthesis.

Proteins with major biological effects are often synthesized as inactive precursors that are converted to the active forms only as needed. There are also rare cases in which more than one active protein is generated from a single large polypeptide chain. A variety of types of modification of residues, addition of prosthetic groups, cross-linking, and interactions among subunits are also important in generation of the final differentiated forms of the proteins coded for by the genome.

Although the individual contributions of these many steps to the expression of differentiation may be indirect, they are all important components of the overall process of gene expression in living cells. Neither the final expression of differentiation nor the complex feedback loops that appear to be involved in its control could function properly without these many component steps.

ROLE OF CYTOPLASMIC PROCESSES IN CONTROL OF GENE EXPRESSION

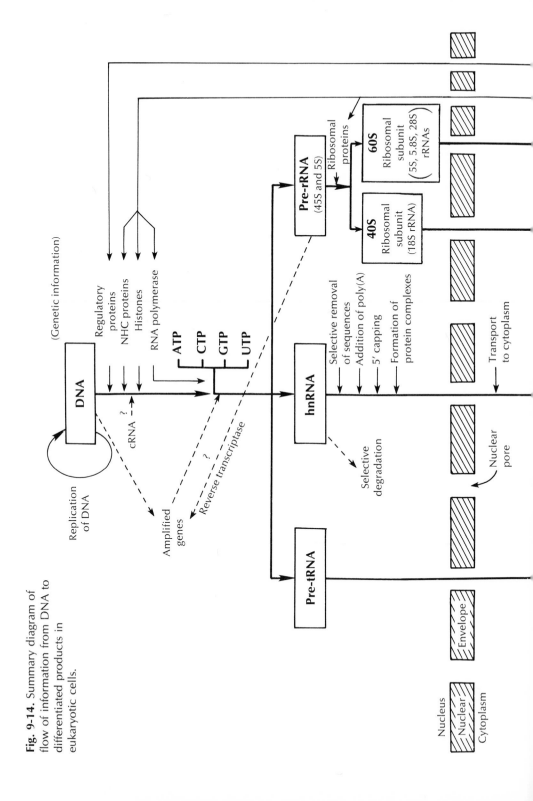

Fig. 9-14. Summary diagram of flow of information from DNA to differentiated products in eukaryotic cells.

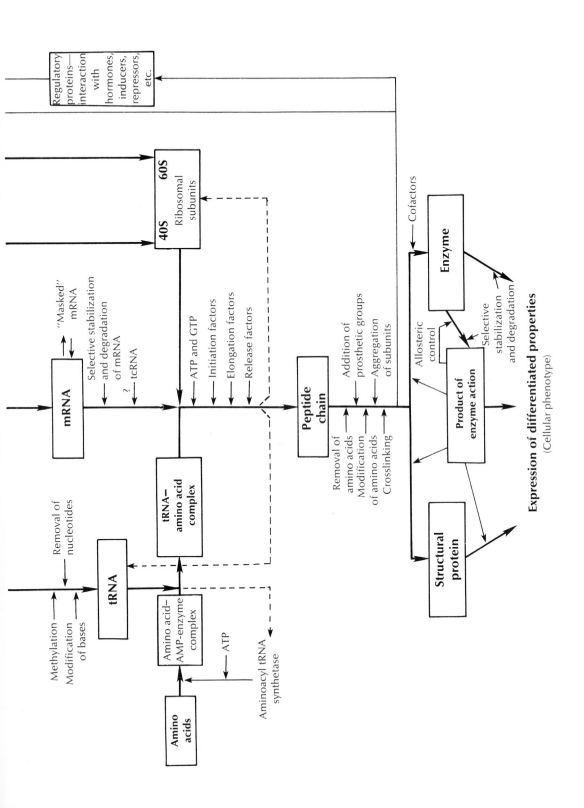

SUMMARY OF REGULATION
OF GENE EXPRESSION IN
EUKARYOTIC CELLS

Major aspects of the overall flow of information from DNA to biochemical differentiation are summarized in Fig. 9-14. To present this complex picture as clearly as possible, we have dealt with each individual step separately. However, we must emphasize that living cells do not work that way. Everything in Fig. 9-14 plus many more details, including some that have not yet been discovered, must be regarded as a single delicately balanced and integrated system. Although the major controls over differentiation are probably exercised only at a limited number of places in this overall scheme, no one part of the system can be considered to be acting alone.

A hint of the overall complexity that is probably involved can be gleaned from what is currently known about hemoglobin synthesis. The types of chains that are synthesized are probably determined at the transcriptional level, and it is likely that regulatory proteins in the NHC protein fraction are involved, although we know little about what they do or what controls them. The α- and β-genes occur in the human genome in a ratio of $2:1$. Somewhere between the genes and cytoplasmic mRNA, the ratio of α- to β- is changed to $1.5:1$. A delicate balance relationship involving the relative affinity of the two messages for initiation, the availability of hemin, and the overall rate of initiation further reduces the ratio to $1.04:1$, which is almost exactly what is needed for synthesis of the types of hemoglobin found in adult human blood. Any adjustment in the amount of hemoglobin synthesis that involves only the supply of mRNA or only the supply of initiation factors will alter the ratio of synthesis of the two chains. Perhaps there exist integrating controls of which we are not yet aware, but at the present time it appears that the control systems which have evolved involve a complex, balanced relationship among products coded for by many different genes.

Our goal in this section of the book has been to summarize current understanding of those molecular mechanisms which either have been proved to be important to the process of cellular differentiation or are likely to be found important to it by future research. Obviously this represents only one aspect of development. We still have to deal with major questions such as generation of shape and integrated biological function at the cellular, tissue, and organismic levels. As we proceed to those topics, our emphasis will shift abruptly to phenomena observable at supramolecular levels of organization, which in many cases cannot yet be related precisely to molecular mechanisms.

The lack of a smooth transition between these two areas accurately reflects the current state of knowledge. Many phenomena that occur at the level of intact cells cannot yet be explained in molecular terms, and many known molecular phenomena cannot yet be related precisely to the behavior of intact cells or organisms. As we proceed, we will keep in mind our goal of ultimately explaining all developmental phenomena at the molecular level, and we will provide molecular explanations when they are available. However, there will be many cases in which we will simply have to say "this is an interesting phenomenon that needs to be investigated at the molecular level, but no data are yet available."

BIBLIOGRAPHY
Books and reviews

Anderson, W. F., et al. 1977. International Symposium on Protein Synthesis: summary of Fogarty Center–NIH workshop held in Bethesda, Maryland on 18-20 October, 1976. F.E.B.S. Lett. **76**:1.

Blobel, G. 1977. Synthesis and segregation of secretory protein: the signal hypothesis. *In* B. R. Brinkley and K. R. Porter. eds. International cell biology 1976-1977. Rockefeller University Press, New York.

Ilan, J., and J. Ilan. 1975. Regulation of mRNA translation during insect development. Curr. Top. Dev. Biol. **9**:89.

Kafatos, F. C., and R. Gelinas. 1974. mRNA stability and the control of specific protein synthesis in highly differentiated cells. *In* J. Paul, ed. International review of biochemistry. Vol. 9. Biochemistry of cell differentiation. University Park Press, Baltimore.

Lodish, H. F. 1976. Translational control of protein synthesis. Annu. Rev. Biochem. **45**:39.

Neurath, H., and K. A. Walsh. 1976. Role of proteolytic enzymes in biological regulation. Proc. Natl. Acad. Sci. U.S.A. **73**:3825.

Paul, J. 1974. Macromolecular synthesis in sea urchin development. *In* J. Paul, ed. International review of biochemistry. Vol. 9. Biochemistry of cell differentiation. University Park Press, Baltimore.

Weissbach, H., and S. Ochoa. 1976. Soluble factors required for eukaryotic protein synthesis. Annu. Rev. Biochem. **45**:191.

Selected original research articles

Bandman, E., and T. Gurney. 1975. Differences in the cytoplasmic distribution of newly synthesized poly (A) in serum-stimulated and resting cultures of BALB/c 3T3 cells. Exp. Cell Res. **90**:159.

Bester, A. J., D. S. Kennedy, and S. M. Heywood. 1975. Two classes of tcRNA: their role in regulation of protein synthesis. Proc. Natl. Acad. Sci. U.S.A. **72**:1523.

Heywood, S. M., D. S. Kennedy, and A. J. Bester. 1974. Separation of specific initiation factors involved in the translation of myosin and myoglobin mRNA's and the isolation of a new RNA involved in translation. Proc. Natl. Acad. Sci. U.S.A. **71**:2428.

Lassam, N. J., H. Lerer, and B. N. White. 1976. A re-examination of the leu-tRNA's and the leu-tRNA synthetase in developing *Tenebrio molitor*. Dev. Biol. **49**:268.

Lodish, H. F. 1974. Model for the regulation of mRNA translation applied to haemoglobin synthesis. Nature **251**:385.

Mains, R. E., and B. A. Eipper. 1978. Coordinate synthesis of corticotropins and endorphins by mouse pituitary tumor cells. J. Biol. Chem. **253**:651.

Palmiter, R. D., and N. H. Carey. 1974. Rapid inactivation of ovalbumin mRNA after acute withdrawal of estrogen. Proc. Natl. Acad. Sci. U.S.A. **71**:2357.

Palmiter, R. D., J. Gagnon, and K. A. Walsh. 1978. Ovalbumin: a secreted protein without a transient hydrophobic leader sequence. Proc. Natl. Acad. Sci. U.S.A. **75**:94.

Roberts, J. L., and E. Herbert. 1977. Characterization of a common precursor to corticotropin and β-lipotropin: identification of β-lipotropin peptides and their arrangement relative to corticotropin in the precursor synthesized in a cell-free system. Proc. Natl. Acad. Sci. U.S.A. **74**:5300.

Steinberg, R. A., B. B. Levinson, and G. M. Tomkins. 1975. "Superinduction" of tyrosine aminotransferase by actinomycin D: a re-evaluation. Cell **5**:29.

SECTION THREE

Control signals in differentiation

☐ In the first section of this book some of the basic concepts of development and cellular differentiation were introduced, primarily at a descriptive level. The second section dealt with genes and the molecular mechanisms of gene expression, with emphasis on those mechanisms which are currently considered likely to be involved in the selective gene expression that occurs during differentiation. At this point we are ready to begin examining development and differentiation from the perspective of the cell, with special emphasis on the control signals that are responsible for differentiation.

In Chapter 3 we examined a variety of definitions of cellular differentiation and found it virtually impossible to arrive at an all-encompassing definition that did not also include many other phenomena not commonly regarded as differentiation. For our purposes in this section, differentiation can be defined simply as the summation of all processes within a multicellular organism that give rise to cells that are different from one another. We are concerned with how cells become and remain different, and our attention will be focused primarily on the control signals and mechanisms, both intracellular and extracellular, that are responsible for the differences.

We begin in Chapter 10 by examining the behavior of individual differentiated cells that have been isolated from the rest of the organism in an artificial culture medium. Extracellular influences on the expression of differentiated properties are examined, together with the apparent inheritance of tissue specificity over many cell generations in culture, whether or not that specificity is actively expressed. Chapter 11 deals with experimental inquiries into the nature of intracellular mechanisms that control expression of differentiated properties. Topics that receive special attention include the role of cell division in the establishment of new states of differentiation, and various experimental manipulations such as cellular hybridization and treatment with 5-bromodeoxyuridine that are beginning to provide valuable clues to the nature of the intracellular processes involved in regulating differentiation.

In Chapter 12 attention is shifted to differentiation in the intact embryo, with special emphasis on the first appearance of factors that influence subsequent differentiation and the extent to which they are asymmetrically distributed in the egg cytoplasm and cortex at the time of fertilization and during early cleavage divisions. In Chapter 13 the process of embryonic induction, in which the course of differentiation of one type of tissue is determined by intimate interactions with a second type of tissue, is examined in detail. Finally, in Chapter 14 hormonal signals that are transmitted chemically through the body fluids and have their effects at locations remote from their tissue of origin are examined, including current views of the molecular mechanisms responsible for hormonal action and an analysis of the elaborate feedback networks that regulate hormonal action.

In each of these chapters emphasis is given both to the nature of the external signals received by the cells and to the nature of the intracellular changes that occur in response to those signals. In some cases our approach is by necessity mostly descriptive, but whenever molecular mechanisms are known, they are included. The goal of this section is to bridge the gap, insofar as present knowledge permits, between the detailed mechanisms for information flow from gene to protein described in Chapters 4 to 9 and the purely descriptive approach to development that was outlined in Chapters 2 and 3. Obviously our knowledge is still incomplete in many of these areas. However, most of the experimental tools needed to obtain answers to the major questions appear to be available, and rapid progress toward the understanding we seek can be expected during the next few years. It is our hope that this section will provide students with the foundation needed for an understanding of that progress as it is reported in the professional literature.

CHAPTER 10

Growth and behavior of differentiated cells in culture

☐ If it were technically feasible, the ideal place to study cell differentiation would be in the developing embryo, where all the many cellular interactions could be examined in their natural setting. However, embryos of higher organisms are so complex that it often is not practical to ask precise experimental questions in the confines of the embryonic system. Consequently, many workers studying cellular differentiation have opted for one of two alternative approaches.

The first approach is to study differentiation in primitive organisms, which have a simple organization and contain only a few cell types. An example of such an organism is the cellular slime mold *Dictyostelium discoideum*, whose development is discussed in Chapter 23. The disadvantage of this approach, however, is the uncertainty as to whether the answers that are obtained will be applicable to the more complex differentiation that occurs among cells of more advanced organisms.

The second alternative approach, which has already been discussed briefly in Chapter 3, is to study cells from a complex organism in a simplified environment such as cell culture. The complex interactions that occur among different cell types in the intact embryo are eliminated in culture, and it is possible, at least in theory, to analyze cellular responses under precisely controlled environmental conditions. However, one must be cautious in extending results obtained with cultured cells to intact organisms, since the simplified culture system may not contain all the components that affect differentiation in the intact organism.

Before we examine experiments that have been done with differentiated cells in culture, it is desirable to review some of the historical background of the development of cell culture techniques. The early techniques of "tissue" culture were far from precisely defined. For

MODEL SYSTEMS FOR THE STUDY OF DIFFERENTIATION

DEVELOPMENT OF CELL CULTURE TECHNIQUES

319

growth of cells from vertebrates, a culture vessel was typically coated with a clot of blood plasma (the gelatinous material left when the fluid portion of blood is allowed to clot after its cells have been removed by centrifugation). A tissue fragment was placed on the clot and was covered by a liquid medium consisting of a mixture of physiological saline solution, serum (the clear fluid formed when whole blood is allowed to clot and the clot and entrapped cells are removed), and embryo extract (a clear fluid obtained by homogenizing chicken embryos in saline solution and centrifuging out the debris).

In such a culture set-up, cells migrate out from the central fragment of tissue onto the surface of the plasma clot and undergo proliferation. The dividing cells tend to lose their specialized appearance and assume a simplified morphology (Fig. 10-1). Cellular multiplication is slow under such conditions, and typically the outgrowths were subcultured by being cut in two and placed on plasma clots in fresh culture vessels about once a month. A major breakthrough in tissue culture was made in the early 1940s, when Wilton Earle of the National Cancer Institute treated a fragment of mouse subcutaneous connective tissue that had been in culture for about a year with a carcinogenic agent, 20-methylcholanthrene, for a period of nearly 4 months. As a result of this treatment the cells became malignant, growing much more rapidly in culture and also producing tumors when injected into mice. This was the origin of the well-known mouse L cell line. A few years later, one of Earle's co-workers, Katherine Sanford, placed a single mouse L cell in a plasma-coated capillary tube and initiated a clonal culture (a population of cells derived from a single cell) designated strain L-929 (Fig. 10-2).

In the early 1950s, an established line of human cells was initiated by George Gey and his associates at Johns Hopkins University. As described in Chapter 3, this line, designated HeLa, was derived from a highly malignant cervical carcinoma surgically removed from a black woman. This cell line was also initiated on a plasma clot under a natural medium composed of saline solution, serum, and embryo extract, comparable to that used to initiate the mouse L line.

These two established lines of malignant cells were shipped to laboratories throughout the world and dominated the study of cell culture for a number of years. Rapid progress was made in developing improved media and culture conditions, and by the end of the 1950s a number of chemically defined media were available that would support growth of these cells either without undefined supplements or with only small amounts of serum or serum proteins. Many additional established lines with similar properties were adapted to grow either in the original natural media or in the defined media that had been developed for mouse L and HeLa.

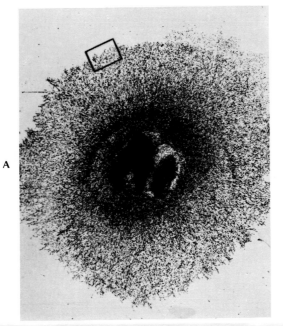

Fig. 10-1. Outgrowth of fibroblast-like cells from an explant of mouse subcutaneous connective tissue on a plasma clot. **A,** Broad halo of outgrowth surrounds the explant after 4 days in culture. **B,** Enlargement of the area in the rectangle in **A,** showing elongated shape of individual cells. (**A,** ×14; **B,** Harris hematoxylin stain, ×260.) (From Bloom, W., and D. W. Fawcett. 1975. A textbook of histology, 10th ed. W. B. Saunders Co., Philadelphia.)

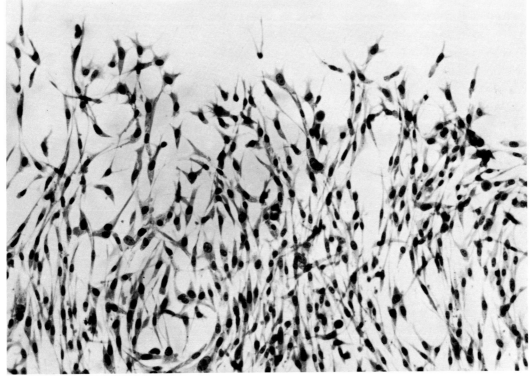

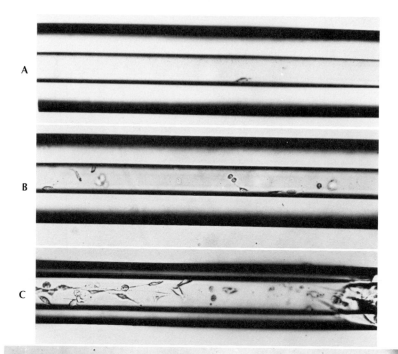

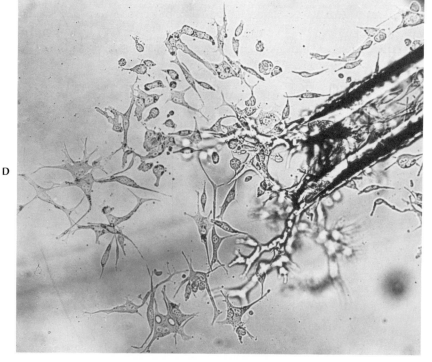

Fig. 10-2. Cloning of mouse L cells in a capillary tube. **A,** Single cell isolated in a plasma-coated capillary tube. **B,** Multiplication within the capillary tube. **C,** Multiplying cells have reached the end of the capillary tube and are ready to begin migrating out. **D,** Outgrowth of cells from the end of the capillary tube onto the plasma clot, 28 days after beginning of experiment. (From Sanford, K. K., W. R. Earle, and G. D. Likely. 1948. J. Natl. Cancer Inst. **9:**229.)

Other advances in culture technique occurred during the same period. The plasma clot was eliminated, and satisfactory growth of cells was obtained directly on glass surfaces (and later on specially prepared polystyrene plastic surfaces). The enzyme trypsin was used to release the cells from the culture surface so that they could be subcultured easily. Replicate culture techniques were devised so that precisely controlled experiments could be done comparing the responses of identical cultures to various conditions. Clonal growth techniques were developed whereby single isolated cells developed into large colonies, allowing the recovery of homogeneous populations of cells with specialized properties and allowing quantitative measurement of cellular survival.

The development of cell culture methods has made it possible for investigators to work with cells from complex organisms in a manner similar to that used for growth of microorganisms. Indeed, it can be argued that cells in culture *are* independent microorganisms, since they grow, reproduce, and evolve independently of the organism from which they were originally derived. In fact, one of the major arguments against use of established cell lines such as mouse L and HeLa as research tools is that they may have evolved so extensively in culture, because of their genetic instability and the selective pressures that they were exposed to early in their culture histories, that they are no longer representative of cells of the organisms from which they were originally derived.

During the period when modern cell culture techniques were being developed, the bulk of the research that was done with cultured cells employed established cell lines such as mouse L and HeLa. Although these cells were highly abnormal in many respects, they were far more reliable and easier to work with than normal cells. A few investigators worked with primary cultures of normal cells (cultures of cells taken directly from animal tissues) or with subcultures of these cells. However, such cultures required large amounts of serum or other undefined supplements for growth and failed to exhibit any significant differentiated properties. When cultures were deliberately initiated from differentiated tissues, one of three things tended to happen: (1) nothing grew, (2) the cultures were overrun with fibroblast-like cells (cells similar in appearance to relatively undifferentiated connective tissue cells), or (3) the differentiated cells began to multiply in culture but lost their differentiated phenotype and became fibroblast-like in their properties.

Culture media and conditions have been improved over the years, and, as summarized briefly in Chapter 3, two major classes of differ-

REQUIREMENTS FOR THE GROWTH OF DIFFERENTIATED CELLS IN CULTURE

entiated cells can now be grown well in culture. The first are cells derived from malignant tumors that express differentiated properties in the intact animal and continue to express similar properties when placed in culture (Table 10-1). Such cells are generally less demanding in their requirements for growth in vitro than normal cells, and

Table 10-1. Examples of cells that exhibit differentiated properties in culture*

Type of cell	Differentiated properties expressed in culture
Differentiated tumor cells	
Adrenal tumor	Synthesis of steroids in response to ACTH
Chorionic carcinoma	Synthesis of chorionic gonadotropin
Erythroleukemia	Synthesis of hemoglobin
Glioma	Synthesis of S-100 protein characteristic of glial cells
Hepatoma	Synthesis of serum albumin and various liver-specific enzymes; induction of tyrosineaminotransferase by glucocorticoids
Mammary tumor	Synthesis of milk proteins
Melanoma	Presence of melanosomes with or without synthesis of melanin pigment
Muscle cell tumor	Fusion of fibroblast-like forms to yield multinucleated myotubes
Myeloma	Synthesis of immunoglobins
Neuroblastoma	Formation of neuronlike cells with long axons; presence of acetylcholine esterase, neurotransmitters, and action potentials
Renal adenocarcinoma	Synthesis of kidney-specific form of esterase
Teratocarcinoma	Formation of embryoid bodies; multiple pathways of differentiation
Differentiated normal cells	
Adrenal cortex	Synthesis of steroid hormones
Cartilage	Synthesis of chondroitin sulfate and cartilage-specific collagen
Heart muscle	Spontaneous rhythmic contraction; muscle proteins
Keratinocytes	Keratinization
Lens	Synthesis of crystallins (lens protein)
Liver	Synthesis of serum protein
Mammary gland	Synthesis of milk proteins and fats
Melanocytes	Synthesis of pigmented melanosomes
Myoblasts	Fusion to form multinucleate myotubes; synthesis of muscle-type actin and myosin; contractility
Pituitary	Synthesis of ACTH, somatotropin, and prolactin (from different cultures)
Tendon	Synthesis of large amounts of type I collagen (up to 30% of total protein synthesis)
Thyroid	Synthesis of thyroid hormones

*More extensive listings of differentiated cells in culture can be found in Wigley, C. B. 1975. Differentiation **4:**25, and in Cox, R. P., and J. C. King. 1975. Int. Rev. Cytol. **43:**281.

they are usually also capable of continuing to multiply indefinitely as permanent cell lines. However, since their expression of differentiated properties in culture is often partially or totally divorced from normal control mechanisms, their usefulness as research tools is somewhat limited.

The second class of differentiated cells that can now be grown in culture is derived from normal tissue. The behavior of such cells differs significantly from one tissue to another and also from one species to another, but in general they remain diploid, at least during early passages in culture, and in most cases they do not form permanent lines but instead exhibit cellular senescence after a number of doublings that is characteristic for each cell type and species. Some of the types of differentiated normal cells that have been studied in culture are listed in Table 10-1.

Whether cultures initiated from normal differentiated cells actually exhibit differentiated properties (frequently referred to as differentiated phenotype) is influenced by many different aspects of the culture environment. Among the factors that influence expression of differentiated phenotype are the composition of the culture medium, the nature of the surface the cells are grown on, and what other types of cells are in contact with the cells that are being studied in the culture system.

The culture medium that is used for growth of differentiated cells typically consists of a "chemically defined" basal medium supplemented with serum and sometimes also fortified with various other undefined additives such as embryo extract or pituitary extract. The basal medium consists of an isotonic physiologic saline solution that contains all the major ions needed for cellular survival and growth, plus an assortment of sugars, amino acids, vitamins, trace elements, and various other nutrients. So-called minimal media such as the "minimum essential medium" devised by Harry Eagle may contain less than 30 components, whereas some of the most complex media contain up to 70 or 80 components. The exact composition of a basal medium that is frequently used for growth of differentiated cells in culture is given in Table 10-2.

Although many types of highly adapted permanent cell lines that exhibit little or no differentiation in culture can be grown in defined media without supplementation, all differentiated normal cells, as well as most differentiated tumor cells, still require incompletely defined supplements in their media for multiplication in culture. Reintroduction of undefined components into the culture medium was a price that had to be paid to obtain efficient growth of differentiated cells. The dilemma that is involved was stated well in 1963 by Irwin

Konigsberg, one of the pioneers in the study of muscle cell differentiation in culture: "This [reintroduction of undefined components in media] may seem a retrograde step in view of the important strides made in the elaboration of defined media; however, for the developmental biologist, the choice between ill-defined cells in well-defined media and well-defined cells cultivated in ill-defined media is no choice at all."

Progress toward understanding the nutritional requirements of normal cells in general has been slow. At the present time it is still not possible to grow even a single type of normal cell, differentiated or not, in a completely defined synthetic medium. Most established cell lines that exhibit differentiated properties also require undefined supple-

Table 10-2. Composition of medium F12*

Component	Amount (moles/L)	Component	Amount (moles/L)
Amino acids		*Vitamins — cont'd*	
L-Alanine	1.0×10^{-4}	Pyridoxine · HCl	3.0×10^{-7}
L-Arginine · HCl	1.0×10^{-3}	Riboflavin	1.0×10^{-7}
L-Asparagine · H_2O	1.0×10^{-4}	Thiamine · HCl	1.0×10^{-6}
L-Aspartic Acid	1.0×10^{-4}	Vitamin B_{12}	1.0×10^{-6}
L-Cysteine · HCl · H_2O	2.0×10^{-4}	*Other organic compounds*	
L-Glutamic Acid	1.0×10^{-4}	Choline chloride	1.0×10^{-4}
L-Glutamine	1.0×10^{-3}	D-Glucose	1.0×10^{-2}
Glycine	1.0×10^{-4}	Hypoxanthine	3.0×10^{-5}
L-Histidine · HCl · H_2O	1.0×10^{-4}	*i*-Inositol	1.0×10^{-4}
L-Isoleucine	3.0×10^{-5}	Linoleic acid	3.0×10^{-7}
L-Leucine	1.0×10^{-4}	Putrescine · 2HCl	1.0×10^{-6}
L-Lysine · HCl	2.0×10^{-4}	Sodium pyruvate	1.0×10^{-3}
L-Methionine	3.0×10^{-5}	Thymidine	3.0×10^{-6}
L-Phenylalanine	3.0×10^{-5}	*Inorganic salts*	
L-Proline	3.0×10^{-4}	$CaCl_2$ · $2H_2O$	3.0×10^{-4}
L-Serine	1.0×10^{-4}	$CuSO_4$ · $5H_2O$	1.0×10^{-8}
L-Threonine	1.0×10^{-4}	$FeSO_4$ · $7H_2O$	3.0×10^{-6}
L-Tryptophan	1.0×10^{-5}	KCl	3.0×10^{-3}
L-Tyrosine	3.0×10^{-5}	$MgCl_2$ · $6H_2O$	6.0×10^{-4}
L-Valine	1.0×10^{-4}	NaCl	1.3×10^{-1}
Vitamins		$NaHCO_3$	1.4×10^{-2}
d-Biotin	3.0×10^{-8}	Na_2HPO_4 · $7H_2O$	1.0×10^{-3}
Folic acid	3.0×10^{-6}	$ZnSO_4$ · $7H_2O$	3.0×10^{-6}
DL-α-Lipoic acid	1.0×10^{-6}	*pH indicator*	
Niacinamide	3.0×10^{-7}	Phenol red	3.3×10^{-6}
D-Pantothenic acid (hemicalcium salt)	1.0×10^{-6}		

*Medium F12 was originally developed for clonal growth of established Chinese hamster lines (Ham, R. G. 1965. Proc. Natl. Acad. Sci. U.S.A. **53**:288.) However, it has also been used extensively for growth of differentiated cells, sometimes with its concentrations of amino acids and pyruvate doubled.

ments for growth, although rat glial tumor cells and mouse neuroblastoma cells have both recently been grown successfully in protein-free synthetic media. The current revival of interest in the growth requirements of normal and differentiated cells is encouraging. Hopefully in the near future it will be possible to study differentiated cells in media that are free from unknown substances, and to determine precisely what environmental factors are needed for expression of differentiated properties.

For the present, conclusions regarding the control of cell differentiation in cultured cells must be based on experiments performed with poorly defined additives, such as serum and embryo extract, in the culture medium. Nevertheless, experiments done under these conditions have been able to demonstrate convincingly that the expression of differentiated properties by cultured cells is under environmental control. Thus we are able to distinguish between "permissive" media and culture conditions that support the expression of differentiated properties by cultured cells and "nonpermissive" media and culture conditions that do not.

As an example, we can consider the "ill-defined media" that support muscle cell differentiation. At the time that Konigsberg made his statement, the differentiation of myoblasts (precursors of skeletal muscle cells) in culture was accomplished by growing them in the presence of three types of undefined materials: (1) horse serum (other types of serum, such as fetal calf serum, which is widely used in other culture systems, did not work), (2) embryo extract, and (3) conditioned medium (that is, medium in which cells have been previously grown, but not to the point of exhaustion of the nutrients). The critical component in the conditioned medium was later found to be collagen, and the requirement for conditioned medium can now be satisfied more simply by coating the culture dishes with collagen or gelatin. However, the other requirements remain poorly defined, and the standard procedure in 1979 for obtaining muscle cell differentiation in culture still calls for the addition of horse serum and embryo extract to the medium and for the use of culture vessels coated with collagen or gelatin. In addition, it is now known that the culture medium must contain a high concentration of calcium ion to support myoblast differentiation.

Under these permissive conditions the cells not only multiply, but when they reach an adequate density, they line up end to end and fuse to form multinucleate myotubes that contain cross-striated contractile filaments (Fig. 10-3). If the culture conditions are rendered nonpermissive by use of a lower calcium concentration, by replacement of horse serum with fetal bovine serum, by increasing the concentration of embryo extract, or by eliminating the collagen or gelatin coating, reasonably good growth of the myoblasts still occurs, but there is little cellular fusion or formation of differentiated myotubes (Fig. 10-4).

Two other types of differentiated cells that have been studied extensively in culture are cartilage cells and pigmented retina cells. Differentiated retina cells grown under permissive conditions in vitro contain large numbers of pigment granules so that the cultures look black to the naked eye. The cartilage cells secrete large amounts of

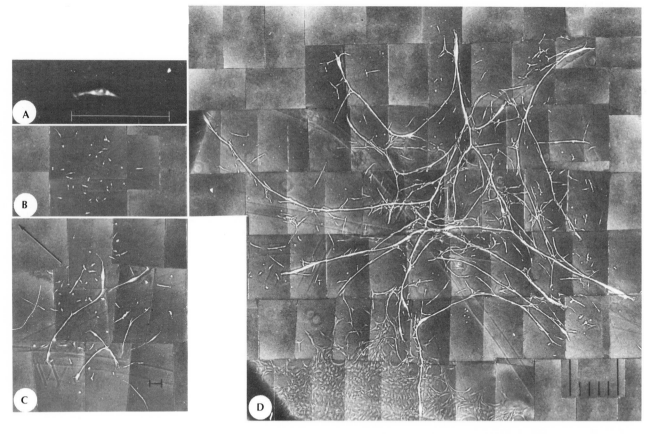

Fig. 10-3. Clonal growth of muscle-forming cells. **A,** Single myoblast from embryonic chicken muscle, showing typical bipolar appearance. **B,** Small colony of mononucleated cells, derived from cell in **A. C,** Cellular fusion has begun to yield multinucleated myotubes. Mononucleated cells continue to divide. **D,** Colony has continued to grow and now contains very long multinucleated myotubes.in addition to proliferating mononucleated myoblasts. An adjacent colony of fibroblast-like cells is seen at the lower left. (From Konigsberg, I. R. 1963. Science **140:**1273. Copyright 1963 by the American Association for the Advancement of Science.)

extracellular matrix, rich in chondroitin sulfate, which can easily be detected by staining techniques or by uptake of radioactively labeled sulfate. Differentiation in culture of either of these cell types derived from chicken embryos requires the presence of a low molecular weight fraction of embryo extract; the high molecular weight fraction is inhibitory to expression of differentiated properties (Fig. 10-5).

Another aspect of the culture system found to be of importance in obtaining cell differentiation in vitro is growth of the cells under clonal conditions. If cells are cultured as monolayers (i.e., at a high enough

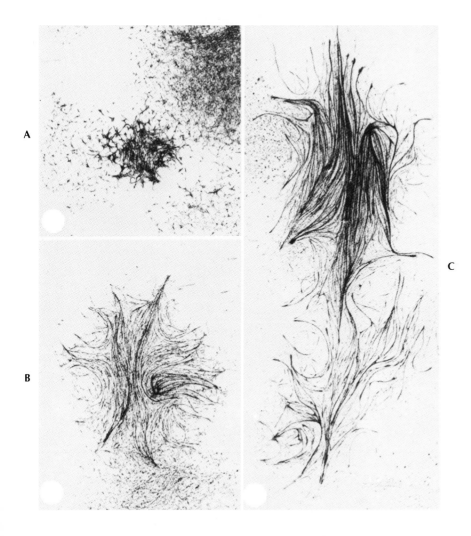

Fig. 10-4. Effect of conditioned medium and collagen on muscle cell differentiation in culture. **A,** Colonies grown without the medium being conditioned or the dishes being treated with collagen. The small colony in the center contains a few multinucleated myotubes, but the neighboring colonies are entirely undifferentiated. **B,** Colony grown in conditioned medium on an untreated surface. Myotubes are longer and more numerous than in **A. C,** Two muscle cell colonies typical of those which develop without conditioned medium on a collagen-coated surface. (From Hauschka, S. D., and I. R. Konigsberg, 1966. Proc. Natl. Acad. Sci. U.S.A. **55:**119.)

concentration in the culture vessel so that the cells form a continuous layer on the culture surface), contact with interfering cell types tends to suppress expression of differentiation by the cell types of interest. It has been shown conclusively that the presence of a small percentage of fibroblast-like cells in a monolayer of cartilage or pigment cells will effectively suppress expression of differentiation by the latter. Under clonal growth conditions, where each colony develops without contact with "foreign" cells, no such interference occurs, and cells that have the ability to express differentiated properties do so freely.

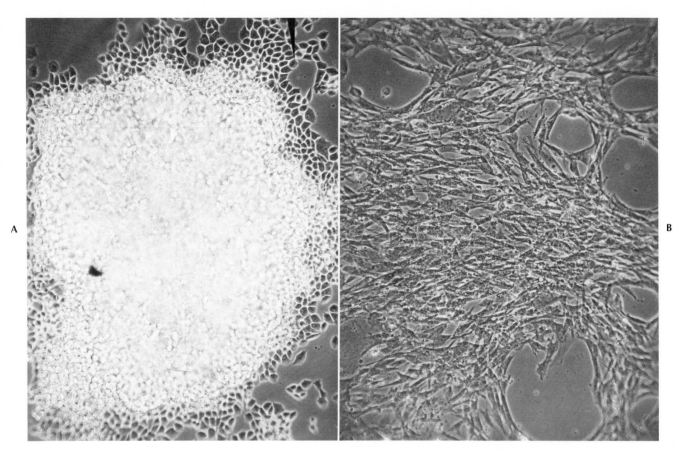

Fig. 10-5. Effect of embryo extract fractions on differentiation of embryonic chicken cartilage cells in clonal culture. **A,** Well-differentiated colony grown in medium with low molecular weight fraction of embryo extract for 18 days. The cells in the center of the colony are piled into multiple layers and surrounded by a refractile matrix that gives them a very bright appearance in this phase contrast micrograph. **B,** A portion of the edge of a sibling colony grown for 18 days in a medium containing the high molecular weight fraction of embryo extract. Cartilage-like differentiation has been entirely suppressed, and the cells have assumed a fibroblast-like morphology. Cells in the centers of these colonies pile up but do not secrete matrix. (From Coon, H. 1966. Proc. Natl. Acad. Sci. U.S.A. **55:**66.)

The behavior of differentiated cells in permissive and nonpermissive media has made it clear that there are two distinct aspects of differentiation. The display of differentiated properties by cultured cells is clearly under the control of environmental influences. Yet cells that are committed to a specific differentiated pathway can be grown in nonpermissive media for many generations in culture and they will still "remember" what type of cell they are. When they are returned to a permissive medium, they always express the same type of differentiated properties to which they were originally committed. For example, when retinal pigment cells and cartilage cells are grown as clonal cultures in nonpermissive media, both form similar-appearing colonies of fibroblast-like cells with no obvious pigment or cartilage matrix. If those colonies are dispersed into suspensions of single cells with trypsin and subcultured in permissive media, they again exhibit differentiated properties. Cells derived originally from cartilage form only cartilage-making colonies and never pigmented colonies in permissive media. Likewise, cells from pigmented tissue form only pigmented colonies and never cartilage-making colonies. Each of these cells "inherits" the ability to express a particular differentiated property even though environmental conditions have prevented that expression during a large number of cell doublings.

The inherited potential to express a particular type of differentiation has been termed the epigenotype of the cell. The implication of the term "epigenotype" is that the predisposition of each cell to differentiate in a particular direction is a heritable characteristic that has somehow been superimposed on the basic genetic program of that cell. A corollary term, "epiphenotype," is used to denote the differentiated properties that are actually expressed by a cell of a particular epigenotype. (The terms "phenotype" or "phenotypic expression" are also sometimes used somewhat loosely to express the same concept.) As an example to illustrate the concepts of epigenotype and epiphenotype, a cell whose epigenotype is cartilage may express a cartilage epiphenotype (that is, it may act like a cartilage cell and secrete a matrix rich in chondroitin sulfate), or it may exhibit an undifferentiated epiphenotype (that is, it may look like a nondescript fibroblast-like cell and not produce any cartilage matrix), but it will never exhibit a pigment cell epiphenotype (that is, it will never form melanosomes or synthesize melanin pigment) (Fig. 10-6).

The distinction between epigenotype, or commitment to a particular pathway of differentiation, and epiphenotype, or actual expression of differentiated properties, does not appear to be an artifact of cell culture. In developing embryos, commitment to a particular pathway of differentiation often occurs long before any overt expression of dif-

COMMITMENT OF CELLS TO
SPECIFIC DIFFERENTIATED
STATES

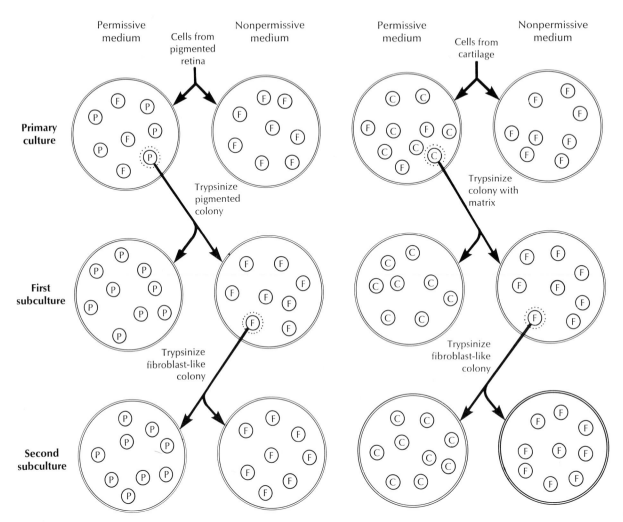

Fig. 10-6. Inheritance of epigenotype. Parallel experiments are diagrammed for cells from cartilage and pigmented retina of chicken embryos. C = Colonies making cartilage matrix; P = pigmented colonies; and F = fibroblast-like colonies that make neither matrix nor pigment. Primary cultures in permissive medium exhibit a mixture of differentiated and undifferentiated phenotypes due to heterogeneity of the inoculum. When cells from differentiated colonies are subcultured, they exhibit a differentiated epiphenotype in permissive medium and a fibroblast-like epiphenotype in nonpermissive medium. When subcultured again, this time in permissive medium, cells that had seemingly dedifferentiated in nonpermissive medium during the first subculture once again exhibit their characteristic differentiated properties. Cells derived originally from a pigmented colony again synthesize pigment, and those from a colony that made cartilage matrix again synthesize matrix components. (Based on data from Coon, H. G., and R. D. Cahn. 1966. Science **153:**1116.)

ferentiated properties. This phenomenon will be discussed in greater detail in Chapters 12 and 13.

It is generally assumed that environmentally caused changes in epiphenotype are due to regulatory effects on gene expression, and in at least some cases, such as muscle cell differentiation, definite changes in mRNA synthesis can be observed. However, there is one unusual case in which the change in epiphenotype may be entirely due to posttranslational processes. Treatment of cartilage cells in clonal culture with an excess of vitamin A results in their failure to accumulate the extracellular matrix that is characteristic of cartilage cell differentiation. It is currently believed that the primary effect of vitamin A is to increase the lability of lysosomes, causing leakage of lysosomal enzymes into the cytoplasm. The major component of the matrix secreted around cultured cartilage cells is a chondromuco-protein, which consists of several molecules of a complex polysaccha-ride, chondroitin sulfate, attached to a protein backbone. That back-bone protein is highly susceptible to digestion by cathepsin D, which is one of the enzymes released from the lysosomes. Chondroitin sulfate is normally synthesized in situ on the intact backbone protein. In addi-tion, even if it is still synthesized, without the backbone, chondroitin sulfate alone is highly soluble and unlikely to produce a firm matrix around the cells. Thus, although definitive studies have not been done, it currently appears that the inhibition of cartilage differentiation in clonal culture in the presence of excess vitamin A is due primarily to destruction of a key component of the extracellular matrix, rather than to inhibition of its synthesis.

The ability to control differentiation in culture by alteration of growth conditions has made possible a clarification of the relationship between cell multiplication and differentiation. In primitive culture systems, expression of differentiation by growing cells generally did not occur, and many investigators concluded that differentiation and growth in culture were mutually exclusive events. For example, carti-lage cells in an actively growing monolayer lost their cartilage-specific properties. These properties were expressed in culture only when the cells were centrifuged into a dense pellet, which inhibited cellular multiplication. In retrospect, it now appears that the very crowded conditions somehow permitted the cells to overcome the inhibitory effects of foreign cell types. However, at that time many investigators believed that it was not possible, perhaps for energetic reasons, for a cell to make the large amounts of the basic cellular materials neces-sary for proliferation and at the same time to synthesize large amounts of the "luxury molecules" characteristic of differentiation.

CELL DIVISION VERSUS DIFFERENTIATION IN CULTURED CELLS

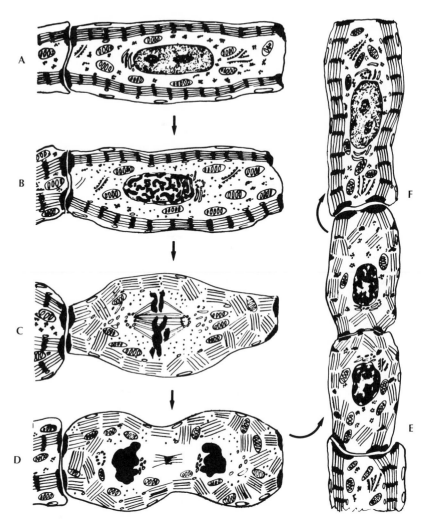

Fig. 10-7. Schematic representation of the cell cycle in embryonic heart muscle cells. **A,** Interphase cell with well-developed contractile filaments and junctions to adjacent cells. **B,** Prophase, showing condensation of chromosomes. **C,** Metaphase. A selective intracellular protease has distrupted the Z discs of the contractile fibers. **D,** Anaphase. Cytokinesis is in progress. **E,** Telophase. Desmosomes are being established between the newly divided cells, and the Z bands are beginning to form again. **F,** Interphase. The daughter cells (only one shown here) have fully organized contractile fibers and intercellular junctions, comparable to **A. G,** Electron micrograph showing mitosis of a well-differentiated heart muscle cell from an 11-day chicken embryo. Condensed chromosomes are seen as dark masses in the left center. Well-developed myofibrils are present (indicated by *MF* and also by the box, which outlines the area shown at higher magnification in the insert). Extensive deposits of glycogen and other features characteristic of the ultrastructure of heart muscle cells are also visible. (**A** to **F** from Rumyantsev, P. P. 1977. Int. Rev. Cytol. **51:**187; **G** from Manasek, F. J. 1968. J. Cell Biol. **37:**191.)

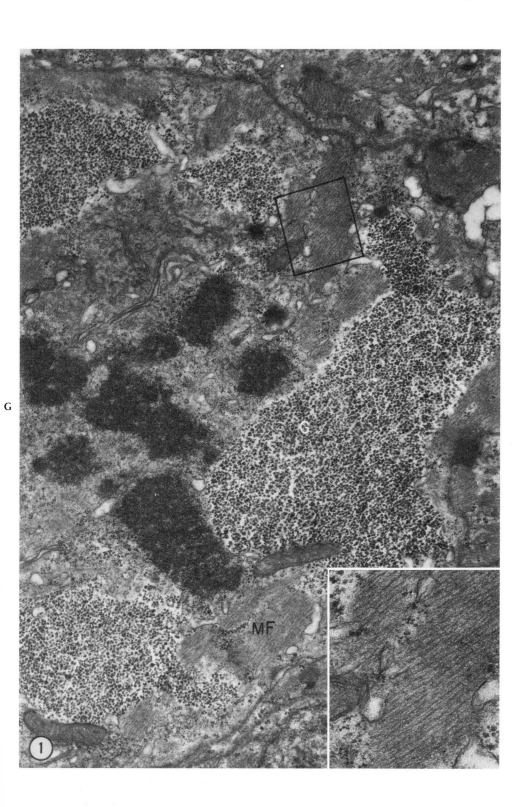

(The term "luxury molecules" has been used by a number of investigators to refer to specialized substances such as cartilage matrix, pigment, or contractile proteins that are found in large amounts in many types of differentiated cells but are not required for growth and replication of cells in general. It should be pointed out, however, that such molecules represent a "luxury" only for individual types of cells in the highly artificial environment of cell culture. They are generally absolutely necessary for survival of the intact organism.)

With the introduction of improved media and clonal growth conditions, it became possible by the mid-1960s to obtain growing colonies of several different types of cells that exhibited specific differentiated phenotypes. However, it was still necessary to demonstrate that the colonies were not composed of two distinct cell populations, one that expressed differentiated properties but had ceased to divide and another which proliferated but did not express differentiated properties. Experiments designed to test these alternatives indicated that the simultaneous occurrence of cell division and differentiation was possible in some types of differentiated cells but not in others.

Cartilage cells were shown by the following experiment to multiply and to synthesize differentiated products simultaneously. Colonies containing differentiated cartilage cells were incubated for a short period of time with two different radioactive labels: ^{3}H-thymidine, which would be incorporated into DNA molecules, and ^{35}S-sulfate, which would be incorporated primarily into a cartilage-specific product, chondroitin sulfate. On autoradiography of doubly labeled cultures, cells were found that contained label both over the nucleus and over the cytoplasm and extracellular matrix (the expected locations for chondroitin sulfate). These labels could be removed by deoxyribonuclease and hyaluronidase, respectively, demonstrating that they did indeed represent incorporation of labeled precursors into DNA and into chondroitin sulfate within the same cell. In heart muscle, concurrent differentiation and cell division have been demonstrated by electron micrographs that show contractile fibers and condensed mitotic chromosomes in the same cell (Fig. 10-7).

Not all types of cells, however, are capable of division in the terminally differentiated state. An extreme case is the mammalian erythrocyte, which no longer even contains a nucleus. Among cultured cells the classical example of a nondividing cell type is skeletal muscle. As described earlier, mononucleate myoblast cells in a permissive medium undergo a period of rapid cell division, then line up end to end and fuse to form multinucleated myotubes. After fusion, muscle-specific gene products are synthesized, contractile fibers appear in the cytoplasm, and there is no further DNA synthesis or cell division.

It has been clearly demonstrated by a number of experimental approaches both in culture and in intact embryos that multinucleated myotubes arise strictly by cell fusion and not by nuclear division without concurrent cytoplasmic division. One experiment in which this was shown is summarized in Table 10-3. Chick myoblasts taken from thigh muscle of 10-day embryos were cultured for 18 hours, and then two groups of cells were incubated for a short period of time in a medium containing ^{3}H-thymidine. One group of cells was then immediately fixed and autoradiographed; the second group was transferred to unlabeled medium and incubated until the mononucleated myoblasts had fused and formed myotubes (3 to 5 days), after which it was fixed and autoradiographed. A third group of cells from the same original source was allowed to begin forming myotubes in unlabeled medium and then was transferred to a medium containing ^{3}H-thymidine after myotube formation had begun. These cells were then fixed and autoradiographed immediately after incubation with the labeled thymidine. The first group of cells contained label over the nuclei of single myoblasts, indicating that DNA synthesis was occurring in mononucleate cells. The second group exhibited label over some of the nuclei in the myotubes, indicating that labeled mononucleated cells were later incorporated into myotubes. The third group of cells showed incorporation of label only into mononucleated cells and not into nuclei of myotubes, demonstrating that myotube nuclei do not synthesize DNA. Measurement of the DNA content of myotube nuclei indicated that all such nuclei were diploid and thus that they had not been formed by division in the absence of DNA synthesis. These data together indicate that the sequence of events required to obtain a labeled myotube nucleus is DNA synthesis (shown by incorporation of labeled thymidine) in a mononucleate cell, cell division (to return the amount of DNA per

Table 10-3. Incorporation of ^{3}H-thymidine into mononucleated myoblasts and multinucleated myotubes as a function of time after labeling*

	Experiment I	Experiment II	Experiment III
Time in culture before labeling	18 hours	18 hours	4 to 6 days
Exposure to label	30 minutes	30 minutes	30 minutes
Time after label before fixing	0	3 to 5 days	0
Cell types labeled			
Mononucleated	Yes	Yes	Yes
Multinucleated	No	Yes	No

*Based on data from Stockdale, F. E., and H. Holtzer, 1961. Exp. Cell Res. **24:**508.

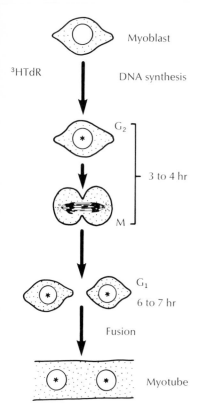

Fig. 10-8. Proposed sequence of events leading to formation of multinucleated myotubes from mononucleated myoblasts. Details of the time sequence are discussed in Chapter 11.

nucleus to a diploid amount), and finally, fusion of the labeled daughter cells into new or existing myotubes (Fig. 10-8).

Evidence that this sequence of events occurs during myotube formation in the intact animal and is not an artifact of cell culture has been obtained from experiments with "allophenic" mice, which are formed by fusing two genetically different embryos at the eight-cell stage of development. Such animals, which are described in greater detail in Chapter 12, contain two genetically different types of cells in their tissues. When the two embryos that are mixed are each homozygous for an electrophoretically distinct form of the enzyme isocitrate dehydrogenase (IDH), the resulting individuals contain two populations of cells, each able to make one or the other form of the enzyme. No single cell that has been derived by mitosis from the original population contains both types of genes, and therefore no single cell can produce both types of enzyme. However, if multinucleated skeletal muscle cells are formed as the result of cell fusion, some may contain nuclei derived from both types of mononucleated precursor cells and thus may be able to produce both types of enzymes. Since IDH is a dimer, the production of both types of monomers within the cytoplasm of a single cell can be detected by the appearance of a hybrid form of the enzyme (ab), which is distinguishable electrophoretically from either of the two parental forms (aa and bb) (Fig. 10-9). Such a hybrid enzyme is normally found, together with the parental forms, in cells of genetically heterozygous mice, which contain both genes within every diploid nucleus. When IDH is isolated from different tissues of allophenic mice formed by fusion of two homozygous embryos of different genotypes, those tissues in which all cells have arisen by mitosis contain only the pure aa and bb forms expected from the parental cells. However, the skeletal muscle tissue also contains the ab form of the enzyme, indicating that both monomers are being produced within a common cytoplasm and therefore that nuclei derived from the two parental cell populations are present in the same myotube, presumably as a result of cellular fusion during myotube formation.

It appears that no broad generalizations are possible concerning the relationship between growth and expression of differentiation either in cultured cells or in intact organisms. Neuroblastoma, a nerve cell tumor, is an example of cells that show intermediate behavior in culture. When the cells are actively growing in a serum-rich medium, they do not appear differentiated. When serum is removed, growth stops and the cells extend long axonlike processes and exhibit nervelike biochemical and physiological properties. When serum is restored, the processes are retracted and growth of seemingly undifferentiated cells again proceeds. However, during the transition period between active growth and differentiation, cells are seen to extend processes and look differentiated but then to retract them, divide, and reextend them. Thus growth and the expression of differentiation by neuroblastoma cells in culture shows some overlap, although one state or the other generally predominates.

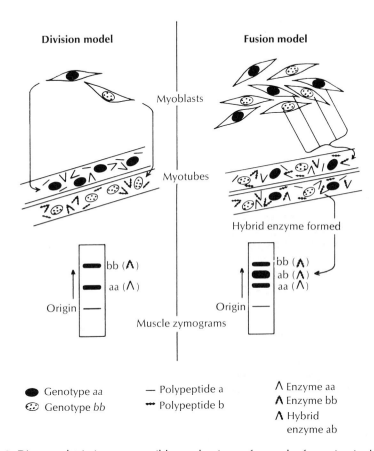

Fig. 10-9. Diagram depicting two possible mechanisms of myotube formation in the development of skeletal muscle and the experimental procedure for distinguishing between them. Allophenic mice are formed by fusing two early embryos, each of which is homozygous for a different electrophoretic variant of the enzyme isocitrate dehydrogenase. If myotubes arise by repeated nuclear division within a single cytoplasm, each myotube will contain only one kind of gene, since each mononucleated cell is homozygous. Thus myotubes arising by division will be able to make only pure aa and pure bb enzyme. However, if myotubes arise by fusion, both kinds of nuclei will be incorporated within a single cytoplasm. Since both a and b subunits for the dimeric enzyme will be synthesized in that cytoplasm, they will be able to combine to form hybrid enzyme (ab) in addition to the parental types (aa and bb). The presence of hybrid enzyme in skeletal muscle but not other tissues of these allophenic mice confirms the fusion model. (From Mintz, B., and W. W. Baker. 1967. Proc. Natl. Acad. Sci. U.S.A. **58:**592.)

Fig. 10-10. Changes in cell surface morphology during the cell cycle. These scanning electron micrographs show cells from synchronized cultures of an established Chinese hamster ovary line at various stages of the cell cycle. **A,** S phase. Note the extent of flattening and the relative absence of cell surface features. **B,** Late G_2 and M phases. The two lower cells are in late G_2 and are just beginning to round up at their margins. The cell at the right is in early prophase and is in the process of rounding up. The fully rounded cell at top center is in late prophase or perhaps metaphase.

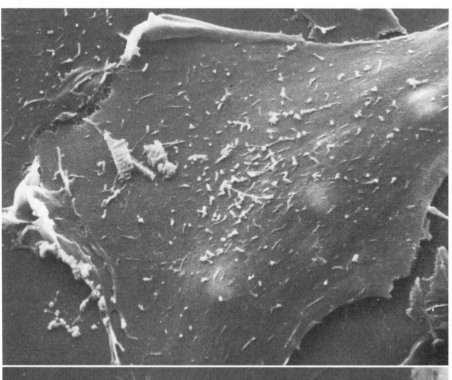

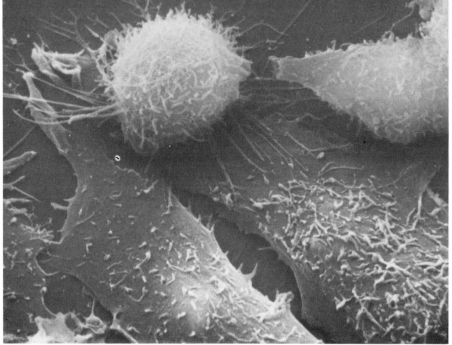

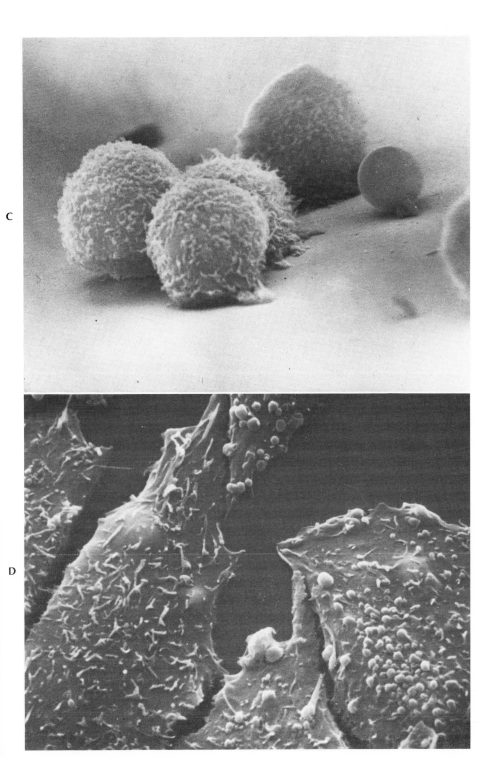

Fig. 10-10, cont'd. C, Early G$_1$ phase cells. These cells are shown 1 hour after they were collected as mitotic cells by being shaken loose from their substrate. Reattachment is in progress, but the cells are not yet fully flattened (undisturbed telophase cells would show a similar rounded appearance.) **D,** Late G$_1$ phase cells. Note surface features, including many rounded blebs, particularly on the cell at the right. (From Porter, K., D. Prescott, and J. Frye. 1973. J. Cell Biol. **57:**815.)

RELATIONSHIP BETWEEN
DIVISION AND
DIFFERENTIATION IN
INTACT TISSUES

A broad range of relationships between growth and differentiation is also apparent in intact animals. Classical cytologists tended to believe that most differentiated cells did not normally divide. One of the reasons for this was the fact that mitotic figures are not usually seen in cytological preparations of fully differentiated cells. However, studies of dividing cells in culture have shown that major morphological changes accompany mitosis. A cell that is extended and flattened in culture during all other phases of the cell cycle rounds up and assumes a totally different morphology during mitosis (Fig. 10-10). If similar changes also occur in vivo, a differentiated cell in mitosis should not be expected to have the typical morphological appearance exhibited by that kind of differentiated cell during other parts of the cell cycle.

The question of whether differentiated cells in intact tissues can divide has been approached by the use of autoradiographic techniques to determine which cells are synthesizing DNA. Through administration of ^{3}H-thymidine to test animals over a moderately long period of time, it is possible to detect even those cells which enter the cell cycle only occasionally. The results of such studies indicate a full range of cellular behavior, including rapidly cycling stem cells, a variety of cells that divide occasionally, and some cells that do not divide at all such as myotubes in skeletal muscle and neurons in the central nervous system.

WHEN IS A CELL
DIFFERENTIATED?

Finally, we can ask about those cells which do not express obvious differentiated properties while they are multiplying in culture under nonpermissive conditions. Should we consider them to be differentiated? The answer is a matter of semantics. The cells have ceased to exhibit the characteristic properties of the differentiated tissue that they were derived from, but they have not lost their epigenotype. They still have a commitment to a specific type of differentiation, and when they are returned to a proper environment, they will again express their characteristic type of differentiation and no other. Thus, in one sense, they remain "differentiated" at all times, even while they are growing in a nonpermissive medium. We have already indicated in Chapter 3 that many investigators avoid problems of this type by viewing all cells as "differentiated" and classifying them according to their level of differentiation. This concept of multiple levels of differentiation will be developed further during the discussion of the quantal cell cycle hypothesis in Chapter 11.

The development of techniques whereby cells can be grown in culture and made to express their differentiated characteristics is a challenging task that is still only partially accomplished. Work in this field has provided some understanding of the relationship of cell division to differentiation and of the commitment of cells to particular pathways of differentiation. The importance of environmental factors in the expression of differentiation has also become apparent, but we still know relatively little about the complex interactions between environment and the inherent properties of the cell that are inferred from current data. This is a challenging area in which we can expect rapid progress during the next few years as cell culture media and systems are refined and the recently developed techniques of molecular biology are increasingly applied to the study of gene expression by cells cultured under defined environmental conditions.

The following chapter describes pioneering studies into the control of differentiation in cultured cells and some of the particularly interesting questions that remain to be answered.

BIBLIOGRAPHY
Books and reviews

Cahn, R. D. 1968. Factors affecting inheritance and expression of differentiation: some methods of analysis. *In* A. Ursprung, ed. Results and problems in cell differentiation. Vol. 1. The stability of the differentiated state. Springer-Verlag, Inc., New York.

Cameron, I. L. 1971. Cell proliferation and renewal in the mammalian body. *In* I. L. Cameron, and J. D. Thrasher, eds. Cellular and molecular renewal in the mammalian body. Academic Press, Inc., New York.

Cameron, I. L., and J. R. Jeter, Jr. 1971. Relationship between cell proliferation and cytodifferentiation in embryonic chick tissue. *In* I. L. Cameron, G. M. Padilla, and A. M. Zimmerman, eds. Developmental aspects of the cell cycle. Academic Press, Inc., New York.

Cox, R. P., and J. C. King. 1975. Gene expression in cultured mammalian cells. Int. Rev. Cytol. 43:281.

Ham, R. G., and W. L. McKeehan. 1978. Development of improved media and culture conditions for clonal growth of normal diploid cells. In Vitro 14:11.

Harris, M. 1964. Cell culture and somatic variation. Holt, Rinehart & Winston, Inc., New York.

Hauschka, S. D. 1972. Cultivation of muscle tissue. *In* G. H. Rothblat, and V. J. Cristofalo, eds. Growth, nutrition and metabolism of cells in culture. Vol. 2. Academic Press, Inc., New York.

Holtzer, H., et al. 1972. The cell cycle, cell lineages, and cell differentiation. Curr. Top. Dev. Biol. 7:229.

Jakoby, W. B., and I. H. Pastan, eds. 1979. Methods in enzymology. Vol. 58. Cell culture. Academic Press, Inc., New York. (Entire volume is dedicated to cell cultures and includes many chapters on differentiated cells in culture.)

Konigsberg, I. R. 1963. Clonal analysis of myogenesis. Science 140:1273.

Lasher, R. 1971. Studies on cellular proliferation and chondrogenesis. *In* I. L. Cameron, G. M. Padilla, and A. M. Zimmerman, eds. Developmental aspects of the cell cycle. Academic Press, Inc., New York.

Paul, J. 1975. Cell and tissue culture. 5th ed. Churchill Livingstone, Edinburgh.

Puck, T. T. 1972. The mammalian cell as a microorganism; genetic and biochemical studies in vitro. Holden-Day, Inc., San Francisco.

Rumyantsev, P. P. 1977. Interrelations of the proliferation and differentiation processes during cardiac myogenesis and regeneration. Int. Rev. Cytol. 51: 187.

Wigley, C. B. 1975. Differentiated cells in vitro. Differentiation 4:25.

Selected original research articles

Bryan, J. 1968. Studies on clonal cartilage strains. I. Effect of contaminant non-cartilage cells. Exp. Cell Res. 52:319.

Buonassisi, V., G. Sato, and A. L. Cohen. 1962. Hormone-producing cultures of adrenal and pituitary tumor origin. Proc. Natl. Acad. Sci. U.S.A. 48:1184.

Cahn, R. D., and M. B. Cahn. 1966. Heritability of cellular differentiation: clonal growth and expression of differentiation in retinal pigment cells in vitro. Proc. Natl. Acad. Sci. U.S.A. 55:106.

Coon, H. G. 1966. Clonal stability and phenotypic expression of chick cartilage cells in vitro. Proc. Natl. Acad. Sci. U.S.A. 55:66.

Coon, H. G., and R. D. Cahn. 1966. Differentiation in vitro: effects of Sephadex fractions of embryo extract. Science 153:1116.

Ham, R. G., L. M. Murray, and G. L. Sattler. 1970. Beneficial effects of embryo extract on cultured rabbit cartilage cells. J. Cell Physiol. 75:353.

Hauschka, S. D., and I. R. Konigsberg. 1966. The influence of collagen on the development of muscle clones. Proc. Natl. Acad. Sci. U.S.A. 55:119.

Holtzer, H., et al. 1960. The loss of phenotypic traits by differentiated cells in vitro. I. Dedifferentiation of cartilage cells. Proc. Natl. Acad. Sci. U.S.A. 46:1533.

Yasumura, Y., A. H. Tashjian, and G. Sato. 1966. Establishment of four functional, clonal strains of animal cells in culture. Science 154:1186.

CHAPTER 11

Analysis of controls of differentiation in cultured cells

☐ Techniques for growth of differentiated cells in culture, described in the previous chapter, have made possible many types of experiments designed to provide an understanding of the basic mechanisms responsible for cellular differentiation. Three different approaches to the study of differentiation in cultured cells are presented in this chapter as examples. The first approach involves analysis of the temporal relationship of DNA synthesis and cell division to the appearance of differentiated properties. Results of research along this line have led to the proposal that cells must pass through one or more "quantal" cell cycles to become programmed to express new functions. The second approach stems from the discovery that the drug 5-bromodeoxyuridine (BrdU) selectively and reversibly suppresses expression of differentiated properties in a large variety of cell types. The mechanism of action of this drug has been studied intensively in the hope that an understanding of what BrdU does will shed light on the molecular mechanisms responsible for differentiation. The third approach makes use of the fact that specific and reproducible patterns of expression of differentiated properties occur in hybrid cells formed by fusion of differentiated and undifferentiated cell types. Although few firm conclusions are yet possible, such studies are providing many new insights into the complex set of phenomena that we refer to collectively as cellular differentiation.

Cell differentiation, as it occurs both in culture and in the developing embryo, is usually associated with cell proliferation. Cells that have reached a terminal stage of differentiation may or may not cease dividing, as discussed in Chapter 10. However, the first appearance of differentiation always seems to occur within populations of dividing cells. Many workers in the field of development have asked whether cell division and cell differentiation are unconnected events whose

QUANTAL CELL CYCLE
HYPOTHESIS

345

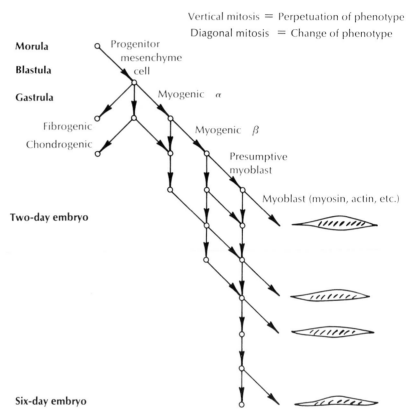

Fig. 11-1. Hypothetical scheme showing proliferative and quantal cell cycles in the formation of terminally differentiated myoblasts (mononucleated cells that are ready to fuse and begin synthesis of muscle protein without further cell cycles). In this scheme a series of quantal cell cycles leads to a succession of transitory myogenic precursor cells prior to definitive myoblast formation. It is postulated that myogenic cells ancestral to the myoblast are not capable of fusing or synthesizing muscle-specific proteins without undergoing the requisite numbers of additional quantal cycles. Each quantal cycle is assumed to segregate a cytoplasmic factor and/or alter expression of a portion of the genome. That change is then transmitted to the cell's progeny. The developmental potential of the primitive mesenchyme cell, which initially can become many things besides muscle, is progressively narrowed during early cycles. Each intermediate stage has a well-defined set of biochemical properties and possible kinds of differentiation that it can undergo. Proliferative cycles in which there is no change in these properties can occur at any level prior to the final quantal cycle. The postmitotic definitive myoblast is assumed to be essentially the same whether it undergoes the final quantal cycle early in development or later. (From Holtzer, H. 1970. Symp. Int. Soc. Cell Biol. **9:**69.)

parallel occurrence merely reflects the requirement for both during the normal course of development or whether there is a causal connection between division and differentiation. The latter view holds that passage through a cell cycle, or more specifically DNA synthesis followed by karyokinesis (nuclear division), is an essential prerequisite for cell differentiation. This viewpoint has been strongly supported by Howard Holtzer and his associates who have presented extensive evidence for their theory of the quantal cell cycle.

The quantal cell cycle theory proposes that during development two qualitatively different types of cell cycles can occur. The first is the proliferative cell cycle, which results in the production of two daughter cells that are identical to their parent cell in all respects, including their state of differentiation. The second is the quantal cell cycle, which results in an altered state of differentiation in one or both of the resulting daughter cells. The term "quantal" refers to a quantum jump from one type of differentiated expression to another. A hypothetical scheme showing the occurrence of proliferative and quantal cycles in the development of terminally differentiated muscle cells is presented in Fig. 11-1. Current versions of the quantal theory, based on studies with inhibitors, suggest that both DNA synthesis and nuclear division are essential for expression of the muscle phenotype, whereas cytoplasmic division is not. Specific molecular mechanisms whereby a cell's differentiated state is altered are not yet known, but any major reprogramming of a cell's activity would be likely to require a redistribution of the regulatory protein molecules associated with its DNA. This might be accomplished either as DNA polymerase traverses the DNA during periods of DNA synthesis, or in conjunction with the complex folding of chromosomal material that accompanies chromosomal condensation in mitosis, or both.

Muscle differentiation

The concept of the quantal cell cycle has not been rigorously proved for any developing system, but data from several experimental systems strongly support the hypothesis. One of the systems in which the concept of quantal cell cycles has been pursued most vigorously is that of skeletal muscle development. Fusion of myoblasts to form functional myotubes is prevented by a number of techniques that inhibit the cells from undergoing the final cell cycle that normally occurs prior to fusion. If muscle from a 10-day chicken embryo is placed in culture, about 25% of the myoblasts are capable of rapid fusion. The rest will not fuse without undergoing DNA synthesis. This has been demonstrated in a number of ways. If the cells are divided and one group is cultured at a low population density and the other at a high density for approximately 24 hours, only those at the lower cell density undergo much cell division during this period. (Density-dependent regulation of growth and the failure of cells in crowded cultures to multiply are discussed in Chapter 17.) When both groups are subsequently plated at equal densities, either high or low, only those cells previously cultured at low density (and thus allowed to divide) exhibit increased ability to fuse. Similarly, if the myoblasts are placed in culture in the presence of inhibitors of DNA synthesis, the fraction that is able to fuse is small and the rest of the cells remain unfused for long periods of time. However, when the inhibition of DNA synthesis is released, fusion resumes after a few hours.

The following experiment suggests that the cell cycle that occurs immediately prior to myoblast fusion is in some respect different from normal proliferative cell cycles. Myoblasts taken from an animal and cultured in a nonpermissive medium undergo cell division but do not fuse. If they are subsequently shifted to a medium that is permissive for muscle differentiation, fusion occurs, but only after a time lag during which DNA synthesis and mitosis both occur. It thus appears to be essential for the cells to undergo DNA synthesis and division under conditions permissive for their differentiation before they are able to express that differentiation.

The timing of biochemical differentiation of muscle cells in relationship to DNA synthesis and mitosis has been analyzed. Myoblast cells were placed in culture, and DNA synthesis was monitored by incorporation of ^{3}H-thymidine while cell differentiation was determined by the ability of cells to bind fluorescent antibody to myosin. At very early times after the beginning of the experiment some cells were found that had incorporated ^{3}H-thymidine (those already in the S phase) and other cells were found that bound fluorescent antimyosin (those already differentiating), but no cells were found with both labels. It was only after the cultures had been grown in the presence of ^{3}H-thymidine for 10 hours that cells began to be detected that had both taken up thymidine into DNA and synthesized myosin. These results imply that after being labeled with thymidine in the S phase of the cell cycle, the cells must complete that cycle (G_2 + M in these cells takes 3 to 4 hours) and then spend 6 to 7 hours in the G_1 phase of the next cycle prior to detectable biochemical differentiation (myosin synthesis). (Readers unfamiliar with the phases of the cell cycle should read Appendix C.)

Under most conditions the synthesis of muscle myosin does not occur until after fusion of myoblasts into myotubes. When cells labeled with ^{3}H-thymidine are examined at different intervals after addition of the isotope, they are found to fuse only after they have finished their cell cycles, including mitosis, and have spent approximately 5 hours in the G_1 phase of the next cycle. After fusion, initiation of biochemical differentiation occurs within a short time. After fusion the myotube nuclei remain permanently in the G_1 (or G_0) phase of the cell cycle. Each myotube nucleus contains a diploid amount of DNA and undergoes no DNA synthesis, other than small amounts of repair synthesis that can be induced by agents such as ultraviolet irradiation.

Data from a number of recent experiments suggest that the critical events that must precede biochemical differentiation in muscle cells are DNA synthesis and karyokinesis (nuclear division), but not cytokinesis (cytoplasmic division) or cell fusion. When cytokinesis of myoblasts is blocked with cytochalasin B (Chapter 15), nuclear division proceeds normally and binucleate cells are obtained that synthesize muscle proteins and become contractile. If fusion of myoblasts that have completed the quantal cycle but not yet fused is blocked by the calcium-binding agent EGTA, or other agents, muscle proteins are synthesized by the mononucleate cells. On removal of the inhibitors, fusion occurs without an intervening period of DNA synthesis. Holtzer claims that cells that have been kept from fusing do not reenter the cell cycle when placed in a growth-promoting medium at densities too low to permit fusion. This finding has been offered as evidence that myoblasts withdraw permanently from the cell cycle before they fuse.

However, not all investigators agree. A recent study of rat myoblast L_6E_9, a permanent cell line that retains the ability to fuse and undergo musclelike biochemical differentiation, has shown clearly that a quantal cell cycle is not always needed for differentiation. In growth medium, nearly all L_6E_9 cells are capable of multiplication and the formation of colonies. When transferred to a medium that favors differentiation and simultaneously subjected to conditions that inhibit DNA synthesis, the vast majority of cells from an actively growing L_6E_9 culture will fuse and exhibit musclelike differentiation with no intervening DNA synthesis or mitosis.

Additional experimentation is needed to reconcile this new observation with earlier data. The data that have been presented appear to show clearly that L_6E_9 cells do not need a quantal cell cycle to switch from multiplication to terminal differentiation. However, until comparable experiments have been done with fully normal cells in primary culture, it can be argued that the observed results simply represent abnormal control mechanisms in the genetically altered permanent myoblast line.

Another developmental system for which strong claims of a quantal cell cycle have been made is the differentiating red blood cell in chicken embryos. In this case the purported quantal cell cycle is not

Other systems with proposed quantal cycles

the final cell division but rather the one at which the cells change their program from self-renewal to a stepwise sequence of terminal differentiation that includes both the synthesis of hemoglobin and a precisely limited amount of additional proliferation. In chicken embryos between 30 and 60 hours of incubation, primitive red blood cell precursors called hematocytoblasts undergo a quantal cycle to form primitive erythroblasts that synthesize detectable amounts of hemoglobin. These erythroblasts then go through exactly six additional cell cycles before ceasing division permanently. As in the case of myoblasts, the quantal cell cycle is considered to include a "decision" to cease DNA synthesis, but in the case of the primitive erythroblast, that decision is not acted on until six cell generations later. During the intervening period hemoglobin is made by the dividing primitive erythroblasts, and its synthesis is not affected by inhibitors of DNA synthesis. However, if DNA synthesis is inhibited in the hematocytoblast, which has not yet been programmed to produce hemoglobin, the initiation of hemoglobin synthesis is prevented. Thus the reprogramming of these cells for a new kind of synthetic activity appears to require DNA synthesis.

A major corollary that has been inserted into recent versions of the quantal cell cycle theory is that a cell cannot choose from more than two alternatives at any given quantal cycle. Thus to generate more than two different cell types, a precursor cell must go through sequential quantal cycles, each of which is a binary choice and each of which determines the options that are available to the progeny cells at the next quantal cycle.

The alternative possibility is that stem cells are pluripotent and can enter any of several different pathways of differentiation provided that they receive the correct inductive stimulus. For example, as will be discussed in greater detail in Chapter 13, lung bud endoderm can form structures typical of lung, liver, stomach, intestine, or trachea, depending on which type of mesoderm it is cultured with. Current methods of analysis are not yet sophisticated enough to determine whether such transitions can be achieved in a single step or whether a series of binary choices, each of which involves completion of a quantal cell cycle, are required.

The data in favor of the quantal cell cycle concept are persuasive but not totally conclusive. The primary criticism of this work has been its heavy reliance on the use of DNA synthesis inhibitors and other drugs whose full effects are not completely understood. The argument for the necessity of specific cell divisions to initiate cell differentiation is also weakened somewhat by the fact that cell differentiation occurs in the cellular slime mold *Dictyostelium discoideum* in the absence of

DNA synthesis and cell division (Chapter 23). Also, in the development of certain types of complex organisms such as ascidians differentiation-like changes have been found to occur in localized regions of early embryos whose normal sequence of cell division has been inhibited. The occurrence of DNA synthesis and accompanying rearrangements of chromatin structure cannot be ruled out by the experiments that have been done with these latter systems, but such findings must nonetheless be reconciled with the quantal cell cycle concept if it is to be considered to be of general applicability.

The drug 5-bromodeoxyuridine (BrdU) is a structural analog of thymidine that has a bromine atom replacing the methyl group that occupies the 5 position in the pyrimidine ring of thymidine (Fig. 11-2). The presence of BrdU in the culture medium selectively inhibits the expression of differentiated properties in vitro by many different types of cells. In addition, there are certain cases in which BrdU treatment induces the expression of genes that are not expressed in its absence. Table 11-1 lists the effects of BrdU on a variety of types of cells, and Fig. 11-3 illustrates the reversible effect of BrdU on pigment formation by melanoma cells.

EFFECT OF BrdU ON CELL DIFFERENTIATION

Thymine

5-Bromouracil

Fig. 11-2. Structural formulas of thymine and 5-bromouracil. The two molecules are identical in all respects except that the methyl group of thymine is replaced by bromine in 5-bromouracil.

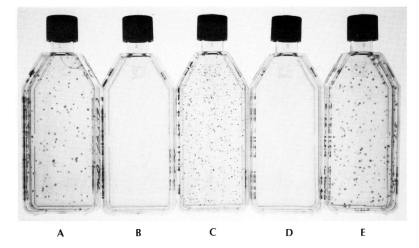

A   B   C   D   E

Fig. 11-3. Reversibility of effect of 5-bromodeoxyuridine on pigment formation. **A,** Untreated B$_5$59 mouse melanoma cells (control). **B,** Clone C471 after long-term growth with 1 μg/ml of BrdU. **C,** A subculture of **B** grown in normal medium. **D,** A subculture from **C** returned to BrdU medium. **E,** A subculture from **D** returned to normal medium. The colonies in **A, C,** and **E** are visible because of their content of melanin pigment. Flasks B and D contain similar numbers of colonies, which are not visible in these unstained preparations. Inhibition of the pigmented epiphenotype by treatment with BrdU in **B** and **D** has no effect on the epigenotype, as seen in **C** and **E** after return to permissive conditions. (From Silagi, S. 1971. In Vitro **7:**105.)

Table 11-1. Effects of BrdU on gene expression[*]

System	Parameter measured
Repression of differentiated properties	
Pancreatic acinar cells	Exocrine enzymes
Pancreatic B cells	Insulin
Chondrocytes	Chondroitin sulfate
Myoblasts	Myotube formation, myosin
Pigmented retina cells	Melanin
Erythroblast precursors	Hemoglobin
Mammary gland	Casein, α-lactalbumin
Lymphocytes (primed)	Antibodies
Amnion cells	Hyaluronic acid
Liver cells (avian)	Estrogen induction of phosvitin
Hepatoma cells (HTC)	Glucocorticoid induction of tyrosine aminotransferase
Mouse lung—mammary carcinoma hybrid cells	Hyaluronic acid
Melanoma	Tumorigenicity, pigments
Induction of gene expression	
Mouse lung—mammary carcinoma hybrid cells	Alkaline phosphatase
Mouse mammary carcinoma cells	Alkaline phosphatase
Pancreatic exocrine cells	Alkaline phosphatase
Neuroblastoma	Neurite formation
	Cell membrane glucoprotein
Lymphoid cells (Burkitt lymphoma clones and NC37 line)	Epstein-Barr virus
Pituitary tumor line	Prolactin

[*]Modified from Rutter, W. J., R. L. Pictet, and P. W. Morris. 1973. Annu. Rev. Biochem. **42:**601.

BrdU is readily incorporated into DNA in place of thymidine, and many investigators believe that its selective effects on gene expression result from such incorporation. However, as will be discussed later, there are some data that are difficult to understand on that basis. Large amounts of BrdU also inhibit cellular multiplication, but in most cases the amount of BrdU that is needed to alter gene expression is not enough to affect multiplication significantly.

BrdU-induced mutation   Incorporation of BrdU into DNA causes mispairing of bases and therefore increases the probability of mutation. In addition, the presence of BrdU in DNA shifts its maximum absorption of light from around 260 nm in the ultraviolet spectrum to longer wavelengths that are near the violet end of the visible light spectrum. Thus DNA that contains BrdU is susceptible to radiation damage from ordinary fluorescent lights. This property is sometimes used to select for auxo-

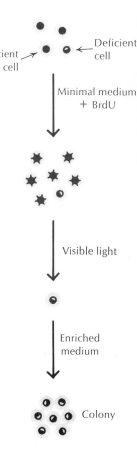

Fig. 11-4. Selection of auxotrophic mutants with BrdU and visible light. After exposure to mutagenesis, cells are grown in a minimal medium plus BrdU. Wild-type cells, which possess all the enzymes needed for growth in the minimal medium, multiply and incorporate BrdU into their DNA. Auxotrophic mutants, which are deficient in one of the enzymes needed for growth in the minimal medium, do not multiply and hence do not incorporate BrdU. The cultures are then exposed to ordinary fluorescent light, which selectively kills the cells that have BrdU in their DNA. The survivors are then transferred to an enriched medium that contains the factors needed for growth of the nutritionally more demanding auxotrophic mutants. This procedure yields populations enriched in auxotrophic mutants, which can then be isolated in pure form by selection of clones. (Modified from Puck, T. T. 1972. The mammalian cell as a microorganism. Holden-Day, Inc., San Francisco.)

trophic mutants (i.e., those which have additional requirements for multiplication beyond the minimal nutrient requirements of wild-type cells [Fig. 11-4]). If a mutagen-treated population of cells is placed in a medium that contains BrdU and lacks the special growth factors required by auxotrophic mutants, the wild-type cells that are capable of multiplying in that medium will incorporate BrdU. They can then be killed by exposure to fluorescent light. Mutant cells that do not incorporate BrdU into their DNA because they are unable to multiply without the special growth factors are not killed by the fluorescent light treatment. When the survivors are placed in an enriched medium that contains the factors that are needed for growth by the auxotrophic mutants, multiplication is resumed and clones with the desired genetic properties can be selected. Because of the light sensitivity of BrdU-treated cells, it is important to minimize exposure of such cells to short wavelength visible light while studying the effects of BrdU on the expression of differentiated properties.

Although BrdU can increase mutation rates both by mispairing and by making the DNA sensitive to radiation damage from short wavelength visible light, several lines of evidence suggest that the effects of BrdU on the expression of differentiated properties are not due to mutational changes:

1. The level of mutagenesis caused by BrdU at the concentrations used is low, whereas the effect of BrdU on differentiation is uniform over the entire population of treated cells.
2. BrdU selectively inhibits differentiation at concentrations that have little or no effect on growth.
3. In most cases the effects of BrdU on differentiation are fully reversible.

Nucleoside diphosphate sugar metabolism

Since simultaneous administration of thymidine with BrdU prevents the effects of BrdU on differentiation, it is probable that some aspect of thymidine metabolism is involved. DNA is generally considered to be the most likely site of BrdU action, but the possibility that BrdU may affect some other aspect of the metabolism of thymidine or other pyrimidine derivatives, such as formation of nucleoside diphosphate sugar complexes, cannot be ruled out entirely. Thymidine diphosphate (TDP) complexes are known to occur as intermediates in rhamnose metabolism in plants and bacteria, but there are no known TDP-sugars in animals. However, the possibility that metabolites of BrdU might interfere with normal metabolism of other nucleoside diphosphate sugar complexes (e.g., UDP-sugars or CDP-sugars) is interesting in view of the role of such intermediates in the formation of complex polysaccharides that are found on cell surfaces. In addition to selective inhibition of gene expression, BrdU frequently alters the pattern of cellular interactions with the culture substrate. BrdU-treated cells are typically flattened and tend to adhere unusually strongly to the underlying substrate.

Is incorporation of BrdU into DNA required?

If DNA is involved in the effects of BrdU on differentiation, which is currently the most generally accepted possibility, the effect could be due either to a generalized replacement of thymidine by BrdU throughout the genome or, alternately to selective incorporation of BrdU into special target areas that are particularly susceptible to replacement of thymidine by BrdU. It has been firmly established that BrdU at high concentrations replaces thymidine throughout the entire genome. By a process of adaptation, cultured cell strains have even been obtained in which it is not only possible to obtain 100% substitution of BrdU for thymidine but in which growth is actually dependent on the presence of BrdU in the culture medium. Such cells cease to grow when they are transferred to a medium containing thymidine instead of BrdU.

However, the effect of BrdU on differentiation can be obtained with amounts of BrdU that cause only partial replacement of thymidine by BrdU. There is one report claiming that differentiation is inhibited under conditions that result in as little as 2% substitution (one BrdU per 50 normal thymidine sites) in the DNA of the affected cells, and there is fairly general agreement that inhibition of differentiation need not be accompanied by more than 20% substitution (one BrdU per five sites) in many different types of cells.

Most investigators have been unable to find evidence for a special BrdU-sensitive target DNA, although there have been several recent claims that low concentrations of BrdU are preferentially incorporated into certain moderately repeated DNA fractions. There is also a recent report that during cartilage differentiation the number of sequences that preferentially incorporate BrdU is increased. Although they are interesting, these findings are too preliminary to analyze further until they have been verified or rejected by additional research groups.

A number of investigators have attempted to obtain experimental evidence that BrdU must be incorporated into DNA to have its effect. Unfortunately, the data that have been gathered all have alternate interpretations. Two such classes of evidence and their criticisms are presented here to give the student a better understanding of the difficulties involved in such experiments. The two lines of evidence are as follows:

1. If BrdU is administered to cells for a limited period of time, it is found to be effective only if present during the S phase of the cell cycle.

2. When inhibitors of DNA synthesis are administered simultaneously with BrdU, no suppression of cell differentiation occurs.

Both these findings are compatible with the hypothesis that BrdU must enter DNA to have its effect. However, neither is conclusive, since both are also compatible with the following alternative mechanisms that do not involve DNA synthesis:

1. Since BrdU is a structural analog of thymidine, the enzymes that are responsible for transport of thymidine into the cell and its phosphorylation to form thymidine monophosphate (TMP), TDP, and thymidine triphosphate (TTP) also transport and phosphorylate BrdU. Those enzymes are present in much greater amounts during the S phase than during other parts of the cell cycle. If BrdU must be taken into the cell and phosphorylated to be effective (which would be true even if it affected some other aspect of thymidine metabolism such as nucleoside diphosphate sugar metabolism rather than DNA synthesis), the S phase would still be the time when the enzymes needed to convert it to the active form would be at their highest concentrations. Therefore BrdU would have its greatest effect when administered during S phase.

2. The lack of effect of BrdU in the presence of DNA inhibitors could result from a buildup of thymidine or thymidine nucleotides in the absence of their use in DNA synthesis. These unused precursors might then have the same reversing effect on BrdU action as does simultaneous administration of thymidine. Most of the so-called inhibitors of DNA synthesis actually block the formation of some DNA precursor rather than the synthesis of DNA itself and act in a manner that could allow accumulation of unused thymidine-containing com-

pounds. One exception among the commonly used inhibitors of DNA synthesis is 5-fluorodeoxyuridine, which acts by preventing synthesis of TMP from deoxyuridine monophosphate. Unlike the other inhibitors, it actually enhances the effect of BrdU by reducing the amount of endogenously produced thymidine-containing compounds competing with BrdU for incorporation. Thus, although it currently seems likely that DNA is involved in the effect of BrdU on differentiation, that likelihood cannot yet be considered to be a scientifically proved fact.

One finding that must be considered in regard to the mechanism of action of BrdU on cell differentiation is that BrdU needs to be present for only a single cycle of DNA replication to be effective. In view of the generally accepted semiconservative mode of DNA replication, one round of replication in the presence of BrdU should result in production of DNA molecules that contain BrdU in only one strand of the double helix. In half the daughter chromosomes the BrdU would be in the "sense," or coding, strand for any particular gene, whereas in the other half it would be in the complementary strand. This result has been interpreted by many investigators to imply that BrdU inhibits the expression of genes for differentiated cell products both when it is incorporated into the strand that serves as a template for transcription and when it is incorporated into the complementary strand that does not participate directly in transcription.

Effects of BrdU on DNA-protein interactions

If incorporation of BrdU into either strand of the DNA is sufficient to block expression of differentiation, it seems probable that the mode of action of BrdU involves something other than direct interference with transcription. One of the most likely candidates at the present time appears to be modification of the interaction between DNA and specific regulatory proteins.

Proteins that are capable of recognizing base sequences in double-stranded DNA are thought to do so without disrupting the double helical structure of the DNA. Since the structures of the two deoxyribose phosphate backbones, which are on the outside of the DNA double helix, are essentially independent of the base sequence of the DNA, proteins that read base sequence presumably must recognize the edges of the flat base pairs that project into the "grooves" between the deoxyribose phosphate strands.

The double helical structure of DNA has one wide and one narrow groove, due to the angle at which deoxyribose is attached to the base pairs. The methyl group of thymine is positioned in AT base pairs so that it projects conspicuously into the wide groove of double helical DNA, where it could easily interact with sequence-reading regulatory molecules (Fig. 11-5). Substitution of bromine in place of the methyl

Fig. 11-5. Projection of methyl and bromo groups into the wide groove of DNA. **A,** Thymidine paired with deoxyadenosine. The methyl group projects outward from the planar structure of the base pair into the wide groove between the deoxyribose phosphate backbones. dR = Deoxyribose. **B,** 5-Bromodeoxyuridine paired with deoxyadenosine. Base pairing is unaltered, with the bromine projecting into the space in the wide groove normally occupied by the methyl group of thymidine. **C,** Space-filling model of DNA. The wide and narrow grooves are labeled. The circles identify two methyl groups of adjacent thymidines that project conspicuously into the wide groove. Bromine atoms occupy the same positions in BrdU-substituted DNA. (**C** courtesy Professor M. H. F. Wilkins, Biophysics Department, King's College, London.)

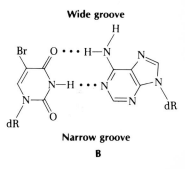

group of thymine would have little effect on base pairing or formation of a double helix (Fig. 11-5, *B*). However, it could significantly alter the charge distribution and perhaps also the shape of a localized area within the wide groove. BrdU substitution could thus have a major effect on the interaction between double-stranded DNA and specific regulatory proteins that respond to thymine-rich coding sequences within that DNA.

Recent data from several sources suggest that DNA-protein interactions are in fact altered significantly by the presence of BrdU in the DNA and thus that such interactions may play an important role in the effects of BrdU. In the lactose operon of *E. coli*, for example, binding of the repressor protein to the operator locus is much tighter when the DNA contains BrdU in place of thymidine. This may reflect the fact that the *lac* operator is rich in AT base pairs (Fig. 11-6). Studies with chromatin from eukaryotic cells have shown a greater thermal stability for chromatin that contains BrdU-substituted DNA. Other studies have shown a reduced binding of NHC proteins to DNA-histone complexes in chromatin reconstitution experiments using BrdU substituted DNA. Since recent experiments suggest that transcription can be activated by specific NHC protein molecules, BrdU substitution might inhibit that activation at certain sites. Consistent with such a hypothesis is the recent finding from DNA:RNA hybridization studies that a greater portion of the genome is normally transcribed in myotubes than in myoblasts, and that in myoblasts whose differentiation has been blocked with BrdU, the number of sequences transcribed is even less than that in normal myoblasts prior to their differentiation (Fig. 11-7).

Studies with hematocytoblasts also suggest that the effect of BrdU is at a regulatory level. After the quantal cycle leading to hemoglobin synthesis has been completed, extensive BrdU substitution, replacing

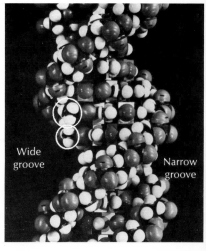

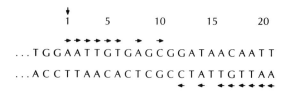

Fig. 11-6. The *lac* operator sequence of *E. coli*. The starting point for transcription of *lac* mRNA is indicated by a vertical arrow. Fourteen of the first 21 base pairs in the transcribed sequence are AT, and six of the eight symmetrically arrayed base pairs (horizontal arrows) are AT. Substitution of BrdU in place of thymidine in this sequence increases substantially the strength of binding of the lactose repressor protein to the operator.

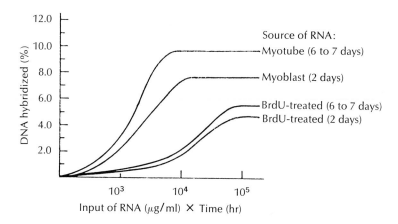

Fig. 11-7. Reduced diversity of transcription in BrdU-inhibited myoblasts. A small amount of radioactively labeled DNA was annealed with an excess of RNA from chicken embryo myoblasts (2 days in culture), myotubes (6 to 7 days in culture), or BrdU-treated myoblasts (either 2 or 6 to 7 days in culture). Percentage of hybridization was plotted against the product of initial RNA concentration multiplied by time of hybridization (analogous to cot curves, Chapter 5 and Appendix A), and hybridization conditions were maintained until all sites on the DNA capable of hybridization with the RNA were occupied (indicated by no further rise in the percentage of the DNA found as a hybrid). Myotube RNA has slightly greater species diversity than that from myoblasts (based on fraction of the genome hybridized). BrdU-treated cells, which continue to multiply but do not differentiate, exhibit substantially less RNA diversity than either myoblasts or myotubes. (Modified from Colbert, D. A., and J. R. Coleman. 1977. Exp. Cell Res. **109**:31.)

up to 80% of the thymidine residues, has little effect on the transcription of specific hemoglobin RNA sequences. However, during the quantal cell cycle in which hematocytoblasts are converted to primitive erythroblasts, as little as 20% substitution is sufficient to block all subsequent hemoglobin synthesis.

If these observations regarding the effect of BrdU on transcription are valid, it is still necessary to explain the selective effect that moder-

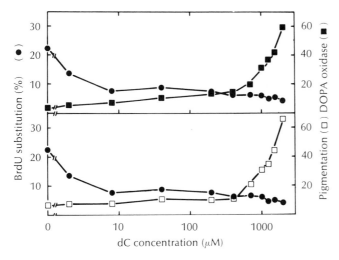

Fig. 11-8. Effect of deoxycytidine on the BrdU content of DNA and differentiation of pigmented melanoma cells. A pigmented Syrian hamster melanoma line was grown in medium containing 0.5 μM BrdU, which suppressed pigmentation. The indicated amounts of deoxycytidine (dC) were added to replicate cultures, and the cells were assayed for percentage incorporation of BrdU into their DNA (●———●), and for DOPA-oxidase, which is one of the enzymes involved in synthesis of melanin (■———■), and pigmentation (□———□). DOPA-oxidase activity and pigmentation are both expressed as percentages of the amounts found in controls without BrdU. A major reduction in amount of BrdU substitution occurs at low levels of deoxycytidine that have little effect on the expression of differentiation, and very little change in the amount of BrdU in the DNA occurs over the range of concentrations of deoxycytidine that have a major effect on the expression of differentiation. Top, Changes in BrdU substitution and DOPA-oxidase activity in response to changes in deoxycytidine concentration. Bottom, Changes in BrdU substitution and pigmentation in response to changes in deoxycytidine concentration. (Modified from Davidson, R. L., and E. R. Kaufman. 1977. Cell **12**:923. Copyright © MIT; published by the MIT Press.)

ate doses of BrdU have on differentiation, with relatively little effect on cell survival or multiplication. One possibility is that genes for differentiated products are subject to a different level of regulation than those for cellular survival and "housekeeping" functions.

New data are still appearing in the literature regularly, and it is too early to construct a final model of BrdU action. At present, the most widely accepted models involve altered interactions between DNA and regulatory proteins, but other possibilities, such as a special differentiation-specific DNA that behaves differently than the bulk of the chromosomal DNA with regard to BrdU incorporation, or a mechanism that does not involve DNA at all, cannot be fully ruled out. For example, one recent report claims that reversal of the effects of BrdU on melanoma differentiation by deoxycytidine does not significantly change the amount of BrdU incorporated into cellular DNA (Fig. 11-8).

In the same report the authors show that deoxycytidine has its effect only after being converted metabolically to thymidine nucleotides. They favor the view that sugar nucleoside diphosphate metabolism and the synthesis of complex carbohydrates are being affected by BrdU and that effects of BrdU on differentiation are secondary consequences of such changes.

The widespread and selective effects of BrdU on cellular differentiation strongly suggest that a full understanding of its mechanisms of action will provide valuable clues to the overall mechanisms responsible for selective gene expression in cellular differentiation. However, as we have seen, the research literature is still filled with confusing and contradictory data concerning the effects of BrdU. Extensive research is still needed to resolve current conflicts and to provide final answers about how BrdU achieves its effects and about how they are related to the overall phenomenon of cellular differentiation.

EXPRESSION OF DIFFERENTIATED PROPERTIES BY CELLULAR HYBRIDS

The formation of viable "hybrid" cells by fusion of different kinds of whole cells (or their component parts) provides a powerful tool for analyzing differentiation in terms of dominance or recessiveness of the differentiated phenotype and possible mechanisms that might be responsible for the observed behavior. Several techniques have been developed in recent years for the formation of fully viable cells that contain components derived originally from two or more parental cells of different phenotype (Fig. 11-9). These include (1) fusion of two complete cells to yield a hybrid that is essentially the sum of the two parental cells, (2) reconstitution of a viable cell that has the bulk of its cytoplasm from one parent and its nucleus from another, (3) fusion of an enucleated cytoplasm from one cell type with a complete cell of another type, and (4) the addition of one or a few chromosomes from one cell to another by fusion of a "microcell" containing a fragment of a normal nucleus with a normal cell. These techniques are all being applied to the study of controlled gene expression in differentiation, but at present, extensive data are available only for studies in which two complete cells have been fused to form a viable hybrid cell.

Cellular hybridization was originally discovered as a spontaneous phenomenon occurring with a low frequency in mixed cultures of established cell lines. When two different types of permanent cell lines with identifiable chromosomal and genetic markers were grown together, a new type of cell was occasionally found that contained cytologically distinct marker chromosomes from both of the original cell types and also exhibited genetic properties characteristic of both of them. Although interesting, these spontaneous hybrids occurred too rarely to be of much experimental value. However, the subsequent

A. Hybridization

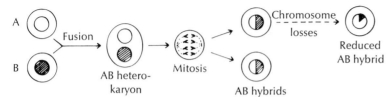

B. Reconstitution

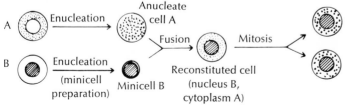

C. Cytoplasmic hybrids ("Cybrids")

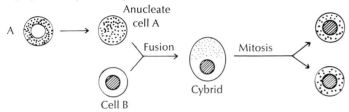

D. Microcell hybrids

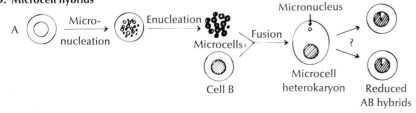

Fig. 11-9. Schematic summary of cellular hybridization and reconstruction techniques. **A,** Hybridization. Two complete cells are fused to yield a heterokaryon (cell with two different types of nuclei). A single spindle forms at the first mitosis, and two daughter cells are formed, each containing chromosomes from both parents (see Fig. 11-11 for details). Subsequent loss of chromosomes derived from one of the parents may occur. **B,** Reconstitution. Treatment with cytochalasin B and centrifugation is used to generate an anucleate cytoplasm (cytoplast) and a nucleus surrounded by a thin ring of cytoplasm (minicell or karyoplast). The cytoplasm from cell A is then fused to the nucleus from cell B, yielding a reconstituted cell that has its nucleus from one parental cell and essentially all its cytoplasm from another. **C,** "Cybrids" (cytoplasmic hybrids). An enucleated cytoplasm is fused to a complete cell, yielding a cell whose cytoplasm is a composite from two parental cells. **D,** Microcell hybrids. Micronuclei are induced in one of the parental cells by disrupting the mitotic spindle and allowing nuclear membrane to form around single chromosomes or small groups of chromosomes. These micronuclei are isolated with cytochalasin B and centrifugation, and the resulting "microcells" are fused to complete cells with Sendai virus. The resulting hybrid has all the chromosomes of one parent plus one or more chromosomes derived from the other. (From Ege, T., et al. 1976. *In* N. Muller-Bernat et al., eds. Differentiation research. Elsevier/North Holland Biomedical Press, Amsterdam.)

development of experimental methods for increasing the frequency of hybrid formation and also for separation of the hybrids from their parental cells has overcome that problem.

It was found in the late 1950s that increased numbers of hybrid cells could be obtained by briefly exposing mixed cultures to inactivated Sendai virus (an RNA virus with hemagglutinating properties, named for Sendai, Japan, where it was first studied). By careful treatment of the virus with radiation or chemicals, it is possible to destroy its ability to infect cells without destroying its agglutinating properties. Controlled exposure to the inactivated virus preparation causes cells to stick together and fuse into binucleate (and some multinucleate) forms without loss of viability. The exact mechanism by which cells fuse is not certain. The virus particles appear to attach to the membranes of both cells, after which surface glycoproteins disappear

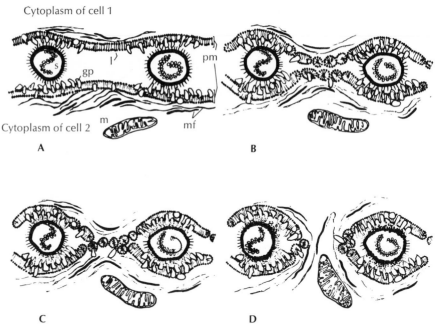

Fig. 11-10. Hypothetical steps in cell fusion induced by Sendai virus. **A,** Local agglutination of two cells by two virus particles, with glycoproteins displaced from the adjacent cell surface membranes by their attachment to adsorbing sites on the virus particles. **B,** Formation of micelles by the exposed lipids of the membranes. This process is enhanced by high pH and calcium ion concentration. Both cells have begun to endocytose the virus particles. **C,** Membrane coalescence. Microfilaments are beginning to move into the affected area, which now has ionic continuity between the cells. **D,** A nascent cytoplasmic bridge has been formed and is being stabilized and enlarged by microfilament activity. pm = Plasma membrane; l = lipid molecules; gp = glycoprotein molecules; mf = microfilaments; s = Sendai virus particles; m = mitochondrion. (From Ringertz, N. R., and R. E. Savage. 1976. Cell hybrids. Academic Press, Inc., New York.)

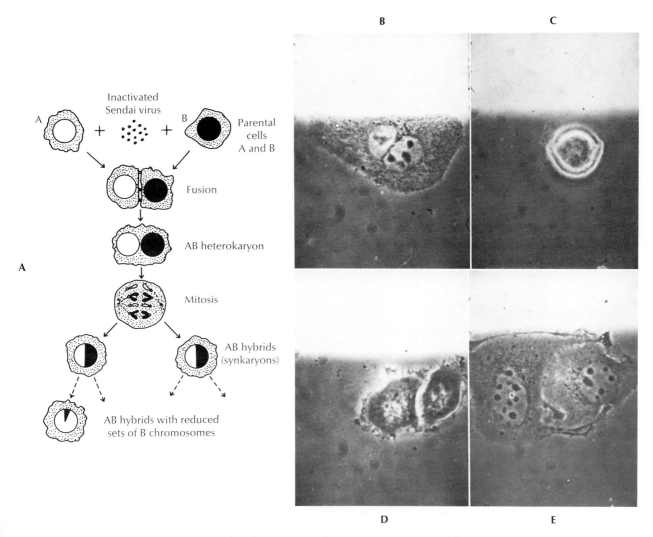

Fig. 11-11. Formation of heterokaryons and synkaryons. **A,** Schematic representation of fusion of two mononucleate cells induced by inactivated Sendai virus. Both nuclei in the resulting heterokaryon undergo DNA synthesis and preparation for mitosis synchronously. The resulting mitosis is organized around a single spindle and results in approximately equal distribution of chromosomes from both parental nuclei into the two daughter cells, which are referred to as synkaryons or simply as hybrid cells. Selective loss of chromosomes (in this case those derived from cell B) may occur in the synkaryons. **B** to **E,** Frames from a time-lapse movie sequence showing division of a binucleate heterokaryon to yield two synkaryons. **B,** Heterokaryon prior to mitosis. **C,** Late prophase or metaphase. **D,** Telophase, showing synkaryon nuclei beginning to form. **E,** Closely adjacent synkaryons. Note the larger number of nucleoli per nucleus in the synkaryons as compared to the heterokaryon. (**A** from Ringertz, N. R., and R. E. Savage. 1976. Cell hybrids. Academic Press, Inc., New York: **B** to **E** from Harris, H. 1970. Cell fusion. Clarendon Press, London.)

and lipoprotein membranes coalesce and break down in the space between the virus particles (Fig. 11-10). It is significant that not all agents that agglutinate cells cause them to fuse. Recently cellular fusion has also been achieved by treatment of cells with concentrated solutions (about 45% to 50%) of polyethylene glycol.

Fusion of two different kinds of cells results in the formation of a binucleate cell called a heterokaryon. The term "heterokaryon" refers specifically to cells that contain more than one type of nucleus. In viable hybrids, both nuclei pass through DNA synthesis and enter mitosis simultaneously. A single spindle is formed, and mononucleate daughter cells containing essentially complete sets of chromosomes from both parental cells are formed (Fig. 11-11). The term "synkaryon" is sometimes used to describe such cells.

Selective media for hybrid cells

Even when hybrid formation is enhanced by use of Sendai virus or polyethylene glycol, it is still necessary to be able to separate the hybrid cells and their progeny from parental type cells (including the products of fusion of two or more cells of the same type.) Most such separation procedures involve the use of culture media or conditions that will not support the multiplication of either parent but will allow hybrids that contain genes from both parents to multiply. One of the most frequently used selective systems is the hypoxanthine-aminopterin-thymidine (HAT) medium. Selection of hybrids by the HAT medium is based on the use of parental strains carrying mutations that render them unable to multiply in HAT. The aminopterin in HAT prevents the cells from synthesizing purines or thymidine de novo. One of the parental lines lacks the enzyme hypoxanthine-guanine phosphoribosyltransferase (HGPRT), which is needed for use of hypoxanthine, a purine source that is included in the HAT medium. The other parental line lacks the enzyme thymidine kinase (TK), which is needed for use of thymidine from the HAT medium. Thus, when in HAT medium, neither parental cell can obtain all the precursors needed for nucleic acid synthesis and growth. However, hybrid cells that have received a functional *TK* gene from the *HGPRT*⁻ parent and a functional *HGPRT* gene from the *TK*⁻ parent can use both hypoxanthine and thymidine and thus can multiply in the presence of the aminopterin contained in the HAT medium. The theoretical basis for use of the HAT medium is presented in greater detail in Appendix B.

It is also possible to use HAT medium in another procedure called half-selection. In this system, only one of the parents lacks an enzyme required for growth in HAT medium. The other parental cell is selected against by virtue of some other biological property. For example, a human lymphocyte cell that does not attach to the culture surface might be fused with a mouse cell line that attaches but lacks HGPRT.

In this case hybrid cells can be selected for their ability to grow in HAT medium and to remain attached to the culture dish when it is washed with medium.

One interesting recent application of the half-selection technique has been the isolation of clonal lines of hybrid cells that produce a single type of antibody. Hybrids are constructed that have as one parent a myeloma line that synthesizes incomplete antibodies and is unable to grow in the HAT medium and as the other parent freshly isolated spleen cells from a mouse that has been immunized with a particular antigen. The basal medium to which hypoxanthine, aminopterin, and thymidine are added does not support good growth of primary spleen cultures, and the selective effect of the HAT medium prevents growth of the myeloma parent. By testing clones that grow in the HAT medium for antibody production, it is possible to select clonal cultures that specifically produce a single class of antibodies directed against the antigen originally used to immunize the mouse. Fusion of the spleen cells with myeloma cells does not result in extinction of antibody production in the hybrid and allows the antibody-producing "hybridoma" to be grown for an extensive time in culture. This relatively new technique is proving to be useful for preparing highly purified antibodies against a single antigen, such as those used in fluorescent antibody studies of the type described in Chapter 15.

In theory, any pair of properties that prevent cellular multiplication by two different mechanisms can form the basis for selection against parental cell types as long as both are recessive (not expressed) in hybrid cells. Also, if no suitable selective conditions can be found, it is possible, although tedious and time-consuming, to isolate a large number of clones and to test each individually for the combination of properties that are expected in the desired hybrid.

At least initially, hybrid clones generally contain most of the chromosomes from both parental strains. This has been demonstrated both by direct analysis of karyotypes (Fig. 11-12) and by showing that genes from both parents are present in the hybrid cells. Over a period of time, some hybrids will preferentially lose chromosomes derived from one of the parents. For example, in crosses between established mouse lines and human diploid fibroblasts the human chromosomes tend to be lost preferentially.

Selective loss of human chromosomes from hybrids has proved to be a powerful tool in human genetic analysis. Clones of cells can be selected that have retained only one or a few human chromosomes, and the specific human chromosomes that they contain can be identified through use of staining techniques that give each chromosome a characteristic banded appearance. The human genes that are present

Selective loss of chromosomes from hybrid cells

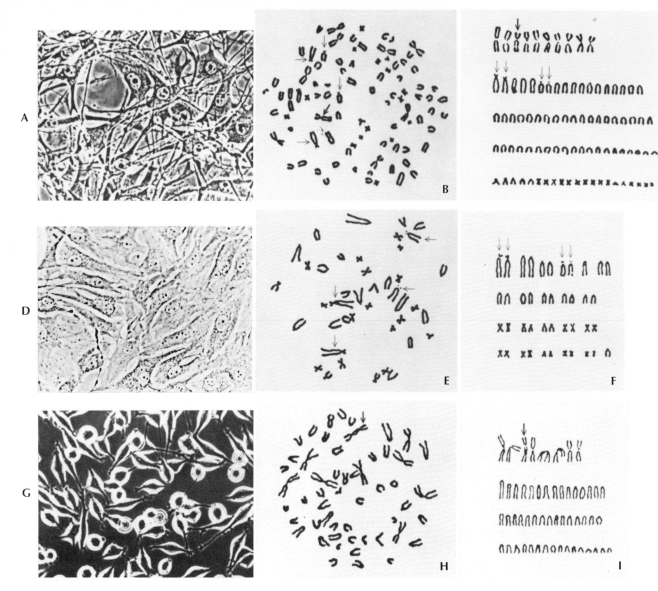

Fig. 11-12. Comparison of morphology and karyotype of a cellular hybrid and its parental strains. **A,** Appearance of hybrid formed between a diploid rat fibroblast and an established mouse line. **B,** Chromosomes of hybrid. **C,** Karyotype of hybrid. Thin arrows in **B** and **C** point to marker chromosomes from the diploid rat parent cell and thick arrow to a mouse marker chromosome. **D,** Appearance of the parental rat line. **E,** Chromosomes of the rat cells. **F,** Karyotype of the rat cell. A normal diploid number of 42 chromosomes is seen. All are matched in pairs except the sex chromosomes. **G,** Appearance of the established mouse line. Note the spindle-shaped appearance of the cells in contrast to the flattened shape of the rat fibroblasts and the intermediate morphology of the hybrid. **H,** Chromosomes of the mouse line. **I,** Karyotype of the mouse line. This line has an abnormal number of chromosomes (54) and several chromosomes with characteristic "marker" appearances, including the one indicated with the thick arrow, both here and in the karyotype of the hybrid. (From Ephrussi, B., and M. C. Weiss. 1969. Sci. Am. **220:**26 [April].)

in such cells can be identified by their biochemical expression or in some cases in which no expression occurs (e.g., globin chains of hemoglobin) by hybridization of total cellular DNA with cDNA prepared with reverse transcriptase from purified human mRNA for the gene in question. Table 11-2 contains a partial list of genes that have been as-

Table 11-2. Assignment of genes to selected human chromosomes*

Gene symbol	Name	Status†	Gene symbol	Name	Status†
Chromosome 1			*Chromosome 6*		
A12-1	Adenovirus 12: chromosome 1 modification site	P	HL-A region		C
Ak-2	Adenylate kinase 2	C	Bf	Properdin factor B	C
Amy-1	Amylase, salivary	C	C2		P
Amy-2	Amylase, pancreatic	C	Chi	Chido	C
Aod	Auriculo-osteodysplasia	P	Ir	Immune response locus	C
Cae	Zonular pulverulent cataract	P	MLC loci		C
El-1	Elliptocytosis, Rh linked	C	P	P blood group	C
Fh	Fumarate hydratase	C	Pg-5	Urinary pepsinogen	C
Fy	Duffy blood group	C	Sd-1		C
Guk-1	Guanylate kinase	C	Sd-2		C
Pep-c	Peptidase C	C	Sd-3		C
Pgd	Phosphogluconate dehydrogenase	C	Me-1	Malic enzyme, cytoplasmic	C
Pgm-1	Phosphoglucomutase 1	C	Pgm-3	Phosphoglucomutase 3	C
Pph	Phosphopyruvate hydratase	C	Sod-2	Superoxide dismutase, mitochondrial	C
Rh	Rhesus blood group antigen	C	*Chromosome 17*		
Rn-5	5S ribosomal RNA	C	A12−17	Adenovirus-12: chromosome 17 modification site	C
Udgp	Uridyl diphosphate glucose pyrophosphorylase	P	Gk	Galactokinase	C
Uk	Uridine kinase	P	Tk	Thymidine kinase	C
Umk	Uridine monophosphate kinase	C	*X Chromosome*		
Chromosome 2			Aga	α-Galactosidase A	C
Acp-1	Acid phosphatase, red-cell type	C	Dhtr	Dihydrotestosterone receptor	C
Gac	Galactose enzyme activator	P	Gpd	Glucose-6-phosphate dehydrogenase	C
Gput	Galactose-1-phosphate uridyltransferase	I	Pgk	Phosphoglycerate kinase	C
Hba	Hemoglobin α	C	Hprt	Hypoxanthine-guanine phosphoribosyltransferase	C
Idh-1	Isocitrate dehydrogenase, cytoplasmic form	C	Tar	Tyrosine aminotransferase regulator	P
If-1	Interferon 1	P	Xg	Xg blood group	C
Mdh-1	Malate dehydrogenase, cytoplasmic form	C			
MNSs	MNSs blood group antigen	P			
Tys	Sclerotylosis	P			

*From Ruddle, F. H., and R. P. Creagan. 1975. Annu. Rev. Genet. **9**:407. Reproduced, with permission, from the *Annual Review of Genetics.* © 1975 by Annual Reviews, Inc.

†C=Confirmed; P=provisional; I=inconsistent or conflicting reports.

signed to specific human chromosomes by this technique in combination with data on known linkage groups. The two genes that are selected for by the HAT medium, *TK* and *HGPRT*, are located on human chromosome 17 and the human X chromosome, respectively.

Expression of differentiated properties in hybrid cells

Thus far the bulk of studies on the expression of differentiated properties in cellular hybrids have been done with permanent lines of differentiated tumor cells. These cells are generally not diploid, and there is a real risk that their regulation of expression of differentiated properties may be abnormal. However, because they do not have a finite lifetime in culture, and because they are easier to grow, they have proved to be more useful in hybridization studies than normal diploid differentiated cells.

Because they are relatively easy to manipulate genetically, the differentiated tumor lines are useful in studying the effects of gene dosage on expression of differentiated phenotypes. It is, for example, easy to obtain differentiated lines with double the usual number of chromosomes, either spontaneously because of failure to complete mitosis or because of fusion of identical cells with Sendai virus. Such variants with double the usual number of chromosomes are referred to as 2s, and the parental cells (which generally are not diploid to start with) are referred to as 1s. The designations N, 2N, and 4N, which refer to haploid, diploid, and tetraploid, respectively, are normally not used to describe cellular hybrids or their parental cells.

The patterns of gene expression that are observed in hybrid cells can generally be described in terms of various combinations of codominance, extinction, and cross activation.

Codominance refers to equal expression of genes from both parental cells in the hybrid. Codominant expression is observed for most enzymes of general metabolism that are not unique to differentiated cells. In hybrids involving two different species, codominance is often easy to demonstrate by differences in electrophoretic mobility or immunological properties of the two parental enzymes (Fig. 11-13).

Extinction refers to the frequently observed inhibition of expression of differentiated properties in hybrids between differentiated and undifferentiated cells. An example is the absence of pigment synthesis in hybrids between 1s hamster pigmented melanoma cells and 1s mouse fibroblast cells.

Cross activation refers to the activation of genes from an undifferentiated cell as the result of hybridization with a differentiated cell. An example is the synthesis of mouse serum albumin in a hybrid between a rat hepatoma and a mouse fibroblast-like cell that does not normally make serum albumin. In some special cases cross activation

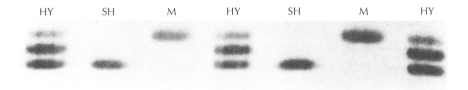

Fig. 11-13. Codominant expression of genes in a hybrid. The electrophorectic mobilities of NAD-dependent malate dehydrogenase *(MDH)* from mouse and Syrian hamster cells are sufficiently different so that the two enzymes can be separated easily. This zymogram shows MDH from a Syrian hamster melanoma *(SH)*, a mouse fibroblast *(M)*, and a hybrid prepared by fusing the two cells *(HY)*. The mouse and Syrian hamster genes are both expressed in the hybrid, and since MDH is a dimer, three electrophoretically distinct bands of MDH activity are seen. One corresponds to the normal mouse enzyme (two mouse subunits), the second corresponds to the normal Syrian hamster enzyme (two Syrian hamster subunits), and the third has intermediate mobility and is composed of one mouse subunit plus one Syrian hamster subunit. (From Davidson, R. L. 1973. Symp. Soc. Dev. Biol. **31**:295.)

may be accompanied by extinction (e.g., synthesis of mouse serum albumin, but not of rat serum albumin in certain hybrids between rat hepatomas and mouse fibroblasts).

In the simplest cases, in which a 1s differentiated cell is fused with a nondifferentiated cell, expression of the differentiated phenotype is frequently partially or completely absent in the resulting hybrid (Table 11-3). In the example just cited, no pigment is synthesized by 1:1 hybrids between hamster melanomas and mouse fibroblasts, even though codominance of hamster and mouse forms can be demonstrated easily for general metabolic enzymes (Fig. 11-13). Likewise, in crosses between 1s rat hepatomas and mouse lymphoid lines, expression of a variety of liver-specific properties of the hepatoma line is inhibited. Detailed studies by Mary Weiss and her co-workers have shown that there is usually partial or complete extinction in such hybrids of synthesis of serum albumin, aldolase B, liver-specific alcohol dehydrogenase, and both basal synthesis and glucocorticoid-inducible synthesis of tyrosine aminotransferase and alanine aminotransferase.

When 1s hybrids are constructed so that chromosomes from the genome of the nondifferentiated parent are preferentially lost, hybrid clones in which expression is initially suppressed frequently give rise to subclones exhibiting reexpression of the original epiphenotype of the differentiated parent (see "Remarks" column of Table 11-3). This was first seen for a kidney-specific esterase in hybrids between a mouse renal adenocarcinoma line and human diploid fibroblasts. More recently reexpression after chromosome loss has also been observed for several of the liver-specific properties described. The reex-

Table 11-3. Extinction of differentiated markers in hybrid cells*

Marker	Differentiated parental cell	Undifferentiated parental cell	Remarks†
Melanin	Melanoma (mouse)	L cell (mouse)	
	Melanoma (Syrian hamster)	L cell (mouse)	Gene dosage effect
Dopa-oxidase	Melanoma (Syrian hamster)	L cell (mouse)	
TAT, high baseline activity	Hepatoma (rat)	3T3 (mouse)	
	Hepatoma (rat)	SV40 transformed cells (human)	
	Hepatoma (rat)	L (mouse)	
TAT inducibility	Hepatoma (rat)	3T3 (mouse)	
	Hepatoma (rat)	Epithelial (rat)	Reappearance
	Hepatoma (rat)	3T3 (mouse)	
	Hepatoma (rat)	WI-38 (human)	Reappearance linked to loss of human X
Albumin	Hepatoma (rat)	3T3 (mouse)	Gene dosage effect and cross activation
Aldolase B	Hepatoma (rat)	3T3 (mouse)	
	Hepatoma (rat)	L (mouse)	
	Hepatoma (rat)	BRL-1 (rat)	Reappearance
	Hepatoma (rat)	DON (hamster)	Reappearance
Alcohol dehydrogenase (ADH)	Hepatoma (rat)	3T3 (mouse)	
	Hepatoma (rat)	BRL-1 (rat)	
Alanine aminotransferase	Hepatoma (rat)	BRL-1 (rat)	
Growth hormone	Pituitary cells (rat)	L cell (mouse)	
Protein S 100	Diploid glia cell (rat)	L cell (mouse)	
	"Near tetraploid" glia cell	L cell (mouse)	
GPDH inducibility	Glia cell (rat)	L cell (mouse)	
Immunoglobulins	Plasmocytoma (mouse)	L cell (mouse)	Incomplete extinction
	Myeloma (mouse)	3T3 (mouse)	
	Lymphoblast (human)	L cell (mouse)	
	Myeloma (mouse)	Lymphoma (mouse)	Light chain synthesis not extinguished
Esterase 2	Kidney cell (mouse)	L cell (mouse)	
	Kidney cell (mouse)	WI-38 (human)	Reappearance
Macrophage-specific surface receptors	Macrophage (mouse)	L cell (mouse)	
Steroid sulfatase	Neuroblastoma (mouse)	L cell (mouse)	
Sensitivity to 6-hydroxydopamine	Neuroblastoma (mouse)	Fibroblast (human)	

*Modified from Ringertz, N. R., and R. E. Savage. 1976. Cell hybrids, Academic Press, Inc., New York.
†"Gene dosage effect" indicates that the degree of extinction observed is dependent on the ratio of undifferentiated to differentiated genomes in the hybrid. "Reappearance" indicates that the differentiated trait has been observed to reappear after loss of chromosomes derived from the undifferentiated parent. "Cross activation" indicates that synthesis of the differentiated product by the undifferentiated genome can be activated at favorable gene dosage ratios.

pression of liver functions is particularly interesting in that individual liver-specific properties can be reexpressed independently. Thus, rather than a single switching mechanism that turns liver functions on and off coordinately, it appears that there is a separate control mechanism that can act independently for each of the liver-specific properties.

When hybrids are constructed from 2s differentiated cells and diploid or 1s nondifferentiated cells, the expression of differentiated properties is usually not suppressed. For example, many, but not all, of the hybrids obtained by fusing 2s hamster melanomas and 1s mouse fibroblasts synthesize pigment and the enzymes involved in pigment formation. Also, in hybrids involving 2s differentiated cells, genes from the nondifferentiated parent may be cross activated (Table 11-4). Thus in crosses of 2s rat hepatoma with 1s mouse lymphoid cells, rat liver properties continue to be expressed, and, in addition, synthesis of mouse liver proteins such as mouse serum albumin, mouse tyrosine aminotransferase, and mouse aldolase B can be detected. At the present time no satisfactory means exists to distinguish between mouse and rat forms of alanine aminotransferase and liver-specific alcohol dehydrogenase, so it is not known if the mouse forms of these enzymes are also cross activated. Even in cases in which there is partial extinction of liver-specific properties in 1s hybrids of rat hepatoma and mouse lymphoid cells, there is activation of the mouse genome. The reduced level of liver specific products that are synthesized con-

Table 11-4. Appearance of new activities not present in parental cells*

Product or property	Parental cells
Cross activation	
Mouse serum albumin	Rat hepatoma + mouse 3T3 cells
	Rat hepatoma + mouse lymphoblasts
Human immunoglobulins	Mouse myeloma + human lymphocytes
Human serum albumin	Mouse hepatoma + human leukocytes
Human complement factor (C4)	Guinea pig macrophages (C4⁻) + human HeLa cells
Possible cross activation	
Complement factor (C5)	Mouse spleen cells (C5⁻) + chick erythrocytes
Other new activities	
Hyaluronic acid	Mouse cell line + Chinese hamster cell line
Esterases	Chinese hamster ovary cells + human fibroblasts
Radioresistance	Rat pituitary cells + mouse L cells
Choline acetyltransferase	Mouse neuroblastoma + human fibroblasts
	Mouse neuroblastoma + rat glioma
Morphine receptors	Mouse neuroblastoma + rat glioma

*From Ringertz, N. R., and R. E. Savage. 1976. Cell hybrids. Academic Press, Inc., New York.

tains mouse as well as rat forms, although less mouse than rat product is made. There have also been a few interesting cases in which mouse, but not rat, albumin is made. However, these exceptional cases may have been due to loss of the chromosomes coding for rat albumin from the hybrids.

Positive versus negative controls of differentiation

These data show clearly that the expression of differentiated properties is under the control of diffusible substances. Such substances are referred to as "transactive" because they can affect chromosomes other than the ones that code for their synthesis. It is of interest that such regulatory substances have been conserved during evolution and are able to act on genes from other species. The data also show clearly that both the structural genes needed for differentiated functions and the tissue-specific determination (epigenotype) associated with those functions are retained in hybrid cells during extinction of their expression.

Beyond these points, however, the data are open to a multitude of interpretations, and it is not yet possible to say with certainty whether the controls that are involved act in a positive or a negative manner, or both. The presence of specific inhibitors for each differentiated function is suggested by extinction of expression in 1s hybrids. If such inhibitors exist, reexpression could reflect loss of the specific chromosomes coding for them. Continued expression in 2s hybrids would be caused by dilution of a limited amount of inhibitor among too many genes, which could also account for cross activation of genes from the nondifferentiated parent. However, it is also possible to explain all these phenomena with models based on positive activators. In the latter case, extinction would occur when limited supplies of the activator were diluted too greatly by nondifferentiated genomes. Partial or complete activation of genes from both parental genomes would occur when there was a more favorable ratio of activator molecules to genes. Reexpression could reflect a loss of some of the genes from the nondifferentiated parent that were competing for the activator molecules.

The data discussed thus far are all based on the fusion of two complete cells to yield a synkaryon that is essentially the sum of the two parental cells. Only a few studies of differentiation have been undertaken with cytoplasmic hybrids (cybrids) or with reconstituted cells (Fig. 11-9). Two recent experiments are of interest, although at present the results reported are difficult to reconcile with the interpretations of the whole-cell hybridization studies described. In the first, mouse erythroleukemia cells were fused with cytoplasms from nonerythroid cells such as mouse neuroblastoma or mouse L. In the parental erythroleukemia cell, hemoglobin synthesis can be induced by treatment with dimethyl sulfoxide. Extinction of that property occurred in the cybrids and persisted even after extended periods of continuous culture. These findings suggest the presence of a long-lived inhibitory factor in the cytoplasm that continues to be inherited through multiple rounds of cellular multiplication, or induction by a cytoplasmic factor of an epigenetic change in the genome of the erythroleukemia cell.

In the other experiment, minicells consisting of a nucleus and a thin adherent layer of cytoplasm were prepared from rat L6 myoblasts, a permanent line that retains the ability to undergo differentiation into myotubes in culture. Those minicells were fused with anucleate cytoplasms prepared from mouse fibroblasts. Viable cells were obtained that were able to fuse and form normal-appearing myotubes. Thus a period of exposure of the rat myoblast nucleus to mouse fibroblast cytoplasm did not alter its viability or its program of differentiation. Superficially, this and the previous experiment appear to be in direct conflict with each other. In the one case, hybridization with a cytoplasm permanently inactivated differentiation, whereas in the other it appeared to have no effect on the differentiated program. The ability to fuse parts of cells opens many new and exciting possibilities for studies of the controls over expression of differentiation. It is reasonable to expect many different combinations to be tested in the near future, including the mixing of cytoplasts from one type of differentiated cell with nucleated minicells from another type.

Experimentally induced mutation of cultured cells can also be used to study the nature of controls over gene expression. For example, a program of treatment of CHO cells (an established line derived from the ovary of a Chinese hamster) with mutagens followed by selection of clones and testing of them for changes in the isozyme distributions of various enzymes has resulted in the isolation of several clones that appear to have mutations in regulatory genes. These include a clone that expresses six additional electrophoretically distinct bands of esterase activity not found in the parental culture. All the additional bands appear to correspond precisely to forms of esterase that are found in one or more tissues in Chinese hamsters. The simplest interpretation of this observation is that a gene coding for a negative-acting control substance has been rendered nonfunctional by mutation, thereby permitting additional types of esterase to be formed. It is particularly interesting that one of the additional esterase bands is similar to the "kidney-specific" esterase of mouse renal adenocarcinoma that undergoes extinction when the mouse line is hybridized with human diploid fibroblasts and later reappears as human chromosomes are lost from the hybrid. However, other interpretations of the CHO mutants are possible, including mutational activation of a positive regulatory system and generation of esterase isozymes by posttranslational modification rather than by activation of additional esterase genes.

Data available at the present time are not adequate to distinguish with certainty between positive and negative control mechanisms. Experiments involving transcription from reconstituted chromatin seem to favor positive controls (Chapter 7), whereas many of the hybridization studies presented here seem to favor negative controls, although both sets of data are also capable of being interpreted in the opposite way. Ultimately, the distinction between positive and negative controls may prove to be unnecessary. The controls over differentiation are obviously complex. It is possible that both positive and negative elements are involved, and that either may be observed experimentally, depending on the nature of the particular experiment.

ANALYSIS OF DIFFERENTIATION IN CULTURED CELLS

As can be seen from the work described in this chapter, the use of cultured cells to study differentiation has made a variety of different experimental approaches possible. Examination of the relationship of DNA synthesis and cell division to differentiation has led to the hypothesis that special quantal cell cycles are required for differentiation. Study of the action of BrdU on cell differentiation has provided results consistent with the idea that differentiation requires the interaction of regulatory proteins with specific DNA sequences. Finally, the ability to construct hybrid cells from two different types of cells has made it possible to assign the structural genes for differentiated characteristics to specific chromosomes and has shown that diffusible regulatory molecules are important in control over the expression of cellular differentiation. Although these approaches have thus far raised more questions than they have answered, they are almost certainly leading us in the direction of profitable new areas of study.

In the following chapter we will direct our emphasis away from the study of cells that are already differentiated and begin exploring the apparent origins of cellular differentiation in developing embryos.

BIBLIOGRAPHY
Books and reviews

Bernhard, H. P. 1976. The control of gene expression in somatic cell hybrids. Int. Rev. Cytol. **47**:289.

Davidson, R. L. 1973. Somatic cell hybridization: studies on genetics and development. Addison-Wesley Modules in Biology No. 3. Addison Wesley Publishing Co., Inc., Reading, Mass.

Davidson, R. L. 1974. Gene expression in somatic cell hybrids. Annu. Rev. Genet. **8**:195.

Davis, F. M., and E. A. Adelberg. 1973. Use of somatic cell hybrids for analysis of the differentiated state. Bacteriol. Rev. **37**:197.

Dienstman, S. R., and H. Holtzer. 1975. Myogenesis: a cell lineage interpretation. In J. Reinert and H. Holtzer, eds. Cell cycle and cell differentiation. Springer-Verlag, Inc., New York.

Ephrussi, B. 1972. Hybridization of somatic cells. Princeton University Press, Princeton, N.J.

Ephrussi, B., and M. C. Weiss. 1969. Hybrid somatic cells. Sci. Am. 220(4):26.

Hamprecht, B. 1977. Structural, electrophysiological, biochemical and pharmacological properties of neuroblastoma-glioma cell hybrids in cell culture. Int. Rev. Cytol. 49:99.

Harris, H. 1970. Cell fusion. Harvard University Press, Cambridge, Mass.

Holtzer, H. 1970. Proliferative and quantal cell cycles in the differentiation of muscle, cartilage, and red blood cells. Symp. Int. Soc. Cell Biol. 9:69.

Holtzer, H., et al. 1972. The cell cycle, cell lineages, and cell differentiation. Curr. Top. Dev. Biol. 7:229.

Puck, T. T. 1972. The mammalian cell as a microorganism – genetic and biochemical studies in vitro. Holden-Day, Inc., San Francisco.

Ringertz, N. L., and R. E. Savage. 1976. Cell hybrids. Academic Press, Inc., New York.

Ruddle, F. H., and R. P. Creagan. 1975. Parasexual approaches to the genetics of man. Annu. Rev. Genet. 9:407.

Ruddle, F. H., and R. S. Kucherlapati. 1974. Hybrid cells and human genes. Sci. Am. 231(1):36.

Rutter, W. J., R. L. Pictet, and P. W. Morris. 1973. Toward molecular mechanisms of developmental processes. Annu. Rev. Biochem. 42:601.

Silagi, S. 1976. Effects of 5-bromodeoxyuridine on tumorigenicity, immunogenicity, virus production, plasminogen activator, and melanogenesis of mouse melanoma cells. Int. Rev. Cytol. 45:65.

Wilt, F. H., and M. Anderson. 1972. The action of 5-bromodeoxyuridine on differentiation. Dev. Biol. 28:443.

Selected original research articles

Bick, M. D., and R. L. Davidson. 1974. Total subsitution of bromodeoxyuridine for thymidine in the DNA of a bromodeoxyuridine-dependent cell line. Proc. Natl. Acad. Sci. U.S.A. 71:2082.

Biswas, D. K., J. Lyons, and A. H. Tashjian. 1977. Induction of prolactin synthesis in rat pituitary tumor cells by 5-bromodeoxyuridine. Cell 11:431.

Brown, J. E., and M. C. Weiss. 1975. Activation of production of mouse liver enzymes in rat hepatoma-mouse lymphoid cell hybrids. Cell 6:481.

Buckley, P. A., and I. R. Konigsberg. 1977. Do myoblasts in vivo withdraw from the cell cycle? A reexamination. Proc. Natl. Acad. Sci. U.S.A. 74:2031.

Colbert, D. A., and J. R. Coleman. 1977. Transcriptional expression of nonrepetitive DNA during normal and BUdR-mediated inhibition of myogenesis in culture. Exp. Cell Res. 109:31.

David, J., J. S. Gordon, and W. J. Rutter. 1974. Increased thermal stability of chromatin containing 5-bromodeoxyuridine-substituted DNA. Proc. Natl. Acad. Sci. U.S.A. 71:2808.

Davidson, R. L., and E. R. Kaufman. 1977. Deoxycytidine reverses the suppression of pigmentation caused by 5-BrdUrd without changing the amount of 5-BrdUrd in DNA. Cell 12:923.

Gilbert, W., and A. Maxam. 1973. The nucleotide sequence of the lac operator. Proc. Natl. Acad, Sci. U.S.A. 70:3581.

Gopalakrishnan, T. V., E. B. Thompson, and W. F. Anderson. 1977. Extinction of hemoglobin inducibility in Friend erythroleukemia cells by fusion with cytoplasm of enucleated mouse neuroblastoma or fibroblast cells. Proc. Natl. Acad. Sci. U.S.A. 74:1642.

Lapeyre, J. N., and I. Bekhor. 1976. Chromosomal protein interactions in chromatin and with DNA. J. Mol. Biol. 104:25.

Lin, S. Y., and A. D. Riggs. 1972. Lac operator analogues: bromodeoxyuridine substitution in the lac operator affects the rate of dissociation of the lac repressor. Proc. Natl. Acad. Sci. U.S.A. 69:2574.

Nadal-Ginard, B. 1978. Commitment, fusion and biochemical differentiation of a myogenic cell line in the absence of DNA synthesis. Cell 15:855.

Peterson, J. A., and M. C. Weiss. 1972. Expression of differentiated functions in hepatoma cell hybrids: induction of mouse albumin production in rat hepatoma-mouse fibroblast hybrids. Proc. Natl. Acad. Sci. U.S.A. 69:571.

Puck, T. T., and F. T. Kao. 1967. Genetics of somatic mammalian cells. V. Treatment with 5-bromodeoxyuridine and visible light for isolation of nutritionally deficient mutants. Proc. Natl. Acad. Sci. U.SA. 58:1227.

Ringertz, N. R., U. Krondahl, and J. R. Coleman, 1978. Reconstitution of cells by fusion of cell fragments. Exp. Cell Res. 113:233.

Schubert, D., and F. Jacob. 1970. 5-Bromodeoxyuridine induced differentiation of a mouse neuroblastoma. Proc. Natl. Acad. Sci. U.S.A. 67:247.

Schwartz, S. A., D. Horio, and W. H. Kirsten. 1974. Non-random incorporation of 5-bromodeoxyuridine in rat cell DNA. Biochem. Biophys. Res. Commun. 61:927.

Siciliano, M. J., M. R. Bordelon, and R. M. Humphrey. 1978. Genetics of regulation in cultured mammalian cells. In G. F. Saunders, ed. Cell differentiation and neoplasia. Raven Press, New York.

Strom, C. M., and A. Dorfman. 1976. Distribution of 5-bromodeoxyuridine and thymidine in the DNA of developing chick cartilage. Proc. Natl. Acad. Sci. U.S.A. 73:1019.

Strom, C. M., M. Moscona, and A. Dorfman. 1978. Amplification of DNA sequences during chicken cartilage and neural retina differentiation. Proc. Natl. Acad. Sci. U.S.A. 75:4451.

Walther, B. T., et al. 1974. On the mechanism of 5-bromodeoxyuridine inhibition of exocrine pancreas differentiation. J. Biol. Chem. 249:1953.

Whittaker, J. R. 1973. Segregation during Ascidian embryogenesis of egg cytoplasmic information for tissue specific enzyme development. Proc. Natl. Acad. Sci. U.S.A. 70:2096.

CHAPTER 12

Cell differentiation in the embryo—mosaic and regulative patterns of development

☐ In the previous chapters we discussed the behavior of differentiated cells in culture and the factors that affect the expression of differentiation in cultured cells. In this chapter we turn to cellular differentiation within the embryo. As is the case in cultured cells, cellular differentiation in the embryo seems to involve two distinguishable steps. The first is the determination or commitment of a cell to a particular differentiated phenotype, a step that tends to occur prior to morphological or biochemical specialization of the cells. The second is the actual expression of that differentiated phenotype. Just as is seen in differentiated cells in culture or in hybrid cells formed by cell fusion, embryonic cells that are not expressing differentiated characteristics may still have an inherited determination or epigenotype whereby they and their progeny are able to express only a specific differentiated phenotype.

The process of determination often occurs gradually and in a stepwise manner during embryonic development. In some invertebrate organisms cell determination occurs very early in development, and cells formed by even the earliest cleavage divisions are not totipotent. In higher organisms, however, cells of very early embryos are frequently totipotent, that is, unrestricted as to their developmental potential. But as embryogenesis proceeds, the developmental potential of cells inevitably becomes more and more restricted. In the mammalian embryo, for example, restriction of developmental potential probably occurs first with the separation of inner cell mass and trophoblast cells. Only inner cell mass cells can develop into embryonic structures; trophoblast cells form only extraembryonic/supporting structures, for example, the fetal part of the placenta. As development proceeds, a particular cell in a mammalian embryo might undergo sequential determinations to become, for example, inner cell mass, then endoderm, then midgut, then pancreas, then pancreatic exocrine cell. Cell determination does not occur in a single step by which a totipotent cell becomes directly determined to form a specific terminally differentiated cell type. Rather, it occurs in a series of steps during which the cell's developmental fate becomes progressively narrower.

PRESUMPTIVE FATE VERSUS DETERMINATION IN EMBRYONIC DEVELOPMENT

In following the progressive determination of cells during embryonic development, it is important to distinguish between the presumptive fate of a cell and its actual determination to a specific cell type. The presumptive fate of a cell is the type of cell into which it will normally develop if left undisturbed. A cell's presumptive fate can usually be accurately predicted, even at early stages, from its position within the embryo. By staining different cells of an embryo and following their subsequent development, investigators have constructed fate maps for different kinds of embryos that indicate the developmental fate of cells in all areas of the embryo. Fig. 12-1 is a fate map showing the presumptive fate of cells in different regions of the frog embryo at the gastrula state. A cell's presumptive fate can be changed in embryos in which cell determination is not an early event simply by moving the cell to a different location within the embryo. Determination of a cell, on the other hand, is an irrevocable event and is not altered by moving a cell to a new environment. Once a cell is determined to develop in a specific direction, it develops in that direction or not at all.

Transplantation experiments are frequently used to elucidate the time in development at which cells undergo specific determining events. Cells from a donor embryo, which are distinguished by vital staining, natural pigmentation, or some other marking device, are transplanted to a new location in a host embryo. Young embryos of a variety of organisms can easily tolerate this kind of surgical manipulation and develop normally. In the amphibian embryo, for example,

Fig. 12-1. Fate map of the early frog gastrula showing presumptive developmental fate of regions on the surface of the embryo. (Modified from Watterson, R. L., and R. M. Sweeney. 1973. Laboratory studies of chick, pig and frog embryos, 3rd ed. Burgess Publishing Co., Minneapolis, Minn.)

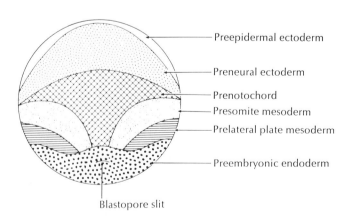

- Preepidermal ectoderm
- Preneural ectoderm
- Prenotochord
- Presomite mesoderm
- Prelateral plate mesoderm
- Preembryonic endoderm

Blastopore slit

presumptive neural ectoderm from a late blastula or early gastrula can be transplanted to a region of presumptive epidermis. The presumptive neural ectoderm then develops in accordance with its new location and becomes epidermal (Fig. 12-2, *A*). If the reverse procedure is done, presumptive epidermis develops into neural structures. Even more dramatically, presumptive ectodermal cells that are transplanted to regions of presumptive mesoderm or presumptive endoderm develop in accordance with their new locations. If transplantation is done at a later stage in gastrulation, however, the transplanted tissues differentiate autonomously. For example, presumptive neural ectoderm that has been transplanted to a region of presumptive epidermis sinks into the interior of the embryo, folds itself into a vesicle, and forms a neural tube (Fig. 12-2, *B*). Thus during the time span from early gastrulaton to late gastrulation a determining event has occurred in these cells such that their developmental potential has been restricted to the formation of neural tube derivatives.

Transplantation experiments have also been done to demonstrate cell determination in the embryos of higher vertebrates. In the chick embryo the presumptive wing-forming region can be localized prior to appearance of the wing bud by its position along the anterior-posterior axis of the embryo. If this presumptive wing bud region is transplant-

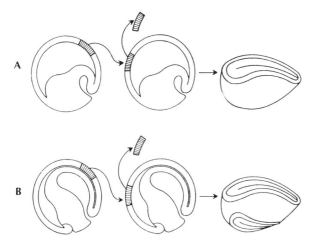

Fig. 12-2. Transplantation of presumptive neural ectoderm in the newt. **A,** Presumptive neural ectoderm of an early newt gastrula is transplanted to a region of presumptive epidermis in a second embryo of the same age. The transplanted material develops in accordance with its new location and becomes epidermis. **B,** Presumptive neural ectoderm of a late newt gastrula develops autonomously after transplantation and forms a secondary neural tube in the host embryo. (From Saxèn, L., and S. Toivönen. 1962. Primary embryonic induction. Prentice-Hall, Inc., Englewood Cliffs, N.J.)

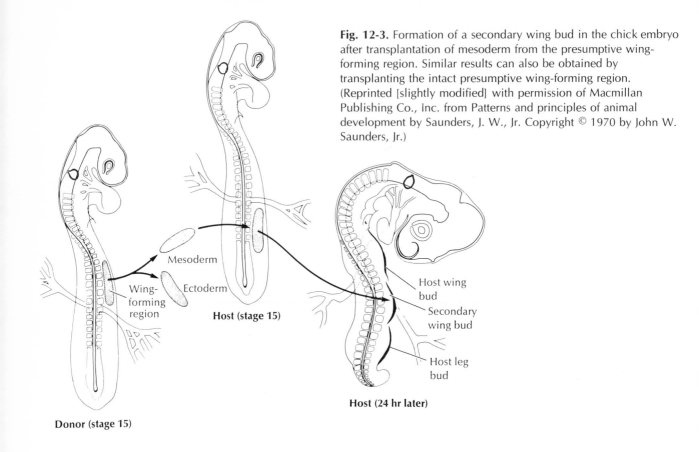

Fig. 12-3. Formation of a secondary wing bud in the chick embryo after transplantation of mesoderm from the presumptive wing-forming region. Similar results can also be obtained by transplanting the intact presumptive wing-forming region. (Reprinted [slightly modified] with permission of Macmillan Publishing Co., Inc. from Patterns and principles of animal development by Saunders, J. W., Jr. Copyright © 1970 by John W. Saunders, Jr.)

Mesoderm

Wing-forming region   Ectoderm

Host (stage 15)

Host wing bud

Secondary wing bud

Host leg bud

Host (24 hr later)

Donor (stage 15)

Fig. 12-4. Experimental procedures involved in the production of allophenic mice. The cells of two genetically different eight-cell stage embryos are combined after removal of their outer envelope and allowed to develop to the blastocyst stage in vitro. They are then transferred to the uterus of a foster mother where they complete their development. (Modified from Mintz, B. 1967. Proc. Natl. Acad. Sci. U.S.A. **58**:344.)

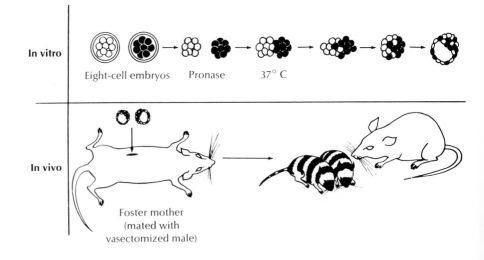

In vitro

Eight-cell embryos   Pronase   37° C

In vivo

Foster mother (mated with vasectomized male)

ed to a different location along the body wall of the embryo, it nonetheless develops into a wing bud at the usual time (Fig. 12-3). The cells in this region were thus determined to become wing structures prior to any detectable morphological or biochemical specialization in the direction of wing development.

The timing of cell determination in mammalian embryos has been studied by a different approach that involves the use of allophenic embryos. These embryos, which contain two genetically different populations of cells, are produced in the following manner (Fig. 12-4). At the eight-cell stage of development, the zona pellucida (the noncellular envelope that surrounds the cells) is removed from two different mouse embryos, and the cells are pushed together. The cell masses fuse and mix to form a single composite embryo. The composite embryos are allowed to develop to the blastocyst stage in vitro and are then transferred to the uterus of a suitably prepared foster mother, where they continue normal embryonic development. The embryo eventually regulates its size and develops into a normal, often fertile, mouse containing two distinct cell types. If the two embryos that are fused are from strains of mice with different pigmentation, the allophenic offspring normally exhibit both pigment types. From the patterns of pigmentation seen in these allophenic mice, one can calculate the approximate time in embryonic development at which pigment cell determination occurred. A salt and pepper pattern of pigmentation, which would indicate that pigment cell determination occurred at a relatively late embryonic stage, is never seen in allophenic animals. Instead, cells expressing the two different pigment types are usually located in large transverse patches on the animal along each side of a middorsal line (Fig. 12-5). It is believed that each patch represents a clone of cells derived from a single pigment cell precursor. The number and pattern of pigment patches seen suggest that they arise from about 34 primordial melanoblasts that become determined at about 5 to 7 days of development. The design of this experiment, however, does not allow one to distinguish whether these cells are irreversibly determined at the time they become melanoblast precursors or if their fate might be altered in a different environment.

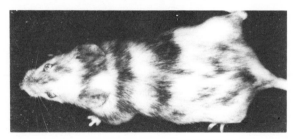

Fig. 12-5. Adult allophenic mouse exhibiting transverse patches of black and white pigment in its coat. (From Mintz, B. 1967. Proc. Natl. Acad. Sci. U.S.A. **58**:344.)

MOSAIC AND REGULATIVE EMBRYOS

The time in embryonic development at which cell determination first occurs varies considerably with different organisms. Embryos can be roughly classified into two types on this basis—mosaic embryos, in which cell determination occurs in the earliest stages of embryonic development, and regulative embryos, in which the developmental fate of early embryonic cells can be altered by experimental manipulation. In general terms, separation of cells at the four-cell stage in a mosaic embryo would result in each cell producing a specific quarter-part of the embryo, since each cell is already irreversibly programmed to a specific developmental fate. In contrast, separation of cells at the four-cell stage in a regulative embryo would result in each cell producing a complete, although possibly smaller, embryo. The cells are able to regulate the direction of their development depending on the environmental circumstances. This distinction between mosaic and regulative embryos is useful in contrasting the two types, but it is simplified and idealized and the behavior of most embryos lies somewhere along the spectrum between these two extremes.

Mosaic and regulative aspects of development in invertebrates

Mosaic embryos occur primarily among mollusks, annelids, and arthropods. The eggs of these animals frequently contain cytoplasmic components that become distributed to specific cells as a result of cleavage. A classic example of such compartmentalization is seen in formation of the polar lobe in the embryos of some mollusks. In the

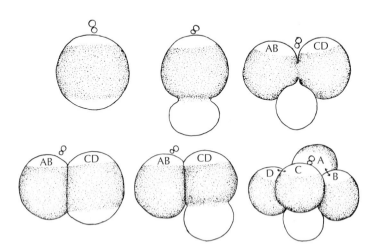

Fig. 12-6. Cleavage and formation of the polar lobe in the mollusk *Dentalium*. After the first cleavage division, cell *CD* contains all the polar lobe material. After the second cleavage division, cell *D* contains all the polar lobe material. (From Balinsky, B. I. 1975. An introduction to embryology, 4th ed. W. B. Saunders Co., Philadelphia.)

egg of the tusk shell mollusk *Dentalium*, for example, a protrusion (the polar lobe) forms at the vegetal end just prior to the first cleavage division (Fig. 12-6). Following the first cleavage, which cuts the egg parallel to the animal-vegetal axis, the polar lobe is attached to one of the two daughter cells. The polar lobe is then retracted into this cell but reappears prior to the second cleavage division. After this division it is associated with only one of the four blastomeres. Thus as early as the two-cell stage the cells of the embryo differ regarding their cytoplasmic composition, with one of the two cells containing the mass of the vegetal cytoplasm. This apparent difference in cytoplasmic composition reflects an actual difference in developmental potential of the cells. This has been demonstrated by experiments in which cells of the early embryo are separated or in which the polar lobe is removed. If the cells of a *Dentalium* embryo are separated at the two-cell stage, the cell with the polar lobe produces a complete, though distorted, larva. The other cell develops into a defective larva that lacks mesodermal components (Fig. 12-7). If the blastomeres are separated at the four-cell stage, again only the blastomere containing the polar lobe forms a complete, though distorted, larva; the other three form defective larvae lacking mesoderm. These findings suggest that some component present in the cytoplasm of the polar lobe is essential for the formation of mesoderm. The same conclusion is drawn from experiments in which the polar lobe is removed. If the polar lobe is

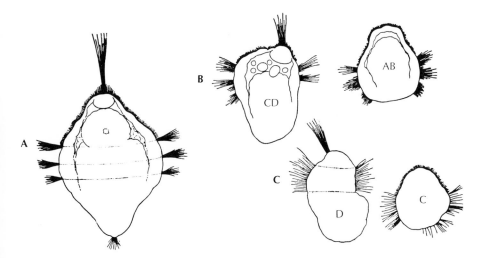

Fig. 12-7. Larvae of *Dentalium* that have developed, **A,** from a complete egg, **B,** from cells separated at the two-cell stage, and **C,** from cells separated at the four-cell stage. Letters on larvae in **B** and **C** correspond to those in Fig. 12-6 and indicate the cell from which the larva has developed. (From Balinsky, B. I. 1975. An introduction to embryology, 4th ed. W. B. Saunders Co., Philadelphia.)

amputated at the time when it protrudes, embryo development proceeds but the resulting larva lacks mesoderm.

Some attempts have been made to elucidate the cytoplasmic determinants that distinguish the cytoplasm of the polar lobe from the remaining cytoplasm of the embryo. Differences have been found in the proteins synthesized by the two cells at the two-cell stage and also in the proteins synthesized by normal and delobed embryos. Differences in protein synthesis in normal and delobed embryos occurred even in the absence of RNA synthesis. This suggests either that mRNA molecules are differentially segregated into polar and nonpolar lobe cytoplasm or that there is differential segregation of specific translational factors that distinguish among messages common to all the cells of the embryo.

Another clear example of a mosaic embryo is that of the ascidian *Styela*. The fertilized egg of *Styela* contains four regions of cytoplasm that differ in their pigmentation (Fig. 12-8). The animal half of the egg contains clear cytoplasm; the vegetal half contains two crescent-shaped regions opposite each other just below the equator, one light gray, the other yellow. The remainder of the vegetal hemisphere is a slaty gray and laden with yolk. During cleavage of the egg the four cytoplasmic regions are distributed to different cells. In later development the cells containing clear cytoplasm develop into ectoderm; cells with yellow cytoplasm develop into mesoderm; cells with light gray cytoplasm develop into neural system and notochord; and cells with slaty gray cytoplasm develop into endoderm. The results of experiments in which the cells of ascidian embryos are separated have demonstrated that the developmental fates of the different cytoplasmic regions are determined at an early stage. If an eight-cell embryo is

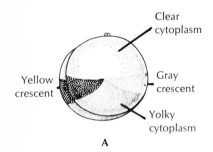

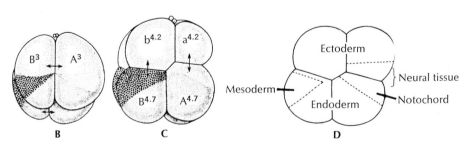

Fig. 12-8. Distribution of cytoplasmic regions in cleavage-stage embryos of the ascidian *Styela* at the, **A,** two-cell stage **B,** four-cell stage, and, **C,** eight-cell stage. **D** shows the presumptive fate of regions in the eight-cell embryo. The process of cytoplasmic streaming at the time of fertilization that leads to the distribution of cytoplasmic regions in **A** is shown in Fig. 2-3. (From Kuhn, A. 1971. Lectures on developmental physiology. Springer-Verlag, Inc., New York.)

separated into four pairs of blastomeres, each pair is able to produce only those structures which it would normally produce or fewer (Fig. 12-9). Localization of different cytoplasmic substances is also suggested by the results of experiments in which fertilized ascidian eggs are centrifuged. Embryos that develop from such centrifuged eggs have tissues arranged in a chaotic manner, implying that cytoplasmic substances have been displaced from their usual positions by centrifugation. However, if the eggs are centrifuged and enough time is allowed between centrifugation and cleavage, normal development occurs. Thus the cytoplasmic determinants tend to return to their original positions, perhaps under the influence of cortical regions that are not affected by centrifugation.

Compartmentalization of cytoplasmic materials in the *Styela* egg probably occurs concurrently with the rearrangement of pigment granules that occurs at fertilization (Fig. 2-3). Prior to fertilization the egg does not show mosaic characteristics. The unfertilized egg can be

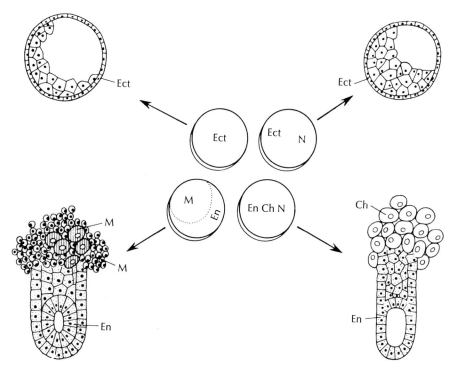

Fig. 12-9. Development of pairs of cells separated at the eight-cell stage in the ascidian embryo. The presumptive fate of each pair of cells is indicated, as well as the actual structure formed. *Ch* = Notochord; *Ect* = ectoderm; *En* = endoderm; *M* = mesoderm; *N* = neural structures. (Modified from Kuhn, A. 1971. Lectures on developmental physiology. Springer-Verlag, Inc., New York.)

separated into two halves either by dissection or centrifugation. Whether the plane of bisection is along or across the animal-vegetal axis, each half can be fertilized and will develop into a normal larva. Some of these embryos are, however, haploid due to the absence of the egg nucleus. If an intact egg is fertilized and the blastomeres are separated after the first cleavage division, each will form only a right or left half-embryo.

There is some evidence that the molecular species segregated by cleavage in ascidian embryos may be specific activators of transcription rather than mRNA molecules, translational factors, or proteins themselves. If cleavage is arrested at different points in the development of the ascidian *Ciona*, the activities of two tissue-specific enzymes nonetheless appear at the normal time. The enzymes are acetylcholinesterase, which is made in large amounts by nerve and muscle cells, and dopa oxidase, which is made in large amounts by pigment cells. These enzymes have been histochemically detected only in those cells which are precursors to the cells in which the enzymes are normally found, suggesting that their differential synthesis may be correlated with the segregation of cytoplasmic determinants. The appearance of enzyme activity is blocked both by inhibitors of RNA synthesis and inhibitors of protein synthesis, indicating that their expression results from differential gene expression in the cells of the developing embryo.

Cell determination in insect embryos, especially *Drosophila*, has been studied extensively in recent years, and evidence indicates the occurrence of both mosaic and regulative properties in these embryos. As discussed in Chapter 6, at one pole of some insect eggs there is a cytoplasmic region called the pole plasm, which is required for the determination of germ cells. In *Drosophila* the distinctive granules associated with the pole plasm are first seen during oogenesis. When the pole plasm from unfertilized eggs or from oocytes in later stages of oogenesis is implanted into the anterior region of cleavage-stage embryos, it is integrated into anterior cells that morphologically resemble normal pole cells. These pole cells have been shown to be capable of forming functional germ cells when transplanted into the polar region of genetically distinct host embryos. Some of the progeny produced by these host embryos expressed the markers characteristic of the transplanted cells. Thus it appears that, in *Drosophila*, cytoplasmic factors that influence germ cell development are present even prior to fertilization. Further evidence for the existence of cytoplasmic factors that influence cell determination in *Drosophila* comes from the study of maternal effect mutants, which are discussed in Chapter 22.

The results of the pole plasm transfer experiments also indicate that the nuclei of developing embryos are not irreversibly determined but are capable of multipotential development. In these experiments anterior nuclei that would not normally form pole cells do so under the influence of pole plasm. Other nuclear transplantation experiments have indicated that nuclei of cleavage-stage *Drosophila* embryos can support the development of a wide variety of adult structures.

Nuclei from *Drosophila* embryos that have undergone cellularization are also not determined as to their developmental potential. This was demonstrated in an experiment done by Karl Illmensee, which is diagrammed in Fig. 12-10. Single nuclei from cells in different regions of early gastrula embryos were implanted into genetically different unfertilized eggs. Most of these eggs began development and devel-

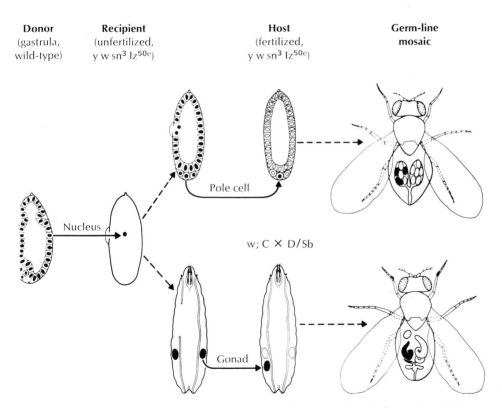

Fig. 12-10. Experimental procedure involved in transplantation of nuclei from *Drosophila* gastrulae. Nuclei of donor embryos at the gastrula stage were transplanted into unfertilized eggs of genetically different flies. These recipient eggs were allowed to develop to the cellular blastoderm stage (top) or the larval stage (bottom). Pole cells from the cellular blastoderms or gonads from the larvae were then implanted into hosts of the same developmental stage. The hosts developed into adult flies whose gonads contained gametes of both donor and host genotype. (From Illmensee, K. 1973. Wilhelm Roux Archiv. **171**:331.)

oped to a larval stage. Fertile adults were obtained from descendants of the donor nuclei in two ways. Some recipient eggs were allowed to develop to the cellular blastoderm stage, and then their presumptive germ cells (pole cells) were transplanted into host blastoderms. Other recipients were allowed to develop to a larval stage, and their gonads were then transplanted into host larvae. Both the hosts receiving transplanted germ cells and those receiving transplanted gonads produced functional gametes with the genotype of the original transplanted nucleus. Although *Drosophila* nuclei are still totipotent after formation of the cellular blastoderm, whole cells taken from embryos after cellularization do show a restriction in their developmental potential. When aggregates of genetically marked cells from anterior or posterior halves of cellular blastoderm embryos are cultured in adult flies and then induced to differentiate by being injected into larvae undergoing metamorphosis, the anterior and posterior cells form only anterior and posterior adult structures, respectively.

The sea urchin provides a classic example of a regulative embryo among the invertebrates. Isolated cells from two- or four-cell-stage embryos develop into complete, though small, larvae. However, the sea urchin egg also shows some mosaicism. This has been shown by experiments in which the unfertilized sea urchin egg is cut into halves, each of which is then fertilized and allowed to develop (Fig. 12-11). In describing these experiments, we will visualize sea urchin eggs oriented as shown in Figs. 12-11 and 12-12 with the animal pole at the top and the vegetal pole at the bottom. If the cut is made in the vertical direction (along the animal-vegetal axis), two small but normal larvae develop from the fertilized halves. However, if the cut is horizontal, two incomplete larvae are formed. The animal half forms a ball of ciliated cells, called a dauerblastula; the vegetal half forms an embryo containing a gut, ectoderm, and some skeleton. Similar results have been obtained in experiments with more advanced sea urchin embryos. Half-embryos formed by vertical cuts develop into normal small larvae; half-embryos formed by horizontal cuts form incomplete embryos similar to those which develop from half-eggs. The findings of these experiments strongly suggest that there is a differential distribution of cytoplasmic determinants along the animal-vegetal axis of the sea urchin egg.

In normal sea urchin development the first two cleavages are vertical, producing four equivalent cells. The third cleavage is horizontal and occurs just above the equator, resulting in four smaller animal cells and four larger vegetal cells. The fourth cleavage is unusual—the upper cells divide vertically to form a ring of eight cells, the mesomeres; and the lower cells divide obliquely and produce four large cells, the macromeres, atop four small, centrally located cells, the micromeres (Fig. 12-12). Each of these three types of cells gives rise to different parts of the developing embryo. The mesomeres produce most of the embryo's ectoderm; the macromeres develop into endoderm and some ectoderm; the micromeres develop into mesenchyme, which eventually forms the skeleton of the larva. These developmental fates do not appear to be strictly determined, however. If the micromeres are removed from an embryo, the remaining cells regulate and develop into a normal larva. The combining of micromeres or

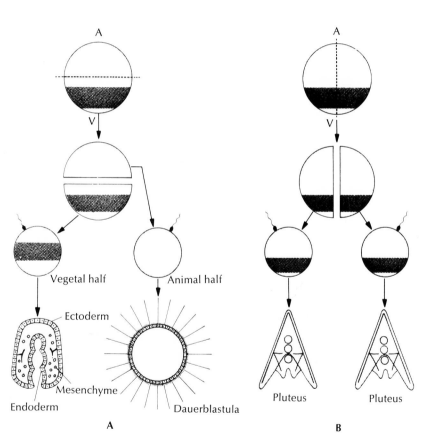

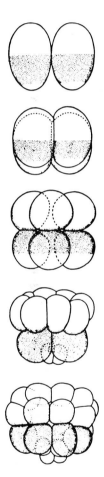

Fig. 12-11. Development of halves of sea urchin eggs that have been, **A,** bisected horizontally or, **B,** bisected vertically and then fertilized and allowed to develop. In **A** animal and vegetal halves form incomplete embryos. In **B** halves formed by a vertical cut form complete, though small, embryos. (From Barth, L. G. 1953. Embryology. Holt, Rinehart & Winston, Inc., New York.)

Fig. 12-12. Diagram of the first five cleavage divisions in the sea urchin *Paracentrotus lividus*. (From Balinsky, B. I. 1975. An introduction to embryology, 4th ed. W. B. Saunders Co., Philadelphia.)

macromeres with mesomeres likewise results in the development of a nearly normal larva. However, mesomeres alone develop into a highly animalized embryo, and macromeres plus micromeres develop into a highly vegetalized embryo. The results of such experiments suggest that both animal and vegetal regions of the embryo are required for normal development. It has been postulated that sea urchin embryos contain two gradients of activities, one of which has its high point at the animal pole, the other at the vegetal pole. Development of normal structures is postulated to require the interaction of these two opposing gradients. A gradient of reducing activity has been found along the animal-vegetal axis in sea urchin embryos. However, the mechanism by which this gradient is produced and its significance with regard to development of animal and vegetal structures is unclear.

Mosaic and regulative aspects of development in amphibian embryos

Although frog embryos are usually considered regulative, they show both regulative and mosaic characteristics. If the cells of the two-cell embryo are separated, each can develop into a normal embryo. However, this occurs only if both cells contain a portion of the gray crescent (Chapter 2). If one of the cells completely lacks gray crescent material, it does not develop normally, since the gray cres-

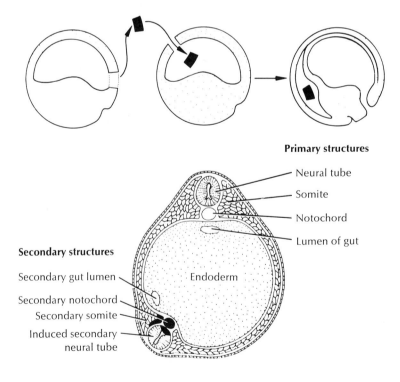

Fig. 12-13. Formation of a secondary embryo in the amphibian. The dorsal lip of the blastopore (the region derived from the gray crescent) is transplanted into the blastocoel of a second embryo. The transplanted tissue forms a secondary center of gastrulation and induces formation of a second set of embryonic structures. (Reprinted [modified] with permission of Macmillan Publishing Co., Inc. from Patterns and principles of animal development by Saunders, J. W., Jr. Copyright © 1970 by John W. Saunders, Jr.)

Primary structures

- Neural tube
- Somite
- Notochord
- Lumen of gut

Secondary structures

Endoderm

- Secondary gut lumen
- Secondary notochord
- Secondary somite
- Induced secondary neural tube

cent is essential for gastrulation to occur. The importance of the gray crescent region for development is also apparent from experiments in which pieces of this region are transplanted to different locations on a second embryo or placed within the blastocoel of a blastula-stage embryo. The result is that a new center of gastrulation forms at the site of the transferred gray crescent tissue, and a secondary embryo develops (Fig. 12-13).

The results of experiments done with *Xenopus laevis* by A. S. G. Curtis in the early 1960s indicate that the cellular determinants present in the gray crescent region are localized in the cortex, at least initially. The cortex in the amphibian egg is a thin, nonfluid layer of cytoplasm located beneath the plasma membrane. It is free of yolk granules and contains the pigment characteristic of the animal pole and gray crescent region. Curtis found that when the cortex of the gray crescent region of fertilized eggs is excised, the egg undergoes cleavage but fails to gastrulate (Fig. 12-14, *A*). Removal of cortex from other regions of the fertilized egg did not impede development to such a de-

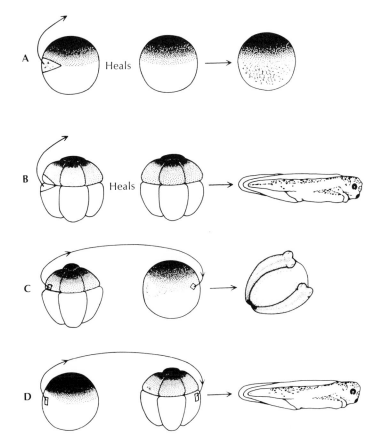

Fig. 12-14. Gray crescent cortex excision and transplantation experiments done by Curtis in *X. laevis.* **A,** Excision of the gray crescent cortex from a fertilized egg does not impede cleavage but prevents gastrulation. **B,** Excision of the gray crescent cortex from an eight-cell embryo does not alter its development. **C,** Transplantation of gray crescent cortex from an eight-cell embryo to the ventral region of a fertilized egg results in formation of a secondary embryonic axis. **D,** Transplantation of gray crescent cortex from a fertilized egg to the ventral region of an eight-cell embryo does not result in formation of a secondary embryonic axis. (From Curtis, A. S. G. 1962. J. Embryol. Exp. Morphol. **10:**410.)

gree. Removal of the gray crescent cortex of eight-cell eggs, however, did not prevent gastrulation (Fig. 12-14, *B*). This suggests that by this stage the information in the cortex has already been transmitted to the appropriate cells or cytoplasmic regions of the embryo. However, the gray crescent cortex of the eight-cell embryo is still able to transmit its information. This was shown by the results of technically difficult grafting experiments in which the gray crescent cortex was transplanted from an eight-cell embryo to the ventral region of a fertilized egg. The grafted cortex induced the formation of a secondary embryonic axis in the host embryo (Fig. 12-14, *C*). A final interesting result of these experiments is that gray crescent cortex from a fertilized egg was not able to induce formation of a secondary embryonic axis when transplanted to the ventral side of an eight-cell embryo (Fig. 12-14, *D*). This suggests that at this stage the cells of the embryo are not able to respond to the stimulus present in the gray crescent cortex.

Attempts to repeat the experiments of Curtis have failed because of the difficulty of the cortical grafting technique. However, the results have been verified in general by other experimental approaches. For example, other workers have shown that implantation of gray crescent cortex from a fertilized egg or an eight-cell embryo into the blastocoel of a midblastula embryo results in induction of a secondary embryonic axis. These results conflict to some degree with those of Curtis, since he was unable to induce secondary axis formation by grafts onto eight-cell embryos. It is possible that this difference may be related to the difference in age of the responding embryos or to the difference in location of the transplanted cortex.

A simple but elegant experiment done by Hans Spemann in 1928 showed that at least the nuclei of the amphibian embryo can regulate after a number of cleavage divisions. Spemann constricted fertilized eggs of the newt *Triturus* with a baby hair so that the two halves were connected by only a fine bridge of cytoplasm (Fig. 12-15). The side of the egg containing the nucleus underwent normal cleavage, whereas the nonnucleated half remained uncleaved. At about the 16-cell stage,

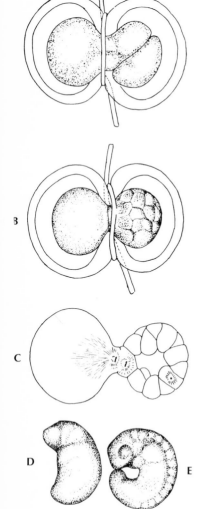

Fig. 12-15. Spemann's experiment demonstrating the regulative ability of cleavage-stage nuclei of the newt embryo. **A,** A fertilized egg is constricted with a baby hair. The nucleated half begins cleavage. **B,** A nucleus has crossed the cytoplasmic bridge between the two halves, and a cell membrane now separates the halves. **C,** Same as **B** but in section. **D,** Embryo formed by originally nonnucleated half is delayed in development compared to **E,** which is an embryo from the originally nucleated half. (From Spemann, H. 1938. Embryonic development and induction. Yale University Press, New Haven, Conn.)

daughter nuclei were small enough that it was possible for one to cross the thin bridge of cytoplasm into the uncleaved half of the egg. As a result, this half also began cleavage. The two halves were then separated completely and allowed to develop. In those cases in which the nonnucleated half of the egg contained some of the gray crescent material, it was able to develop into a complete embryo after being supplied with a nucleus. This finding indicates that even at the 16-cell stage, the cleavage nuclei of the amphibian embryo are all equivalent to the egg cell nucleus and have not undergone any irreversible specialization. More recently, nuclear transplantation experiments (described in Chapter 6) have shown that nuclei from amphibian cells retain the full potential to support normal development even after cellular differentiation has occurred. This suggests that mosaic signals such as determination of the site of gastrulation by the gray crescent are probably cytoplasmic or cortical in nature rather than nuclear.

Regulative ability of the mammalian embryo

The mammal provides our final example of an embryo showing considerable regulative ability. A variety of experimental techniques have been used to demonstrate the regulative capacity of cleavage-stage cells of mammalian embryos. For example, if one of two cells of a mouse or rabbit embryo is destroyed, the remaining cell will usually form a normal blastocyst and after transfer to the uterus of a foster mother, it can develop fully (Fig. 12-16). If the two cells of a mouse embryo are separated, some develop into normal embryos; others, however, form merely a trophoblastic vesicle containing no inner cell mass. Such trophoblastic vesicles are also frequently formed when three cells of a four-cell mouse embryo are destroyed and are almost invariably formed when seven cells of an eight-cell embryo are destroyed. Although these results might be interpreted to indicate that localized cytoplasmic determinants, critical to development, are lost by cell destruction, formation of trophoblastic vesicles by isolated mouse blastomeres has an alternative explanation. Blastocyst formation in the mouse occurs after four or five cleavage divisions when only 20 to 30 cells are normally present, and formation of an inner cell mass may not be possible if this number is significantly reduced by cell destruction. (The mechanism of determination of the inner cell mass of mammalian embryos is discussed later in this chapter.) In the rabbit embryo, in which blastocyst formation occurs after more cleavages have taken place, normal development can occur following destruction of seven cells of an eight-cell embryo.

The regulative ability of mammalian embryos is also supported by the results of embryo fusion experiments. As mentioned earlier in this

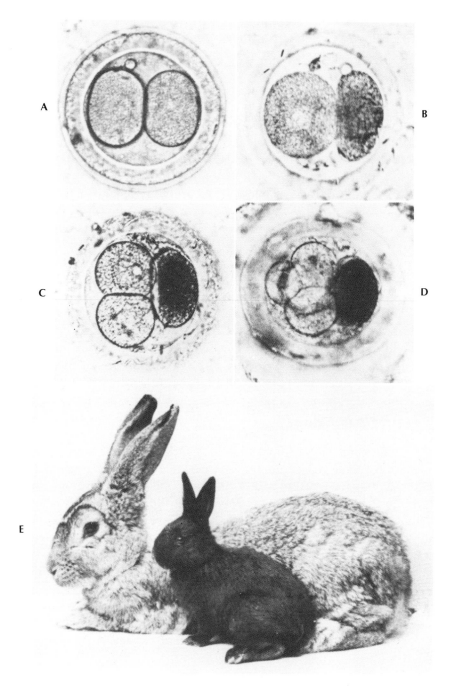

Fig. 12-16. Development of rabbit embryo after destruction of one cell at the two-cell stage of cleavage. **A,** Two-cell–stage embryo in vitro. **B,** Two-cell–stage embryo after destruction of the cell on the right with a fine glass needle. **C** and **D,** Cleavage of the surviving cell into two and then four cells. **E,** Black offspring that developed after transfer of a treated egg to the uterus of a foster mother (large gray rabbit). (From Seidel, F. 1952. Naturwissenschaften **39:**355.)

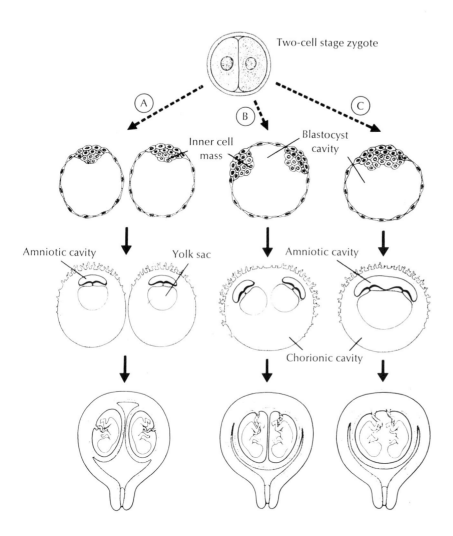

Two-cell stage zygote

Inner cell mass

Blastocyst cavity

Amniotic cavity

Yolk sac

Amniotic cavity

Chorionic cavity

Fig. 12-17. Diagram illustrating different types of monozygotic twinning in humans and other mammals. Intermediate rows show inner cell mass and primitive streak stages of development. Lower row shows advanced fetuses in relationship to the amnion, chorion, and placenta for each case. *A,* Cells separate fully during cleavage and form two separate blastocysts. Each twin has its own amnion, chorion, and placenta. *B,* Two inner cell masses form in a single blastocyst. The resulting twins share a common chorion and placenta, but each has its own amnion. *C,* Two primitive streaks form within a single inner cell mass, or one streak forms and subsequently splits longitudinally. In either case both twins develop within a single amnion and also share a common chorion and placenta. Incomplete separation may lead to conjoined twins. (From Langman, J. 1975. Medical embryology, 3rd ed. The Williams & Wilkins Co., Baltimore.)

chapter, it is possible to mix the cells of two eight-cell mouse embryos and obtain normal embryo development. There is no evidence of any cell sorting occurring after embryo fusion; the cells appear to mix at random. It has been reported that as many as 16 such embryos have been mixed and gone on to develop into a normal blastocyst.

Finally, evidence for the regulative nature of mammalian embryos is found by examining the types of identical twinning that occur in these embryos. Identical or monozygotic twins are thought to arise in at least five different ways (Fig. 12-17).

1. Cells may become separated during cleavage, resulting in the formation of two blastocysts. Twins of this type have separate placentas, since the fetal part of the placenta develops from the trophoblast.

2. Two inner cell masses may form in a single blastocyst. In this case the twins will share a common placenta, but each has its own amniotic sac, since this develops from the inner cell mass.

3. Two primitive streaks, and consequently two embryonic axes, may form within a single inner cell mass. In this case the twins share a common placenta and a common amniotic sac.

4. A single primitive streak may undergo longitudinal division at some point. This type of twinning tends to result in some degree of conjoining. Again the twins share a common placenta and a common amniotic sac.

5. A final type of identical twinning occurs normally in the armadillo. It is actually a combination of the third and fourth types. The armadillo embryo forms two primitive streaks within a single inner cell mass, and each of these primitive streaks then undergoes longitudinal division, resulting in the production of four identical individuals.

Formation of twins by development of two primitive streaks or the splitting of a single primitive streak indicates that at least the inner cell mass cells remain regulative for many cell divisions in the mammalian embryo.

The regulative ability of inner cell mass cells has been further demonstrated by the results of experiments in which one or more inner cell mass cells from one embryo are injected into a host blastocyst. Such injected cells are subsequently incorporated into a variety of adult tissues. Injected trophoblast cells, on the other hand, are not incorporated into embryonic tissues of the host. Although the regulative ability of inner cell mass cells at the blastocyst stage is well established, it appears that inner cell mass and trophoblast cells are not interchangeable at this stage and do not have equivalent developmental potential.

Differentiation of cells into either trophoblast or inner cell mass appears to be the first occurrence of cell determination in the mammalian embryo. Trophoblast and inner cell mass cells are distinguished basically by their locations—trophoblast cells are on the outside of the blastocyst; inner cell mass cells are on the inside (Fig. 12-18). However, the two cell types are also distinguishable morphologically. Trophoblast cells are large and flat and have tight cell junctions, whereas inner cell mass cells are smaller, rounded, and not tightly joined.

Blastocysts can be separated into trophoblast and inner cell mass cells microsurgically, and the properties of the two isolated cell types can then be examined (Fig. 12-18). An isolated group of trophoblast cells will tend to assume a spherical shape and secrete fluid interiorly to form a fluid-filled vesicle. Different groups of trophoblast cells do not interact with each other to form a common vesicle. In contrast, inner cell mass cells remain packed together and do not form vesicles, but they readily stick to other groups of inner cell mass cells to form aggregates. The two cell types also differ in their ability to evoke an implantation response in the uterus of a host animal. Trophoblast cells are able to elicit this response; inner cell mass cells are not. Finally, different proteins are synthesized by isolated trophoblast and inner cell mass cells when they are incubated in medium containing labeled amino acids.

CELL DETERMINATION IN THE EARLY MAMMALIAN EMBRYO

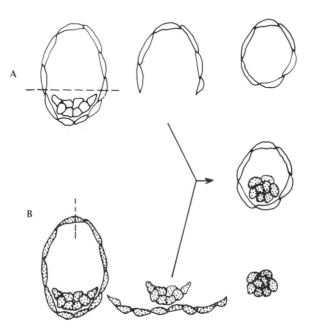

A

B

Fig. 12-18. Microsurgical separation and reconstitution of outside (trophoblast) and inside (inner cell mass) cells of the mouse embryo. **A,** Isolated outside cells form a fluid-filled vesicle that can induce an implantation reaction but develops only as trophoblast. **B,** Inside cells remain as a clump and do not implant in the uterus or develop further. If the inside cell mass is placed within the vesicle of outer cells, normal development can take place. (From Herbert, M. C., and C. F. Graham. 1974. Curr. Top. Dev. Biol. **8:**151.)

Extensive work has been done to elucidate the timing of determination and the mechanism whereby the developmental potential of trophoblast and inner cell mass cells is determined. It is clear from cell destruction and embryo fusion experiments that all cells of the mammalian embryo are totipotent up to at least the eight-cell stage. Determination of inner cell mass and trophoblast cells is likely to be due to factors intrinsic to the embryo rather than to outside influences, since normal embryonic development to beyond the blastocyst stage can occur in vitro in complete isolation from the female reproductive tract.

Two hypothetical mechanisms whereby cell determination might occur in the mammalian embryo are diagrammed in Fig. 12-19. The first theory suggests that determination to become trophoblast or inner cell mass results from compartmentalization of special cytoplasmic factors during cleavage of the egg. This model seems unlikely in view of the evidence for totipotency of cells of the early embryo. The second theory proposes that cells become determined to develop into inner cell

First mechanism

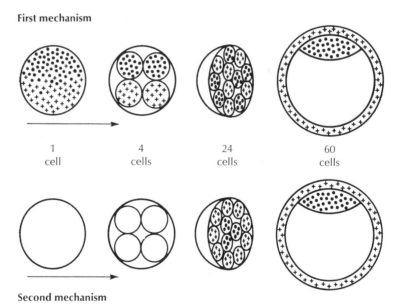

Second mechanism

Fig. 12-19. Two possible mechanisms whereby cells of the early mammalian embryo become determined to form the inner cell mass or trophoblast. In the first mechanism, cells differentiate as a result of differential distribution of cytoplasmic factors. Cells receiving trophoblast determinants assume an exterior location. Cells receiving inner cell mass determinants assume an interior location. In the second mechanism, cell determination occurs as a result of cell position. Exterior cells differentiate into trophoblast. Interior cells differentiate into inner cell mass. (From McClaren, A. 1969. Proc. R. Inst. Great Britain. III. no. 196. p. 156.)

mass or trophoblast by virtue of their position within the embryo. That is, outside cells with an exposed surface become determined to form trophoblast, whereas inside cells, surrounded on all sides by other cells, become determined to form inner cell mass. Up to about the eight-cell stage, all cells have an exposed surface. But from the 16-cell stage onward some cells are enclosed in the interior of the cell aggregate.

Christopher Graham and co-workers have tried to test the theory that determination to become inner cell mass or trophoblast results from cell position within the embryo. They constructed composite mouse embryos containing radioactively labeled cells from one embryo and unlabeled cells from a second embryo combined in specific arrangements. They then followed the subsequent development of each type of cell to determine whether it was influenced by the cell's position. In one set of experiments, labeled four-cell embryos were dissociated, and single blastomeres were placed either on the outside or the inside of other unlabeled four-cell embryos (Fig. 12-20). When 64-cell blastocysts that had developed from these composite embryos were examined, daughter cells of the labeled blastomeres placed in an outside position had developed only into trophoblast. Daughter cells of the labeled blastomeres placed in an inside position developed into both trophoblast and inner cell mass, but the percentage of cells developing into trophoblast was less than would be predicted from a random distribution of these cells into trophoblast and inner cell mass.

The results from a variety of other similar experiments support the general conclusion that inner cells contribute primarily to embryonic structures, whereas outer cells contribute mainly to trophoblast-derived structures. This theory of cell determination as a result of cell position has been used to explain the inability of a single cell from an eight-cell mouse embryo to develop into anything other than a trophoblastic vesicle. Too few cells would be present at the time of determination for any cells to be enclosed; thus none would be signaled to develop into inner cell mass. This same theory would also explain normal development of one cell from an eight-cell rabbit embryo.

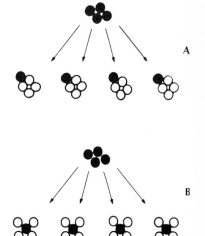

Fig. 12-20. Embryo reconstitution experiments designed to test the effect of cell position on determination to become inner cell mass or trophoblast. **A,** Single cells from a radioactively labeled four-cell embryo were placed outside the cells of unlabeled four-cell embryos. **B,** Single cells from a labeled four-cell embryo were surrounded by the dissociated cells of an unlabeled embryo. (From Herbert, M. C., and C. F. Graham. 1974. Curr. Top. Dev. Biol. **8:**151.)

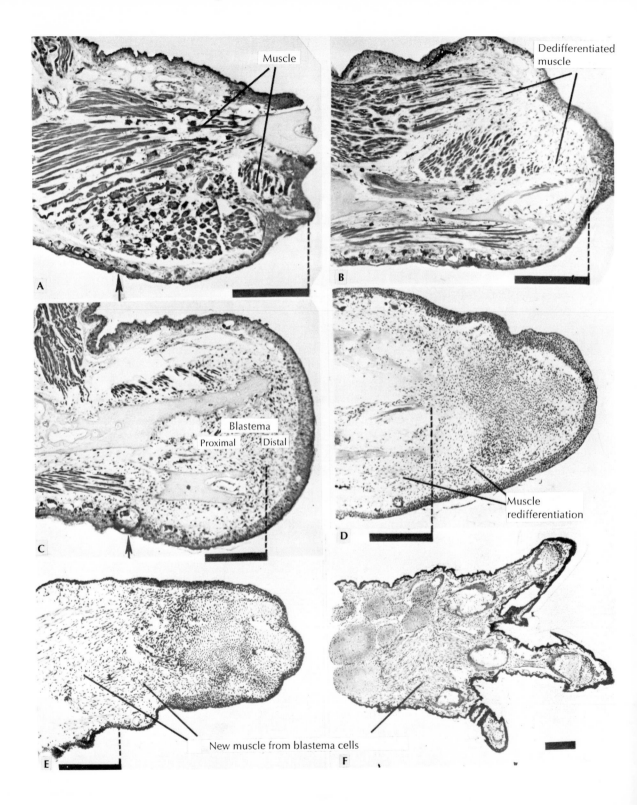

Muscle

Dedifferentiated muscle

Blastema

Proximal   Distal

Muscle redifferentiation

New muscle from blastema cells

A

B

C

D

E

F

Since the rabbit normally has a large number of cells at the time of blastocyst formation, it is possible even with only one eighth the normal number of cells for one or a few cells to be completely enclosed at the time of determination.

As mentioned in Chapter 2, it has been suggested that the signal for blastocyst formation may be the size of the cleavage cells in the morula, measured in terms of nuclear:cytoplasmic ratio. If such is the case, determination of inner cell mass versus trophoblast may simply reflect "inside" versus "outside" cells at the time that cytoplasmic volume is reduced to a critical size by cleavage.

Another developmental phenomenon pertinent to the subject of cell determination and differentiation in vivo is the tissue regeneration that occurs in mature animals after wounding. Although mammals show some capacity for tissue regeneration, the most dramatic examples of vertebrate tissue regeneration are observed in amphibians. For example, following limb amputation, the salamander is able to regenerate a complete, functional appendage (Fig. 12-21). The first event after amputation is closing over of the wound by epithelial cells that border it. Collagenase is released in the region underlying the wound epithelium and helps to remove differentiated limb cells from their normal matrix. These cells undergo a dedifferentiation in which they take on the appearance of embryonic mesenchyme cells. Multiplication of the mesenchymal cells produces a mound of cells called the blastema, which lies beneath the wound epithelium. The blastema enlarges by cell proliferation, then slows in growth as its component cells undergo redifferentiation and morphogenesis to form new limb structures. Severed nerves regrow into the wound area, and their presence is essential for regeneration to occur.

There has been controversy as to whether regeneration is accomplished by undifferentiated reserve cells present in the adult organism or by dedifferentiation and redifferentiation of cells. Recent evidence tends to favor the latter possibility. For example, in the case of

REGENERATION IN VERTEBRATES

Fig. 12-21. Light micrographs depicting regeneration of newt limb following amputation. The black bar indicates the same region of limb stump in **A** to **E. A,** Limb stump at time of amputation showing differentiated muscle. Arrow indicates level to which muscle dedifferentiation later occurs. **B** and **C,** Limb stump at 18 and 21 days after amputation showing progressive cell dedifferentiation. Proliferating cells have grown out from the level of amputation. **D** to **F,** Successive stages of cell redifferentiation in the limb. (From Hay, E. D. 1974. *In* J. Lash and J. R. Whittaker, eds. Concepts of development. Sinauer Associates, Inc., Sunderland, Mass.)

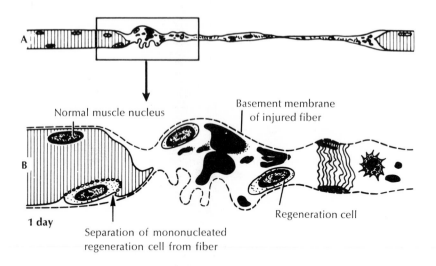

Normal muscle nucleus

Basement membrane
of injured fiber

B

Regeneration cell

1 day

Separation of mononucleated
regeneration cell from fiber

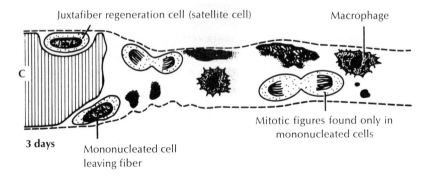

Juxtafiber regeneration cell (satellite cell)

Macrophage

C

Mitotic figures found only in
mononucleated cells

3 days

Mononucleated cell
leaving fiber

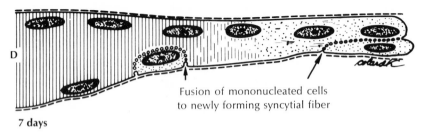

D

Fusion of mononucleated cells
to newly forming syncytial fiber

7 days

Fig. 12-22. Diagram of events in mammalian muscle regeneration. Mononucleated cells arise from muscle fibers and become separated from the fibers by small vesicles that fuse. The mononucleated cells proliferate and later fuse in the same manner as occurs in embryonic muscle development. **A,** Muscle fiber 1 day after injury. **B** to **D,** Enlargement of boxed area from **A** at 1, 3, and 7 days after injury. (From Hay, E. D. 1974. *In* J. Lash and J. R. Whittaker, eds. Concepts of development. Sinauer Associates, Inc., Sunderland, Mass.)

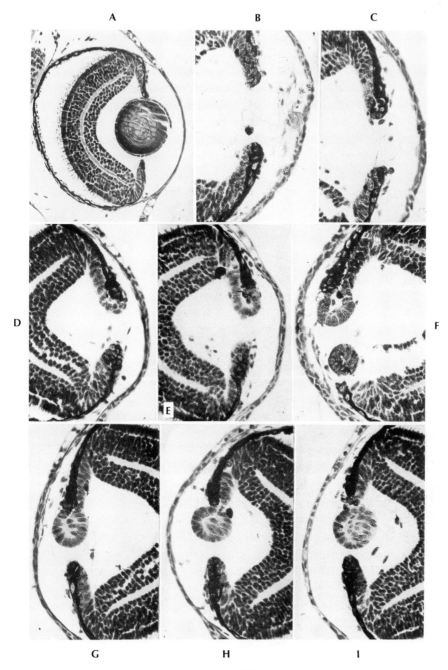

Fig. 12-23. Light micrographs showing stages of lens regeneration from the dorsal iris in larvae of the newt *Triturus viridescens*. **A,** Normal eye. **B** to **E,** Treated eye at 4, 5, 6, and 7 days after lens removal. **F** and **G,** Eye 9 days after lens removal. **H** and **I,** Eye 11 days after lens removal. (From Reyer, R. W. 1954. Q. Rev. Biol. **29:**1.)

regenerating mammalian muscle, one can observe mononucleated myoblasts arising from the muscle fiber in a manner that essentially reverses the fusion process by which the muscle fiber arises (Fig. 12-22). During amphibian limb regeneration, muscle fibers near the blastema lose some of their myofibrils. Cell membranes form in these areas and delimit mononucleated cells, which are able to make DNA and divide. These cells resemble the mesenchymal cells formed by cartilage dedifferentiation, and it is not known whether mesenchymal cells derived from muscle fibers can redifferentiate into cartilage cells or vice versa.

Only one clear case is known in which regeneration involves transformation of one cell type into a completely different cell type. This occurs in the regeneration of lens from the dorsal iris in the salamander. If the lens is removed from the eye of an adult salamander, a new one is produced by the adjacent dorsal iris. The normally pigmented cells of the dorsal iris begin to lose their pigment granules and start cell division. After the necessary amount of proliferation, some of the cells differentiate into lens epithelium and lens fibers while others regain their original pigmented phenotype (Fig. 12-23). Transformation of the iris into lens can also occur in vitro but only in the presence of neural retina, which apparently provides a necessary inductive stimulus.

PROGRESSIVE DETERMINATION DURING CELL DIFFERENTIATION

The change in cellular determination that occurs in amphibian lens regeneration appears to be an exceptional case in development. Most cells do not seem to be multipotential by the time they express their terminally differentiated phenotype. A progressive narrowing of developmental potential from the totipotent egg cell through a variety of multipotential cell types to the final unipotential terminally differentiated cell is the characteristic pathway of cell differentiation in the embryo. Determination of a cell to a particular type of differentiation may occur long before that differentiation is overtly expressed, but one can detect it by observing the fate of the cell after transplantation or isolation. The time in development at which cell determination first occurs varies considerably with different types of embryos, beginning at the earliest stages of development in highly mosaic embryos and as late as gastrulation in more regulative embryos. In the next chapter we will examine the most well-known mechanism whereby cell determination occurs during embryonic development, that is, inductive interactions between cells.

BIBLIOGRAPHY
Books and reviews

Austin, C. R., and R. V. Short, eds. 1972. Reproduction in mammals. Book 2. Embryonic and fetal development. Cambridge University Press, New York.

Balinsky, B. I., 1975. An introduction to embryology. 4th ed. W. B. Saunders Co, Philadelphia.

Gardner, R. L. 1971. Manipulations on the blastocyst. In G. Raspe, ed. Advances in the biosciences 6. Shering Symposium on Intrinsic and Extrinsic Factors in Early Mammalian Development. Pergamon Press, Inc., Elmsford, NY.

Hamburgh, M. 1971. Theories of differentiation. American Elsevier Publishing Co., Inc., New York.

Hay, E. M. 1974. Cellular basis of regeneration. In J. Lash and J. R. Whittaker, eds. Concepts of development. Sinauer Associates, Inc., Sunderland, Mass.

Herbert, M. C., and C. F. Graham, 1974. Cell determination and biochemical differentiation of the early mammalian embryo. Curr. Top. Dev. Biol. 8:151.

Kuhn, A. 1971. Lectures on developmental physiology. Springer-Verlag Inc., New York.

Mintz, B. 1965. Experimental genetic mosaicism in the mouse. In G. E. W. Wolstenholme and M. O'Conner, eds. Preimplantation stages of pregnancy. Little, Brown & Co., Boston, Mass.

Reyer, R. W. 1954. Regeneration of the lens in the amphibian eye. Quart. Rev. Biol. 29:1.

Saunders, J. W., Jr. 1970. Patterns and principles of animal development. Macmillan Publishing Co., Inc., New York.

Stearns, L. 1974. Sea urchin development: cellular and molecular aspects. Dowden, Hutchison & Ross, Inc., Stroudsburg, Pa.

Wilson, E. B. 1925. The cell in development and heredity. 3rd ed. Macmillan Publishing Co., Inc., New York.

Selected original research papers

Chan, L. N., and W. Gehring. 1971. Determination of blastoderm cells in Drosophila melanogaster. Proc. Natl. Acad. Sci. U.S.A. 68 : 2217.

Curtis, A. S. G. 1962. Morphogenetic interactions before gastrulation in the amphibian Xenopus laevis —the cortical field. J. Embryol. Exp. Morphol. 10:410.

Donohoo, P., and F. C. Kafatos. 1973. Differences in proteins synthesized by the progeny of the first two blastomeres of Ilyanassa, a "mosaic" embryo. Dev. Biol. 32:224.

Hillman, N., M. I. Sherman, and C. Graham. 1972. The effect of spatial arrangement on cell determination during mouse development. J. Embryol. Exp. Morphol. 28:263.

Illmensee, K. C. 1973. The potentialities of transplanted early gastrula nuclei of Drosophila melanogaster. Production of their imago descendents by germ-line transplantation. Wilhelm Roux Arch. 171:331.

Illmensee, K. 1976. Ontogeny of germ plasm during oogenesis in Drosophila. Dev. Biol. 49:40.

Newrock, K. M., and R. A. Raff. 1975. Polar lobe specific regulation of translation in embryos of Ilyanassa obsoleta. Dev. Biol. 42:242.

Van Blerkom, J., S. C. Barton, and M. H. Johnson. 1976. Molecular differentiation in the preimplantation mouse embryo. Nature 259:319.

Whittaker, J. R. 1973. Segregation during ascidian embryogenesis of egg cytoplasmic information for tissue-specific enzyme development. Proc. Natl. Acad. Sci. U.S.A. 70:2096.

CHAPTER 13

Embryonic induction

☐ In the previous chapter we saw that cell determination can begin very early in development and that it has begun in almost all organisms by the time of gastrulation. The movement of cells that occurs in gastrulation produces an arrangement of cells that is basically retained throughout the rest of embryogenesis. It also brings cells that were previously widely separated into close association. As a result of these new associations, cellular interactions known as embryonic inductions take place that have an important effect on cellular determination.

We will define embryonic induction as a process whereby two tissues interact in such a way that one or both become permanently and irreversibly changed. According to this definition, two components participate in the interaction; one tissue supplies or transmits a stimulus, and the other tissue responds to this stimulus. Embryonic inductions differ in a number of ways from other cellular interactions that meet these criteria (e.g., some hormonal interactions). First, the tissue that transmits the stimulus may be able to do so for only a short period of time. Likewise, the tissue responding to the stimulus may be capable of responding for only a restricted period of time (the period during which the cells can react is called the period of *competence* in the older literature). Moreover, embryonic inductions result in the gradual restriction of the developmental potential of the responding cell. In the terminology we have used previously, the interaction results in establishment of an epigenotype, or, in other words, the *determination* of the competent cell. Inductive interactions differ from tissue interactions found in mature animals in that the latter usually result in transitory changes which occur in response to immediate stimulation and are maintained on a long-term basis only by continuous stimulation. Inductive interactions, on the other hand, trigger new developmental programs that, once initiated, proceed to completion relatively autonomously.

Although many examples of embryonic induction are known, the precise cellular and molecular mechanisms involved in the interaction are not known even for a single example. This is a result of the complexity of the process and of our failure to ask the right experimental questions about it. One noteworthy gap in our knowledge regarding induction is a lack of understanding of the physical basis for the epigenotype. This lack of understanding of the basis of a cell's developmental potential makes it difficult to ask meaningful questions about the mechanisms whereby that potential is altered. A further problem in studying induction is that one encounters all the problems found in the study of cell differentiation, with the added complexity that an inductive interaction normally occurs long before its results can be recognized. The alteration of the epigenotype can be recognized by alterations that occur at a later time in the morphology of cells, in their adhesive properties, in their reproductive capacity, or in their cytodifferentiation. The two most general properties examined in studies of induction are tissue morphogenesis and cytodifferentiation.

The formation of the central nervous system (CNS) and the establishment of the anterior-posterior* axis has been one of the most intensely studied examples of embryonic induction. As was discussed in Chapter 12, if a portion of the prospective neural tube ectoderm of an amphibian embryo (as determined from the fate map) is transplanted to a region of prospective epidermis or even prospective endoderm shortly before gastrulation begins, the cells can regulate and normal development will proceed. However, if the same experiment is performed near the end of gastrulation, the transferred cells develop into a portion of a neural tube (Fig. 12-2). In the period between early and late gastrulation the ability of the presumptive neural cells to interact with their neighbors has been altered, and a neural epigenotype has been imprinted into the cells; that is, they have been induced to form neural tissue.

Over half a century ago Hans Spemann and Hilde Mangold demonstrated that the inducing tissue in neural tube induction in the amphibian embryo is the chordamesoderm. As mentioned in Chapter 12, if the dorsal lip of the blastopore (the location of the presumptive notochord) is transplanted to a different site on the surface of an early gastrula or is placed within the blastocoel, it interacts with host tissue and causes the development of a secondary neural tube and eventually a fairly complete secondary embryo (Fig. 13-1). This dorsal lip material was consequently called the "primary organizer" because it induces

CLASSICAL EXAMPLES OF EMBRYONIC INDUCTION
Primary induction

---

*For consistency with the research literature, the terms "anterior" and "posterior" are used in this chapter to refer to cranial and caudal locations.

Fig. 13-1. Secondary embryo induction following transplantation of dorsal lip of blastopore in the newt. **A** and **B,** Dorsal lip of the blastopore from an early gastrula is transplanted into the blastocoel of another early gastrula. **C,** Transplanted material induces formation of secondary structures in the host embryo. Cross section shows tissues derived from grafted material in black and host tissues in white. **D,** Secondary embryo formed as a result of transplantation. (Modified from Holtfreter, J., and V. Hamburger. 1955. *In* B. H. Willier, P. A. Weiss, and V. Hamburger, eds. Analysis of development. W. B. Saunders Co., Philadelphia.)

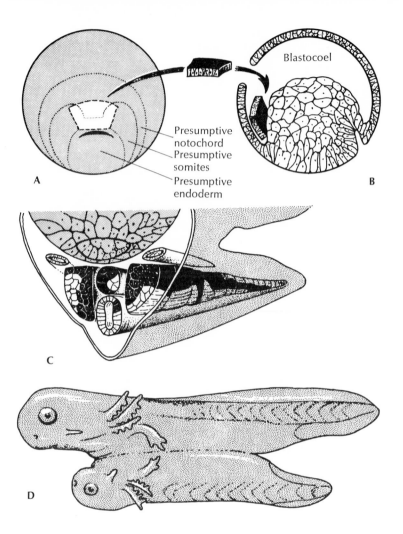

Blastocoel

Presumptive
notochord
Presumptive
somites
Presumptive
endoderm

A

B

C

D

formation of the basic axial organization of the embryo and because it is among the first in the chain of inducing events that results in the formation of the entire embryo.

The initial interaction of the chordamesoderm and ectoderm in induction of the neural tube is followed by other inductive interactions that progressively narrow the fate of specific neural tube areas. For example, specific interactions are involved in formation of the forebrain, midbrain, hindbrain, and spinal cord. Chordamesoderm from an anterior position tends to induce forebrain, whereas more posterior pieces of chordamesoderm induce more posterior neural tube structures. The different brain regions in turn may be involved in secondary or higher order inductions. These are inductions in which tissues formed as a result of an inductive interaction act as the inducing tissue for a subsequent induction. This is the case, for example, in the induction of the lens by the optic vesicles of the forebrain and the induction of the auditory vesicles by the hindbrain.

The basis for the differentiation of the embryonic CNS into the distinct regions along an anterior-posterior axis of the animal has been investigated in numerous different experiments. The results of many of these have been consistent with the existence of gradients of what have been termed "neuralizing" and "mesodermalizing" determinants within the embryo that influence the development of CNS structures. The neuralizing gradient is postulated to be highest in the mid-dorsal region and to decrease ventrally. The mesodermalizing gradient is postulated to be highest in the posterior region of the embryo and decrease anteriorly. It is thought that cells receive different concentrations of these factors and differentiate in a manner that is determined by the relative levels of the two factors.

Lens induction

The basic features of the inductive process are apparent in another classical example of embryonic induction – the formation of the lens of the eye under the influence of the optic vesicle. As shown in Fig. 13-2, the forebrain produces two lateral bulges, the optic vesicles, which come into contact with the overlying head ectoderm. This contact induces the head ectoderm to thicken into a lens placode, which then invaginates and detaches from the overlying ectoderm to form the lens vesicle and ultimately the lens. If the optic vesicle is placed beneath the ectoderm in a different part of the animal where the ectoderm would normally form skin or hair, this ectoderm is now induced to form a lens. Likewise, if the optic vesicle is removed or otherwise prevented from contacting the head ectoderm, generally no lens is formed. However, sometimes in amphibians a lens forms in the complete absence of the optic vesicles. Further investigation of this phenomenon has revealed that lens formation in the absence of the optic vesicles is dependent on the presence of accessory tissues in the normal vicinity of the eye. Therefore several tissues may be involved in this inductive interaction, and other stimuli may operate in conjunction with that from the optic cup.

Primary embryonic induction and lens induction are only two examples of an almost endless list of inductive interactions involved in organogenesis in vertebrate embryos. For example, the lens induces the ectoderm that grows over it to become cornea. Mosoderm and ectoderm interact in the formation of limbs (Chapter 21) and also skin derivatives (e.g., hair, feathers, teeth). Endoderm and mesoderm inter-

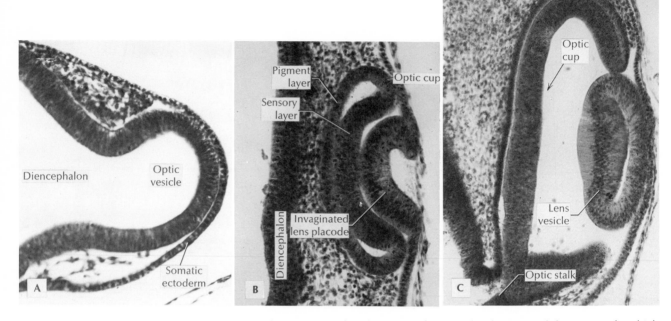

Fig. 13-2. Photomicrographs of sections showing development of the eye in the chick embryo. **A,** Optic vesicle in the 33-hour chick. **B,** Optic cup and lens placode in the 48-hour chick. **C,** Optic cup and lens vesicle in the 72-hour chick. The diencephalon is the portion of the forebrain adjacent to the developing eye. (Modified from Torrey, T. W. 1971. Morphogenesis of the vertebrates, 3rd ed. John Wiley & Sons, Inc., New York.)

act in the development of the lungs and various derivatives of the gut such as the liver and pancreas. Mesoderm-mesoderm interactions are found in the formation of some structures in the urinary and genital systems. We will not attempt to describe here the large variety of inductive systems that have been studied, but instead we will concentrate on a few representative systems that illustrate some of the interactions which occur. The student interested in pursuing particular systems can consult the bibliography at the end of the chapter.

INSTRUCTIVE VERSUS PERMISSIVE INTERACTIONS IN INDUCTION

A question of considerable importance that arises in analyzing the mechanisms of embryonic induction is the relative roles of the stimulating and responding tissues. Theoretically, at least two different types of tissue interaction are possible. The first is an instructive interaction in which tissue A specifies the direction in which tissue B develops; that is, a signal from tissue A specifies the epigenotype of tissue B. The second type of interaction is often called permissive interac-

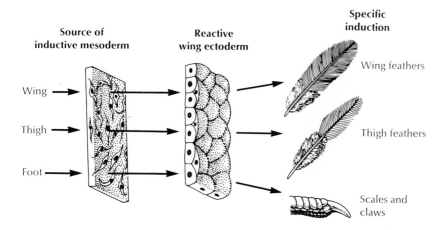

Source of inductive mesoderm

Reactive wing ectoderm

Specific induction

Wing →

Thigh →

Foot →

Wing feathers

Thigh feathers

Scales and claws

Fig. 13-3. Formation of ectodermal derivatives by wing ectoderm in combination with wing, thigh, or foot mesoderm. The mesoderm induces the ectoderm to form derivatives characteristic of the origin of the mesoderm. (Reprinted with permission of Macmillan Publishing Co., Inc. from Patterns and principles of animal development by Saunders, J. W., Jr. Copyright © 1970 by John W. Saunders, Jr.)

tion. This is analogous to the "permissive conditions" in cell culture but occurs in vivo. In this type of interaction, tissue A provides a suitable environment in which an intrinsic epigenotype in tissue B can be displayed. An instructive interaction implies a lability in the developmental potential of the responding tissue — its course of action is determined by the type of inducing stimulus it receives. A permissive interaction, on the other hand, implies that the responding tissue is already predetermined. Although expression of this predetermination requires some stimulus, no other type of expression is possible.

An examination of the types of embryonic inductions that have been studied provides evidence for both types of interactions during development. We shall first describe a few systems in which instructive interactions seem to occur. One is the induction of ectodermal structures of the wing, thigh, and foot of the chick embryo. Specific ectodermal derivatives are formed in each of these three limb regions. Wing and thigh ectoderm each produce a specific type of feather; foot ectoderm forms scales and claws. These ectodermal derivatives are produced in response to induction by the underlying mesoderm, and the source of the mesoderm determines the type of derivative that is formed. For example, if thigh mesoderm is combined with wing ectoderm, the wing ectoderm forms thigh-specific feathers. If mesoderm from the foot is combined with wing ectoderm, the wing ectoderm produces scales and claws (Fig. 13-3). The inducing tissue appears to be supplying specific information that instructs the responding tissue to develop in a particular direction.

The development of the lung also appears to involve an instructive interaction in which the mesenchyme exerts a specific influence on the direction of epithelial development. A number of vertebrate glands and organs (e.g., lung, pancreas, salivary gland, mammary gland, kidney) have a similar basic morphology and developmental sequence. Early in development an epithelial sheet of cells is surrounded by a mass of loose mesenchymal cells. The epithelial part of the structure typically develops into the functional units and the duct system of the organ, whereas the mesenchymal cells form the organ's connective tissue and vascular system. In recent years considerable attention has

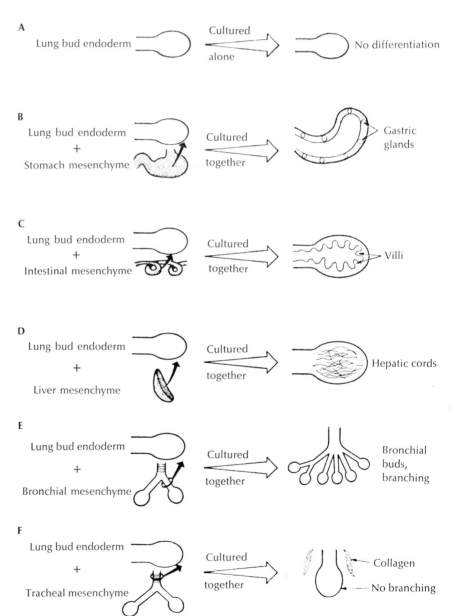

Fig. 13-4. Development of lung endoderm cultured alone or in combination with mesenchyme from different sources. **A,** Lung endoderm does not differentiate in isolation, but when combined with mesenchyme, **B** to **F,** it differentiates in accord with the source of the mesenchyme. (Modified from Deuchar, E. M. 1975. Cellular interactions in animal development, London, Chapman & Hall, Ltd. © 1975 Elizabeth M. Deuchar.)

been focused on the epithelial-mesenchymal interaction that takes place in the development of these organs.

Cultured lung endoderm differentiates only when combined with mesenchyme, but the type of structure it forms varies with the type of mesenchyme that is used (Fig. 13-4). For example, lung endoderm combined with stomach mesoderm forms gastric glands; combined with intestinal mesoderm, it forms intestinal villi; and combined with liver mesoderm, it forms hepatic cords. Furthermore, different regions of lung bud mesenchyme determine just which type of lung structure will develop from the lung endoderm. Lung endoderm combined with bronchial mesoderm forms branching bronchi, whereas lung endoderm combined with tracheal mesoderm exhibits no branching and causes the mesoderm to deposit collagen.

An examination of the development of the pancreas provides a transition from instructive interactions to permissive interactions. The pancreas develops as an evagination of the gut tube into the surrounding mesenchyme (Fig. 13-5). The endodermal layer forms ducts

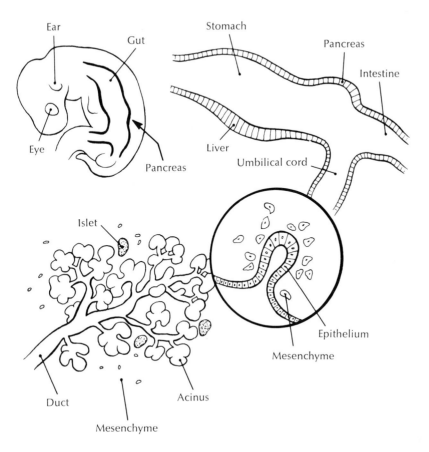

Fig. 13-5. Pancreas development in the mouse embryo. The pancreas develops at about midgestation as an evagination of the gut endoderm into the surrounding mesenchyme. The epithelium later develops into the ducts, acini, and islets of the pancreas. (From Wessells, N. K. 1973. Tissue interactions in development. Addison-Wesley module in biology no. 9. Addison-Wesley Publishing Co., Inc., Reading, Mass.)

with spherical clusters of secretory cells (acini) at their ends. Most of the cells in these clusters differentiate into exocrine cells that produce the digestive enzymes of the pancreas. Others bud off and form islets that contain different types of endocrine cells such as the β-cells, which produce insulin, and the α-cells, which produce glucagon.

Pancreas development can be followed biochemically because of the availability of sensitive assays for pancreas-specific enzymes and hormones. This allows the investigator to correlate cell differentiation and morphogenesis in pancreas development. Pancreas-specific proteins are not detectable prior to the bulging of the pancreatic primordium. But at the time that the bulge first becomes apparent, low levels of these proteins can be detected. The levels of these materials remain low during the next 5 days while the pancreas is undergoing considerable cell proliferation and morphogenesis. Then, as some acinar cells cease division and acquire the characteristics of differentiated pancreas cells (complex endoplasmic reticulum, Golgi apparatus, and zymogen granules), the levels of pancreatic proteins increase dramatically (Fig. 13-6).

Studies of the epithelial-mesenchymal interactions involved in these phases of pancreas development have indicated that the specificity of mesenchymal requirement varies at different stages. If gut endoderm is placed in culture about 18 hours before outgrowth of the pancreatic rudiment, it will not develop into pancreas in combination

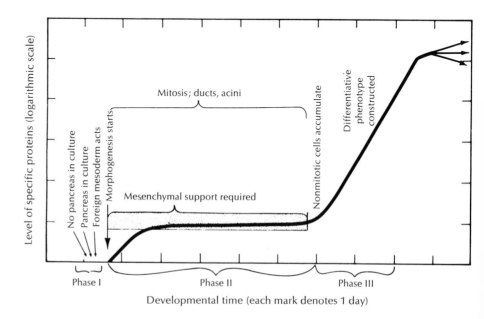

Fig. 13-6. Levels of pancreas-specific proteins at different stages in pancreas development. (From Wessells, N. K. 1973. Tissue interactions in development. Addison-Wesley module in biology no. 9. Addison-Wesley Publishing Co., Inc., Reading, Mass.)

with foreign mesenchyme but requires interaction with its homologous mesenchyme and adjacent tissues. (Gut tissue taken from embryos at an even earlier age does not develop into pancreas in culture at all, although it forms other gut derivatives such as liver and stomach.) However, when the endoderm of the midgut region from which the pancreas develops is dissected out at about 8 hours prior to the time at which the pancreatic rudiment first bulges and is then grown in culture in combination with mesenchymal tissue, the source of the mesenchyme is not critical. Pancreatic development from the endodermal epithelium still requires the presence of mesenchyme, but at this point the presumptive pancreatic endoderm requires only a nonspecific support from mesenchyme. Thus, although pancreas induction appears to involve a permissive interaction in that pancreatic epithelium develops only into pancreas, the interaction is specific at least early in pancreas development and occurs only in response to homologous mesenchyme.

Pancreas development does not require actual contact between mesenchyme and epithelium. Pancreas development occurs normally in culture when pancreatic epithelium and mesenchyme are separated by a thin filter that allows the passage of molecules but not cell processes. In fact, the requirement for mesenchyme is also met by a particulate cell-free extract of whole chick embryos or of mesodermal tissue or by a partially purified glycoprotein preparation extracted from mesenchyme. This material, which has been termed mesenchymal factor, need not enter the epithelial cells to exert its influence. The partially purified factor is completely active when it is bound to Sepharose beads that are too large to enter the cells and therefore must exert its influence by acting at the cell surface.

A second example of an inductive interaction that is permissive yet requires a specific inducing tissue occurs in the development of the salivary gland. Pioneering experiments in epithelial-mesenchymal interactions were done by Clifford Grobstein in the 1950s on the development of the salivary gland. When the epithelium and mesenchyme of the salivary primordium were separated and grown in culture, neither developed in a manner indicative of salivary gland development. However, if the two tissues were placed near each other, the mesenchyme cells surrounded the epithelium and the epithelium underwent branching. This interaction is specific; only homologous mesenchyme causes salivary epithelium to branch in a manner typical of the salivary gland. If mouse salivary epithelium is combined with salivary mesenchyme on one side and lung mesenchyme on the other, the epithelium in contact with the salivary mesenchyme undergoes branching; the region associated with the lung mesenchyme does not branch

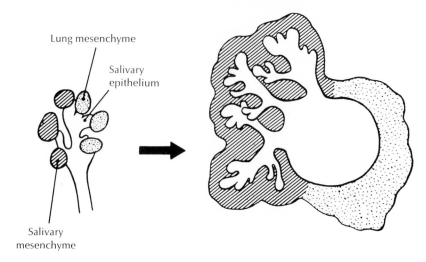

Fig. 13-7. Development of salivary epithelium cultured with salivary mesenchyme on one side and lung mesenchyme on the other. The epithelium in contact with the salivary mesenchyme undergoes branching characteristic of the salivary gland. The epithelium in contact with lung mesenchyme does not develop at all. (Modified from Wessells, N. K. 1973. Tissue interactions in development. Addison-Wesley module in biology no. 9. Addison-Wesley Publishing Co., Inc., Reading, Mass.)

at all (Fig. 13-7). The same result is obtained with mesenchyme derived from a variety of other sources. (Salivary gland morphogenesis is also discussed in Chapter 15.)

An example of a highly permissive inductive system is found in the formation of kidney tubules in the nephric mesoderm. This inductive system is somewhat different from the others discussed, since the inductive tissue and the responding tissue are both mesodermal. Normally kidney tubule formation is induced in the relatively unorganized mesoderm by the outgrowth of the ureter, which was previously organized from the mesoderm. The development of the nephric system in higher vertebrates occurs as a progression through three stages (see Chapter 18). The pronephros, a primitive excretory system, is the first to organize; the pronephric duct induces the intermediate form of kidney, the mesonephros. The mesonephric duct forms a bud, the future ureter, which then induces the mesoderm to form the metanephric tubules and kidney.

The ureteric bud and mesoderm are able to form metanephric tubules in organ culture. This interaction is not specific, however, since a large number of other tissues will induce the nephric mesoderm to form tubules. For example, spinal cord, brain, salivary mesenchyme, head mesenchyme, and a variety of other tissues interact with the

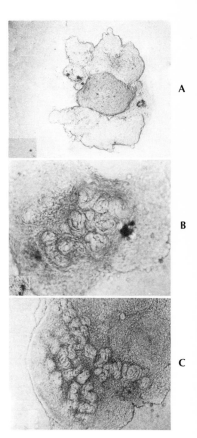

Fig. 13-8. Development of metanephrogenic mesenchyme cultured with different tissues. **A,** Metanephrogenic mesenchyme surrounding salivary mesenchyme at the beginning of the culture period. **B,** Metanephric tubule formation by metanephrogenic mesenchyme after 6 days of culture with salivary mesenchyme. **C,** Tubule formation by metanephrogenic mesenchyme after 3 days of culture with forebrain. (From Unsworth, B., and C. Grobstein. 1970. Dev. Biol. **21:**547.)

nephric mesoderm in the absence of the ureteric bud to permit the formation and organization of tubules in the nephric mesoderm (Fig. 13-8). It therefore appears as if the nephric mesoderm is already determined (i.e., its epigenotype has been established) but that this mesoderm must interact with other tissues to display that determination. In other words, the nephric mesoderm must participate in a permissive induction to differentiate.

One obvious problem that arises in considering permissive interactions in induction is determining the agent that initially establishes the epigenotype of the responding tissue. We discussed this earlier in the chapter when we noted that the problems encountered in the study of differentiation are also encountered in the study of induction. The epigenotype is established before cytodifferentiation is observed. Therefore the mechanisms that establish the epigenotype in systems which show permissive interactions must be sought at earlier developmental stages.

TWO-WAY INTERACTIONS

The transfer of an inductive stimulus, whether instructive or permissive, does not always occur in one direction. Numerous examples are known in which development requires the transmission of signals in both directions; that is, each of two interacting tissues is an inducing tissue and each is a responding tissue. One example of such a system is seen in the formation of digits by the chick limb. (Limb development is also considered in Chapters 18 and 21.) Two distinct mutations affecting digit formation are known in the chick. Each mutation results in the same phenotype—polydactyly, or an excess of digits. In animals expressing one of the mutations, which is called *eudiplopodia*, the defect has been shown to reside in the ectodermal component of the digit-forming region. In the other mutant, called *polydactylous*, the defect resides in the mesoderm. A combination of ectoderm from a *eudiplopodia* mutant with mesoderm from a normal embryo results in the development of a limb with supernumerary digits, whereas combination of mutant mesoderm and normal ectoderm does not. Mixing of mesoderm from a polydactylous mutant with ectoderm from a normal embryo causes supernumerary digit formation, but the reverse procedure does not. Finally, if mesoderm from a polydactylous mutant is combined with ectoderm from a eudiplopodia mutant, the effect on digit formation is compounded so that limbs develop with an even greater excess of digits than is seen with either mutant alone. Clearly, in this case, information regarding the production of digits resides in both the mesodermal and ectodermal components of the developing limb.

INDUCTIVE INTERACTIONS BETWEEN TISSUES OF DIFFERENT SPECIES

Inductive interactions can be obtained between tissues taken from different species and even different phyla. The fact that such interactions can occur indicates that the signals generated by the inductive tissues are conserved during evolution and that information on inductive mechanisms derived from the study of one species may be applicable to many other species. The precise response of the responding tissues in many cases is determined by the genetic constitution of that species. The types of information specified by the inducing tissue tend to be positional and tissue specific. For example, the oral epithelium in frog tadpoles forms suckers, whereas the same region in the

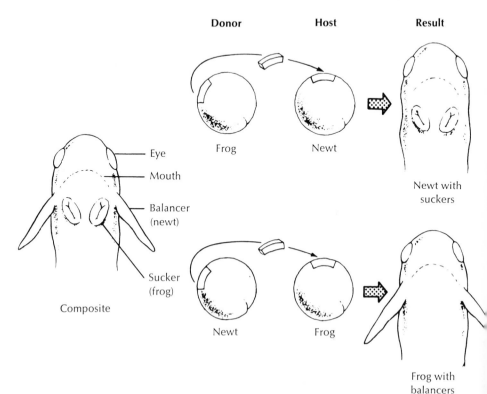

Fig. 13-9. Development of the oral epithelium in frog and newt embryos. Transplantation of frog oral epithelium to newt results in formation of suckers, which are characteristic of the frog mouth region. Transplantation of newt oral epithelium to frog results in formation of balancers, which are characteristic of the newt mouth region. (From Hamburgh, M. 1970. Theories of differentiation. American Elsevier Publishing Co., Inc., New York.)

newt embryo forms club-shaped balancers. When the ectoderm from an early frog embryo is transplanted to the future oral region of a newt embryo, this ectoderm is induced to form frog suckers. When the reverse experiment is done and newt ectoderm is transplanted to the future oral region of a frog embryo, balancers are formed (Fig. 13-9).

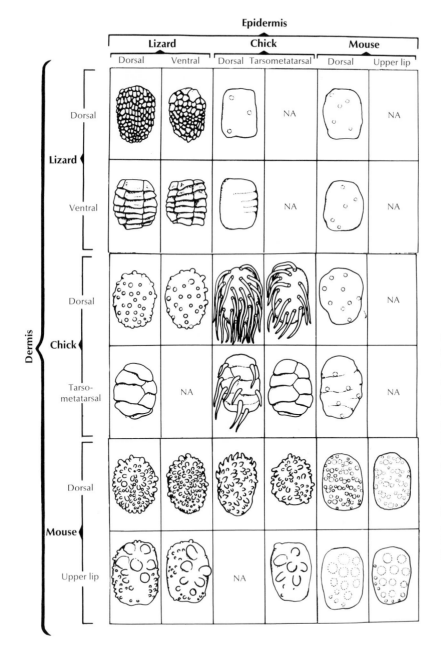

Fig. 13-10. Formation of epidermal derivatives (scales, feathers, and hair follicles) after various combinations of dermis and epidermis from lizard, chick, and mouse are cultured together. The size, number, and pattern of the derivatives is determined by the dermis in most cases. *NA* = Not available. (Modified from Sengel, P. 1975. *In* Cell Patterning, CIBA Foundation Symposium 29. Associated Scientific Publishers, Amsterdam. Additional combinations have been added courtesy Professor P. Sengel.)

Thus, although development of oral structures by the ectoderm requires an inductive stimulus from the oral mesoderm, the precise response to this stimulus is determined by the responding tissue itself. In other words, in this example the host oral mesoderm transmits the information "oral" and also transmits the information for "epithelial structure" to the foreign ectoderm. However, the foreign tissue responds by using its genetic program and forms the oral structures appropriate for its species.

Another slightly more complicated system in which positional information is translated appropriately by responding tissue of a different species is the interaction of dermis and epidermis in the formation of epidermal derivatives such as feathers, hairs, and scales. Primordia of these epidermal derivatives are produced when the dermis and epidermis from two different species of vertebrates are cultured together. The type of structure that is produced under these conditions is determined by the origin of the epidermis. However, the topographical distribution of the epidermal derivatives is under dermal control (Fig. 13-10). For example, when dermis from mouse upper lip (which normally contains large and small hairs) is combined with lizard epidermis, the epidermis produces large and small scale placodes. Mouse upper lip dermis combined with chick epidermis results in the production of large and small feather follicles. Similarly, if dermis from mouse dorsal skin (which contains only small hairs) is combined with lizard or chick epidermis, the result is production of a dense pattern of small scale placodes or feather follicles. (The epidermal derivatives do not develop fully in these experiments. Their complete development seems to require an additional stimulus from dermis that is not supplied in the heterospecific combinations. Formation of mature epidermal structures occurs only if a small amount of dermis from the species providing the epidermis is present, although the dermis need not come from an area that produces epidermal derivatives.)

MORPHOGENESIS AND CYTODIFFERENTIATION ARE SEPARABLE

In some permissive interactions the responding tissue may show normal cytodifferentiation without assuming the normal tissue morphology; morphogenesis and cytodifferentiation are separable processes, at least in a few systems. For example, the mammary gland epithelium requires mesenchymal tissue to develop. Mammary gland epithelium combined with salivary gland mesenchyme forms a duct system similar to the duct system of the salivary gland (Fig. 13-11). However, the epithelial cells respond to lactogenic hormones and produce at least some proteins found in milk. The mammary gland epithelium epigenotype is not altered by salivary gland mesenchyme, although the morphology of the epithelium is altered.

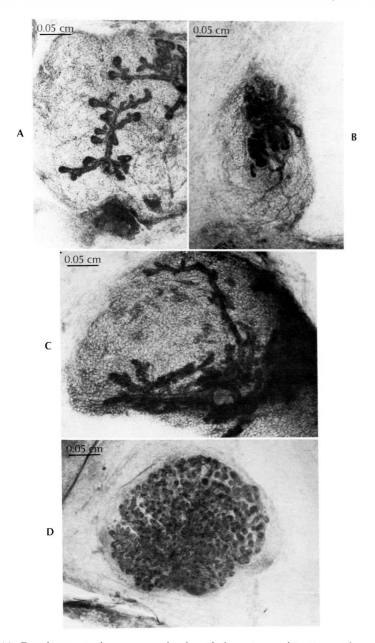

Fig. 13-11. Development of mammary gland epithelium in combination with mammary gland mesenchyme and salivary gland mesenchyme. Epithelium and mesenchyme were combined in vitro and later transplanted beneath the kidney capsule. **A,** and **B,** Sixteen-day mammary epithelium combined with mammary, **A,** and salivary, **B,** mesenchyme 7 days after transplantation. **C** and **D,** Sixteen-day mammary epithelium combined with mammary, **C,** and salivary, **D,** mesenchyme 14 days after transplantation. **A** and **C** show typical mammary branching pattern, whereas **B** and **D** show dense branching characteristic of the salivary gland. (From Sakakura, T., Y. Nishizuka, and C. Dawe. 1976. Science **194:** 1439. Copyright 1976 by the American Association for the Advancement of Science.)

Similarly pancreatic epithelial cells can display cytodifferentiation in the absence of normal morphogenesis. When dissociated pancreatic cells that are determined but undifferentiated are grown in cell culture, they undergo little cellular proliferation and form a two-dimensional monolayer. The cells, however, produce pancreatic enzymes after several days in cell culture. Since the cells remain as a flat sheet and do not develop into the typical acinar structures, the pancreatic epithelium can undergo cytodifferentiation in the absence of normal morphogenesis. The relative proportion of endocrine and exocrine cells that is seen during culture of dissociated pancreatic cells is dependent on the presence of mesenchymal factor. In the absence of mesenchymal factor, the proportion of endocrine to exocrine cells is much greater than normal. Addition of mesenchymal factor to the culture medium results in a more normal ratio of endocrine to exocrine cells.

TYPES OF INTERACTIONS INVOLVED IN INDUCTION

As discussed earlier, inductive interactions can be classified generally as instructive or permissive. The mechanisms involved in the two types of interactions need not be similar. Progress toward defining the molecular mechanisms involved in instructive interactions will probably be limited until the physical basis for the epigenotype is understood. Progress toward defining the molecular mechanisms of permissive interactions may be more rapid, since, as discussed below, some of the molecules involved are apparently being identified.

As several workers have noted, there appear to be three types of possible cellular interactions (Fig. 13-12). In the first type of interaction the inducing cells produce a product that acts at a distance; the inducing agent diffuses from its site of origin to the target tissue. The interaction of the diffusible product and the competent tissue constitutes the inductive event. This type of interaction is similar in some respects to hormonal action. In the second type of interaction the inducing cells produce an extracellular material with limited ability to diffuse. The interaction of the competent cells with this matrix constitutes the inductive event. The third type of interaction is a direct cell-cell interaction mediated by surface contact and/or the formation of junctional complexes. There is evidence that all three types of interactions occur during inductive events.

Diffusible agents

One inductive system that appears to involve diffusible factors is primary induction, which was discussed earlier. In the half century since Spemann and Mangold described primary induction, intensive research has been undertaken to find the inducer of neural tube formation. However, this research has shown that environmental factors (e.g., pH, monovalent cations), defective or heterologous tissue (e.g., dead tissue, guinea pig liver, hydra), and completely foreign chemicals (e.g., methylene blue) can induce neural tube formation. The list of such inducing agents explains why some workers consider primary induction as a "can of worms." In reality the list exemplifies both the complexity of induction and the fact that, at the present time, only the competent cells know the "correct" mechanism of induction.

However, work has continued on the isolation and purification of cellular agents that are capable of reproducing events typical of primary induction. Although the agents studied have been from heterologous sources, it is thought that by determining their mechanism of action, it will be possible to elucidate the molecular responses of competent cells that occur in normal induction. In recent years purified protein fractions from guinea pig bone marrow and guinea pig liver have been isolated that induce mesodermal and forebrain structures, respectively. These fractions are highly active inducers. When combined in the same experiment (e.g., when early gastrula ectoderm is treated with both inducers simultaneously), intermediate (e.g., hindbrain) structures are formed. The structure along the anterior-posterior axis that is formed is dependent on the ratio of neuralizing to mesodermalizing factor—the higher the ratio, the more anterior the structure. This has reinforced the gradient hypothesis for establishing the anterior-posterior embryonic axis mentioned earlier in our discussion of primary induction.

Primary neural induction appears to occur at a distance in the embryo. Electron micrographs of gastrulae indicate that the chordamesoderm and ectoderm are never contiguous but are always separated by at least 500 Å. Nevertheless, proteins and other cellular agents emanating from the chordamesoderm can be detected in the ectoderm. Although there is no direct evidence that the agents that have been shown to diffuse to the ectoderm are inducers, these observations are consistent with a diffusible inducer.

Another example of a diffusible factor involved in inductive interactions is the mesenchymal factor required for development of the pancreas. Earlier work on the epithelial-mesenchymal interactions in this system made extensive use of the transfilter technique introduced by Grobstein. The epithelium of the pancreas was found to react in vitro to mesenchyme that was separated from the epithelium by a Millipore filter. These filters (which consist of a meshwork of nitrocellulose fibers) can be constructed in such a manner that particles of a given size are excluded. For example, 0.45 μm filters exclude particles larger than 0.45 μm, and 0.2 μm filters exclude particles larger than 0.2 μm. The filters used in these experiments appeared to exclude cellular pro-

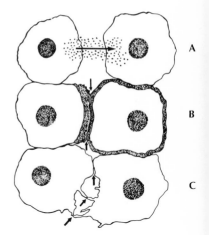

Fig. 13-12. Diagrammatic representation of three possible mechanisms for inductive interactions between cells. **A,** Long-range diffusion. **B,** Matrix interaction. **C,** Contact interaction. (From Saxen, L., and J. Kohonen. 1969. Int. Rev. Exp. Pathol. **8:**57.)

cesses so that the two reacting tissues were never in direct contact. Nonetheless, induction of the pancreatic epithelium occurred, and therefore the inductive interaction had to occur by means of an extracellular material. As discussed earlier, a distinct glycoprotein has since been isolated that is capable of substituting for the mesenchymal tissue. This factor, which acts as a mitogen stimulating DNA synthesis in the pancreatic epithelial cells, can exert its effects when bound in such a manner that it cannot enter the cell.

Extracellular matrix   The fact that the mesenchymal factor can exert its effects on pancreatic epithelium from the outside of the cell demonstrates that cellular surface interactions are important in inductive events. Cells synthesize and secrete a number of different types of molecules. Recent experiments indicate that three classes of these compounds might be important in inductive interactions: (1) collagen, (2) glycosaminoglycans (GAG) such as chondroitin sulfate, heparan sulfate, and keratan sulfate, and (3) proteoglycans (proteins to which a large number of GAG molecules are covalently attached). Theoretically these compounds might constitute an "alphabet" similar to the alphabet of bases in DNA. The particular environment of extracellular matrix might be unique for each tissue and cell type and theoretically might account for the specificity of some inductive interactions. Many of the permissive inductive interactions might use such an extracellular matrix. The interaction of the responding cells with this matrix might be analogous to the interaction of myoblasts with collagen. As discussed in Chapter 10, myoblasts in cell culture require a complex medium to fuse into myotubes. One condition that facilitates their fusion is a collagen-coated substrate which can replace "conditioned" medium.

Results from work done in several systems of inductive interactions in recent years have substantiated the view that the extracellular matrix is involved in induction. For example, it has long been known that the notochord interacts with somitic mesoderm and induces the mesoderm to form cartilage. This is the basic interaction that leads to the formation of the axial skeleton. It has been found, however, that collagen and proteoglycans can replace the notochord in this inductive interaction—collagen and proteoglycans act more or less as a primer for the chondrogenic differentiation of the somitic mesoderm. In agreement with this concept is the observation that notochord from which the extracellular matrix has been removed is a poorer inducer of chondrogenesis of the mesoderm.

In the next few years the role of the extracellular matrix in the development of a number of structures (e.g., in somitic chondrogenesis, development of the cornea of the eye, development of the limb) is expected to become clearer and may allow a better description of the mechanisms involved in their induction. Interested students are advised to watch the literature.

Direct cellular interactions can take two forms – simple direct contact and contact with the formation of specialized junctions. In simple contact, alterations in a restricted area of the plasma membrane may be propagated throughout the surface of the cell and subsequently affect internal metabolic properties of the cell. Contact with the formation of specialized junctions allows more direct intercellular communication. For example, cells joined by gap junctions can exchange low molecular weight material and are electrically coupled.

**Direct cellular
interactions**

Because of the possibilities for direct exchange of information, contact between cells was originally considered the most likely explanation for induction. Until recently, however, no example of an inductive interaction involving direct cellular contact was known. The transfilter technique discussed earlier seemed to prove that induction in many systems could occur without direct cellular contact. Recently new types of filters have been used in place of Millipore filters. The new membrane Nuclepore filters differ from Millipore filters in that a rather uniform membrane possesses straight channels of various sizes rather than a meshwork of fibers. With these filters it has been found that at least one inductive interaction earlier thought to occur without cell contact does require direct cellular contact. As discussed earlier, kidney mesenchyme can be induced to form tubules when supplied with many different types of tissues. No cellular processes were seen to penetrate Millipore filters when kidney mesenchyme was induced with the transfilter system. When similar experiments were performed with Nuclepore filters, however, membrane processes were observed to penetrate the filters (Fig. 13-13). As the size of the pores was reduced, it was found that the minimum pore size that permitted tubule induction was the same as the minimum pore size that allowed the cytoplasmic processes to penetrate the filter. The ability to observe the penetrating cellular extensions resulted from the use of the straight-pore membrane filters and improved cellular fixation techniques. It thus appears that direct cellular contact is essential for the formation of the tubules, although no specialized cell junctions have been observed. Further details of the mechanisms involved in this inductive interaction are not known.

In this chapter we have described the interaction of different cell types that produce stable alterations in the developmental capacities of at least one of those cell types. The interaction that produces these

OVERVIEW

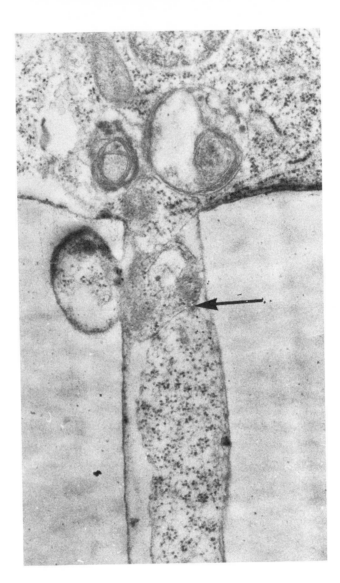

Fig. 13-13. Electron micrograph showing cell contact in pore of a Nuclepore filter separating metanephric mesenchyme (above) from inducing spinal cord (below). (From Saxen, L. 1975. Clin. Obstet. Gynecol. **18:**149.)

alterations can transmit precise instructions to the responding cell and somehow limit its developmental capacity, or it can permit the responding cell to express its own predetermined potential. Although the precise molecular events whereby these interactions occur are not known, there is evidence that induction occurs through diffusible factors, through the production of extracellular material, and through direct cellular contact.

In the next chapter we shall consider cellular interactions that tend to be more transient and that involve the transmission of chemical messages—hormonal interactions.

BIBLIOGRAPHY

Books and reviews

Balinsky, B. I. 1975. An introduction to embryology, 4th ed. W. B. Saunders Co., Philadelphia.

Deuchar, E. M. 1975. Cellular interactions in animal development. Halsted Press, New York.

Fleischmajer, R., and R. E. Billingham, eds. 1968. Epithelial-mesenchymal interactions. The Williams & Wilkins Co., Baltimore.

Hamburgh, M. 1971. Theories of differentiation. American Elsevier Publishing Co., Inc., New York.

Lash, J. W., and N. S. Vasan. 1977. Tissue interactions and extracellular matrix components. *In* J. W. Lash and M. M. Burger, eds. Cell and tissue interactions. Raven Press, New York.

Lehtonen, E. 1976. Transmission of signals in embryonic induction. Med. Biol. **54**:108.

Newlon, C., G. Gussin, and B. Lewin. 1975. Molecular information in developmental genetics. Cell **5**:213.

Saunders, J. W., Jr. 1970. Patterns and principles of animal development. Macmillan Publishing Co., Inc., New York.

Saxèn, L. 1975. Embryonic induction. Clin. Obstet. Gynecol. **18**:149.

Saxèn, L., and S. Toivönen. 1962. Primary embryonic induction. Academic Press, Inc., New York.

Sengel, P. 1975. Feather pattern development. *In* Cell patterning. Ciba Foundation Symposium 29 (new series). Associated Scientific Publishers, Amsterdam.

Spemann, H. 1938. Embryonic development and induction. Yale University Press, New Haven, Conn.

Tiedemann, H. 1975. Substances with morphogenetic activity in differentiation of vertebrates. *In* R. Weber, ed. The biochemistry of animal development. Vol. III. Academic Press, Inc., New York.

Wessells, N. K. 1973. Tissue interactions in development. Addison-Wesley module in biology no. 9. Addison-Wesley Publishing Co., Inc., Reading, Mass.

Wessells, N. K., and W. J. Rutter. 1969. Phases in cell differentiation. Sci. Am. **220**:36 (March).

Selected original research articles

Grobstein, C. 1953. Epitheliomesenchymal specificity in the morphogenesis of mouse submandibular rudiments *in vitro*. J. Exp. Zool. **124**:383.

Ronzio, R. A., and W. J. Rutter, 1973. Effects of a partially purified factor from chick embryos on macromoelcular synthesis of embryonic pancreatic epithelia. Dev. Biol. **30**:307.

Rutter, W. J., N. K. Wessells, and C. Grobstein. 1964. Control of specific synthesis in the developing pancreas. Natl. Cancer Inst. Monograph **13**:51.

Sakakura, T., Y. Nishizuka, and C. J. Dawe. 1976. Mesenchyme-dependent morphogenesis and epithelium-specific cytodifferentiation in mouse mammary gland. Science **194**:1439.

Spemann, H., and H. Mangold. 1924. Induction of embryonic primordia by implantation of organizers from a different species. *In* B. H. Willier and J. M. Oppenheimer, eds. 1964. Foundations of experimental embryology. Prentice-Hall, Inc., Englewood Cliffs, N.J.

Spooner, B. S., H. I. Cohen, and J. Faubion. 1977. Development of the embryonic mammalian pancreas: the relationship between morphogenesis and cytodifferentiation. Dev. Biol. **61**:119.

Spooner, B. S., and N. K. Wessells. 1970. Mammalian lung development: interactions in primordium formation and bronchial morphogenesis. J. Exp. Zool. **175**:445.

CHAPTER 14

Hormonal mechanisms in cellular differentiation

☐ Cellular interactions that influence differentiation of cells during the early phases of embryonic development tend to be short-range interactions in which one type of cell acts on an immediately adjacent cell of another type. Later in development, however, cellular interactions over greater distances, mediated by diffusible chemical substances known as hormones, become increasingly important. A hormone can be defined as a chemical substance that is carried by the circulation or some other body fluid and that exerts an influence on target cells or organs at a point remote from the cells that synthesize and release it. This definition is rather broad and includes a number of substances not classically thought of as hormones. In particular, there exist a number of "tissue-specific growth factors" such as nerve growth factor and erythropoietin that stimulate the growth and function of remote target tissues but are not usually called hormones (although some investigators are beginning to include them in the hormone category). Since these factors will be dealt with separately in Chapter 17 in the discussion of growth control, the current chapter is restricted primarily to those molecules which are widely accepted as hormones. However, we wish to point out that the distinction between hormones in the classical sense and tissue-specific growth factors is probably artificial.

It is useful to think of hormones as chemical messages that carry information from one type of cell to another in the body. The concept of a hormone as a message implies that there is a preestablished code that is used by the receiving cell to interpret the message. Just as with human gestures and words, chemical messages do not have an intrinsic meaning, and the message can be (and often is) interpreted in different ways by different groups of cells. The nature of the response to the hormone, both in development and in other physiological processes, is ultimately determined by the preprogrammed reaction of the target cell to the presence of the hormone (and also sometimes to its amount).

Thus one type of cell might respond to a particular hormone by proliferation, whereas another would respond by secretion, and a third by contraction. Other cell types might not respond at all. In some cases this would be because of their lack of the appropriate receptor molecules for the hormone in question, but there are also many cases known in which apparently normal hormone receptors are present in cells that do not respond in any obvious manner to the hormones that interact with those receptors. Although hormones frequently alter the functional differentiated state of a cell, the ability to respond to a hormone must, in itself, be viewed as a form of differentiation.

Most studies of hormonal effects on development are focused either on relatively late stages of embryogenesis when the organism is already in a fairly advanced stage of development or on developmental processes that occur in juvenile or adult stages of the life cycle. In some cases it becomes difficult to distinguish between normal physiological function and true developmental processes, but most of the processes discussed here fall reasonably well within the broad definition of development that we adopted in Chapter 1.

In most cases cellular responses to hormones involve the expression of specific differentiated properties rather than the determination of cell type. There are some hormones that have been shown to be present in early stages of embryogenesis of some types of plants and animals, but in most cases they have not been shown to be causative factors in the determination of cell type. One organism in which cell determination may be hormonally influenced is the cellular slime mold *Dictyostelium discoideum,* whose development is discussed in detail in Chapter 23.

The hormonal interactions that occur in vertebrate development need to be viewed in terms of the overall organization of the vertebrate endocrine system. That system is highly complex and cannot be described in full detail here. However, we will summarize it briefly, including some of the major feedback circuits that serve to integrate its function.

Two major classes of hormones and their functions are summarized in Table 14-1 — those which originate in the pituitary gland and those which originate from peripheral endocrine glands. We will be concerned with selected examples from these two classes of hormones and also with a third class, the regulatory hormones produced by the hypothalamus (or adjacent areas of the brain) that control the release and/or synthesis of pituitary hormones. Table 14-2 describes the

VERTEBRATE ENDOCRINE SYSTEM

Table 14-1. Major hormones of the pituitary and the peripheral endocrine glands*

Hormones	Cellular source	Principal actions
Hormones of the pituitary gland		
Anterior pituitary (pars distalis, adenohypophysis)		
Somatotropin (STH, growth hormone)	Somatotrophs	Growth of bone and muscle; promotes protein synthesis; affects lipid and carbohydrate metabolism
Adrenocorticotropin (ACTH)	Corticotrophs	Stimulates secretion of adrenal cortical steroids; certain extra-adrenal actions
Thyrotropin (TSH)	Thyrotrophs	Stimulates thyroid gland to form and release thyroid hormones
Gonadotropins		
Luteinizing or interstitial cell–stimulating hormone (LH or ICSH)	Gonadotrophs (luteotrophs or interstitiotrophs)	*Ovary:* formation of corpora lutea; secretion of progesterone; probably acts in conjunction with FSH; *Testis:* stimulates the interstitial cells of Leydig, thus promoting the secretion of androgen
Follicle-stimulating hormone (FSH)	Gonadotrophs (folliculotrophs)	*Ovary:* growth of follicles; functions with LH to cause estrogen secretion and ovulation; *Testis;* possible action on seminiferous tubules to promote spermatogenesis
Prolactin (PRL, lactogenic hormone, luteotropin)	Lactotrophs (mammotrophs)	Initiation of milk secretion; acts on crop sacs of some birds; maternal behavior in birds; resembles STH in affecting many tissues
Pars intermedia		
Melanophore-stimulating hormone (intermedin, MSH)	Melanotrophs	Expansion of amphibian melanophores; contraction of iridophores and xanthophores; melanin synthesis; darkening of skin
Posterior pituitary (neurohypophysis)		
Vasopressin (ADH, antidiuretic hormone)	Hypothalamic neurons	Elevates blood pressure through action on arterioles; promotes reabsorption of water by kidney tubules
Oxytocin	Hypothalamic neurons	Affects postpartum mammary gland, causing ejection of milk; promotes contraction of uterus; possible action in parturition and in sperm transport in female tract
Hormones of the peripheral endocrine glands		
Thyroxine (T_4); triiodothyronine (T_3)	Thyroid	Growth; amphibian metamorphosis; molting; metabolic rate in birds and mammals
Thyrocalcitonin (calcitonin)	Thyroid; ultimobranchial bodies; parathyroids(?)	Lowers blood calcium and phosphate
Parathyroid hormone (PTH)	Parathyroids	Elevates blood calcium; lowers blood phosphate

*Modified slightly from Turner, C. D., and J. T. Bagnara, 1976. General endocrinology, 6th ed. W. B. Saunders Co., Philadelphia.

Table 14-1. Major hormones of the pituitary and the peripheral endocrine glands — cont'd

Hormones	Cellular source	Principal actions
Epinephrine	Adrenal medulla; other chromaffin cells	Mobilization of glycogen; increased blood flow through skeletal muscle; increased oxygen consumption; heart rate
Norepinephrine	Adrenal medulla; other chromaffin cells; adrenergic neurons	Adrenergic neurotransmitter; elevation of blood pressure; constricts arterioles and venules
Insulin	Pancreatic islets (β-cells)	Lowers blood glucose; increases utilization of glucose and synthesis of protein and fat; decreases gluconeogenesis
Glucagon	Pancreatic islets (α-cells)	Increases blood glucose; stimulates catabolism of protein and fat
Androgens (e.g., testosterone)	Testis; adrenal cortex; ovary	Male sexual characteristics
Estrogens (e.g., estradiol)	Ovary; placenta; testis; adrenal cortex	Female sexual characteristics
Progestogens (e.g., progesterone)	Ovary; placenta; adrenal cortex	Maintenance of pregnancy; inhibition of reproductive cycles
Relaxin	Ovary; uterus; placenta	Enlargement of birth canal by relaxation of uterine cervix and pelvic ligaments
Glucocorticoids (e.g., cortisol, corticosterone)	Adrenal cortex	Promote synthesis of carbohydrate; protein breakdown; anti-inflammatory and antiallergic actions
Mineralocorticoids (e.g., aldosterone)	Adrenal cortex	Sodium retention and potassium loss through kidneys
Melatonin	Pineal gland	May mediate systemic responses to environmental lighting; effects on pigment cells
Human chorionic gonadotropin (HCG)	Placenta	Various effects on gonads; some actions similar to pituitary LH
Pregnant mare serum gonadotropin (PMSG)	Placenta	Ovulation in immature rats; various effects on gonads
Placental lactogen	Placenta	Simulates activities of pituitary growth hormone and prolactin
Gastrin	Stomach	Secretion of gastric juice
Secretin	Small intestine	Secretion of fluid by acinar pancreas
Cholecystokinin	Small intestine	Contraction of gallbladder; enzyme secretion by pancreas
1,25-$(OH)_2$-cholecalciferol	Kidney	Elevates serum calcium; promotes intestinal absorption of calcium and mobilizes bone calcium

Table 14-2. Structures of some mammalian hormones*

Hormone	Structure
Hypothalamus	
Thyrotropin releasing hormone	L-(pyro) glu-his-pro-NH$_2$
Luteinizing hormone–releasing hormone (LH-RH; gonadotropin-releasing hormone)	L-(pyro) glu-his-trp-ser-tyr-gly-leu-arg-pro-gly-NH$_2$
Growth hormone releasing hormone	$\leq$ decapeptide
Posterior pituitary	
Oxytocin, vasopressin	Peptides, 9 amino acids, cyclic
Anterior pituitary	
Thyrotropin, FSH, LH	Glycoproteins, MW 30,000
ACTH	Peptide, 39 amino acids
Growth hormone	Protein, 188 amino acids (human)
Prolactin	Protein, MW 25,000 (sheep, ox)
Parathyroid	
Parathyroid hormone	Protein, 84 amino acids (bovine)
Thyroid	
Calcitonin	Peptide, 32 amino acids
Thymus	
Thymosin	Protein, 105 amino acids
Pancreas	
Glucagon	Peptide, 29 amino acids
Insulin	Protein, 51 amino acids

ADRENAL CORTEX

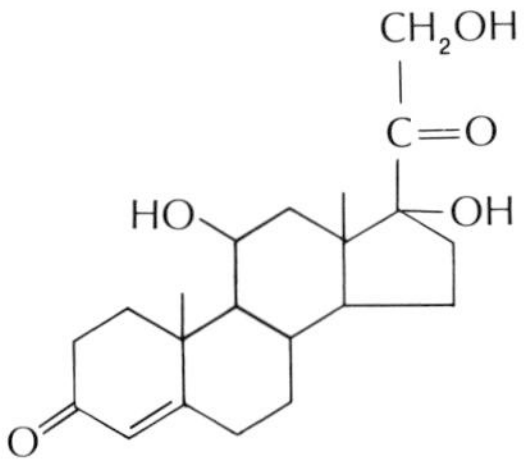

Hydrocortisone

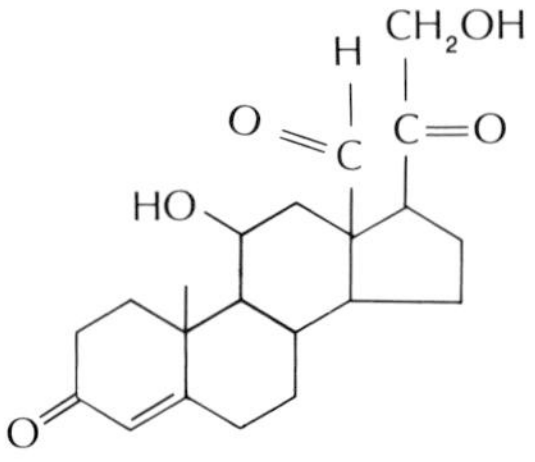

Aldosterone

TESTIS

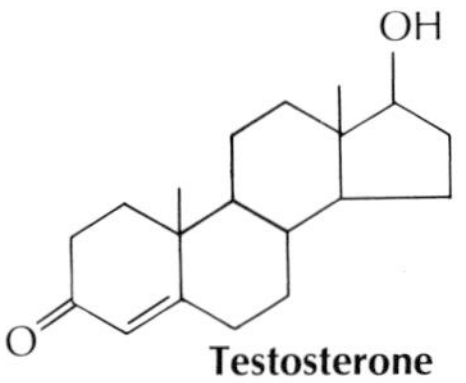

Testosterone

OVARY

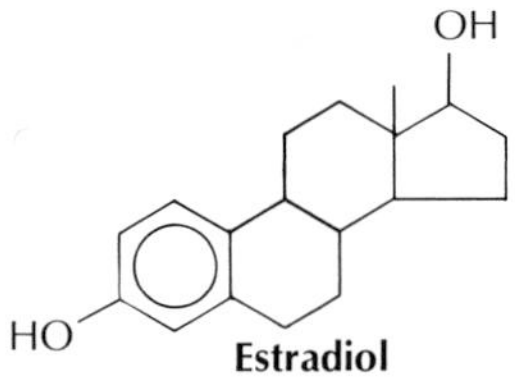

Estradiol

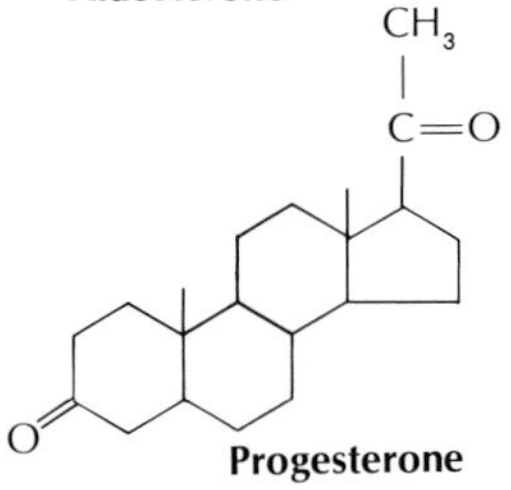

Progesterone

THYROID

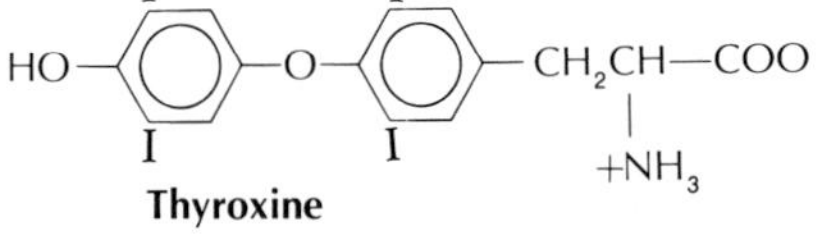

Thyroxine

ADRENAL MEDULLA

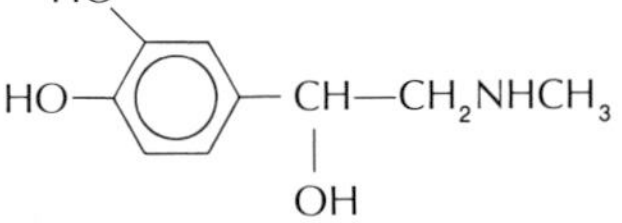

Epinephrine

*Modified slightly from Filburn, C. R., and G. R. Wyatt. 1974. *In* J. Lash and J. R. Whittaker, eds. Concepts of development. Sinauer Associates, Inc., Sunderland, Mass.

chemical nature of a number of hormones from Table 14-1, as well as several of the hypothalamic hormones. On a chemical basis, these hormones can be divided into two major classes, peptide and nonpeptide hormones. The peptide hormones are small protein molecules that range from as few as three amino acids in the case of thyrotropin-releasing hormone (TRH) to several hundred amino acids in the case of some of the larger hormones of the anterior pituitary. Many of the nonpeptide hormones are steriods (e.g., those of the gonads and adrenal glands), and the remainder are generally modified derivatives of amino acids.

Endocrinologists find it convenient to subdivide the vertebrate hormones into three major classes: (1) those produced by the hypothalamus, (2) those produced by the pituitary gland, and (3) those produced by peripheral endocrine tissues such as the gonads, adrenals, pancreas, and thyroid. A complex set of negative and positive feedback loops, involving both the nervous and endocrine systems, controls endocrine function and permits hormones to influence not only their own secretion but also that of other hormones (Fig. 14-1, A).

The hypothalamus is a part of the brain, and as such it receives input from the entire central nervous system. It also contains hormone receptors that allow it to receive endocrine input of many types, and it appears to serve as the main integrator of the body's hormonal activities. Neurosecretory cells of the hypothalamus produce releasing hormones and inhibiting hormones that determine the release and, in at least some cases, the synthesis of hormones from the anterior pituitary. The hypothalamic hormones travel down the axons of their neurosecretory cells of origin and are released into portal blood vessels that travel directly to the anterior pituitary (Fig. 14-1, B). The hypothalamic hormones are typically very small peptides. For example, luteinizing hormone releasing hormone (LH-RH) contains only 10 amino acids. Recent evidence suggests that these hypothalamic hormones also have peripheral effects, for example, induction of mating behavior in rats by LH-RH.

The pituitary gland (or hypophysis as it is sometimes called) is composed of two discrete lobes, the anterior pituitary (adenohypophysis) and the posterior pituitary (neurohypophysis), plus an intermediate region known as the pars intermedia. Only the anterior pituitary will be considered in detail here, since the hormones of the pars intermedia and the posterior pituitary do not appear to be of major importance in development. The hormones of the anterior pituitary are all proteins or glycoproteins. Two of these hormones, prolactin and growth hormone (and also melanophore-stimulating hormone [MSH], which is produced by the pars intermedia), are under both posi-

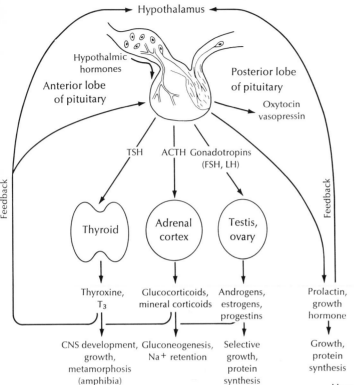

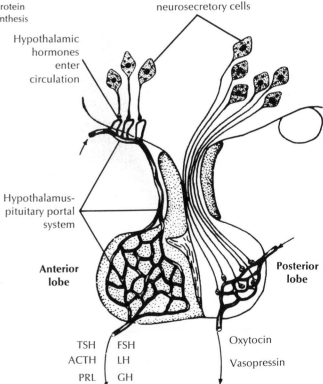

Fig. 14-1. Hypothalamic and pituitary control over peripheral endocrine glands. **A,** Schematic diagram showing functional relationship among the hypothalamus, pituitary gland, and peripheral endocrine glands. Neurosecretory release and inhibitory hormones from the hypothalamus are carried by the circulation to the anterior lobe of the pituitary, controlling its secretion into the circulation of tropic hormones that stimulate the peripheral endocrine glands. Hormonal and neural feedbacks to the hypothalamus maintain the entire system in a precisely controlled balance relationship. (Modified from Filburn, C. R., and G. R. Wyatt. 1974. *In* J. Lash and J. R. Whittaker, eds. Concepts of development. Sinauer Associates, Inc., Sunderland, Mass.)

B, Details of the relationship between the hypothalamus and the pituitary gland. The arrows show the pathway of blood circulation through the anterior and posterior pituitary. Note that neurosecretory products are carried to the anterior lobe of the pituitary via the circulation and are thus true hormones, whereas neurosecretory products that are stored in and released from the posterior lobe are carried directly to it by axons of the neurosecretory cells. (Modified from Hickman, C. P., et al. 1974. Integrated principles of zoology, 5th ed. St. Louis, The C. V. Mosby Co.)

tive and negative hypothalamic controls. These hormones, in contrast to the other anterior pituitary hormones, have nonendocrine target tissues. The remaining anterior pituitary hormones (thyroid-stimulating hormone [TSH], adrenocorticotropic hormone [ACTH], luteinizing hormone [LH] and follicle-stimulating hormone [FSH]) are under positive control by the hypothalamus and negative control by their endocrine target tissues.

The third group of hormones, those produced by peripheral endocrine glands, is more diverse. These include steroid hormones from the gonads and the adrenal cortex, as well as thyroid hormones, all of which are under pituitary control. Other peripheral gland hormones (e.g., insulin, epinephrine) are not under pituitary control but are secreted in response to neural or metabolic stimuli.

As we discuss roles and mechanisms of action of various hormones, it is important to remember that the same hormone may interact with many different target tissues and that their responses may be very different from one another. For example, in chickens, estrogen stimulates synthesis of the egg white protein ovalbumin in the oviduct and synthesis of the yolk protein precursor vitellogenin in the liver. There are also a number of cases in which the same or very similar hormones have been conserved during evolution but have become associated with different functions in different species. Prolactin, which was named for its ability to stimulate lactation in female mammals, is a good example. Prolactin-like hormones occur in most types of vertebrates, where they have such diverse functions as osmoregulation and salt balance in certain fishes (needed for survival of saltwater species in fresh water), stimulation of pigment formation in other fishes, delay of metamorphosis in amphibia, promotion of nesting behavior in a variety of species ranging from fishes to birds, and formation of "crop milk" in the crops of pigeons that are rearing young. Even in mammals, prolactin has other effects, including an essential role in normal development and function of the testes of male mice and effects similar to luteinizing hormone on the ovaries of female mice. There is considerable, although not total, cross reactivity among the prolactin-like hormones from different species. Thus, for example, amphibian prolactin will stimulate mammalian milk production. This suggests that lactation and its control must have evolved around a preexisting hormonal control system.

Another circumstance to keep in mind is the frequent practice of naming hormones for the first function that they were shown to have. Most of the functions of prolactin in nonmammalian species, for example, have nothing to do with the promotion of lactation. A similar situation exists for the gonadotropins produced by the anterior pituitary. They are called follicle-stimulating hormone and luteinizing hormone on the basis of their roles in promoting follicle maturation and ovulation and in maintenance of the corpus luteum after ovulation, respectively, in the female. However, subsequent research has shown that the same two hormones are also indispensable in male reproductive physiology. LH stimulates the androgen-secreting interstitial cells (Leydig cells) of the testis. Because of this role, it is sometimes referred to as "interstitial cell-stimulating hormone" (ICSH), but ICSH and LH are in fact the same. The exact role of FSH in the male is less clearly defined, but it is known to be essential for normal spermatogenesis.

We make no attempt in this chapter to describe in detail the multitude of diverse hormonal effects that are known to occur in developing organisms. Instead, we have chosen to present the mechanisms currently believed to be responsible for the action of some of the major types of hormones and then to concentrate on a few selected examples of hormonal function during development. Two additional well-known examples of hormonal action in development, the role of steroid hormones in mammalian sexual differentiation and the action of juvenile hormone and ecdysone during insect development, are presented in Chapters 20 and 22, respectively.

MECHANISMS OF HORMONE ACTION

The molecular mechanisms by which the protein and steroid hormones influence their target tissues are beginning to be reasonably well understood. Although not all the details are yet known, it is clear that these two classes of hormones accomplish their effects by very different means. Receptors for peptide hormones are located on the outer surfaces of cells, and there is substantial evidence suggesting that such hormones do not need to enter the cells at all to have their effects. Receptors for steroid hormones, on the other hand, are located in the cytoplasm, and the preponderance of evidence indicates that steroid hormone—receptor complexes enter the nucleus and interact directly with the genomes of cells.

Peptide hormones

All the peptide hormones and related peptide growth factors that have been examined have been found to interact with receptors localized on the outer surface of the cell membrane. It is generally assumed that the hormones and growth factors do not have to enter the cells to have their effects. This has been demonstrated in a number of cases in which the active substances were chemically bound to large particles such as Sepharose beads and were shown to retain their biological activity. Although the data supporting external receptor action are in general convincing, there remains a possibility that alternate mechanisms may also exist. For example, careful studies with a peptide growth factor for cultured cells have shown that small amounts of radioactively labeled factor that may be enough to account for the activity are slowly released from Sepharose beads during the course of a typical experiment. In addition, recent experiments have shown that insulin can be taken into cells in small amounts and that complexes of insulin plus insulin receptors even enter the nucleus.

Although these recent findings raise some interesting questions that deserve further study, interactions with cell surface receptors appear to be the major mechanism of peptide hormone action. This leaves the problem of how such interactions can affect intracellular events. For a number of the peptide hormones it has been shown that a "second messenger" functions inside the cell in response to signals transmitted through the cell membrane, as shown in Fig. 14-2. In the case shown, which is valid for several different hormones, including ACTH, TSH, MSH, LH, and glucagon, the second messenger is 3′, 5′-cyclic adenosine monophosphate (cyclic AMP, or cAMP). Hormone binding to receptors on the cell surface results in stimulation of the activity of the membrane-bound enzyme adenyl cyclase, which causes

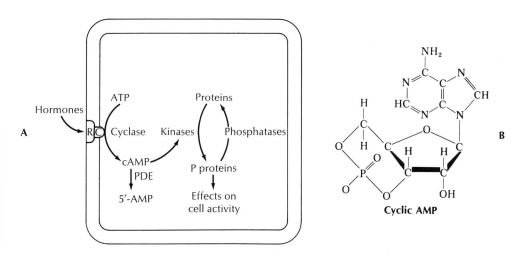

Fig. 14-2. Mechanism of action of peptide hormones that use cAMP as a second messenger. **A,** The hormone (first messenger) binds to a surface receptor *(R)*, transmitting a stimulus through the membrane that causes the enzyme adenyl cyclase *(C)* to convert ATP to cAMP. The increased level of intracellular cAMP activates cAMP-dependent kinase enzymes that phosphorylate regulatory or other proteins in the cell and thereby alter their biological activity. Excess cAMP is degraded by phosphodiesterase *(PDE)* to keep the system responsive to changes in extracellular hormonal signals. **B,** Structural formula of cAMP. (**A** from Filburn, C. R., and G. R. Wyatt. 1974. *In* J. Lash and J. R. Whittaker, eds. Concepts of development. Sinauer Associates, Inc., Sunderland, Mass.)

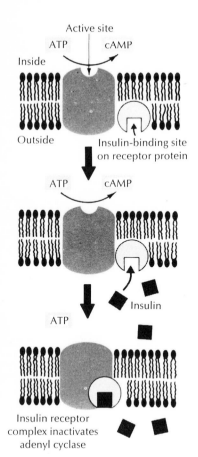

Active site

ATP   cAMP

Inside

Outside

Insulin-binding site
on receptor protein

ATP   cAMP

Insulin

ATP

Insulin receptor
complex inactivates
adenyl cyclase

Fig. 14-3. Schematic representation of the inactivation of adenyl cyclase by insulin. Insulin binds to an extracellular receptor, which causes an allosteric change in adenyl cyclase that inactivates the intracellular enzymatic site. The receptor and the enzyme are both embedded in the lipid bilayer of the cell membrane. (Modified from James D. Watson, 1976. Molecular biology of the gene, 3rd ed., copyright © 1976, 1970, 1965 by James D. Watson and The Benjamin/Cummings Publishing Co., Inc., Menlo Park, Calif.)

the conversion of ATP to cAMP. The cAMP in turn activates kinases that phosphorylate and thereby alter the activity of specific cell proteins. Other hormones such as insulin inhibit adenyl cyclase activity when they bind to membrane receptors (Fig. 14-3), thereby reducing the intracellular level of cAMP and the amount of cAMP-dependent phosphorylation that occurs within the hormone-treated cell.

The initial actions of hormones that act through cAMP take place rapidly and do not require the synthesis of new RNA or protein species, since they occur through modification of preexisting molecules. Changes in intracellular cAMP levels appear to be responsible for many different types of hormone actions. It should be emphasized, however, that cAMP is simply one more link in a message-carrying system and does not contain specific information itself. Thus the particular response that cAMP elicits is determined entirely by the type of cell that is responding to the hormone. For some of the peptide hormones and growth factors, the nature of the intracellular second messenger is not yet known.

Steroid hormones

Steroid hormones act on their target tissues by an entirely different type of mechanism (Fig. 14-4). Steroids are lipid soluble and readily enter cells, presumably by diffusion through the cell membrane, although it is also possible that their transport may be protein mediated. They do not become biologically active, however, until they become bound to specific receptor proteins, which generally are found only in the cells of appropriate target tissues. These receptor proteins (thousands per cell) are initially located in the cytoplasm, but on binding to the steroid hormone, they undergo a structural change and enter the nucleus. The structural transformation may occur within the receptor molecule itself, or it may involve the addition of a second protein molecule. (It has been argued by some workers in the field that this change occurs after transfer to the nucleus.) Once in the nucleus, the hormone-receptor complex binds to the chromatin and causes a selective stimulation of transcription of certain genes. This occurs only in

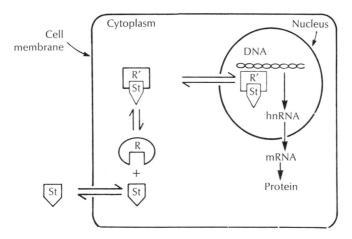

Fig. 14-4. Schematic representation of the control of gene expression by steroid hormones. The steroid *(St)* enters the cytoplasm and binds to a specific receptor *(R)*, which becomes modified *(R')*. The steroid-receptor complex then enters the nucleus and interacts with the genome, stimulating the transcription of specific genes. (Modified from Baxter, J. D., et al. 1973. *In* J. K. Pollack and J. W. Lee, eds. The biochemistry of gene expression in higher organisms. D. Reidel Publishing Co., Inc., Hingham, Mass.)

target tissues; no effect is seen if steroid-receptor complex is added to the nuclei of nontarget tissues. Thus the ability to respond to a particular steroid is determined not only at the cytoplasmic level by the presence or absence of receptors but also at the nuclear level by the ability to respond to the hormone-receptor complex.

Some of the earliest work on steroid receptors was done with the estradiol receptor found in uterine tissue, where it was shown that a soluble protein isolated from uterine cytosol has the ability to bind radioactively labeled estradiol. This protein is found in large amounts only in estrogen target tissues (e.g., vagina, uterus, mammary gland, hypothalamus). When uterine tissue is treated with ³H-estradiol, the labeled hormone accumulates in the nucleus. However, if nuclei are first isolated from the cells of target tissues and then incubated with labeled estradiol, no hormone-receptor complex is found, indicating that unbound receptor molecules occur only in the cytoplasm. Transfer of the hormone-receptor complex from the cytoplasm to the nucleus does not occur at low temperatures, which suggests that it requires metabolic energy. It has been reported that the structural transformation of the estrogen receptor that occurs after estrogen binding is also temperature dependent.

The binding of the hormone-receptor complex to chromatin has been suggested by the results of experiments in which treatment with deoxyribonuclease releases the complex from nuclei. Direct binding of complexes to DNA has also been demonstrated in homogenates of target cells. The majority of the roughly 10,000 receptors present in each target cell move to the nucleus and become bound to the genome when they are complexed with steroids. This amount of binding seems large in view of the selective effects of hormones on their target tissues. However, it appears that much of the binding to chromatin is of low affinity and rather nonspecific. Such low-affinity binding may make it difficult to detect the presence of smaller numbers of high-affinity binding sites. One recent proposal suggests that the cumulative effect of the low-affinity binding is to alter the structure of a relatively large segment of chromatin and thereby make it available for transcription. Clustered low-affinity binding sites for receptors are postulated to occur in regions that also contain specific DNA sequences that bind receptor-hormone complexes with high affinity. If this view is correct, a combination of specific and nonspecific receptor binding would be responsible for activation of specific genes by steroid hormones.

The affinity of chromatin for the progesterone-receptor complex appears to reside at least in part in the NHC proteins. This has been demonstrated by experiments in which DNA and chromatin proteins from target and nontarget tissues are reconstituted. When chicken oviduct chromatin is reconstituted with oviduct histone and erythrocyte NHC protein, binding of the progesterone-receptor complex occurs at a low level characteristic of native erythrocyte chromatin. However, if erythrocyte DNA is reconstituted with erythrocyte histone and oviduct NHC protein, binding of the progesterone receptor complex occurs at a level similar to that found in native oviduct chromatin. The binding capacity has been localized in a specific fraction of NHC protein designated AP_3.

A genetic approach to studying the mechanism of steroid hormone action has been pursued by Gordon Tomkins and co-workers. They have examined mutants of a line of mouse lymphoma cells that are normally killed by exposure to physiological doses of glucocorticoids. (Similar mechanisms are probably responsible for the mildly immunosuppressant effects of hydrocortisone and similar glucocorticoids.) Several classes of steroid-resistant mutants have been isolated and analyzed. One class of steroid-resistant mutants (r^-) is defective in steroid binding to cytoplasmic receptors. A second class (nt^-) is defective in transfer to the nucleus, and the hormone-receptor complexes from this class do not bind to DNA efficiently. The third class (nt^i) shows increased nuclear transfer, and the receptors from these mutants show a higher affinity for DNA than those of wild-type cells, although there is still no killing of the cells. The fourth class (d^-) is defective in reactions that occur subsequent to nuclear transfer of the steroid-hormone complex. The existence of these classes of mutants further supports the model of steroid hormone action shown in Fig. 14-4.

During amphibian metamorphosis, a long-tailed, herbivorous, aquatic, fishlike animal becomes a tailless, carnivorous, land-based animal with four long legs (Fig. 14-5). In addition to the obvious morphological changes occurring in metamorphosis (Table 14-3), a num-

HORMONE ACTION IN DEVELOPING SYSTEMS
Amphibian metamorphosis

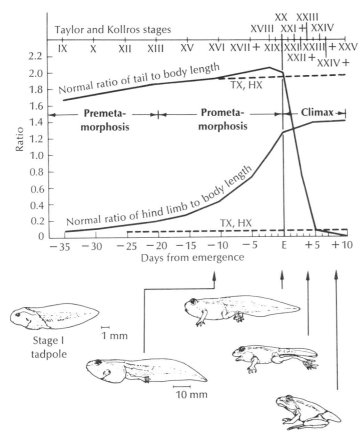

Fig. 14-5. Metamorphosis of a frog, *Rana pipens*. Time is in days before or after the emergence of the second forelimb *(E)*. Roman numerals are a standardized set of "stages" used to describe development and metamorphosis. Ratios of length of tail to length of body, and of length of hind limb to length of body are plotted. After emergence of the forelimbs, the metamorphic climax proceeds rapidly. The broken lines demonstrate the lack of metamorphosis that occurs in animals that have been either thyroidectomized *(TX)* or hypophysectomized *(HX)*. The drawings show morphological appearance at the indicated stages. (From Beckingham-Smith, K., and J. R. Tata. 1976. *In* C. F. Graham and P. F. Wareing, eds. The developmental biology of plants and animals. W. B. Saunders Co., Philadelphia.)

Table 14-3. Anatomical changes during anuran metamorphosis*

Tissue, organ, or system	Changes	Function
External features		
Head	Loss of horny beaks to mouth; widening of mouth	Adaptation to new diet
	Development of tympanum; repositioning of eyes	Accommodation to aerial sensory input
Limbs	Complete growth of forelimbs and hindlimbs	Locomotion on land
Tail	Complete resorption	Loss of swimming as major mode of locomotion
Skin	Pigmentation changes	Protective coloration
	Hardening	Protection against water loss on land
Internal features		
Digestive system	Development of muscular tongue; major shortening of gut; repositioning of anus	Change from vegetarian to carnivorous diet
Respiratory system	Resorption of gills; development of lungs; development of hyoid cartilages and muscles for respiration	Change from water- to air-based respiratory system
Reproductive system	Development of gonads	Sexual maturity only in adult form
Nervous system	Degeneration of Mauthner cells	Denervation of degenerating tissue
	Growth of new neurons and nerves	Innervation of new structures

*From Beckington-Smith, K., and J.·R. Tata. 1976. *In* C. F. Graham and P. F. Wareing, eds. The developmental biology of plants and animals. W. B. Saunders Co., Philadelphia.

ber of associated biochemical changes also occur (Table 14-4). These include the appearance of a new form of hemoglobin, changes in the digestive enzymes of the intestine, synthesis of enzymes required for the excretion of nitrogen as urea rather than ammonia, and many others.

The primary control over all the diverse processes that occur in metamorphosis is exerted by a single type of hormone, thyroid hormone.* The coordinated sequence of events that occur in normal metamorphosis appears to require a progressive increase in the concentration of circulating thyroid hormone. If a large dose of thyroxine or

*The term "thyroid hormone" is used to describe the mixture of thyroxine (T4) and triiodothyronine (T3) that is released from the thyroid gland in response to TSH.

Table 14-4. Biochemical changes during anuran metamorphosis*

Tissue or organ	Change induced	Function
Eye	Change in visual pigment from porphyropsin to rhodopsin	— — —
Liver	Synthesis of urea cycle enzymes	Excretion of urea
	Synthesis of serum albumin	Maintenance of homeostasis
	Synthesis of ceruloplasmin	Connected with changed iron utilization(?)
Erythropoietic tissue	Change from synthesis of larval hemoglobin to adult hemoglobin	Lower affinity oxygen carrier for air-based respiration
Gut	Synthesis of hydrolytic enzymes	Resorption of tissue
	Appearance of peptic activity in foregut	Change to digestion of animal tissue
Skin	Melanin synthesis	Protective coloring
	Induction of Na^+-K^+-ATPase	Maintenance of electrolyte balance
	Serotonin synthesis	— — —
	Changes in collagen deposition and breakdown	Changes in mechanical properties of skin for terrestrial life
Tail	Synthesis of hydrolytic enzymes	Resorption of tissue

*From Beckington-Smith, K., and J. R. Tata. 1976. In C. F. Graham and P. F. Wareing, eds. The developmental biology of plants and animals. W. B. Saunders Co., Philadelphia.

triiodothyronine is given to a tadpole, the final events of metamorphosis occur before the first are completed, and an abnormal "adult" is formed (Fig. 14-6).

The precise mechanism responsible for the controlled increase is not completely understood, but it is thought to involve a temporary positive feedback effect of thyroid hormone on its own production. At a certain point during tadpole development (perhaps as a result of environmental factors such as light and temperature) the normal negative feedback effect of thyroid hormone on the hypothalamus appears to become temporarily reversed so that the low level of thyroid hormone already present in the circulation begins to stimulate the release of TRH from the hypothalamus. TRH stimulates the anterior pituitary to secrete thyrotropin, or TSH, which in turn causes the release of thyroid hormone from the thyroid gland. The increased level of circulating thyroid hormone stimulates further production of TRH, and the cycle continues until high levels of circulating thyroid hormone are gradually established. (This positive feedback loop contrasts sharply with the negative feedback system that normally operates both in younger tadpoles and in the adult).

Fig. 14-6. Discoordinate metamorphosis induced by abrupt increase in thyroid hormone. The top drawing shows a normal Florida swamp tadpole, *Rana heckscheri,* during early stages of metamorphosis. The lower drawing shows results 5 days after injection of enough triiodothyronine to initiate metamorphic climax. Tail regression has occurred prior to normal rearrangement of body contours and limb development. (From Freiden, E. 1963. Sci. Am. **209**:111 [Nov.]. Copyright © 1963 by Scientific American, Inc. All rights reserved.)

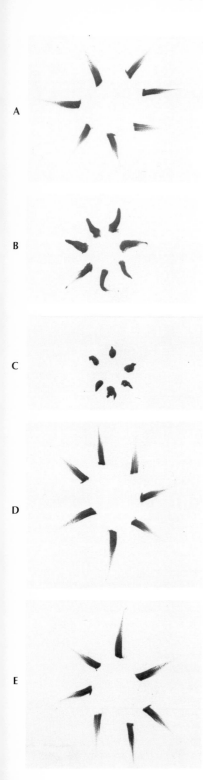

Thyroid hormone acts directly on its target tissues and not through the mediation of any other systemic factor. This is demonstrated by the finding that thyroid hormone is active in organ culture (Fig. 14-7) and that it also acts locally when pellets are implanted into specific tissues (Fig. 14-8). Administration of thyroid hormone to thyroidectomized animals has shown that thyroid hormone must be present continuously for metamorphosis to proceed to completion.

Although virtually all aspects of amphibian metamorphosis can be attributed to changes in the level of circulating thyroid hormone, the overall picture is considerably more complex than this. Different target tissues become responsive to thyroid hormone at different times in tadpole development and also show differing degrees of sensitivity to thyroid hormone. Furthermore, the effects of thyroid hormone are inhibited by prolactin (and possibly also by growth hormone). The inhibitory effect of prolactin on metamorphosis is pronounced during early growth of the tadpole. However, at the beginning of metamorphosis, secretion of prolactin ceases at the same time that the major release of TSH from the anterior pituitary begins. The orderly sequence of events that occurs in metamorphosis thus appears to result from at least four different factors—the progressive increase in the level of circulating thyroid hormone, the withdrawal of prolactin, development of target tissue responsiveness, and the relative sensitivity of each target tissue to thyroid hormone.

Amphibian metamorphosis provides a dramatic example of the specificity of tissue response to hormonal stimulation. Each type of target tissue is preprogrammed to respond to thyroid hormone in a unique manner. For example, limb muscle responds by growth, but tail muscle responds by degeneration. The type of response is intrinsic to the tissue and is not influenced by its location within the animal. For example, if an eyecup is transplanted to the tail, it remains completely healthy while the tail regresses and it can eventually fuse to the animal's body after the tail is totally resorbed (Fig. 14-9, *A* to *C*). On the other hand, a second tail transplanted onto the body undergoes atrophy at the same time as the host animal's own tail (Fig. 14-9, *D* to

Fig. 14-7. Regression of tadpole tails in organ culture. **A,** Appearance of freshly amputated tails at beginning of experiment. **B,** Appearance after 4-day exposure to 1 μg/ml of triiodothyronine (T_3). **C,** Appearance after 8 days in T_3. **D,** Appearance of controls without T_3 after 8 days in culture. **E,** Same as **C,** but with RNA synthesis inhibited by actinomycin D. (Modified from Tata, J. R. 1969. *Gen. Comp. Endocrinol.* [Supp.] **2:**385.)

Fig. 14-8. Localized regression in a tadpole tail 20 days after implantation of a pellet containing tetraiodothyropropionic acid, an active synthetic analog of thyroxin. (From Kaltenbach, J. C. 1970. J. Exp. Zool. **174:**55.)

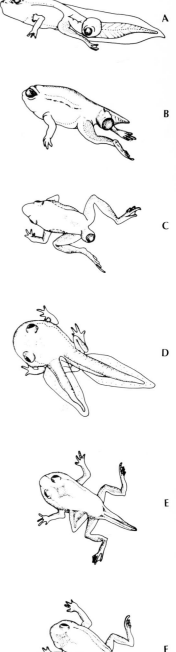

F). Also, as was shown in Fig. 14-7, tail tissue can be maintained in a healthy condition in culture but will degenerate when thyroxine is added to the culture medium, whereas other tissues in culture do not respond in this manner.

The molecular mechanism of thyroid hormone action in amphibian metamorphosis is not fully understood. Studies on tadpole liver cells have shown that extensive changes in protein synthesis occur in response to thyroid hormone. The changes that occur in enzymes involved in urea synthesis are shown in Fig. 14-10. Circumstantial evidence suggests that thyroid hormone acts at least in part by affecting transcription. Labeled thyroid hormone accumulates in the nuclei of tadpole liver cells and binds tightly to chromatin. Actinomycin D inhibits many aspects of metamorphosis, including autolysis of cultured tails (Fig. 14-7) and synthesis of carbamyl phosphate synthetase, one of the enzymes of the urea cycle that increases sharply in the liver during metamorphosis. The increase in carbamyl phosphate synthetase activity has been shown to result from de novo protein synthesis both by immunoassay of the enzyme and by prevention of the increase with inhibitors of protein synthesis. However, it has also been shown that carbamyl phosphate synthetase is made as an inactive precursor molecule and that conversion from the inactive to the active form is stimulated by thyroid hormone. Thus the overall effect of thyroid hormone may result from a combination of effects at different levels.

Fig. 14-9. Organ specificity of metamorphosis. **A** to **C,** Successive stages in the metamorphosis of a tadpole with an eye grafted into the tail. As the tail regresses, the eye persists and becomes fused to the trunk. **D** to **F,** Metamorphosis of a tadpole with a second tail grafted to the trunk between the fore and hind limbs. Although that region supports active limb development, the grafted tail regresses synchronously with the host tail. (From Berrill, N. J., and G. Karp. 1976. Development. McGraw-Hill Book Co., New York. Copyright © 1976 by McGraw-Hill Inc. Used with permission of McGraw-Hill Book Co.)

Fig. 14-10. Changes in urea excretion and activities of enzymes of the urea cycle during tadpole metamorphosis. Roman numerals refer to developmental stages shown in Fig. 14-5. Values for "arginine synthetase" are for the overall synthesis of arginine from citrulline and aspartate, which involves two different enzymes. (From Brown, G. W., and P. P. Cohen. 1958. *In* W. D. McElroy and B. Glass, eds. The chemical basis of development. The Johns Hopkins University Press, Baltimore. Copyright 1958 by the Johns Hopkins University Press.)

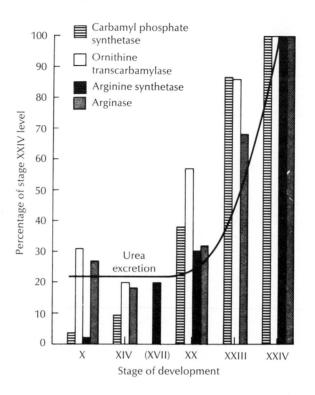

Mammary gland development and function

The mammary gland is a good example of a developmental system that is regulated by a number of different hormones that interact in a complex manner. Lactation is critically important to the continuity of mammalian species but is not required continuously. Typically the mammary gland of the female mammal remains relatively undeveloped until sexual maturity is reached, and then its function develops and regresses in response to complex hormonal signals related to reproduction.

The major phases of mammary gland development are shown diagrammatically in Fig. 14-11. Mammary glands arise as ingrowths of ectodermal tissue into the underlying mesoderm during embryonic development. A relatively simple duct system is formed at that time (Fig. 14-11, *A*), both in the male and in the female of most species (the mouse is an exception, as described in Chapter 18). Little more happens until the time of sexual maturation, when in the female increased estrogen secretion stimulates development of the fat pad underlying the glandular tissue and also promotes extensive proliferation and branching of the ductal system (Fig. 14-11, *B*). Normal background levels of insulin, thyroxine, somatotropin (growth hormone), and glucocorticoids also appear to be needed for the response to estrogen.

During pregnancy there is a rise in the circulating levels of progesterone and lactogenic hormones (prolactin from the anterior pituitary and a lactogen produced by the placenta). These hormones, in combination with estrogen, insulin, thyroxine, glucocorticoids, and somatotropin, stimulate extensive development of secretory alveoli at the ends of the highly branched ducts (Fig. 14-11, C). However, little actual synthesis of milk proteins occurs until after delivery of the young. It is currently believed that progesterone, possibly acting in combination with estrogen, inhibits synthesis of milk proteins and that one of the major changes that initiates milk production after parturition (birth of the young) is the abrupt decrease in the levels of circulating sex steroids, which had previously been supplied in large amounts by the placenta. Recent studies suggest that progesterone affects both the amount of casein mRNA that accumulates in the cytoplasm of the alveolar cells and the extent to which that mRNA is translated.

The maintenance of active lactation during the period while the young are nursing is an interesting example of interactions between the nervous system and the endocrine system. Lactation is maintained only as long as adequate amounts of prolactin are released from the anterior pituitary. That release is controlled by prolactin inhibiting hormone (PIH) from the hypothalamus, which in turn is controlled by tactile stimulation of sensory nerves in the nipples. Thus each time the young nurse (or each time a dairy animal is milked), a neurosecretory feedback mechanism temporarily inhibits the output of PIH from the hypothalamus and causes a surge of prolactin release, thereby stimulating further synthesis of milk proteins and lipids. Suckling also has other neuroendocrine effects. Release of milk from the mammary gland is triggered by the hormone oxytocin, which is released from the posterior pituitary in response to suckling. The milk is actually ejected from the alveoli into the ductal system by contraction of myoepithelial cells, and when the neuroendocrine reflex is blocked by anesthesia,

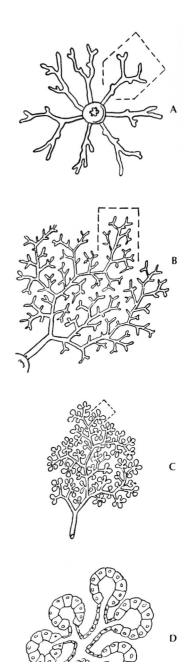

Fig. 14-11. Diagrams illustrating functional differentiation of the mammary gland. **A,** Prior to sexual maturity the gland consists of relatively simple ducts radiating from the nipple. **B,** Enlargement of a section of the duct system in **A** in an adult female prior to pregnancy. Estrogen has induced extensive growth and branching of the duct system, but secretory alveoli have not yet developed. **C,** Further enlargement of a segment of the mammary gland in **B** in a pregnant female. Extensive development of secretory alveoli at the ends of the branched ducts has taken place in preparation for lactation. **D,** Enlargement of a small area of **C,** showing the arrangement of cells in the alveoli and ducts. (From George W. Corner, The hormones in human reproduction [copyright 1942, © 1970 by Princeton University Press] Rev. edn. 1947, sections A, B, C, and D, p. 209. Reprinted by permission of Princeton University Press.)

the young are unable to obtain enough milk to survive. Suckling also inhibits the release of LH-RH, resulting in an inhibition of ovulation that in many cases lasts until the time of weaning.

Two mechanisms appear to be responsible for cessation of milk production when the young stop nursing. One is disruption of the positive neuroendocrine feedback loop that leads to continued release of prolactin and continued synthesis of milk as long as there is demand for it, and the other is a buildup of pressure within the mammary gland, which reduces the circulation to the gland. It has been shown that the second mechanism can stop lactation even when the prolactin level remains high. When lactation ceases, the mammary gland undergoes a major involution and reverts to a state similar to the prepregnant condition shown in Fig. 14-11, B.

The specific hormonal signals required for the synthesis of milk proteins and milk fats by mammary gland tissue have been studied extensively in organ culture systems, both from rodents and from dairy cattle. Such studies are normally performed with tissue either from mature virgin animals or from midpregnant animals and thus reflect primarily those events involved in the actual initiation of synthesis of milk, rather than those which occur during formation of the mammary gland or its maturation at puberty. Functional differentiation of mammary gland organ cultures is usually measured either in terms of synthesis of specific milk proteins (e.g., casein or α-lactalbumin) or milk fats (incorporation of acetate into specific fatty acids).

In these organ culture studies three types of hormones—insulin, glucocorticoids (e.g., hydrocortisone), and prolactin—are found to be essential for initiation of milk protein synthesis. Insulin alone maintains the tissue in a healthy state and stimulates cell division but by

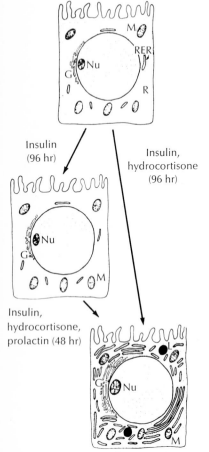

Insulin
(96 hr)

Insulin,
hydrocortisone
(96 hr)

Insulin,
hydrocortisone,
prolactin (48 hr)

Insulin, hydrocortisone,
prolactin (48 hr)

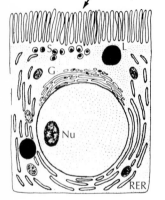

Fig. 14-12. Schematic representation of ultrastructural changes in the alveolar epithelial cells of explants from midpregnant mouse mammary glands in a synthetic culture medium with added hormones. Insulin alone causes proliferation but no morphological change. The addition of hydrocortisone causes buildup of rough endoplasmic reticulum and the Golgi zone but no secretion. Addition of prolactin with the other two hormones leads to secretion. When prolactin and hydrocortisone are added simultaneously to insulin-treated cells, changes that appear to be due to hydrocortisone are observed before those which appear to be due to prolactin. G = Golgi zone; L = lipid droplet; M = mitochondria; Nu = nucleolus; R = "free" ribosomes; RER = rough endoplasmic reticulum; S = secretory granules. (Modified from Mills, E. S., and Y. J. Topper. 1970. J. Cell Biol. 44:310.)

itself does not promote functional differentiation (Fig. 14-12). The addition of a glucocorticoid promotes rRNA synthesis and the development of extensive rough endoplasmic reticulum, giving the cell the characteristic appearance of a secretory cell, but little or no synthesis of milk protein occurs. It is only when prolactin is added to the organ cultures in combination with the other two hormones that significant amounts of milk protein synthesis occur. Any one or combination of two of these hormones is inactive.

Extensive studies have been made of the time sequence of action of these three hormones in the stimulation of lactation, but the conclusions remain controversial. It appears to be necessary to have insulin present at all times to obtain good stimulation of lactation. There are some reports indicating that insulin can be replaced at least partially by epidermal growth factor (Chapter 17) and by factors from serum that have insulin-like properties but are antigenically distinct from insulin. There is also good evidence that prolactin can be added to the cultures relatively late, after the buildup of rough endoplasmic reticulum has already been induced by insulin and hydrocortisone, as shown in Fig. 14-12.

The exact roles of insulin and hydrocortisone and their temporal relationship are less clear. Early data suggested that the primary role of insulin was to induce DNA synthesis and cell division and that it was necessary for hydrocortisone to be present at the time the cells were induced to enter the cell cycle to obtain responsiveness to prolactin. This led to the proposal that a critical mitosis or quantal cell cycle (Chapter 11) triggered by insulin with hydrocortisone present as an inducing agent was necessary to obtain responsiveness to prolactin. However, more recent experiments have raised questions about that interpretation. Experiments with inhibitors of DNA synthesis have shown that milk protein synthesis can be induced in tissue from pregnant, but not from virgin, animals without the need for a cell cycle. In addition, they have shown that when a cycle is needed, it can be triggered by insulin alone, and the subsequent induction of lactation by the three hormones can then be achieved in the presence of cytosine arabinoside, which effectively blocks all DNA synthesis. These data have been interpreted to indicate that some sort of quantal cycle is necessary but that it does not have to be accomplished in the presence of hydrocortisone, provided that one waits long enough for the subsequent induction. Induction without a cell cycle in cultures from mid-pregnant animals is viewed as being due to completion of the critical cycle in vivo prior to the initiation of cultures, although not all investigators agree. Attempts to induce milk protein synthesis in individual cultured cells, as opposed to organized tissue in organ culture, have

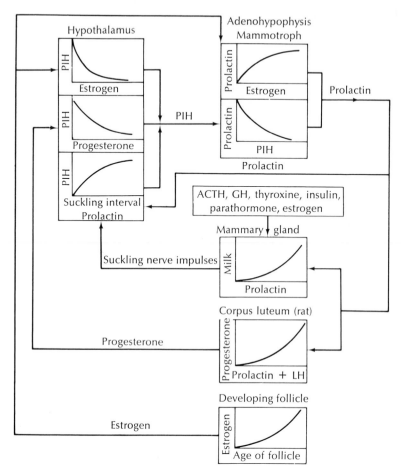

Fig. 14-13. Schematic representation of some of the major controls over lactation. The small graphs indicate the nature of responses. Thus prolactin stimulates milk production (near center of chart). Also, in combination with LH, it stimulates formation of progesterone, which in turn depresses prolactin inhibiting hormone (PIH). Thus prolactin builds to a rather high level during pregnancy, although lactation is inhibited until after birth of the young by a direct effect of progesterone on the mammary gland (not shown). Suckling inhibits release of PIH; thus the longer the interval between episodes of suckling, the higher the level of PIH. Prolactin also acts directly on the hypothalamus to increase PIH. Estrogen reduces PIH and thus increases prolactin. Estrogen also acts directly on the prolactin-secreting cells of the adenohypophysis (anterior pituitary) to enhance prolactin production. Adrenocorticotropic hormone (ACTH), growth hormone (GH), thyroxine, insulin, parathyroid hormone, and estrogen are also shown to have direct effects on lactation at the level of the mammary gland. (Modified from Earl Frieden and Harry Lipner. Biochemical endocrinology of the vertebrates, 1971, p. 31. Reprinted by permission of Prentice-Hall, Inc., Englewood Cliffs, N.J.)

not succeeded well, suggesting that not all the variables that control lactation are yet understood.

Although the short-term organ culture experiments are normally done in serum-free defined media, there is undoubtedly a substantial carryover of hormones from the donor animals into the cultures. Data from experiments performed in vivo and some longer term culture experiments suggest that, as a minimum, all the following play a role in the control of lactation in the intact animal: estrogen, progesterone, LH, and FSH (both of which control production of estrogen and progesterone), placental gonadotropin, insulin, glucocorticoids, ACTH (controlling glucocorticoid production), thyroid hormone, thyrotropin (controlling thyroid hormone), somatotropin, placental lactogen, prolactin, oxytocin, releasing and inhibiting factors for all the pituitary hormones involved directly or indirectly, and neuroendocrine responses to tactile stimuli. Some of the feedback loops that are involved are shown in Fig. 14-13.

The oviduct of the hen is a convenient system in which to study hormone-mediated developmental changes because, as in the case of the mammary gland, it produces large amounts of specific assayable products in response to hormonal stimulation. The major proteins of the egg white are produced by the magnum of the oviduct and secreted around the egg yolk as it passes (Fig. 14-14). Treatment of the immature oviduct with estrogen induces proliferation and differentia-

Hen oviduct development

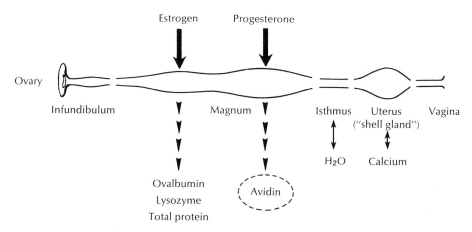

Fig. 14-14. Schematic representation of the functional segments of the chicken oviduct. The magnum is the site of synthesis of the major components of egg white. Estrogen stimulates a large increase in synthesis of ovalbumin, lysozyme, and total egg white protein. Synthesis of avidin is stimulated by progesterone in the estrogen-primed oviduct. (Modified from O'Malley, B. W., et al. 1969. Rec. Prog. Horm. Res. **25:**105.)

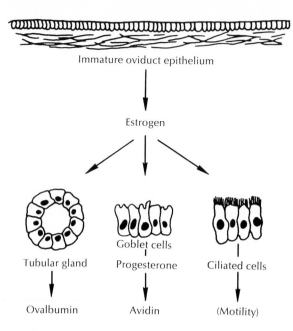

Fig. 14-15. Diagram showing cytological and biochemical actions of steroid hormones on the oviduct of the immature chicken. Morphogenesis of the tubular glands is presented in greater detail in Chapter 15. (Modified from O'Malley, B. W., et al. 1969. Rec. Prog. Horm. Res. **25:**105.)

tion of tubular gland cells, goblet cells, and ciliated cells (Fig. 14-15). The tubular gland cells produce a number of egg white proteins, including lysozyme, conalbumin, and ovalbumin. Goblet cells produce the egg white protein avidin, and ciliated cells are concerned with motility.

Ovalbumin comprises 50% to 60% of the total protein synthesized by the differentiated oviduct. When estrogen treatment is discontinued, tubular gland cells persist, but ovalbumin synthesis declines. During primary estrogen stimulation of the immature oviduct, a lag of about 36 hours occurs before ovalbumin synthesis is detectable. However, once tubular gland cells have differentiated, secondary stimulation with estrogen results in ovalbumin synthesis with a lag of only 3 to 4 hours.

The large-scale synthesis of ovalbumin by oviduct cells has made it possible to isolate ovalbumin mRNA and its nuclear precursor RNA, as well as to determine the number of ovalbumin-coding genes present in the genome of oviduct cells. It has thus been possible to study in detail the molecular mechanisms responsible for ovalbumin synthesis in response to estrogen treatment. The absence of differential amplification of ovalbumin genes was mentioned in Chapter 6. The primary mechanism responsible for the increase in ovalbumin synthesis in estrogen-stimulated oviduct cells appears to be an increase in the total amount of ovalbumin mRNA contained by the cells. The amount of functional ovalbumin message that can be detected by translation in cell-free systems is directly proportional to the rate of ovalbumin synthesis occurring in the intact tissue. This remains true at all times, including primary estrogen stimulation, withdrawal, and secondary stimulation. The amount of ovalbumin message present in oviduct tissue has also been determined by hybridization of oviduct mRNA to cDNA prepared from ovalbumin mRNA. This technique allows the detection of ovalbumin messages that may not be in a translatable form. The results of these experiments also indicate that the number of ovalbumin mRNA sequences present in oviduct cells is directly proportional to the rate of ovalbumin synthesis. Production of avidin in response to progesterone also appears to result from an increase in translatable avidin message, as detected in cell-free protein-synthesizing systems.

Estrogen clearly causes a major increase in the transcription of ovalbumin-specific RNA sequences. However, this is not the only mechanism involved in increasing the total amount of ovalbumin mRNA found in estrogen-treated oviduct cells. The half-life of ovalbumin mRNA in the estrogen-treated cells is increased to about 24 hours, as compared to only about 3 hours in untreated tissue. When estrogen is withdrawn, the half-life of the ovalbumin mRNA quickly falls to that found in untreated oviduct cells, whereas that of most other mRNAs remains unchanged. Thus during the withdrawal period there is a selective loss of ovalbumin mRNA from the cytoplasm.

The prodigious output of ovalbumin by the estrogen-stimulated hen oviduct (6×10^5 molecules/cell/minute or 3×10^{19} molecules/hen/day) is the cumulative result of a number of hormone-mediated changes that go beyond increased transcription and increased stability of ovalbumin mRNA. Other effects of estrogen on the oviduct include an increase in the number of ductal gland cells due to stimulation of their proliferation and a general increase in the efficiency of the translational machinery in the ductal gland cells.

The presence of noncoding intervening sequences within the coding region of the ovalbumin gene (Chapter 5) has led to speculation that such intervening sequences might be related in some way to hormonal activation of transcription of the ovalbumin gene. However, recent studies appear to show that the intervening sequences are unchanged between transcriptionally active and inactive ovalbumin genes and between oviduct and other tissues, including gametes.

Interestingly, however, the intervening sequences seem to differ from one chicken to another, suggesting that they are subject to a lower level of evolutionary selection than the ovalbumin coding sequence.

HORMONES IN PLANT DEVELOPMENT

Hormones also play critical roles in morphogenesis and cell differentiation in plants. Although the major focus of this book is on animal development, the importance of hormones in plant development warrants a brief discussion. The action of hormones on plant cells shows some similarities to the action of hormones on animal cells and also some differences. Both plant and animal hormones initiate expression of particular differentiated characteristics in already predetermined cells. Plant hormones, however, generally exhibit a lesser degree of target cell specificity than most animal hormones. Also, the molecular mechanisms responsible for plant hormone action are not yet well understood.

The patterns of growth and development in plants are totally different from those in animals. In particular, in all higher plants, centers of cell division known as meristems and immediately adjacent regions of active cellular differentiation play a major role in development. The meristems, which are located at the tips of all branches and roots, consist of small, relatively undifferentiated cells and are the site of most of the cell division in the plant. Initial stages of differentiation generally occur soon after division. The meristem can be viewed as a cluster of stem cells that remains at the tip of the growing root or shoot and lays down differentiated structures behind itself as growth proceeds.

Cell division and increase in mass are largely separated in plants. Most of the division occurs among the very small cells in the meristems. Most of the increase in mass is accomplished in the region behind the meristem, primarily by synthesis of new cell wall material and by the expansion of large fluid-filled vacuoles that are characteristic of plant cells. Thus in plant growth most of the increase in cell number is accomplished in the meristems, with little increase in mass, and most of the increase in mass is accomplished with little further cell division. It is not uncommon for the term "growth" to be used to describe the latter process in books on plant development.

The meristem located at the top (apex) of a growing plant is known as the apical meristem. Meristems located at the tips of lateral branches (or buds for potential lateral branches) are known as lateral meristems, and those located at the tips of roots are called root meristems. Both a shoot meristem and a root meristem typically become established early in embryonic development of a plant. Lateral meristems that give each kind of plant its characteristic pattern of shoot and root branching begin to be established soon after and continue to arise

Table 14-5. Involvement of hormones in plant growth and development*†

Type of hormone‡	Cell elongation	Cell division	Induction of primary vascular tissue	Induction of secondary vascular tissue	Root and shoot initiation	Breaking of seed dormancy	Senescence	Abscission of flowers, fruits, and leaves	Fruit growth	Sex expression	Control of stomatal aperture
Auxins (IAA)	✔	✔	✔	✔	✔	✔	✔	✔	✔	✔	
Gibberellins (GA₃)	✔	✔		✔	✔	✔	✔	✔	✔	✔	
Cytokinins (Zeatin)		✔	✔	✔	✔	✔	✔	✔	✔		✔
Inhibitors (ABA)	✔				✔	✔	✔	✔			✔
Ethylene	✔	✔			✔	✔	✔	✔	✔	✔	

*Modified from Hall, M. A. 1976. *In* C. F. Graham and P. F. Wareing, eds. The developmental biology of plants and animals. W. B. Saunders Co., Philadelphia.
†A check indicates that the hormone in question has been shown to affect the process in some way, although not necessarily in all instances.
‡Examples in parentheses.

throughout the period of active growth of the plant. In addition to their roles as centers of cell division and differentiation, meristems also function as endocrine organs in plants, as will be described later.

There are five major classes of plant hormones: auxins, gibberellins, cytokinins, growth inhibitors, and ethylene. Each class of hormone can be identified by its characteristic effects, but most developmental phenomena in plants seem to involve complex interactions among the various classes of hormones (Table 14-5).

Auxins

Auxin is the generic name for a group of plant hormones produced by shoot meristems and traditionally associated with stem elongation in the regions immediately behind the meristem. The principal naturally occurring auxin is indole-3-acetic acid (IAA), whose structure is shown in Fig. 14-16 together with that of 2,4-dichlorophenoxyacetic acid (2,4-D), a synthetic auxin widely used as a weed killer because of its ability to stimulate unbalanced growth that is lethal to many kinds of plants with broad leaves.

The effect of auxin on stem elongation is primarily the result of cell expansion due to enlargement of vacuoles and synthesis of new cell wall. Auxin is produced by the meristem, and all elongation stops when the tip bearing the meristem is cut off (Fig. 14-17). Auxin from an external source stimulates elongation in the absence of a meristem. Such experiments are generally done with coleoptiles (the sprout, or more properly the outer sheath that forms the first true leaf, of seedlings of grasslike plants such as oats).

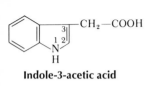

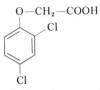

Fig. 14-16. Structural formulas of indole-3-acetic acid *(IAA)* and 2,4-dichlorophenoxyacetic acid *(2,4-D)*.

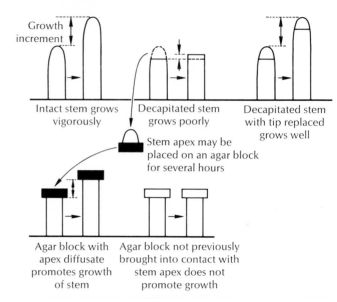

Fig. 14-17. Experiments demonstrating production by the stem apex of substances (auxins) that promote stem elongation. The active material produced by the apex can be collected in an agar block and used to stimulate elongation of a decapitated stem. Pure auxins such as indoleacetic acid will fully replace the material produced by the stem apex. (From Galston, A. W., and P. J. Davies. 1970. Control mechanisms in plant development. Prentice-Hall, Inc., Englewood Cliffs, N.J.)

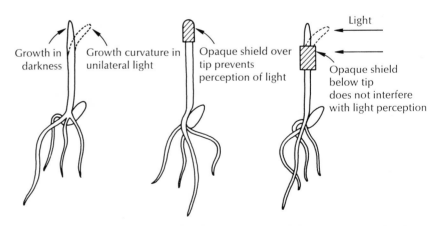

Fig. 14-18. Demonstration that the shoot apex controls curvature toward light. Covering the apex with an opaque shield prevents curvature, whereas an opaque shield placed over the stem below the tip does not. (From Galston, A. W., and P. J. Davies. 1970. Control mechanisms in plant development. Prentice-Hall, Inc., Englewood Cliffs, N.J.)

Fig. 14-19. Asymmetrical distribution of auxin caused by unilateral light. The numbers refer to the relative amounts of auxin recovered. Auxin transported into agar blocks by intact and vertically split coleoptiles is approximately equal, both in darkness and in unilateral light (**A** to **D**). However, if the lower extremity of the coleoptile is split and auxin is collected into two separate agar blocks, more auxin emerges from the side of the coleoptile away from the illumination than from the side toward it (**E**). If the entire coleoptile is split, an equal amount of auxin is collected from each half, showing that synthesis in both halves remains equal (**F**). The conclusion from this experiment is that auxin moves laterally across the coleoptile in a direction away from the source of the light as it is transported downward. (From Galston, A. W., and P. J. Davies. 1970. Control mechanisms in plant development. Prentice-Hall, Inc., Englewood Cliffs, N.J.)

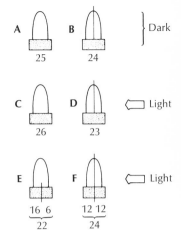

Auxins are also well known for their role in the phototropic curvature that occurs when seedlings are exposed to unilateral light. The auxin responsible for the curvature is produced in the apical meristem, and removal of the tip of the shoot prevents the curvature, as does covering the tip with an opaque cap (Fig. 14-18). The curvature is caused by differential elongation as a result of auxin accumulation on the side of the shoot opposite or distal to the impinging light (Fig. 14-19). Growth curvature in the absence of unilateral light can be achieved by applying auxin selectively to one side of a shoot tip (Fig. 14-20).

In combination with gibberellin (see following discussion) auxin stimulates cell division in the cambium (a layer of tissue just below the bark) of deciduous woody shoots and causes differentiation of its cells into xylem and phloem (the hollow vascular structures of plants). In the spring, apical buds produce increased amounts of the hormones, which travel downward toward the base of the plant. Cell division and differentiation in the cambium occur in the same downward pattern. Removal of apical buds prevents cambial division and differentiation. Application of IAA and gibberellic acid to disbudded shoots reinitiates cambial division and differentiation. Neither hormone alone is able to bring about growth in diameter or formation of new vascular tissue.

Auxin traveling downward also stimulates root development. The formation of roots by cuttings can be enhanced dramatically by treatment with auxin (Fig. 14-21). Auxin is also involved in the negative geotropism of plant shoots (ability to grow away from gravity) and the positive geotropism of roots (ability to grow toward gravity). The exact mechanisms are not known, but in shoots, auxin becomes distributed

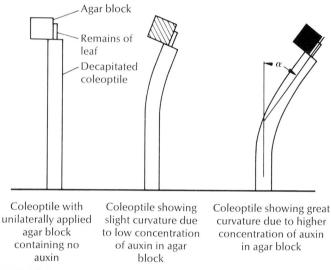

Fig. 14-20. Bending of decapitated coleoptile in response to asymmetrical application of auxin. The bending of coleoptiles can be used as a bioassay to determine auxin concentration. The angle α is related to the concentration of auxin in the agar block. (From Galston, A. W., and P. J. Davies. 1970. Control mechanisms in plant development. Prentice-Hall, Inc., Englewood Cliffs, N.J. Courtesy Boyce Thompson Institute.)

Coleoptile with unilaterally applied agar block containing no auxin

Coleoptile showing slight curvature due to low concentration of auxin in agar block

Coleoptile showing great curvature due to higher concentration of auxin in agar block

Not treated   Treated

Fig. 14-21. Induction of root formation by auxin. The basal end of the camelia cutting on the right was dipped in an auxin solution. (From Galston, A. W., and P. J. Davies. 1970. Control mechanisms in plant development. Prentice-Hall, Inc., Englewood Cliffs, N.J.)

in a gravitational field in such a way that the lower side of the shoot is disproportionately stimulated to elongate, causing curvature upward. It has been clearly shown with radioactively labeled IAA that there is a displacement of auxin to the lower side of stems (Fig. 14-22). In roots the effect is reversed, with displacement of auxin to the upper side, which results in downward bending. Thus no matter what position a seed is planted in, its shoot will grow upward and its root downward.

Another example of the action of auxin is found in the development of strawberries. The achenes ("seeds") on the fleshy part of the berry contain large amounts of IAA. Their removal prevents the normal process of swelling of the berry as it develops (Fig. 14-23). Replacement of the action of the achene with auxin dissolved in lanolin restores the normal process of swelling.

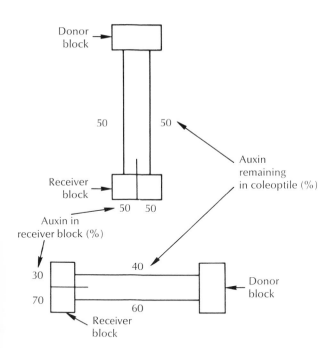

Fig. 14-22. Effect of gravity on distribution of symmetrically applied auxin in a segment of corn coleoptile. When the segment is vertical, the auxin is transported through it without lateral displacement. However, when it is horizontal, there is an asymmetrical accumulation of the auxin on the lower side. (From Galston, A. W., and P. J. Davies. 1970. Control mechanisms in plant development. Prentice-Hall, Inc., Englewood Cliffs, N.J.)

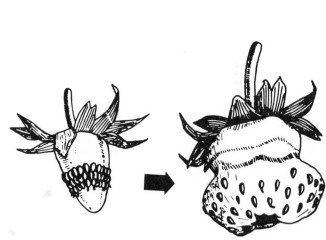

Fig. 14-23. Effect of removing all but three rows of achenes from an immature strawberry. Only the region containing the achenes swells during subsequent development. (From Nitsch, J. P. 1950. Am. J. Bot. 37:211.)

Gibberellins   The gibberellins are named for a fungus, *Gibberella fujikuroi*, which causes abnormal growth of rice seedlings. The infected seedlings grow very rapidly, forming long, spindly plants that fail to produce grain. The structure of gibberellic acid (GA_3), which was isolated from the fungus, is shown in Fig. 14-24. More than 20 different gibberellins, all with related chemical structures and hormonal effects, have since been isolated from a variety of sources, both in fungi and in higher plants.

The most striking characteristic of the gibberellins is their ability to promote stem elongation in intact plants. The differences between short, dwarf strains of plants and tall, spindly ones (e.g., bush beans versus pole beans) can usually be traced to gibberellins (Fig. 14-25).

Gibberellins also have a number of other roles. For example, when a plant that requires long days for flowering is kept under short-day conditions and is treated with GA_3, the plant undergoes stem elongation and flowering. The amounts and types of gibberellins present change when untreated plants are moved from short- to long-day conditions. Similarly, biennial plants such as the carrot, which require 2 years to complete their life cycle, normally do not flower until they have been exposed to low temperatures (winter). However, treatment with gibberellin causes them to flower without such cold treatment (Fig. 14-26). Increases in endogenous gibberellins are also associated with the breaking of dormancy, and exogenous application of gibberellin can artificially break dormancy in the buds of many kinds of plants.

The balance between gibberellins and auxins influences sexual expression in plants such as cucumber and squash that produce separate male and female flowers. High levels of gibberellin increase the frequency of male flowers, whereas high levels of auxin shift the ratio in favor of female flowers. In dioecious species (those with male and female flowers on separate plants) such as hemp, application of auxin to the male plant leads to the formation of female flowers. In studies of endogenous hormone levels, plants with predominantly female expression have been found to contain high levels of auxin, whereas those with male expression have high levels of gibberellin.

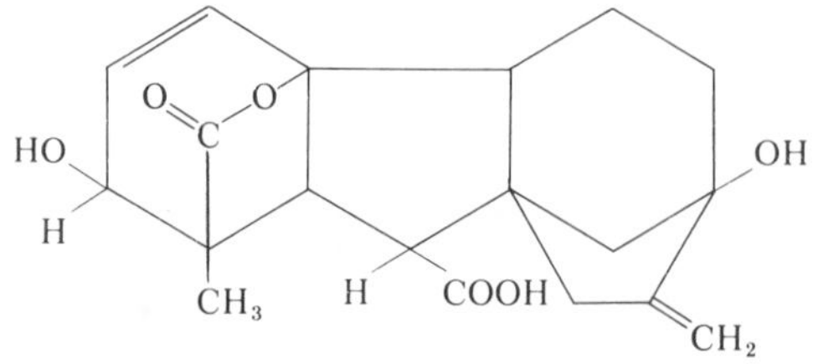

Fig. 14-24. Structural formula of gibberellic acid (GA_3).

Gibberellic acid

Fig. 14-25. Effect of gibberellin on the growth of dwarf bean plants. The plant on the right was treated with gibberellin. (From Wittwer, S. H., et al. 1957. Plant Physiol. **32:**39.)

Fig. 14-26. Use of gibberellin to stimulate flowering in a biennial plant (carrot) without cold treatment. The control plant on the left received no treatment. The plant in the center received 10 μg of gibberellin daily. The plant on the right was exposed to low temperatures for 8 weeks. The results of gibberellin treatment without exposure to cold are almost identical to those from cold treatment without gibberellin. (From Land, A. 1957. Proc. Natl. Acad. Sci. U.S.A. **43**:709.)

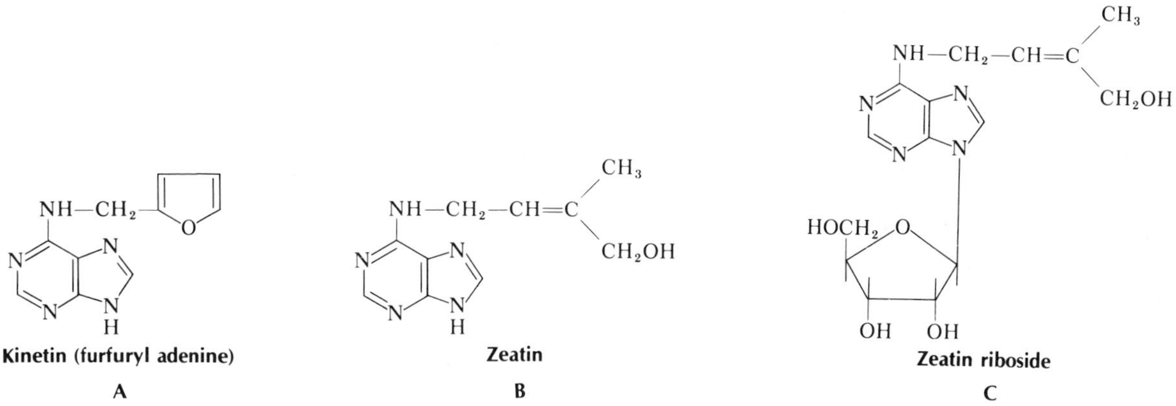

Kinetin (furfuryl adenine)

A

Zeatin

B

Zeatin riboside

C

Fig. 14-27. Structural formulas of kinetin, zeatin, and zeatin riboside.

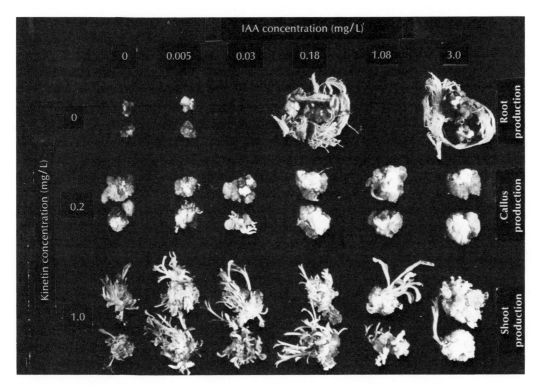

Fig. 14-28. The interactions of kinetin and IAA in cultures of tobacco cells. Both hormones are required for maximum growth. High ratios of auxin to kinetin produce roots (upper right), whereas at low auxin-kinetin ratios shoots predominate (lower left). At equivalent levels, relatively undifferentiated callus growth predominates (upper left to lower right). (From Skoog, F., and C. O. Miller. 1957. Symp. Soc. Exp. Biol. **11:**118.)

Cytokinins

The class of plant hormones known as cytokinins was discovered through studies of the growth of plant cells in culture. Prior to the discovery of cytokinins, coconut milk had been shown to contain an unidentified ingredient that was essential for rapid division of many types of plant cells in culture. The first purified chemical substance capable of replacing coconut milk effectively was kinetin (Fig. 14-27, *A*), which was isolated initially from an aged preparation of DNA. Subsequent attempts to isolate kinetin from fresh DNA were not successful, and it was found to be necessary to "age" the DNA artificially by autoclaving to obtain good yields of kinetin. Subsequently a large number of synthetic analogs of kinetin were prepared and found to be biologically active, but it was nearly 10 years before the first naturally occurring cytokinin, zeatin (Fig. 14-27, *B*), was isolated from young maize seeds. More recently it has been shown that the cytokinin activity in coconut milk is due to zeatin riboside (Fig. 14-27, *C*).

Cytokinins and auxins interact in a complex manner in plant cell cultures. For example, in cultures of tobacco stem pith, both hormones are needed for cellular multiplication. When the concentrations of

Fig. 14-29. Apical dominance. **A,** In a normally growing young pea plant, auxin produced by the apical meristem inhibits growth of lateral buds. **B,** When the apex is removed, the lateral buds begin to grow. **C,** An agar block containing auxin inhibits lateral growth in a plant with the apex removed. (From Biology of developing systems by Philip Grant. Copyright © 1978 by Holt, Rinehart & Winston. Reprinted by permission of Holt, Rinehart & Winston.)

A   B   C

both are either high or low, the primary form of growth is an undifferentiated "callus." However, when the ratio of auxin to cytokinin is high, the culture tends to begin differentiating and forming roots. When the ratio of auxin to cytokinin is low, differentiation also occurs, but in this case shoots are formed (Fig. 14-28).

In contrast to auxins, cytokinins appear to be made in the roots and travel upward in the plant. Although both cytokinins and auxins appear to be needed for some types of processes, their actions are frequently antagonistic. One of the consequences of the antagonism seems to be the maintenance of balance between root and shoot growth, since each produces a hormone that favors the growth of the other when present in excess. Negative as well as positive control mechanisms appear to be involved in maintaining balanced growth. For example, one of the effects of the auxin that is produced by the apical meristem is the inhibition of development of lateral buds less than a certain distance from the apex (Fig. 14-29). This effect, which is called apical dominance, can be counteracted by cytokinin treatment without removal of the apical meristem or reduction of the auxin

Fig. 14-30. The release of lateral buds from apical dominance by cytokinin. The plant on the right was treated with a local application of kinetin to the bud 3 days previously. (From Galston, A. W., and P. J. Davies. 1970. Control mechanisms in plant development. Prentice-Hall, Inc., Englewood Cliffs, N.J.)

Fig. 14-31. Witches'-broom formation on a willow tree due to growth of many lateral branches caused by production of cytokinin by an invading plant pathogen. The normal pattern of branching can be seen at the upper right. (From Galston, A. W., and P. J. Davies. 1970. Control mechanisms in plant development. Prentice-Hall, Inc., Englewood Cliffs, N.J. Courtesy Dr. W. A. Sinclair, Cornell University.)

concentration at the level of the bud (Fig. 14-30). Witches'-broom, an abnormal pattern of growth in trees characterized by excessive formation of lateral branches, is a naturally occurring example of reversal of apical dominance (Fig. 14-31). It is caused by excess cytokinins produced by an invading plant pathogen.

Growth inhibitors   The growth inhibitors are a class of plant hormones that tend to promote abscission (release of fruit and leaves) and dormancy. They were discovered independently by groups studying the control of abscission in cotton plants and bud dormancy in deciduous trees (those which lose their leaves each year). Chemically they consist of abscisic acid (ABA) and a group of related compounds. Some investigators prefer the term "dormin" rather than "abscisic acid" to emphasize the role of this compound in promoting dormancy. The level of ABA has been shown to increase in cotton fruits during premature abscission and also during dehiscence (bursting open) of mature fruits. Treatment of cotton plants with ABA causes abscission of the fruits. In deciduous trees the level of ABA in leaves increases in the autumn and is correlated with a decreased rate of shoot growth and with the formation of dormant buds. Formation of dormant buds on the shoots of actively growing seedlings can also be induced by application of ABA.

Abscisic acid

Among the plant hormones, ethylene is unique in that it is a gas at the temperatures at which plants are viable. Ethylene is involved in a variety of developmental processes, including suppression of leaf growth until the seedling emerges from the ground into the light, stimulation of flowering, and abscission, but its best known role is in the ripening of fruit. Endogenously produced ethylene appears to be the trigger stimulus that leads to the softening and removal of acid that characterizes the ripening process in many fruits. Progress of the ripening reaction can be hastened by exposure of unripe fruit to ethylene.

Ethylene

Ethylene

We have seen in this chapter how hormones function during development as signals that trigger the expression of preprogrammed responses in target tissues. In animal cells the specificity of the target tissue response is related both to the presence of specific receptors for the hormone on the cell surface or in the cytoplasm and to the specific differentiated properties of the cell. Some of the molecular mechanisms involved in interactions between hormones and their receptors and their subsequent effects on cellular metabolism, such as changes in phosphorylation of regulatory molecules or changes in transcription of RNA at specific gene loci, are beginning to be understood. However, the molecular basis for the different responses that different target cells can have to a particular hormonal signal remains obscure, although it is undoubtedly related to the molecular events that are responsible for determination of epigenotype (Chapter 10).

RELATIONSHIP OF HORMONAL ACTION TO THE OVERALL PROCESS OF DEVELOPMENT

The special value of hormones in development is that they spread throughout the entire organism and make it possible to coordinate the timing of many different processes during the development of that organism. For example, in amphibian metamorphosis, thyroid hormone is the coordinating stimulus that induces a controlled pattern of widely differing changes in essentially every part of the organism. Similarly, hormones can function to switch the activities of specialized cells on and off in cases in which their activities are only needed periodically, as occurs in the mammary gland and the oviduct. The actions of hormones in plant development are not as fully understood, but it is clear that nearly all aspects of plant development are regulated by complex interactions among several different classes of hormones. Just how complex and widespread such interactions are is shown in Table 14-5, which lists overall hormonal effects, positive and negative, on a variety of developmental and physiological processes in plants.

In our discussion of the role of hormones in development, we have already begun to shift our attention away from a narrow focus on selective gene expression and toward the question of how shape arises, which will be the principal theme of the next four chapters. The role of hormones in the shaping of plants is clear, and it is also obvious that processes such as metamorphosis in a tadpole and the formatin of tubular glands in the chicken oviduct involve changes in shape as well as changes in patterns of protein synthesis. During the integrated discussion of selected developmental systems, which occurs toward the end of this book, we will observe dramatic effects of hormones on morphology, including metamorphosis in insects, as well as sexual differentiation in mammals, where virtually every structural difference between male and female, other than the primary organization of the gonads, is due to the effects of hormones.

BIBLIOGRAPHY
Books and reviews

Banerjee, M. R. 1976. Responses of mammary cells to hormones. Int. Rev. Cytol. 47:1.

Beck, J. C., ed. 1976. Polypeptide hormones: molecular and cellular aspects. Ciba Foundation Symposium 41 (new series). Elsevier-North Holland, Inc., New York.

Beckingham-Smith, K., and J. R. Tata. 1976. The hormonal control of amphibian metamorphosis. In C. F. Graham and P. F. Wareing, eds. The developmental biology of plants and animals. W. B. Saunders Co., Philadelphia.

Cohen, P. P. 1970. Biochemical differentiation during amphibian metamorphosis. Science 168:533.

Czech, M. P. 1977. Molecular basis of insulin action. Annu. Rev. Biochem. 46:359.

Deuchar, E. M. 1975. Cellular interactions in animal development. Chapman & Hall Ltd., London.

Evans, M. L. 1974. Rapid responses to plant hormones. Annu. Rev. Plant Physiol. 25:195.

Filburn, C. R., and G. R. Wyatt. 1974. Developmental endocrinology. In J. Lash and J. R. Whittaker, eds. Concepts of development. Sinauer Associates, Inc., Sunderland, Mass.

Frantz, A. G. 1978. Prolactin. N. Engl. J. Med. 298:201.

Frieden, E., and H. Lipner. 1971. Biochemical endocrinology of the vertebrates. Prentice-Hall, Inc., Englewood Cliffs, N.J.

Galston, A. W., and P. J. Davies. 1970. Control mechanisms in plant development. Prentice-Hall, Inc., Englewood Cliffs, N.J.

Gorski, J., and F. Gannon. 1976. Current models of steroid hormone action: a critique. Annu. Rev. Physiol. **38**:425.

Hall, M. A. 1976. Hormones and differentiation in plants. *In* C. F. Graham and P. F. Wareing, eds. The developmental biology of plants and animals. W. B. Saunders Co., Philadelphia.

Hill, T. A. 1973. Endogenous plant growth substances. Edward Arnold Ltd., London.

Larson, B. L., and V. R. Smith, eds. 1974. Lactation: a comprehensive treatise. Academic Press, Inc., New York.

Litwak, G., ed. 1970-1975. Biochemical actions of hormones. 3 Vols. Academic Press, Inc., New York.

Mepham, B. 1976. The secretion of milk. Edward Arnold Ltd., London.

O'Malley, B. W., and A. R. Means. 1974. Female steroid hormones and target cell nuclei. Science **183**:610.

O'Malley, B. W., H. C. Towle, and R. J. Schwartz. 1977. Regulation of gene expression in eukaryotes. Annu. Rev. Genet. **11**:239.

Palmiter, R. D. 1975. Quantitation of parameters that determine the rate of ovalbumin synthesis. Cell **4**:189.

Robison, G. A. 1971. Cyclic AMP. Academic Press, Inc., New York.

Schally, A. V., A. Arimura, and A. J. Kastin. 1973. Hypothalamic regulatory hormones. Science **179**:341.

Schimke, R. T., G. S. McKnight, and D. J. Shapiro. 1975. Nucleic acid probes and analysis of hormone action in oviduct. *In* G. Litwack, ed. Biochemical actions of hormones. Vol. III. Academic Press, Inc., New York.

Topper, Y. J., and T. Oka. 1974. Some aspects of mammary gland development in the mature mouse. *In* B. L. Larson and V. R. Smith, eds. Lactation: a comprehensive treatise. Vol. I. Academic Press, Inc., New York.

Topper, Y. J., and B. K. Vonderhaar. 1974. The role of critical cell proliferation in differentiation of mammary epithelial cells. *In* B. Clarkson and R. Baserga, eds. Control of proliferation in animal cells. Cold Spring Harbor Laboratory, Cold Spring Harbor, N.Y.

Yamamoto, K. R., and B. M. Alberts. 1976. Steroid receptors: elements for modulation of eukaryotic transcription. Annu. Rev. Biochem. **45**:722.

Selected original research papers

Bartke, A., et al. 1977. Effects of prolactin (PRL) on pituitary and testicular function in mice with hereditary PRL deficiency. Endocrinology **101**:1760.

Cox, R. F. 1977. Estrogen withdrawal in chick oviduct. Selective loss of high abundance classes of polyadenylated messenger RNA. Biochemistry **16**:3433.

Griswold, M. D., M. S. Fischer, and P. P. Cohen. 1972. Temperature-dependent intracellular distribution of thyroxine in amphibian liver. Proc. Natl. Acad. Sci. U.S.A. **69**:1486.

Palmiter, R. D., and N. H. Carey. 1974. Rapid inactivation of ovalbumin mRNA after acute withdrawal of estrogen. Proc. Natl. Acad. Sci. U.S.A. **71**:2357.

Rosen, J. M., et al. 1978. Progesterone-mediated inhibition of casein mRNA and polysomal casein synthesis in the rat mammary gland during pregnancy. Biochemistry **17**:290.

Schwind, J. L. 1933. Tissue specificity at the time of metamorphosis in frog larvae. J. Exp. Zool. **66**:1.

Sibley, C. H., and G. M. Tomkins. 1974. Mechanisms of steroid resistance. Cell **2**:221.

Weinstock, R., et al. 1978. Intragenic DNA spacers interrupt the ovalbumin gene. Proc. Natl. Acad. Sci. U.S.A. **75**:1299.

Morphogenesis

☐ With the exception of the introductory survey of descriptive embryology in Chapter 2, we have until now concentrated almost entirely on events at the cellular and subcellular levels. In addition, most of our discussion of specific mechanisms has been focused on chemical and molecular aspects of development and on the signals that pass from one cell to another in development. Although it is our firm conviction that all aspects of development are the result of interactions among molecules and therefore ultimately explainable at the molecular level, it is also clearly obvious that we must discuss the mechanisms of development at higher levels of structural organization than we have yet undertaken. This section is therefore devoted to a discussion of morphogenesis, or the generation of shape and form, which occurs at all levels of development from the assembly of molecular aggregates to the molding of the body contours of a complete organism.

Morphogenetic processes play many important roles in the development of multicellular organisms. They are involved in such diverse phenomena as the sorting out of germ layers, the formation of a tube from a sheet of cells, and the branching of glandular tissues. Morphogenesis is responsible for the specific pattern or organization of parts within tissues, organs, and ultimately organisms. This organization of parts is basic to the very identity of biological structures. A pancreas, for example, that has been disaggregated into its component cells is really no longer a pancreas. The organ is more than the simple sum of its parts; those parts must be arranged in a specific and organized

471

manner to perform the functions required of the pancreas in the vertebrate body.

Organization is a fundamental characteristic of biological entities, and organization is evident at all levels in biological systems. Atoms are organized into molecules, molecules into organelles, organelles into cells, cells into tissues, etc. At the most fundamental levels, organization is often the result of a self-assembly process. The component parts of particular structures are capable of spontaneously forming those structures under biological conditions without the intervention of enzymes, templates, or any other promoters of organized assembly that are not part of the final structure itself. In those cases, formation of the structure occurs simply because that structure is the thermodynamically favored configuration of its components.

As illustrations of this point, self-assembly is responsible for the folding of polypeptide chains into three-dimensional structures, for the double helical configuration of DNA chains, and for the association of polypeptide subunits into multimeric proteins. Some self-assembling structures, such as collagen fibrils, are aggregates of large numbers of identical or very similar molecules, whereas others such as bacterial ribosomes or chromatin are composed of highly diverse molecular species. Some virus particles with relatively simple structures appear to be formed entirely by self-assembly processes. Tobacco mosaic virus, for example, can be dissociated into protein subunits and nucleic acid, which reassociate readily to form infectious virus particles. Many of the steps in the assembly of far more complex viruses, such as bacteriophage T4, can also occur spontaneously, but at that level of organization we also begin to encounter a few processes that require specific catalytic assembly factors. In all these cases of self-assembly, structural organization is achieved simply by molecules assuming energetically favorable configurations. No direct input of energy is required; the molecules need only to be brought into reasonable proximity to one another under environmental conditions (e.g., ionic strength, temperature) that are favorable for their interaction. Molecular self-assembly and model systems that have been used to study it are described in greater detail in Appendix D.

Although self-assembly clearly must be involved in certain aspects of the development of complex multicellular organisms, one could not hope to obtain an organism such as a mouse merely by mixing its molecular or even cellular components in a flask and adjusting conditions until they are suitable for self-assembly. Nevertheless, mice are somehow routinely assembled during the course of embryonic development, as are also all other types of organisms. This section attempts to deal with the mechanisms currently believed to be involved in this assembly process. A higher organism begins as a single cell, the fertilized egg, from which it progresses to what we call a mouse, or a frog, or a fruit fly. At the same time that the cells of an embryo are multiplying and undergoing determination and differentiation, they are also involved in a variety of morphogenetic processes that collectively result in the production of a complete and fully organized organism.

During the discussions of cellular interactions such as embryonic induction and hormonal effects in the previous section, we pointed out that morphology as well as chemistry is affected when regulatory signals interact with target tissues. However, discussion of the specific mechanisms responsible for the morphological responses was deferred until now.

In this section we begin by discussing morphogenesis at the cellular level in Chapter 15. In addition to the obvious question of how any particular cell acquires its characteristic shape, we will also discuss related topics, including cellular motility, which is very important to many aspects of development, and the effects of changes in shapes of individual cells on the shaping of tissues and the morphogenesis of the entire developing embryo. In Chapter 16 we will shift our attention to interactions among cells and the specific roles played in development by selective cellular adhesion and by the sorting out of cells into relatively homogeneous tissues. In Chapter 17 we will examine the control of cellular multiplication, which is important not only in determining the overall size of the organism but also the relative proportions of its component parts. We will also examine briefly some of the cellular characteristics associated with the loss of control over growth that occurs during malignancy. Finally, in Chapter 18 we will examine the processes that are employed by developing embryos to eliminate tissues that are not needed for further development, with particular emphasis on the role played by cellular death. Although we commonly think of development as a dynamic process of growth and the formation of new structures, selective cellular death is an integral part of the overall developmental program, and if it fails to occur or is prevented, major abnormalities will be produced in the embryo.

This section will complete our discussion of individual mechanisms. After we have added the basic principles of morphogenesis to the information concerning gene expression and cellular differentiation that we have already discussed, we will devote the remainder of this book to an examination of a few selected developmental phenomena at multiple levels of organization ranging from the molecule to the intact organism.

CHAPTER 15

Cellular movement and changes in cellular shape

☐ The importance to morphogenesis of cellular movements and changes in the shapes of cells was well known to early observers of embryonic development. Classical descriptive embryology abounds with accounts of such dynamic processes as the spreading, folding, and rounding up of sheets of cells and the migration of individual cells (Chapter 2). In addition, the study of histology provides many examples of the diversity of cell shapes generated during development (Chapter 3). However, it has been only in recent years that biologists have begun to understand the underlying mechanisms responsible for cellular movement and for changes in cellular shape. This understanding has come largely from the discovery of two subcellular structural elements of higher cells, the microtubules and the microfilaments, and from a study of their behavior. The first part of this chapter is therefore devoted to a review of some of the properties of microtubules and microfilaments.

MICROTUBULES   The widespread occurrence of microtubules was not adequately appreciated during early studies of the ultrastructure of cells because of the fact that their structure was not preserved by the fixatives (e.g., osmic acid) generally used by early electron microscopists. However, with the introduction of cross-linking fixatives such as glutaraldehyde it became evident that microtubules are universal components of eukaryotic cells. Without exception, they have been found in all types of nucleated plant and animal cells, including unicellular protozoa and algae, that have been adequately examined for their presence.

The main structural component of microtubules is a protein called tubulin. Gentle dissociation of microtubules yields a dimer with a molecular weight of 115,000, sometimes referred to as 6S tubulin, which can be further dissociated by denaturing solvents into two similar, but not identical, subunits called α-tubulin and β-tubulin. The two sub-

474

units occur in equal amounts in microtubules, and the stable 6S dimer, which is often referred to simply as "tubulin," is generally viewed as the basic building block for assembly of microtubules. There are also smaller amounts of several other proteins, generally referred to as microtubule-associated proteins (MAPs), that appear to be integral parts of the microtubule structure. At least one such protein may play a role in the regulation of microtubule assembly.

A   B   C

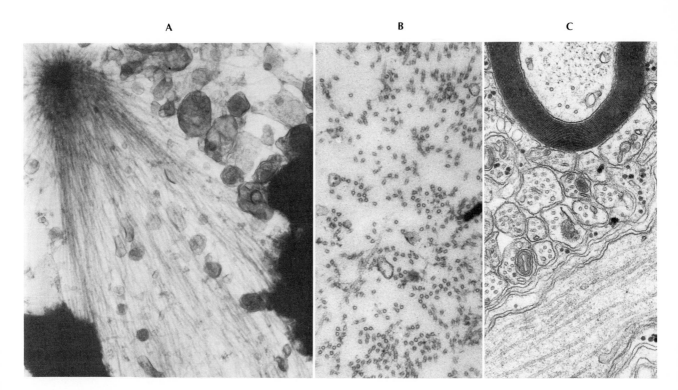

Fig. 15-1. Microtubules. **A,** Longitudinal section through the mitotic spindle of a cultured rat kangaroo cell at prometaphase. A spindle pole is seen at the upper left, with spindle microtubules extending toward the lower right. Condensed chromosomes are seen as dark masses on either side of the bundle of microfilaments. **B,** Cross section of a mitotic spindle similar to that in **A.** Microtubules cut perpendicular to the plane of section exhibit a circular cross section, and those cut obliquely appear elongated. **C,** Thin section of frog cerebellum. Microtubules are seen in cross section in both the myelinated axon at the top and in the nonmyelinated nerve processes (dendrites) below it. Longitudinally sectioned microtubules are present in the dendrite running diagonally across the lower part of the picture. (**A** from McIntosh, J. R., et al. 1975. *In* S. Inoue and R. Stephens, eds. Molecules and cell movement. Raven Press, New York; **B** courtesy Dr. J. R. McIntosh; **C** from Porter, K. R., and M. A. Bonneville. 1968. Fine structure of cells and tissues, 4th ed. Lea & Febiger, Philadelphia.)

A

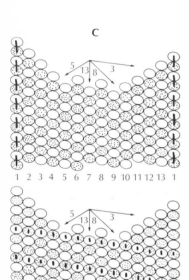

B

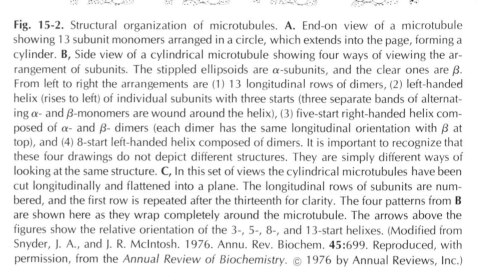

C

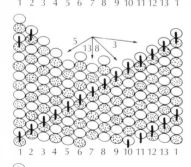

Fig. 15-2. Structural organization of microtubules. **A.** End-on view of a microtubule showing 13 subunit monomers arranged in a circle, which extends into the page, forming a cylinder. **B,** Side view of a cylindrical microtubule showing four ways of viewing the arrangement of subunits. The stippled ellipsoids are α-subunits, and the clear ones are β. From left to right the arrangements are (1) 13 longitudinal rows of dimers, (2) left-handed helix (rises to left) of individual subunits with three starts (three separate bands of alternating α- and β-monomers are wound around the helix), (3) five-start right-handed helix composed of α- and β- dimers (each dimer has the same longitudinal orientation with β at top), and (4) 8-start left-handed helix composed of dimers. It is important to recognize that these four drawings do not depict different structures. They are simply different ways of looking at the same structure. **C,** In this set of views the cylindrical microtubules have been cut longitudinally and flattened into a plane. The longitudinal rows of subunits are numbered, and the first row is repeated after the thirteenth for clarity. The four patterns from **B** are shown here as they wrap completely around the microtubule. The arrows above the figures show the relative orientation of the 3-, 5-, 8-, and 13-start helixes. (Modified from Snyder, J. A., and J. R. McIntosh. 1976. Annu. Rev. Biochem. **45:**699. Reproduced, with permission, from the *Annual Review of Biochemistry.* © 1976 by Annual Reviews, Inc.)

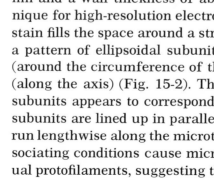

In the electron microscope, microtubules are seen as long, cylindrical structures with a circular cross section and an apparently hollow center (Fig. 15-1). Their length varies from a fraction of a micron to several microns, and in certain special cases they can be much longer. They are uniform in appearance with an outer diameter of about 25 nm and a wall thickness of about 5 nm. Negative staining (a technique for high-resolution electron microscopy in which a heavy metal stain fills the space around a structure without penetrating it) reveals a pattern of ellipsoidal subunits, each of which is about 5 nm wide (around the circumference of the microtubule) and about 4 nm high (along the axis) (Fig. 15-2). The volume of each of these structural subunits appears to correspond to one α- or β-tubulin molecule. The subunits are lined up in parallel rows, forming "protofilaments" that run lengthwise along the microtubules (Fig. 15-2, *B*). Certain mild dissociating conditions cause microtubules to separate into the individual protofilaments, suggesting that the stable α-β dimer bonds are oriented lengthwise along the protofilaments.

Most microtubules are composed of 13 protofilaments, joined side to side to form a cylinder, as shown in Fig. 15-2. However, variants are also known with 15 subunits around their circumference (crayfish sperm arms and cockroach epidermis) and with only 12 (crayfish nerve cord). In addition to being arranged into longitudinal protofilaments, the tubulin subunits are also offset slightly from one protofilament to the next so that they form a spiral or helical pattern, as shown in cylindrical form in Fig. 15-2, *B*, and in flattened form in Fig. 15-2, *C*. The α- and β-subunits are arranged in a strictly alternating pattern, as indicated in these figures. Several ways of viewing the organization of the subunits are shown in Fig. 15-2, but it must be emphasized that these are simply different mental devices for viewing a structure that remains the same no matter how we look at it. The easiest arrangements to visualize are either 13 parallel protofilaments offset from each other just enough so that by following one row of subunits completely around the microtubule, one moves three subunits along its length, or else a helical structure composed of three separate rows of subunits wound around the microtubule (referred to as a three-start helix). The three-start helix does not, however, take into account the fact that microtubules appear to be assembled from dimers. Therefore, if we are going to view the microtubule as a helical structure, it is better to view it either as a five-start right-handed helix or an eight-start left-handed helix, both of which are also shown in Fig. 15-2.

A definite polarity, with all the α-ends of the tubulin dimers pointing one way and all the β-ends pointing the other way is inherent in the structure of microtubules. With an odd number of protofilaments arranged into a cylinder, it is impossible to construct a simple model (such as adjacent protofilaments pointed in opposite directions) that does not have an inherent polarity. Recent ultrastructural and x-ray crystallographic studies verify the polarized arrangement of dimers, as shown in the three-dimensional model in Fig. 15-3.

Microtubules occur as structural units in the cytoplasm and as major components of the mitotic and meiotic spindles. They are also major components of cilia and flagella, which typically contain nine fused microtubular doublets surrounding a central pair of single microtubules (Fig. 15-4). Animal cells also contain complex microtubular structures known as centrioles, which appear to serve as foci for microtubular growth during formation of the mitotic and meiotic spindles (Fig. 15-5) and also as "seeds" for the polymerization of tubulin to form flagella and cilia (Fig. 15-6). Centrioles are composed of fused microtubular triplets, typically with nine triplets arranged in a cylindrical shape, as shown in Fig. 15-5, *C*. Some protozoa contain even more complex microtubular structures, such as axopodia and axostyles, which will be discussed later in this chapter.

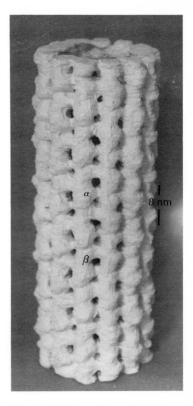

Fig. 15-3. Three-dimensional model of the microtubule structure described in Fig. 15-2. The α- and β-subunits are depicted as similar but not identical, giving the structure an 8 nm periodicity along the axis of the microtubule. (From Amos, L. A., and A. Klug. 1974. J. Cell Sci. **14:**523.)

The mechanisms involved in assembly of microtubules from tubulin and the means by which the timing and location of microtubule assembly are controlled are currently areas of active research. It has been known for some time that cytoplasmic microtubules, as well as the microtubules in certain more ordered structures such as the mitotic spindle and the axopodia of *Echinosphaerium,* undergo disassembly at low temperature and reassemble when the temperature is returned to normal. These findings indicate that the microtubules are formed by protein-protein interactions (e.g., hydrophobic bonds) that depend on the state of the solvent. Guanosine triphosphate (GTP) must be bound to the tubulin to obtain assembly. Hydrolysis to GDP and phosphate normally occurs during assembly, but it has been shown with a nonhydrolyzable analog of GTP (5' guanylylmethylene diphosphonate, GMPPCP) that GTP hydrolysis is not essential for assembly and that microtubules containing GMPPCP can be dissociated and reassociated several times by cycles of low and high temperatures.

The ability of microtubules to assemble and dissociate in response to changes in the intracellular environment may have important implications for the control of microtubule involvement in various developmental processes, as will be discussed later in this chapter. One recent series of studies has shown that under equilibrium conditions,

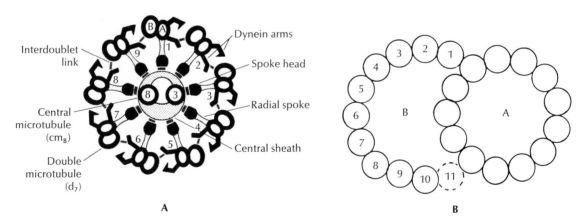

Fig. 15-4. Arrangement of microtubules in flagella and cilia. **A,** Schematic cross section of a cilium showing the arrangement of microtubules and other associated macromolecular components. A ring of nine microtubule doublets surrounds two single microtubules associated with the central core of the cilium. The dynein arms associated with the *A* tubule of each doublet contain ATPase activity that permits the cilia to convert biochemical energy to motion. **B,** Schematic representation of protofilaments in a doublet, based on negative-staining studies. The portion of the doublet designated *A* is a normal microtubule composed of 13 protofilaments, whereas *B* consists of 10 or 11 protofilaments attached to it to form a second partial tubule. (**A** from DeRobertis, E. D. P., et al. 1975. Cell biology, 6th ed. W. B. Saunders Co., Philadelphia; **B** from Tilney, L. G., et al. 1973. J. Cell Biol. **59:**267.)

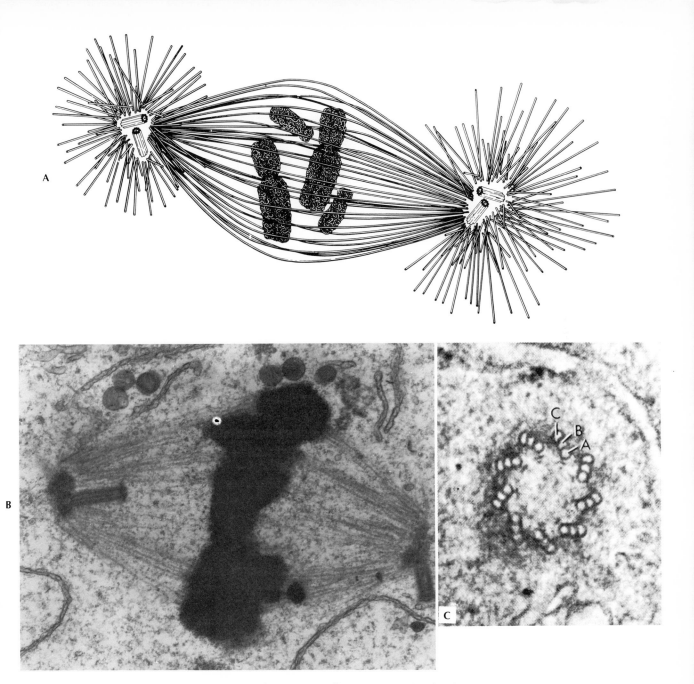

Fig. 15-5. Relationship of centrioles to mitotic and meiotic spindles. **A,** Interpretive drawing of mitotic apparatus, showing centriole pairs, characteristically oriented at right angles, at poles, with microtubules radiating out from them. **B,** Electron micrograph of second meiotic metaphase in a chicken testis. Centrioles are plainly visible at both poles. **C,** Transverse section through a centriole from a cell in a chicken embryo. The fused microtubule triplets *(A, B, C)* can be seen clearly. (**B,** ×18,000.) (**A** from Dupraw, E. J. 1968. Cell and molecular biology. Academic Press, Inc., New York; **B** from McIntosh, J. R. 1974. J. Cell Biol. **61:**166; **C** from DeRobertis, E. D. P., et al. 1975. Cell biology, 6th ed. W. B. Saunders Co., Philadelphia.)

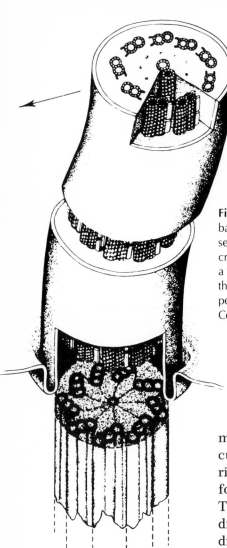

Fig. 15-6. Interpretive drawing of the basal region of a flagellum, showing the underlying basal body (centriole) with its characteristic arrangement of nine microtubule triplets. In serial sections an orderly transition can be seen from the nine doublets plus two single microtubules of the flagellum to the nine triplets of the centriole. Note that the basal body has a "wagon wheel" configuration with a single central microtubule inside the nine triplets at the level shown here. The arrows indicate the direction of flagellar motion, which is perpendicular to the plane defined by the two central microtubules. (From DuPraw, E. J. 1968. Cell and molecular biology. Academic Press, Inc., New York.)

microtubule assembly may occur at one end while disassembly is occurring at the other end. The energy for this process is apparently derived from hydrolysis of GTP. Although GTP hydrolysis is not essential for microtubule assembly, it normally does occur at the growing end. The resultant GDP remains tightly bound to the subunits and favors disassembly at the end opposite the growing point but does not lead to disintegration of the central portion of the microtubule. Reassembly occurs only when GTP, which has a slightly higher affinity for tubulin, displaces GDP from the dissociated tubulin dimers. Adenosine triphosphate (ATP) will also promote assembly, provided that GDP is first stripped from the tubulin, but the affinity of ATP for the subunits is not strong enough to displace the GDP competitively. Thus ATP will not cause spontaneous assembly of GDP-tubulin into microtubules Speculative models have been proposed in which growth of microtubules at one end and their dissociation at the other end might provide a kind of biological "conveyor belt" for movement of intracellular structures such as the chromosomes in the mitotic spindle.

Assembly of microtubules in vitro is highly dependent on ring-shaped aggregates of tubulin that serve as nucleation centers for growth of the microtubules. It is not clear, however, that microtubule assembly in vivo depends on these tubulin aggregates. Microtubule formation in cells is usually initiated at defined structures such as the poles of the mitotic spindle or the basal bodies of flagella, although some foci of microtubule initiation are more amorphous. The sites of microtubule formation are known collectively as microtubule organizing centers (MTOCs). In animal cells a substantiated part of the MTOC activity appears to be closely related to centrioles, as described earlier.

There is suggestive evidence for specific control systems that regulate microtubule assembly, but they are not yet well understood. Unfertilized clam eggs contain tubulin in an aggregated form that is different from microtubules. These aggregates appear to be storage forms for tubulin that will be used in spindle formation during cleavage. MTOCs isolated from activated clam eggs support formation of asterlike arrays of microtubules from tubulin that has been isolated either from activated or unactivated eggs, whereas MTOCs from unactivated eggs do not. This observation suggests that specific changes in the MTOCs induced at or soon after fertilization or artificial activation give them the ability to initiate microtubule growth and aster formation. Similarly, MTOCs isolated from cultured mammalian cells at the time that mitotic spindles normally are formed initiate microtubule growth more efficiently than MTOCs isolated at earlier stages of mitosis. Divalent cations, especially Mg^{2+}, are essential for polymerization of microtubules, but an excess, particularly of Ca^{2+}, will inhibit the process. One of the MAPs, designated τ (tau), also appears to be essential for assembly of microtubules.

Microtubules, as seen in the electron microscope, appear to be rigid with no sharp bends. They are believed to function as a type of supporting structure, commonly referred to as a "cytoskeleton," giving shape to cells. However, over a longer distance, microtubules are clearly able to move and bend in relation to each other, as is seen, for example, in flagella and cilia.

The generation of motion by structures that contain microtubules is typically associated with cross bridges between the tubules. An interesting example occurs in *Saccinobaculus*, an anaerobic flagellate protozoan that lives in the gut of a wood-feeding roach and helps its host to digest cellulose. That organism possesses a motile ribbonlike intracellular structure known as an axostyle, which is composed of microtubules with conspicuous cross bridging that is believed to be involved in motility (Fig. 15-7). Similarly, in cilia and flagella, arms attached to the "A" subfiber of each doublet consist of an ATP-hydrolyzing enzyme called dynein. These arms bind transiently to the neighboring tubule and are believed to provide the energy needed for movement (Fig. 15-4, *A*). In flagella it is well established that a sliding motion of microtubules past one another, driven by hydrolysis of ATP, is responsible for movement. Formation of cross bridges and sliding of microtubules has also been suggested as a possible cause of chromosomal movement in the mitotic spindle, but alternate models such as the "conveyor belt" that has already been described must also be considered.

Study of the functions of microtubules has been greatly facilitated by use of drugs such as colchicine, Colcemid (demecolcine), and vinblastine, which bind specifically to tubulin dimers. These drugs prevent assembly of new microtubules and in most cases also cause disaggregation of existing microtubules by reducing the concentration of soluble tubulin and thereby disrupting the normal equilibrium that exists between the microtubules and free tubulin. As an example, colchicine is frequently used to arrest cells at the mitotic phase of the cell cycle for karyotypic analysis of the condensed mitotic chromosomes. Completion of mitosis is prevented by disruption of the mitotic spindle.

The extreme length of microtubules and the fact that they usually do not lie entirely within the plane of a single thin section makes an analysis of their morphological role by standard ultrastructural tech-

Fig. 15-7. Cross-links between microtubules in the axostyle of *Saccinobaculus ambloaxostylus*. **A,** Cross section of the axostyle showing the regular pattern of links between microtubules within each row. **B,** Enlarged area showing occasional links between rows. **A,** ×85,000; **B,** ×310,000. (From McIntosh, J. R., et al. 1973. J. Cell Biol. **56:**304.)

niques difficult. Although microtubules are too small to resolve directly with a light microscope, the use of indirect immunofluorescence techniques has made it possible to study their intracellular distribution at light microscope magnifications. In this technique the cells to be examined are first treated with antibody to tubulin and then with a fluorescent antiglobulin that binds to the antibody directed against tubulin. This technique permits a clear visualization of the location and configuration of microtubules in an entire cell (Fig. 15-8). Colchicine can be used to demonstrate that the structures visualized are actually microtubules (Fig. 15-9, *A* and *B*).

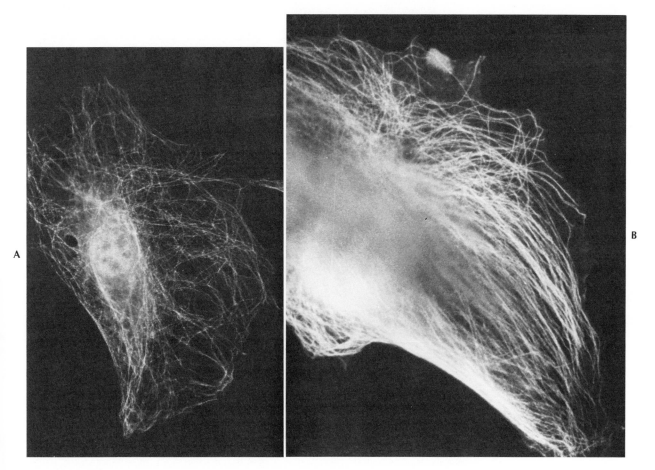

Fig. 15-8. Visualization of microtubules by indirect immunofluorescence. **A,** Extensive network of cytoplasmic microtubules in an untransformed mouse 3T3 cell. **B,** Higher magnification of a portion of a 3T3 cell showing microtubules terminating at the cell surface. (**A** from Brinkley, B. R., et al. 1977. Cancer Bull. **29**:13; **B** courtesy B. R. Brinkley.)

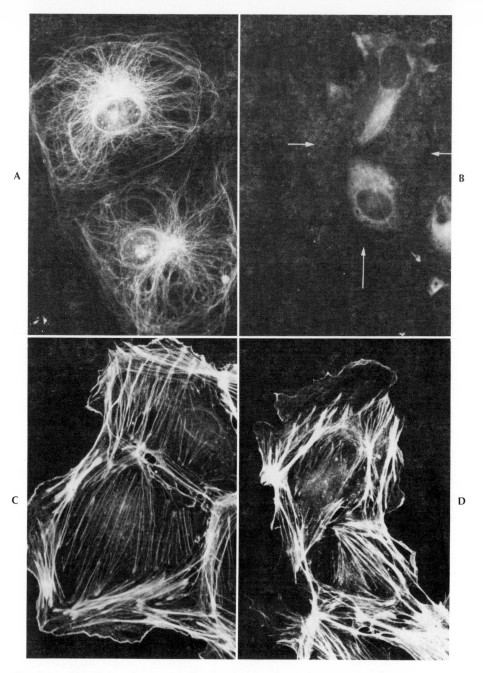

Fig. 15-9. Specificity of microtubule visualization by indirect immunofluorescence. **A,** Pt K2 cells stained with antibody against tubulin. The characteristic radiating network of cytoplasmic microtubules is clearly evident. **B,** Colchicine-treated Pt K2 cells stained with antibody against tubulin. The cytoplasmic network is no longer visible. White arrows indicate the cell boundaries. Nuclear fluorescence in **A** and **B** is believed not to be due to tubulin, since it is not seen with other species of cells. **C,** Pt K2 cells stained with antibody against actin. Cytoplasmic fibers consisting of bundles of microfilaments are clearly visible. **D,** Colchicine-treated Pt K2 cells stained with antibody against actin. Colchicine has not significantly altered the distribution of actin filaments. (From Osborn, M., W. W. Franke, and K. Weber. 1977. Proc. Natl. Acad. Sci. U.S.A. **74:**2490.)

Microtubules in normal cells show a branching pattern, often radiating from one or two centers near the nucleus. This pattern, which is evident in Fig. 15-8, can be seen even more clearly in cells that are recovering from treatment with agents that disrupt microtubules, such as Colcemid (Fig. 15-10). As recovery proceeds, the cytoplasmic microtubule complex can be seen to spread outward from the organizing centers until it once again fills the entire cytoplasm.

In addition to their roles in processes involving motility that have been discussed and their roles in morphogenesis, which will be discussed, microtubules have also been shown to be involved in such processes as protein and hormone secretion and in determining the arrangement of molecules on cell surfaces. Procedures that affect the cell surface, such as treatment of cultured cells with trypsin, have a major disruptive effect on the cytoplasmic microtubule complex.

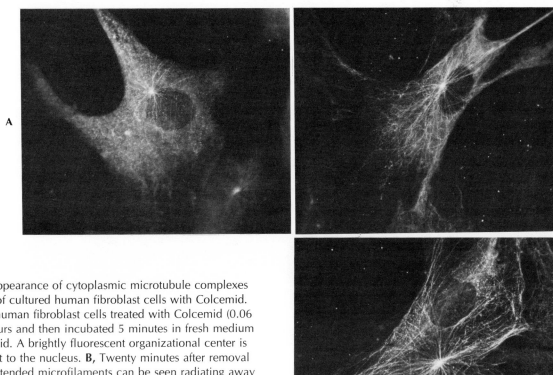

Fig. 15-10. Reappearance of cytoplasmic microtubule complexes after treatment of cultured human fibroblast cells with Colcemid. **A,** Strain PA-2 human fibroblast cells treated with Colcemid (0.06 μg/ml) for 2 hours and then incubated 5 minutes in fresh medium without Colcemid. A brightly fluorescent organizational center is evident adjacent to the nucleus. **B,** Twenty minutes after removal of Colcemid, extended microfilaments can be seen radiating away from the bright center. **C,** Sixty minutes after removal of Colcemid the cytoplasm is nearly filled with the cytoplasmic microtubule complex, including long radiating microtubules spreading to the edges of the cell. (From Brinkley, B. R., et al. 1978. *In* G. F. Saunders, ed. Cell differentiation and neoplasia. Raven Press, New York.)

MICROFILAMENTS   Microfilaments are one of several types of filamentous structures commonly found in plant and animal cells. They have a diameter of about 6 nm and appear solid in cross section. These characteristics permit them to be distinguished both from microtubules and from other types of cytoplasmic filaments, which are generally larger in diameter. Microfilaments are usually found in parallel array in groups or bundles beneath the cell membrane. Recent evidence indicates that microfilaments are polymers of an actinlike protein. Although slightly different than muscle actin, the protein of microfilaments is classified as an actin, since it can be "decorated" with heavy meromyosin either in situ or after isolation (Fig. 15-11). (Heavy meromyosin is a fragment of myosin that is specifically bound by actin to give a characteristic "arrowhead" pattern.) Detailed amino acid sequence studies suggest that the actins of microfilaments are coded for by at least two different genes and that they differ in primary structure from skeletal muscle actin in at least 25 positions. (There is also evidence suggesting that heart muscle actin, smooth muscle actin, and skeletal muscle actin are all different from one another.)

Microfilament bundles can be visualized in the light microscope by indirect immunofluorescence with antibody to actin, as shown in Fig.

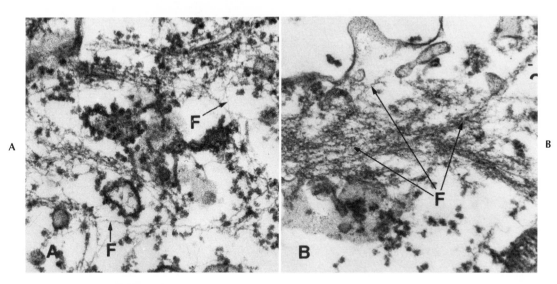

Fig. 15-11. Reaction of microfilament actin with heavy meromyosin. **A,** Microfilaments *(F)* of a control salivary gland cell showing normal smooth appearance. **B,** A similar cell treated with heavy meromyosin, which has interacted specifically with the actin of the microfilaments to "decorate" them. At higher magnifications the complexes between actin and heavy meromyosin exhibit a characteristic "arrowhead" pattern, which is just perceptible at the tip of the middle arrow. Both cells were treated with glycerin to make them permeable to large molecules. (From Spooner, B. S. 1973. Am. Zool. **13:**1007.)

15-9, C, and D. The pattern of filaments seen with this technique is the same as the microfilament pattern seen with EM. Immunofluorescence techniques have also shown that other proteins involved in muscle contraction, including α-actinin, myosin, and tropomyosin, are closely associated with microfilaments. As is the case with actin, the myosin of nonmuscle cells appears to be different from that of muscle cells.

Microfilaments have been implicated in contractile activities of cells, both from cytological studies and as a result of experiments using the fungal metabolite cytochalasin B. This drug inhibits a variety of cellular processes that involve contractility. It is presumed to operate by disruption of microfilaments, an effect that can be demonstrated ultrastructurally. Questions have been raised about the drug's specificity of action, since it also affects a number of membrane properties and could thus act primarily by alteration of a membrane structure or function that is essential to microfilament action. However, the fact that microfilaments disappear ultrastructurally when certain processes are disrupted by the drug is strong circumstantial evidence for their involvement in those processes. Cellular motility probably also involves the other musclelike proteins that are found in the cytoplasm of nonmuscle cells. Those proteins are known to interact with the actin of microfilaments, but the mechanisms by which the interactions induce motility in the nonmuscle cells are almost totally unknown at present.

Structures rich in microfilaments are associated with a variety of types of cellular movement and cellular shape change. One example is the process of cytokinesis (cytoplasmic division) that occurs immediately after karyokinesis (nuclear division) during both mitosis and meiosis. A beltlike ring of microfilaments encircles the equatorial region of the anaphase cell and begins to draw up tighter and tighter, literally pinching the cell in two. Cytochalasin B disrupts the belt of microfilaments, and if applied at the right time during mitosis, can result in the formation of a binucleate cell. Such a treatment has been used, for example, to show that cytokinesis is not necessary for skeletal muscle cell differentiation following a "quantal" cycle of DNA synthesis and karyokinesis, both of which are claimed to be essential (Chapter 11). Similarly, cytochalasin B has been used to show that in ascidian embryos with mosaic development, blockage of further cleavage does not prevent the appearance on schedule of specific enzymatic activities in the precursor cells for tissues that would normally acquire the activities only after further cleavage (Chapter 12).

We have already seen that normal cell division (karyokinesis plus cytokinesis) requires the participation of both microtubules and microfilaments. The examples that follow illustrate the variety of roles that these two subcellular structures play in morphogenesis as a result of their involvement in cellular movement and cellular shape change. Some writers have attempted to form a simple analogy in which microfilaments are viewed as cellular "muscles" and microtubules as a sort of cellular "skeleton." However, the total picture is probably far more complex.

FORMATION OF THE NEURAL TUBE AND LENS VESICLE

Formation of the neural tube and formation of the lens vesicle are analogous morphogenetic processes in vertebrates (Fig. 15-12). Both processes begin with cells arranged in an epithelial layer. The cells undergo elongation to a more columnar shape, resulting in the formation of the neural plate and lens placode, respectively. Next, these structures constrict at their apical surfaces to produce a curvature in the epithelium, which forms the neural groove and lens vesicle, respectively. Finally, cells at the edges of the invagination meet and

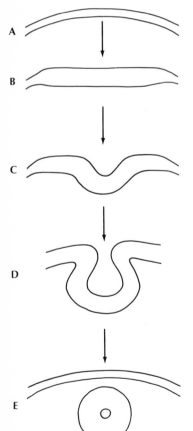

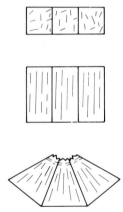

Fig. 15-12. Morphological changes involved in formation of the neural tube and lens vesicle. **A,** Ectodermal layer prior to overt differentiation. Microtubules (and microfilaments) are random. **B,** Elongation of cells to form the columnar epithelium of the neural plate or lens placode. Microtubules are seen oriented along the long axis of the cells. **C,** Formation of the neural groove or lens vesicle. Contraction of microfilaments occurs at the apex of each cell. **D,** Continuation of the processes in **C. E,** Formation of a closed tube or vesicle by fusion of the edges. (From Spooner, B. S. 1974. *In* J. Lash and J. R. Whittaker, eds. Concepts of development. Sinauer Associates, Inc., Sunderland, Mass.)

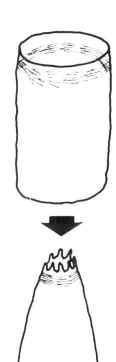

Fig. 15-13. Role of microfilaments in cell shape change occurring during invagination. A "purse-string" contraction of microfilaments causes the apex of each cell to contract. The collective effort of such contraction in a sheet of adjacent cells is the invagination process shown in Fig. 15-12. (Modified slightly from Wessells, N. K., et al. 1971. Science **171**:135. Copyright 1970 by the American Association for the Advancement of Science.)

form a closed structure that becomes separate from the overlying epithelium, the result being the definitive neural tube or lens.

The first step in these analogous processes, cellular elongation, has been correlated with the presence of microtubules oriented along the long axis of the cells. Microtubules in cells that are not undergoing elongation are more randomly oriented. In addition, colchicine has been shown to inhibit cellular elongation in neural tube formation. The precise role of the microtubules in cell elongation is not clear. Directed polymerization of microtubules may cause the cell to be pushed outward in the direction of polymerization. Or elongation may result from the sliding of microtubules in relationship to one another. Once cell elongation has occurred and invagination begins, however, intact microtubules are apparently no longer needed for these processes, since colchicine does not inhibit further invagination.

The cell shape change that produces invagination seems to be due primarily to the microfilaments. Microfilaments are seen at the apex of each cell just prior to invagination. In the optic vesicle they encircle the apical ends of the cells and are believed to narrow the apexes by a kind of purse-string action (Fig. 15-13). The process is similar in the neural tube, except that it occurs in only one direction, producing a cylinder rather than a closed vesicle. Treatment of the invaginating neural tube with cytochalasin B causes it to open up and flatten. Formation of tubular glands in the chicken oviduct (Chapter 14) proceeds by a similar mechanism.

The roles of microtubules and microfilaments have been examined in the morphogenesis of a number of glandular tissues, including the salivary gland. Microfilaments are responsible for the changes in cell shape that cause the epithelial layers of the glands to form complex patterns of branched outpocketings. Epithelial cells in the developing salivary gland have bands of microfilaments both at their apical ends (next to the lumen) and at their basal ends (next to the basement membrane). Branching is thought to occur by contraction of microfilaments at the apical and basal ends of different groups of cells (Fig. 15-

SALIVARY GLAND MORPHOGENESIS

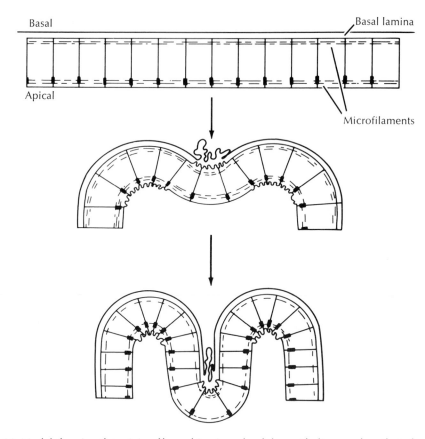

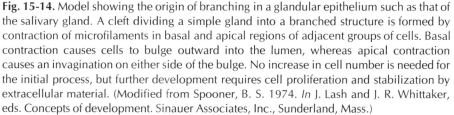

Fig. 15-14. Model showing the origin of branching in a glandular epithelium such as that of the salivary gland. A cleft dividing a simple gland into a branched structure is formed by contraction of microfilaments in basal and apical regions of adjacent groups of cells. Basal contraction causes cells to bulge outward into the lumen, whereas apical contraction causes an invagination on either side of the bulge. No increase in cell number is needed for the initial process, but further development requires cell proliferation and stabilization by extracellular material. (Modified from Spooner, B. S. 1974. *In* J. Lash and J. R. Whittaker, eds. Concepts of development. Sinauer Associates, Inc., Sunderland, Mass.)

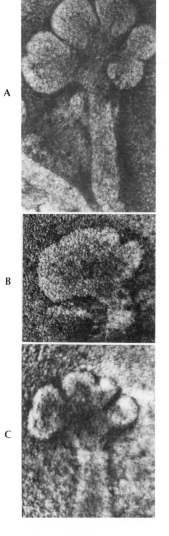

Fig. 15-15. Inhibition of morphogenesis of cultured mouse salivary gland with cytochalasin B. **A,** Control gland after 24 hours of culture exhibits numerous deep clefts. **B,** Gland treated with cytochalasin B for 24 hours has a simple rounded shape with essentially no clefts. **C,** Twenty-three hours after transfer to medium without cytochalasin B, the gland shown in **B** has formed numerous clefts. (From Spooner, B. S. 1974. *In* J. Lash and J. R. Whittaker, eds. Concepts of development. Sinauer Associates, Inc., Sunderland, Mass.)

14). When developing salivary glands are cultured in the presence of cytochalasin B, morphogenesis ceases and the epithelium flattens and loses its clefts. If the cytochalasin B is removed, the cells recover and morphogenesis continues (Fig. 15-15). Microfilaments are not seen in electron micrographs of these cells after cytochalasin B treatment, but they reappear in the cells during recovery from the drug. The reappearance of microfilaments does not require protein synthesis, which suggests that they are reassembled from preexisting components that are not destroyed by the drug treatment.

Although microfilaments are necessary for the formation of clefts in glandular tissue and their initial maintenance, they are not required for maintaining the shape of mature clefts, which appear to be stabilized by extracellular materials such as collagen and mucopolysaccharides. In addition, the morphology that is assumed by the epithelium is largely determined by the type of mesoderm that is used, as was discussed in Chapter 13 (Figs. 13-4, 13-7, and 13-11).

The role of microtubules in salivary gland morphogenesis is more indirect than that of microfilaments. Colchicine treatment of developing glands does not cause flattening of the clefts nor does it inhibit the progress of cleft formation once it has begun. However, the drug does prevent the formation of new clefts. This effect is likely to be due to the inhibition of mitosis by colchicine, which would suggest a critical role for cellular proliferation in this morphogenetic process.

The formation of axons during development of a neuroblast into a neuron is an example of a morphogenetic process that occurs entirely within an individual cell, as opposed to the previous examples, which have involved coordinated changes in many cells to achieve morphogenesis at the tissue level. The basic mechanisms involved are similar, however. The axon of a developing neuroblast undergoes all its elongation at the distal tip, which is called the growth cone. This region undulates and produces projections termed microspikes that protrude and withdraw as the axon moves over its substratum (Fig. 15-16). The microspikes are filled with a fine network of microfilaments, and the body of the axon contains numerous microtubules aligned with the long axis of the axon. If cultured neuroblasts are treated with colchicine, their axons start to collapse after about 30 minutes of treatment, and eventually the axon is completely withdrawn into the cell body (Fig. 15-17). Treatment of neuroblasts with cytochalasin B, on the other hand, has little effect on the length of the axon but causes retraction of the microspikes, which results in a cessation of axon elongation (Fig. 15-17). From these results it appears that the microtubules, which run longitudinally through the axon (Fig. 15-1, C), func-

AXON FORMATION IN NEUROBLAST DEVELOPMENT

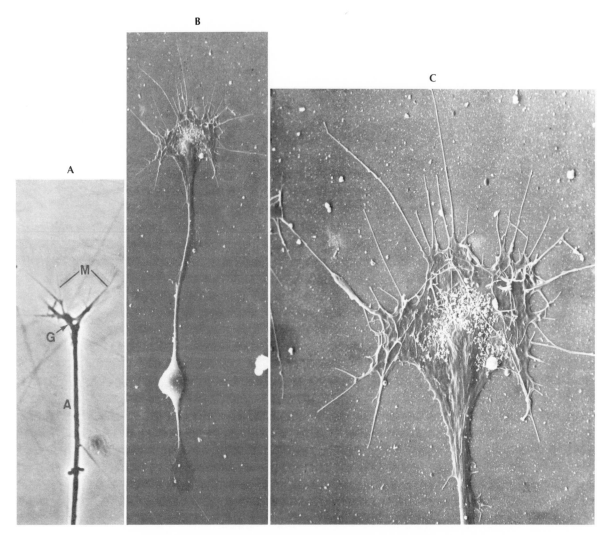

Fig. 15-16. Growth cone of a nerve axon. **A,** Phase contrast micrograph showing growth cone and microspikes of a dorsal root ganglion cell. A = Axon; G = growth cone; M = microspikes. **B,** Scanning electron micrograph of a single parasympathetic neuron. **C,** Enlarged view of the growth cone in **B.** Long, thin microspikes can be seen clearly. (**A** from Yamada, K. M., et al. 1971. J. Cell Biol. **49**:614; **B** and **C** from N. K. Wessels. Tissue interactions and development. Benjamin/Cummings Publishing Co., Menlo Park, Calif., 1977.)

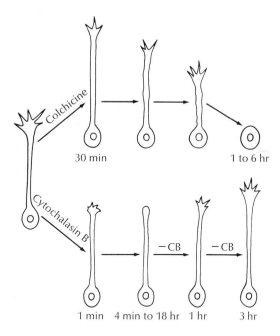

Fig. 15-17. Schematic diagram showing effects of colchicine and cytochalasin B on axon growth in cultured nerve cells. Colchicine treatment causes the axon to retract into the cell body. Cytochalasin B *(CB)* treatment causes retraction of microspikes without collapse of the axon and results in cessation of axon elongation. Removal of cytochalasin B results in a reappearance of microspikes and resumption of axon growth. (From Spooner, B. S. 1974. *In* J. Lash and J. R. Whittaker, eds. Concepts of development. Sinauer Associates, Inc., Sunderland, Mass.)

tion as cytoskeletal elements maintaining the elongated shape of the axon and that microfilaments play a role in the actual elongation process, possibly by providing the contractile activity necessary for movement of the growth cone over the substratum.

CELL MOVEMENT IN CULTURE

Microtubules and microfilaments have also been implicated in the movement of cells growing in culture. Although cells undergoing migration in vivo do not typically show the same morphology as those in vitro, it is still possible that some of the same mechanisms may be involved in cell movement in both environments.

Migrating cells in culture assume a roughly triangular form. Each cell has a broad area of ruffled membranes, or lamellipodia, at its leading edge and tapers to a point at the back (Fig. 15-18). The undulating membranes at the front of the cell are filled with actin-containing microfilaments. In addition, heavy actin-containing fibers can be seen extending from the leading edge to the point at the rear of the cell (Fig. 15-18, *C*). When migrating cells in culture are treated with cytochalasin B, membrane ruffling and locomotion both cease (Fig. 15-19). Both activities are resumed after removal of the cytochalasin B.

Microtubules are also seen in migrating cells and are aligned roughly along the direction of cell movement, particularly in the pointed rear portion of the cell, which has been likened by some investigators to a cellular rudder. Treatment of migrating cells with colchicine

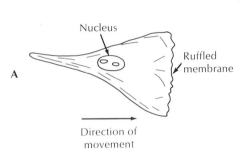

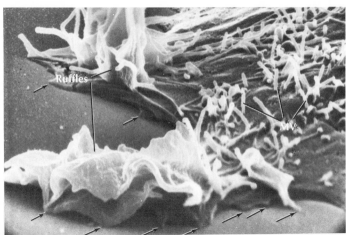

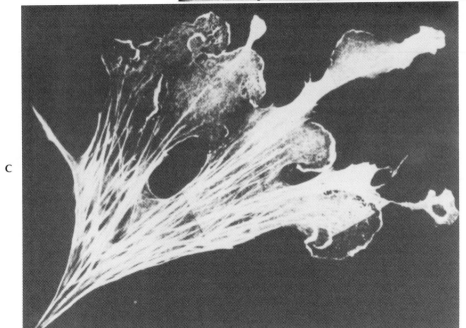

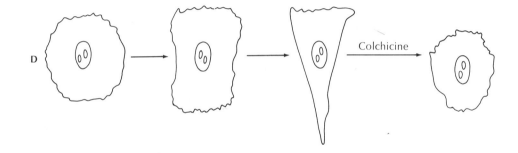

Fig. 15-18. Morphological characteristics of migrating cells in culture. **A,** Schematic diagram showing ruffled membrane at the leading edge and the pointed trailing portion of the cell. **B,** Scanning electron micrograph showing a portion of the leading edge of a cultured mouse cell. Arrows indicate points of attachment to the substrate. Lifting of the ruffled membranes above the culture surface is clearly evident. MV = Microvilli. **C,** A single mouse cell moving on a glass slide visualized by indirect immunofluorescence of antibody to actin. The ruffled membranes are rich in actin, and actin-containing filaments can be seen to run to the point at the rear of the cell. **D,** Sequential steps in the spreading of a trypsinized cell on a culture surface. Initially, ruffled membranes are seen around the entire periphery of the cell, but they soon become concentrated at the leading edge. Colchicine causes a loss of directionality and resumption of ruffling around the entire periphery. (**A** and **D** modified from Spooner, B. S. 1974. *In* J. Lash and J. R. Whittaker, eds. Concepts of development. Sinauer Associates, Inc., Sunderland, Mass.; **B** from Revel, J. P. 1974. Symp. Soc. Exp. Biol. **28:**447; **C** from James D. Watson, Molecular biology of the gene, 3rd ed., copyright © 1976, 1970, 1965 by James D. Watson and The Benjamin/Cummings Publishing Co., Inc., Menlo Park, Calif.)

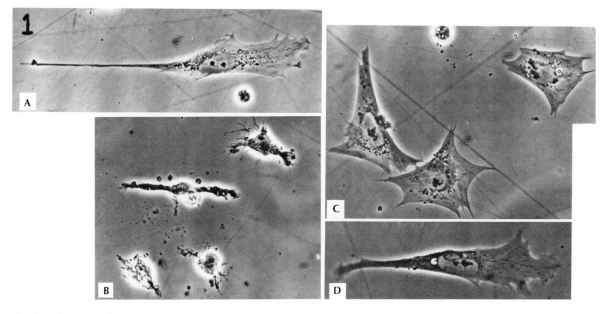

Fig. 15-19. Effect of cytochalasin B on cell motility. **A,** Phase contrast micrograph of a migrating glial cell in culture with an undulating membrane at the leading edge (right) and extreme elongation of the cytoplasm. **B,** After 18 hours in cytocholasin B, glial cells are rounded with some long, thick processes. These cells show no migratory behavior. **C,** Three hours after removal of cytochalasin B the cells have again flattened onto the substratum and are beginning to resume migratory activity. **D,** Twenty-two hours after removal of cytochalasin B the cells are again elongated with undulating membranes and are migratory. (From Spooner, B. S., et al. 1971. J. Cell Biol. **49:**595.)

does not inhibit either ruffling or locomotion in general, but it does affect the directionality of cell movement. Cells normally move in approximately straight lines, altering their direction only after contact with another cell or an obstacle. In the presence of colchicine, however, cells move more randomly, and ruffling is seen around the entire periphery of the cell instead of at a single edge (Fig. 15-18, *D*). Microtubules thus seem to play a role in the polarity of cell movement, whereas microfilaments appear to be basic to cell motility itself.

GASTRULATION   Gastrulation is one of the most spectacular and important morphogenetic processes in embryonic development. The basic process of gastrulation and its role in shaping the embryo were described in Chapter 2. The following discussion is focused on specific mechanisms of cell movement and cell shape change that occur during gastrulation in representative species.

Sea urchin   Gastrulation in the sea urchin is usually considered to begin after the formation of the mesenchyme blastula. It is composed of two major phases, which can be seen in Fig. 15-20. In the first phase of gastrulation the portion of the blastula surface that will become the gut undergoes a partial invagination, moving about a third of the total distance that it will eventually move. This movement is the result of forces within the cells at the vegetal end of the embryo. The second phase completes the invagination of the gut and is due to the interaction of gut cells with cells on the roof of the blastocoel.

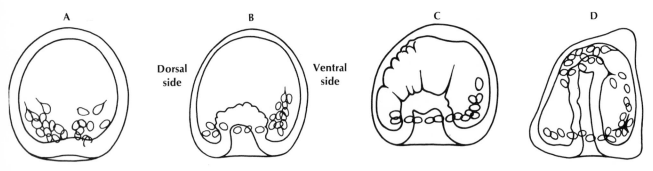

Fig. 15-20. Diagram of some stages seen in sea urchin gastrulation. **A,** Mesenchyme blastula at the start of gastrulation. The vegetal pole has thickened, and cells that will form the primary mesenchyme are moving inward. **B,** Early gastrula at the end of the first phase of gastrulation. The gut has partially invaginated. **C,** Gastrula at the start of the second phase of gastrulation showing filopodia extending from the gut toward the dorsal end of the blastocoel. **D,** Late gastrula after completion of invagination. (From Gustafson, T., and L. Wolpert. 1963. Int. Rev. Cytol. **15:**139.)

The mechanisms involved in the first stage of gastrulation are not completely understood. By time-lapse cinematography, cells in the vegetal region can be seen to pulsate, round up, and lose some contact with each other at the time of invagination. However, they retain their contact with the extracellular hyaline layer. Such a rounding up of cells would normally cause a sheet of cells to increase in length. However, in this case it has been proposed that the cells at the perimeter of the invaginating archenteron are tightly attached to the nonelastic hyaline layer so that as the cells round up, the sheet is forced to curve (Fig. 15-21). The forces involved in this process are confined to the vegetal plate, since isolated vegetal plates continue invagination when separated from the remainder of the embryo.

The mechanisms responsible for the second stage of gastrulation in the sea urchin are more clearly defined. In this stage, mesenchyme cells produced by the invaginating gut extend fine pseudopodia, called filopodia, which contact the roof of the blastocoel and then contract to pull the gut inward to its full extent (Fig. 15-20, C and D). The filopodia contain microtubules at their bases. Treatment of embryos with colchicine prevents the formation of filopodia and thereby blocks the second phase of gastrulation. Microfilaments are also found in the filopodia and are thought to be associated with their contractile activity.

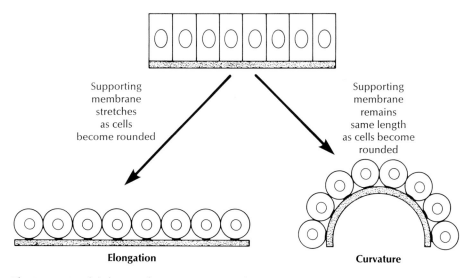

Supporting membrane stretches as cells become rounded

Supporting membrane remains same length as cells become rounded

Elongation

Curvature

Fig. 15-21. Model showing how curvature can be generated by rounding of epithelial cells firmly attached to a supporting membrane. On the left the membrane stretches as the cells round up, resulting in elongation. On the right the membrane does not stretch, and the rounding of the cells causes a bulging curvature away from the membrane-bound surface. (Modified from Gustafson, T., and L. Wolpert. 1963. Int. Rev. Cytol. **15:**139.)

Amphibia   As described in Chapter 2, amphibian gastrulation begins with the movement of cells from the surface of the embryo through the lips of the blastopore into the interior of the embryo. As cells in the blastopore region elongate and burrow toward the interior through the presumptive endoderm, they remain partially in contact with the surface of the embryo for a while. This causes them to assume a shape that has been described by the terms "bottle cell" or "flask cell" (Figs. 15-22 and 2-21). The long, narrow neck of each bottle cell is attached to the acellular membrane that surrounds the embryo and also to the necks of its neighboring cells. As the early stages of gastrulation proceed, the main body of each bottle cell migrates to the interior, causing the neck region to elongate considerably. The necks of bottle cells are filled with microtubules running parallel to the long axis of the cell. The blasto-

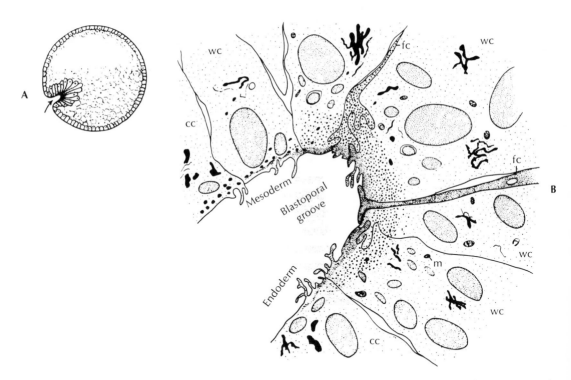

Fig. 15-22. Formation of bottle cells during amphibian gastrulation. **A,** Schematic sagittal section showing elongated bottle-shaped cells attached to the surface of the blastoporal groove (arrow). The upper surface of the groove is composed of invaginating mesoderm and the lower surface of endoderm. **B,** Drawing prepared from a montage of electron micrographs of the deepest part of the blastoporal groove in **A.** The heavy shading indicates a densely staining area containing many fine filaments adjacent to the blastoporal groove. *fc* = Neck region of an elongated flask or bottle cell; *wc* = wedge-shaped cell not sufficiently elongated to be classified as bottle-shaped; *cc* = cuboidal cell. (Modified from Baker, P. C. 1965. J. Cell Biol. **24:**95.)

poral end of the neck appears dense in the electron microscope and is filled with a network of microfilaments. The exact role of microfilaments in amphibian gastrulation remains controversial. The contraction of the necks of blastoporal cells looks very much like the "purse-string" action described in Fig. 15-13, and a model has been put forth postulating microfilament contraction as a possible mechanism for bottle cell formation (Fig. 15-23). Experiments with cytochalasin B have not yielded clear results. It is possible to inhibit gastrulation with that drug, but only by using concentrations high enough to cause extensive disaggregation of cells. However, amphibian embryos are notoriously impermeable to externally applied agents. Therefore the negative experimental results are not necessarily informative.

Experiments with isolated dorsal lip appear to indicate that the

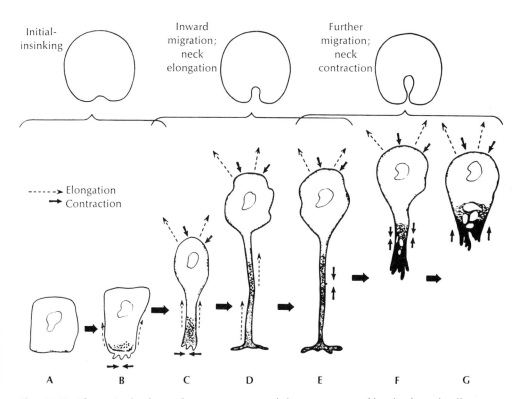

Fig. 15-23. Theoretical scheme for appearance and disappearance of bottle-shaped cells during amphibian gastrulation. **A,** Cell at surface before gastrulation. **B,** Contraction of filamentous "dense layer" adjacent to the surface of the embryo initiates the process of elongation leading to bottle cell formation. **C** and **D,** As elongation proceeds, the main body of the cell moves to the interior, leaving a long thin neck. **E** to **G,** Contractile forces shorten the neck of the bottle cell as it becomes part of the lining of the developing archenteron. (From Baker, P. C. 1965. J. Cell Biol. **24:**95.)

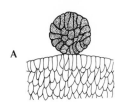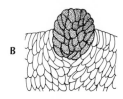

Fig. 15-24. Invagination of an isolated blastoporal fragment. **A,** Graft of blastoporal cells partly covered by the "surface coat" is placed on the endoderm. **B,** Blastoporal cells sink into the endoderm. **C,** Blastoporal graft begins to invaginate and form bottle-shaped cells. (From Holtfreter, J. 1944. J. Exp. Zool. **95:**171.)

characteristic shape of bottle cells is due to properties intrinsic to the cells themselves and not to forces exerted on them by other parts of the embryo. When groups of cells attached to the surface coat of the blastoporal region are separated from the remainder of the embryo, the cells retain their bottle-shaped configuration. Also, if an isolated group of blastoporal cells with its surface coat in place is grafted onto endodermal tissue, a process of invagination similar to that which occurs in the blastopore is initiated (Fig. 15-24).

Mammals

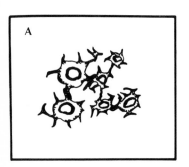
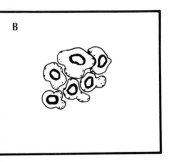

During gastrulation in the mouse embryo, mesoderm cells that migrate to the interior through the primitive streak normally have numerous filopodia that contain microfilaments. There exist in mice a number of so-called T locus mutants that are characterized by defective cellular adhesion and/or movement at various stages of early embryonic development (Chapter 16). In one class of these mutants (homozygous t^9/t^9), presumptive mesodermal cells fail to migrate through the primitive streak but instead pile up in the region adjacent to it. In the affected embryos the presumptive mesodermal cells lack the filopodia found on the normal cells and instead possess broad pseudopodia that contain few microfilaments (Fig. 15-25). The mutant cells also fail to form the intercellular junctions that are found between the filopodia of adjacent normal cells. The mutation is lethal, presumably because of failure to form adequate mesodermally derived organs and tissues, particularly the circulatory system. Although the exact role of microfilaments in mammalian gastrulation has not

Fig. 15-25. Diagram of mesoderm cells from normal and t^9/t^9 8-day embryos. **A,** Cells from the normal embryo have fine filopodia, which have a network of microfilaments and are the site of intercellular junctions. **B,** Cells of the t^9/t^9 embryo have broad pseudopodia with few microfilaments, and they form few junctions with each other. (From Bennet, D. 1975. Cell **6:**441. Copyright © MIT; published by the MIT Press.)

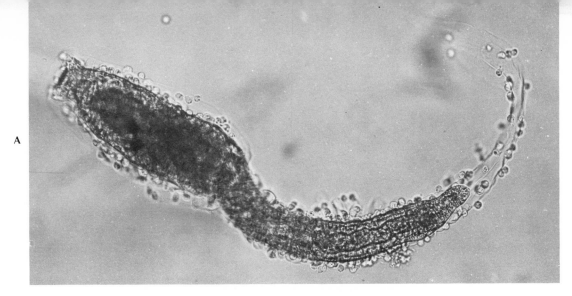

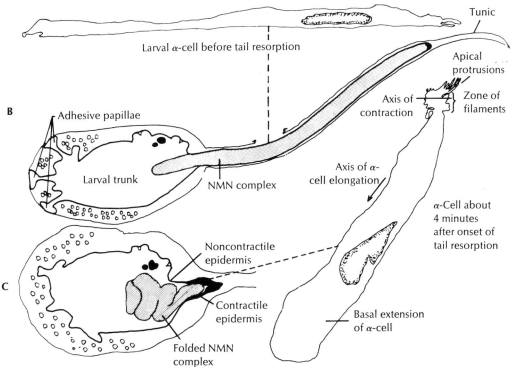

Fig. 15-26. Tail retraction during ascidian metamorphosis. **A,** Early stages of metamorphosis in *Ciona intestinalis*. The tail has already withdrawn from a portion of the acellular membranous "tunic" that surrounds it. **B,** Schematic diagram of *Amaroecium constellatum* before metamorphosis. The notochord-muscle-nerve (NMN) complex is fully extended, as are the epithelial α-cells. **C,** Same as **B,** 4 minutes after onset of tail resorption. Contraction of the epidermis has forced the NMN complex to fold up inside the larval trunk. The α-cells have contracted to about a twentieth of their original length with a resultant protrusion of their cytoplasm at right angles to the axis of contraction and a major thickening of the contractile epidermis. (**A** from N. K. Wessells, Tissue interactions and development. Benjamin/Cummings Publishing Co., Menlo Park, Calif., 1977; **B** and **C** from Cloney, R. A. 1966. J. Ultrastruct. Res. **14:**300.)

been established, they appear to be essential both for the extensive cellular migration that occurs in gastrulation and for specific cellular interactions.

ASCIDIAN METAMORPHOSIS

Ascidians or tunicates are distant relatives of the vertebrates that develop as free-swimming tadpolelike larvae and then undergo a rapid metamorphosis into sessile adults with a totally different appearance. Metamorphosis includes a dramatic retraction of the tail into the body that is completed within 6 minutes after its onset. The tail contains notochord, muscle, nervous tissue, and elongated epithelial cells known as α-cells. During metamorphosis the α-cells contract to approximately a twentieth of their original length and literally force the notochord-muscle-nerve complex of the tail to fold up inside the body of the organism (Fig. 15-26). The rapid and extreme contraction of the α-cells is caused by bands of microfilaments that run the length of each cell prior to contraction and are found in the crumpled apex of the cell after contraction. Contraction of the tail is reversibly inhibited by cytochalasin B. However, if the drug is left in contact with the organisms for a prolonged time, reversibility is lost and the adult permanently retains a larval tail.

RELATIONSHIP OF CELL SHAPE AND MOTILITY TO THE OVERALL PROCESS OF MORPHOGENESIS

The examples cited, although by no means exhaustive, provide a general picture of the types of cellular movements and cellular shape changes that occur during morphogenesis. Microtubules and microfilaments play active, important roles in these processes, which provide, at least in part, the mechanical basis for morphogenesis. However, many other processes, including changes in cellular surfaces, selective cellular adhesiveness, differential cellular proliferation, and differential cellular death, also play important roles in the generation of form that occurs during development. These mechanisms will be examined in subsequent chapters.

BIBLIOGRAPHY
Books and reviews

Arnold, J. M. 1976. Cytokinesis in animal cells: new answers to old questions. *In* G. Poste and G. L. Nicolson, eds. The cell surface in animal embryogenesis and development. Elsevier North-Holland, Inc., New York.

Balinsky, B. I. 1975. An introduction to embryology, 4th ed., W. B. Saunders Co., Philadelphia.

Bennett, D. 1975. The T-Locus of the mouse. Cell **6:** 441.

Clarke, M., and J. A. Spudich. 1977. Non-muscle contractile proteins: the role of actin and myosin in cell motility and shape determination. Annu. Rev. Biochem. **46:**797.

Deuchar, E. M. 1975. Cellular interactions in animal development. Chapman & Hall Ltd., London.

Fuller, G. M., and B. R. Brinkley. 1976. Structure and control of assembly of cytoplasmic microtubules in normal and transformed cells. J., Supramol. Struct. **5:**497.

Gustafson, T., and L. Wolpert. 1963. The cellular basis of morphogenesis and sea urchin development. Int. Rev. Cytol. **15:**139.

Hepler, P. K., and B. A. Palevitz, 1974. Microtubules and microfilaments. Annu. Rev. Plant Physiol. **25:**309.

Korn, E. D. 1978. Biochemistry of acto-myosin-dependent cell motility (a review). Proc. Natl. Acad. Sci. U.S.A. **75:**588.

Lazarides, E., and J. P. Revel. 1979. The molecular basis of cell movement. Sci. Am. **240:**100 (May).

Lu, R., and M. Elzinga. 1976. Comparison of amino acid sequences of actins from bovine brain and muscles. *In* R. Goldman, T. Pollard, & J. Rosenbaum, eds. Cell motility. Cold Spring Harbor Conferences on Cell Proliferation. Vol. 3. Book B. Cold Spring Harbor Laboratory, New York.

Porter, K. R. 1966. Cytoplasmic microtubules and their functions. *In* G. E. W. Wolstenholme and M. O'Conner, eds. Principles of biomolecular organization. Churchill Livingstone, London.

Snyder J. A., and J. R. McIntosh. 1976. Biochemistry and physiology of microtubules. Annu. Rev. Biochem. **45:**699.

Spooner, B. S. 1974. Morphogenesis of vertebrate organs. *In* J. Lash and J. R. Whittaker, eds. Concepts of development. Sinauer Associates, Inc., Sunderland, Mass.

Stearns, L. W. 1974. Sea urchin development: cellular and molecular aspects, Dowden, Hutchinson & Ross, Inc., Stroudsburg, Pa.

Tilney, L. G. 1968. The assembly of microtubules and their role in the development of cell form. *In* M. Locke, ed. The emergence of order in developing systems. Academic Press, Inc., New York.

Trinkaus, J. P. 1976. On the mechanism of metazoan cell movements. *In* G. Poste Hall, Inc., Englewood Cliffs, N.J.

Trinkaus, J. P. 1976. On the mechanism of metazoan cell movements. In G. Poste and G. L. Nicolson, eds. The cell surface in animal embryogenesis and development. Elsevier North-Holland, New York.

Vasiliev, J. M., and I. M. Gelfand. 1977. Mechanisms of morphogenesis in cell cultures. Int. Rev. Cytol. **50:**159.

Wessells, N. K. 1971. How living cells change shape. Sci. Am. **225:**77 (Oct.).

Wessells, N. K. 1977. Tissue interactions and development. W. A. Benjamin, Inc., Menlo Park, Calif.

Selected original research articles

Amos, L. A., and A. Klug. 1974. Arrangement of subunits in flagellar microtubules. J. Cell Sci. **14:**523.

Baker, P. C. 1965. Fine structure and morphogenic movements in the gastrula of the treefrog *Hyla regilla*. J. Cell Biol. **24:**95.

Byers, B., and K. R. Porter. 1964. Oriented microtubules in elongating cells of the developing lens rudiment after induction. Proc. Natl. Acad. Sci. U.S.A. **52:** 1091.

Cloney, R. A. 1966. Cytoplasmic filaments and cell movements. Epidermal cells during ascidian metamorphosis. J. Ultrastruct. Res. **14:**300.

David-Pfeuty, T., J. Laporte, and D. Pantaloni. 1978. GTPase activity at ends of microtubules. Nature **272:**282.

Fujiwara, K., and T. D. Pollard. 1978. Simultaneous localization of myosin and tubulin in human tissue culture cells by double antibody staining. J. Cell Biol. **77:**182.

Lazarides, E. 1976. Actin, α-actinin and tropomyosin interaction in the structural organization of actin filaments in nonmuscle cells. J. Cell Biol. **68:**202.

Lazarides, E., and K. Weber. 1974. Actin antibody: the specific visualization of actin filaments in non-muscle cells. Proc. Natl. Acad. Sci. U.S.A. **71:**2268.

Lockwood, A. H. 1978. Tubulin assembly protein: immunochemical and immunofluorescent studies on its function and distribution in microtubules and cultured cells. Cell **13:**613.

Margolis, R. L., and L. Wilson. 1978. Opposite end assembly and disassembly of microtubules at steady state in vitro. Cell **13:**1.

Margolis, R. L., L. Wilson, and B. I. Kiefer. 1978. Mitotic mechanism based on intrinsic microtubule behaviour. Nature **272:**450.

Miller, C. L., J. W. Fuseler, and B. R. Brinkley, 1977. Cytoplasmic microtubules in transformed mouse × nontransformed human cell hybrids: correlation with in vitro growth. Cell **12:**319.

Spiegelman, M., and D. Bennett. 1974.

Fine structural study of cell migration in the early mesoderm of normal and mutant mouse embryos (T-locus:t^9/t^9). J. Embryol. Exp. Morphol. **32**:723.

Spooner, B. S., and N. K. Wessells. 1972. An analysis of salivary gland morphogenesis: role of cytoplasmic microfilaments and microtubules. Dev. Biol. **27**:38.

Spudich, J. A., and S. Lin. 1972. Cytochalasin B, its interaction with actin and actomyosin from muscle. Proc. Natl. Acad. Sci. U.S.A. **69**:442.

Tilney, L. G., and J. R. Gibbins. 1969. Microtubules and filaments in the filopodia of the secondary mesenchyme cells of *Arbacia punctulata* and *Echinarachnius parma.* J. Cell Sci. **5**:195.

Vanderckhove, J., and K. Weber. 1978. Mammalian cytoplasmic actins are the products of at least two genes and differ in primary structure in at least 25 identified positions from skeletal muscle actins. Proc. Natl. Acad. Sci. U.S.A. **75**:1106.

Weber, K., R. Pollack, and T. Bibring. 1975. Antibody against tubulin: the specific visualization of cytoplasmic microtubules in tissue culture cells. Proc. Natl. Acad. Sci. U.S.A. **72**:459.

Yamada, K. M., B. S. Spooner, and N. K. Wessells. 1970. Axon growth: roles of microfilaments and microtubules. Proc. Natl. Acad. Sci. U.S.A. **66**:1206.

CHAPTER 16

Cellular adhesion and sorting

☐ Cellular adhesion, or the tendency of cells to stick to one another, plays an important role in the generation of form during development. Various types of cells exhibit different degrees of adhesion both to their own cell type (homotypic adhesion) and to different cell types (heterotypic adhesion). These differences are presumed to reflect differences in their cell surface properties. Differential patterns of cellular adhesion play important roles in embryonic morphogenesis from the earliest cleavage stages through the final precise patterning of cells into complex structures such as the brain.

The primary mediator of cell affinity appears to be the cell membrane. Cell membranes vary greatly with regard to the details of their composition and structure, but the gross aspects of these properties are similar for most cells and allow one to make many useful generalizations about cell membranes. The adhesion of cells to one another has been studied experimentally by dissociating tissues or entire organisms into free cells and then observing the reaggregation of these cells. Cells separated by mechanical and chemical procedures reaggregate and arrange themselves into specific patterns, including the sorting out of different types of cells from each other. The effects of various environmental conditions on aggregation and sorting have been analyzed in studies seeking to understand the mechanisms responsible for the specificity of these processes.

The major components of the cell membrane are proteins and lipids. In addition, membranes also contain a small percentage (1% to 5%) of carbohydrate in the form of oligosaccharides, which may be very important to the specificity of cellular interactions. The primary lipid components of the plasma membrane are cholesterol and phospholipids. Proteins, some of which are conjugated with carbohydrates or lipids to form glycoproteins or lipoproteins, are a major component of most cell membranes. Many of these proteins are membrane-bound enzymes, but others probably have equally important roles, including

COMPOSITION AND STRUCTURE OF THE CELL MEMBRANE

Extracellular water

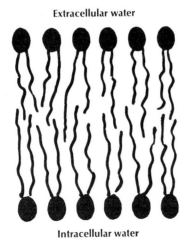

Intracellular water

Fig. 16-1. A schematic cross-sectional view of a phospholipid bilayer. The dark circles represent the polar groups of the phospholipids, which are exposed to the aqueous environment. The lines represent fatty acid chains that are sequestered in the interior of the bilayer. (From Singer, S. J., and G. L. Nicolson. 1972. Science **174:**720. Copyright 1972 by the American Association for the Advancement of Science.)

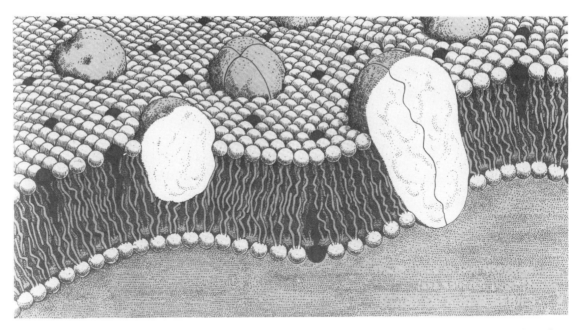

Fig. 16-2. Fluid mosaic model of the cell membrane. Phospholipids are depicted as white hydrophilic balls with gray hydrophobic tails (fatty acids) in a lipid bilayer similar to that in Fig. 16-1. Cholesterol molecules (black) are intermingled. Globular proteins (some composed of more than one subunit) "float" in the lipid bilayer with their relatively hydrophilic areas protruding. Some proteins pass entirely through the bilayer, whereas others protrude on only one side. The proteins are able to move about freely within the plane of the bilayer. (Drawing by B. Tagawa. Singer, S. J. Architecture and topography of biologic membranes. Hosp. Practice **8**(5), and from Singer, S. J. 1975. *In* G. Weissmann and R. Claiborne, eds. Cell membranes: biochemistry, cell biology, and pathology. H. P. Publishing Co., Inc., New York. Reprinted with permission.)

structural, antigenic, transport, and receptor functions. Proteins and attached polysaccharides are responsible for the cell surface specificity that distinguishes different cell types and different tissues and also for the antigenic sites recognized by antibodies.

Present evidence regarding the structure of the cell membrane can be summarized by presenting the widely accepted fluid mosaic model of membrane structure put forth in 1972 by S. J. Singer and Garth Nicolson. The model postulates that the lipids of the cell membrane are arranged in a phospholipid bilayer (Fig. 16-1), with the polar groups of the phospholipids exposed to the aqueous environment both on the extracellular and the cytoplasmic surfaces of the membrane. The nonpolar fatty acid chains from the two layers are in contact with each other in the interior of the bilayer. Membrane proteins, which have been found to be largely globular, are postulated to be randomly distributed in the plane of the bilayer (Fig. 16-2), but with each oriented so that its polar groups are exposed to the aqueous phase and its nonpolar groups are buried in the hydrophobic interior of the membrane. Some proteins are thought to span the entire membrane and have two exposed surfaces, whereas others are believed to be restricted to either the outer surface or the cytoplasmic surface of the membrane.

The entire structure is assumed to be fluid so that proteins can move around freely in the plane of the membrane. This dynamic, fluid nature of the cell membrane has been demonstrated by a variety of methods. One of the most dramatic demonstrations is found in an experiment in which mouse and human cells are fused with Sendai virus (Chapter 11) and the distribution of their antigenic components is followed by immunofluorescent methods. Within 40 minutes after cell fusion, a total mixing of the antigens from the two parental cells occurs in 90% of the heterokaryons (Fig. 16-3). This mixing occurs in the presence of metabolic inhibitors that would prevent the synthesis of new surface antigens and therefore appears to be the result of diffusion of preexisting proteins within the plane of the membrane. Al-

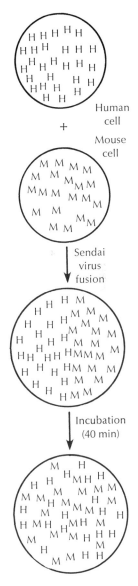

Fig. 16-3. Mixing of cell surface antigens after cell fusion. Indirect immunofluorescence (Chapter 15) was used to visualize species-specific antigens before and after fusion of human and mouse cells with Sendai virus. Mouse antigens were visualized with fluorescein (green fluorescence) and human antigens with rhodamine (red fluorescence) so that the positions of both could be detected within the same cell. Immediately after fusion (5 to 10 minutes) separate areas of red and green fluorescence were seen. However, within 40 minutes both types of antigens were uniformly mixed over the entire cell surface, even in the presence of inhibitors that would block synthesis of new antigens. (Modified from Singer, S. J. 1975. *In* G. Weissmann and R. Claiborne, eds. Cell membranes: biochemistry, cell biology, and pathology. H. P. Publishing Co., Inc., New York.)

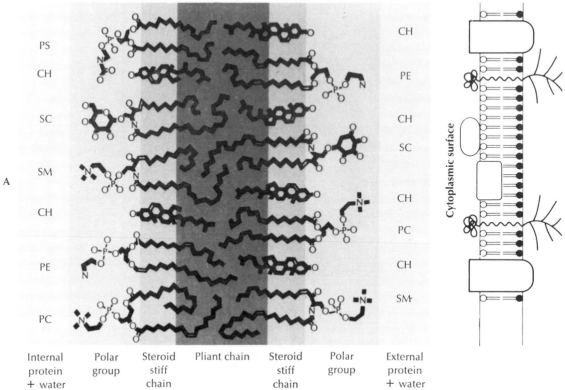

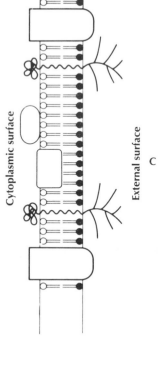

		Steroid		Steroid		
Internal protein + water	Polar group	stiff chain	Pliant chain	stiff chain	Polar group	External protein + water

Cytoplasmic surface

External surface

Phosphatidyl-choline

(Phosphatidyl)-ethanol amine

(Phosphatidyl)-serine

Sphingomyelin

though proteins appear to have translational mobility within the membrane, they probably cannot rotate freely because of constraints placed on them by hydrophobic and hydrophilic interactions.

A detailed analysis of membrane structure is far beyond the scope of this book. However, our readers should be aware that membranes are actually far more complex than indicated in the brief summary just presented. Some of these new levels of complexity that are just beginning to be understood may ultimately prove to be important to cellular interactions in development. Additional aspects of cell membrane structure that need to be considered include the following:

1. The lipid bilayer is a complex mixture of many different types of molecules, including cholesterol and various types of phospholipids and glycolipids (Fig. 16-4, *A* and *B*). In addition, the fatty acids associated with each class of molecule can vary both in length and in degree of unsaturation. Thus different membranes can contain lipid bilayers that differ from one another.

2. The types of phospholipids on the inside and outside of the plasma membrane tend to differ from one another. The external surface is enriched in phosphatidylcholine and sphingomyelin, whereas the internal surface contains more phosphatidylserine and phosphatidylethanolamine (Fig. 16-4, *C*).

3. Membranes contain significant amounts of carbohydrates. These vary from single sugar residues in simple glycolipids to multiple oligosaccharide side chains in large glycoproteins. Es-

---

Fig. 16-4. Details of membrane structure. **A,** Schematic drawing depicting the variety of types of lipid molecules that make up a membrane lipid bilayer. *CH* = Cholesterol; *PC* = phosphatidylcholine; *PE* = phosphatidylethanolamine; *PS* = phosphatidylserine; *SC* = sphingosine cerebroside; *SM* = sphingomyelin. This drawing is intended to show only how the various molecules might fit together and does not reflect spatial distribution or relative frequencies of components in any particular type of membrane. **B,** Chemical formulas for the predominant types of phospholipids in membranes of higher organisms. Wavy lines indicate fatty acids of unspecified length and degree of unsaturation. **C,** Asymmetrical distribution of components in a cellular plasma membrane. Solid circles indicate hydrophilic ends of phosphatidylcholine and sphingomyelin concentrated in the outer leaflet of the lipid bilayer. Open circles indicate phosphatidylserine and phosphatidylethanolamine concentrated on the cytoplasmic surface. Asymmetrical distribution of glycoproteins is also seen. The oligosaccharide side chains (shown as branches) are all on the exterior of the cell. The glycoprotein has a hydrophobic middle region embedded in the lipid bilayer and a second hydrophilic region in the intracellular space. Nonglycosylated proteins either extend through the entire membrane or are found only along the cytoplasmic surface. (**A** by S. Marcus from G. Weissmann and R. Claiborne, eds. 1975. Cell membranes: biochemistry, cell biology, and pathology. HP Publishing Co., Inc., New York. Reprinted with permission; **B** and **C** from Bretscher, M. S. 1974. *In* A. A. Moscona, ed. The cell surface in development. John Wiley & Sons, Inc., New York.)

sentially all the sugars are found attached to the external surface of the cell membrane (Figs. 16-4, *C*, and 16-5). In some cases a heavy glycoprotein coat with a "fuzzy" appearance can be seen in the electron microscope. The term "glycocalyx" is sometimes used to describe such a coat.

4. Complex interactions with intracellular structural elements (particularly microtubules and microfilaments) restrict the mo-

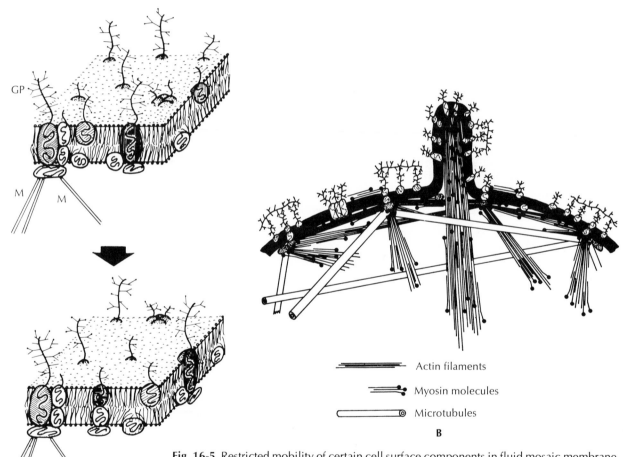

Fig. 16-5. Restricted mobility of certain cell surface components in fluid mosaic membrane. **A,** Three-dimensional model showing glycoproteins in a lipid bilayer. The glycoprotein designated *GP* is restrained from free movement by membrane-associated structures *(M)* in the adjacent cytoplasm, whereas all other glycoproteins shown are able to move about freely in the plane of the membrane. **B,** Schematic drawing showing hypothetical relationships between cell surface glycoproteins and membrane-associated microtubule and microfilament systems. (**A** from Nicolson, G. L. 1974. *In* B. Clarkson and R. Baserga, eds. Control of proliferation in animal cells. Cold Spring Harbor Laboratory, Cold Spring Harbor, N.Y.; **B** from Nicolson, G. L. 1976. Biochim. Biophys. Acta **457:**57.)

bility of certain cell surface glycoproteins (Fig. 16-5). Thus not every cell surface molecule is as free to move around as suggested by the original fluid mosaic model. Such restraints provide mechanisms for control over the spatial distribution of extracellular macromolecules from inside the cell.

5. Membrane proteins that are not glycoproteins appear to be located predominantly on the cytoplasmic surface of the cell membrane.

In the following sections we will consider a series of specific cellular interactions related to development and explore the possible mechanisms that may be involved in these interactions.

The pioneering work in the field of cell reaggregation was done in the early 1900s by H. V. Wilson, who studied the dissociation and reaggregation of sponge cells. Sponges were mechanically dissociated into single cells by being pressed through cloth of fine mesh. When allowed to stand, the cells reaggregated into small, functional sponges. It was later shown that when dissociated cells from two different species of sponges are mixed, the cells sort out by species and reaggregate separately. This is dramatically apparent when cells from *Haliclona occulata*, which is purplish, and *Microciona prolifera*, which is reddish, are mixed. As the reaggregation proceeds, masses of each color emerge from the initial mixture of cells.

Some evidence as to the mechanism whereby species-specific cell sorting occurs in sponges has come from the technique of dissociating cells in seawater that is free of calcium and magnesium. When sponge cells are dissociated in this chemical manner, they must be returned to seawater that contains calcium and magnesium for reaggregation to occur. The cells are normally shaken on a gyratory shaker during reaggregation. Shaking eliminates the variable of differential migratory ability and allows cells to reaggregate strictly on the basis of their adhesive properties.

The reaggregation process occurs more slowly with chemically dissociated cells than with mechanically dissociated cells. In addition, reaggregation does not occur at all at low temperatures (5° C) after dissociation in calcium- and magnesium-free seawater, whereas it occurs slowly at low temperatures after cells have been mechanically dissociated. These findings suggest that chemical dissociation removes something from the cells that is required for reaggregation and that replenishment of this material is a temperature-dependent process. Seawater in which sponges have been chemically dissociated contains a factor that stimulates reaggregation at 5° C. This factor is species specific. Seawater in which *Haliclona* cells have been dissociated

REAGGREGATION OF DISSOCIATED CELLS
Cells from sponges

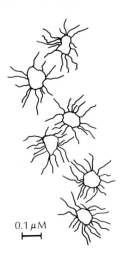

0.1 μM

Fig. 16-6. "Sunburst" configuration of sponge aggregation factor. (Drawing based on electron micrograph of uranium-shadowed preparation of aggregation factor from *Microciona* in Henkart, P., et al. 1973. Biochemistry **12:**3045.)

promotes only *Haliclona* cell reaggregation; seawater in which *Microciona* cells have been dissociated promotes only *Microciona* cell reaggregation.

Recent studies indicate that the factors responsible for species specific reaggregation in sponges are glycoprotein in nature. In *Microciona* the factor appears to be large with a molecular weight of several million. When the factor is visualized in the electron microscope, it is seen to have a sunburst configuration with fibers radiating outward from an inner circle (Fig. 16-6). The overall diameter of each complex is about 160 nm.

Cells from amphibian embryos

The use of reaggregation experiments to examine cellular affinities was extended to vertebrate embryo cells by Philip Townes and Johannes Holtfreter, who studied the ability of cells from gastrula- and neurula-stage amphibian embryos to reassociate and continue differentiating. The tissues were disaggregated to single-cell suspensions by exposure to alkaline solutions. Cells from various tissues were then mixed together and allowed to reaggregate at physiological pH. Initially, the mixed cell populations aggregate randomly, but then they begin to sort out and form tissue-specific associations. In many cases they ultimately enter into spatial arrangements similar to those seen in a normal embryo (Fig. 16-7). For example, when epidermal cells are combined with either neural plate or mesodermal cells, the epidermal cells remain on the surface while the other cell type moves to the interior. In such mixed aggregates the neural plate cells tend to form hollow vesicles that resemble neural tube histologically, whereas mesodermal cells assume the configuration of connective tissue surrounding coelomic cavities. When epidermal cells are combined with both mesodermal and endodermal cells, the mesoderm becomes sandwiched between the epidermis and the endoderm. Some of the endodermal cells, however, usually remain exposed on the surface of the mass, a position similar to their exposed location in the lumen of the primitive gut. These results demonstrate not only an affinity of each type of cell for other cells of its own kind but also a complex pattern of relationships with other types of cells that leads naturally to histotypic interactions similar to those which occur in intact embryos.

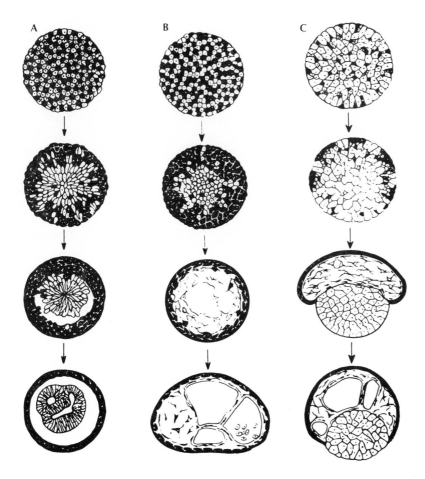

Fig. 16-7. Cell arrangements seen with different combinations of vertebrate embryo cells following dissociation and reaggregation. **A,** Mixture of epidermal and neural plate cells. **B,** Mixture of epidermal and mesodermal cells. **C,** Mixture of epidermal, mesodermal, and endodermal cells. (From Townes, P. L., and J. Holtfreter. 1955. J. Exp. Zool. **128:**53.)

Extensive reaggregation studies have been done with cells from warm blooded vertebrates and particularly from chicken and mouse embryos. Questions that have been studied include the relative importance of tissue specificity versus species specificity, the nature and strength of the attractive forces between cells, and the molecular basis for cellular adhesion.

The dissociation technique that is used has a major effect on the results that are obtained. In most cases disaggregation of tissues is accomplished by some combination of treatment with trypsin and/or the calcium-binding agent EDTA, together with mechanical agitation. Such treatment generally removes glycoproteins from the cell surface and renders the cells dependent on synthesis of proteins and polysaccharides for tissue-specific cell sorting. Inhibition of the synthesis of either protein or glucosamine (a component of cell surface glycopro-

Cells from avian and mammalian embryos

teins) prevents normal sorting, but it does not prevent an initial random aggregation. Even when normal synthesis of glycoproteins is allowed, an initial period of random aggregation is generally observed before any specific sorting begins. However, recent experiments appear to indicate that aggregation can be highly specific from the beginning if the dissociation conditions are sufficiently mild.

Four basic approaches have been widely used to measure cellular aggregation. In the first, the disaggregated cells are placed in a stationary culture and allowed to find one another by their own random motility. As was described for sponge cells, the disadvantage of this approach is that it is dependent on the motility as well as the adhesiveness of the cells under study. The second approach involves the use of a rotary shaker to increase the probability of collisions between the cells in question. This approach increases the rate of aggregation and the sizes of aggregates that can be obtained, provided that the shaking is not too vigorous to prevent cells from sticking together after they make contact. In these first two approaches, quantitative measurement is based on the sizes of the aggregates formed or the decrease in total particle number that accompanies aggregation.

The third approach involves the collection of radioactively or fluorescently labeled single cells on the surfaces of preformed aggregates or a monolayer of cells on a culture dish during a period of incubation on a rotary shaker. Unattached single cells are then washed away, and the amount of label that has accumulated on the aggregates or monolayer provides a quantitative measurement of the number of single cells that have attached to them. This method allows not only for precise quantification but also for measurement of heterotypic aggregation of one type of cell on a preformed aggregate containing cells of a different type. In addition, by use of two different isotopes that can be counted independently in the same sample with a liquid scintillation counter (e.g., phosphorus 32 and carbon 14), it is possible to measure the relative affinities of two different types of cells for the same type of cell.

The fourth approach involves the use of a Couette viscometer, which consists of two concentric cylinders about 1.5 mm apart with culture medium between them. A precisely measured shearing force can be generated in the medium by rotating the outer cylinder at a controlled speed. This method has the advantage of measuring directly the strength of the adhesive forces between cells, but it is generally unable to distinguish among cell types in mixed aggregation experiments.

Collection onto preformed aggregates has shown that in most cases homotypic adhesion (i.e., collecting type A cells on type A aggregates and type B cells on type B aggregates) is substantially greater than heterotypic adhesion (collecting type A cells on type B aggregates or vice versa). Both tissue and species specificity are observed, but in many cases tissue specificity appears to be of greater importance. Current data suggest that there has been a substantial evolutionary conservation of the cell surface signals that are responsible for tissue-specific aggregation. In some cases, such as cells from the neural retina, mixtures of cells from chicken and mouse embryos (whose nuclei can be distinguished histologically) combine freely to form mosaic tissues that do not subsequently sort out. However, for certain other tissues such as heart and liver there is also species specificity, and a definite sorting out will occur if the cell suspensions are prepared in such a way that they are not damaged.

Even in cases in which an initial random aggregation takes place, the degree of sorting out that will occur as the cells regenerate surface molecules is extensive. In the case of cells derived from a single tissue, much of the original histotypic organization is regenerated. For example, dispersed myocardial cells from chicken embryos first aggregate into small, loose clumps, which then grow to larger masses by the addition of more single cells and small clusters (Fig. 16-8). The structure becomes compact as the cells make associations appropriate for the tissue, and by 24 to 48 hours a dense, pulsating sphere is formed. The aggregation of cells from the retina or brain is even more striking with regard to reconstruction of cell patterns characteristic of the tissue. Early in reaggregation of dissociated retina cells, different cell types adhere randomly. Later they become aligned into distinct layers. It is not known, however, to what degree the alignment represents a true sorting out of different cell types or a positionally dependent differentiation of the same cell type. Retinal cell aggregates that are retained in culture for about a week undergo considerable differentiation and morphogenesis to produce a stratified structure typical of advanced retinal development. Cells dissociated from different regions of the brain such as the cerebrum and the optic tectum form aggregates of characteristic size, shape, and histological differentiation (Fig. 16-9). When cultured for long periods of time, the cells in these aggregates show distinctive patterns of cell layering and specialization of cell types.

In cases in which cells from two different tissues initially form a mixed aggregate, the process of sorting out generally involves one tissue moving to the center of the aggregate while the other forms an enveloping outer layer around it. In any given combination of two different types of cells, one cell type will consistently assume an internal position and the other will surround it. For example, a mixture of cartilage and liver cells always forms a composite aggregate in which cartilage is central and liver is peripheral (Fig. 16-10). Examination of a

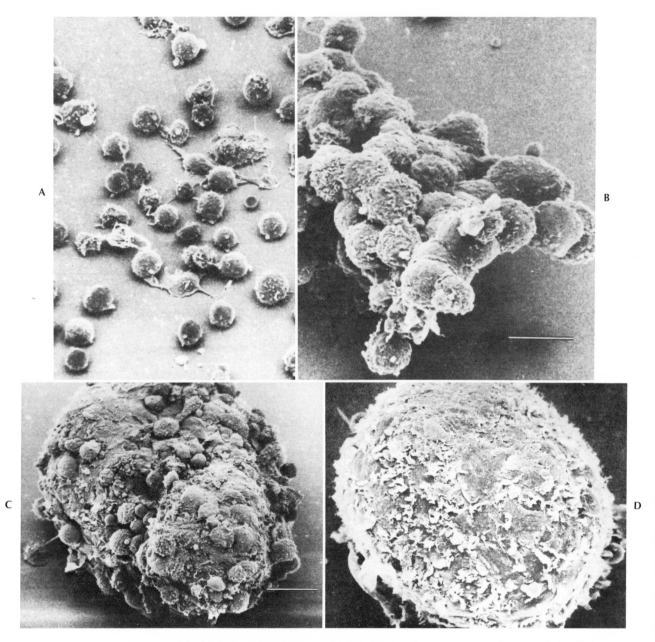

Fig. 16-8. Scanning electron micrographs showing aggregation of myocardial cells from 7-day chick embryo. **A,** Dissociated cells. **B,** A 3-hour cell aggregate. **C,** A 9-hour cell aggregate. **D,** A 48-hour cell aggregate. (From Shimada, Y., A. A. Moscona, and D. A. Fischman. 1974. Dev. Biol. **36:**428.)

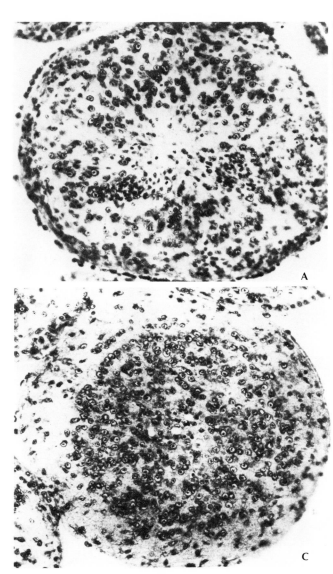

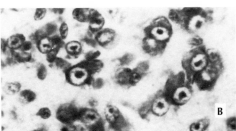

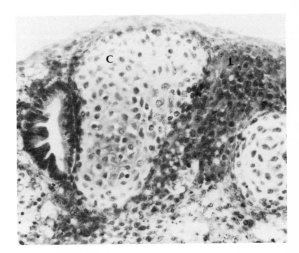

Fig. 16-9. Sections of 14-day aggregates from brain regions of 6-day chick embryos. **A,** Cerebrum. **B,** Cerebrum at higher magnification, showing differentiation of neurons. **C,** Optic tectum. (From Garber, B. B., and A. A. Moscona. 1972. Dev. Biol. **27:**217.)

Fig. 16-10. Histological section through part of a composite aggregate formed by a mixture of embryonic cartilage-forming and liver-forming cells after 48 hours in culture. Cells from the two tissues have segregated and formed nodules of cartilage (C) and masses of hepatic tissue (L) with a bile duct on the far left. (From Moscona, A. A. 1974. In A. A. Moscona, ed. The cell surface in development. John Wiley & Sons, Inc., New York.)

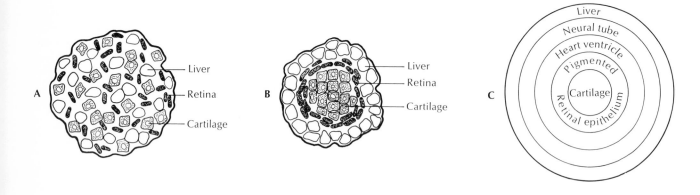

Fig. 16-11. Schematic representation of the hierarchy of relative positions assumed by reaggregating cells from chick embryos. **A,** A fresh aggregate of liver, retina, and cartilage that has not yet sorted out. **B,** Sorting out of **A** with the most cohesive cells (cartilage) at the center with maximized area of contact between cells and the least cohesive cells (liver) at the periphery. **C,** Relative positions assumed by five different types of cells. In mixed aggregates of two different kinds of cells, any type of cell indicated on this diagram will surround all types of cells shown in smaller circles and will be enveloped by any type of cell shown in a larger circle. Among these five tissues, cartilage always sorts to the center of aggregates and liver always sorts to the outside. (From Spratt, N. T., Jr. 1971. Developmental biology. Wadsworth Publishing Co., Inc., Belmont, Calif.)

variety of cell combinations suggests that there is a hierarchical order in cell segregation in which any one cell type is always found external to those above it in the hierarchy and internal to those below it (Fig. 16-11). Those cells which move to the more interior positions are assumed to have a greater degree of adhesiveness, as will be discussed.

POSTULATED MECHANISMS OF CELL ADHESION AND SORTING

A number of different theories have been put forward regarding the mechanism by which cells sort themselves out from one another in a specific manner. Several theories that are of special interest and for which there is some evidence will be discussed here. These theories are not mutually exclusive, and the mechanisms involved in them, as well as others, may play roles in specific cell adhesion.

Differential cell adhesion

The differential adhesion hypothesis has been put forth by Malcolm Steinberg to explain the cell sorting behavior observed when different types of cells are mixed and allowed to aggregate. The theory postulates that each kind of cell has a characteristic adhesive energy and that in a mixed population of cells, rearrangement into concentric layers results from differences in the strengths of intercellular adhesion in a manner similar to the behavior of immiscible liquids (e.g., oil and water), which separate on the basis of differences in intermolecular adhesion.

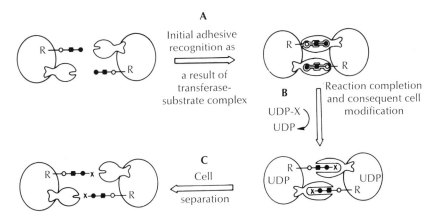

Fig. 16-12. Model of cell recognition and adhesion by means of interaction between glycosyl transferases and their substrates. **A,** Recognition and adhesion occur as a result of the binding of a complex carbohydrate chain (—○—■—●) to its corresponding glycosyl transferase. **B,** Sugar *X* is provided to the enzyme by a sugar nucleotide (UDP-X) and is added to the carbohydrate chain. **C,** After completion of the reaction the enzyme dissociates from the reaction product and the cells separate. *R* = Lipid or protein. (From Roseman, S. 1974. *In* A. A. Moscona, ed. The cell surface in development. John Wiley & Sons, Inc., New York.)

According to the differential adhesion theory, when two types of cells are mixed, the cell type with greater adhesive strength maximizes its area of contact with similar cells by forming the core of the composite aggregate. In the process it becomes surrounded by the cell type with the lesser degree of adhesion. The theory is supported by the hierarchical pattern of layering previously described, in which one member of a mixed aggregate becomes a roughly spherical core while the other forms a surrounding shell. The hierarchy of relative positions assumed by different types of cells in mixed aggregates (Fig. 16-11) is also seen when whole fragments of different tissues are combined and cultured together.

The differential adhesion hypothesis does not deal with the physicochemical mechanism of cell adhesion but is concerned only with the thermodynamics of the adhesion process. Steinberg has suggested that adhesion may result from a general mechanism that is identical for all cells plus specific mechanisms that are unique for each kind of cell. The latter are necessary to explain the fact that homotypic adhesions (A-A and B-B) are usually stronger for both cell types than is the heterotypic adhesion (A-B).

Saul Roseman and co-workers have offered an intriguing theory whereby cell recognition and adhesion are accomplished by means of an enzyme-substrate type of recognition (Fig. 16-12). The postulated

Enzyme-substrate recognition

enzymes and substrates are glycosyl transferases and complex carbohydrates, respectively. Binding between a carbohydrate and its corresponding glycosyl transferase would result both in cell recognition and in cell adhesion. Cell separation could be achieved simply by completing the enzymatic reaction through transfer of the appropriate sugar to the carbohydrate. Such a mechanism is especially attractive to explain cell adhesions that are transitory in embryonic development, such as those of migratory cells. Although direct evidence in support of this theory is lacking, it has been experimentally demonstrated that carbohydrates are important in cell-cell adhesion and also that glycosyl transferases are present on cell surfaces.

Cell surface adhesive molecules

Several interrelated theories have been developed suggesting that surface molecules play an important role in cell adhesion. Aaron Moscona and co-workers have developed the cell ligand theory, which postulates that cell recognition and adhesion are mediated by specific cell surface molecules termed "cell ligands." In discussing this theory, it is necessary to distinguish between two different levels of specificity

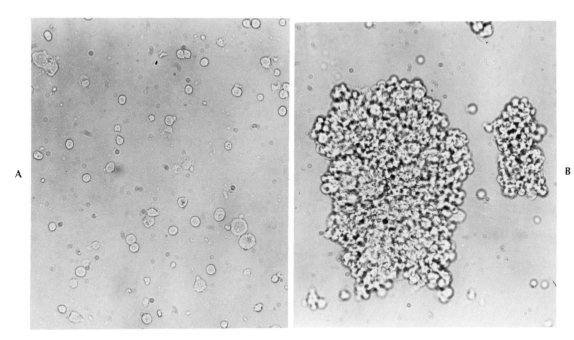

A   B

Fig. 16-13. Reaction of suspension of neural retina cells from 10-day chick embryo with tissue-specific antiserum. **A,** Control cells exposed to preimmunization rabbit serum show no agglutination. **B,** Cells exposed to rabbit antiretina serum absorbed with embryonic liver cells show massive agglutination. (From Goldschneider, I., and A. A. Moscona. 1972. J. Cell Biol. **53:**435.)

in cell aggregation. There is fairly broad agreement that specific cell surface molecules are involved in the tissue specificity of aggregation, such as occurs in the sorting out of liver and neural cells. However, the possible role of specific cell surface ligands in the sorting out of different cell types to generate normal histotypic organization within a single tissue remains more controversial.

The results of a variety of experiments have suggested that cell specific surface molecules may play an important role in cell adhesion. The existence of tissue-specific molecules on the surfaces of embryonic cells has been demonstrated conclusively by immunological methods. Antiserum prepared against neural retina cells and then completely absorbed with other types of cells from the same species reacts specifically with neural retina cells (Fig. 16-13). Similarly, antiserum prepared against liver cells reacts specifically with this cell type.

The first evidence for specific cell surface molecules important to cellular adhesion was obtained from cell aggregation experiments. Supernatants taken from cultures of neural retina cells specifically stimulate the reaggregation of other neural retina cells (Fig. 16-14). It has been reported that other cell types, even those of other neural tissues, do not respond to the factor. The retina-specific cell-aggregating factor has been purified from materials released into the medium

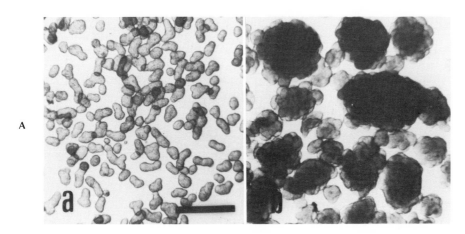

Fig. 16-14. Effect of retina-specific cell-aggregating factor on aggregation of a suspension of neural retina cells from a 10-day chick embryo. **A,** Twenty-four hour control aggregates without factor. **B,** Twenty-four hour aggregates with 0.5 μg/ml of retina-specific cell-aggregating glycoprotein (II-B-2) added at time 0. Length of bar is 0.5 mm. (From Moscona, A. A. 1975. *In* D. McMahon and C. F. Fox, eds. Developmental biology: pattern formation, gene regulation. ICN UCLA Symposia on Molecular and Cellular Biology. Vol. 2. W. A. Benjamin, Inc., Menlo Park, Calif.)

by cultured neural retina cells and has been found to be a glycoprotein with a molecular weight of approximately 50,000. A factor has also been isolated from cerebrum cells that specifically enhances their reaggregation. Recently there have also been reports that low molecular weight factors (< 1000 daltons) specifically promote liver cell reaggregation.

A different approach to the identification of cell surface molecules that are involved in cellular aggregation has been taken in recent studies by Urs Rutishauser, Gerald Edelman, and their co-workers. They started by preparing antibodies against the cell surface antigens of neural retinal cells. Fragments prepared from those antibodies block the aggregation of neural retinal cells, presumably by attaching to the cell surface molecules that are needed for aggregation. (Antibody fragments are used because whole antibody molecules would cause the cells to clump together.) Cell surface molecules isolated from neural tissue neutralize the effects of the antibody fragments. On fractionation, it was found that the bulk of the neutralizing activity was associated with a single type of molecule with a molecular weight of 140,000. This molecule, which has been designated "cell

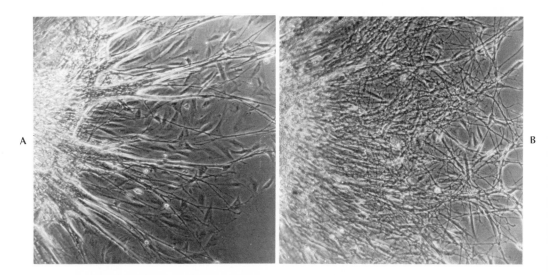

Fig. 16-15. Effect of anti-CAM antibody fragments on outgrowth of neurites. **A,** Control. Thoracic ganglion *(G)* from 10-day chicken embryo cultured for 24 hours in medium containing antibody fragments prepared from serum of unimmunized rabbits. **B,** Similar culture, except that the medium contained antibody fragments prepared from serum of rabbits immunized against CAM. The presence of anti-CAM antibody fragments has resulted in a tangled outgrowth of fine processes rather than the thick and relatively straight fascicles formed in cultures without this antibody. (From Rutishauser, U., W. E. Gall, and G. M. Edelman. 1978. J. Cell Biol. **79:**382.)

adhesion molecule" (CAM), has been shown to play an important role in aggregation and histotypic sorting of all types of neural cells.

Indirect immunofluorescence studies with highly specific antibody prepared against purified CAM have shown that CAM is uniformly distributed over the surfaces of all parts of all types of neural cells, including the filamentous neurites (dendrites and axons) that extend from the nerve cell bodies. The finding that antibody fragments specific for CAM block aggregation of all types of nerve cells raises questions about previous reports suggesting that different ligands are involved in the aggregation of cells from different parts of the nervous

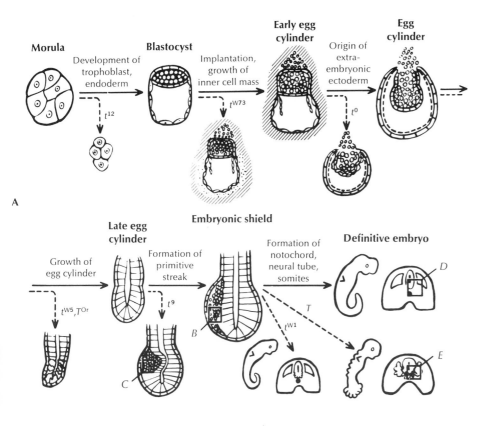
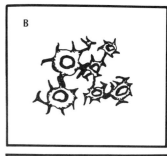
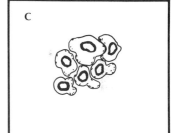
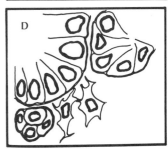
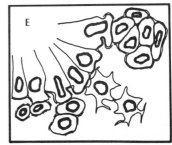

Fig. 16-16. Diagrammatic representation of effects of homozygous T locus mutants on development of mouse embryos. **A,** Summary of embryonic development in the mouse showing the points at which various homozygous mutations have their effects. In every case abnormalities of cellular associations appear to be involved. Squares *B, C, D,* and *E* indicate cellular interactions that are shown in greater detail in the small drawing. **B,** Normal pattern of association of mesodermal cells at the time of primitive streak gastrulation. **C,** Abnormal pattern of association of t^9/t^9 mesoderm. **D,** Normal relationships of neural tube, notochord, somite, and lateral mesenchyme cells in a wild-type embryo. **E,** Abnormal cellular interactions in a *T/T* embryo at the same gestational age as **D.** (From Bennett, D. 1975. Cell **6:**441. Copyright © MIT; published by the MIT Press.)

system. Studies with antibodies to CAM have also provided clues about its possible role in normal morphogenesis. Antibody fragments that are highly specific for CAM block the normal process of histotypic sorting that occurs in cultured aggregates of neural tissue (e.g., clustering of cell bodies in one area of the aggregate and neurites in another). CAM also appears to be involved in the gathering of neurites together into organized bundles (fascicles). Anti-CAM antibody fragments prevent that process and cause the neurites surrounding a cultured ganglion to form a tangled web of individual processes rather than an organized pattern of relatively thick and organized fascicles (Fig. 16-15).

Variations in the mechanisms that have been discussed may be important in generating the diverse examples of cell recognition and adhesion that occur in development. For example, the topographical arrangement of cell surface molecules may differ with cell type in a manner that could influence cell-to-cell interactions. Possible control of such arrangements from inside the cell has already been discussed (Fig. 16-5). Also, the timing of appearance and disappearance of different cell surface moieties in development could have a profound influence on cellular interactions in a system as dynamic as a developing embryo. In support of the latter concept is the finding that the capacity for tissue-specific reaggregation of trypsin-dissociated cells declines with the age of the embryo. Likewise, the retina-specific aggregating factor cannot be isolated from older embryos; neither will cells from such embryos respond to the factor obtained from younger cells.

T LOCUS IN MICE   A genetic approach to the role of cell surface molecules in development has been made possible by study of the T locus system in the mouse embryo. The T locus system is a chromosomal region that consists of at least two genetic loci and is characterized by a group of dominant and recessive mutations that affect embryonic development (Fig. 16-16). Dominant T locus mutations (designated by a capital letter) result in short tails in heterozygous mice ($T/+$) and are lethal between the seventh and eleventh days of gestation in homozygous embryos (T/T). Recessive alleles (designated by a lower case letter) have no effect in heterozygotes with the wild-type allele ($t/+$) but interact with T to produce a tailless phenotype in the heterozygote (T/t). Many of the recessive alleles are also lethal in homozygous embryos (t/t) at different points in development, as shown in Fig. 16-16.

Examination of different classes of homozygous T locus mutants suggests that lethality is due to failures in cellular interactions that are essential for normal morphogenesis, such as failure to establish connections of trophoblast and maternal cells at implantation, defective migration of mesodermal cells into the primitive streak (discussed in Chapter 15), and lack of normal cellular associations by cells of the notochord and somites. Recent evidence suggests that the T locus genes code for cell surface molecules. One allele in particular, the wild-type allele of t^{12}, has been shown to code for a cell surface antigen that is present only during early cleavage and morula stages, which corresponds to the time at which homozygous t^{12}/t^{12} embryos exhibit their defective phenotype (failure to organize into a normal blastocyst). These findings strongly support the contention that cell surface components play an important role in embryonic morphogenesis. Similar conclusions have also been reached for the lethal talpid[3] mutation in chicken embryos, which results in abnormally tight association of mesodermal cells in culture, and multiple developmental anomalies, particularly in limb morphogenesis, in affected embryos.

From the experiments described in this chapter, it is apparent that cell surface interactions that mediate cell recognition and adhesion are important in morphogenesis. However, these cell associations probably also play another significant role in embryonic development by influencing cell differentiation. Evidence for such a role has been found in the neural retina system. During development of the neural retina the enzyme glutamine synthetase (GS) becomes inducible by hydrocortisone at a particular developmental stage. This induction can be monitored in retina tissue grown in organ culture by immunoprecipitation of newly synthesized GS, as well as by assay of GS activity.

Inducibility of GS in embryonic chick retina has been shown to be dependent on the specific pattern of cell associations normally found in this tissue. For example, if neural retina is dissociated into single cells and grown in culture as a monolayer, the inducibility of GS is extremely low or nonexistent, even though receptors for hydrocortisone are present and the hormone is taken up by the cells. In contrast, if dissociated neural retina cells are allowed to reaggregate and form histotypic cell contacts, the aggregates are inducible for GS. If aggregation is accomplished with a plant agglutinin, which causes random cell agglutination and immobilizes the cells so that histotypic cell contacts fail to be made, GS induction is much lower than in normal aggregates.

It is clear from the foregoing discussion that cell adhesion plays a major role in determining the arrangement of cells in the developing organism. The ability of dissociated cells from a variety of sources to reassociate in a specific manner indicates that cell adhesion does not occur randomly but is an ordered process. The failure of normal cell

CELL SURFACE INTERACTIONS AND DIFFERENTIATION

CELLULAR AGGREGATION AND MORPHOGENESIS

associations to occur when the cell surface is altered experimentally or by mutation illustrates the importance of the cell surface in mediating specific cell adhesion. The precise physicochemical mechanisms involved in cell adhesion are just beginning to be uncovered. We have discussed a few of the postulated mechanisms and examined evidence in support of their roles in bringing about specific cell adhesion in various systems. Other mechanisms of cell adhesion have been proposed and may well be equally important. It seems likely that a number of different mechanisms may operate together or in different circumstances to bring about the multitude of highly specific cell adhesions that are required for normal embryonic development.

Although the processes of shaping and aggregation of cells both play important roles in morphogenesis, they alone are inadequate to explain the entire phenomenon. As we shall see in the next two chapters, selective control over cellular multiplication and selective cellular death also play major roles in the generation of adult body shape.

BIBLIOGRAPHY
Books and reviews

Bennett, D. 1975. The T-locus of the mouse. Cell **6**:441.

Bosmann, H. B. 1977. Cell surface enzymes: effects on mitotic activity and cell adhesion. Int. Rev. Cytol. **50**:1.

Maslow D. E. 1976. In vitro analysis of surface specificity in embryonic cells. *In* G. Poste and G. L. Nicholson, eds. The cell surface in animal embryogenesis and development. Cell Surface Reviews. Vol. 1. Elsevier/North-Holland Biomedical Press, Amsterdam.

McLaren, A. 1976. Genetics of the early mouse embryo. Annu. Rev. Genet. **10**:361.

Moscona, A. A., ed. 1974. The cell surface in development, John Wiley & Sons, Inc., New York.

Nicolson, G. L. 1976. Transmembrane control of the receptors on normal and tumor cells. 1. Cytoplasmic influence over cell surface components. Biochim. Biophys. Acta **457**:57.

Poste, G., and G. L. Nicolson, eds. 1976. The cell surface in animal embryogenesis and development. Cell Surface Reviews. Vol. 1. Elsevier/North-Holland Biomedical Press, Amsterdam.

Poste, G., and G. L. Nicolson, eds. 1977. Dynamic aspects of cell surface organization. Cell Surface Reviews. Vol. 3. Elsevier/North-Holland Biomedical Press, Amsterdam.

Sherman, M. I., and L. R. Wudl. 1977. T complex mutations and their effects. *In* M. I. Sherman, ed. Concepts in mammalian embryogenesis. MIT Press, Cambridge, Mass.

Singer, S. J., and G. L. Nicolson. 1972. The fluid mosaic model of the structure of cell membranes. Science **175**:720.

Steck, T. L. 1974. The organization of proteins in the human red blood cell membrane—a review. J. Cell Biol. **62**:1.

Steinberg, M. S. 1964. The problem of adhesive selectivity in cellular interactions. In M. Locke, ed. Cellular membranes in development. Academic Press, Inc., New York.

Steinberg, M. S. 1975. Adhesion-guided multicellular assembly: a commentary upon the postulates, real and imagined, of the differential adhesion hypothesis, with special attention to computer simulations of cell sorting. J. Theor. Biol. **55**:431.

Trinkaus, J. P. 1969. Cells into organs. Prentice-Hall, Inc., Englewood Cliffs, N.J.

Weissmann, G., and R. Claiborne, eds. 1975. Cell membranes: Biochemistry, cell biology, and pathology. H. P. Publishing Co., Inc., New York.

Selected original research articles

Burdick, M. L. 1972. Differences in the morphogenetic properties of mouse and chick embryonic liver cells. J. Exp. Zool. 180:117.

Ede, D. A., and O. P. Flint. Intercellular adhesion and formation of aggregates of normal and talpid³ mutant chick limb mesenchyme. J. Cell Sci. 18:97.

Frye, L. D., and M. Edidin. 1970. The rapid intermixing of cell surface antigens after formation of mouse-human heterokaryons. J. Cell Sci. 7:319.

Grady, S. R., and E. J. McGuire. 1976. Intercellular adhesive selectivity. III. Species selectivity of embryonic liver intercellular adhesion. J. Cell Biol. 71:96.

Henkart, P., S. Humphreys, and T. Humphreys. 1973. Characterization of sponge aggregation factor. A unique proteoglycan complex. Biochemistry 12:3045.

Humphreys, T. 1963. Chemical dissolution and in vitro reconstruction of sponge cell adhesions. I. Isolation and functional demonstration of the components involved. Dev. Biol. 8:27.

Moscona, A. A. 1968. Cell aggregation properties of specific cell-ligands and their role in the formation of multicellular systems. Dev. Biol. 18:250.

Roth, S., E. J. McGuire, and S. Roseman. 1971. An assay for intercellular adhesive specificity. J. Cell Biol. 51:525.

Roth, S., E. J. McGuire, and S. Roseman. 1971. Evidence for cell-surface glycosyltransferases. J. Cell Biol. 51:536.

Rutishauser, U., W. E. Gall, and G. M. Edelman. 1978. Adhesion among neural cells of the chick embryo. IV. Role of the cell surface molecule CAM in the formation of neurite bundles in cultures of spinal ganglia. J. Cell Biol. 79:382.

Steinberg, M. S. 1970. Does differential adhesion govern self-assembly processes during histogenesis? Equilibrium configurations and the emergence of a hierarchy among populations of embryonic cells. J. Exp. Zool. 173:395.

Steinberg, M. S., and L. L. Wiseman. 1972. Do morphogenetic tissue rearrangements require active cell movements? The reversible inhibition of cell sorting and tissue spreading by cytochalasin B. J. Cell Biol. 55:606.

Thiery, J. P., et al. 1977. Adhesion among neural cells of the chick embryo. II. Purification and characterization of a cell adhesion molecule from neural retina. J. Biol. Chem. 252:6841.

Townes, P., and J. Holtfreter. 1955. Directed movements and selected adhesion of embryonic amphibian cells. J. Exp. Zool. 128:53.

Wilson, H. V. 1907. On some phenomena of coalescence and regeneration in sponges. J. Exp. Zool. 5:245.

CHAPTER 17

Growth control

☐ Control of growth in the developing organism (and in the adult) is obviously a process of crucial importance. A single average-sized cell of about 10^{-9} grams could produce a 90 kg (200-pound) adult in approximately 42 cell doublings if one assumes that there is no cell loss, that all cells continue to divide at the same rate, and that cell size is constant. None of these assumptions (including the size of the initial cell) is fully valid for embryonic development, but it is nevertheless true that only a relatively small number of cell divisions is required to produce an adult from a fertilized egg. Furthermore, most of that cellular proliferation occurs prior to birth or hatching. In humans, for example, only four to five doublings of mass are needed to generate a normal-sized adult from a 3 kg (7-pound) newborn baby. There is essentially no net growth in the normal, healthy adult. Cell division and cell loss balance each other, and the total mass remains fairly constant. Our discussion of growth control in this chapter will be focused on the control of cellular reproduction. Except where specified otherwise, we assume that cellular mass and cellular volume remain relatively constant.

The regulation of growth during development is also apparent from the diversity of reproductive patterns of different types of cells. Different cells vary in the length of their generation times and in the point in the animal's life cycle at which proliferation normally slows or ceases. Some cells even contain an inherent death program, which, as we shall see in the next chapter, results in their demise during embryonic development. Different types of cells also respond differentially to factors that stimulate or inhibit growth.

Although control of growth in multicellular organisms is not totally understood, current research is providing a number of clues as to possible mechanisms. Current investigation in the area of growth control is focused in three general areas. The first area of research deals with external factors that stimulate or inhibit growth. Some of these factors have general effects on many types of cells, whereas others are highly selective for specific types of target cells. The second area of research concerns events that take place at the cellular membrane. The cellular membrane provides the interface between external factors and the internal cellular milieu and appears to play an important role in growth regulation. The final area of current research deals with intracellular factors that influence growth. Since cellular growth, whether by cell division or an increase in cell size, occurs within the cell itself, it can be expected that all mechanisms of growth control have an intracellular facet.

The actual control of growth is undoubtedly a complex process, that may involve all the mechanisms to be discussed plus others that are yet to be discovered. Likewise, control of proliferation in any individual cell probably involves overlap among several mechanisms. Certain cases of overlap have already been suggested by research data. The separate discussion of each proposed mechanism on the following pages reflects our current lack of understanding of the complex interactions that must be involved in the overall phenomenon of growth control. We can describe individual pieces of the puzzle in considerable detail, but at the present time we can only speculate on how they fit together into a complete picture.

There are two main types of factors that have been studied with regard to growth stimulation—hormones and growth factors. Although we cover them separately, the distinction between them may be somewhat artificial, as will be discussed. A single general class of growth inhibitory factors has been studied. These factors are called chalones from the Greek word *chalan*, to slacken.

EXTERNAL FACTORS THAT STIMULATE OR INHIBIT GROWTH

Hormones

The hormone classically associated with growth regulation in vertebrates is growth hormone (GH), or somatotropin, a protein hormone produced by the anterior pituitary. The magnitude of its effect is dramatically apparent in cases in which the amount present is abnormal. In the human, for example, excessive prepuberal production of GH results in giantism. Excessive GH production after puberty produces a condition called acromegaly in which there is no increase in height because the epiphyses (growth areas) of the long bones have become closed over, but there is a disproportionate growth of skeletal tissues of the extremities (lower jaw, nose, feet, hands) and a few other organs. A deficiency of GH results in dwarfism in the preadolescent; it has no immediately obvious effects in the adult. Numerous metabolic responses to GH by its target tissues have been delineated, such as increases in protein synthesis, fat mobilization, and output of glucose from the liver. The mechanism of action of GH is unknown, but it appears to act on cells by means of cell surface receptors.

Although GH acts directly on some of its target tissues, its effect on skeletal growth is indirect and is mediated by a plasma growth factor called somatomedin. The role of somatomedin was discovered from the inability of GH to stimulate the uptake of sulfate into cartilage in vitro, which it normally does in vivo. Sulfate uptake was found to be stimulated in vitro by serum of normal rats but not by serum from hypophysectomized rats, which lack GH. It was thus proposed that GH action on skeletal tissue was due to a factor released in response to GH and carried in plasma. This factor was termed somatomedin or sulfation factor, and it was subsequently shown that GH stimulates output of the factor from the liver.

Somatomedin activity in plasma is found in a number of distinct molecular species in the same animal. All the species are GH dependent. Three have been isolated so far and are proteins with molecular weights of 7000 or less. Thus somatomedin is not a single substance but a group of substances. All of them are insulin-like in their activity and belong to a class of plasma substances termed nonsuppressible insulin-like activity (NSILA) that mimic the activity of insulin but are unaffected by antibody to pancreatic insulin. The NSILA peptides are partially responsible for the growth-promoting effect of serum on certain types of cultured cells (e.g., chicken embryo fibroblasts).

Somatomedin has the following biological activities:
1. It mediates the actions of GH on cartilage, including increased proliferation of cartilage cells, stimulation of the uptake of sulfate, and the synthesis of DNA, RNA, and proteoglycans of the cartilage matrix.
2. It mimics the action of insulin in every tissue studied so far and competes with insulin for membrane receptor sites.
3. It stimulates the division of certain cultured cells in the absence of serum.

Numerous hormones other than GH are also important for the growth of various tissues. Among these are thyroxine, hydrocortisone, insulin, testosterone, and estradiol. Some of these act at least in part by influencing either the secretion of GH or the response of tisssues to GH. Other developmental effects of hormones are discussed in Chapters 14, 20, and 22.

Growth factors

Most of the substances called growth factors are small polypeptide molecules with hormonelike actions. Some have highly specific target tissues, and others affect a variety of tissues. The major distinction between growth factors and hormones lies in their cell-type of origin rather than in their mechanism of action, and some investigators use

Table 17-1. Tissue-specific growth factors

Growth factor	Action stimulated
Nerve growth factor	Proliferation and neurite outgrowth of sympathetic ganglia (also sensory ganglia in embryo)
Epidermal growth factor	Growth and development of epithelial tissues
Erythropoietin	Production of red blood cells
Colony-stimulating factor	Proliferation of granulocytes and macrophages
Thrombopoietin	Production of blood platelets
Pancreatic mesenchymal factor	Proliferation of pancreatic epithelial cells
Somatomedin	Proliferation of cartilage cells
Tumor angiogenesis factor	Development of blood vessels
Fibroblast growth factor	Growth of fibroblasts and mesodermally derived cells in culture

the two terms interchangeably. The list of specific growth factors has grown increasingly long in recent years. Table 17-1 presents a partial listing of these factors. This section will describe as examples several growth factors that have been studied extensively. (The somatomedins were described in the discussion of hormones on p. 530.)

Nerve growth factor (NGF). NGF has been studied extensively since its discovery more than 25 years ago, and consequently we will discuss it at greater length than the other growth factors. The story of the discovery and isolation of NGF is interesting. In the original investigation, transplantable mouse tumors were grafted to the limb bud region of 3-day chicken embryos to study the effect of rapidly growing tissue on innervation. Some of the grafted tumors became highly innervated by nerves from the embryo. Crude extracts prepared from the tumors were able to stimulate the growth of nerves, especially those of sympathetic ganglia. During experiments designed to determine whether the factor contained nucleic acid, snake venom was used as a source of phosphodiesterase. Treatment with snake venom was found to enhance rather than destroy the activity, and control preparations that contained only the snake venom and no tumor extract were found to have a potent stimulatory activity on nerve growth. Since the snake venom gland is a modified salivary gland, workers tried to obtain NGF activity from salivary glands of mice, which are easier to obtain and work with than cobras. High NGF activity was found in the submaxillary glands of male (but not female) mice, and NGF from this source has been used for most subsequent studies.

NGF isolated either from mouse submaxillary glands or from snake venom is a protein, and NGF from both these sources has recently been sequenced. Native NGF from the mouse is a dimer composed of two identical subunits of 118 amino acids. Cobra venom NGF is similar in size, and about 60% of its amino acid residues are identical to those in mouse salivary NGF. The amino acid sequence of mouse NGF shows considerable homology with the human proinsulin molecule. Proinsulin is synthesized as a single chain, which is subsequently cleaved to yield the two chains (A and B) that form insulin (Fig. 9-10). Cleavage results in the removal of a sequence of 36 amino acids designated C. The sequence of proinsulin starting at the N-terminal end is BCA. Mouse NGF has a sequence that bears a resemblance to BCAB (Fig. 17-1). Some investigators believe that the NGF gene could have arisen by duplication of the proinsulin gene, followed by a second duplication of only one of the two proinsulin genes and deletion of the second CA sequence. There is some evidence from in vitro synthesizing

Mouse NGF	Ser	—	Ser	Thr	His	Pro	—	Val	Phe	His	Met	Gly	Glu	—	—	Phe	Ser	Val	Cys	Asp	Ser
Human proinsulin	Phe	Val	**Asn**	**Gln**	**HIS**	Leu	Cys	Gly	Ser	**HIS**	**Leu**	Val	**GLU**	Ala	Leu	**Tyr**	**Leu**	**VAL**	**CYS**	Gly	Glu
Guinea pig insulin	Phe	Val	**SER**	Arg	**HIS**	Leu	Cys	Gly	Ser	**Asn**	**Leu**	Val	**GLU**	Thr	Leu	**Tyr**	**SER**	**VAL**	**CYS**	Gln	Asp

B peptide

Mouse NGF	Val	Ser	Val	Trp	Val	Gly	Asp	Lys	Thr	Thr	Ala	Thr	Asn	Ile	Lys	Gly	Lys	Glu	Val	Thr	Val
Human proinsulin	Arg	**Gly**	Phe	**Phe**	Tyr	Thr	Pro	**LYS**	**THR**	Arg	Arg	Glu	**Ala**	Glu	Asp	Leu	**Gln**	Val	Gly	Gln	**VAL**
Guinea pig insulin	Asp	**Gly**	Phe	**Phe**	Tyr	Ile	Pro	**LYS**	Asp												

B peptide   C peptide

Mouse NGF	Leu	Ala	Glu	Val	Asn	Ile	Asn	Asn	Ser	Val	Phe	Arg	Gln	Tyr	Phe	Phe	Glu	Thr	Lys	Cys	Arg
Human proinsulin	Glu	Leu	**Gly**	Gly	**Gly**	Pro	**Gly**	**Ala**	**Gly**	Ser	**Leu**	**Gln**	**Pro**	Leu	Ala	**Leu**	**GLU**	Gly	Ser	Leu	**Gln**
Guinea pig insulin																					

C peptide

Mouse NGF	Ala	Ser	Asn	Pro	Val	Glu	Ser	Gly	Cys	Arg	Gly	Ile	Asp	Ser	Lys	His	—	Trp	Asn	Ser	Tyr
Human proinsulin	Lys	Arg	**Gly**	Ile	**VAL**	**GLU**	Gln	Cys	**CYS**	Thr	**Ser**	**ILE**	Cys	**SER**	Leu	Tyr	Gln	Leu	**Glu**	**Asn**	**TYR**
Guinea pig insulin			**Gly**	Ile	**VAL**	Asp	Gln	Cys	**CYS**	Ala	**GLY**	Thr	Cys	**Thr**	**Arg**	**HIS**	Gln	Leu	**Glu**	**SER**	**TYR**

C peptide   A peptide

Mouse NGF	Cys	Thr	Thr	Thr	His	Thr	Phe	Val	Lys	Ala	Leu	Thr	Thr	Asp	Glu	Lys	Gln	Ala	Ala	Trp	Arg
Human proinsulin	**CYS**	**Asn**	COOH			NH₂	**PHE**	**VAL**	**Asn**	Gln	His	Leu	Cys	**Gly**	**Ser**	**His**	Leu	Val	**Glu**	Ala	Leu
Guinea pig insulin	**CYS**	**Asn**	COOH			NH₂	**PHE**	**VAL**	Ser	Arg	His	Leu	Cys	**Gly**	**Ser**	**Asn**	Leu	Val	**Glu**	Thr	Leu

A peptide   B' peptide

Mouse NGF	Phe	Ile	Arg	Ile	Asn	Thr	Ala	Cys	Val	Cys	Val	Leu	Ser	Arg	Lys	Ala	Thr	Arg	COOH
Human proinsulin	**Tyr**	**Leu**	Val	Cys	**Gly**	Glu	Arg	Gly	Phe	Phe	Tyr	Thr	**Pro**	—	**LYS**	**Thr**	Arg	**ARG**	C-peptide
Guinea pig insulin	**Tyr**	Ser	Val	Cys	**Gln**	Asp	Asp	Gly	Phe	Phe	Tyr	**Ile**	**Pro**	—	**LYS**	Asp	COOH		

B' peptide

Fig. 17-1. Comparison of amino acid sequences of mouse NGF, human proinsulin, and guinea pig insulin. In the insulin and proinsulin sequences, boldface capitals indicate amino acids identical to those in NGF, and boldface lower case type indicates "favored" amino acid substitutions whose probability is greater than random (e.g., one-base mutations). Addition and deletion have been allowed as shown for best sequence match. A and B refer to the A and B subunits of insulin, and C refers to the connecting peptide. B' refers to the repeated B subunit. The guinea pig C sequence was not available for this comparison. Substantial evolutionary divergence occurs even among insulin and C peptide sequences. Thus the similarities with NGF may be highly significant. (Modified from Frazier, W. A. et al. 1972. Science **176**:482.)

systems that NGF is translated as a pro-NGF of about 170 amino acids, which can be cleaved to 118 amino acids by an extract from submaxillary glands. The 170 amino acid molecule is about the size of a double-length proinsulin gene, suggesting that the second CA sequence may be retained and removed posttranslationally from the carboxy-terminal end of pro-NGF.

A relationship between NGF and proinsulin has also been suggested by other techniques. Certain procedures used to determine the three-dimensional structure of proteins suggest a resemblance between NGF and proinsulin. Also, both have specific cell surface receptors, although there is apparently only limited cross reaction of either protein with the other's receptors. It should be pointed out, however, that some investigators believe the differences between NGF and proinsulin are greater than their similarities and that they probably do not have a common evolutionary origin.

NGF may be released by a variety of cell types. It has been shown to be present in mouse heart, spleen, kidney, normal fibroblasts, mouse 3T3 and L cells (both of which are fibroblasts), glial tumors, and submaxillary gland. In some of these cases it is unclear whether the cells actually make NGF or merely store it. Although the submaxillary glands are the richest source of NGF, they are not likely to be the primary source, since after their removal the plasma level of NGF drops only temporarily. The most likely candidates for the synthesis of NGF are considered to be fibroblasts and/or glial cells. Cultured fibroblasts secrete a material that is immunologically similar to, if not identical with, NGF. Neurons cultured in vitro without NGF are viable only in the presence of other nonneural ganglion cells, which are probably glial cells. If antibody to NGF is added to glial cells before they are combined with cultured neurons, their ability to maintain survival of the neurons is greatly reduced.

NGF has a number of biological activites. For example, it promotes the viability of neurons grown in culture. Outgrowth of axons from sympathetic ganglion tissue is frequently used to measure NGF activity (Fig. 17-2). In vivo, NGF stimulates growth of sympathetic neurons and sensory neurons in the embryo, and it stimulates growth of sympathetic neurons in the adult. NGF may also have effects on the central nervous system. Injection of NGF into the brains of experimental animals has been shown to stimulate regeneration of severed axons of neurons of the central nervous system.

The biological activities of NGF found by treating cells or animals with the factor do not necessarily indicate its role within the animal. However, experiments done with antibody to NGF indicate that it does have a critical role in both the development and maintenance of neu-

Fig. 17-2. Biological assay for NGF. These micrographs show sensory ganglia dissected out of an 8-day chicken embryo and cultured for 12 hours either in the absense (**A**) or the presence (**B**) of purified NGF. (**A** from Levi-Montalcini, R., and P. Calissano. 1979. Sci. Am. **240**:68 [June]. Copyright 1979 by the American Association for the Advancement of Science; **B** from Levi-Montalcini, R. 1964. Science **143**. Cover. [Jan.])

B

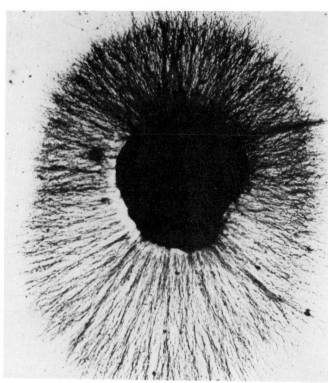

A

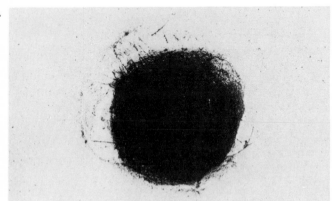

Fig. 17-3. Micrographs showing effect of NGF bound to Sepharose beads on sensory ganglion from 8-day chick embryo. **A,** Ganglion after 18 hours of incubation in NGF-Sepharose. **B,** Higher magnification view of periphery of ganglion in **A.** Both views show dense outgrowths of nerve fibers stimulated by the NGF on the beads. (From Frazier, W. A., L. F. Boyd, and R. A. Bradshaw. 1973. Proc. Natl. Acad. Sci. **70:** 2931.)

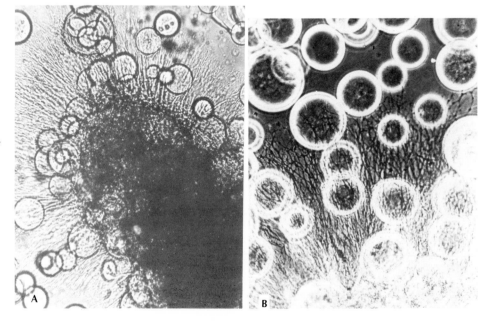

ral tissues. When antibody to NGF is given to newborn mice, 90% to 95% of the sympathetic nervous system is destroyed, producing what are called immunosympathectomized mice. If the antibody is administered to an adult, the size of sympathetic ganglia is reduced.

The primary action of NGF on cells is unknown, but it is known to interact with cells by virtue of specific cell surface receptors. NGF apparently does not need to enter its target cells for activity, since NGF is fully active in promoting neurite extension when bound to Sepharose beads too large to enter cells (Fig. 17-3). When such beads are added to neurons in culture, the cells orient themselves around the beads. Only those cells in contact with the beads are stimulated, suggesting that NGF is not being released from the beads in a soluble form. However, as is the case with peptide hormones (Chapter 14), there is some lingering doubt that small amounts of factor could be released directly to the cells in closest proximity to the beads.

Recent studies have shown that small amounts of NGF normally enter the cytoplasm of cells stimulated with free NGF. Specific recognition by cell surface receptors appears to be necessary for the internalization, which also results in internalization of the receptors. NGF taken up at the tips of axons of sympathetic neurons undergoes retrograde axonal transport to the cell body, sometimes over a considerable distance. It has been suggested that the internalization of the receptor may serve to render the stimulated cell temporarily unresponsive to further stimulation until new receptors are synthesized and transported to the cell surface. Some of the internalized NGF appears to be transported to the nucleus and to interact with nuclear receptors. (For further discussion of this phenomenon, see the review by Bradshaw cited at the end of this chapter). Similar internalization phenomena have been reported for insulin (Chapter 14) and epidermal growth factor.

The specificity of binding of NGF to cells has been studied by using NGF labeled with a radioactive isotope of iodine (^{125}I). Some investigators report that NGF binds specifically to neurons of sympathetic and sensory ganglia and is not competed with by other proteins, including proinsulin, suggesting a high degree of receptor specificity. Others, however, report limited competition by insulin and proinsulin for NGF binding sites.

In rat and chick embryos, NGF also binds to a number of tissues other than sensory and sympathetic ganglia, including brain and organs that are innervated by the sympathetic system, such as the heart. The significance of these interactions is unclear, but it has been postulated that the attachment of NGF to such cells during development could attract growing sympathetic nerves.

The binding of NGF to sensory and sympathetic ganglia has also been studied as a function of time in the development of the chick embryo. Binding to sensory ganglia is greatest at about the seventh and eighth days of development and then drops sharply. Binding to sympathetic ganglia is fairly constant from the eighth day on. These binding patterns correspond well with the timing of ganglion sensitivity to NGF, suggesting that the responsiveness of the cells to NGF is probably controlled by the relative presence or absence of receptors.

Epidermal growth factor (EGF). EGF was originally found as a contaminant in NGF preparations from mouse submaxillary glands. Human EGF has also been isolated recently from the urine of pregnant women. The EGFs from these two sources differ slightly in size, but both are small, with molecular weights between 5000 and 6000. The two EGF molecules have some similarity in amino acid composition, show some immunological cross reaction, and compete for the same cell surface receptors.

EGF appears to be involved in the development of epithelial tissues. When injected into newborn mice, it causes precocious eyelid opening and tooth eruption. It likewise stimulates the growth of a variety of epithelial tissues in organ culture. In these two cases it promotes increase in cell size rather than cell division. However, it is mitogenic for mouse 3T3 and normal diploid fibroblasts. Finally, it has been shown to stimulate the structural and biochemical differentiation of keratinocytes in culture, but only when mesodermally derived fibroblastic cells are also present in the culture. This recent observation suggests the interesting possibility that "epidermal" effects of EGF may actually be mediated by its effects on underlying mesodermal cells. It also should be emphasized that there is still no direct evidence for a role of EGF in the normal development of epithelial cells in the animal, although such a role seems likely from the experiments that have been described.

Tumor angiogenesis factor. Many types of cancerous growth are quickly invaded by actively growing blood vessels after they become established in the body. Early investigations showed that outgrowth from neighboring blood vessels was stimulated even when the tumor was isolated from contact with neighboring tissue by Millipore filter membranes. Detailed studies by Judah Folkman and associates have shown that a specific factor, which has been termed tumor angiogenesis factor, is responsible for the vessel outgrowth. The factor has been isolated from tumor cells, and the activity is sensitive to both ribonuclease and proteolytic enzymes. A similar activity can be isolated from placental tissue, which is a site of extensive vascular growth.

Tumor angiogenesis factor has attracted considerable attention, since tumors that fail to become vascularized do not grow beyond a small size and appear not to be harmful to their host animals. If a means of selectively blocking the effect of tumor angiogenesis factor can be found, it could prove to be of major clinical value in inhibiting the development of secondary cancer as a result of metastasis (cells being freed from a primary tumor and establishing new tumors in other parts of the body). The normal role of this growth factor is not known, but it is possible that it is produced by rapidly growing tissues during normal development to stimulate their vascularization.

Chalones are molecules that specifically inhibit cell division. Although they remain controversial, a substantiated body of evidence supporting their existence has been gathered in recent years. Chalones provide an autoregulatory type of growth control in that they inhibit cell proliferation in the tissue where they are made. Following are some of the systems for which tissue-specific chalones have been described:

Chalones

Eccrine (sweat) gland	Granulocyte	Lymphocyte
Epidermis	Hair follicle	Melanocyte
Erythrocyte	Kidney	Sebaceous gland
Fibroblast	Liver	

Chalones have been postulated to act in cases of compensatory hyperplasia. For example, if one kidney is removed from an animal, the other kidney compensates for the loss by growing. Or if part of the liver is removed, the remaining part undergoes cell proliferation. The chalone hypothesis postulates that the kidney and liver produce labile mitotic inhibitors that regulate their own growth. When the amount of tissue is decreased, the concentration of chalone in the animal's bloodstream is also decreased, and this results in a stimulation of cell division.

Chalones are also postulated to operate in the normal loss and replacement of cells in tissues. It has been proposed, for example, that keratinizing cells in epidermis produce a chalone that diffuses into the basal epidermis to inhibit cell division (Fig. 17-4). As the thickness of the epidermis waxes and wanes, the amount of chalone produced increases and decreases correspondingly and thus regulates proliferation by basal cells.

It has been demonstrated experimentally that this type of negative feedback system operates in a variety of tissues in different vertebrates. For example, a small peptide isolated from rabbit liver decreases DNA synthesis in the remaining lobe of liver after partial hepatectomy. Extracts taken from adult amphibian kidney specifically inhibit cell division by the embryonic kidney, whereas extracts from other

Fig. 17-4. Diagram of proposed scheme for chalone action in epidermal cells. Postmitotic, differentiating cells produce an inhibitor that feeds back to actively dividing basal cells and controls their proliferation. (Modified from Bullough, W. S. 1969. Science J. **5:**71 [April]).

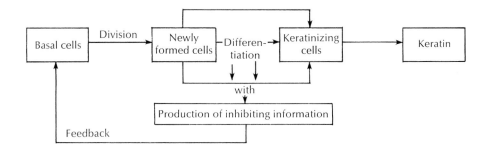

adult tissues have no such effect. A number of chalones have been isolated and chemically characterized. All are proteins or glycoproteins with molecular weights varying from less than 4000 to up to 50,000.

One of the more extensively studied chalone systems is that involving the epidermal chalone. Epidermal chalone inhibits mitosis of epidermal cells when injected in vivo or when added to cultured cells. It is tissue specific, but does not appear to be particularly species specific. In fact, an extract from cod skin is able to inhibit division in mouse and human epidermal cells. The activity of epidermal chalone requires the synergistic action of stress hormones. The chalone is effective in vitro only if small amounts of adrenalin are also present.

One of the major problems involved in studying any factor claimed to have a negative effect on cell proliferation is proving that a natural inhibitor and not a toxic by-product of the fractionation procedure has been isolated. Although this objection has not yet been met completely for chalones, the data that are currently emerging make it appear increasingly likely that the chalone concept is valid.

RELATIONSHIP OF THE CELLULAR MEMBRANE TO GROWTH CONTROL

The results of numerous studies have indicated that the cellular membrane is intimately connected with the control of growth. Although these studies have typically been done with cultured cells, it is likely that the results are applicable to some extent to growth control in the developing animal. For example, it is possible that changes during development in the composition or arrangement of molecules on the surfaces of cells may alter the growth potential of cells. It is also possible that cells in a tissue are able to monitor their density by mechanisms involving the cell surface and cease growing when a certain density is reached.

When normal diploid cells are grown in culture, most types cease division at a saturation density that is characteristic of the type of cell, and to some extent, of the culture conditions. This phenomenon is commonly called contact inhibition of division, although some investigators prefer to refer to it as density-dependent inhibition of growth, since environmental conditions other than cell contact also appear to affect it.

Most established cell lines generally do not exhibit contact inhibition but continue to grow with cells piling on top of one another after they reach confluency. This difference from normal cells is presumably a consequence of the techniques used to establish cell lines. Typically, the cultures are maintained at high cell densities, and therefore they quickly reach a degree of crowding that inhibits further multiplication of all cells that are sensitive to contact inhibition. However, if any variants arise in the culture that are not sensitive to contact inhibition, they can continue to multiply until the nutrient medium is depleted. This provides a large selective advantage for variants that can continue growing under crowded conditions, and such cells soon dominate the cultures.

If the selective advantage of cells that are not sensitive to contact inhibition is eliminated by inoculating cultures at a lower density and subculturing frequently so that all cells are constantly multiplying, permanent cell lines can be obtained that continue to exhibit contact inhibition. The mouse 3T3 line is an example. It has an altered karyotype and is capable of indefinite growth, but it has remained highly contact inhibited. In fact, it ceases growth at a lower cell density than most types of diploid cells. Transformation of 3T3 cells by cancer viruses causes them to lose their contact-inhibited properties. A number of other cell lines comparable to 3T3 have also been produced by similar culture procedures.

The term "contact inhibition" implies that the inhibitory effects on cell division result from direct physical contact rather than from other aspects of cell crowding. This remains a controversial point. It is clear that the phenomenon is influenced by factors other than contact, but there is also considerable evidence that some degree of cell-to-cell contact is necessary to achieve the inhibition of cell division. For example, "wounding" a contact-inhibited monolayer of cells by scraping away some of the cells results in growth of cells into the wounded area while the rest of the culture remains inhibited. Furthermore, the presence of a dense, contact-inhibited culture of cells on one side of a Millipore filter does not inhibit multiplication of a sparse population of cells on the other side. The sparse population shares the same medium and culture conditions as the dense population but lacks the factor of cell contact.

Serum factors play a major role in determining the cell density at which growth inhibition occurs. The final density achieved by a sparsely seeded culture is determined rather directly by the amount of serum protein in the culture medium (Fig. 17-5). The addition of more serum to a culture that has reached its final density will cause additional growth if other nutrients in the medium have not been depleted. A detailed analysis of this phenomenon has not yet been achieved because of the multiplicity of roles that serum factors play in the culture system. One or more factors from serum are needed just for survival of 3T3 cells in conventionally used culture media. A second factor from serum is needed for cell migration into a wound area, and still another factor is required for multiplication of sparse cultures. The factor or factors that stimulate further proliferation in crowded cultures are different from all these other factors and at present must be studied against a background containing all of them. Hopefully, new, better-defined media currently under development will simplify such studies in the near future.

The mechanisms that cause certain serum factors to become growth limiting only when the cultures are crowded are not well understood. One interesting but controversial theory suggests that essential factors are depleted from the thin layer of medium immediately overlying confluent cultures and that diffusion is not adequate to transfer such factors from the bulk of the medium to that boundary layer adjacent to the cells. Support for this theory is derived from experiments showing that shaking or stirring of cultures stimulates density-inhibited 3T3 cells to divide, presumably by redistribution of

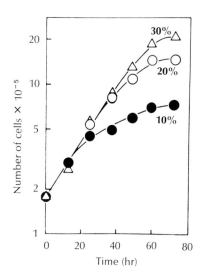

Fig. 17-5. Growth curves of mouse 3T3 cells in media containing 10%, 20%, and 30% calf serum. (From Holley, R. S., and J. A. Kiernan. 1968. Proc. Natl. Acad. Sci. U.S.A. **60**:300.)

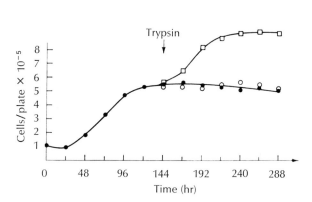

Fig. 17-6. Effect of trypsin treatment on growth of mouse 3T3 cells that have been grown to density where their division is contact inhibited. ● = Untreated control cells; □ = trypsin-treated cells; ○ = cells treated with inactivated trypsin. (Modified from Burger, M. M. 1970. Nature **227**:170.)

growth factors through the medium. The same theory could possibly also explain why medium taken from dense cultures still supports the growth of sparse cultures and why wounding stimulates growth. In the latter case the process of scraping would disrupt the boundary layer of medium immediately over the culture surface, and the reduced cell population of the scraped area would reduce the rate of depletion of growth factors in that area. Extreme proponents of the diffusion boundary theory argue that the entire phenomenon of density-dependent inhibition can be explained by it with no need to consider cell contact. However, opponents of this theory have recently shown that density-dependent inhibition is not enhanced by increases in viscosity of the medium that are sufficient to reduce diffusion rates substantially.

Although the exact role of cell contact in contact inhibition remains uncertain, cell surface and membrane properties appear to be important to the phenomenon. Transformed cells that do not exhibit contact inhibition have glycoprotein receptors on their surfaces that permit them to be agglutinated readily with plant lectins such as concanavalin A and wheat germ agglutinin. Normal cells also possess the receptors, but they are not as agglutinable. The cells become more agglutinable briefly during mitosis, and after protease treatment, which also sometimes stimulates division of contact-inhibited cells (Fig. 17-6). It was originally thought that receptors on normal cells

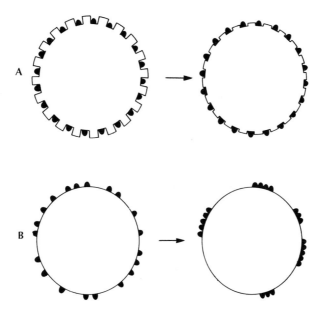

Fig. 17-7. Two mechanisms to explain the findings that transformed cells and protease-treated normal cells are more readily agglutinable by several plant agglutinins. **A,** Agglutinin-binding sites are "masked" on the surfaces of normal cells but become exposed after proteolysis or transformation. **B,** Agglutinin-binding sites of normal cells are dispersed in the membrane, but after proteolysis or transformation they diffuse and aggregate in clusters. The probability of agglutination of two such modified cells is enhanced by the clustering. (From Singer, S. J., and G. L. Nicolson. 1972. Science **175:**720.)

were "masked" by proteins and became "unmasked" by protease treatment. However, many investigators now believe that the mobility of molecules in the cell membrane is increased as a result of mitosis, protease treatment, and transformation, and that the surface receptors become clustered into more reactive groups. These two mechanisms to explain the increased agglutinability of transformed or protease-treated cells are diagrammed in Fig. 17-7.

Restoration of growth in contact-inhibited cultures involves a number of rapid physiological and biochemical changes, including reduction in the level of intracellular cyclic AMP (cAMP), increases in transport of phosphate, uridine, and amino acids, and initiation of synthesis of RNA and proteins. One research group has used the term "pleiotypic activation" to refer to the multiplicity of apparently concurrent (rather than sequential) changes that occur and has suggested that a single master control mechanism must be involved. Integrated theories of growth control are discussed in greater detail at the end of this chapter.

INTRACELLULAR FACTORS THAT INFLUENCE GROWTH

Much of the research that has been done relating intracellular changes to the control of cell growth has centered on the cyclic nucleotides, especially cAMP. An effect of cyclic nucleotides on cell proliferation has been demonstrated in a variety of different experimental systems. The rate of cell division in cultured fibroblast cells is related to the level of cAMP in these cells. Cell division is slowed by the addition of theophylline, a drug that inhibits cAMP phosphodiesterase and thereby increases the intracellular level of cAMP. Cell division is likewise slowed by analogs of cAMP such as dibutyryl cAMP (Bt_2cAMP), which mimics cAMP actions but is more readily taken up by cells and is less rapidly metabolized to 5' AMP. Finally, agents such as certain prostaglandins also slow cell division by stimulating adenyl cyclase, the enzyme that synthesizes cAMP from ATP.

Measurements of cAMP levels in cells growing at different rates also suggest a correlation between cAMP levels and cell proliferation (Fig. 17-8). In general, cAMP levels are lower in dividing cells. In cells that exhibit contact inhibition of division the level of cAMP increases as the cells become confluent. Such an effect is not seen in transformed cells that do not show contact inhibition. Many of the growth factors and other agents that stimulate cell division in cultured cells, such as serum, insulin, and proteases, cause a decrease in cAMP levels. Most of these agents act at least in part by inhibiting the activity of adenyl cyclase (Fig. 14-3). The stimulation of cell division caused by such agents can be prevented by the addition of Bt_2cAMP.

Regulation of cell proliferation by cyclic nucleotides has been extensively studied in the lymphocyte system. Lymphocytes can be in-

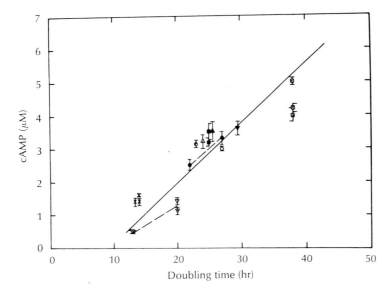

Fig. 17-8. Comparison of cAMP levels with the growth rate in 10 different lines of fibroblasts. The dashed lines connect values for cell lines whose doubling times (and cAMP levels) changed during the course of the study. (From Otten, J., G. C. Johnson, and I. Pastan. 1971. Biochem. Biophys. Res. Commun. **44:** 1192.)

duced to divide by exposing them to antigens or mitogenic plant lectins. Although the results vary considerably with different experimental conditions, the general picture seems to be that cyclic guanosine 3′, 5′-monophosphate (cGMP) mediates the action of mitogens that stimulate lymphocyte proliferation, whereas cAMP antagonizes the actions of these mitogens. It has been recently demonstrated that these effects of cAMP and cGMP are paralleled by their influence on phosphorylation of nuclear acidic proteins—cGMP stimulates phosphorylation of these proteins while cAMP inhibits their phosphorylation—and it has been suggested that these phosphorylations may have a role in the stimulation of proliferation.

The effect of cAMP on cell division has also been studied by examining cAMP regulation of the passage of cells through the cell cycle.* When the cAMP levels of synchronized cells of the Chinese hamster ovary (CHO) cell line are examined at different times in the cell cycle, it is found that cAMP is low during mitosis and high in early G_1 (Table 17-2). Large and rapid changes in cAMP levels are particularly dramatic during the cleavage stage of developing sea urchin embryos (Fig. 17-9). If the drop in cAMP that normally accompanies division is prevented by the addition of caffeine (which is thought to block cAMP phosphodiesterase), cell division is inhibited. These and similar results found with other types of cells suggest that cAMP levels may regulate the passage of cells through the different phases of the cell cycle, a low

---

*Details of the cell cycle and its analysis are described in Appendix C.

Table 17-2. cAMP concentration in CHO cells*

Phase of cell cycle	cAMP (μM)	Picomoles of cAMP per milligrams of protein
Mitosis	2.9 ± 0.4	16 ± 2
Early G_1	8.4 ± 0.9	44 ± 8
Late G_1	4.0 ± 0.5	24 ± 4
S	5.2 ± 0.4	28 ± 4

*From Sheppard, J. R., and D. M. Prescott. (1972). Exp. Cell Res. **75**:293.

level being required for division and a high level for holding cells in the G_1 period prior to DNA synthesis.

Cyclic AMP regulation of the cell cycle has been studied in a line of mouse lymphoma cells that carries a mutation altering the effect of cAMP on growth. In the parental line, cAMP inhibits growth and subsequently kills the cells, but some of the mutant lines have escaped sensitivity to the cAMP effect on growth. When the parental lymphoma cells are treated with Bt_2cAMP, cells are blocked in the G_1 phase of the cell cycle, and other phases of the cell cycle are unaltered in length. When one of the mutant lines is treated with Bt_2cAMP, its cell cycle is totally unaltered. This mutant has a lowered amount of cytoplasmic cAMP-binding protein and decreased cAMP-dependent protein kinase activity. These results thus suggest that cAMP control of cell proliferation operates in the G_1 phase of the cell cycle and that it may operate at least in part through protein phosphorylation.

The basis for the transition from the G_1 phase to the S phase of the cell cycle is currently an area of active research regarding intracellular regulation of cellular proliferation. Some event (possibly more than one) in G_1 is clearly critical to the regulation of cellular reproduction. This has been deduced from a variety of observations. For example, the variability in cellular generation times can be accounted for primarily by variation in the length of G_1. In addition, most agents or conditions that inhibit cellular reproduction (e.g., unfavorable growth conditions) cause cells to become arrested in G_1. Also, nonreproducing terminally differentiated cells are normally in G_1.

Another approach to the study of intracellular regulation of cellular reproduction is the study of mutations that affect cellular growth properties. For example, a recessive mutation has been identified in *Drosophila* in which neuroblasts of the brain primordia fail to differentiate in the larva and continue reproducing. These cells also continue to reproduce if transplanted to an adult fly, and they eventually kill the host. The gene affected by this mutation must be critical to regulation of cell reproduction in neural cells. It is possible that some genes

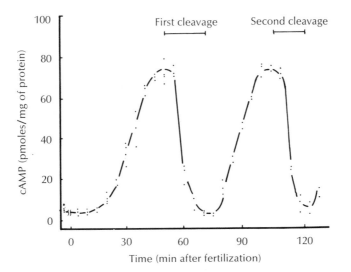

Fig. 17-9. Change in the content of cAMP in the sea urchin egg after fertilization. (From Yasumasu, I., A. Fujiwara, and K. Ishida. 1973. Biochem. Biophys. Res. Commun. **54:**628.)

which regulate cellular reproduction are specific for different cell types and act during differentiation of their respective cell types.

Many gaps still exist in our knowledge of how cellular proliferation is controlled and how that control is lost in malignant cells. Nonmultiplying cells are generally arrested in the G_1 phase of the cell cycle. Specific new protein synthesis appears to be needed to initiate the S phase. Marginally deficient levels of certain amino acids such as leucine, isoleucine, and glutamine permit cells to continue limited protein synthesis and to complete the cycle they are in but do not support the initiation of a new round of DNA synthesis. When normal levels of amino acids are returned to the inhibited cells, new protein synthesis is initiated quickly, but the cells do not enter S for several hours.

Contact-inhibited or serum-starved cells arrested in G_1 tend to have low levels of active transport of nutrients through their membranes and reduced rates of synthesis of RNA and protein. They also have high intracellular levels of cAMP and enhanced breakdown of RNA and protein. When such cells are stimulated to enter active proliferation, all these properties are rapidly reversed.

The large number of essentially simultaneous changes that accompany initiation of active growth led Avram Hershko and associates to propose a theory of pleiotypic activation, in which some kind of master control mechanism is responsible for many simultaneous changes, as opposed to a sequential mechanism in which each change in turn triggers the next. The mechanisms responsible for each pleiotypic activation have not been identified, but at least in theory such a system is

THEORIES OF GROWTH CONTROL

compatible with the Britten and Davidson model for gene regulation, in which interaction between an environmental stimulus and a single sensor gene can trigger changes in expression of many different structural or producer genes via a network of multiple integrator and receptor genes (Chapter 7).

The factors that influence cellular multiplication tend to be either nutrients that must be transported into the cells or regulatory substances (and possibly direct contact) that act on cell surface receptors. A theory that attempts to unify such observations has been proposed by Arthur Pardee and co-workers. They propose that the cellular membrane is the key element in growth control and that growth control is achieved primarily as a result of the actions of extracellular agents such as serum proteins and tissue-specific growth factors on receptors located in the membrane. Interactions at the cellular surface influence the activity of adenylcyclase and thus the intracellular level of cAMP. This in turn influences transport through the membrane of compounds that are critical for growth. All agents that have an effect on cell proliferation are considered to act by altering the state of the cell membrane either directly or indirectly.

Although this theory undoubtedly contains elements of truth, it is probably an oversimplification. In particular, it gives little importance to nuclear events such as changes in phosphorylation of NHC proteins, which probably also play an important role in activation of growth. When we finally understand fully what is involved in control of cellular multiplication, we will probably find that it includes a complex network of communication involving all parts of the cell and that there are many points in the chain of information flow where the growth state of a cell can be altered.

BIBLIOGRAPHY
Books and reviews

Angeletti, R. H., P. U. Angeletti, and R. Levi-Montalcini. 1973. The nerve growth factor. In J. Lobue and A. S. Gordon, eds. Humoral control of growth and differentiation. Vol. I. Academic Press, Inc., New York.

Boyd, L. F., et al. 1974. Nerve growth factor. Life Sci. 15:1381.

Bradshaw, R. A. 1978. Nerve growth factor. Annu. Rev. Biochem. 47:191.

Brooks, R. F. 1976. Growth regulation in vitro and the role of serum. In A. C. Allison, ed. Structure and function of plasma proteins. Vol. 2. Plenum Press, New York.

Bullough, W. S. 1973. Chalone control systems. In J. Lobue and A. S. Gordon, eds. Humoral control of growth and differentiation. Vol. I. Academic Press, Inc., New York.

Clarkson, B., and R. Baserga, eds. 1974. Control of proliferation in animal cells. Cold Spring Harbor Laboratory, Cold Spring Harbor, New York.

Cohen, S., and C. R. Savage, Jr. 1974. Recent studies on the chemistry and biology of epidermal growth factor. Recent Prog. Horm. Res. 30:551.

Cohen, S., and J. M. Taylor. 1974. Epidermal growth factor: chemical and biological characterization. Recent Prog. Horm. Res. 30:533.

Folkman, J. 1974. Tumor angiogenesis. Adv. Cancer Res. 19:331.

Hershko, A., et al. 1971. Pleiotypic response. Nature (New Biol.) 232:206.

Holley, R. W. 1975. Control of growth of mammalian cells in culture. Nature **258**: 487.

Mobley, W. C., et al. 1977. Nerve growth factor. Parts I to III. N. Engl. J. Med. **297**:1096, 1149, 1211.

Nicolson, G. L. 1976. Trans-membrane control of the receptors of normal and tumor cells. II. Surface changes associated with transformation and malignancy. Biochim. Biophys. Acta **458**:1.

Pardee, A. B., L. Jimenez de Asua, and E. Rozengurt. 1974. *In* B. Clarkson and R. Baserga, eds. Control of proliferation in animal cells. Cold Spring Harbor Laboratory, Cold Spring Harbor, New York.

Pastan, I. H., G. S. Johnson, and W. B. Anderson. 1975. Role of cyclic nucleotides in growth control. Annu. Rev. Biochem. **44**:491.

Prescott, D. M. 1976. The cell cycle and the control of cellular reproduction. Adv. Genet. **18**:99.

Prescott, D. M. 1976. Reproduction of eukaryotic cells. Academic Press, Inc., New York.

Reich, E., D. B. Rifkin, and E. Shaw, eds. 1975. Proteases and biological control. Cold Spring Harbor Laboratory, Cold Spring Harbor, New York,

Server, A. C., and E. M. Schooter. 1977. Nerve growth factor. Adv. Protein Chem. **31**:339.

Van Wyk, J. J., et al. 1974. The somatomedins: a family of insulinlike hormones under growth hormone control. Recent Prog. Horm. Res. **30**:259.

Selected original research articles

Cohen, S., and G. Carpenter. 1975. Human epidermal growth factor: isolation and chemical and biological properties. Proc. Natl. Acad. Sci. U.S.A. **72**:1317.

Daniel, V., G. Litwack, and G. M. Tomkins. 1973. Induction of cytolysis of cultured lymphoma cells by adenosine 3':5'-cyclic monophosphate and the isolation of resistant variants. Proc. Natl. Acad. Sci. U.S.A. **70**:76.

Pardee, A. B. 1974. A restriction point for control of normal animal cell proliferation. Proc. Natl. Acad. Sci. U.S.A. **71**: 1286.

Peehl, D. M., and R. G. Ham. 1979. Growth and differentiation of human keratinocytes without a feeder layer or conditioned medium. In Vitro. (In press.)

Stoker, M., and D. Piggott. 1974. Shaking 3T3 cells: further studies on diffusion boundary effects. Cell **3**:207.

Whittenberger, B., and L. Glaser. 1978. Cell saturation density is not determined by a diffusion-limited process. Nature **272**:821.

CHAPTER 18

Cell death and elimination of unneeded tissues

☐ From our perspective as vital organisms with a strong will to survive, we tend to view death as an abnormal or pathological process in all circumstances except extreme old age. Most individuals who are not already familiar with the evidence are surprised to learn of the major role that cell death plays in the shaping of organisms. It seems more natural to view morphogenesis as a dynamic formative process occurring in embryos whose cells are actively multiplying and becoming differentiated. However, the paradoxical fact is that death plays a constructive role in the orderly development of individuals, just as it does in the development of ecosystems.

The death of cells in embryonic development is not a random process. It occurs at specific locations within the embryo and at a definite time during development for each location. Therefore cell death must be regarded as an integral part of a precisely controlled developmental program.

This chapter describes a number of developmental events in which cell death plays a significant role. In many cases an important part of the molding of final body contours is achieved by selective death of cells. Some of the best-studied cases occur in vertebrate limb development, where processes such as the separation of digits are accomplished by cell death. Another important role of cell death is the removal of epithelial coverings so that underlying mesodermal structures can fuse. An example is the formation of the palate or roof of the mouth in mammals. A third important role is elimination of embryonic tissues and organs that are of no value to the adult. Structures that are eliminated include (1) vestigial reminders of evolutionary history (e.g., pronephric and mesonephric kidneys in mammalian embryos); (2) developmental options that are never exercised (e.g., unused sex ducts left over from the indifferent stage of sex differentiation); and (3) structures that are functionally important for a limited time during development and then degenerate (e.g., larval structures in an organism that subsequently undergoes metamorphosis).

Some investigators believe that death (or loss of reproductive capacity) in individual cells in the body may be a causative factor in aging. This concept remains controversial and will be discussed separately in Chapter 24.

Development of the limbs in chicken embryos between the third and ninth days of incubation includes a well-defined pattern of cell death. Cell death begins at the junction between the limb bud and the trunk and subsequently spreads to more distal areas along both the anterior and posterior margins of the growing limb, appearing last in the interdigital regions (Fig. 18-1). The areas that contain dying cells can be identified readily by staining fresh unfixed preparations with basic dyes of high molecular weight, such as trypan blue. The degenerating cells are rapidly stained, but the dyes do not readily enter living cells whose membranes are intact. Alternatively, one can distinguish living and dead cells by use of vital stains such as neutral red and Nile blue, which preferentially stain living cells.

Cell death sharpens the delineation between the trunk and the limb and removes excess tissue in the regions of future joints. It is also responsible for the removal of tissue between the digits. Careful comparisons have been made between the feet of chickens and ducks,

LIMB MORPHOGENESIS

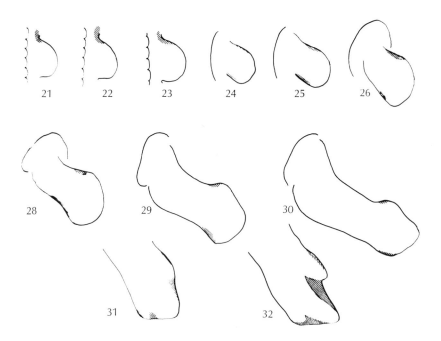

Fig. 18-1. Sketches of the developing chick wing bud. Stippled areas are regions of massive cell death as revealed by vital staining with Nile blue. Numbers refer to developmental stages of the chick embryo according to Hamburger, V., and H. L. Hamilton. 1951. J. Morphol. **88**:49. (From Saunders, J. W., M. T. Gasseling, and L. C. Saunders. 1962. Dev. Biol. **5**: 147.)

which are similar except for webbing between digits II and III and digits III and IV in the duck. Extensive cell death occurs in all the spaces between digits in the chicken, resulting in a complete separation of digits without webbing. In the duck, on the other hand, necrosis is minimal between the digits that remain connected by webs (Fig. 18-2). Recombination experiments have shown that it is necessary to have both mesoderm and ectoderm from chickens to obtain the pattern of cell death and separation of digits that is characteristic of chickens. The presence of either mesoderm or ectoderm from ducks reduces substantially the amount of necrosis that occurs. In sharp contrast to the differences observed between the feet of chicken and duck embryos, their wings show similar patterns of cell death (Fig. 18-3), which are reflected in similarities in adult morphology.

A region of cell death in the developing chicken wing that has been studied extensively is the posterior necrotic zone (PNZ), a mesodermal

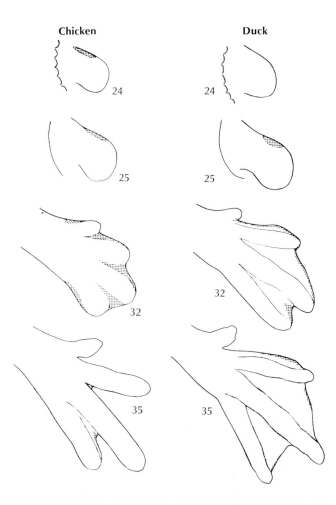

Fig. 18-2. Sketches of developing chick and duck leg buds. Differences in the pattern of cell death (shown by stippling) account for much of the difference in final morphology of the feet. Numbers refer to stages of Hamburger and Hamilton. (From Saunders, J. W., and J. F. Fallon. 1966. Symp. Soc. Dev. Biol. **25**:289.)

region located on the posterior edge of the wing bud near its junction with the body wall (Fig. 18-4). The cells in this region undergo programmed death at stage 24 of development* (about 4 1/2 days of incubation). Within a period of 8 to 10 hours, about 2500 cells in this region die and are subsequently engulfed by macrophages.

There is substantial evidence that cells in the PNZ become programmed or "determined" at an earlier stage to die at stage 24. Cells can be removed from the PNZ and placed in culture at stage 17 of development (about 48 hours' incubation), which is the earliest time that the posterior outline of the limb is visible. These cells will die on

---

*Developmental stages based on the overall anatomy of the embryo are a far more accurate measure of progress of development than incubation time, since rate of development is also dependent on temperature and the amount of development that has occurred before the timing of incubation begins. The stages used in this book are those described by Hamburger, V., and H. L. Hamilton. 1951. J. Morphol. **88**:49.

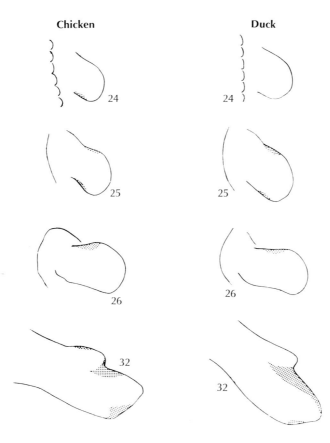

Fig. 18-3. Sketches of developing chick and duck wing buds showing similarities in pattern of cell death. Numbers refer to stages of Hamburger and Hamilton. (From Saunders, J. W., and J. F. Fallon. 1966. Symp. Soc. Dev. Biol. **25**:289.)

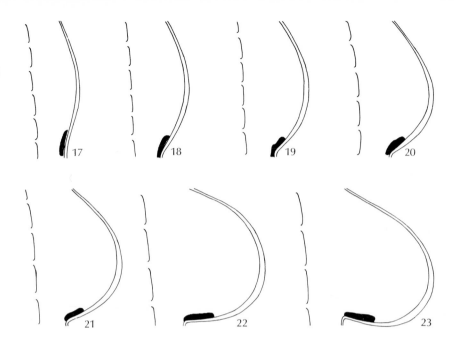

Fig. 18-4. Diagram of developing chick wing bud. Shaded region indicates the approximate location of prospective degenerating cells of the PNZ. Numbers refer to stages of Hamburger and Hamilton. (From Saunders, J. W., M. T. Gasseling, and L. C. Saunders. 1962. Dev. Biol. **5**:147.)

schedule at the time that the embryo from which they were taken would have reached stage 24. Similarly, if cells from the PNZ are transplanted to the somites of another embryo, they die when the donor embryo would have reached stage 24, quite independently of whether the host embryo into which they have been transplanted is older or younger. When stage 17 wing bud mesoderm from regions other than the PNZ is placed in culture or transplanted onto the somites of another embryo, it does not die.

The nature of the stimulus that is originally responsible for programming the cells of the PNZ to die when they reach stage 24 is not known. However, it appears to be gone by stage 17. If other cells are grafted to the region of the PNZ at stage 17, they do not become programmed to die. Between stages 17 and 22, the cells of the PNZ give no hint that they have been programmed to die. No signs of degeneration are seen prior to stage 22. The ultrastructural appearance of the cells remains normal, and they do not show the increase in lysosomal enzymes that often occurs in cells that are destined to degenerate. It has been shown autoradiographically that the number of cells that incorporate ^{3}H-thymidine into DNA begins to decrease at stage 22. Likewise, the level of protein synthesis in the cells begins to decrease at about stage 22 to 23.

Although it is determined by stage 17 that cells of the PNZ will die when they reach stage 24, the determination is reversible up to stage 22, when the first signs of degeneration begin to appear. If cells are taken from the PNZ at any time between stages 17 and 22 and placed deep in the mesoderm in the central region of a limb bud, they change their program and do not die at stage 24. In addition, when cells from these same stages are placed in culture with central limb mesoderm, their death is prevented. The ability of central limb mesoderm to prevent death of the PNZ cells is apparently due to a diffusible factor, since PNZ cells do not die when separated from the limb mesoderm by a Millipore filter that prevents cell contact. The central limb bud mesoderm must, however, be present as a block of tissue. Dissociated cells from limb mesoderm do not prevent cell death, although if the dissociated cells are centrifuged into a pellet and then grown in organ culture as a pellet for 24 hours, they regain their ability to prevent death of the PNZ cells.

The change in the program of PNZ cells brought about by central limb mesoderm is initially reversible, but it can become permanent. When PNZ cells from embryos at stage 19 of development are placed in culture transfilter to central mesoderm and left until stage 24, no death occurs. However, if the PNZ cells are then removed and incubated in the absence of central limb mesoderm, they undergo massive death within 12 hours. This pattern continues until 6 days past the normal time of death. However, when the PNZ cells have been kept in culture with central limb mesoderm for a full 6 days past the time of stage 24, they can be separated from the limb mesoderm without losing their viability.

The fact that the cells of the PNZ die on schedule, even in culture, indicates that the death program is intrinsic to the cells and not due to the influence of some external signal. However, this intrinsic program can be altered either temporarily or permanently in response to external signals from central limb mesoderm cells.

It has been suggested by John Saunders that many different cell types may contain similar programs, or "death clocks," and that they may be rescued from programmed death by the influence of signals from their cellular environment. Evidence in support of this concept is found in the development of the sensory neurons of the spinal ganglia in the chicken embryo. If the fibers from these ganglia come in contact with a peripheral field whose size has been diminished, they undergo a commensurate degree of degeneration. Massive destruction of sensory neuroblasts occurs on removal of the limb bud primordium. Similarly, when certain embryonic sense organs are destroyed, neurons in the corresponding brain centers undergo degeneration.

DEVELOPMENT OF THE PALATE

Another morphogenetic event in which cell death has been shown to be important is formation of the palate or roof of the mouth in mammalian embryos. During palate development, the palatal processes grow vertically from the maxillary (upper jaw) processes, then swing into a horizontal position where they meet and fuse in the midline, thus separating the oral and nasal cavities (Fig. 18-5). Failure of this process of fusion results in the congenital defect known as cleft palate. The palatal processes are largely mesenchymal, but are encased in an epithelial layer. During movement of the processes from a vertical to a horizontal position, the epithelium along the presumptive midline region acquires a glycoprotein coat, ceases DNA synthesis, and begins to degenerate. When the medial edges meet, they adhere, and with the death of the epithelial layer, the underlying mesenchyme of the two palatal processes fuses and becomes continuous.

The biochemical and physiological events associated with fusion of the palatal processes (glycoprotein synthesis, cessation of DNA synthesis, cell death) also occur when palatal processes are cultured in vitro; and when portions of two processes are juxtaposed in culture,

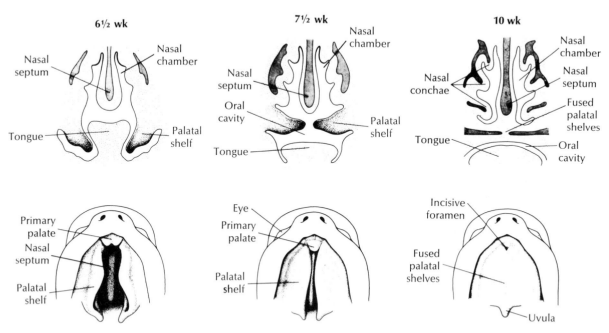

Fig. 18-5. Diagrammatic representation of palate formation in the human. Top row, frontal sections of embryos showing movement of the palatal processes from the vertical to the horizontal position and their fusion, forming separate nasal and oral cavities. Bottom row, ventral views of the palatal region showing the same stages of palate formation. (From Langman, J. 1975. Medical embryology, 3rd ed. The Williams and Wilkins Co., Baltimore.)

they adhere and fuse. Epithelial contact is not necessary for epithelial death, however, since it also occurs in palatal processes that are cultured singly. Although the signal that triggers these events is unknown, it has recently been demonstrated that the level of cAMP increases dramatically in palatal processes at the time these events normally occur. In addition, the shift to increased glycoprotein synthesis and cessation of DNA synthesis is accelerated in immature palatal processes when dibutyryl cAMP is added to their culture medium.

Recent experimental evidence has indicated that the synthesis of glycoproteins which occurs in cells of the palatal epithelium prior to their death is necessary for cell death to occur. When palatal shelves are cultured in the presence of 6-diazo-5-oxo-L-norleucine (DON), a glutamine analog that inhibits glycoprotein synthesis, the epithelial cells fail to undergo programmed cell death (Fig. 18-6). The effect of the inhibitor is prevented by adding metabolites to the medium that overcome the effect of DON on glycoprotein synthesis but not by the addition of metabolites that overcome its effect on other synthetic pathways. It is possible that inhibition of glycosylation may limit the

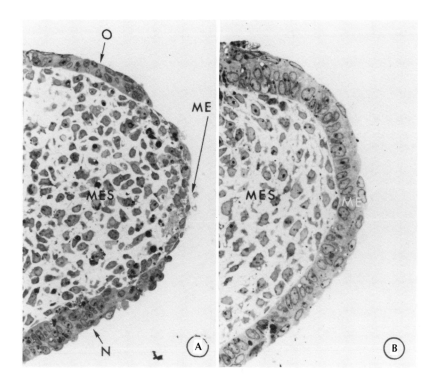

Fig. 18-6. A, Light micrograph of a rat palatal shelf cultured for 48 hours. Complete autolysis of the medial-edge epithelium *(ME)* has occurred, whereas oral *(O)* and nasal *(N)* epithelia and mesenchyme *(MES)* are intact. **B,** Light micrograph of a rat palatal shelf cultured for 48 hours in the presence of DON, an inhibitor of glycoprotein synthesis. Cells of the medial-edge epithelium have not undergone cell death. (From Pratt, R. M., and R. M. Greene, 1976. Dev. Biol. **54:**135.)

release of lysosomal enzymes, since lysosomal enzyme activity in DON-treated cells is found only within lysosomes while activity is located throughout the cytoplasm of untreated cells.

ELIMINATION OF UNUSED STRUCTURES

Development of mammals and other advanced vertebrates is characterized by the transient appearance of many structures that are eliminated or greatly reduced in size and function in the adult. This is particularly true for the circulatory, urinary, and reproductive systems. The structures that are eliminated probably reflect evolutionary history, since similar structures often remain fully functional in the adult forms of lower vertebrates. In many cases evolutionary change appears to have been accomplished by making additions rather than substitutions to preexisting developmental patterns. Structures characteristic of the adult stage in more advanced organisms often arise relatively late in development. Prior to their disappearance, the more primitive organs and tissues that are ultimately eliminated frequently play an essential role in directing the formation of more advanced organs and tissues that are retained. For example, the pronephros is a primitive kidney, which arises early in vertebrate development and later degenerates in all vertebrates except primitive fishes. Although the pronephros itself is discarded as development proceeds, its duct system is retained and plays essential inductive and structural roles in the development of both the urinary and the reproductive systems of adult mammals (described later in greater detail). It is likely that such roles are the reason why a variety of ancestral structures continue to make at least transitory appearances during development.

One of the prominent changes that occurs in the development of the circulatory system in higher vertebrates is the elimination of aortic arches. Early embryos of all vertebrates develop a set of branchial clefts analogous to the gill clefts of fish embryos. A set of six paired arteries called the aortic arches provide major circulation around the clefts. These persist into the adult stage in fishes (Fig. 18-7, A), whereas several of the arches degenerate in adult mammals (dotted lines in Fig. 18-7, C). A similar loss of aortic arches also occurs in frogs during metamorphosis as they switch from gill breathing to lung breathing (Fig. 18-7, B).

Major changes also occur in the venous system as developing mammalian embryos switch from use of the mesonephric kidneys to the metanephric kidneys (described later) and as an adult pattern of circulation through the liver is established (Fig. 18-8). The posterior vena cava assumes a major role in returning blood from the posterior half of the body to the heart, and the postcardinal and renal portal veins, which are no longer needed, degenerate.

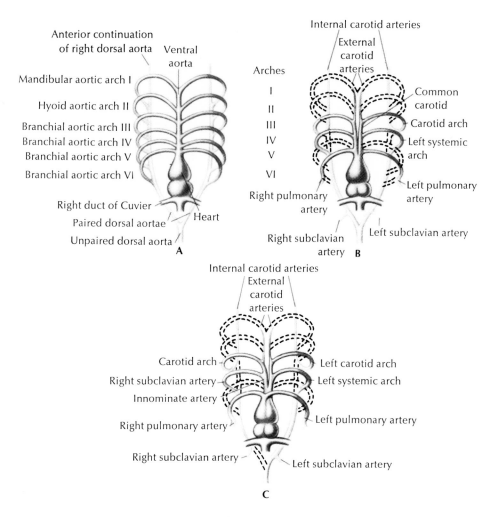

Fig. 18-7. Diagram showing ventral views of the aortic arches in the A, fish, B, frog, and, C, mammal. Dotted lines indicate arches that are eliminated during development. (From Balinsky, B. I. 1975. An introduction to embryology, 4th ed. W. B. Saunders Co., Philadelphia.)

A total of three different types of kidneys are present at various times during embryonic development in the more advanced vertebrates. (The patterns of venous circulation to all three are indicated in Fig. 18-8.) The most anterior, and the first to appear in development, is the pronephros, which is retained as the functional kidney in primitive fishes. In all more advanced vertebrates a second kidney, the mesonephros, forms slightly later in a more posterior position, and the pronephros degenerates. The mesonephros is retained as the adult kidney in advanced fishes and amphibia. In reptiles, birds, and mammals a third kidney, the metanephros, is formed later in a still more posterior position, and the mesonephros degenerates.

Although the pronephros degenerates very early in development, its duct system is retained and plays important roles in development of

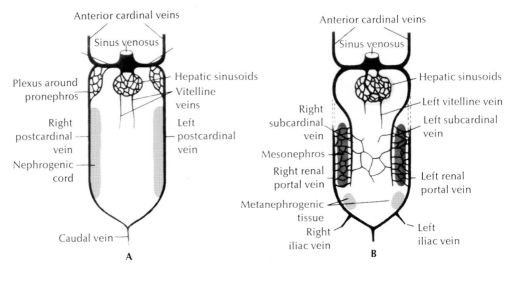

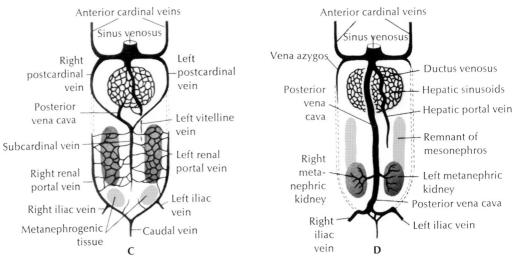

Fig. 18-8. Diagram showing stages in development of posterior venous system in terrestrial vertebrates. Dotted lines indicate areas of blood vessel degeneration. **A,** Vitelline veins have given rise to the hepatic sinusoids. **B,** Postcardinal veins have begun to degenerate anteriorly and to form the renal portal veins posteriorly. The subcardinal veins have formed from the venous network of the mesonephros. **C,** Subcardinal veins have fused to form a single vein and the posterior vena cava has formed from the right subcardinal and the hepatic system. **D,** Postcardinal veins have completely degenerated and the posterior vena cava is fully developed as the major channel for collecting blood from the posterior part of the body. (Modified from Balinsky, B. I. 1975. An introduction to embryology, 4th ed. W. B. Saunders Co., Philadelphia.)

the mesonephros and metanephros and also in the development of reproductive ducts. Concomitant with the formation of the pronephros, a pronephric duct is formed and begins to grow in a posterior direction from the pronephros. The mesonephros forms from an elongated band of undifferentiated nephrogenic mesoderm in response to an inductive stimulus from the advancing pronephric duct. As the mesonephros is formed, its tubules connect to the elongating pronephric duct, which is then called the mesonephric duct. Differentiation of the mesonephros proceeds from anterior to posterior. If growth of the developing duct system is disrupted, no further differentiation of the mesonephros occurs posterior to the point of disruption (Fig. 18-9).

Formation of the metanephric kidney is also dependent on the mesonephric (extended pronephric) duct. A bud forms on the mesonephric duct near its posterior end and grows away from the duct and toward the nephrogenic mesoderm. This branch becomes the metanephric duct or ureter, and when it makes contact with the mesoderm, it induces formation of the metanephric kidney from that mesoderm (Chapter 13). When growth of the mesonephric duct in a posterior direction is disrupted, there is no formation of the ureter, and thus differentiation of the metanephric kidney is not induced (Fig. 18-9).

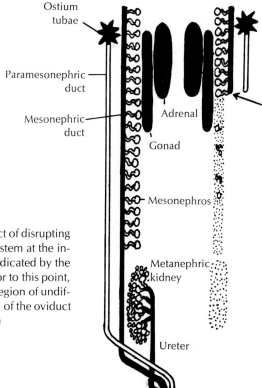

Fig. 18-9. Diagram of the urogenital system in an avian embryo showing effect of disrupting growth of the mesonephric duct. The left side shows a normal urogenital system at the indifferent stage. The duct on the right side has been disrupted at the point indicated by the arrow. The mesonephros and paramesonephric duct do not develop posterior to this point, and the ureter and metanephric kidney do not form. Stippling indicates the region of undifferentiated nephrogenic mesenchyme. The ostium tubae becomes the funnel of the oviduct in females. (Modified from Gruenwald, P. 1952. Ann. N.Y. Acad. Sci. **55**:42.)

The development of still another set of ducts, called the paramesonephric or müllerian ducts, is dependent on induction by the mesonephric (pronephric) ducts. The paramesonephric ducts first appear in the region where the pronephros has degenerated, and grow in a posterior direction parallel to the mesonephric ducts. If development of the mesonephric duct is disrupted, growth of the paramesonephric (müllerian) duct also stops (Fig. 18-9).

The mesonephric and paramesonephric ducts both have important roles in sexual differentiation. In males the mesonephric duct, which is often referred to as the wolffian duct when its sexual role is discussed, becomes the vas deferens (the tube that conveys sperm from the testis to the urethra). A portion of the mesonephros becomes epididymis, and the remainder disintegrates. In females, the paramesonephric or müllerian ducts become the oviducts and the upper portion of the uterus.

There exists during sexual differentiation an "indifferent" stage during which both müllerian and wolffian ducts are present. As definitive sexual differentiation proceeds, either the müllerian or the wolffian ducts degenerate, depending on the sex of the individual, while the other set of ducts grows and differentiates into mature sexual ducts. Different mechanisms are responsible for the degeneration of the two kinds of ducts.

The müllerian ducts are programmed to grow and differentiate into oviducts and uterine tissue unless they are stopped by the action of a protein factor from the testis, commonly referred to as "müllerian-inhibiting hormone," which specifically induces their degeneration, both in intact embryos and in culture. The wolffian ducts, on the other hand, must receive a specific external stimulus in the form of testosterone to grow and differentiate. In female embryos and in embryos whose gonads have been removed, the wolffian ducts degenerate from lack of specific stimulus, and the müllerian ducts develop because they have not been inhibited. Thus development of the ductal system proceeds automatically in the female direction except when altered by hormonal signals from the developing testes in male embryos. These relationships and other aspects of sexual differentiation are discussed in greater detail in Chapter 20.

The mammary glands of mice provide another interesting example of sex-linked tissue degeneration. The adult male mouse, unlike the adult male human, does not possess rudimentary mammary glands. Early in development, mammary glands are formed by an inward budding of the epithelium in both male and female mouse embryos. However, at a slightly later stage of development in the males the mesoderm surrounding the mammary bud constricts and appears to literally pinch off the primitive gland from the overlying ectoderm. The glandular tissue subsequently degenerates.

When organ cultures of mammary gland rudiments from mouse embryos of either sex are treated with testosterone, they undergo mesodermal constriction and epithelial degeneration in vitro, comparable to that which occurs in intact male embryos. Explanted mammary rudiments are sensitive to androgen treatment from late day 13 until early day 15 of gestation. Male (XY) embryos that carry the X-linked

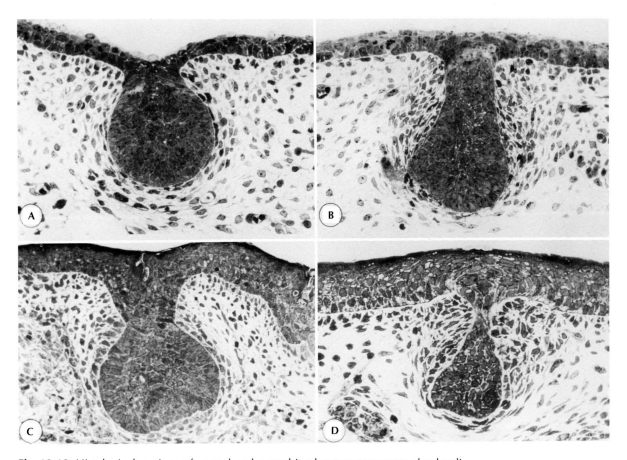

Fig. 18-10. Histological sections of normal and recombined mouse mammary gland rudiments. **A,** Mammary gland rudiment of 14-day female embryo. **B,** Mammary gland rudiment of 14-day male embryo showing beginning of mesodermal condensation and constriction of the epithelial bud. **C,** Mammary gland rudiment formed by combining normal epithelium with *Tfm* mesoderm and cultured in the presence of testosterone. No androgen response is visible. **D,** Mammary gland rudiment formed by combining *Tfm* epithelium and normal mesoderm and cultured in the presence of testosterone. Condensation of mesoderm and constriction at the neck of the epithelial bud are apparent. (From Kratochwil, K., and P. Schwartz. 1976. Proc. Natl. Acad. Sci. U.S.A. **73:**4041.)

gene for testicular feminization *(Tfm)* fail to respond to the effects of testosterone (presumably because of defective hormone receptor molecules). Such embryos do not lose their mammary glands as they develop. It is possible to separate mammary epithelium and mesoderm and to recombine them in organ culture without altering subsequent developmental patterns. Recombination experiments with tissue from normal and *Tfm* embryos have revealed that the mesodermal component is the only one that responds directly to testosterone (Fig. 18-10). When *Tfm* ectoderm, which is insensitive to testosterone, is combined with normal mesoderm and then treated with testosterone, mesodermal constriction and degeneration of the epithelial bud occur. However, when *Tfm* mesoderm, which is insensitive to testosterone, is combined with normal ectoderm, no constriction and no degeneration of the ectodermal component occur. Thus in this case we are not dealing with intrinsically programmed death of the mammary epithelium but rather with hormone-induced "killing" of the ectodermal tissue of the mammary rudiment by the surrounding mesoderm.

MECHANISMS OF TISSUE ELIMINATION

In this chapter we have examined a number of cases in which tissue that is no longer useful is eliminated during the course of development. The process of elimination is very selective. Targeted cells degenerate at specific times during development while the organism as a whole remains vigorous and continues to grow.

Cell death is unquestionably the mechanism responsible for elimination of some of the tissues we have discussed. However, in other cases the mechanism is less clear and it is dangerous to draw broad generalizations in the absence of specific data. For example, the separation of digits during limb development in birds, mammals, and reptiles has been shown beyond any reasonable doubt to involve cell death. It would therefore seem reasonable to predict a similar mechanism in amphibia. Nevertheless, a recent study of the separation of digits in the forelimbs of *Xenopus laevis* by Joanne Cameron and John Fallon found absolutely no evidence for cell degeneration in the interdigital spaces. Their data indicate that the separation is entirely due to differences in cell division, which stops early in the future interdigital spaces while continuing unabated in the future digits. In the hindlimbs, which have webbed digits, rapid cellular multiplication continues in the interdigital spaces as well as in the digits.

Most of the examples of cell death and tissue degeneration that have been presented fall into one of two broad general patterns. In the first, an intrinsic "death clock" is set, and the cells subsequently die on schedule unless they are "rescued" by an overriding signal from the environment. The posterior necrotic zone of the embryonic chicken wing bud is probably the best-studied example of such a mechanism. However, degeneration of sensory neurons in the absence of an adequate peripheral field and the degeneration of the wolffian duct in the absence of specific stimulation by androgenic hormones are also similar in principle.

The second general pattern requires an active signal to tell the tissues when to die. The action of the müllerian-inhibiting hormone in male embryos is a clear example of destruction of a specific tissue triggered by an extrinsic agent. The degeneration of mammary gland epithelial buds induced by the surrounding mesoderm in response to testosterone in male mouse embryos is also similar in principle. Although the signal to die comes from an external source in these cases, that signal is effective only in tissues that have been programmed to respond specifically to it. An extreme example of such specificity of response can be seen in amphibian metamorphosis (Chapter 14). A single hormonal signal in the form of an increased amount of thyroid hormone appears to be responsible for all the changes that occur in metamorphosis. These include both the degeneration of larval structures and the growth of adult structures. For example, muscle, cartilage, and nervous tissue are being destroyed in the tail in response to thyroid hormone at the same time that muscle, cartilage, and nervous tissue are being formed in developing limbs in response to thyroid hormone.

Although some of the trigger stimuli responsible for cell death and its prevention are beginning to be identified, we still have little or no understanding of the actual intracellular mechanisms that are involved. Modern cell and organ culture techniques have made it possible to study both cell death and the rescue of cells from programmed death in controlled environments. The time is right for a precise analysis of the intracellular molecular mechanisms that ultimately decide whether a cell lives or dies. An understanding of how to control such mechanisms would open exciting possibilities for future biomedical research, particularly on the control of malignancies.

BIBLIOGRAPHY
Books and reviews

Balinsky, B. I. 1975. An introduction to embryology, 4th ed. W. B. Saunders Co., Philadelphia. (Chapter 13.)

Langman, J. 1975. Medical embryology. 3rd ed. Williams & Wilkins Co., Baltimore. (Chapters 11 and 12.)

Lockshin, R. A., and J. Beaulaton. 1975. Programmed cell death. Life Sci. **15:** 1549.

Saunders, J. W. 1966. Death in embryonic systems. Science **154:**604.

Saunders, J. W., and J. F. Fallon. 1966. Cell death in morphogenesis. *In* M. Locke, ed. Major problems in developmental biology. Academic Press, Inc., New York.

Selected original research articles

Cameron, J., and J. F. Fallon. 1977. The absence of cell death during development of free digits in amphibians. Dev. Biol. **55:**331.

Kratochwil, K., and P. Schwartz. 1976. Tissue interaction in androgen response of embryonic mammary rudiment of mouse: identification of target tissue for testosterone. Proc. Natl. Acad. Sci. U.S.A. **73**:4041.

Pollak, R. D., and J. F. Fallon. 1974. Autoradiographic analysis of macromolecular syntheses in prospectively necrotic cells of the chick limb bud. I. Protein synthesis. Exp. Cell Res. **86**:9.

Pratt, R. M., and R. M. Greene. 1976. Inhibition of epithelial cell death by altered protein synthesis. Dev. Biol. **54**:135.

Pratt, R. M., and G. R. Martin. 1975. Epithelial cell death and cAMP increase during palatal development. Proc. Natl. Acad. Sci. U.S.A. **72**:874.

Saunders, J. W., M. T. Gasseling, and L. C. Saunders. 1962. Cellular death in morphogenesis of the avian wing. Dev. Biol. **5**:147.

SECTION FIVE

Analysis of selected developmental processes

☐ At this point we have completed our survey of the fundamental mechanisms that are involved in cellular differentiation and in morphogenesis, and we are ready to begin examining selected developmental processes in terms of the variety of mechanisms that are collectively involved. By necessity in the previous sections we have had to focus on one type of mechanism at a time—first, on controls over the flow of information from coded nucleotide sequences in DNA to functional proteins; next, on control signals that influence the course of cellular differentiation; and, finally, on mechanisms involved in the generation of shape, first at a cellular level and then at the level of tissues, organs, and whole-body contours. Now, for the first time, we shall begin to think of development in a way that is closer to what actually happens, with a multitude of different mechanisms at many different levels of organization all operating simultaneously.

It is important, however, even in this section, for our readers to keep in mind the fact that we will discuss only key points and that development in real life is far more complex than can reasonably be depicted in a coherent manner here. Within every cell thousands of different genes are continually being transcribed and translated, some constitutively and others under specific differentiated control. The entire process of metabolism, which in itself can fill a good-sized textbook, is occurring in all cells in a pattern that is partially constitutive and partially modulated by differentiation. Within each tissue there is a balance of biosynthesis and degeneration and of cellular multiplication and cellular loss that may result either in a net increase in tissue mass and cell number, a stationary state, or regression, depending on the system and its developmental stage. Superimposed on this background, and involving many complicated interactions with it, are the basic processes of differentiation and morphogenesis that we are primarily concerned with.

To have any hope of seeing the forest as an organized unit rather than as a multitude of individual trees, we must step back and look at generalities while ignoring many of the details (except as they may be important for the generalities we are attempting to develop). In addition, there are many cases in which the details that we would like to examine are simply not available because research in the particular area has not yet progressed that far. Thus, frequently the conclusions that we shall draw will by necessity be provisional. However, they are still valuable in that they represent current thinking in those areas, as well as the starting points for understanding new advances as they are published in the professional literature.

In the six chapters that follow, we will consider six diverse aspects of development. These will involve organisms ranging from cellular slime molds to humans and developmental stages ranging from gametogenesis and fertilization to postreproductive senescence. These are but a brief sampling from an almost limitless number of possibilities that could fill many books of this size. We make no pretense of a comprehensive analysis of development. Our goal is simply to illustrate how the various individual mechanisms that we have studied fit together into an integrated system and to fill in a few of the most glaring gaps in the limited analysis that we have made up to this point of developmental phenomena (as opposed to underlying mechanisms). Readers who wish to acquire a more comprehensive background in the phenomena and processes of development may wish to supplement this book with some of the textbooks on embryology and developmental biology listed at the end of Chapter 2.

We began our survey of development in Chapter 2 somewhat arbitrarily with a fertilized egg ready for the first cleavage division. In our discussions up to this point we have focused mostly on events between the first cleavage division and the formation of a new functional individual. In Chapter 19 the life cycle is completed with an analysis of the origin of germ cells, the formation of gametes, and the process of fertilization. Emphasis is given to the separation of germ and somatic lines, to the intense level of specialization for the reproductive functions that occur in sperm and egg cells, and to the events that occur as sperm and egg cells fuse to initiate the development of a new diploid individual.

Chapter 20 deals with the existence of two alternative pathways of development in higher organisms, one leading to formation of an egg-producing female and the other to the formation of a sperm-producing male. Emphasis is on the regulatory signals involved in the choice of one pathway or the other and on the ways in which individuals that are phenotypically different can be obtained from developmental pathways that are indistinguishable from one another until their point of divergence (which does not occur until most other organ systems are already well developed).

In Chapter 21 attention is shifted to the vertebrate limb, a complex structure whose development, once initiated, proceeds with almost total autonomy. A complex set of morphological determinants, including genetic properties of the tissue, inductive and feedback relationships, positional and time sequence information, selective outgrowth, and selective cellular death, all influence the final shape that is generated.

In Chapter 22 we turn away from the dominant vertebrate theme of this book and examine selected aspects of insect development as it occurs in *Drosophila*. Particular attention is given to genetic control of early development, hormonal regulation of gene expression and metamorphosis, control over the developmental fate of imaginal discs (larval precursors of adult structures), and the role of positional information in determining adult morphology.

Chapter 23 analyzes developmental mechanisms in the cellular slime mold *Dictyostelium discoideum*, which at different stages of its life cycle can exist either as a unicellular amoeba-like organism or as a relatively complex cellular aggregate that is capable of coordinated movement and differentiation into two distinctly different types of cells, each of which is also different from the cells of the unicellular stage.

Finally, in Chapter 24 we return to vertebrates and focus on yet another aspect of the life cycle that we have not previously considered—postreproductive senescence or aging. Although not part of the continuing life cycle of the species as a whole, aging can nevertheless be viewed as a continuation of development from the point of view of the individual. Adulthood is not a static condition but rather a time of continuing changes that are no different in principle from those which occur earlier in life. Development-like changes continue to occur with advancing age, and although not everyone agrees, it is becoming increasingly common both for developmental biologists and for gerontologists to view aging as an uninterrupted continuation of the developmental program that begins at fertilization (or perhaps even during oogenesis).

CHAPTER 19

Germ cells, gametes, and fertilization

☐ As discussed in Chapter 1, development of higher organisms that reproduce sexually can be viewed as a continuous life cycle in which gametes derived from germ cells unite to form a zygote, which in the course of its development gives rise both to somatic cells that eventually die and to new germ cells that are capable of initiating another cycle of development. This chapter focuses on the cells responsible for the continuity of life from one generation to the next—the germ cells and the gametes that are derived from them—and on the mechanisms involved in their formation and function. Specifically, we will examine the origin of the primordial germ cells during embryonic development, the movement of primordial germ cells from their point of origin to the gonads, the conversion of germ cells into haploid gametes (sperm and egg cells), and the union of male and female gametes in the process of fertilization to give rise to a new diploid individual.

ORIGIN AND MIGRATION OF GERM CELLS

Primordial germ cells can often be distinguished from precursors of somatic cells early in embryonic development. Frequently the primordial germ cells are first seen in locations distant from the area in the embryo that is destined to become the gonads in which the germ cells will ultimately reside. We have already examined some extreme cases of early origin of germ cells in the discussion of chromatin diminution in Chapter 6. For example, in the nematode *Ascaris* at the two-cell stage of embryonic development, one of the two cells is already committed exclusively to somatic cell differentiation, whereas the other retains the potential to form both germ and somatic cells. The descendants of the totipotent cell retain a full complement of chromosomal material after the second cleavage division, whereas the future somatic cells undergo a substantial loss of chromatin during that division (Fig. 6-7). Similar processes in subsequent divisions leave two cells at the 32-cell stage that are destined to become germ cells and 30 that will form somatic cells. A similar loss of chromosomes from all except the future germ cell nuclei also occurs in some, but not all, insects during the fourth or fifth nuclear division (Fig. 6-8).

Germm plasm   In certain insects and in some frogs and toads, a region of the oocyte cytoplasm, known as the germ plasm, which is rich in RNA-containing granules, is associated with the formation of primordial germ cells. Since germ cell determination in insects has already been discussed in Chapters 6 and 12, it will be reviewed only briefly here. In insect embryos, where nuclear division occurs without cytoplasmic division during early stages of development, the germ plasm is localized at the extreme posterior end or pole of the elongated egg (Fig. 6-8). Nuclei that enter this region prior to the partitioning off of individual cells are incorporated into large "pole cells" (Fig. 19-1), which are direct precursors of the germ cells and gametes. If nuclei are prevented from coming into contact with the pole plasm by constriction of the egg, or if the pole plasm is inactivated by ultraviolet irradiation, the resulting insects are devoid of germ cells and therefore sterile. In the latter case, injection of pole plasm from the posterior cytoplasm of unirradiated eggs reverses the effects of irradiation. Similarly, if pole plasm is injected into the anterior region of an embryo, apparently normal pole cells are formed in that region. Such cells are capable of forming functional gametes when transferred to the posterior region of a genetically different embryo. The pole plasm is thus a mosaic determinant of germ cell formation. This is true not only for species that undergo chromatin loss in their somatic cells but also for species such as *Drosophila melanogaster*, in which the germ and somatic cells appear to have identical genomes.

In frog embryos the germ plasm can be identified on the basis of its basophilic staining properties (affinity for basic dyes) even before the first cleavage division has occurred (Fig 19-2). If the germ plasm is destroyed by ultraviolet irradiation of the ventral surface of the egg or is mechanically removed, few or no germ cells develop in the embryo.

Fig. 19-1. Germ plasm and origin of germ cells in insects. **A,** Diagrammatic representation of the egg of the beetle *Calligrapha bigsbyana* 4 hours after deposition. Male and female pronuclei are fusing deep in the yolk mass, but no nuclear division has occurred. **B,** Same as **A** 16 hours after deposition of the egg. Many nuclei have been formed, including several future germ cell nuclei that have entered into close association with the germ plasm. **C,** Scanning electron micrograph of the egg of the fly *Drosophila melanogaster* at the start of gastrulation. Large pole cells (primordial germ cells) are prominently visible at the posterior end (right). **D,** Schematic diagram of pole cell formation in the fly *Compsilura concinnata*. The germ plasm (dense shading) is being incorporated into the pole cells as they are being formed. **E,** Scanning electron micrograph of pole cells of *Drosophila melanogaster* at a stage somewhat earlier than in **C.** Note that some of the pole cells (arrows) appear to be dividing. (**A, B,** and **D** reprinted with permission of Macmillan Publishing Co., Inc. from *Patterns and principles of animal development* by Saunders, J. W., Jr. Copyright © 1970 by John W. Saunders, Jr.; **C** and **E** from Turner, F. R., and A. P. Mahowald. 1976. Dev. Biol. **50:**95.)

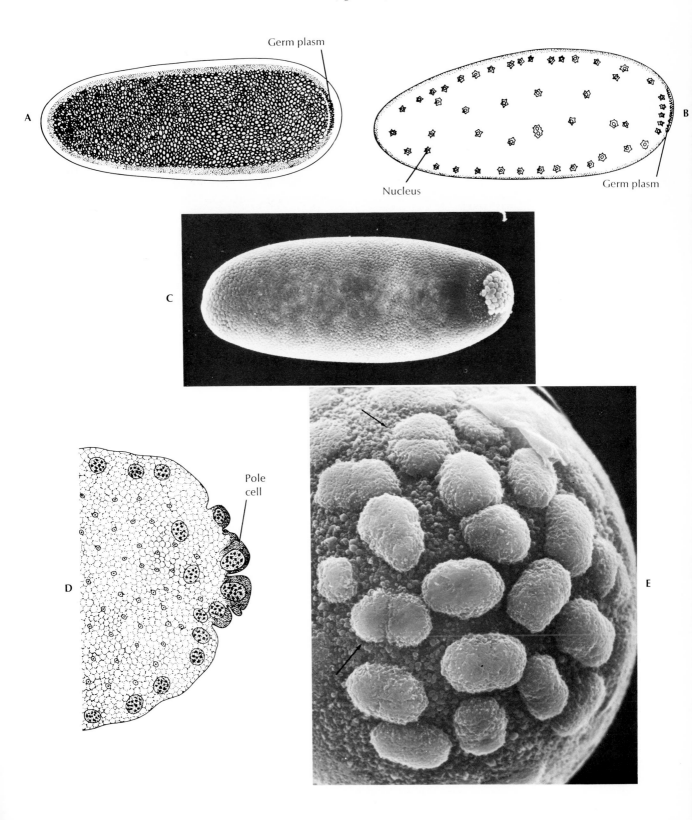

Germ plasm

Nucleus

Germ plasm

Pole cell

A

B

C

D

E

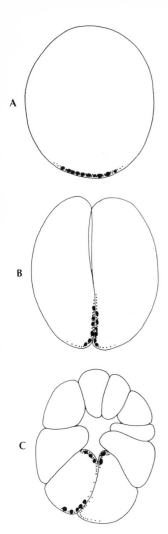

Fig. 19-2. Distribution of RNA-rich germ plasm in the frog embryo. **A,** Germ plasm is at the vegetal pole in the uncleaved egg. **B,** As the egg cleaves, some of the germ plasm is carried toward the interior of the embryo. **C,** By the 16- to 32-cell stage some germ plasm is located in the floor of the future blastocoel. (Reprinted with permission of Macmillan Publishing Co., Inc. from Patterns and principles of animal development by Saunders, J. W., Jr. Copyright © 1970 by John W. Saunders, Jr.)

Because of the size of the amphibian egg, one side can be treated with ultraviolet irradiation with little effect on the other. Thus, for example, it is possible to destroy the germ plasm, which is located near the vegetal pole of the egg, with little or no damage to the nucleus, which is located near the animal pole.

As cleavage progresses, the germ plasm is segregated into endodermal cells that are first located in the floor of the blastocoel and then after gastrulation in the floor of the gut. These cells will subsequently become primordial germ cells and migrate to the embryonic gonad.

Cytologically the germ plasm of insect and frog embryos is similar. The light microscope shows it to be an intensely staining basophilic area (i.e., it reacts strongly with basic dyes, a property that is characteristic of areas rich in nucleic acids). The ultraviolet absorption spectrum (based on the effectiveness of various wavelengths for inactivating the germ plasm) and removal of stainable material with ribonuclease both suggest that the germ plasm is rich in RNA, although it may also contain other components. With the electron microscope, characteristic dense granules are seen (Fig. 19-3), often in close association with polyribosomes and mitochondria. Mature germ cells of many different species are known to contain densely staining regions of basophilic material known as "nuage" (from the French word for "cloud"), which ultrastructurally appear to be similar to the germ plasm of insects and frogs. Current data are not adequate to permit a positive statement that nuage is derived directly from germ plasm, or that it represents a retention in mature germ cells of the mosaic determinant for their formation, but there is currently much speculation to that effect.

In the frog, as in insects, injection of egg cytoplasm from areas rich in germ plasm reverses the effect of ultraviolet irradiation of the germ plasm and restores fertility. Extirpation and transplantation experiments have demonstrated that the cells in the endoderm that receive the germ plasm are the precursors of the definitive germ cells in frogs. For example, the ventral endoderm of the gut can be removed from early neurula-stage embryos. The wound heals quickly, and development is sufficiently regulative that it appears to be normal. However, the frogs that develop are sterile and have no germ cells. Replacement of the ventral endoderm with the same region from another embryo restores normal germ cell formation (Fig. 19-4). Through use of nucleolar markers, it can be shown that the germ cells formed in such cases are from the donor endodermal tissue. For example, when the host has one nucleolus and the donor of the endoderm has two nucleoli, the germ cells will all have two nucleoli (in the diploid state prior to doubling of chromosomal material at the beginning of meiosis). Simi-

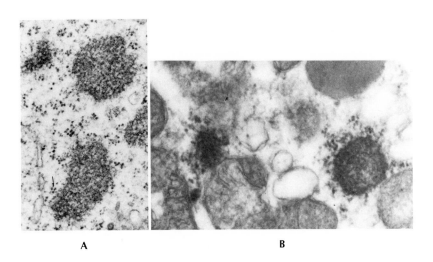

Fig. 19-3. Dense bodies in germ plasm. **A,** Electron micrograph of polar granules in germ plasm of *Drosophila hydei.* **B,** Electron micrograph of dense bodies in germ plasm of two-cell *Rana pipiens* embryo. Note similarity of structure and clusters of polysomes surrounding both. (**A,** ×29,500; **B,** ×33,000.) (**A** from Mahowald, A. P. 1968. J. Exp. Zool. **167:**237; **B** from Mahowald, A. P., and S. Hennen. 1971. Dev. Biol. **24:**37.)

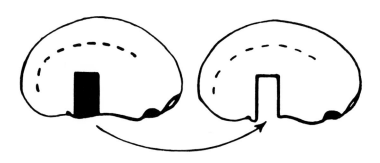

Fig. 19-4. Diagram showing scheme of germ cell transfer experiment in *Xenopus* neurulae. The germ cell region from the donor embryo is transplanted to a host embryo from which this region has been removed. (From Blackler, A. W., and M. Fishberg. 1961. J. Embryol. Exp. Morphol. **9:**634.)

lar transplantations have also been done between two subspecies of *Xenopus.* In such cases the germ cells have the genetic properties of the donor subspecies rather than of the host.

In salamanders the origin of germ cells appears to be very different than in frogs. Transplantation experiments between related species indicate that in urodeles (salamanders and their close relatives) the primordial germ cells originate from the lateral mesoderm.

It is characteristic of the vertebrates in general that the germ cells migrate into the presumptive gonads from other areas of the body. The gonads are typically formed from germinal ridges that protrude slightly into the abdominal cavity from the dorsal body wall on either side of the dorsal mesentery. In the frog the germ cells migrate from the ventral endoderm to the germinal ridges, whereas in the salamander the migration is from the lateral mesoderm to the germinal ridges.

Germ cell migration

Presumptive germ cells or regions of germ plasm have not been localized at extreme early stages of development in the embryos of birds and mammals. The chicken egg at the time of laying already contains an extensive cellular blastoderm, but no cytologically distinct germ cellls can be found. In the intact egg, primordial germ cells are first detected at about 20 hours of incubation, a time at which the body axis has been established, but the brain and somites are not yet formed. The germ cells are identified by their large size and prominent nuclei and also by cytochemical staining for the abundant glycogen granules that they contain in their cytoplasm. When they first become detectable, the germ cells are localized in a cresent-shaped area of the yolk sac endoderm anterior to the head fold of the embryo called the germinal crescent (Fig. 19-5). Destruction of this area does not alter embryonic development in general, but the gonads of the resulting individuals contain no germ cells.

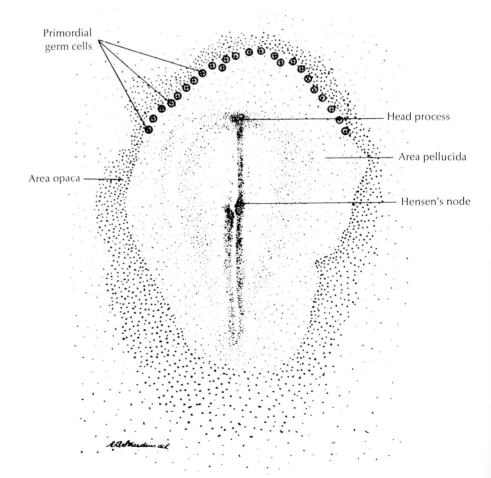

Fig. 19-5. Schematic diagram of chick embryo at the head-process stage showing location of the primordial germ cells in a crescent-shaped region anterior to the embryo proper. (From Swift, C. H. 1914. Am. J. Anat. **15**:483.)

It is not certain whether germ cells originate only in this area in the intact embryo or whether they migrate there from other parts of the blastoderm. If the cellular blastoderm of a newly laid egg is cut into pieces and incubated in culture, primordial germ cells are formed by all parts of it. However, no further formation of germ cells occurs when the germinal crescent is removed from a slightly older embryo.

The primordial germ cells of chicken embryos reach the gonads by traveling through the blood vessels of the extraembryonic membranes and into the embryo. Initially the germ cells are found throughout the embryo, but by about 48 hours of incubation, they have become concentrated in the gonads. If the posterior part of a second embryo is grafted onto the extraembryonic membranes in such a manner that it becomes vascularized, the common circulation also carries germ cells to its gonads. Migration of germ cells through the circulatory system appears to be unique to birds. In mammals the primordial germ cells do not enter the circulation but instead migrate by ameboid motion, as described later.

The primordial germ cells of mammals are thought to originate within the endoderm of the yolk sac at the posterior end of the embryo, although this point has been disputed. In human embryos, primordial germ cells are seen in this region at about 3 weeks of development (Fig. 19-6). From here, primordial germ cells, which can be identified by their high level of alkaline phosphatase, have been shown in the mouse to migrate into the mesoderm ventral to the gut and then move around the gut and into the dorsal mesentery from which the gut is suspended. Finally, they move laterally into the gonad-forming region (Fig. 19-7). Human primordial germ cells reach the gonad by about 4 to 5 weeks of development.

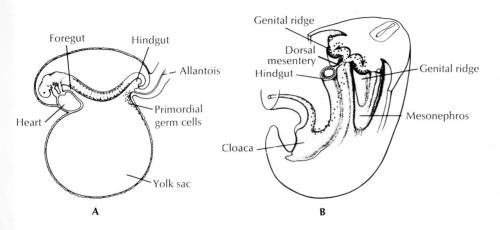

Fig. 19-6. Schematic drawings showing, **A,** the location of primordial germ cells in a 3-week human embryo in the yolk sac at the posterior end of the embryo and, **B,** the path of their migration around the gut and dorsal mesentery and into the genital ridge. (Modified from Langman, J. 1975. Medical embryology, 3rd ed. The Williams & Wilkins Co., Baltimore.)

Fig. 19-7. Light micrograph of a cross section of mouse embryo at about 10 days. Primordial germ cells stained for alkaline phosphatase can be seen entering the genital ridges (GR) from the dorsal mesentery (DM). (From Mintz, B. 1957. J. Embryol. Exp. Morphol. **5**:396.)

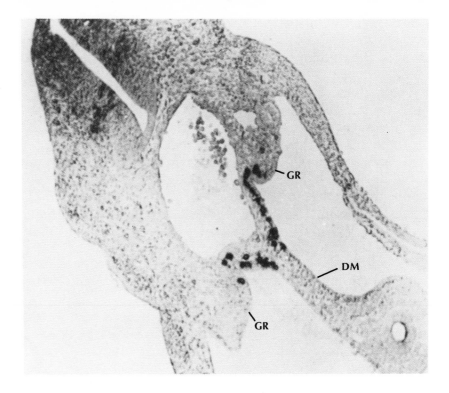

Germ cell migration in most animals appears to involve an active ameboid type of movement by the germ cells. What causes germ cells to migrate to the gonadal region is not known, but there is some evidence that chemotactic factors may be involved. When tissue of the chick gonadal ridge is placed in culture, it attracts primordial germ cells. Likewise, if gonadal ridge tissue is transplanted to an unusual site within the embryo, germ cells still migrate to it. If such an attractant is involved in germ cell migration, it may not be species specific. This is suggested by an experiment in which hindguts from mouse embryos were transplanted to the posterior part of the coelomic cavity of chick embryos. Primordial germ cells that left the grafted mouse tissue and entered chick tissue migrated toward the gonadal region, although they did not actually enter the gonad and eventually degenerated. Primordial germ cells from chicken and duck embryos freely enter the gonads of the other species.

The primordial germ cells enter the gonad at a time when the gonad is undifferentiated with regard to sex. As shown in Fig. 19-8, the immature gonad of higher vertebrates consists of an outer cortex and

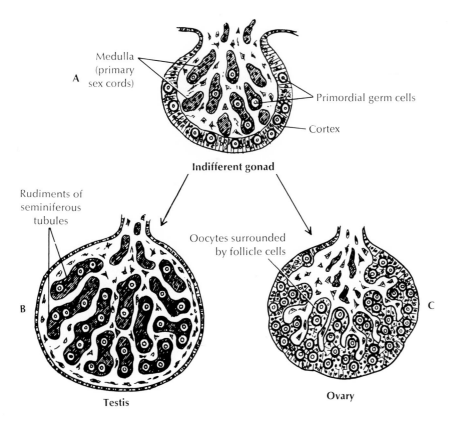

Medulla
(primary
sex cords)

A

Primordial germ cells

Cortex

Indifferent gonad

Rudiments of
seminiferous
tubules

Oocytes surrounded
by follicle cells

B

C

Testis

Ovary

Fig. 19-8. Events in gonadal development in higher vertebrates. **A,** Indifferent gonad containing primordial germ cells in the cortex and medulla. **B,** In the testis, germ cells enter the primary sex cords, which develop into the seminiferous tubules, and the cortex is reduced. **C,** In the ovary the germ cells are located in the cortex, which forms the bulk of the gonad, and the primary sex cords are reduced. (From Balinsky, B. I. 1975. An introduction to embryology, 4th ed. W. B. Saunders Co., Philadelphia.)

an inner medulla. The germ cells initially enter the cortex. If the gonad differentiates in the female direction, the primordial germ cells remain concentrated in the cortex, which becomes the predominant structure of the ovary. In the case of male differentiation the germ cells move from the cortex into the medulla, which eventually forms the seminiferous tubules of the testis. Germ cells that end up in the "wrong" layer of mammalian gonads appear to undergo degeneration. Models of gonadal differentiation and the fates of germ cells whose chromosomal sex does not match the type of gonad they are in are discussed in Chapter 20.

In nonmammalian vertebrates, germ cell differentiation is not determined by the genetic constitution of the germ cell itself but occurs in response to the sex of the gonad in which the germ cell resides. In the female chick, for example, the left ovary normally develops, and the right ovary remains rudimentary and its primordial germ cells degenerate. However, removal of the left ovary of the young chick causes the right ovary, which is composed primarily of medullary tissue, to develop into a testis. If the left ovary is removed before the primordial germ cells in the right ovary have degenerated, these germ cells, which are genetically female, develop into mature functional spermatozoa. Sex reversal in amphibians is readily induced by temperature changes. High temperature causes degeneration of the cortex and can cause ovaries to become transformed into testes. Low temperature has the opposite effect. In both cases germ cell differentiation is in accord with the phenotypic sex of the gonad. Sex reversal can also be obtained in frogs and chicks by exposure to steroid sex hormones.

Germ cells increase in number during their migratory period, but their greatest proliferation occurs after they enter the gonad. In the human female, as shown in Fig. 19-9, the number of germ cells undergoes about a tenfold increase between the second and sixth months of development and rises to about 7 million. At this point the germ cells begin to degenerate; their number declines dramatically at first and then in a more gradual manner. In most mammals, including the human, female germ cells begin meiosis prior to birth, and once they have entered meiosis, they are incapable of further proliferation. Still, the number of germ cells present in the human female at birth seems far more than adequate for her reproductive needs. In an estimated 30 to 35 years of fertility with approximately 13 ovulations per year (this implies that there are no pregnancies), a woman would require fewer than 500 germ cells for her entire reproductive life span. This leaves more than 999,000 excess germ cells. What becomes of them? For reasons that are unknown, the vast majority of these germ cells undergo

Fig. 19-9. Changes in the number of germ cells in the human ovary with increasing age. (From Baker, T. G. 1972. *In* C. R. Austin and R. V. Short, eds. Reproduction in mammals. Book 1. Germ cells and fertilization. Cambridge University Press, Cambridge.)

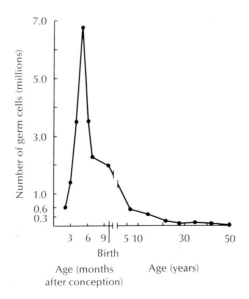

degeneration at various phases in their development. Many are lost prior to puberty, and by the time of menopause the ovaries are essentially devoid of germ cells. The mechanisms whereby some germ cells complete their development and are ovulated whereas others degenerate are completely unknown.

Germ cells in the male are capable of mitosis for an indefinite period of time. The adult male retains a population of germ cells that continually proliferates and replaces itself as well as providing cells for meiosis and spermatogenesis.

GAMETOGENESIS

Prior to their entry into the gonad the development of male and female germ cells is essentially identical. After entry into the gonad, germ cells of both sexes undergo meiosis to form haploid gametes capable of fusing with gametes of the opposite sex. However, the details and timing of the differentiation of male and female sex cells are different. It is therefore more convenient to discuss male and female gametogenesis separately. In addition, since the details of the development and organization of gametes vary considerably from one species to another, our discussion will be restricted to a limited number of

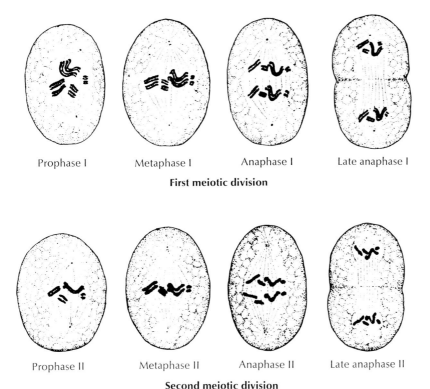

| Prophase I | Metaphase I | Anaphase I | Late anaphase I |

First meiotic division

| Prophase II | Metaphase II | Anaphase II | Late anaphase II |

Second meiotic division

Fig. 19-10. Schematic diagram showing separation of homologous chromosomes in the first meiotic division and separation of chromatids in the second meiotic division to form haploid gametes. (From Patten, B. M., and B. M. Carlson. 1974. Foundations of embryology, 3rd ed. McGraw-Hill Book Co., New York. Copyright © 1974 by McGraw-Hill, Inc. Used with permission of McGraw-Hill Book Co.)

animal species whose egg and sperm cells have been especially well studied, either because of the ease of obtaining and working with them (e.g., sea urchins, amphibians) or because they are closely related to humans (e.g., various mammals).

Oogenesis

The female germ cell, prior to the initiation of meiosis, is called an oogonium. Oogonia are formed from primordial germ cells after they colonize the ovary and are distinguished from the primordial germ cells by their larger size and differing distribution of cytoplasmic organelles. When oogonia cease mitosis and are ready to begin meiosis, they are called primary oocytes. The primary oocyte undergoes DNA synthesis and then enters the first meiotic division (Fig. 19-10). In this division the duplicated chromosomes, each consisting of two strands (sister chromatids), undergo homologous pairing to form "tetrads," and then separate in such a way that each daughter cell receives one full set of the duplicated chromosomes (which still consist of paired sister chromatids). The product of the first meiotic division is the secondary oocyte, which then undergoes the second meiotic division without further DNA synthesis. This results in separation of the sister

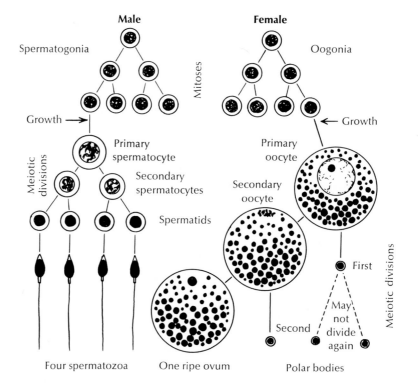

Fig. 19-11. Comparison of gametogenesis in the male and female. Two meiotic divisions occur in both sexes. In the male each primary spermatocyte forms four spermatids, which all differentiate into functional spermatozoa. In the female, however, each oocyte produces only one functional ovum, which contains essentially all the original oocyte cytoplasm. The small polar bodies degenerate. (Modified from Balinsky, B. I. 1975. An introduction to embryology, 4th ed. W. B. Saunders Co., Philadelphia.)

chromatids and the formation of haploid daughter cells. In the female, as shown in Fig. 19-11, both meiotic divisions are unequal, forming one large cell and one much smaller cell, the polar body. Thus the end result of the two meiotic divisions is one large cell, the egg or ovum, and two or three polar bodies, depending on whether the first polar body undergoes the second meiotic division.

The prophase of the first meiotic division typically lasts for a very long time—up to 50 years in human females. Generally, meiosis proceeds as far as the diplotene stage prior to birth and then stops. At this stage chromosomal pairing has already occurred, as has crossing over, if it is going to take place. However, the chromosomes are still only partially condensed. In many mammals all the oocytes are at the diplotene stage at birth. Meiosis does not proceed further until puberty and then only in a few oocytes at a time in preparation for ovulation.

Although these primary oocytes are "resting" with regard to their meiotic activity, they are very active biosynthetically. This period coincides with the lampbrush chromosome stage, which has been studied extensively in amphibian oocytes and which also occurs in the oocytes of mammals. At the lampbrush stage the paired homologous chromosomes are in an extended form and are joined at points of crossing over (chiasmata) (Fig. 19-12). Each homolog contains two duplex DNA strands and has pairs of lateral loops extending symmetrically from the main chromosomal axis. Under the light microscope these loops have a fuzzy brushlike appearance from which the name "lampbrush" derives. The loops consist of extended DNA fibers and have been shown by autoradiography to be active in RNA synthesis, whereas DNA in the main axis is transcriptionally inactive. The fuzzy matrix associated with the loops is newly synthesized RNA in the form of ribonucleoprotein.

During oogenesis, oocytes typically accumulate large reserves of materials, such as ribosomes, tRNAs, mRNAs and nutrients, that are needed during the rapid divisions that follow fertilization. The amplification of rRNA genes, which was described in Chapter 6, occurs during the extended prophase of the first meiotic division. There is substantial evidence, particularly in sea urchins and amphibians, that the mRNA necessary for early development of the embryo is synthesized prior to fertilization and is stored in the egg.

Experiments with inhibitors of protein synthesis have demonstrated that protein synthesis is required after fertilization for normal early development, as would be expected. However, experiments in which RNA synthesis has been prevented, have shown that no new RNA synthesis is needed until relatively late in development. As an extreme example, many aspects of early development of sea urchin embryos

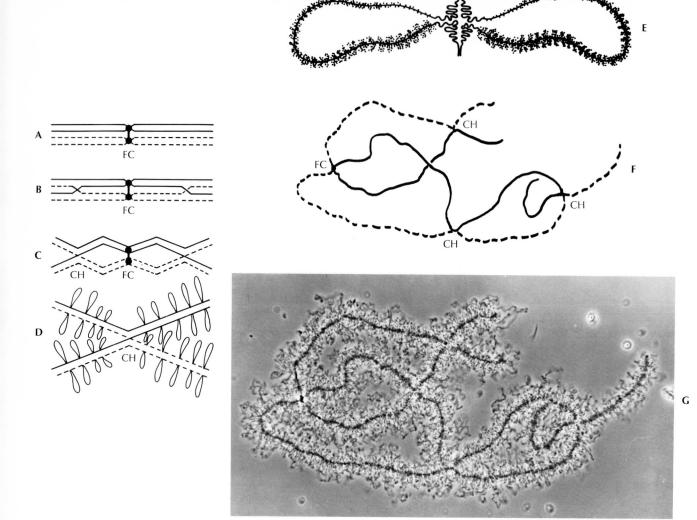

Fig. 19-12. Structure and origin of lampbrush chromosomes. **A,** In preparation for the first meiotic division each chromosome duplicates, forming two sister chromatids, and homologous chromosomes pair to yield a structure with four closely apposed chromatids and fused centromeres *(FC)*. **B,** Crossing over between adjacent chromatids occurs during the pachytene stage of meiosis. **C,** Partial separation of homologous chromosomes occurs during the diplotene stage of meiosis, with the chromosomes remaining attached only at their fused centromeres and at chiasmata *(CH)* resulting from crossing over. **D,** Symmetrically paired loops of fully extended DNA radiate out from the partially condensed chromosomes, giving them a brushlike appearance. Each chromosome consists of two sister chromatids, both of which extend loops at the same point. **E,** Drawing depicting the proposed structure of one pair of extended loops on a lampbrush chromosome. The fuzzy material consists of newly synthesized RNA and associated proteins. **F,** Interpretive drawing of lampbrush chromosome shown in **G.** One of the paired chromosomes (consisting of two chromatids) is shown as a solid line and the other dashed. The two chromosomes are attached at their fused centromeres and three chiasmata. **G,** Phase contrast photograph of a lampbrush chromosome from an oocyte of *Triturus*. (**E** from Gall, J. D. 1963. Symp. Soc. Study Growth Dev. **21:**119; **G** from Gall, J. D. 1966. *In* D. Prescott, ed. Methods in cell physiology. Vol. 2. Academic Press, Inc., New York.)

will take place in the complete absence of a nucleus. This has been shown by parthenogenetically activating enucleate fragments that have been produced from sea urchin eggs centrifugation. Such fragments undergo a normal increase in protein synthesis and "cleave" into smaller fragments. Alternately, RNA synthesis can be prevented by treating intact fertilized eggs with actinomycin D. In the sea urchin, protein synthesis occurs at normal rates for several hours in treated embryos and development proceeds normally up to the blastula stage, using only the so-called maternal mRNA that is already present in the egg prior to fertilization. Evidence for the oogenetic origin of specific types of mRNA has been obtained from experiments showing that tubulin and histones are synthesized both by sea urchin embryos treated with actinomycin D and by enucleated, parthenogenetically activated sea urchin eggs. Eggs from a variety of other species also proceed normally through early stages of development after treatment with actinomycin D.

In the case of mammalian embryos, inhibition of RNA synthesis disrupts development early in cleavage. However, recent studies have shown that some of the proteins that are synthesized for brief periods at specific stages during cleavage are coded for by mRNA molecules that are already present in the oocyte prior to fertilization. Comparison of patterns of protein synthesis in normally fertilized embryos and in oocytes that have been ovulated but not fertilized reveal two distinct classes of proteins (Fig. 19-13). One class is synthesized only in the fertilized embryos and appears to require new RNA synthesis, whereas the other class is synthesized at the same chronological time after ovulation with or without fertilization and does not require new RNA synthesis (essentially no RNA synthesis occurs in the unfertilized oocyte, whose nucleus remains arrested in the second meiotic metaphase). In vitro translation experiments have demonstrated that the mRNA for synthesis of the second class of proteins is present in freshly ovulated oocytes, even though it is not normally translated at that time. Current data suggest that the critical event during ovulation that initiates the timed pattern of synthesis of various proteins from "maternal" mRNAs is the breakdown of the germinal vesicle (the large oocyte nucleus) in preparation for the first meiotic division.

The developing oocytes of many species accumulate large amounts of stored nutrients for use during development. These food reserves are given the general name "yolk," and the process of yolk production is frequently called vitellogenesis. In humans and other mammals the developing embryo is largely dependent on an external supply of maternal nutrients rather than on stored reserves, and the oocyte contains little yolk and remains small. At the other extreme the large eggs

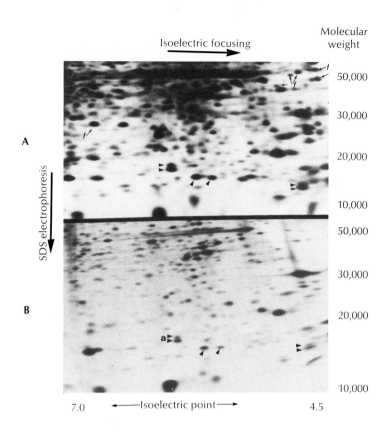

Fig. 19-13. Delayed translation of mRNA that is present in rabbit oocytes prior to fertilization. Proteins synthesized at specific times after ovulation were labeled radioactively by exposing oocytes or embryos to ^{35}S-methionine for brief time periods. Two dimensional "O' Farrell gels" were used to separate the proteins in the horizontal dimension according to their isoelectric points (4.5 to 7.0) and in the vertical dimension by their molecular weights (10,000 to 50,000). Autoradiography was then used to determine where the radioactive proteins that were synthesized during the labeling period are located on the gels. **A,** Proteins synthesized by normal 16-cell rabbit embryos (about 36 hours after ovulation and fertilization). The small arrows *(f)* identify easily recognizable proteins that are synthesized at this stage only in eggs that have been fertilized. **B,** Proteins synthesized by unfertilized oocytes between 36 and 48 hours after ovulation. The triangular arrowheads here and in **A** identify easily recognized proteins that are synthesized for a limited time beginning about 36 hours after fertilization, both in unfertilized oocytes and in cleaving embryos. Their synthesis appears to begin at a specific time after breakdown of the germinal vesicle (rupture of the oocyte nuclear membrane), whether or not fertilization has occurred. Since no significant RNA synthesis occurs in the unfertilized oocytes at any time after ovulation, the mRNAs that code for these proteins must already be in the oocytes in "masked" form. This has been verified by in vitro translation of mRNA isolated from freshly ovulated oocytes. Those proteins indicated by **a** have been specifically identified among the products of such translation. (From Van Blerkom, J. 1979. Dev. Biol. **72:**188.)

of birds and reptiles contain enough stored nutrients for the development of large and essentially self-sufficient individuals.

The nutrients that are stored in developing oocytes are often synthesized by other cells. For example, in amphibia and birds substantial amounts of yolk proteins are synthesized in the liver in response to estrogen (Chapter 14). These are transported through the circulation to the ovary, where they are taken up by the growing oocyte. In amphibia the liver makes a protein called vitellogenin or serum lipophosphoprotein (SLPP), which moves into the bloodstream and is taken up by the ovary, specifically by vitellogenic oocytes. SLPP is made in vitro by liver slices but is not made by isolated oocytes. SLPP is converted by oocytes into two major yolk proteins, phosvitin and lipovitellin (Chapter 9). This conversion has been demonstrated by injecting animals with radioactively labeled SLPP and noting the appearance of label in phosvitin and lipovitellin. A similar process also occurs in birds.

Female germ cells in the ovary are typically surrounded by accessory cells, which contribute in various ways to their development. In

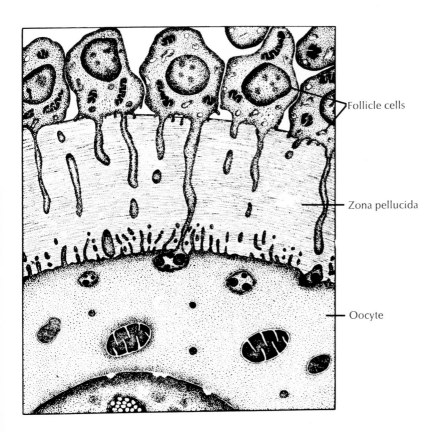

Follicle cells

Zona pellucida

Oocyte

Fig. 19-14. Diagram showing microvilli of follicle cells interdigitating with those of an oocyte across the zona pellucida. (From Baker, T. G. 1972. *In* C. R. Austin and R. V. Short, eds. Reproduction in mammals. Book 1. Germ cells and fertilization. Cambridge University Press, Cambridge.)

some types of insects, yolk material is made in "fat bodies" and transported through the hemolymph to the ovaries. The oocytes of certain species of insects and some other invertebrates are surrounded by nurse cells that are derived from the oogonium and are directly connected to the oocyte by cytoplasmic bridges. Nurse cells contribute to oocyte development by synthesis of macromolecules, such as RNA, that are transferred to the oocytes through the cytoplasmic bridges. Nurse cells also absorb substances from the circulation that are then transferred to the growing oocytes. Comparison of rates of oocyte growth in insect species that do and do not have nurse cells in their ovaries has shown that the first meiotic prophase is much shorter in those species which do have nurse cells than in those which do not.

Fig. 19-15. Life cycle of the female germ cell in the mammal. (From Baker, T. G. 1972. In C. R. Austin and R. V. Short, eds. Reproduction in mammals. Book 1. Germ cells and fertilization. Cambridge University Press, Cambridge.)

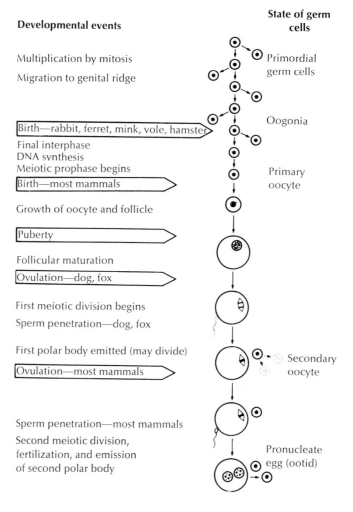

Developmental events

State of germ cells

Multiplication by mitosis

Migration to genital ridge

Primordial germ cells

Birth—rabbit, ferret, mink, vole, hamster

Oogonia

Final interphase
DNA synthesis
Meiotic prophase begins

Birth—most mammals

Primary oocyte

Growth of oocyte and follicle

Puberty

Follicular maturation

Ovulation—dog, fox

First meiotic division begins

Sperm penetration—dog, fox

First polar body emitted (may divide)

Ovulation—most mammals

Secondary oocyte

Sperm penetration—most mammals

Second meiotic division, fertilization, and emission of second polar body

Pronucleate egg (ootid)

In many animals, particularly vertebrates, oocytes are surrounded by cells called follicle cells, which are derived from gonadal tissue. In contrast to nurse cells, follicle cells do not play a major role in synthesis of materials to be stored in the egg. However, materials entering the oocyte from the bloodstream pass through the follicle cells on their way to the oocyte. Microvilli projecting from the surfaces of the egg and follicle cells may be involved in the transport of these materials (Fig. 19-14). In addition, follicle cells contribute to the synthesis of some of the protective coverings that surround oocytes, for example, the zona

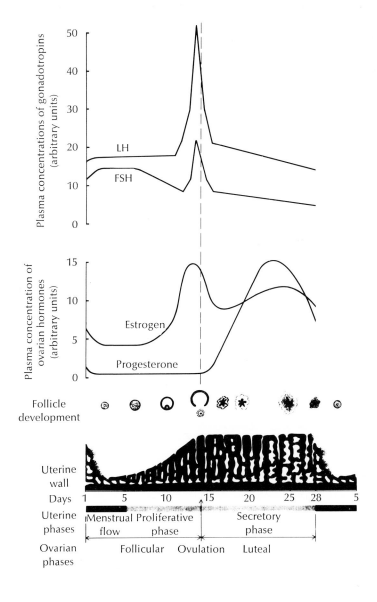

Fig. 19-16. Diagram showing correlation among gonadotropic hormones, ovarian hormones, follicle development, ovulation, and uterine phases in the human female. (From Vander, A. J., et al. 1975. Human physiology: the mechanisms of body functions. McGraw-Hill Book Co., New York. Copyright © 1975 by McGraw-Hill, Inc. Used with permission of McGraw-Hill Book Co.)

pellucida of mammalian oocytes. Several layers of follicle cells remain attached to the mammalian egg after it is ovulated, forming the cumulus oophorus, which the spermatozoa must penetrate. The layer of follicle cells immediately adjacent to the egg is called the corona radiata.

The female germ cells of many types of animals do not complete meiosis prior to their union with the male germ cell. Mammalian oocytes normally remain in the prophase of the first meiotic division throughout their growth phase (Fig. 19-15). Cyclic increases in gonadotropic hormones stimulate a few oocytes to undergo maturation and ovulation at each period of fertility (Fig. 19-16). In most mammals, with the exception of dogs and foxes, the oocyte completes the first meiotic division and proceeds to metaphase of the second meiotic division during the period of follicle maturation just prior to ovulation. The egg then remains arrested at the second meiotic metaphase until it is penetrated by a sperm, after which meiosis is completed. The stages of meiosis at which sperm penetration occurs in other species vary from young primary oocytes in certain worms to after completion of both meiotic divisions in the sea urchin, as shown in Table 19-1.

Table 19-1. Stage of egg maturation at which sperm penetration occurs in different animals*

Young primary oocyte	Fully grown primary oocyte	First metaphase	Second metaphase	Female pronucleus
The annulate worm *Dinophilus*	The round worm *Ascaris*	The nemertine worm *Cerebratulus*	The lancelet *Amphioxus*	Coelenterates (e.g., anemones)
The polychaete worm *Histriobdella*	The mesozoan *Dicyema*	The polychaete worm *Chaetopterus*	The amphibian *Siredon*	Echinoids (e.g., sea urchins, starfish)
The flatworm *Otomesostoma*	The sponge *Grantia*	The mollusk *Dentalium*	Most mammals	
The onychophoran *Peripatopsis*	The polychaete worm *Myzostoma*	The cone worm *Pectinaria*		
The annulate worm *Saccocirrus*	The clam worm *Nereis*	Many insects		
	The clam *Spisula*			
	The echiuroid worm *Thalassema*			
	Dog and fox			

*Modified from Austin, C. R. 1965. Fertilization. Prentice-Hall, Inc., Englewood Cliffs, N.J.

Spermatogenesis

The process of spermatogenesis is fairly similar for many animals; so for the sake of simplicity we will describe only mammalian spermatogenesis except where noted. In contrast to the situation in the female, primordial germ cells in the male do not differentiate into spermatogonia until after birth. In the human, formation of spermatogonia occurs between birth and puberty, during which time the testis is relatively undifferentiated and the spermatogonia are located in solid cords of tissue. At puberty these cords become hollowed out to form the seminiferous tubules and some of the spermatogonia begin spermatogenesis.

The primordial germ cells of male vertebrates are associated with smaller supporting cells, which differentiate into the Sertoli cells of the sexually mature male. Mature Sertoli cells are large and triangular in shape, with their bases at the periphery and their apexes at the lumen of the seminiferous tubule (Fig. 19-17). Although the role of the Sertoli cells is not entirely clear, they appear to be involved in the nutrition and protection of germ cells, release of spermatozoa into the lumen of the tubule, and removal of cytoplasm that is cast off by the developing spermatozoa. They also have been shown to produce an androgen-binding protein in response to follicle-stimulating hormone. Recent studies with mixed populations of normal cells and those which lack testosterone receptors suggest that the androgen dependence of spermatogenesis is mediated by the Sertoli cells and does not require androgen receptors in the germ cells themselves.

Spermatogonia proliferate throughout the reproductive life span of the male, both replacing themselves and producing primary spermatocytes. Conversion of the stem cell spermatogonia into primary sper-

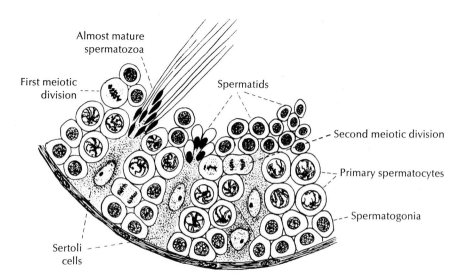

Fig. 19-17. Diagram of part of a seminiferous tubule in a mammal, showing the relationship of Sertoli cells and germ cells at different stages of development. (From Balinsky, B. I. 1975. An introduction to embryology, 4th ed. W. B. Saunders Co., Philadelphia.)

matocytes involves several mitotic divisions that produce intermediate classes of spermatogonia (which are distinguishable by their sizes and by cytological characteristics) prior to formation of primary spermatocytes. The primary spermatocyte, in turn, enters the first meiotic division and produces two secondary spermatocytes, which then undergo the second meiotic division to yield a total of four haploid spermatids (Fig. 19-11). In contrast to the situation in oogenesis, in which an oocyte produces only one ovum, each of the four products of meiosis in the male can form a functional gamete.

Spermatogenesis occurs in an orderly pattern within the seminiferous tubule. As shown in Fig. 19-17, which illustrates the typical ar-

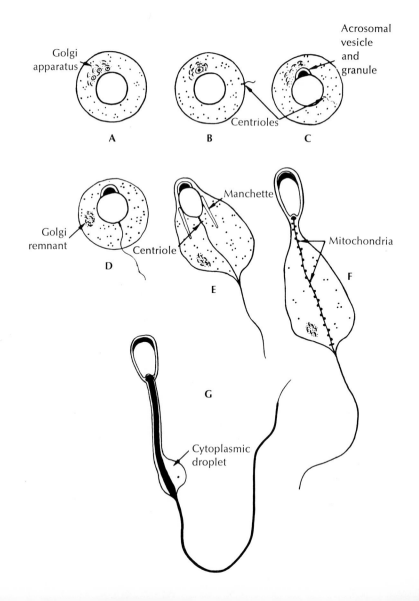

Fig. 19-18. Diagram showing morphogenetic changes occurring during spermiogenesis in the rabbit. **A,** Proacrosomal vesicles appear within the Golgi body, each containing a granule. **B to D,** Vesicles coalesce and form a single acrosomal vesicle and granule, which forms a cap over the anterior end of the nucleus. At the same time, one of the centrioles forms the basal body for the flagellum. **C to F,** Nucleus moves to the edge of the cell, and the cytoplasm is displaced posteriorly. Mitochondria surround the anterior portion of the flagellum, and a cylindrical structure called the manchette is formed. **G,** Nucleus and acrosome assume their final shapes, and most of the cytoplasm is discarded. (From Austin, C. R. 1965. Fertilization. Prentice-Hall, Inc., Englewood Cliffs, N.J. Reprinted by permission of Prentice-Hall, Inc.)

rangement of cells in the seminiferous tubule of a mammal, spermatogonia are located at the periphery of the tubule, and the more developed stages are found at progressively more interior positions, with the most mature spermatozoa next to the lumen of the tubule. The meiotic divisions of spermatocytes occur much more rapidly than those of oocytes. Nevertheless, there is abundant RNA synthesis during the prophase of meiosis I, just as there is in female germ cells. The RNA produced during the first meiotic prophase in the male is thought to be required for the differentiation of spermatids into spermatozoa.

Although the spermatid has the same genetic constitution as a mature spermatozoon, it must undergo extensive morphological and biochemical specialization before it becomes a functional gamete (Fig. 19-18). The process of specialization is called spermiogenesis. Little, if any, RNA synthesis occurs during spermiogenesis, and protein synthesis is thought to be directed primarily by RNA molecules that were synthesized during the meiotic phase of sperm formation. The chromatin condenses into a dense form in which the DNA is tightly complexed with histones. In certain species of fish (e.g., trout, salmon) the DNA becomes associated with a special class of small arginine-rich histones called protamines. The nucleus also becomes compacted and assumes a shape that is characteristic for the species, such as a sickle shape in the rat and mouse and a pearlike shape in the human.

Some of the morphological changes that occur during spermiogenesis are illustrated in the formation of the acrosome, a membrane-bound vesicle that plays an important role in fertilization. The acrosome is formed by the Golgi body, which produces proacrosomal vesicles containing proacrosomal granules. The proacrosomal vesicles coalesce to form a single cap over the nucleus at the anterior end of the developing sperm cell. The acrosome is thought by some workers to be a modified lysosome. It contains a number of enzymes, including hyaluronidase and proteases, which are used by the sperm to pass through the protective coverings that surround the egg cell.

In specializing for fertilization the sperm discards all cytoplasmic components not essential to that function. The bulk of its cytoplasm is displaced posterior to the nucleus and is cast off as a "residual body" prior to the release of the sperm into the lumen of the seminiferous tubule. The two centrioles move to the end of the cell opposite the acrosome, and a flagellum develops from the more posterior centriole. Mitochondria, which provide energy for flagellar motion, are located at the proximal end of the flagellum. Thus after completion of spermiogenesis the sperm cell consists primarily of a head containing the acrosome and the nucleus and a tail containing centrioles, mitochondria, and a flagellum, all of which are surrounded by a plasma membrane.

The final steps in differentiation of the sperm occur after its release into the lumen of the seminiferous tubule. In the epididymis the acrosome assumes its final shape and the sperm loses its last bit of cytoplasm and becomes motile. Sperm that have not passed through most of the epididymis are not fertile when collected for artificial insemination. (In humans frequent intercourse sometimes results in infertility for similar reasons, since sperm are not allowed sufficient maturation time within the epididymis.) In the human the process of spermatogenesis, from spermatogonium to release of the sperm into the lumen of the seminiferous tubules, lasts about 74 days. Sperm are produced continuously after puberty in the male; in the absence of sexual activity, they are released slowly in the urine. In some species of mammals with a well-defined breeding season, spermatogenesis is a cyclic process whose peak of activity coincides with the breeding season.

FERTILIZATION   Gametogenesis results in the formation of male and female gametes with the potential of fusing, combining their genetic information, and forming a new individual. However, realization of this potential requires a complex sequence of events to occur:

1. Release of the gametes from the gonads or storage organs
2. Transport of the gametes to the site of fertilization (external for some species and in the female reproductive tract for others)
3. Maturation of the gametes to render them capable of undergoing fertilization (including capacitation of sperm in mammals)
4. The acrosome reaction, a release of enzymes and change in morphology by the sperm in response to contact with outer coats surrounding the egg
5. Penetration of the sperm through outer coats surrounding the egg (aided by acrosomal enzymes)
6. Fusion of egg and sperm plasma membranes when they make contact
7. Rapid modification of the membrane potential of the plasma membrane of the egg, resulting in an "early" block to polyspermy (fertilization by more than one sperm) within a few seconds
8. The cortical reaction, a complex series of changes that spread over the egg surface to provide a slower, but more absolute, block to polyspermy
9. Engulfment of the sperm by the egg cytoplasm
10. Completion of meiosis by the oocyte nucleus (except in species like the sea urchin in which meiosis is completed prior to ovulation)
11. Fusion of the haploid sperm nucleus (male pronucleus) with the haploid egg nucleus (female pronucleus)

12. Initiation of the developmental program, including synthesis of protein, RNA, and DNA; cleavage; and eventually differentiation, morphogenesis, and growth.

This orderly sequence of events has been studied extensively in recent years. Particular attention has been focused on the acrosomal reaction of the sperm and on "activation" of the egg. The early responses of the sperm and egg, respectively, share a number of properties in common. Both are initiated by contact with the complementary gamete (or its surrounding coats), and both appear to be under the control of transient changes in ionic flow through the plasma membrane and in intracellular concentrations of common inorganic ions, as will be discussed.

The male and female gametes are produced by two separate individuals in most species of higher animals. A large portion of the physiology and behavior of such animals is related to the specific goal of successful union of male and female sex cells. In this chapter we will focus on the process of fertilization itself and on some of the activities of the gametes that occur between their release from the gonad and their actual union at fertilization. Somatic differentiation related to the bringing together of gametes will be considered in the next chapter.

In animals with external fertilization, male and female gametes are usually deposited in the same vicinity and at approximately the same time. Although most animal sperm are motile and therefore potentially capable of moving toward an attractant, chemotactic attraction does not seem to be a primary factor in fertilization of most animals. Gametes come together for the most part by random collision. However, once the germ cells are in close proximity, sticky gelatinous coats surrounding the egg can serve to trap the sperm.

In animals with internal fertilization, male and female gametes come together in specific locations to which they both must travel. In mammals the ovulated oocyte is normally transported from the surface of the ovary into the oviduct (fallopian tube) by fluid flow, with the help of fingerlike projections called fimbria that are located at the ovarian end of the oviduct. It is possible for the oocyte to remain in the abdominal cavity in some mammals, including the human, and if it is fertilized by a spermatozoon that has traveled completely through the fallopian tube, it may undergo embryonic development. Such abdominal pregnancies can even go to term if a suitable site for formation of a placenta is found. Obviously in such cases delivery must be by means of cesarean section. Fertilization normally occurs in the ampulla, a swollen region of the oviduct near the ovary. The oocyte is capable of

Release and meeting of male and female gametes

being fertilized for only about 24 hours after ovulation, and successful fertilization and development are more likely early in this period.

The seminal fluid containing the sperm is deposited in the vagina in the case of humans or directly into the uterus in some other mammals. The volume varies considerably with the species (a few milliliters in the human, 250 ml in the pig). The sperm move rapidly through the female reproductive tract, reaching the ampulla in an average time of 30 minutes in the human. Their progress is most efficient at the time of ovulation, when estrogen stimulates muscular contractions in the female tract and causes changes in the cervical mucus that help sperm to penetrate into the uterus. The release of oxytocin by the female at orgasm further facilitates muscular contraction and sperm transport, as do prostaglandins that are present in the semen. Sperm movement results primarily from these contractile processes of the female organs and from ciliary action in the uterus. However, the motility of the sperm is important for keeping the sperm in suspension as well as in helping them to traverse passages between divisions of the reproductive tract (i.e., the cervix and the uterotubal junction) and in the actual penetration of the oocyte and its surrounding coats. Several hundred million sperm are deposited in one ejaculation in the human, but only a few hundred reach the region of the egg. Human sperm are capable of fertilization for about 48 hours.

Final maturation of the gametes

Following their release from the gonads, male and female gametes frequently must undergo further changes before they are capable of participating in fertilization. For example, the oocytes of some animals complete meiosis in the period between ovulation and fertilization. Sperm also frequently undergo changes necessary to their fertilizing ability after their release from the gonads. As mentioned earlier, mammalian sperm undergo a final morphological change in the epididymis. Following their release from the male reproductive tract, sperm from many species of mammals must undergo a change called capacitation. Capacitation takes a number of hours in some animals (5 to 6 in the human), and although the events that take place during this time are not fully understood, they must occur for the sperm to be able to undergo the acrosome reaction, which is essential to fertilization in many different types of animals. Capacitation normally occurs in the female reproductive tract, but for many mammals it can occur also in vitro in a saline solution, which has led to the suggestion that it may involve the dilution of some inhibitory component of semen.

Acrosome reaction and penetration of egg coats

The acrosome reaction is easily seen with light and electron microscopes and has been well characterized. It occurs in the presence of oocytes, and it can be induced in mammalian sperm by a particular fraction of follicular fluid and in sea urchin sperm by a component of the egg jelly coat. Details of the structural changes involved in the acrosome reaction vary with different animals, but the reaction basically involves an opening up of the acrosomal vesicle and consequent release of its contents.

The acrosome reaction has been well studied in *Saccoglossus* (a wormlike marine invertebrate) and the process seems to be similar in the sea urchin. As shown in Fig. 19-19, the plasma membrane of the sperm fuses with the apical membrane of the acrosome and the two membranes break at the point of fusion, forming an opening. The contents of the acrosome, which are thought to be enzymes that aid the sperm in passing through the egg coverings, are released. A portion of the acrosome membrane adjacent to the nucleus then evaginates and forms a tube called the acrosomal tubule (or acrosomal filament), which plays a major role in sperm-egg fusion in those species in which it occurs. The tubule makes contact with the egg surface and fuses with the egg plasma membrane. This forms a funnel between the egg and sperm through which the spermatozoan nucleus enters the egg cytoplasm.

The regulatory mechanisms believed to be responsible for the acrosome reaction in sea urchin sperm are summarized in Fig. 19-20. The jelly coat of the egg (or possibly a substance produced at the egg surface that diffuses through the jelly coat) initiates the acrosome reaction when the sperm comes into contact with the jelly. The first identifiable event is an influx of calcium ions into the sperm cell, which appears to activate all the other events. Exposure of sperm to a calcium ionophore (a substance that allows calcium to penetrate cell membranes readily) will evoke a full acrosome reaction, but only if an adequate amount of calcium is present in the extracellular environment. The increased intracellular concentration of calcium appears to cause exocytosis (release of contents to the exterior of the cell) of the acrosomal granule. The substances that are externalized include the enzymes needed to digest through the jelly coat and the vitelline layer of the egg and also specific receptor substances (bindins) that remain

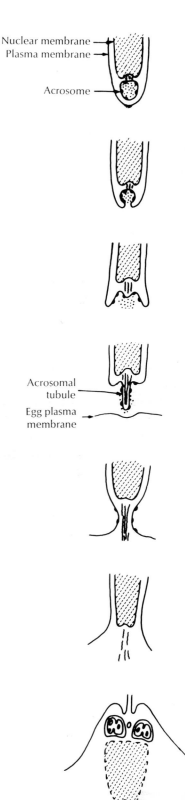

Fig. 19-19. Diagram showing acrosome reaction and sperm penetration in *Saccoglossus*. The acrosome membrane fuses with the plasma membrane of the sperm tip and releases its contents. The acrosome membrane evaginates and forms the acrosomal tubule, which contacts the egg plasma membrane and forms a funnel through which the sperm nucleus enters the egg cytoplasm. (From Guidice, G. 1973. Developmental biology of the sea urchin embryo. Academic Press, Inc., New York.)

attached to the tip of the acrosomal tubule and interact with receptor sites on the vitelline layer of the egg.

The influx of calcium ions also triggers a movement of hydrogen ions out of the cell. By analogy with egg activation (described later), the efflux of H^+ is thought to be balanced by an influx of Na^+ and to result in a rise in intracellular pH. The increased pH is believed to be responsible for the polymerization of actin, which in turn is known to be responsible for extension of the acrosomal process. It is speculated, but not yet fully proved, that increased intracellular pH causes dissociation from the actin of a protein that binds to it and prevents spontaneous polymerization when the pH is lower.

The acrosome reaction in a typical mammalian spermatazoon is diagrammed in Fig. 19-21. In this case the sperm plasma membrane overlying the acrosome fuses at several points with the membrane of the acrosome itself. Small holes are formed that allow the release of enzymes from the acrosome. These enzymes are able to dissolve the gelatinous matrix between the follicle cells that surround the egg (cumulus oophorus), allowing the sperm to pass through this cell mass. By the time the sperm has penetrated the follicle cells and reached the zona pellucida, the outer acrosome membrane and adjoining plasma membrane have usually been lost and the sperm head is enclosed by the inner acrosome membrane. This inner acrosome membrane is believed to carry another enzyme that is needed for penetration of the noncellular zona pellucida. Mammalian sperm do not form any structure comparable to the acrosomal tubule of *Saccoglossus* or sea urchins. Instead, the plasma membrane of the sperm head adjacent to the acrosomal region fuses with the egg plasma membrane to provide a pathway for entry of the sperm nucleus into the egg cytoplasm.

Fig. 19-20. Summary of major events in the acrosome reaction in sea urchin sperm. When receptors on the surface of the sperm cell contact the jelly coat of the egg, a series of events is triggered that culminates in the acrosome reaction, as shown. The regulatory mechanisms that are involved are similar to those responsible for activation of the egg (Fig. 19-23). Steps shown in parentheses are postulated on the basis of similarity to egg activation but have not yet been proved to occur during the acrosome reaction. (Modified from Epel, D. 1978. Curr. Top. Dev. Biol. **12**:185.)

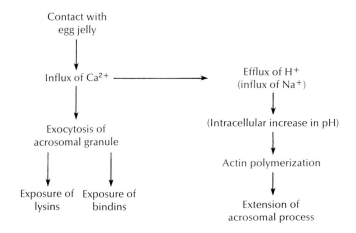

The enzyme activities associated with the acrosome have been well studied in the mammal. Among the activities found in extracts of acrosomal vesicles are hyaluronidase, which is responsible for penetration of the cumulus oophorus and acrosin, a trypsinlike protease that is involved in penetration of the zona pellucida. Hyaluronidase also appears to be located on the sperm plasma membrane, since intact rabbit sperm react with antibody to rabbit hyaluronidase.

Fig. 19-21. Schematic representation of mammalian fertilization. **A** and **B,** Diagram showing acrosome reaction and egg sperm fusion. *1,* Intact sperm; *2,* acrosome reaction; *3,* penetration of sperm through the corona radiata and zona pellucida, facilitated by acrosomal enzymes; *4,* fusion of sperm and egg plasma membranes. **C** to **H,** Detailed representation of changes in sperm during fertilization. **C,** Intact sperm. **D,** Fusion of the acrosome membrane with the sperm plasma membrane. **E,** Release of acrosomal contents. **F,** Sperm approaching the egg plasma membrane. **G,** Fusion of the egg and sperm plasma membranes. **H,** Sperm nucleus sinking into the egg cytoplasm. (**A** and **B** from Moore, K. L. 1974. Before we are born. Basic embryology and birth defects. W. B. Saunders Co., Philadelphia; **C** to **H** from Austin, C. R. 1972. *In* C. R. Austin and R. V. Short, eds. Reproduction in mammals. Book 1. Germ cells and fertilization. Cambridge University Press, Cambridge.)

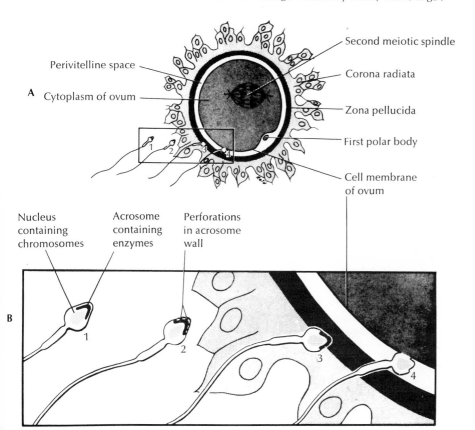

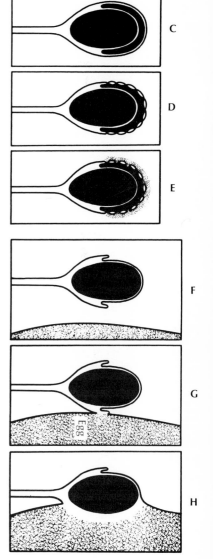

Engulfment of the sperm

The sperm plasma membrane does not enter the egg. In the sea urchin, microinjection experiments done with intact sperm have shown that sperm enclosed in their plasma membranes are incapable of initiating development. The microinjection technique does not harm the eggs, since they are still capable of normal fertilization. Other parts of the sperm—for example, the mitochondria, centrioles, and flagellum—may or may not enter the egg, depending on the species involved. When they do enter, they have been shown in many cases to degenerate. For example, when two species of *Xenopus*, whose mitochondrial DNAs are distinguishable by hybridization, are crossed, only the mitochondrial DNA of the female is found in the developing embryos. Male centrioles persist in the egg at least for a while in some animals and are thought by some investigators to participate in cleavage of the fertilized egg. However, they are not essential for cleavage, since parthenogenetically activated eggs of many species undergo cleavage with no exposure to sperm at all.

Fusion of pronuclei

After the sperm nucleus has entered the egg cytoplasm, its chromatin becomes less densely compacted. In sea urchins the sperm and egg nuclei, which are now called male and female pronuclei, enlarge, move toward each other, and fuse to form a single zygote nucleus. DNA replication is normally not initiated until the fusion has oc-

Fig. 19-22. Schematic drawing of mammalian oocyte, **A,** at the time of ovulation; **B,** after sperm penetration; **C,** after formation of male and female pronuclei; **D** and **E,** during the first cleavage division; and, **F,** at the two-cell stage. (From Langman, J. 1969. Medical embryology, 2nd ed. The Williams & Wilkins Co., Baltimore.)

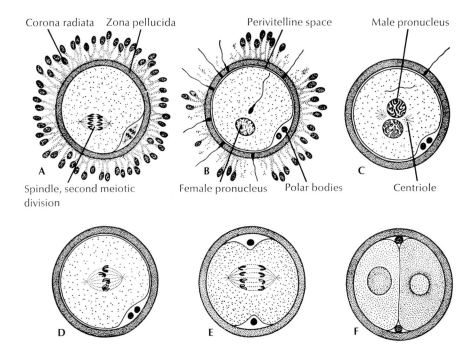

curred, although in some cases replication can begin before pronuclear fusion. Any extra sperm nuclei present in the egg cytoplasm initiate DNA synthesis at the same time as the pronuclei, suggesting that DNA synthesis is induced by a cytoplasmic factor.

In most species of animals, unlike the sea urchin, the oocyte nucleus does not complete meiosis until after the sperm nucleus has entered the egg cytoplasm (Table 19-1). In those species, formation of a new polar body (which may be either the first or second depending on the species) is one of the earliest signs of a successful fertilization. Fusion of male and female nuclei does not occur until meiosis is completed and a haploid female pronucleus has been formed. In mammals, DNA synthesis occurs during the enlargement of the pronuclei that follows the completion of meiosis, and actual fusion of the pronuclei does not occur until their nuclear envelopes break down in preparation for the first cleavage division. At that time the maternal and paternal chromosomes are incorporated into a single mitotic spindle (Fig. 19-22) and distributed equally to both blastomeres.

The response of the egg to sperm penetration (and to various artificial stimuli) is called egg activation. Egg activation can be defined most simply as the beginning of embryonic development. Key events that we will focus on include a rapid change in membrane potential, which is responsible for the early block to polyspermy; the cortical reaction, which is responsible for the late block to polyspermy; and the actual initiation of development. We will also examine changes in intracellular concentrations of specific inorganic ions that appear to regulate those processes. None of these events require direct participation by the male gamete, and all can be observed in parthenogenetically activated eggs, as well as in normally fertilized eggs.

The major regulatory steps currently believed to be involved in the overall process of activation of the sea urchin egg are summarized in Fig. 19-23. The most rapid response that has been observed is a minor influx of sodium ions, with no corresponding counterflow of ions. This leads to a rapid change in membrane potential. Initially the inside of the egg is somewhat negatively charged relative to the outside, but within a few seconds after sperm contact, it becomes slightly positive. The mechanisms are not fully understood, but the change in membrane potential appears to contribute significantly to "early" resistance to polyspermy, although other factors may also be involved. In the eggs of *Strongylocentrotus purpuratus*, for example, artificially modifying the membrane potential from its normal prefertilization value of -60 mV to a value of $+10$ mV (a normal value during the first minute after fertilization) prevents fusion of sperm and egg plasma

EGG ACTIVATION

membranes, although it does not prevent the initial attachment phase. Also, eggs of species that show a rapid rise in membrane potential appear to be far more resistant to polyspermy than those of species in which the rise is slow.

The second response that is observed is an increase in the intracellular concentration of calcium ions, which appears to be due to release of intracellular stores of bound Ca^{2+} rather than transport of calcium into the egg, since it will occur in calcium-free seawater. Once triggered, the release of Ca^{2+} is self-catalyzed and spreads as a wave across the egg from the point of fertilization. The spreading wave of Ca^{2+} release is very closely associated with the spreading cortical reaction. However, detailed studies have shown that the Ca^{2+} release occurs just ahead of the cortical exocytosis, and that Ca^{2+} release occurs even when the cortical granules have been displaced by centrifugation. The source of the stored calcium is thus in the cytoplasm and not in the cortical granules. The period of enhanced cytoplasmic concentration of free Ca^{2+} lasts for only about 60 seconds, after which the level falls rapidly to the prefertilization level. The normal source for

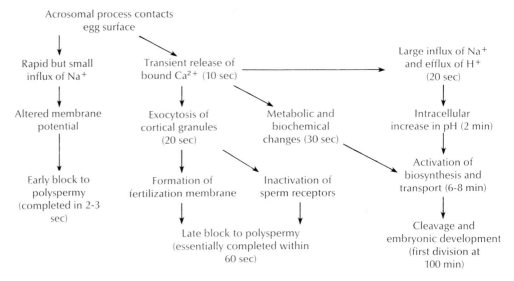

Fig. 19-23. Summary of major events in activation of the sea urchin egg. Three major sequences are triggered by contact of the acrosomal process of the sperm with receptors on the vitelline layer of the egg and the subsequent fusion of egg and sperm membranes. The first, shown on the left, is a rapid block to polyspermy brought about by a change in membrane potential. The second, in the center, is the late block to polyspermy that results from the cortical reaction. The third, on the right, is activation of embryonic development. These three sequences are described in detail in the text. The times indicated are from sperm-egg contact to the beginning of each event. (Modified from Epel, D. 1978. Curr. Top. Dev. Biol. **12:**185.)

the initial Ca^{2+} that triggers the spreading wave of Ca^{2+} release and the subsequent cortical reaction is still unknown. However, the process can be triggered artificially by a calcium ionophore, just as was the case with the acrosome reaction.

In the cortical reaction the cortical granules, which are small membrane-bound vesicles lying just under the egg plasma membrane, fuse with the plasma membrane and release their contents into the space between the plasma membrane and the overlying noncellular vitelline membrane (Fig. 19-24). As described, the cortical reaction spreads as a self-propagating wave, which typically begins 20 to 30 seconds after fertilization and then requires about 20 additional seconds to spread completely around the egg. All sperm except the one responsible for fertilization are detached from the egg surface as the cortical reaction spreads (Fig. 19-25, A to F). Within 2 to 3 minutes, fluid accumulates in the space below the vitelline membrane, and that membrane becomes thickened and distinctly raised from the cell surface to form the fertilization membrane (Figs. 19-24 and 19-25, G and H).

Materials released during exocytosis of the cortical granules include the following:

1. A protease called vitelline delaminase, which breaks the bonds between the vitelline layer and the plasma membrane
2. A protease called sperm receptor hydrolase, which alters sperm-binding sites on the vitelline layer and causes release of all bound sperm still on the outside of the vitelline layer

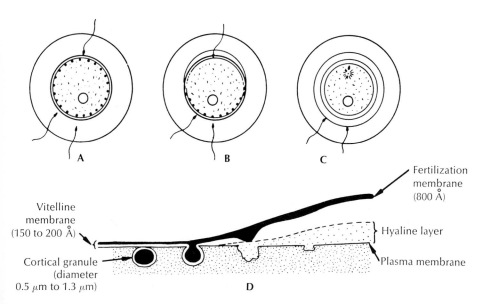

Fig. 19-24. Cortical reaction in sea urchin eggs. A to C, Time sequence of formation of the fertilization membrane. (The outer circle represents the jelly.) D, Exocytosis of cortical granules and elevation of the fertilization membrane. (From Austin, C. R. 1965. Fertilization. Prentice-Hall, Inc., N.J. Reprinted by permission of Prentice-Hall, Inc.)

3. A glycoprotein that draws water into the space between the vitelline layer and the plasma membrane, causing elevation of the vitelline layer to form the fertilization membrane
4. A cross-linking factor (not precisely defined), which hardens the fertilization membrane
5. A structural protein called hyalin, which contributes to formation of the hyaline layer directly over the plasma membrane

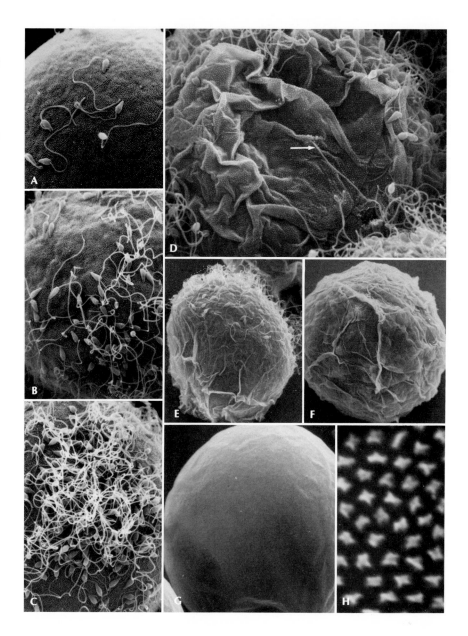

Fig. 19-25. Scanning electron micrographs of the sperm attachment-detachment sequence on the surface of *Strongylocentrotus purpuratus* eggs. The number of sperm bound per egg increases until 25 seconds after insemination, when the cortical reaction and the detachment sequence begin. Times shown after insemination are, **A,** 1 second; **B,** 5 seconds; **C,** 15 seconds; **D,** 30 seconds (arrow points to the tail of the fertilizing spermatazoon and the hole it made in the vitelline layer; the extent of the cortical reactions at this time is marked by the circular area from which sperm have been detached); **E,** 45 seconds (cortical granule breakdown has covered about half the egg); **F,** 55 seconds (cortical reaction is completed, and all but the fertilizing sperm have been detached); **G,** 3 minutes (showing the hardened fertilization membrane). Note that **E** to **G** are at the same magnification. The differences in size and surface wrinkling are related to elevation and hardening of the fertilization membrane. **H,** Enlargement of **G,** showing projections on the surface of the fertilization membrane. (From Tegner, M. J., and D. Epel. 1973. Science **179:**687.)

Collectively these changes result in a complete and permanent block to polyspermy. Full elevation of the fertilization membrane typically requires about 3 minutes. By the end of that time the head of the fertilizing sperm is completely engulfed in the cytoplasm. Fusion of the egg and sperm plasma membranes occurs at approximately the same time as the beginning of the cortical reaction and is probably the stimulus responsible for initiating the release of Ca^{2+} and the beginning of the cortical reaction. (The earlier changes in egg plasma membrane potential appear to be initiated as soon as sperm proteins on the tip of the acrosomal tubule contact receptors on the vitelline layer.) During the engulfment process the membrane of the acrosomal tubule becomes continuous with the egg plasma membrane and the head and tail of the sperm move to the interior of the egg while the membranes are left behind. The subsequent decondensation of the sperm nucleus and the fusion of male and female pronuclei, which normally occurs about 20 minutes after the initial sperm contact, have already been described.

In addition to triggering the cortical reaction, the temporary increase in Ca^{2+} concentration also triggers a series of events that culminates in the initiation of development, as shown in Fig. 19-23. At about a minute after fertilization, when the intracellular Ca^{2+} level is highest, hydrogen ions begin to flow out of the egg cell, balanced electrically by a comparable flow of sodium ions into the cell. This balanced Na^+/H^+ exchange results in a significant accumulation of "fertilization acid" in the seawater surrounding the eggs and a corresponding rise in intracellular pH. Direct measurement of pH with microelectrodes inserted into the eggs of *Lytechinus pictus* has shown that the intracellular pH rises from 6.84 ± 0.02 before fertilization to a value of 7.26 ± 0.06, which is reached within 5 to 6 minutes and then remains stable for at least 30 minutes after fertilization.

Although the increase in pH is normally triggered by the temporary rise in free intracellular Ca^{2+}, both the Na^+/H^+ exchange and the pH rise can be induced artificially by treatment of unfertilized eggs with ammonium salts or with amine anesthetics such as procaine. When this is done, the artificially stimulated increase in pH induces a number of the changes that normally occur during the initiation of development, including activation of protein synthesis and activation of DNA synthesis. The rate of protein synthesis in fertilized and artificially activated eggs is extremely sensitive to small changes in pH, and it has been reported that the effects of the increase in pH are enough to account for the entire increase in protein synthesis that occurs after fertilization.

Increased pH alone does not lead to fully normal development, however, and it currently appears that other metabolic changes, also triggered by the temporary release of free Ca^{2+}, are also needed, as indicated in Fig. 19-23. In the normal development of sea urchin eggs the rate of protein synthesis increases at about 5 minutes after fertilization. Transport systems in the plasma membrane are activated about the same time. The male and female pronuclei fuse about 20 minutes after sperm-egg contact, and DNA synthesis begins almost as soon as the fusion is completed. The first cleavage division normally occurs about a hundred minutes after the initial encounter between egg and sperm cells.

The events associated with fertilization are not known in as much detail for mammalian eggs. However, the basic steps of fast block to polyspermy, slow block to polyspermy, and initiation of development appear to be similar. There is no fertilization membrane as such in mammalian eggs. However, studies with golden hamster eggs have shown that the zona pellucida, a thick noncellular membranous structure that surrounds the mammalian egg, becomes unreceptive to sperm attachment about 15 minutes after attachment of the first sperm and that the loss of receptiveness can be blocked with protease inhibitors. This change, which is known as the zona reaction, involves the release of material from the cortical granules and appears to be similar in principle to the cortical reaction in sea urchin eggs.

ERRORS OF FERTILIZATION

There are a number of situations in which the normal union of one sperm cell and one egg cell does not occur. Most of these are abnormal situations and qualify as "errors" of fertilization, but in a few cases they are part of a normal developmental process.

Polyspermy occurs when two or more sperm fertilize a single egg. In mammals, fertilization of an egg by two sperm results in a triploid individual, which frequently develops normally until about midgestation, whereupon the fetus dies. Polyspermy results in abnormal development from the very earliest stages in the frog and sea urchin due to the development of extra spindles or multipolar spindles. Whereas polyspermy is an unusual occurrence in most kinds of animals and is usually lethal, it occurs rather frequently in animals with very large eggs, such as birds. In these animals, polyspermy is not pathological. Although 50 or more sperm may enter the egg and form pronuclei, only one fuses with the female pronucleus and all others are eventually lost.

Polygyny results from a failure of the oocyte to release the second polar body. The fertilized egg contains a single male pronucleus but two female pronuclei. In mammals the development of these individuals is essentially the same as that of polyspermic individuals, with death occurring at about midgestation.

Gynogenesis is a situation in which the male germ cell acts to stimulate development of the egg but makes no genetic contribution to the resulting individual. There is a species of fish, the Amazon molly, in which there are no males. Males of another species within the same genus activate the eggs of the Amazon molly but make no genetic contribution to the resulting embryo. The eggs develop without releasing the second polar body and form diploid individuals. Although gynogenesis can occur in mammals, development does not progress far.

Androgenesis, the counterpart to gynogenesis, occurs when the female pronucleus is not viable. It is not known to be normal in any animal. Although development may begin in the mammal under the influence of only the male pronucleus, it is short-lived. In some species, however, such as salamanders and sea urchins, it is possible to obtain full development of haploid individuals that contain only a paternal genome.

Double fertilization is possible when the second meiotic division occurs with the spindle in a central position within the egg instead of near the egg surface. The oocyte divides into two cells of fairly equal size, each of which may be fertilized by a sperm. The resulting embryo develops normally in a manner comparable to an allophenic mouse (Chapter 12). Mosaic individuals that appear to have resulted from double fertilization have been observed both in mice and in humans. In cases in which the two sperm that fertilize the egg halves are of different sex type, the individual may show some degree of hermaphroditism (sex differentiation intermediate between normal female and normal male; see Chapter 20).

PARTHENOGENESIS

In parthenogenesis there is no involvement whatever of sperm cells in egg activation or embryonic development. It is a normal method of reproduction in some animals, such as the honeybee, in which males (drones) are haploid individuals produced parthenogenetically and females (workers and queens) result from fertilization. There are also parthenogenetic species of rotifers in which males are completely unknown. Parthenogenesis occurs normally in higher organisms also—turkeys often undergo spontaneous parthenogenesis. The resulting offspring are diploid, presumably because the second polar body is not released, and always male, since the male is the homogametic sex in birds.

Parthenogenesis is easily induced experimentally in amphibian eggs by pricking them with a glass needle contaminated with blood or tissue extract. Normal fertile adult offspring have been obtained in this manner. Parthenogenetic development has also been induced in mammals by such methods as cold shock, and the resulting embryos have developed to the beating-heart stage. There are a few claims in the older literature that live young have been obtained in mammals by parthenogenesis, but these remain controversial because of lack of adequate controls.

IN VITRO FERTILIZATION   The conditions for successful fertilization can be met in vitro for a wide variety of animals. Greater success is achieved with oocytes that have been collected after ovulation, but in some cases oocytes that have been removed from the ovary while still in the prophase of meiosis I and allowed to mature in culture can also be fertilized. Sperm that have undergone capacitation are added to the mature oocytes, and fertilization occurs, followed by embryonic development. In the case of mammals, development past the early embryonic stages requires transfer of the embryo to the reproductive tract of a foster mother that has been suitably prepared hormonally for pregnancy. Normal development to term can readily be achieved from rabbit and mouse eggs fertilized in vitro and has also recently been reported for humans.

GERM CELLS AND THE
CONTINUITY OF LIFE   In this chapter we have concentrated our attention on the cells that are directly involved in the continuity of life from one generation to the next—the germ cells and the gametes that are derived from them. In certain species it is possible to relate the germ cells to a specific cytoplasmic germ plasm that can be detected in the egg cell even before the first cleavage division, and in a wide variety of animal species definitive precursors of the germ cells can be detected early in the process of embryonic development. The end products of germ cell differentiation, the gametes, are highly specialized cells, which when fused together and maintained under appropriate environmental conditions, display the complete developmental program needed for the formation of new individuals. The germ cells thus contain information not only for formation of new germ cells but also for formation of somatic cells, which support and nourish the germ cells and, ultimately, after they have brought appropriate germ cells together to form a new individual, age and die (as outlined in Chapter 1 and Fig. 1-2).

In all higher animals, germ cells occur in two types, male and female. Each type requires specialized somatic organs in which to develop and mature into functional gametes. In addition, the somatic organisms that contain and nurture the germ cells must function in a manner that permits the two types of gametes to come into close proximity at a time when both are mature. Finally, in many species, extended nurturing is required after fertilization before the new individuals achieve self-sufficiency. Since in all except certain relatively primitive species the male and female germ cells reside in separate individuals, a substantive developmental dichotomy (two parallel but different patterns of development) exists, which leads to distinct differences in structure, function, and behavior between the two sexes. The specific mechanisms involved in sexual differentiation during embryonic development and sexual maturation at puberty have been studied extensively and are the subject of the following chapter.

BIBLIOGRAPHY
Books and reviews

Austin, C. R. 1974. Fertilization. *In* J. Lash and J. R. Whittaker, eds. Concepts of development. Sinauer Associates, Inc., Sunderland, Mass.

Austin, C. R., and R. V. Short, eds. 1972. Reproduction in Mammals. Book 1. Germ cells and fertilization. Cambridge University Press, Cambridge.

Balinsky, B. I. 1975. An introduction to embryology, 4th ed. W. B. Saunders Co., Philadelphia.

Eddy, E. M. 1975. Germ plasm and the differentiation of the germ cell line. Int. Rev. Cytol. 43:229.

Epel, D. 1977. The program of fertilization. Sci. Am. 237:129 (Nov.).

Epel, D. 1978. Mechanisms of activation of sperm and egg during fertilization of sea urchin gametes. Curr. Top. Dev. Biol. 12:185.

Fawcett, D. W. 1975. The mammalian spermatozoan. Dev. Biol. 44:394.

Guidice, G. 1973. Developmental biology of the sea urchin embryo, Academic Press, Inc., New York.

Hegner, R. W. 1914. Studies on germ cells. I. The history of germ cells in insects with special reference to the Keimbahm determinants. II. The origin and significance of the Keimbahm determinants in animals. J. Morphol. 25:375.

Longo, F. J., and E. Anderson. 1974. Gametogenesis. *In* J. Lash and J. R. Whittaker, eds. Concepts of development. Sinauer Associates, Inc., Sunderland, Mass.

Mahowald, A. P., et al. 1979. Germ plasm and pole cells of *Drosophila*. Symp. Soc. Dev. Biol. 37:127.

Marx, J. L. 1978. The mating game: what happens when sperm meets egg? Science 200:1256.

McRorie, R. A., and W. L. Williams. 1974. Biochemistry of mammalian fertilization. Annu. Rev. Biochem. 43:777.

Mintz, B. 1961. Formation and early development of germ cells. *In* Symposium on the germ cells and earliest stages of development. Fondazione A. Baselli, Milano.

Monesi, V., et al. 1978. Biochemistry of male germ cell differentiation in mammals: RNA synthesis in meiotic and postmeiotic cells. Curr. Top. Dev. Biol. 12:11.

Monroy, A. 1973. Fertilization and its biochemical consequences. An Addison-Wesley Module, No. 17. Addison-Wesley Publishing Co., Inc., Menlo Park, Calif.

Raff, R. A., et al. 1975. Microtubule protein pools in early development. Ann. N.Y. Acad. Sci. 253:304.

Tarkowski, A. K. 1975. Induced parthenogenesis in the mouse. Symp. Soc. Dev. Biol. 33:107.

Tata, J. R. 1976. The expression of the vitellogenin gene. Cell 9:1.

Tyler, A. 1955. Gametogenesis, fertilization and parthenogenesis. *In* B. H. Willier, P. A. Weiss, and V. Hamburger, eds. Analysis of development. W. B. Saunders Co., Philadelphia.

Van Blerkom, J., and P. Motta. 1979. Cellular basis of mammalian reproduction. Urban and Schwarzenberg, Baltimore.

Wilson, E. B. 1925. The cell in development and heredity, 3rd ed. Macmillan Publishing Co., Inc., New York.

Selected original research articles

Blackler, A. W., and M. Fischberg, 1961. Transfer of primordial germ cells in *Xenopus laevis*. J. Embryol. Exp. Morphol. 9:634.

Eyal-Giladi, H., S. Kochav, and M. K. Menashi. 1976. On the origin of primordial germ cells in the chicken embryo. Differentiation 6:13.

Gilkey, J. C., et al. 1978. A free calcium wave traverses the activating egg of the medaka, *Oryzias latipes*. J. Cell Biol. 76:448.

Grainger, J. L., et al. 1979. Intracellular pH controls protein synthesis rate in the sea urchin egg and early embryo. Dev. Biol. **68:**396.

Gross, P. R., L. I. Malkin, and W. A. Moyer. 1964. Templates for the first proteins of embryonic development. Proc. Natl. Acad. Sci. U.S.A. **51:**407.

Gwatkin, R. B. L., et al. 1973. The zona reaction of hamster and mouse eggs: production in vitro by a trypsin-like protease from cortical granules. J. Reprod. Fertil. **32:**259.

Rogulska, T., W. Ozdzenski, and A. Komar. 1971. Behaviour of mouse primordial germ cells in the chick embryo. J. Embryol. Exp. Morphol. **25:**155.

Shen, S. S., and R. A. Steinhardt. 1978. Direct measurement of intracellular pH during metabolic derepression of the sea urchin egg. Nature **272:**253.

Tegner, M. J., and D. Epel. 1973. Sea urchin sperm-egg interactions studied with the scanning electron microscope. Science **179:**685.

Vacquier, V. D., D. Epel, and L. A. Douglas. 1972. Sea urchin eggs release protease activity at fertilization. Nature **237:**34.

Van Blerkom, J. 1979. Molecular differentiation in the rabbit ovum. III. Fertilization autonomous polypeptide synthesis. Dev. Biol. **72:**108.

Van Blerkom, J., and R. W. McGaughey. 1978. Molecular differentiation of the rabbit ovum. I. During oocyte maturation in vivo and in vitro. Dev. Biol. **63:**139.

Van Blerkom, J., and R. W. McGaughey. 1978. Molecular differentiation of the rabbit ovum. II. During the preimplantation development of in vivo and in vitro matured oocytes. Dev. Biol. **63:**151.

Wallace, R. A., et al. 1972. Studies on amphibian yolk. X. The relative roles of autosynthetic and heterosynthetic processes during yolk protein assembly by isolated oocytes. Dev. Biol. **29:**255.

CHAPTER 20

Sex differentiation

☐ Most multicellular animals, both vertebrate and invertebrate, reproduce sexually. Although there are many exceptions, dioecious species, in which the male and female gonads (and accessory sexual structures) are in separate individuals, are the most common. Since males and females differ significantly from each other, both structurally and functionally, there must exist within dioecious species alternative programs of development for males and females.

Aside from the obvious biological importance of sex differentiation, an exceptional level of interest has been drawn to the subject by sociological influences and psychological factors. We tend to be curious about how and why we became the way we are as either female or male persons, how and why members of the opposite sex developed as they did, and how and why individuals who deviate in any of a huge variety of ways from the accepted sexual norms became as they are. We make no pretense of being able to answer such complex and frequently controversial questions in this book. However, the biological process of sexual differentiation, which is described here, clearly lays down a basic framework on which later physiological, psychological, and cultural influences can act to produce varying degrees of "maleness" and "femaleness" in different individuals.

Our focus in this chapter is primarily on mammalian sex differentiation, with particular emphasis on humans and on experimental animals such as mice, rats, and rabbits that have provided much of the basis for our current level of understanding. Although we remain heavily dependent on data from such experimental animals, care must be exercised in extrapolating that data to humans, since significant differences in physiology and developmental responses are known to occur in certain cases.

Because of the complex interrelationship among the topics that are to be covered in this chapter, it will be helpful to begin with a brief summary of the major concepts:

1. Mechanisms for determination of sex are highly complex and

vary greatly from one type of organism to another. Depending on the species, the primary determinants of sexual differentiation can be genetic, environmental, or a combination of both. In mammals, genetic factors are of primary importance. In humans and most other widely studied mammals, females have two X chromosomes and are referred to as homogametic, since they produce only one kind of gamete, whereas males have one X chromosome and one Y chromosome and are thus heterogametic, producing individual haploid gametes that contain either X or Y chromosomes, but not both.

2. Sexual differentiation occurs rather late in mammalian development and is preceded by a sexually indifferent stage during which precursors for both male and female development are present in the embryo.

3. Differentiation of the sexually indifferent gonads into testes or ovaries is considered to be the primary event in sexual differentiation. In mammals, including humans, a dominant gene, normally carried on the Y chromosome, promotes testicular differentiation. The presence of a specific molecular species, the H-Y antigen, has been shown to be correlated precisely with testicular differentiation in mammals at all times, including special cases in which neither testicular differentiation nor formation of H-Y antigen is under direct control of the Y chromosome.

4. In mammals the fetal testis promotes development of male accessory sexual structures (internal duct systems and external genitalia) and inhibits development of corresponding female structures. Female sexual differentiation prior to birth requires no specific stimulus and proceeds normally in embryos (including genetic males) whose gonads are removed at the sexually indifferent stage. Female differentiation also occurs in embryos whose gonads spontaneously fail to develop.

5. Although male and female are commonly regarded as opposites, they are actually only the extremes of a developmental program that also produces a continuous spectrum of intermediate stages of differentiation when the control signals are ambiguous. It has been estimated that 2 to 3 out of every 1000 persons exhibit some degree of physical intersexuality of the reproductive tract. In addition, wide ranges of variation are considered to be "normal" in sex-related traits such as distribution of body hair and breast development.

6. Certain aspects of the reproductive physiology and behavior of experimental animals can be permanently altered by neonatal treatment with hormones. In mice and rats the so-called "brain sex," which influences both mating behavior and the pattern of release of gonadotropins (cyclic in females versus relatively constant in males) appears

to be determined by the presence or absence of male hormones at a critical time soon after birth. Such effects are much less marked in primates, however.

Although space does not permit a detailed analysis of the diversity of patterns and mechanisms of sexual differentiation in nonmammalian species, we will refer to them from time to time, particularly when they are in a sharp contrast to the pattern of sexual development and its control observed in humans and other mammals.

SEX DETERMINATION

The widespread occurrence among both vertebrates and invertebrates of a relatively uniform pattern of sexual reproduction, in which a female gamete or egg (which is relatively large and contributes, in addition to its nucleus, both the cytoplasm and the nutrients to support development) is fertilized by a motile male gamete or sperm (which is relatively small and contributes little more than its nucleus), makes it easy to assume that fundamental mechanisms of sex determination should be rather similar in highly diverse species. However, this does not appear to be the case. Although there may be unifying principles that have thus far been overlooked, current data make it appear that the actual mechanisms of sex determination are extraordinarily diverse.

First of all, there is diversity in the patterns of sexuality that occur in various organisms. In addition to dioecious species such as ourselves in which the two sexes are in separate individuals, there also exist numerous monoecious or hermaphroditic species, in which both sexes occur in the same individual, either simultaneously or sequentially. The simultaneous hermaphrodites can be further subdivided into those which are self-fertilizing and those in which two identical individuals each fertilize the eggs of the other. In either case the normal developmental program for each such species must call for development both of ovaries and of testes at specific locations within the body of a single organism.

Even in seemingly primitive organisms, complex controls and alternative pathways of sexual differentiation are likely to exist. One example occurs in *Caenorhabditis elegans,* a small nematode worm that is beginning to be used as a model system for the study of genetic controls over development. This organism has several interesting properties that make it particularly useful for such studies. Although it is reasonably complex and possesses a variety of different cell types with different functions, it is very small, with only about 800 somatic cells in the adult, making possible precise tracing of cell lineages during development. It is also a self-fertilizing hermaphrodite, which makes possible the recovery of recessive mutations in a homozygous state in the second generation after mutagenesis without preparing any special crosses. Self-fertilization within progeny of the mutagen-treated worms automatically ensures that a fourth of the second generation progeny will be homozygous for any mutations that have been induced and are viable in the homozygous state. Although the primary mode of reproduction in this organism is by self-fertilization, there also exist genetically determined males with copulatory organs that permit them to fertilize the eggs produced by the hermaphrodites (in competition with the internally produced sperm).

Males have an XO karyotype, with one less chromosome than hermaphrodites, which are XX. Half the progeny of fertilization by males are thus males. However, in addition, a small fraction of the progeny of self-fertilization of XX hermaphrodites are also XO males, and the proportion of males can be increased by certain types of environmental stress such as elevated temperature.

Environmental controls over sex determination

Environmental controls over sex differentiation in sequential hermaphrodites are particularly interesting. Different species of coral reef fish are known that are either first male and then female or first female and later male. In both cases aggressive dominance of the final sex appears to prevent sex change by other individuals within a restricted social unit. For example, within the limited environment of a laboratory aquarium, only one individual of the final sex will exist. If that individual is removed, the most dominant member of the opposite sex will undergo a rapid sex reversal and assume the role of the dominant individual of the final sex.

Although spontaneous sex changes in adult individuals occur in only a limited number of species, there are also many other cases in dioecious species when environmental factors appear to play an important role in sex determination. An interesting example that has already been partially described in Chapter 6 is the mealy bug. In these organisms, females have a diploid set of functional chromosomes, whereas in males the full set of paternal chromosomes remains heterochromatic and is discarded entirely during spermatogenesis. Thus the sperm that are produced all contain a genetically identical set of chromosomes of maternal origin. Although in theory the females could be heterozygous for a dominant female-determining genetic factor (comparable to the WZ pattern of chromosomal sex determination described later), no evidence has been found for such a system, and it currently appears that environmental factors determine the sex ratio among the offspring. This is true not only for the progeny of sexual crosses but also for totally homozygous individuals that develop by a parthenogenetic mechanism which involves duplication of the genome in haploid egg cells after meiosis is completed (Fig. 6-15). In both cases males and females appear to be genetically identical except for the functional haploidy of the males.

There are numerous species of insects that appear to consist only of parthenogenetically reproducing females during the spring and summer, with males being produced only as fall approaches. The males mate with the females, and the resultant fertilized eggs remain quiescent throughout the winter and produce only female offspring the following spring. In a few genera of flies, including *Miastor*, whose chromosome elimination during somatic cell differentiation was described in Chapter 6, the phenomenon of pedogenesis occurs. In this unusual arrangement larval stages produce parthenogenetic eggs that give rise to more larvae. Later in the season some of the larvae undergo metamorphosis into adults, both male and female, that mate and produce fertilized eggs.

Many more examples of environmental controls over sexual differentiation could be cited, including temperature effects on sexual determination in amphibian (and reptilian) embryos, which were described in Chapter 19. However, such controls tend to be the exception, rather than the rule, particularly in the higher vertebrates, in which genetic factors are usually the dominant force in sex determination.

In humans and other mammals the primary (and probably the only) factors determining gonadal sex are specific chromosomal genes (and the products of their transcription and translation). In virtually all mammals, males and females have cytologically distinguishable differences in their chromosomal constitutions. Typically, the female contains two identical sex chromosomes designated by the letter X, whereas the male contains one X, identical in all respects to the female chromosomes, and one Y, which may differ greatly both in morphology and genetic content from the X chromosome. In the human, for example, the X chromosome is of intermediate size and is known to carry many important genes (Chapters 6 and 11), whereas the Y is small, and except for the gene(s) for maleness, appears to be composed primarily of highly repetitive DNA that does not serve any known genetic coding function.

Among animals whose sex determination is primarily genetic or chromosomal, there exist a wide variety of sex chromosome patterns (Table 20-1). Although the XX/XY pattern is by far the most common in mammals, there do exist a number of variations. In certain species, there are two different X chromosomes (X_1, X_2) that exist in one copy in males ($X_1 X_2 Y$) and two copies in females ($X_1 X_1 X_2 X_2$). Similarly there

Genetic control over sex determination

Table 20-1. Relationship between chromosomes and gonadal sex

Sex chromosome karyotype*		Examples
Female	**Male**	**Examples**
XX	XY	Most mammals, including humans, some amphibia (Y chromosome determines maleness)
ZW	ZZ	Birds, reptiles, some amphibia (W chromosome determines femaleness)
$X_1X_1X_2X_2$	X_1X_2Y	Certain mammals with two different X chromosomes
XX	XY_1Y_2	Certain mammals with two different Y chromosomes
XX	$X\overset{*}{X}$	Male gonads without a Y chromosome in humans, mice, and other mammals caused by sex reversal gene ($\overset{*}{X}$)
XO,XXX	XXY,XYY	Nondisjunction of sex chromosomes in humans, other mammals
XX,$\overset{*}{X}$X,$\overset{*}{X}$Y	XY	*Myopus schisticolor* (Scandinavian wood lemming) X-linked gene ($\overset{*}{X}$) prevents Y chromosome from causing maleness
X^FO	X^MO	*Ellobius lutescens* (mole-vole); zygotes with OO and X^FX^M thought to be lethal
XX,XY,YY	XX,XY,YY	Certain fish with environmentally determined sex or with natural sex reversal in a single individual
XX	XY	Certain insects such as *Drosophila melanogaster* (sex determined by ratio of autosomes to sex chromosomes)
ZW	ZZ	Certain insects
XX	XO	Hemiptera (true bugs)
Diploid	Haploid	Honeybee
Diploid	Functional haploid	Mealy bug (all paternal chromosomes remain heterochromatic in male)
Eight somatic cell chromosomes	Six somatic cell chromosomes	*Wachtliella persicariae* (males undergo a second cycle of loss of chromosomes from somatic cells)
XX (hermaphrodite)	XO (true male)	*Caenorhabditis elegans* (nematode)

*Sex chromosomes are defined as those which are cytologically distinguishable between the sexes. Symbols $\overset{*}{X}$, X^F, and X^M are used to denote genetic differences that cannot be detected cytologically.

are species that have one X and two Y chromosomes in the male (XY_1Y_2) and two Xs in the female. Both of these cases result in unequal numbers of chromosomes in males and females. There is one interesting case, the mole-vole, *Ellobius lutescens*, in which the male and female have cytologically indistinguishable XO karyotypes. In this case the male determinant that is normally carried on a Y chromosome appears to have been relocated to another chromosome, possibly the X.

In birds, reptiles, and some amphibia, it is the female rather than the male that is the heterogametic sex. To avoid confusion, the letters W and Z are used to describe the chromosomes in such cases. The karyotype of the male is designated ZZ and that of the female ZW. It is interesting that seemingly rather closely related species of amphibians can have opposite sex determination mechanisms. Thus, for example, the leopard frog, *Rana pipiens,* exhibits an XX/XY pattern with heterogametic males, whereas the South African swimming toad, *Xenopus laevis,* exhibits a ZZ/ZW pattern with heterogametic females.

Among the insects, highly diverse patterns of chromosomal sex determination occur. In the fruit fly *Drosophila melanogaster* an XX/XY mechanism is observed, which superficially seems to be similar to that of mammals. However, on more detailed analysis it turns out that in mammals, sex is determined by the Y chromosome, whereas in *Drosophila,* sex is determined by the ratio of autosomes to X chromosomes. In many of the hemiptera (true bugs) there is no Y chromosome and males (XO) have one chromosome less than females (XX). There also exist ZZ/ZW systems in insects, in which the female is heterogametic. In the honeybee no sex chromosomes as such are evident, and males result from parthenogenetic haploid development of unfertilized eggs, whereas females are derived from fertilized eggs and are diploid. This situation is similar in some respects to that described for the mealy bug, in which the male is functionally haploid, with all paternal chromosomes in a genetically inactive heterochromatic form.*

In view of these complexities, we are left with the conclusion that even in cases when sex determination is primarily or exclusively chromosomal in nature, the factors involved are not necessarily simple or straightforward. Although we will not explore these many alternate forms of sex determination further, it is important for our readers to be aware (1) that there is far more to sex determination than the rela-

---

*Among the insects that lose somatic chromosomes during development, even more complex patterns have been observed. For example, in *Wachtliella persicariae* the somatic cells of males contain six chromosomes, whereas those of females contain eight. At the time of the fourth nuclear division in the embryo, all somatic nuclei in embryos of either sex lose all but eight of their chromosomes, whereas a single germ cell precursor retains the full germ line complement of approximately 40 chromosomes. At the seventh cleavage division in males, but not in females, all somatic cells lose two more chromosomes, leaving the males with a final number of six chromosomes in their somatic cells, as opposed to eight in the females.

tively straightforward XX/XY system that will be examined in detail in this chapter and (2) that even the XX/XY system may be considerably more complex than it superficially appears to be.

SEX CHROMOSOME ABNORMALITIES IN HUMANS
45XO females (Turner's syndrome)

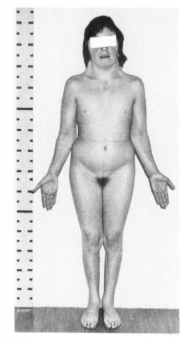

Fig. 20-1. Turner's syndrome. A 17-year-old patient who lacks sex chromatin (Barr body) and exhibits may typical features of Turner's syndrome, including short stature (142 cm, or 4 feet 7½ inches), sparse pubic hair and total absence of breast development or other female secondary sexual characteristics, a broad chest with widely spaced nipples, webbing of the neck, and cubitis valgus (outward bending of the arms at the elbows). (With permission from Hauser, G. A. 1963. In C. Overzier, ed. Intersexuality. Academic Press Ltd., London, Copyright by Academic Press, Inc. [London] Ltd.)

In humans an altered number of sex chromosomes frequently (but not always) results in anomalies of sexual differentiation. Much of the early evidence supporting the concept that the Y chromosome is the determinant of male gonadal differentiation was derived from studies of human individuals with altered numbers of X and Y chromosomes. In describing such individuals, the usual practice is to refer to the total number of chromosomes, followed by a listing of the sex chromosomes that are present. Thus 46XX and 46XY refer to chromosomally normal females and males, respectively.

Individuals with one X and no Y chromosomes (45XO) develop as females but exhibit a pattern of specific defects known as Turner's syndrome (Fig. 20-1). Typically they are very short, exhibit some degree of webbing of the neck, and remain sexually juvenile without the development of female secondary sexual characteristics at the normal age of puberty. Early embryonic development of ovaries and germ cells is relatively normal, but the germ cells degenerate prior to birth, and a major degeneration of the ovaries also occurs so that little more than whitish streaks of ovarian tissue embedded in connective tissue (streak gonads) usually remain. This phenomenon is sometimes referred to as gonadal dysgenesis. Fetuses with a 45XO karyotype frequently abort spontaneously, and among those which survive to term there is an increased incidence of congenital defects. There are also some reports of impaired intelligence, but the majority of affected individuals appear to fall into the normal range, and some are highly intelligent.

Studies of women whose cells contain one normal X chromosome and only part of a second (sometimes translocated onto another chromosome) have shown that a deficiency of the short arm of the X chromosome leads to full expression of Turner's syndrome with both somatic defects and gonadal dysgenesis. Women lacking the distal part (away from the centromere) of the long arm of one X chromosome are fully normal. In cases in which the proximal part of the long arm is disrupted, gonadal dysgenesis occurs without the somatic defects of Turner's syndrome. Such symptoms are similar to those of "pure" gonadal dysgenesis, which can occur in both 46XX and 46XY individuals. In both cases there is failure of gonadal development, resulting in phenotypic females with sexual juvenilism but without other somatic defects associated with the 45XO syndrome.

The cause of somatic defects in the 45XO syndrome is not well understood, particularly in view of the fact that most (but not all) of one of the X chromosomes is inactivated in each somatic cell of normal female mammals, including humans (Chapter 6). In germ cells both X chromosomes are normally active, and there is considerable circumstantial evidence suggesting that in most species two active X chromosomes are needed for long-term survival of oocytes and their full participation in normal oogenesis.* Since germ cell development appears to begin normally in 45XO embryos, it has been suggested that the primary defect may be an inability of the germ cells to survive for prolonged periods in meiotic prophase and that the ovarian dysgenesis that follows may be a secondary consequence of loss of germ cells.

A clear exception to the apparent requirement for two active X chromosomes for oogenesis occurs in the mouse, in which XO animals develop as fully fertile females (although fertility is said to be lost more rapidly with age than in normal mice). In addition, XY germ cells have also been shown to participate in oogenesis in the ovaries of mosaic XX/XY mice that have developed as females. These observations are an example of the danger of extrapolating mouse data to humans or vice versa.

47XXX females

Another sex chromosome abnormality that has been observed in phenotypic females is 47XXX. Individuals with this chromosomal constitution, which can be detected by the presence of two Barr bodies, are generally fertile females with relatively normal sexual differentiation, although there is some tendency toward small breasts, a somewhat juvenile degree of sexual maturation, and menstrual irregularities. There is also a tendency toward mild mental retardation, but many triple X individuals fall within the range of normalcy and go undetected except in special screening programs. When more than three X chromosomes are present, the individual is usually severely retarded.

47XXY males (Klinefelter's syndrome)

Individuals with Klinefelter's syndrome (47XXY) are phenotypic males with small testes that lack germ cells and are consequently sterile. They are usually very tall, and many (but not all) are somewhat mentally retarded. Development until puberty is relatively normal.

---

*An interesting phenomenon occurs in the creeping vole, *Microtus oregoni*, where the X chromosome is routinely eliminated from male germ cells, yielding some sperm that carry no sex chromosomes. This results in the formation of XO female zygotes that develop into embryos whose somatic cells and mitotic germ cells are also XO. However, at the time of transition from oogonia to oocytes, X chromosome nondisjunction occurs so that all functional oocytes become XX.

Some degree of testicular degeneration is usually experienced in early adulthood. However, the level of androgen produced and the consequent sexual phenotype varies considerably from individual to individual (Fig. 20-2). The phenotypic characteristics of Klinefelter's syndrome also appear in individuals with an XXYY chromosome constitution and in individuals that have three, four, or five X chromosomes in combination with a Y chromosome. In these cases all X chromosomes in excess of one are present as heterochromatinized Barr bodies.

Evidence from a number of sources suggests that the presence of more than one X chromosome is incompatible with normal development of male germ cells. In addition to the loss of germ cells in Klinefelter's syndrome, a variety of cases are known in which translocation of X-chromosomal material to an autosome results in sterility in males but not in females. XXY mice exhibit a normal male phenotype but are sterile due to failure of germ cell development. Autosomal sex reversal genes exist in mice and humans that cause XX individuals to develop as males. In both cases there is sterility because of germ cell

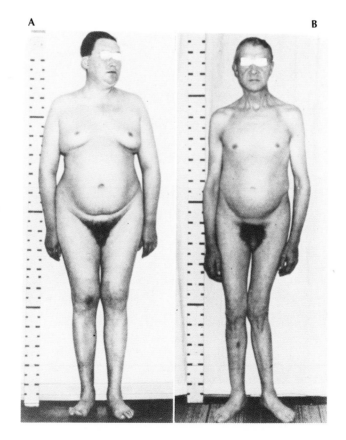

Fig. 20-2. Two men with Klinefelter's syndrome (47XXY) illustrating the phenotypic diversity that occurs. Both have long legs, small testes, and underdeveloped male secondary sexual characteristics.
A, A 38-year-old patient with an underdeveloped penis, gynecomastia (enlargement of breasts), and a generally feminine body shape. **B,** A 50-year-old patient with a normal-sized penis and absence of gynecomastia or other distinctly feminine traits. (With permission from Overzier, C. 1963. *In* C. Overzier, ed. Intersexuality. Academic Press Ltd., London. Copyright by Academic Press, Inc. [London] Ltd.)

failure during meiosis. In XO sex-reversed mice, spermatogenesis is more normal, but abnormal spermatozoa are produced and the animals are still infertile. Thus, with the possible exception of oogenesis in mice, it appears that gametogenesis in mammals requires not only the right kind of gonad but also the correct number of X chromosomes in the germ cells that are involved. Such constraints appear to be less rigorous in the lower vertebrates, in which sex reversal resulting in fully fertile individuals can be achieved rather readily.

Individuals with a 47XYY chromosomal constitution are males who are signficantly taller than average but otherwise physically normal. Surveys of newborn infants suggest that about 0.1% of all male live births are XYY. An increased frequency of XYY individuals is found both in penal institutions and in institutions for the mentally retarded. At one time it was thought that XYY men had a strong predisposition to antisocial and criminal activities. However, this view is at least partially the result of biased sampling. Only a small fraction of XYY births can be accounted for in institutionalized populations, and it is known that at least some XYY men lead normal well-adjusted lives. Unfortunately, there has not yet been an adequate survey of tall men in the population at large to determine accurately the fate of the majority of XYY males.

47XYY males

It is obvious from the sexual phenotypes (male or female) of the individuals with various numbers of X and Y chromosomes just described that the presence of a Y chromosome, and not the number of X chromosomes, determines maleness in humans. Thus 45XO, 46XX, and 47XXX individuals are all female, whereas 46XY, 47XXY, and 47XYY individuals are all male.

ROLE OF THE Y CHROMOSOME IN SEX DETERMINATION

The influence of the Y chromosome on the direction of sex differentiation is also seen from the results of experiments done with allophenic mice (Chapter 12). The genetic sex of the fused embryos is not known when they are mixed, and each of the embryos being combined should have an approximately equal chance of being male or female. It would therefore be expected that 25% of such embryo fusions would combine embryos that are both of the XX genotype and would result in the development of females; that 25% would combine embryos that are both of the XY genotype and would result in the development of males; and the remaining 50% would combine an XX and an XY embryo and result in the production of hermaphrodites (individuals with both male and female traits). However, when the allophenic mice produced by certain strain combinations are examined, it is found that the majority of them are males and fewer than 1% are hermaphrodites. This suggests that many of the animals expected to develop as intersexes instead become males. When such allophenic mice are separated into two groups by virtue of their coat color so that one group contains animals in which one or the other genotype predominates and the other contains animals in which each genotype is about equally represented, the two groups are found to have radically different sex ratios. A normal 1:1 sex ratio is observed in the group in which one genotype predominates, suggesting that sex differentiation in these animals is controlled by the predominating genotype. By contrast, in the group of mice having each genotype represented more equally, the sex ratio approaches 3 males:1 female. The latter result suggests a dominant role of the Y chromosome in directing male sex differentiation.

Although the mechanism whereby the Y chromosome brings about male development is not known with certainty, some interesting theories have been proposed regarding this matter. Ursula Mittwoch and her colleagues have suggested that the Y chromosome causes a rapid rate of growth in cells of the undifferentiated gonad and that rapid growth results in testicular development, whereas slower gonadal growth in the absence of a Y chromosome results in ovarian development. The finding that the gonads of male embryos grow more rapidly than those of female embryos in the same litter is consistent with this theory. Further evidence of a connection between gonadal growth rate and the direction of gonadal development comes from mice that carry the autosomal dominant mutation Sex reversed (Sxr), which causes male development to occur in the absence of a Y chromosome. As described, chromosomal females carrying this mutation are phenotypically male but have small testes that do not contain germ cells. When the gonads of XX Sxr embryos are examined at early developmental stages, they are found to be much larger than those of normal female embryos, although they are slightly smaller than those of normal male embryos in the same litter.

H-Y antigen

Recent studies by Susumu Ohno and his co-workers suggest that the Y chromosome of mammals carries a gene that either codes for or regulates the synthesis of a membrane-bound antigen that promotes testicular differentiation. Specifically, they have postulated that testicular differentiation in male mammals is induced by the H-Y antigen, a well-known male-specific histocompatibility antigen, that is found bound to the plasma membranes of all cells in male mice and is responsible for rejection of male grafts by females of the same inbred strain. Antigens indistinguishable from mouse H-Y have been found in cells of all other male mammals examined, including humans. The relationship between the H-Y antigen and testis determination in mammals has been tested by examining individuals whose gonadal sex is not in accord with their chromosomal sex, such as the sex-reversed XX mice and humans discussed earlier.

In every case examined, individuals that possess testes are H-Y antigen positive, whether or not they have a Y chromosome. Likewise, in cases in which a Y chromosome is present but no testes are formed, no H-Y antigen can be detected. This occurs, for example, in the Scandinavian wood lemming, *Myopus schisticolor,* in which the female to male ratio is 3:1 and an exceptional class of fertile females exists that has an XY somatic cell karyotype and produces only daughters. This unusual situation appears to be due to an X-linked gene that prevents expression or is needed for activation of the Y-linked gene for testicular organization. The XY females also lack H-Y antigen. In their germ cells the Y chromosome is eliminated and the X is duplicated so

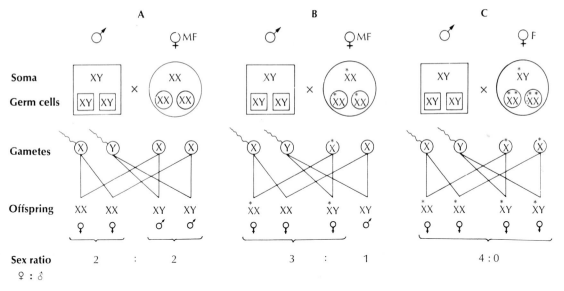

Fig. 20-3. Sex determination in the Scandinavian wood lemming, *Myopus schisticolor.* An X-linked gene (indicated by X̌) causes XY individuals to develop as a special class of fertile females (♀F) that produce only female offspring. During oogenesis in the XY females the Y chromosome is eliminated from the germ cells and the X̌ chromosome is doubled in number by nondisjunction so that all their progeny receive an X̌ chromosome. There also exist XX females (including heterozygotes for X̌) that give rise to both male and female offspring (♀MF). **A,** Cross of a normal XX female with an XY male yields an equal number of male and female progeny. This pattern is typical of mammalian sex determination in general. **B,** Cross of an XX female heterozygous for X̌ with an XY male yields an altered sex ratio consisting of three females (one each of XX, XX̌, and X̌Y) to one male (XY). **C,** Cross of an X̌Y female with an XY male yields only female offspring, two of which are X̌Y and two of which are XX̌. (From Fredga, K., et al. 1977. Hereditas **85**:101.)

that the special anti-male X chromosome is received by all their off-spring, thus resulting in all female litters after fertilization by normal males with presumably normal Y chromosomes (Fig. 20-3).

In bovine freemartins (females virilized by sharing a conjoined placenta with a male twin) the testislike gonads contain as much H-Y antigen as normal bovine male gonads. Freemartins have both XX and XY cells in their gonads, and it has been postulated that the XX cells become coated with H-Y antigen produced by XY cells and as a result participate in testislike differentiation. The testis-organizing ability of the H-Y antigen could also explain other cases in which testes develop from embryonic gonads containing both XX and XY cells, such as in allophenic mice. Ohno suggests that XX cells may also contain anchorage sites for H-Y antigen, which allow them to participate in testis development. Transfer of H-Y antigen from XY to XX cells occurs only between certain strains with appropriate combinations of histocompatibility antigens. The basic model of Ohno is shown in Fig. 20-4.

One curious aspect of the proposed involvement of the H-Y antigen system in sex determination that is difficult to explain on the basis of current data is its consistent association with the heterogametic sex. In all cases in which an XX/XY system of sex determination normally

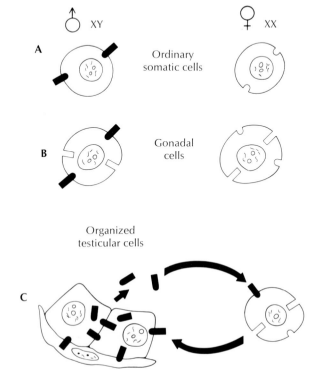

Fig. 20-4. Ohno's model for the role of H-Y antigen in the organization of testicular tissue. **A,** Somatic cells from nongonadal tissue have anchorage sites for H-Y antigen, shown as a rounded depression. The antigen is normally found on male but not female cells. However, the female cells can be coated with free antigen. **B,** Gonadal cells have specific receptors for H-Y antigen (shown as rectangular indentations) in addition to the anchorage sites. **C,** Combination of anchorage sites and receptors permits the H-Y antigen to organize the gonadal cells into the functional morphology of testicular tubules. In mixed aggregates, cellular contact may stimulate production of excess H-Y antigen, which can coat XX gonadal cells and permit them to become incorporated into testes, for example, in allophenic mice produced by fusing embryos of compatible strains. (From Ohno, S. 1976. Cell 7:315. Copyright © MIT; published by the MIT Press.)

operates, the H-Y antigen occurs only on XY male cells. However, in organisms with ZZ/ZW sex determination an antigenic activity that is at least similar, if not identical, to the HY antigen is found on the surface of ZW female cells and not on ZZ male cells. It makes sense for a specific active product to be associated with the dominant heterogametic state in each case, but it is difficult to understand how the same H-Y antigen can promote testicular organization in one case and ovarian organization in the other. Such a dichotomy is even more puzzling when it occurs in two similar species such as *Rana pipiens,* in which H-Y is associated with XY males, and *Xenopus laevis,* in which H-Y is associated with ZW females. One possibility that needs to be explored further is that we may still be dealing with regulatory signals that are remote from the actual molecular mechanisms responsible for testicular and ovarian differentiation. There may be multiple regulatory pathways involved in sex determination that all converge on final steps that are yet to be discovered.

A further question that is not yet fully resolved is whether the H-Y antigen is actually physically responsible for testicular organization in mammals or is merely synthesized in response to the same control signal that also activates testicular differentiation. Recent data from

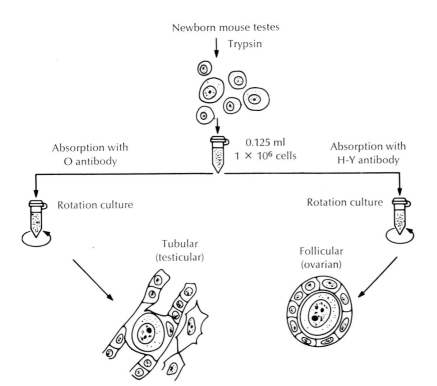

Fig. 20-5. Effect of H-Y antigen on reaggregation of newborn mouse testicular cells. Cells are dispersed with trypsin and treated either with a control antibody preparation (left) or antibody directed against H-Y antigen (right). Treatment with anti–H-Y leads to loss of testislike organization in the reaggregated cells and its replacement with a pattern that resembles ovarian follicles. (From Ohno, S. 1978. J.A.M.A. **239:** 217. Copyright 1978, American Medical Association.)

Ohno's laboratory suggest a direct role. When cells from newborn mouse testis are reaggregated in rotary shaker cultures, they assume a configuration that resembles the organization of testicular tubules. However, if they are first stripped of H-Y antigen by antibody treatment (lysostripping), they reaggregate in a configuration similar to ovarian follicles, with a layer of what had originally been Sertoli cells completely surrounding each germ cell (Fig. 20-5).

The relationship between the H-Y antigen and sex determination has received extensive attention in the scientific literature recently. The correlation between the presence of H-Y antigen and testis formation in mammals appears to be strong, despite the puzzling aspects of H-Y involvement in female differentiation in heterogametic ZW systems. There is probably far more to the whole story than we know at the present time. Interested students (and their instructors) will undoubtedly want to follow the latest developments in this story as they are published. Because of current interest and activity in studies of the role of H-Y antigen in sex determination, we have selected this system for use as an example in the introduction to the scientific literature and methods for literature research presented in Appendix E.

DEVELOPMENT OF THE REPRODUCTIVE SYSTEM
Morphological changes involved in sexual differentiation

Phenotypic sex differentiation occurs at a relatively late stage in mammalian development. In normal sex differentiation in the human, for example, the embryo is sexually indifferent and shows no phenotypic signs of being either male or female until about the seventh week of gestation. By that time the embryo is well formed and human appearing, and most of its organ systems are well differentiated and functional. However, the reproductive structures, which do not need to function until much later in the individual's development, are still in the form of uncommitted precursors with the capacity of differentiating into either male or female forms.

During the sexual differentiation that occurs after the seventh week of gestation in the human, a change from uncommitted precursors to definitive male or female structures occurs in three areas: (1) the gonads, (2) the reproductive ducts and internal sex organs, and (3) the external genitalia and urinary ducts. The final stage of sexual differentiation, involving functional maturation of the reproductive system and development of secondary sexual characteristics, does not occur until puberty.

As discussed in Chapter 19, the primitive gonad is composed of external cortical tissue and internal medullary tissue (Fig. 19-8). In the male the germ cells move from the cortex of the gonad into the medulla, and during the seventh and eighth weeks of development the primary sex cords of the medullary region begin to enclose the germ cells and differentiate into the primitive seminiferous tubules of the testis (which are actually solid cords of tissue, since no lumen forms until much later). In the absence of a positive signal for testicular differentiation (perhaps the H-Y antigen, as discussed), no significant change occurs in the primitive gonad until about the tenth week of development, when the cortex begins to thicken and grow inward until it forms the bulk of the female gonad and contains the majority of the germ cells, each surrounded by a layer of primitive follicle cells. It is not clear whether a specific ovarian organization factor is required. However, normal ovarian development does not take place in the absence of germ cells, and in cases such as Turner's syndrome in which

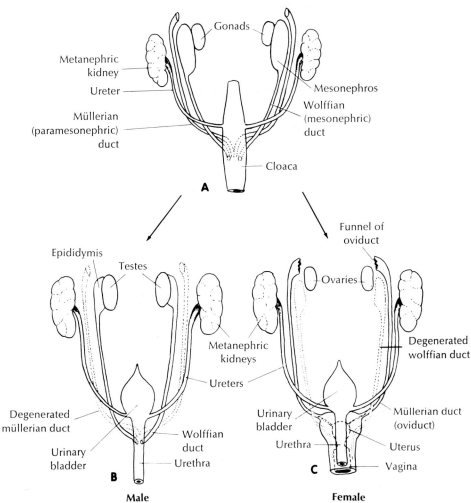

Sexually indifferent

Gonads

Metanephric kidney

Ureter

Müllerian (paramesonephric) duct

Mesonephros

Wolffian (mesonephric) duct

Cloaca

A

Epididymis

Testes

Funnel of oviduct

Ovaries

Metanephric kidneys

Degenerated wolffian duct

Ureters

Degenerated müllerian duct

Urinary bladder

Wolffian duct

Urethra

B

Male

Urinary bladder

Urethra

Müllerian duct (oviduct)

Uterus

Vagina

C

Female

Fig. 20-6. Development of male and female duct systems from the sexually indifferent stage at which precursors of both are present. (Modified from Balinsky, B. I. 1975. An introduction to embryology, 4th ed. W. B. Saunders Co., Philadelphia.)

the germ cells degenerate during the embryonic period, degeneration of the ovaries also occurs.

The sexually indifferent embryo, whether genetically male or female, contains two sets of primitive duct systems, the wolffian ducts, which develop originally as the duct system for the mesonephric kidneys (Chapter 18), and the müllerian or paramesonephric ducts, which are induced by the wolffian ducts and develop in parallel with them. During the course of sexual differentiation, one of the duct systems degenerates while the other develops into sex ducts and internal sex organs of the appropriate sex (Fig. 20-6). In the male the testes induce the wolffian ducts to differentiate into the vasa deferentia and seminal vesicles while the müllerian ducts degenerate in response to a different signal from the embryonic testes. In the female (and also in embryos lacking functional gonads) the müllerian ducts form the oviducts, the uterus, and the upper part of the vagina, and the wolffian ducts regress.

Differentiation of external genital structures is diagrammed in Fig. 20-7. The urinary and genital openings are derived from an embryonic cloaca, which before the end of the sexually indifferent stage becomes separated into the urogenital sinus and the rectum. At the sides of the sexually indifferent urogenital sinus are found the urogenital folds and the labioscrotal swellings. During the fourth week of gestation a genital tubercle develops ventral to the cloaca, which subsequently elongates to become the sexually indifferent phallus. During sexual differentiation in the male the urogenital sinus is closed by fusion of

Fig. 20-7. Diagrammatic representation of differentiation of external genitalia in human embryos. **A,** Cloaca at 4 weeks of gestation. Intestinal and genitourinary tracts are not yet separated. **B,** Urogenital sinus (closed by a thin membrane) at 7 weeks. Male and female are indistinguishable at this stage. **C** and **D,** Male and female at 9 weeks. Appearance remains similar except that the urogenital folds have begun to fuse in the male. Limited fusion of labioscrotal swellings in both sexes has increased the distance between the anus and urogenital sinus. **E** and **F,** Male and female at 11 weeks. In the male, fusion of the genital folds to yield an elongated urethra is complete except near the end of the penis. A cleft has formed in the glans penis for the urethra. Labioscrotal swellings have fused, forming the scrotum. **G** and **H,** Male and female at 12 weeks. Genital differentiation is essentially completed in both sexes. The male urethra extends to the tip of the penis. The prepuce (foreskin) is growing over the glans penis. A raphe (line of fusion) is clearly evident on both the scrotum and penis. In the female the urethra and vagina are fully separated. The genital tubercle has grown less rapidly than the surrounding tissue and has become the clitoris. The urogenital fold and labioscrotal swelling have become the labia minora (singular = labium minus) and the labia majora (singular = labium majus), respectively. (From Moore, K. L. 1973. The developing human — clinically oriented embryology. W. B. Saunders Co., Philadelphia.)

the urogenital folds and elongates to become the urethra, which serves as a common urinary and sexual duct. The urinary bladder is also derived from the urogenital sinus. In the female the urogenital folds do not fuse but rather become the labia minora. In addition, the urogenital sinus becomes divided into separate urinary and sexual ducts, the urethra and the vagina. In the male the labioscrotal swell-

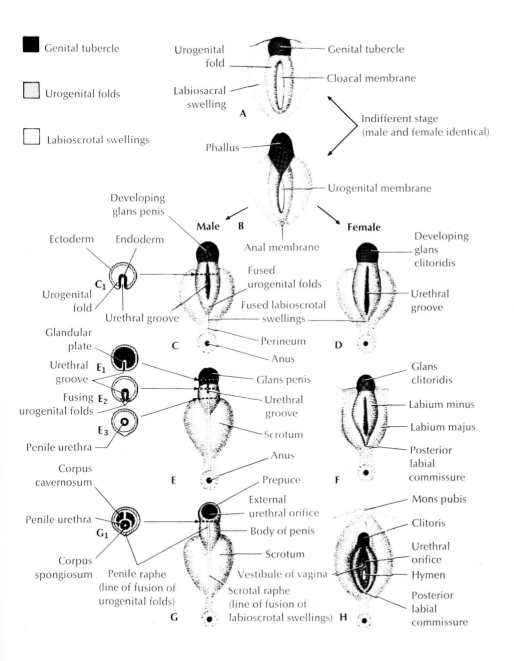

ings fuse to form the scrotum, into which the testes descend from the abdomen at a later stage. In the female the labioscrotal swellings remain separated as the labia majora. In the male the genital tubercle enlarges and elongates and, together with contributions from the urogenital folds, becomes the penis. In the female the genital tubercle remains relatively small and becomes the clitoris. In either case it contains the primary nerve centers for sexual stimulation and satisfaction. Intermediate degrees of enlargement of the genital tubercle and/or partial fusion of the genital folds are frequently encountered forms of intersexuality, as will be discussed.

By about 3 months of development the human embryo is phenotypically either male or female with regard to both internal and external structures. Little additional sexual differentiation then occurs until puberty, when secondary sexual characteristics develop in response to increased levels of steroid sex hormones. These include differences in whole-body growth, differences in relative growth of body parts that result in characteristic male or female body shape, differences in distribution of body hair, differences in larynx development, and functional maturation of previously formed reproductive organs. These changes are all caused by increased production of steroid sex hormones by the gonads and are therefore dependent on proper development of the gonads during embryonic life. It is not unusual, for example, for gonadal dysgenesis in an otherwise healthy phenotypic female to remain undetected until puberty fails to occur at the expected age.

Role of testicular hormones in determining sexual phenotype

The course of differentiation of the internal and external reproductive organs beyond the sexually indifferent stage is entirely dependent on the presence or absence of hormonal factors produced by the fetal testes. In the absence of these hormonal factors the reproductive system develops in a female pattern, regardless of the genetic sex of the embryo. Alfred Jost demonstrated many years ago that castration of male embryos at the sexually indifferent stage results in the development of animals that are phenotypically female with regard both to internal reproductive ducts and external genital structures, although, of course, they lack ovaries (Fig. 20-8). Removal of gonads from a female embryo does not in any way impede development of the rest of the female reproductive system. Thus female development requires no specific stimulus, but merely the absence of testicular factors, whereas male development is totally dependent on positive stimulation by the testicular factors. Such an arrangement makes sense for mammals, in which the embryo develops within the body of a female individual. If female hormones promoted female development, male embryos would need special protection from the effects of maternal

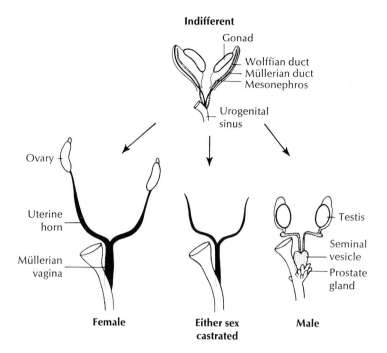

Indifferent

Gonad
Wolffian duct
Müllerian duct
Mesonephros

Urogenital sinus

Ovary

Uterine horn

Müllerian vagina

Female

Either sex castrated

Testis

Seminal vesicle

Prostate gland

Male

Fig. 20-8. Effect of castration on differentiation of the sex ducts in rabbit fetuses. Normal female and male development are shown on left and right. Removal of fetal gonads from either sex leads to a female pattern of duct development, as shown in center. Derivatives of the müllerian ducts are shown in solid black, and derivatives of the wolffian ducts are stippled. (Modified from Jost, A. 1960. Mem. Soc. Endocrinol. **7:**49.)

hormones to develop normally, since steroid sex hormones readily cross the placenta and are even produced by it.

At least two different types of hormonal substances from the fetal testis are needed for normal male development. One is testosterone, a steroid hormone with androgenic activity, which is synthesized by the interstitial (Leydig) cells found between the spermatogenic tubules. Testosterone is converted in target organs to a more active form (5α-dihydrotestosterone) that stimulates male development of the wolffian ducts and the external sex organs. The second hormonelike factor, which appears to be synthesized by the Sertoli cells of the testicular tubules, is a poorly characterized large protein, referred to as the müllerian-inhibiting hormone or antimüllerian factor, whose role is to promote degeneration of the müllerian duct. Preliminary data suggest that the müllerian-inhibiting hormone does not act directly on the epithelial cells of the müllerian duct but rather promotes a "strangling" of the duct by surrounding connective tissue fibroblasts, comparable to the pinching off of the mammary anlage in male mice described in Chapter 18.

Experiments in which castrated animal fetuses are treated with androgen plus experiments in which intact male fetuses are treated with androgen antagonists such as cyproterone acetate demonstrate the roles of the two factors clearly. When a castrated male fetus is treated with androgen, the wolffian duct undergoes development into the vas deferens and its associated structures. However, in the absence of müllerian-inhibiting hormone, the müllerian duct also continues to develop instead of degenerating. Such embryos therefore develop both male and female duct systems. On the other hand, when an intact male is treated with antiandrogenic compounds, the wolffian ducts either degenerate or develop only partially, depending on the species, and the müllerian ducts degenerate. Thus the male embryo treated with antiandrogen contains testes but does not have a fully developed reproductive tract of either sex.

The effect of testicular hormones on the duct systems tends to be localized. For example, if a piece of testis is grafted onto one ovary of a female embryo, the wolffian duct on that side of the animal develops and the müllerian duct regresses. However, normal female structures are formed on the other side of the animal (Fig. 20-9). In contrast, masculinization of the external genital structures is mediated by circulating androgen. Thus any exposure of a female fetus to circulating androgen (which will cross the placenta readily from the mother) typically will cause at least some degree of masculinization of the exter-

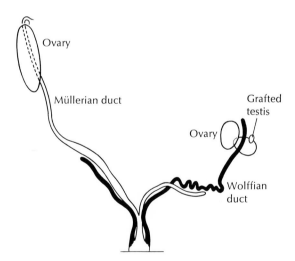

Fig. 20-9. Diagram illustrating localized effect of grafted testis tissue on development of sex ducts in a rabbit fetus. The testicular tissue has produced a localized stimulation of the wolffian duct, regression of the müllerian duct, and partial retention of the mesonephros (which makes a contribution to the epididymis in normal male development). Normal female development has occurred on the opposite side. A similar pattern of differentiation is sometimes observed in true hermaphrodites who have one ovary and one testis. (From Short, R. V. 1972. *In* C. R. Austin and R. V. Short, eds. Reproduction in mammals. Book 2. Embryonic and fetal development. Cambridge University Press, Cambridge.)

nal genitals without altering the female development of internal sex organs. Numerous cases of masculinization of female fetuses have resulted from ill-advised treatment of pregnant women with androgens or with drugs that have androgenic side effects (Fig. 20-10).

It is interesting to note that in the ZZ/ZW systems that have been studied, the heterogametic female gonad appears to have the dominant role in phenotypic sex differentiation. In chickens, for example, the ovary (which normally develops only on the left side) is the fastgrowing gonad whose hormones reverse the innate pattern of male sexual differentiation and actively promote female development. On the basis of data for a limited number of species, it currently appears that the heterogametic gonad may consistently have the dominant role in development, whether it is male or female, and that the homo-

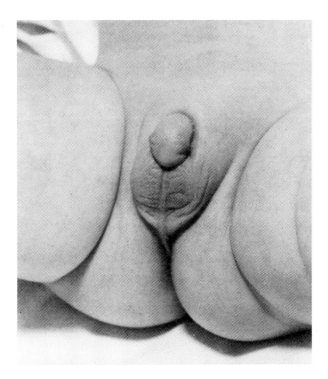

Fig. 20-10. External genitalia of a female infant masculinized by treatment of the mother with methyltestosterone begining about 2 months after conception. The drug was prescribed by a physician to treat alopecia (loss of hair). The labia are fused, and the clitoris is enlarged. The urogenital sinus opens under the clitoris. Similar masculinization has also been reported as a side effect of treatment with certain synthetic progesterone-like drugs to prevent spontaneous abortion. (From Jones, H. W., and Scott, W. W. 1958. Hermaphroditism, genital anomalies, and related endocrine disorders. The Williams & Wilkins Co., Baltimore.)

gametic sex may be the one that develops in the absence of specific gonadal signals. Thus we can speculate that the enigmatic relationship between the H-Y antigen and the heterogametic sex discussed earlier may in some way be related to a dominant role of the heterogametic sex.

Intersexuality

Substantial support for the theory that male development results from the combined actions of androgens and müllerian-inhibiting hormone and that female development occurs in their absence has been derived from studies of abnormalities of human sexual differentiation and the mechanisms that are responsible for them. The terminology used to describe individuals whose sexual differentiation is intermediate between male and female is based on their gonadal sex. The term "hermaphrodite" (or "true hermaphrodite") is used only to describe relatively rare individuals in which both ovarian and testicular tissue are found, either as one gonad of each type or, more commonly, in the form of ambiguous gonads containing both types of tissue. The ambiguity of ductal and genital structures that usually occurs in such cases is presumed to be due to incomplete müllerian duct inhibition and/or incomplete androgen-mediated masculinization resulting from intermediate amounts of testicular tissue. The term "pseudohermaphrodite" refers to individuals with well-defined male or female gonads but with other reproductive organs that are either ambiguous or of the opposite sex from the gonads. Pseudohermaphrodites are defined as "male" or "female" on the basis of their gonads and not on the basis of their external sexual phenotype. Pseudohermaphroditism occurs frequently, and several types that are described in this discussion have helped to confirm the role of testicular factors in sexual differentiation.

In particular, the testicular feminization syndrome *(Tfm)* provides spectacular confirmation of the role of androgens in male sexual differentiation. This form of male pseudohermaphroditism, which occurs both in humans and in mice, results from an X chromosome–linked mutation that renders normally androgen-sensitive target tissues totally unresponsive because of lack of functional androgen receptors. Afflicted individuals develop testes that secrete normal amounts of androgens, both in fetal life and at puberty. However, since their tissues are unable to respond to the androgens, their external sexual differentiation is strikingly female (Fig. 20-11). The male duct system is either absent or underdeveloped, since the wolffian ducts were unable to respond to androgen. However, since production of müllerian-inhibiting hormone by the fetal testis was normal, no female duct system is present either (except for a blind vagina derived from the urogenital sinus). External genital development is totally female, and at puberty enough estrogen is produced by the testes to support fully female secondary sexual differentiation (in the absence of the normally dominant effects of testicular androgens). Breast development and body contours are fully feminine, and except for failure to menstruate, infertility, and sparse or missing pubic and axillary (underarm) hair, these genetic males with undescended abdominal testes generally have no clues that they are not true females. Their psychological outlook is female, they function sexually as women (although corrective surgery is sometimes needed to lengthen a shallow vagina), and they make good adoptive mothers. In extreme cases the condition may not be fully diagnosed until a married "woman" visits a fertility clinic seeking medical advice concerning primary amenorrhea and infertility.

It has recently been demonstrated that total androgen insensitivity in humans can result from either of two distinct genetic defects. In some individuals, androgens fail to act on target cells because of a lack of binding to cellular receptors. However, in other individuals androgen binding and nuclear uptake and retention are normal de-

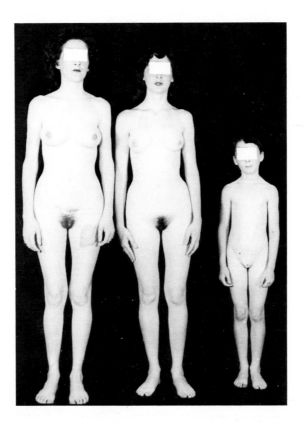

Fig. 20-11. Three siblings, ages, 25, 21, 8, with testicular feminization. All three had 46 XY karyotypes, inguinal or intra-abdominal testes, and blind, shallow vaginas. Except for the presence of more pubic hair than usual, the two older "sisters" exhibit the typical phenotype of sexually mature testicular feminization patients, with very feminine body contours, including well-developed breasts. The oldest has an abdominal scar from surgical removal of the testes, which is generally done after sexual maturation is completed because of the high risk of malignancy. Removal of the testes before puberty prevents development of female secondary sexual characteristics. (From Bartalos, M., and T. A. Baramki. 1967. Medical cytogenetics. The Williams & Wilkins Co., Baltimore.)

spite the absence of androgen-mediated responses. The defect in these latter individuals is thought to occur at the level of receptor-chromatin interaction. In theory, androgen insensitivity can occur at any of the levels described in Chapter 14 for glucocorticoid insensitivity in mutant leukocytes. There is also a less severe syndrome of partial androgen insensitivity in which the target cells lack an enzyme that converts testosterone into its more active derivative dihydrotestosterone. Afflicted individuals exhibit feminizing male pseudohermaphroditism, but not as extreme as in testicular feminization due to lack of functional androgen receptors.

Androgen insensitivity due to defective receptor molecules also occurs in mice, again caused by an X-linked gene. The effects are similar to those observed in human testicular feminization. The existence of androgen-insensitive experimental animals makes possible many types of recombination experiments in which it can be determined which tissue in an interacting pair actually responds to the hormone. We have already seen, for example, in Chapter 18, that degeneration of mammary epithelium in male mice is a secondary consequence of a mesodermal response to testosterone, rather than a direct response of the epithelial tissue. Similar studies on the differentiation of the urogenital sinus have shown that other epithelial responses to testosterone, such as the differentiation of the prostate gland, are also secondary results of mesodermal responses to testosterone rather than primary effects of testosterone action directly on the epithelial "target" tissue.

There is also a form of *female* pseudohermaphroditism that strongly illustrates the effects of androgens on sexual differentiation. It is known as congenital adrenal hyperplasia, and the primary genetic defect that is involved is an inability to produce sufficient amounts of the adrenocortical hormones known as glucocorticoids. The most frequent lesion is in production of an enzyme (21-hydroxylase) that catalyzes an essential step in the synthesis of cortisol and other glucocorticoids. In the absence of adequate production of cortisol, the anterior pituitary releases large amounts of corticotropin (ACTH) in an attempt to stimulate glucocorticoid synthesis. The adrenals enlarge in response to the excessive output of ACTH and produce abnormally large amounts of glucocorticoid precursors. The excess level of precursors (which cannot be converted to glucocorticoids because of the enzyme deficiency) stimulates enhanced production of androgenic steroids, which are normally produced only in small amounts as by-products by the adrenal cortex. These adrenal androgens react with androgen receptors, which are normally present both in male and female cells and exert a masculinizing effect.

Female fetuses with congenital adrenal hyperplasia develop a normal female reproductive tract internally, since no müllerian-inhibiting hormone is present and since there is no local effect of fetal testes on the wolffian ducts. However, they also exhibit varying degrees of masculinization of the external genitalia. In extreme cases this may be so complete that the afflicted individuals are mistaken for males with a normal penis and scrotum but undescended testes. More frequently, however, the afflicted individuals are correctly diagnosed during infancy because of the sexual ambiguity of their genitalia (Fig. 20-12). In most cases the problem can be successfully corrected by a combination of adrenal corticoid treatment and plastic surgery. Since internal reproductive organs are not affected, early treatment usually allows normal development as fertile females. As soon as adequate amounts of glucocorticoids are supplied from external sources, ACTH production by the anterior pituitary and the synthesis of adrenal androgens both fall to normal levels. It is important to start such therapy early in life to minimize masculinization and to permit the afflicted individual to acquire at puberty as feminine a set of secondary sexual characteristics as possible (Fig. 20-13). As long as the condition remains uncorrected, the individual is subjected not only to the direct masculinizing effects of the adrenal androgens but also to their suppressing effect on gonadotropins, which in turn prevents her ovaries from synthesizing normal amounts of female sex hormones.

Congenital adrenal hyperplasia also occurs in males, where its primary effects are rapid early growth, short adult stature due to early

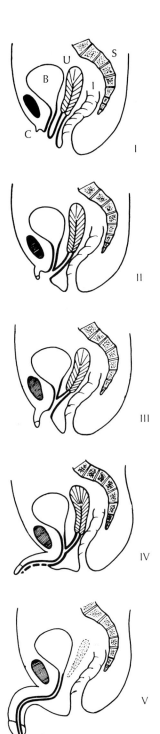

Fig. 20-12. Schematic medial sections illustrating progressive degrees of masculinization of the urogenital tract. Roman numerals refer to designations commonly used to describe the degrees of masculinization. **I,** Normal female urogenital openings with urethra and vagina separate to the surface. Varying degrees of enlargement of the clitoris may occur in mild masculinization, with no other changes. **II,** Enlarged clitoris and hypertrophied labia with some narrowing of the urogenital opening but with the urethra and vagina remaining separate. **III,** Greater enlargement and elongation of the phallus, with a narrow urogenital opening at the base of the phallus leading to both the bladder and the uterus. **IV,** Major elongation of the phallus to form a penislike structure, with the urethra incomplete and opening on the underside (hypospadias). The uterus is connected to the urethra by a narrow tube. The labia are fused to form a scrotum. **V,** Fully male external genitalia. The urethra extends to the tip of the penis. The uterus may or may not be present. In extreme cases of congenital adrenal hyperplasia this pattern is observed with the uterus present and with an empty scrotum. A similar pattern, with the uterus present and with testes in the scrotum, is observed in otherwise normal males whose müllerian ducts do not regress. B = Bladder; C = clitoris; I = intestine; S = spine; U = uterus. (With permission from Overzier, C. 1963. *In* C. Overzier, ed. Intersexuality. Academic Press Ltd., London. Copyright by Academic Press, Inc. [London] Ltd.)

closure of the epiphyses of the long bones, precocious puberty, and precocious maturation of the sex organs. Despite generalized virilization of affected males at a younger than normal age, their testes typically remain small and underdeveloped because of suppression of gonadotropin production. Treatment with glucocorticoids, particularly if it is started at an early age, reverses these effects and permits a more normal pattern of development.

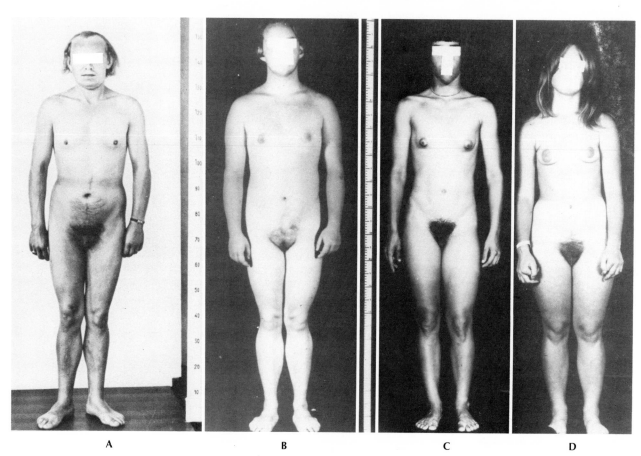

Fig. 20-13. Reversal of masculinization of 46XX females due to congenital adrenal hyperplasia by treatment with glucocorticoids. **A,** Extreme masculinization in an untreated 29-year-old XX female born with ambiguous genitalia. Note the virile physique, hairy body, "male" pattern baldness, and absence of breast development. **B** to **D,** Pubertal girls with congenital adrenal hyperplasia treated with glucocorticoids at progressively earlier ages. Patient in **B** was untreated until age 16; patient in **C** began treatment at age 9, and patient in **D** at age 4. Note progressively more feminine body development with earlier treatment. (**A** from Puschel, L., and H. Nowakowski. 1954. Arch. Ohr. Nas. Kehlk. hK. **166:**255; **B** to **D** from New, M. I., and L. S. Levine. 1973. Adv. Hum. Genet. **4:**251.)

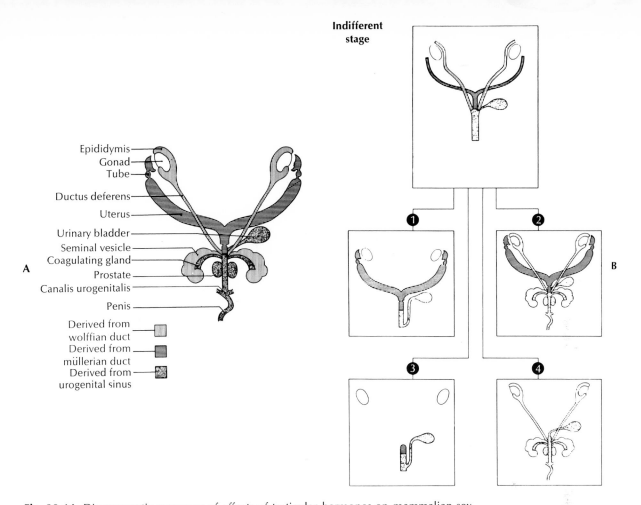

Indifferent stage

Epididymis
Gonad
Tube
Ductus deferens
Uterus
Urinary bladder
Seminal vesicle
Coagulating gland
Prostate
Canalis urogenitalis
Penis

Derived from wolffian duct
Derived from müllerian duct
Derived from urogenital sinus

A

B

Fig. 20-14. Diagrammatic summary of effects of testicular hormones on mammalian sex differentiation. **A,** Composite diagram showing most of the structures of both sexes as they are depicted in the individual diagrams in **B.** These diagrams are based on studies of dog embryos, but except for minor variations such as the size and shape of the uterus and seminal vesicles are also applicable to other mammals. This diagram is the same as the second case represented in **B. B,** Effects of the two testicular hormones that influence sexual differentiation. *1,* Neither hormone active. Derivatives of the wolffian duct are absent. Müllerian ducts have formed the uterus and fallopian tubes. External genitalia are female, with the urethra and vagina separated. This is the normal female pattern of development. *2,* Müllerian-inhibiting hormone inactive, testosterone active. The müllerian duct has formed the uterus and fallopian tubes. The wolffian duct has formed the male duct system. The external genitalia are male, with an elongated urethra that also serves as a sex duct. This pattern may be observed in female or castrated animals that have been completely masculinized by large doses of androgens. It also occurs in males due to failure of müllerian-inhibiting hormone. Variations, often with lateral asymmetry, are also seen in true hermaphrodites. *3,* Müllerian-inhibiting hormone active, testosterone inactive. All müllerian and wolffian duct derivatives are missing. The urogenital sinus has formed a blind vagina and a separate urethra. The external genitalia are female. This pattern is seen in androgen-insensitive males (testicular feminization) and in males treated with androgen antagonists. *4,* Both hormones active. All müllerian duct derivatives are missing. The wolffian ducts have formed the male duct system. The external genitalia are male, with an elongated urethra that also serves as a sex duct. This is the normal male pattern of development. (Modified from Neumann, F., et al. 1975. *In* R. Reinboth, ed. Intersexuality in the animal kingdom. Springer-Verlag, Inc., New York.)

There are also clinical cases known in which the antimüllerian hormone apparently fails to suppress development of derivatives of the müllerian ducts in otherwise normal males. Such defects are normally corrected by surgical removal of the resulting uterus and fallopian tubes, which are frequently found in hernias attached to partially or fully descended testes.

In summary, the difference between male and female differentiation of the urogenital tract lies in the effects of two hormonal activities of the fetal testis—suppression of development of müllerian duct derivatives by the müllerian-inhibiting hormone and stimulation of development of male internal and external structures by androgenic steroids. These relationships are summarized in Fig. 20-14.

SEXUAL DIFFERENTIATION OF THE BRAIN

An additional aspect of sex differentiation that differs between males and females is what has been termed "brain sex." Normal male gamete production and normal output of steroids by male gonads both require that the hypothalamus stimulate the pituitary to release gonadotropins in an essentially continuous manner. Normal gonadal activity in the female, in contrast, involves cyclic stimulation of pituitary gonadotropin output. In rodents the male pattern of gonadotropin secretion appears to develop in response to androgens. Administration of androgen to newborn females results in the development of a continuous pattern of gonadotropin secretion. In addition, newborn males that are castrated or treated with antiandrogen develop the cyclic female pattern of gonadotropin secretion. However, the mechanism is not as simple as it might seem, since estrogen administration is also able to masculinize the hypothalamus of newborn females or castrated males. The hypothalamus contains aromatizing enzymes that can convert testosterone to estradiol, and such a conversion may normally be involved in the sexual imprinting of the brain. Support for this possibility is derived from the fact that *Tfm* mice, which lack androgen receptors, but possess normal estrogen receptors and are able to convert testosterone to estradiol, develop a male pattern of gonadotropin secretion. Sexual differentiation of the hypothalamus may occur by a different mechanism in primates, since neither prenatal nor postnatal androgen treatment alters the pattern of pituitary secretion in these animals. Likewise, human females with adrenal hyperplasia do not develop a male pattern of gonadotropin secretion.

Male sexual behavior in rodents has also been shown to be related to neonatal brain imprinting. Administration of an antiandrogen to newborn male mice prevents the expression of male sexual behavior in the adult. Administration of androgen to newborn females renders them capable of expressing male sexual behavior in adulthood in response to administered androgen. Again, it is unclear whether androgen or estrogen derived from androgen is the immediate imprinting agent. Imprinting of the brain with regard either to sexual behavior or to hypothalamic control of gonadotropin secretion occurs in rodents only during a critical period soon after birth; changes in male and female patterns cannot be induced in adult animals.

"SIMPLE" CONTROLS OVER COMPLEX CHANGES

It is obvious from the material in this chapter that the details of sex determination and sex differentiation involve many highly complex and precisely integrated mechanisms. However, it does not necessarily follow that the basic regulatory signals that control these complex processes must be equally complex. It has been recognized for a number of years that there are two primary points of intervention where an inherently female pattern of development can be disrupted to yield an induced male pattern: (1) induction of a male gonad in the presence of a Y chromosome and (2) induction of accessory male sexual structures in response to testicular hormones (Fig. 20-15, A). Ohno has recently suggested that only two basic regulatory systems are involved and that the regulatory molecules involved in both are now known (Fig. 20-15, B). In agreement with earlier investigators, he assumes that the developmental program for all mammals, including humans, is innately female in the absence of regulatory intervention. A two-step cascade of regulatory control is required for male development. The first step is reorganization of the primitive gonad into a testis, which he considers to be mediated by H-Y antigen. The second step is the switching of all sexually dichotomous developmental pathways from the female program to the male program by testicular hormones. These effects are mediated by testosterone, as shown in the diagram, and also by the müllerian-inhibiting factor, which is not shown.

Thus, although the total differences between male and female are many, it appears that the change from one basic program of development to the other can be achieved rather simply. However, for such a switching system to work effectively, it is necessary for each responding tissue to have built into it two different developmental patterns—one that leads to female structure and function in the absence of a diversionary signal and a second that leads to male structure and function when the signal is received. Thus, although the switches themselves may be simple, the circuits that they are connected to still appear to be highly complex.

In the following chapter we will examine another complex morphogenetic system, the vertebrate limb, whose development is also influenced, and perhaps totally controlled, by a limited number of seemingly rather simple regulatory systems. Although the end results are dif-

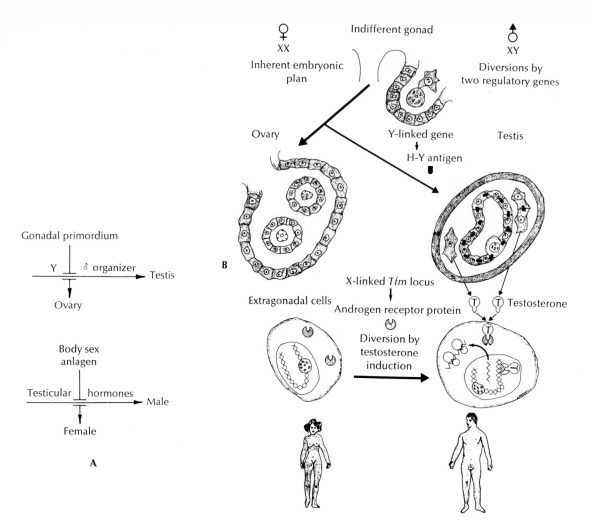

Fig. 20-15. Models for control of mammalian sexual differentiation. **A,** "Asymmetrical" model of A. Jost. Without intervention the normal course of mammalian sex differentiation is entirely in the female direction (represented by vertical lines) both at the gonadal and body levels. The presence of a Y chromosome intervenes to cause formation of a testis rather than an ovary. Testicular hormones in turn cause the body to develop in a male direction. The control system is asymmetrical in the sense that all specific control mechanisms promote male development. There are no comparable mechanisms that actively promote female development. **B,** Model of S. Ohno, in which specific regulatory molecules are identified. H-Y antigen, controlled by a Y chromosome–linked gene, causes the primitive gonads to become testes, whose interstitial cells produce testosterone. All cells (male or female) contain testosterone receptor, coded for by the wild-type allele of the *Tfm* locus. Acting through such receptors, testosterone promotes the development of extragonadal masculine characteristics and thus diverts the organism from its inherent female pattern of development. Although not shown in this diagram for simplicity, the effect of testosterone may be amplified by its conversion in target tissues into 5-α-dihydrotestosterone, which has a higher specific activity. (**A** from Jost, A., et al. 1973. Recent Prog. Horm. Res. **29:**1; **B** from Ohno, S. 1977. J. Steroid Biochem. **8:**585. Reprinted with permission from Pergamon Press, Ltd.)

ferent, many of the detailed mechanisms that are involved in limb development are similar, at least in principle, to those involved in sexual differentiation. The major differences lie in the specific control signals that are involved and in the nature of the overall developmental programs that they regulate.

BIBLIOGRAPHY
Books and reviews

Balinsky, B. I. 1975. An introduction to embryology, 4th ed. W. B. Saunders Co., Philadelphia.

Cunha, G. R. 1976. Epithelial-stromal interactions in development of the urogenital tract. Int. Rev. Cytol. **47**:137.

Editorial. 1974. What becomes of the XYY male? Lancet **2**:1297.

Erickson, R. P. 1977. Androgen-modified expression compared with Y linkage of male specific antigen. Nature **265**:59.

Fredga, K. 1970. Unusual sex chromosome inheritance in mammals. Philos. Trans. R. Soc. Lond. (Biol.) **259**:15.

Gerald, P. S. 1976. Current concepts in genetics: sex chromosome disorders. N. Engl. J. Med. **294**:706.

Harris, G. W., and R. G. Edwards, organizers. 1970. A discussion on determination of sex. Philos. Trans. R. Soc. Lond. (Biol.) **259**:1.

Hoffman, B. F. 1977. Two new cases of XYY chromosome complement and a review of the literature. Can. Psychiatr. Assoc. J. **22**:447.

Josso, N., J.-Y. Picard, and D. Tran. 1977. The antimüllerian hormone. Recent Prog. Horm. Res. **33**:117.

Jost, A., et al. 1973. Studies on sex differentiation in mammals. Recent Prog. Horm. Res. **29**:1.

McEwen, B., et al. 1977. Do estrogen receptors play a role in the sexual differentiation of the rat brain? J. Steroid Biochem. **8**:593.

Mittwoch, U. 1973. Genetics of sex differentiation. Academic Press, Inc., New York.

Mittwoch, U. 1977. H-Y antigen and the growth of the dominant gonad. J. Med. Genet. **14**:335.

New, M. I., and L. S. Levine. 1973. Congenital adrenal hyperplasia. Adv. Hum. Genet. **4**:251.

Northcutt, R. C., and D. O. Toft. 1975. Testicular feminization syndrome: a model of chemical information non-transfer. In J. G. Vassileva-Popova, ed. Physical and chemical bases of biological information transfer. Plenum Publishing Corp., New York.

Ohno, S. 1967. Sex chromosomes and sex-linked genes. Springer-Verlag, Inc., New York.

Ohno, S. 1976. Major regulatory genes for mammalian sexual development. Cell **7**:315.

Ohno, S. 1977. The Y-linked H-Y antigen locus and the X-linked *Tfm* locus as major regulatory genes of the mammalian sex determining mechanism. J. Steroid Biochem. **8**:585.

Ohno, S. 1978. The role of H-Y antigen in primary sex determination. J.A.M.A. **239**:217.

Overzier, C., ed. 1963. Intersexuality. Academic Press, Inc., New York.

Reinboth, R. ed. 1975. Intersexuality in the animal kingdom. Springer-Verlag, Inc., New York.

Sawyer, C. H., and R. S. Gorski, eds. 1971. Steroid hormones and brain function. University of California Press, Berkeley, Calif.

Short, R. V. 1972. Sex determination and differentiation. In C. R. Austin and R. V. Short, eds. Reproduction in mammals. Book 2. Embryonic and fetal development. Cambridge University Press, Cambridge.

Silvers, W. K., and S. S. Wachtel. 1977. H-Y antigen: behavior and function. Science **195**:956.

Wachtel, S. S. 1977. H-Y antigen and the genetics of sex determination. Science **198**:797.

Selected original research articles

Amrhein, J. A., et al. 1976. Androgen insensitivity in man: evidence for genetic heterogeneity. Proc. Natl. Acad. Sci. U.S.A. **73**:891.

Attardi, B., and S. Ohno. 1974. Cytosol androgen receptor from kidney of normal and testicular feminized (Tfm) mice. Cell 2:205.

Bennett, D., et al. 1977. Serological evidence for H-Y antigen in *Sxr*, XX sex-reversed phenotypic males. Nature 265: 255.

Beutler, B., et al. 1978. The HLA-dependent expression of testis-organizing H-Y antigen by human male cells. Cell 13: 509.

Fredga, K., et al. 1977. A hypothesis explaining the exceptional sex ratio in the wood lemming *(Myopus schisticolor)*. Hereditas 85:101.

Fricke, H., and S. Fricke. 1977. Monogamy and sex change by aggressive dominance in coral reef fish. Nature 266:830.

Geyer-Duszynska, I. 1959. Experimental research on chromosome elimination in cecidomyidae (Diptera). J. Exp. Zool. 141:391.

Lyon, M. F., and S. G. Hawkes. 1970. X-linked gene for testicular feminization in the mouse. Nature 227:1217.

Mittwoch, U., and M. L. Buehr. 1973. Gonadal growth in embryos of *Sex reversed* mice. Differentiation 1:219.

Mullen, R. J., and W. K. Whitten. 1971. Relationship of genotype and degree of chimerism in coat color to sex ratios and gametogenesis in chimeric mice. J. Exp. Zool. 178:165.

Nagai, Y., and S. Ohno. 1977. Testis-determining H-Y antigen in XO males of the mole-vole *(Ellobius lutescens)*. Cell 10: 729.

Ohno, S., et al. 1974. *Tfm* mutation and masculinization vs. feminization of the mouse central nervous system. Cell 3: 235.

Ohno, S., et al. 1976. Hormone-like role of H-Y antigen in bovine freemartin gonad. Nature 261:597.

Ohno, S., et al. 1978. Testicular cells lyso-stripped of H-Y antigen organize ovarian follicle-like aggregates. Cytogenet. Cell Genet. 20:351.

Zenzes, M. T., et al. 1978. Studies on the function of H-Y antigen: dissociation and reorganization experiments on rat gonadal tissue. Cytogenet. Cell Genet. 20:365.

CHAPTER 21

Limb development in the chick

☐ The cellular interactions involved in the development of the chick limb have been a subject of intense experimental interest for some time. One of the reasons for the popularity of this system is simply its technical advantages. The limb bud of many organisms is able to withstand considerable surgical manipulation. In the case of the chick, the embryo rests on the yolk with its right wing bud exposed, and it can be treated in situ with little disturbance of normal embryonic development. In addition, the untreated left wing bud serves as a convenient control. In this chapter we will focus on two primary aspects of vertebrate limb development—basic limb outgrowth and the development of limb asymmetry. The role of cell death in limb development was discussed in Chapter 18.

A large number and variety of inductive interactions have been shown to operate in limb development in the chick. The basic induction of limb bud outgrowth from the body wall is the result of an instructive type of interaction (Chapter 13) between mesoderm in the wing-forming region and the overlying ectoderm. The presumptive limb bud region, whose location is determined by its relationship to specific somites, can be removed from the early chick embryo and then dissociated into ectoderm and mesoderm. Dissociation by means of trypsin produces a healthy ectodermal layer (or ectodermal shell in the case of limb buds), and dissociation with ethylenediaminetetraacetate (EDTA, a metal-chelating agent) allows recovery of a healthy mesodermal layer (or core from a limb bud). When mesoderm of the presumptive wing bud is transplanted beneath the ectoderm of a second embryo, it induces the formation of a secondary wing bud in the host embryo. (This experiment is shown in Fig. 12-3, which demonstrates that the fate of the wing-forming area is determined prior to visible formation of a wing bud.) When ectoderm, rather than mesoderm, from the wing-forming region is transplanted, no secondary wing formation occurs. Similar results are seen with the meso-

LIMB OUTGROWTH

643

Fig. 21-1. AER. **A,** Scanning electron micrograph of AER of a chick left wing bud. **B,** Diagrammatic cross section of a 72-hour chick embryo at wing-bud level showing location of AER. **C,** Micrograph of tip of wing bud from chick embryo slightly older than that in **A,** showing morphology of cells in AER and underlying mesoderm. (**A** courtesy K. W. Tasney; from Wessells, N. K. 1977. Tissue interactions in development. The Benjamin/Cummings Publishing Co., Menlo Park, Calif.; **B** from Balinsky, B. I. 1975. An introduction to embryology, W. B. Saunders Co., Philadelphia; **C** from Saunders, J. W. 1948. J. Exp. Zool; **108:**363.)

A

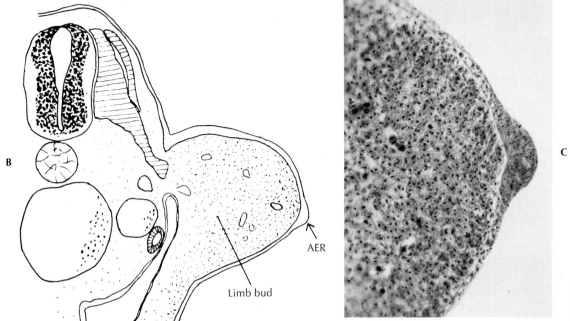

B

AER

Limb bud

C

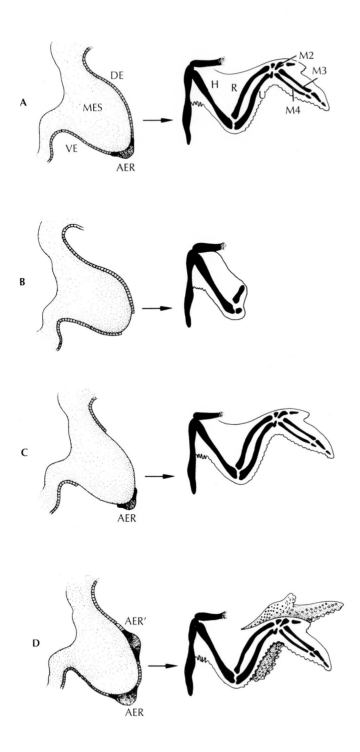

Fig. 21-2. Schematic diagrams showing the influence of the AER on limb outgrowth. In each lettered pair, the left-hand drawing is a schematic cross section of a wing bud comparable in its orientation to Fig. 21-1, *B,* and the right-hand drawing is a dorsal view of the right wing of an older embryo with the tissue partially cleared and the developing bones selectively stained. **A,** Normal limb development in the presence of AER. **B,** Removal of the AER causes cessation of limb outgrowth and lack of terminal limb structures. **C,** Removal of ectoderm from regions of the limb bud other than the AER does not disrupt normal development. **D,** Grafting of the second AER onto the limb bud results in duplication of limb structures distal to the point of graft attachment. *DE* = dorsal ectoderm; *VE* = ventral ectoderm; *MES* = mesoderm; *H* = humerus; *R* = radius; *U* = ulna; *M2, M3,* and *M4* = metacarpals of digits 2, 3, and 4. (Reprinted with permission of Macmillan Publishing Co., Inc. from Patterns and principles of animal development by Saunders, J. W., Jr. Copyright © 1970 by John W. Saunders, Jr.)

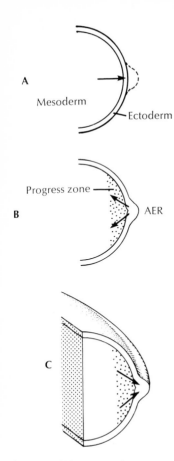

Fig. 21-3. Diagrammatic representation of reciprocal interactions occurring between the mesoderm and ectoderm during limb development. **A,** Mesoderm induces formation of AER. **B,** AER induces growth and differentiation of limb mesoderm in the "progress zone," whose role in limb morphogenesis is explained later in this chapter. **C,** Stimulated mesoderm produces ridge maintenance factor.

DEVELOPMENT OF HANDEDNESS IN THE LIMB

derm and ectoderm of the leg bud. In both cases it is the mesoderm that induces limb bud formation and determines the type of structure formed (wing or leg).

Further outgrowth of the limb involves complex reciprocal interactions between limb mesoderm and a defined region of limb ectoderm known as the apical ectodermal ridge (AER), whose formation is induced by the limb-forming mesoderm. The AER is a ridge of thickened ectodermal cells at the distal end of the limb bud (Fig. 21-1, *A*) that is seen as a nipple-shaped cap in transverse sections (Fig. 21-1, *B* and *C*). If this cap of cells is removed at any stage of limb development, limb outgrowth ceases at that point. The part of the limb that was laid down prior to removal of the AER undergoes normal differentiation, but all distal limb parts fail to develop. In addition, if a second AER is grafted onto a limb bud, one obtains secondary limb formation in which all structures distal to the point of attachment of the second AER are duplicated. If ectoderm is removed from a different region of the limb, the limb develops normally with no loss of distal structures (Fig. 21-2).

The influence of the AER on limb development is not a one-way interaction from ectoderm to mesoderm, however. For the AER to induce outgrowth and differentiation of limb mesoderm, it requires a stimulus from the limb mesoderm. This has been demonstrated by experiments done with the wingless mutant in the chick. Wingless is a recessive mutation, which in homozygous condition results in a failure of wing production. When the wing bud of a wingless embryo is examined, it is seen that the AER is present early on the third day of development but then regresses. If limb bud mesoderm from such an embryo is combined with limb bud ectoderm of a normal embryo, the AER of the normal embryo also undergoes regression. Thus the wingless mutant appears to be deficient in some mesodermal factor that is essential for the maintenance of the AER. This factor has been named apical ectodermal ridge maintenance factor (AERMF). A factor that is capable of maintaining the AER in vitro has been partially purified recently from cell-free preparations of posterior halves of chick limb buds, but it is unclear if the factor corresponds directly to AERMF.

It is apparent from these results that at least three types of ectodermal-mesodermal interactions are required for limb outgrowth. First, mesoderm initiates outgrowth of the limb bud and formation of the AER. Second, the AER stimulates further outgrowth and differentiation of limb bud mesoderm. Third, limb bud mesoderm provides a stimulus essential for the maintenance of the AER (Fig. 21-3).

The orthogonal coordinate system described in Chapter 2 and Fig. 2-17 is used to describe locations within vertebrate limbs. The limb is

A

Anterior

Ventral

Proximal

Distal

Dorsal

Posterior

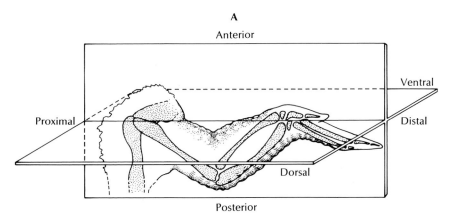

B

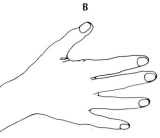

Right hand, dorsal view

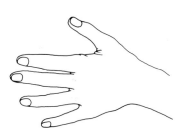

Left hand, dorsal view,
proximal-distal axis reversed

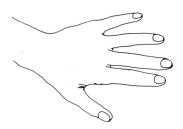

Left hand, dorsal view,
cranial-caudal axis reversed

Left hand, ventral view,
dorsal-ventral axis reversed

Fig. 21-4. A, Schematic representation of the right wing of a chicken embryo viewed from the dorsal aspect, showing the three axes of the basic vertebrate limb and their asymmetry. **B,** Human right hand viewed from the dorsal aspect and demonstration that reversing any one of the three coordinate axes converts it to a left hand.

assumed to be attached to the lateral part of the trunk and extended straight sideways. (Normal bending of the limb to a more ventral position does not alter the terminology.) The names that are most commonly used for the three sets of coordinates are proximal-distal, dorsal-ventral, and anterior-posterior (cranial-caudal in human limb development).

Since the paired limbs are normally asymmetrical with regard to all three axes (Fig. 21-4, *A*), they exhibit handedness (mirror image asymmetry). Reversal of any one of the three axes converts a right-handed limb into a left-handed one and vice versa (Fig. 21-4, *B*). One can verify this by comparing one's own left and right hands. The two hands can be aligned so that any two of the three axes are the same, but the third will always be opposite for the two hands.

The terms "anterior" and "posterior" are widely used in the experimental literature to describe the cranial and caudal regions of limbs in experimental animals such as the chicken embryo. For consistency, they will also be used in this chapter, but with the reservation that they must be changed to "cranial" and "caudal" when used to describe human limb development (p. 34). The terms "preaxial" and "postaxial" are often used to describe the portions of developing limbs that are located anterior (cranial) and posterior (caudal) from the long axis of the developing limb, respectively.

As described earlier (and in greater detail later in this chapter) differentiation along the proximal-distal axis is established by the pro-

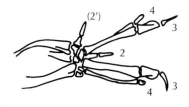

Fig. 21-5. Schematic drawing showing digit formation by a chick right wing bud following 180-degree rotation of the tip. Duplicated wing tips are produced in mirror-image symmetry. Because of the 180-degree rotation of the tip after the dorsal-ventral axis is established, the ventral aspect of the digits is seen in this view. The upper set is right handed and the lower set is left handed. A single digit 2 is shared by both. (Digit 2′ developed from stump tissue.) (From Saunders, J. W., and M. T. Gasseling. 1968. *In* P. Fleischmajer and R. E. Billingham, eds. Epithelial-mesenchymal interactions. The Williams & Wilkins Co., Baltimore.)

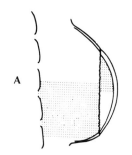

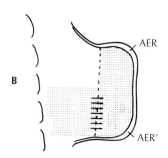

Fig. 21-6. Diagram of chick limb bud showing the postulated distribution of AERMF (stippled region). **A,** Immediately after 180-degree rotation of the tip. **B,** After transmission of AERMF from the stump to the rotated tip. AER′ = Secondary AER. (From Saunders, J. W., and M. T. Gasseling, 1968. *In* P. Fleischmajer and R. E. Billingham, eds. Epithelial-mesenchymal interactions. The Williams & Wilkins Co., Baltimore.)

gressive outgrowth of the limb bud that results from the inductive interactions between the AER and the limb mesoderm. The basis of dorsal-ventral differentiation is not yet known. However, the polarity of this axis becomes stably determined early in limb development and is generally not altered by experimental manipulation.

Polarity along the anterior-posterior axis of the developing limb bud is of particular interest because it is the last determinant of handedness to be established during development, as well as the easiest to modify by experimental manipulation. Interest in the mechanisms that establish anterior-posterior polarity was stimulated by an experiment in which the entire tip of a wing bud was severed, rotated 180 degrees, and then reattached to its stump. This process resulted in duplication of all distal wing parts, and, in addition, the duplicated parts developed as mirror images (i.e., with one right-handed set of digits and one left-handed set) (Fig. 21-5).

Illustrations that are used to demonstrate changes in handedness of embryonic chicken wings tend to be confusing to the uninitiated reader, but much of that confusion can be avoided by remembering two basic rules: (1) it is usually the right wing that has been operated on and (2) the diagram that is shown is normally a dorsal view (relative to the embryo) of the right wing with proximal structures to the left and distal to the right. In the unoperated control the dorsal surface of the wing faces the viewer, with anterior toward the top of the page and posterior toward the bottom. In cases of experimental manipulation the dorsal-ventral or anterior-posterior axes of distal parts of the wing may be reversed.

The "digits" of a chicken wing are generally labeled from anterior to posterior with the numbers 2, 3, and 4 on the basis of developmental homology with mammalian digits. All three of these digits form separate bones, but digits 3 and 4 remain fused with no cleft between them (Fig. 21-2, *A*). The digits normally develop from the postaxial half of the wing bud.

When the tip of a right wing bud is rotated 180 degrees, the part of the tip that was originally posterior, but which is now located in an anterior position, retains its original asymmetry and produces right-handed digits. (Its dorsal-ventral and anterior-posterior axes are both reversed, but its handedness is unchanged, since two axes have been reversed.) The part of the tip that was originally anterior (and not destined to produce digits), but which is now in a posterior position, also produces a set of digits, but these have left-handed asymmetry. Direct cell contact does not seem to be required for these effects, since similar results are obtained when a Millipore filter is placed between the tip and the stump. The effects are prevented, however, by placement of an impermeable film of mylar plastic between stump and tip.

The interpretation of this experiment is more easily understood if the results of wing tip rotation are considered as two separate phenomena—the induction of supernumerary digits and the determination of their asymmetry. The first phenomenon, digit duplication, is thought to be due to the influence of AERMF from the posterior stump region. AERMF is believed to be distributed asymmetrically in limb mesoderm with the greatest concentration in the posterior region (Fig. 21-6, A). This was originally inferred from the fact that the AER is normally thickest in the posterior region and from the fact that the posterior region normally undergoes the greatest degree of outgrowth. Indeed, after the rotation of the wing tip the ectoderm of the originally anterior tip becomes thicker, presumably in response to the increased exposure to AERMF in its new location (Fig. 21-6, B). The factor that is able to maintain the AER in vitro has been detected in both the central and posterior regions of the limb bud. However, it has been shown recently that when an impermeable barrier of mylar film is placed between the center of the limb and its posterior border, the activity normally found in the center of the limb bud disappears. This suggests that the factor may be produced by cells at the posterior border of the limb and then distributed anteriorly through the limb tissue.

The second phenomenon, the determination of asymmetry of the duplicated digits, has been interpreted to occur as follows. Anterior-posterior asymmetry is induced in the posterior region of the limb tip prior to its rotation by a stimulus from the posterior region of the stump. The digits formed by the originally posterior region of the tip develop accordingly—2-3-4 in the originally anterior to posterior direction, which becomes 4-3-2 from anterior to posterior after the tip is rotated. Since both the anterior-posterior and the dorsal-ventral axes are turned 180 degrees during the rotation, the normal right-handed asymmetry is preserved. The originally anterior region of the limb tip comes under the polarizing influence of the posterior mesoderm following rotation, and the digits it produces in response to AERMF develop in a normal anterior-to-posterior sequence (2-3-4). These digits

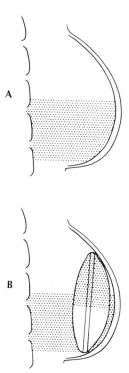

Fig. 21-7. A, Diagram of a normal right chick wing bud. **B,** Similar bud after grafting of the rotated tip of a second wing bud to its dorsal side. Stippling indicates the presumed distribution of AERMF in host and graft tissue at the time of the operation. (From Saunders, J. W., and M. T. Gasseling, 1968. *In* P. Fleischmajer and R. E. Billingham, eds. Epithelial-mesenchymal interactions. The Williams & Wilkins Co., Baltimore.)

show left-handed asymmetry, however, since the dorsal-ventral axis was reversed by rotation of the tip. In the example shown in Fig. 21-5, only a single digit 2 has formed between the "hands," resulting in a 4-3-2-3-4 pattern.

An even more complex set of interactions is observed when a severed wing tip is rotated 180 degrees and grafted to the dorsal side of an intact wing bud (Fig. 21-7). Following this procedure the originally

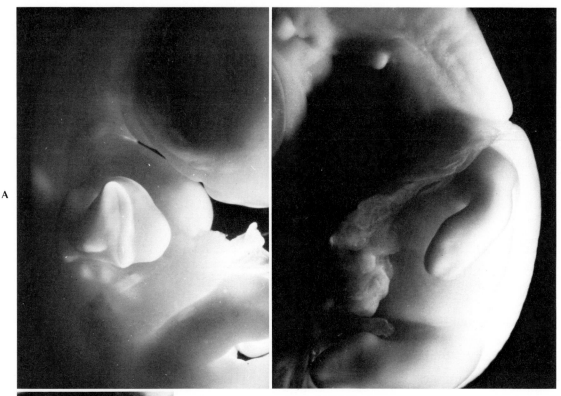

Fig. 21-8. A, Right wing of chick embryo approximately 20 hours after the operation illustrated in Fig. 21-7. **B,** Control left wing of the same embryo. **C,** End-on view of the right wing of an embryo similar to that in **A** sacrificed at 11 days of incubation. The wing contains four sets of hand parts, each the mirror twin of its neighbor anteroposteriorly and dorsoventrally. (From Saunders, J. W., and M. T. Gasseling. 1968. *In* P. Fleischmajer and R. E. Billingham, eds. Epithelial-mesenchymal interactions. The Williams & Wilkins Co., Baltimore.)

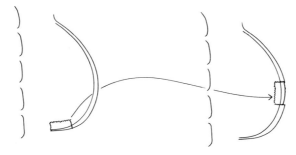

Fig. 21-9. Diagram showing transplantation of mesoderm from the PNZ region of one wing bud to a position beneath the AER at the distal end of a second wing bud. (From Saunders, J. W., and M. T. Gasseling. 1968. *In* P. Fleischmajer and R. E. Billingham, eds. Epithelial-mesenchymal interactions. The Williams & Wilkins Co., Baltimore.)

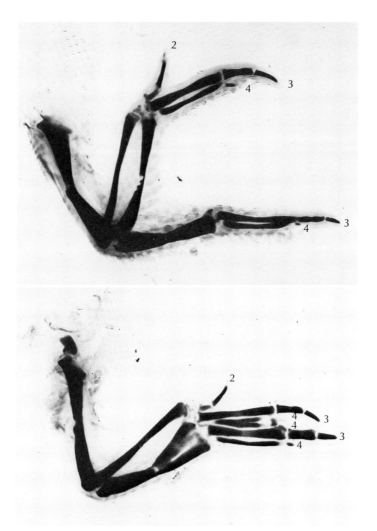

Fig. 21-10. Right wings of 11-day chick embryos showing duplications of distal wing parts as a result of implanting mesoderm from the PNZ region at the apex of the wing bud. (From Saunders, J. W., and M. T. Gasseling. 1968. *In* P. Fleischmajer and R. E. Billingham, eds. Epithelial-mesenchymal interactions. The Williams & Wilkins Co., Baltimore.)

anterior region of the grafted tip comes under the influence of the posterior region of the host, and the anterior region of the host comes under the influence of the originally posterior region of the graft. In some cases four sets of digits are produced. The originally posterior regions of both host and graft develop with right-handed asymmetry, whereas the originally anterior regions develop in a mirror-image pattern and show left-handed asymmetry (Fig. 21-8).

The origin of the mesodermal influence that determines anterior-posterior limb asymmetry became clearer during the course of experiments designed to assess the ability of mesoderm from the posterior necrotic zone (PNZ; see Chapter 18) to maintain the AER. Mesoderm from the PNZ was grafted beneath the AER at the distalmost part of the wing bud (Fig. 21-9). The AER adjacent to the graft flattened, but the AER anterior to the graft thickened and eventually produced supernumerary wing parts. In this case the extra wing parts that developed displayed right-handed asymmetry, the same as those of the host. The normal complement of three digits usually developed in the anterior set of digits, but the posterior set typically had only two digits and was lacking the anterior digit. Thus the order of digits, anterior to posterior, was 2-3-4-3-4. An occasional animal showed duplication of the fourth digit in the medial position to give the pattern 2-3-4-4-3-4, anterior to posterior (Fig. 21-10). If grafts from other regions of the wing bud were transplanted to the apex, no duplications occurred. Only mesodermal grafts from the PNZ and its adjacent area were able to cause duplication of wing parts (Fig. 21-11).

When grafts were done in which mesoderm from the leg bud was transplanted to the apex of the wing bud, similar results were obtained. Duplications were produced only after mesoderm was transplanted from the region of the leg bud whose location was homologous to the PNZ of the wing bud. Because this region of the leg bud is not destined to undergo necrosis, these results demonstrated that the ability of this region of mesoderm to induce duplications does not depend on its necrotic fate. The posterior region of mesoderm that determines anterior-posterior asymmetry does not appear to be identical with the PNZ in regard to location and activity, and it has been given its own title, the zone of polarizing activity (ZPA). It is not yet clear, however, whether the AERMF and the biological activity produced by the ZPA are the same or distinct from each other.

Another set of experiments was done in which the ZPA region was grafted to the anterior region of the limb bud. In this case the duplicated limb parts displayed asymmetry opposite to that of the host parts. For example, digits developed in the order 4-3-2-3-4, anterior to posterior (Fig. 21-12). Further experiments in which the ZPA was trans-

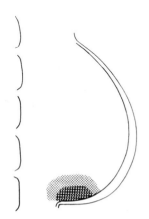

Fig. 21-11. Right chick wing bud showing location of tissue having the ability to induce supernumerary wing parts in anterior tissues when grafted to the apex of the wing bud. This property is restricted to the PNZ (vertical and horizontal cross-hatching) and tissue adjacent to it (diagonal cross-hatching). (From Saunders, J. W., and M. T. Gasseling. 1968. *In* P. Fleischmajer and R. E. Billingham, eds. Epithelial-mesenchymal interactions. The Williams & Wilkins Co., Baltimore.)

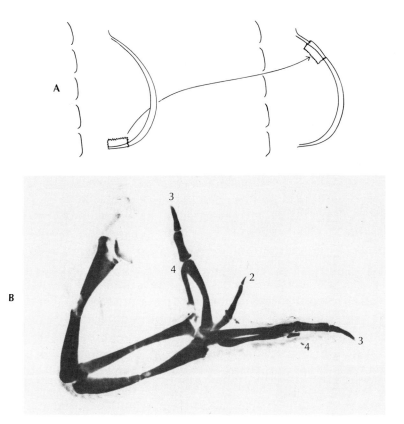

Fig. 21-12. A, Diagram showing transplantation of the ZPA from a donor wing bud to an anterior position in a host wing bud. **B,** Duplication of wing parts following the transplantation experiment diagrammed in **A.** (From Saunders, J. W. and M. T. Gasseling. 1968. *In* P. Fleischmajer and R. E. Billingham, eds. Epithelial-mesenchymal interactions. The Williams & Wilkins Co., Baltimore.)

planted to a variety of sites on the periphery of the wing bud resulted in the following pattern of digit duplications. When the tissue was grafted to a region near the posterior border of the host limb bud, no duplication occurred. If the graft was made in a central location, duplication occurred to produce digits with right-handed asymmetry. If the graft was made near the region of the anterior limb border, duplication occurred to produce digits with left-handed asymmetry. Thus in all cases the order of digits formed showed a consistent relationship to the grafted ZPA, with the normally most posterior digit (digit 4) always developing nearest the graft and the normally more anterior digits always developing at relatively greater distances from the graft.

Lewis Wolpert and his associates have proposed a model to account for the pattern of digit formation found in normal limbs and after grafting of the ZPA to various positions along the anterior-posterior border of the limb bud. They propose that cells receive positional information that differs according to their position along a particular axis of

ROLE OF POSITIONAL SIGNALING IN DEVELOPMENT OF LIMB ASYMMETRY

the limb bud. In the case of the anterior-posterior axis, the signal that cells receive is postulated to depend on their distance from the ZPA. Such a gradient of information could result from the production of a diffusible substance by the cells of the ZPA. Diffusion of the active substance from the ZPA region would result in a gradient of activity that would be highest near to the ZPA and lower at greater distances from the ZPA.

These workers have also postulated that the proximal-distal axis of the limb is laid down by virtue of positional information. In this instance a cell's positional information is proposed to be determined by the amount of time it spends in the "progress zone," a specific region of dividing cells located at the tip of the developing limb. It is postulated that the uniqueness of the progress zone may be related to its proximity to the AER. As cell division occurs in the progress zone, some of the cells leave the zone, and their positional value is determined at that time. Cells that leave the progress zone early receive more proximal positional values; those leaving later receive more distal positional values. Positional information in the proximal-distal dimension is thus considered to result from a clocklike mechanism rather than a physical gradient. One possibility is that time spent in the progress zone could be measured by counting cell divisions—such a mechanism would parallel the concept of quantal mitoses in cell differentiation (see Chapter 11). It is further postulated that only cells in the progress zone are able to respond to the positional information emanating from the ZPA. Thus anterior-posterior determination would be specified by the distance of the cells from the ZPA at the time that they leave the progress zone.

Key elements in the positional signaling model of limb development are diagrammed in Fig. 21-13. In Fig. 21-13, B, cell A, which is located outside the progress zone, is already specified with regard to its positional information and is destined to become part of the radius. Cell B, which is just leaving the progress zone, is in the process of be-

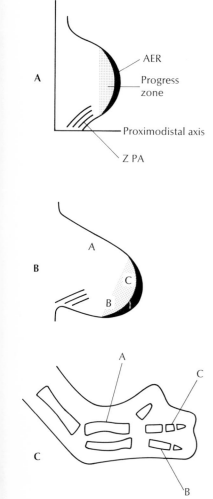

Fig. 21-13. Positional signaling model of chick limb development. **A,** Early limb bud showing location of ZPA, AER, and progress zone. It is postulated that cells receive positional information along the proximal-distal and anterior-posterior axes at the time they leave the progress zone. Anterior-posterior positional information is specified with respect to distance from the ZPA. Proximal-distal positional information is specified by the amount of time spent in the progress zone. **B,** Limb bud at later stage. Cell A has had its positional information specified. Cell B is in the process of being specified, and Cell C is in development. **C,** Limb bud at a late stage in development. Cells A to C have given rise to mature tissue in response to their positional information. (Modified from Tickle, C., D. Summerbell, and L. Wolpert. 1975. Nature **254**:199.)

coming specified proximodistally by the number of cell divisions it has undergone while in the progress zone. It is also being specified in the anterior-posterior dimension by its distance from the ZPA at this time. Cell C, which is still in the progress zone, is multipotential and will remain unspecified until it leaves the progress zone.

The results of numerous transplantation experiments with the chick limb are consistent with this model with regard to the anterior-posterior specification of digits (Fig. 21-14). No matter where the ZPA

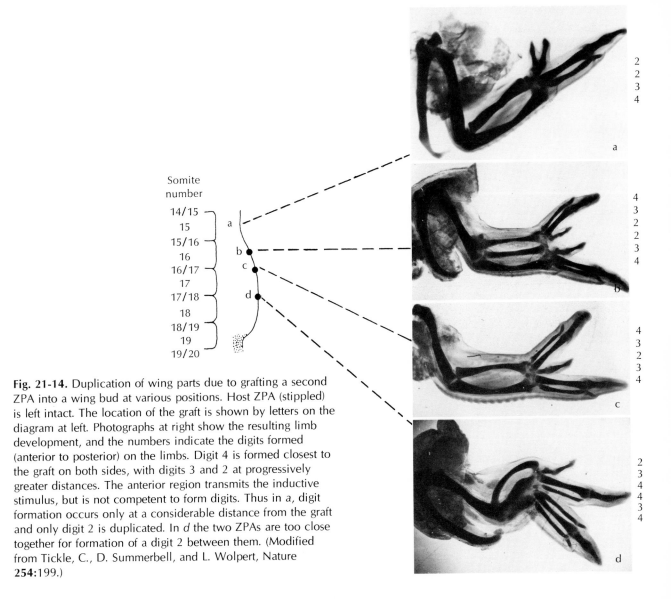

Fig. 21-14. Duplication of wing parts due to grafting a second ZPA into a wing bud at various positions. Host ZPA (stippled) is left intact. The location of the graft is shown by letters on the diagram at left. Photographs at right show the resulting limb development, and the numbers indicate the digits formed (anterior to posterior) on the limbs. Digit 4 is formed closest to the graft on both sides, with digits 3 and 2 at progressively greater distances. The anterior region transmits the inductive stimulus, but is not competent to form digits. Thus in a, digit formation occurs only at a considerable distance from the graft and only digit 2 is duplicated. In d the two ZPAs are too close together for formation of a digit 2 between them. (Modified from Tickle, C., D. Summerbell, and L. Wolpert, Nature **254**:199.)

is grafted, it is always found to be at the same end of the gradient of information after digit formation has been completed. The ZPA seems to direct the adjacent tissue to form digit 4 and relatively more distant tissue to form digits 3 and 2, respectively. Information is propagated from the ZPA in both directions. That is, digits are produced on both sides of the grafted ZPA in the 4-3-2 (posterior to anterior) pattern. The only exceptions are when the ZPA is grafted to the extreme anterior position or left in its natural position in the extreme posterior region of the limb bud. No discontinuities are observed in digit formation (i.e., 4 must be laid down prior to 3, and 3 must be laid down prior to 2). In some cases when the graft is in an extreme anterior position, digits form in sequences such as 2-2-3-4 or 3-2-2-3-4, anterior to posterior. In these cases the high end of the gradient emanating from the grafted ZPA is in tissue that is not competent for digit formation (i.e., body wall tissue), but the noncompetent tissue is still able to transmit the information, and the low end of the gradient that reaches competent limb bud tissue specifies formation of digits 2 and sometimes 3.

The signal that determines positional information along the anterior-posterior axis may be similar in the leg and wing buds, since the ZPA region of the leg bud is able to stimulate formation of wing digits. More dramatically, regions from the forelimb and hind limb buds of the mouse that are homologous to ZPA of the chick limb bud induce the formation of extra digits in the same pattern as the chick ZPA when transplanted to the anterior border of the chick wing bud. The signal from the mouse tissue appears to be somewhat less strong, however, since it induces only digits 3 and 2 and does not induce digit 4 (which according to the theory requires the strongest signal).

Transplantation experiments by Dennis Summerbell strongly support the time-dependent model for determination of proximal-distal positional information by cells in the progress zone. When the tip from a stage-24 limb bud is removed and a tip from stage 19 is grafted in its place, an extra long limb is obtained in which all structures normally laid down between stages 19 and 24 are duplicated (Fig. 21-15, A). The host stage-24 stump contributes an upper wing (humerus) and a "forearm" (radius and ulna). Attached to it at what would normally be the "wrist" level is a complete set of wing parts contributed by the progress zone of the stage-19 graft (humerus, radius and ulna, wrist parts, and digits). In the reciprocal experiment a stage-24 tip grafted onto a stage-19 stump yields only a "wrist" and digits attached directly to the "shoulder" (Fig. 21-15, B). In both cases the progress zone of the grafted tip continues to specify structures that are appropriate for its developmental age with total disregard for environmental signals of any type from the stump onto which it is grafted.

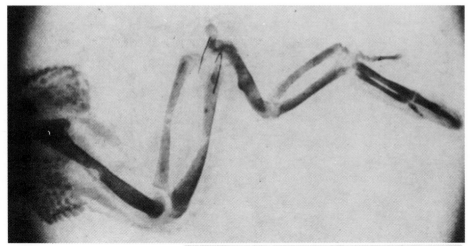

Fig. 21-15. Results obtained by exchanging tips between stage-19 and stage-24 wing buds. **A,** A stage-19 tip was grafted on a stage-24 stump. All structures normally determined between stages 19 and 24 have been duplicated. **B,** A stage-24 tip was grafted on a stage-19 stump. All structures normally determined between stages 19 and 24 are missing. (**A** from Summerbell, D., and J. H. Lewis. 1975. J. Embryol. Exp. Morphol. **33**:621; **B** from Wolpert, L., J. Lewis, and D. Summerbell. 1975. *In* Cell patterning. Ciba Foundation Symposium No. 29. Elsevier North-Holland, Inc., New York.)

Wolpert has suggested that the mechanism of specifying positional information may be the same in all multicellular organisms, with the differences in response to this information being directed by different cellular genotypes. Such a model predicts, for example, that presumptive amphibian neural tissue transplanted to the anterior border of the chick limb bud might develop into forebrain, whereas the same tissue transplanted to the posterior border might develop into hindbrain. It can be expected that such predictions will be tested in the near future by workers interested in mechanisms of positional signaling.

The chick limb provides a fascinating system for the study of mechanisms that determine cell differentiation and spatial patterning of cells within a complex structure. Early development of the limb results from an inductive stimulus of presumptive limb mesoderm on overlying ectoderm. Further limb outgrowth requires reciprocal interactions of limb mesoderm and ectoderm and is relatively autonomous with regard to the rest of the embryo. The specific arrangement of digits in the limb (and the pattern of cell division and differentiation that necessarily brings about this arrangement) appears to result from cell position with respect to a localized zone of the limb. It has likewise been postulated that pattern formation in the proximal-distal axis is specified by a timing mechanism that is active within a region of cell proliferation at the tip of the limb bud. The importance of cellular interactions in bringing about limb development is apparent in all these processes. Elucidation of the precise nature and mode of action of the signals involved in the processes is sure to provide a basis from which the underlying molecular mechanisms of limb development can be studied.

Positional information and cellular interactions are also of major importance in many other aspects of development. Additional examples are provided in the two chapters that follow on the development of the fruit fly, *Drosophila melanogaster,* and the development of the cellular slime mold, *Dictyostelium discoideum,* respectively.

BIBLIOGRAPHY
Books and reviews

Balinsky, B. I. 1975. An introduction to embryology, 4th ed. W. B. Saunders Co., Philadelphia.

Hamburgh, M. 1971. Theories of differentiation. Elsevier North-Holland, Inc., New York.

Saunders, J. W. 1970. Patterns and principles of animal development. Macmillan Publishing Co. Inc., New York.

Saunders, J. W. 1972. Developmental control of three-dimensional polarity in the avian limb. Ann. N.Y. Acad. Sci. **193**:29.

Saunders J. W., and C. T. Gasseling. 1968. Ectodermal-mesenchymal interactions in the origin of limb symmetry. *In* R. Fleischmajer and R. E. Billingham, eds. Epithelial-mesenchymal interactions. The Williams & Wilkins Co., Baltimore.

Wolpert, L., J. Lewis, and D. Summerbell. 1975. Morphogenesis of the vertebrate limb. *In* Cell patterning. Ciba Foundation Symposium No. 29. Elsevier North-Holland, Inc., New York.

Wolpert, L. 1978. Pattern formation in biological development. Sci. Am. **239**: 124 (Oct.).

Zwilling, E. 1956. Genetic mechanisms in limb development. Cold Spring Harbor Symp. Quant. Biol. **21**:349.

Zwilling, E. 1972. Limb morphogenesis. Dev. Biol. **28**:12.

Selected original research articles

Callandra, A. J., and J. A. MacCabe. 1978. The *in vitro* maintenance of the limb-bud apical ridge by cell-free preparations. Dev. Biol. **62**:258.

Fallon, J. F., and G. M. Crosby. 1975. The relationship of the zone of polarizing activity to supernumerary limb formation (twinning) in the chick wing bud. Dev. Biol. **43**:24.

Saunders, J. W. 1948. The proximo-distal sequence of origin of the parts of the chick wing and the role of the ectoderm. J. Exp. Zool. **108**:363.

Tickle, C., et al. 1976. Positional signalling by mouse limb polarising region in the chick wing bud. Nature **259**:396.

Tickle, C., D. Summerbell, and L. Wolpert. 1975. Positional signalling and specification of digits in chick limb morphogenesis. Nature **254**:199.

CHAPTER 22

Development of *Drosophila*

☐ All the information needed for the complex events involved in the development of an organism resides ultimately in the genes of that organism. External factors varying from maternal gene products to conditions of temperature and pH have an influence on development, but the ordered, sequential occurrence of events characteristic of development is based primarily on the ordered, sequential activation and inactivation of specific genes. Interest in a genetic approach to the questions of developmental biology has resulted in a revival of the popularity of insect systems, in particular *Drosophila*, for which a huge fund of genetic information is already available. Aside from its genetic advantages, *Drosophila* has characteristics that are useful for the study of certain aspects of development, such as hormonal actions, cell determination, and pattern formation. Some of the unusual aspects of *Drosophila* development—for example, unusual cleavage pattern, the development of polytene chromosomes in the nuclei of some tissues, and the presence in larvae of discrete blocks of cells (the imaginal discs) that develop into particular adult structures—have proved more beneficial than detrimental for asking questions about developmental mechanisms. Furthermore, some of the answers obtained from studies with insect systems appear to have applicability that extends far beyond the realm of insect development.

Drosophila is an example of a holometabolous insect, that is, one which undergoes a complete metamorphosis during its development. The adult stage is radically different from the larval stage, and the two are separated by a pupal stage during which the transformation from larva to adult occurs. Thus *Drosophila* development can be divided into two distinct phases—embryonic development, which includes the period from fertilization of the egg through hatching of the larva, and postembryonic development, which includes the progression through the larval stages and from larva to pupa to adult.

The eggs of holometabolous insects are typically oval in shape with an average diameter of 1 mm. They are enclosed in a tough protective

**GENETICS OF EARLY
EMBRYONIC DEVELOPMENT**

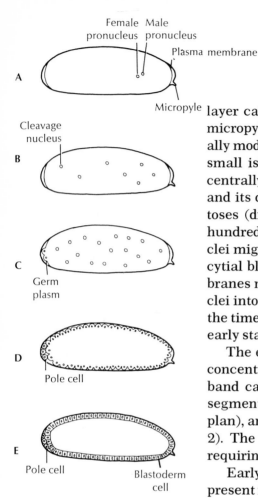

Fig. 22-1. Diagram showing early stages in *Drosophila* development. **A,** Fertilized egg. **B,** Nuclear division. **C,** Entry of nuclei into the germ plasm. **D,** Syncytial blastoderm. **E,** Cellular blastoderm. (Modified from Gehring, W. J. 1973. Symp. Soc. Dev. Biol. **31**:107.)

layer called the chorion, which contains a small opening called the micropyle. The sperm enter through this micropyle. The eggs are usually moderately yolky except for a peripheral region of cytoplasm and a small island of cytoplasm surrounding the nucleus, which is often centrally located (Fig. 2-2). Following fertilization the zygote nucleus and its daughters typically undergo fairly synchronous and rapid mitoses (dividing every 5 to 10 minutes in some dipterans), producing hundreds of nuclei, each surrounded by a halo of cytoplasm. The nuclei migrate to the surface of the egg, forming what is called the syncytial blastoderm. After a number of further mitotic divisions, membranes move inward from the surface of the egg and separate the nuclei into distinct cells. About 4000 nuclei are present in *Drosophila* at the time of cellular blastoderm formation. Fig. 22-1 shows some of the early stages in *Drosophila* development.

The embryo proper is formed from the cellular blastoderm by the concentration of ventral and ventrolateral cells into a broad ventral band called the germ band. The germ band undergoes gastrulation, segmentation (forming the segments characteristic of the larval body plan), and differentiation in an anterior-to-posterior direction (Fig. 22-2). The entire process can be rapid, with *Drosophila*, for example, requiring only 22 hours from fertilization to hatching of the larva.

Early *Drosophila* development is largely controlled by components present in the unfertilized egg. Many of the mutants that are defective in the early stages of development are maternal-effect mutants, that is, mutants in which development depends on the maternal genotype alone or on the maternal genotype as well as that of the offspring. The expression of three of these maternal-effect mutations—deep orange (*dor*), fused (*fu*), and rudimentary (*r*)—all of which are sex-linked recessive mutations, is as follows. Mating of a homozygous mutant female to a hemizygous mutant male (the male has only a single X chromosome) results in the production of offspring that all die as embryos. But mating of a heterozygous female with a hemizygous mutant male produces offspring that all survive (Fig. 22-3). Since half the offspring of the latter mating would contain only the mutant allele, this finding implies that the mother produces some egg component that is essential to embryonic development. The presence of a wild-type allele in the mother allows normal development even of embryos carrying only the mutant allele.

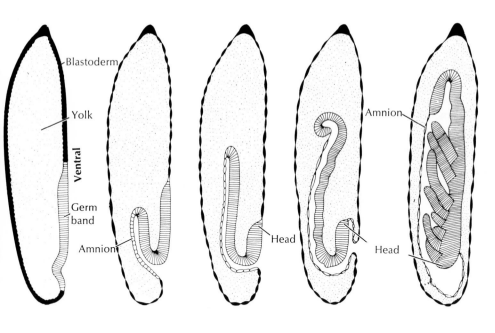

Fig. 22-2. Diagram showing stages in gastrulation and segmentation of the germ band in the dragonfly. At the stages diagrammed the anterior end (head) of the embryo faces the posterior pole of the egg. At a later stage the embryo reverses its orientation so that the anterior of the embryo faces the anterior of the egg. (Reprinted [modified] with permission of Macmillan Publishing Co., Inc. from Patterns and principles of animal development by Saunders, J. W. Jr. Copyright © 1970 by John W. Saunders, Jr.)

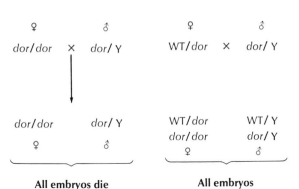

Fig. 22-3. Results of crossing females homozygous or heterozygous for the sex-linked maternal-effect mutation *dor* with hemizygous mutant males. *WT* = Wild type; *Y* = Y chromosome that does not carry the *dor* gene.

Dor, *fu*, and *r* are not strict maternal-effect mutations, since the offspring of homozygous mutant females are rescued when the female is mated to a wild-type male. In this case the hemizygous mutant male offspring all die but about a third of the heterozygous female offspring survive. It is believed that the normal allele present in the sperm is able to supply the factor missing from the egg. Offspring of females homozygous for these maternal-effect mutations have also been rescued by injecting the preblastoderm embryo with cytoplasm from a wild-type embryo. The cytoplasmic components missing in the eggs of *dor*, *fu*, and *r* females appear to be required for normal cleavage, blastoderm formation, gastrulation, and primary organogenesis, since the abnormalities found in mutant offspring encompass all these processes.

Maternal-effect mutants that specifically affect formation of the cellular blastoderm have also been isolated recently. These mutants depend strictly on the maternal genotype and cannot be rescued by wild-type sperm. They appear normal during the first 2 hours of development and form a syncytial blastoderm and primordial germ cells. One mutant is unable to form a cellular blastoderm at all. Two other mutants form partial cellular blastoderms. One forms cells only at the posterior and anterior ends of the embryo; the other forms cells in all but the posterior-dorsal surface. The existence of these mutants demonstrates that certain egg components must be present before fertilization for the transition from syncytial to cellular blastoderm and suggests that different factors are involved in cellularization of different blastoderm regions. One of the mutants is temperature sensitive, and its temperature-sensitive period occurs in the last 12 hours of oogenesis, a finding that correlates with the strict maternal effect of the mutation.

Maternal-effect mutations with specific effects on the development of particular types of tissues have also been found. Females homozygous for the mutation grandchildless (gs), when mated to mutant or wild-type males, produce offspring that are sterile because of a lack of germ cells. It has been found that the primordial germ cells fail to form in the embryos of gs mothers, although the polar granules are present in the germ plasm. Thus the egg must normally contain factors necessary for pole cell formation. Females homozygous for the maternal-effect mutation almondex (amx) produce some viable heterozygous female offspring when mated with wild-type males. However, 80% of the surviving progeny are abnormal, with defects in the thorax and abdomen, such as crippled or missing legs and irregularly shaped abdominal sternites. The development of the thorax and abdomen thus also appears to be influenced by components normally present in the unfertilized egg.

Last, a few maternal-effect mutations are homeotic mutations in which one body part is replaced by another. (Homeotic mutants are discussed in greater detail later in this chapter.) Offspring of homozygous bicaudal females have abnormal heads and die as larvae. The larvae show different degrees of mirror-image duplication of the abdomen, with the most extreme cases having the head completely replaced by an abdomen in mirror-image symmetry to the normal abdomen. These findings suggest that factors necessary for normal head formation are missing or defective in the eggs of homozygous bicaudal females. Ultraviolet irradiation of the anterior region of the egg of the midge *Smittia* results in a similar pattern of bicaudal development, apparently by inactivation of ultraviolet-sensitive anterior determinants, which are thought to contain RNA.

The postembryonic development of holometabolous insects occurs in brief as follows (Fig. 22-4). The insect egg hatches into a wormlike larva that grows and undergoes a series of molts, which are separated by stages called instars. Molts are required for growth of the insect, since its growth is limited by the size of its rigid exoskeleton. After undergoing a number of larval instars (three in *Drosophila*), the larva enters a pupal stage. The pupa is enclosed in a thick cuticle and appears quiescent in that it undergoes no feeding and little movement, but it is actually actively involved in a number of dramatic changes that will result in its metamorphosis into a sexually mature adult, or imago.

Molting and metamorphosis occur under the control of a number of insect hormones (Fig. 22-5). Brain hormone, or ecdysiotropin, is produced by neurosecretory cells of the insect brain in response to certain internal and external signals. Brain hormone stimulates the prothoracic gland to produce a second insect hormone, ecdysone. Ecdysone, or molting hormone, is a steroid that causes the insect to molt, that is, to shed its old cuticle and synthesize a new one. It also stimulates the growth and differentiation associated with a molt. A third insect hormone, juvenile hormone, is produced by the corpora allata, small glands located near the brain. Juvenile hormone controls metamorpho-

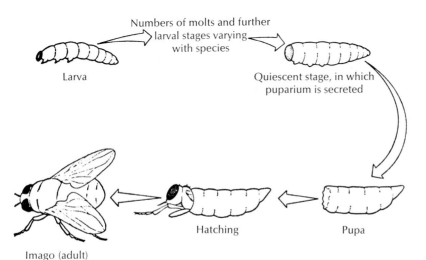

Numbers of molts and further larval stages varying with species

Larva

Quiescent stage, in which puparium is secreted

Hatching

Pupa

Imago (adult)

Fig. 22-4. Diagram showing stages involved in metamorphosis of holometabolous insects. (From Deuchar, E. M. 1975. Cellular interactions in animal development. Chapman & Hall Ltd. © 1975 Elizabeth M. Deuchar.)

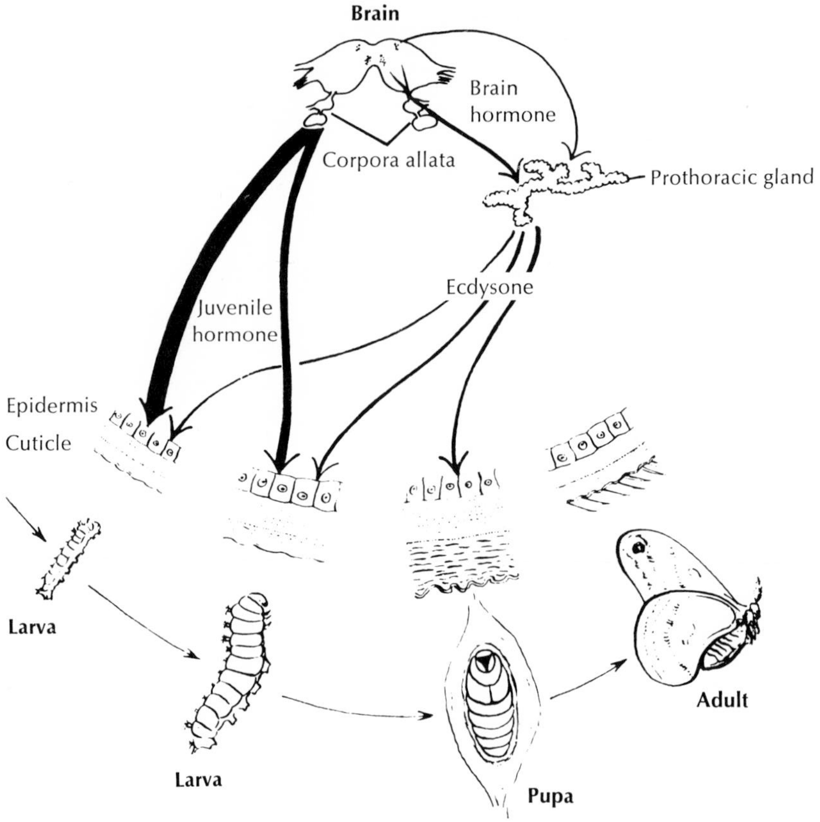

Fig. 22-5. Diagram showing hormonal action in the postembryonic development of the holometabolous silk moth, *Hyalophora cecropia*. The general scheme is the same for *Drosophila*. (From Doane, W. W. 1973. *In* S. J. Counce and C. H. Waddington, eds. Developmental systems: insects. Vol. 2. Academic Press, Inc., New York.)

Fig. 22-6. Diagram of one arm of chromosome 3 in *Drosophila melanogaster* showing the sequence of appearance and regression of seven puffs before the larval-prepupal molt *(L → PP)* and before the prepupal-pupal molt *(PP → P)*. (From Ashburner, M. 1967. Chromosoma **21**:398.)

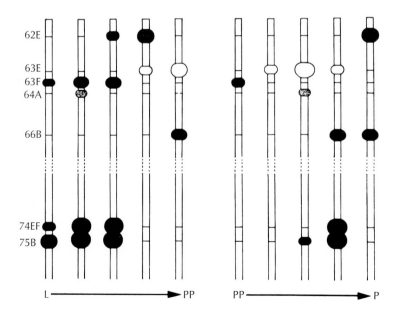

sis by regulating the character of molts. When it is present in sufficient amount, each molt simply results in another larval instar. When the amount of juvenile hormone is reduced, molting results in the formation of the pupal stage. Finally, in the absence of juvenile hormone, the pupal-adult molt takes place. Postembryonic development is thus largely regulated by the relative levels of two hormones—ecdysone, which favors adult differentiation, and juvenile hormone, which favors larval differentiation.

The mechanism of action of ecdysone has been especially well studied. During normal development, specific bands in the polytene chromosomes of *Drosophila* become enlarged and form what are called "puffs." The puffs appear in specific sequential patterns characteristic of the stage of development (Fig. 22-6), and they have been shown to be sites of intense RNA synthesis as was described in Chapter 7. The same patterns of puffing that are seen in normal development also occur when larvae are treated with ecdysone before the usual time of molting and when larval salivary glands are treated with ecdysone in vitro. There has been some debate as to whether ecdysone stimulates transcription directly or indirectly by influencing some other critical factor. For example, chromosome puffing can be specifically induced by heat shock or by altering the ionic environment of salivary glands. The ionic changes that cause puff induction are within the physiological range, and it has been suggested that ecdysone might act primarily by altering the intracellular environment, which then directly influences gene expression. The alternate hypothesis that ecdysone stimulates transcription directly is given some support by the finding that labeled ecdysone accumulates in the nuclei of target cells. The early puffs induced by ecdysone form in the absence of protein synthesis, but later puffs do not, and their formation is believed to require the products of early puffs.

IMAGINAL DISCS IN *DROSOPHILA* DEVELOPMENT

The transformation from the larval to the adult form of *Drosophila* involves radical changes in the insect's architecture. The adult contains a variety of structures that are not present in the larva, such as wings, antennae, and genital structures. However, the pupa does not start from scratch in the elaboration of the adult form. The larva anticipates adult development, and it contains a number of discrete groups of cells called imaginal discs (from imago) that are determined precursors of adult integumentary structures. The larva contains nine pairs of discs that are precursors to paired adult structures such as wings and antennae, plus a single genital disc.

Specific imaginal discs are composed of from a few hundred to many thousands of embryonic cells, and each disc has a specific shape and location within the larva (Fig. 22-7). The discs are first seen in early larvae as thickenings of the epidermis. They later invaginate and form saclike structures that are attached by stalks to the larval epidermis. The cells form a single-layered epithelium, which becomes folded as the cells divide. The disc cells remain in an embryonic state until metamorphosis occurs, whereupon they undergo eversion and the disc epithelium forms the extended structure.

An imaginal disc can be dissected out of the larva and cultured as a larval structure for an extended period of time by placing it in the abdomen of an adult fly. In this environment the disc cells proliferate but do not differentiate. Because of the limited life span of the adult, the discs are subdivided and retransplanted to another adult about every 2 weeks. The cultured imaginal discs are capable of undergoing transformation into adult structures if they are placed in the abdomen of a larva that is then induced to metamorphose. The adult structure appropriate to the disc develops in the larval abdomen and can be identified even though it is not necessarily complete.

Even after prolonged culture in an adult abdomen, discs normally retain their epigenotype and differentiate in their originally determined direction. This can be tested during the successive transfers of disc cells from adult to adult by placing part of a disc into a larval abdomen and inducing metamorphosis. After a number of serial transfers of disc cells, however, cells from one type of disc sometimes undergo a transdetermination and now form a different adult structure.

Fig. 22-7. Diagram showing the shapes, locations, and developmental fates of the imaginal discs of *Drosophila.* The adult eye and antennal structures are both derived from a single pair of large, complex discs. Note that the larva has three pairs of leg discs and also a pair of dorsal thoracic discs that are not shown here. (From Fristrom, J. W., et al. 1969 *In* E. W. Hanly, ed. Problems in biology: RNA in development. University of Utah Press, Salt Lake City.)

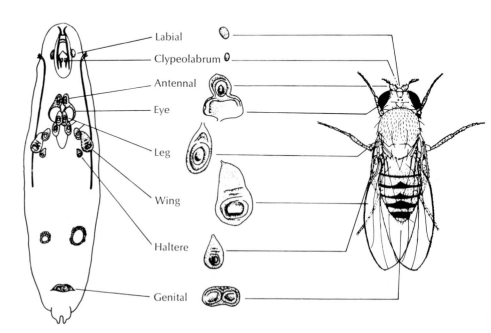

For example, cells from the genital disc may develop into leg cells. Although transdetermination occurs in a group of cells rather than an individual cell, the entire disc need not transform. For example, a genital disc may produce both genital cells and leg cells. The change in determination is inherited, and on successive transfers all the progeny of the transdetermined disc region display the new phenotype.

Transdetermination of imaginal discs tends to occur in a precise and directed sequence. That is, cells of disc type A may change to type B and those of type B may change to type C, but one does not observe changes from type A directly to type C. The types of changes that are observed are diagrammed in Fig. 22-8. It can be seen in the figure, for example, that genital disc cells can change to either leg or antennal cells but not directly to wing cells. Some kinds of changes can occur in either direction, for example, leg to antenna and antenna to leg, whereas others are unidirectional. The change to thorax cell determination appears to be a dead end, and no further transdetermination occurs.

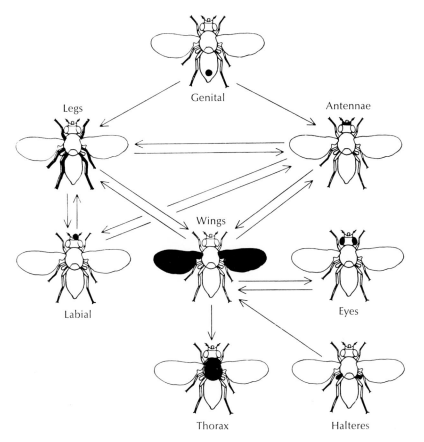

Fig. 22-8. Diagram showing sequence of transdeterminations undergone by seven kinds of imaginal disc cells in *Drosophila*. (From Hadorn, E. 1968. Sci. Am. **219**:110 [Nov.]. Copyright © 1968 by Scientific American, Inc. All rights reserved.)

It is not certain what causes cells to transdetermine, but the probability of transdetermination has been found to be directly correlated with the rate of cell proliferation. The faster a certain type of cell divides, the greater the chance that it will transdetermine. This may be the reason for the lack of transdetermination in cells of the thorax disc, since these cells divide very slowly. It has been postulated that rapid cell division may result in transdetermination because of the dilution of regulatory molecules that bind to the DNA. It is considered probable that transdetermination is the result of a regulatory type of event, since it occurs at too great a frequency to result from mutation. Since the change appears to occur in groups of cells rather than individual cells, it may involve a determinant that is transmitted from cell to cell, for example, via cell surface alteration.

HOMEOTIC MUTATIONS

The Nosobame*

Upon his noses stalketh
around – the Nosobame;
with him, his offspring walketh.
He is not yet in Brehm,

you find him not in Meyer
nor does him Brockhaus cite.
He stepped forth from my lyre
the first time into light.

Upon his noses stalketh
– I will again proclaim –
(with him his offspring walketh),
since then, the Nosobame.

*From Knight, M. E. 1963, translator. Christian Morgenstern's Galgenlieder, a selection. University of California Press, Berkeley.

Single gene mutations occur in *Drosophila* that alter the determination of specific imaginal discs. These mutations are called homeotic mutations, and they result in changes in disc determination that parallel those seen in transdetermination. For example, in the homeotic mutant nasobemia* (literally, leg-on-nose), the antennal disc develops into a leg so that the adult has a leg in the position where the antenna is usually found. In most homeotic transformations, only part of a disc is involved. For example, only some of the antennal segments may be replaced by leg segments so that the adult has basal antennal segments and distal leg segments. Typically, distal antennal segments are replaced only by distal leg segments and proximal antennal segments by proximal leg segments (Fig. 22-9).

Initial determination of the developmental fates of presumptive imaginal discs occurs early in normal development, probably about the time of cellular blastoderm formation. It is uncertain when home-

*Named after a fictional animal, the *Nasobēm*, in a nonsense poem by Christian Morgenstern, which appears at left in translation from the original German.

Fig. 22-9. Homeotic mutants that convert antennal structures to leg structures. **A** to **D,** Photographs showing comparison of antennae of wild-type *Drosophila* with those of various homeotic mutants. **A,** Wild type. **B,** Lethal-(4)29. **C,** Aristapedia. **D,** Nasobemia. **E,** Diagram showing specific segment-by-segment conversion of antennal structures into leg structures. Note that the leg has been drawn upside down for easier comparison with the antenna. Antenna parts: *AI, AII, AIII* = first, second, and third antennal segments; *Ar* = arista. Leg parts: *Cl* = Claw; *Cx* = coxa; *Fe* = femur; *Ta* = tarsus; *Ti* = tibia; *Tr* = trochanter. (**A** to **D** from Gehring, W. J. 1970. *In* E. W. Hanly, ed. Problems in biology: RNA in development. University of Utah Press, Salt Lake City; **E** from Postlethwait, J. H., and H. A. Schneiderman. 1971. Dev. Biol. **25**:606.)

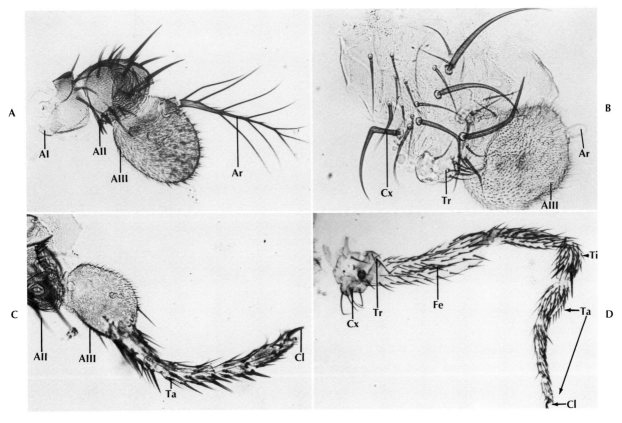

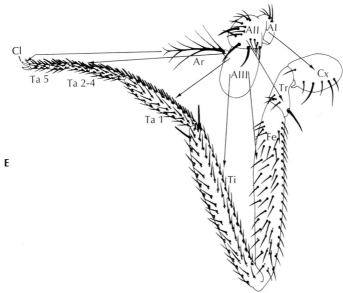

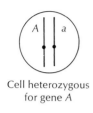

Cell heterozygous
for gene *A*

Duplication of chromatids
and somatic crossing over

Metaphase chromosomes

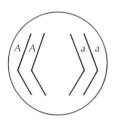

Separation of sister chromatids
at anaphase

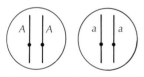

Homozygous daughter cells

Fig. 22-10. Diagrammatic representation showing how somatic crossing-over in a cell heterozygous for a particular gene can result in the production of two daughter cells, each homozygous for one of the parental alleles.

otic mutations normally exert their effect, but there is substantial evidence that they can act much later in development than the time of initial determination. This has been suggested by the results of experiments using the technique of somatic crossing-over. Somatic crossing-over involves chromosome breaks that lead to an exchange of genetic material between individual chromatids, as normally occurs in germ cells during meiosis. Somatic crossing-over occurs in somatic cells and is induced by means of x-irradiation at specific times during development. As a result of somatic crossing-over, a single cell that is heterozygous for a particular gene pair can produce two daughter cells, each homozygous for one allele of the gene pair (Fig. 22-10). The two types of homozygous cells continue division and, if appropriately marked, produce clones of cells that can be distinguished by their differing phenotypes. Such clones have been induced in antennapedia mutants whose antennae normally contain both leg and antennal cells. When crossing-over is induced in these individuals prior to the early third larval instar, clones derived from genetically marked cells can be found that contain both leg and antennal cells. If the cell whose descendants make up such a clone had already been determined with regard to leg or antennal expression, the entire clone would consist of either leg or antennal cells. The fact that both types of cells are formed suggests that the commitment to leg or antennal differentiation was not yet final. When crossing-over is induced at later times in development, the clones that are obtained do contain only one type of cell, indicating that cell determination has occurred. These findings indicate that the commitment to form either normal antennal or homeotic leg tissue is not final until as late as the third larval instar.

Studies with temperature-sensitive homeotic mutations also suggest that homeotic mutations affect gene products required for the expression or maintenance of developmental commitments rather than for the initial determining events that occur early in development. Several temperature-sensitive aristapedia mutants are known that produce a normal arista at the permissive temperature but a distal leg segment in place of the arista at the nonpermissive temperature. By shifting the temperature at which these mutants are grown at different times in development, one can determine the temperature-sensitive period during which the normal gene product is required. The temperature-sensitive period for these mutants has been shown to occur during the third larval instar, rather than at the time of initial disc determination, which occurs much earlier. This suggests that these homeotic mutations alter a gene product that is necessary for the maintenance or expression of a developmental commitment, rather than its initial establishment.

The temperature-sensitive aristapedia mutants are all cold-sensitive mutants. That is, they produce normal structures at 29° C and abnormal structures at 17° C, whereas the opposite situation is seen in most temperature-sensitive mutations that have been studied. In lower organisms, cold-sensitive mutations tend to occur in genes coding for proteins with strict conformational requirements such as ribosomal proteins or allosteric proteins. It has been suggested that the aristapedia gene product might require allosteric specificity for its action, which suggests a regulatory type of role. Most of the homeotic mutations are clustered within a region on one chromosome that is less than 2% of the *Drosophila* genome. This suggests that a distinct chromosomal region may be responsible for gene products that control the stability of disc determination. Finally, the fact that a single gene mutation can prevent the expression of the entire set of genes required for producing one structure and activate another entire set of genes required for the production of a second structure indicates that homeotic mutations occur in regulatory molecules that control large blocks of genes.

A series of intriguing recent observations suggests that a process of cellular compartmentalization occurs during development in *Drosophila*. Clonal analysis of imaginal disc development by the somatic crossover method has shown that there are certain boundary lines in the adult structures that are not crossed by expanding clones of cells after a given moment in development. For example, the mesothoracic area, which gives rise to the wings and the second pair of legs, becomes progressively compartmentalized, as shown schematically in Fig. 22-11. On each side of the larva the area that will give rise to the wing and second leg discs becomes compartmentalized into anterior and posterior halves extremely early, even before wing- and leg-forming potential are separated. Clones generated at the cellular blastoderm stage may involve both the wing and the second leg, but they are always confined either to the anterior halves of both structures or the posterior halves. Thus far, no clones have been generated by somatic crossover that involve both anterior and posterior structures of the wing or second leg.

Slightly later in development, wing and leg become compartmentalized, and their imaginal discs separate physically from each other. Still later the wing disc becomes further compartmentalized into dorsal-versus-ventral surfaces and into the wing blade proper versus the notum (that part of the thorax adjacent to the wing that is also derived from the wing imaginal disc). Further compartmentalization also occurs within the wing blade and within the notum. As somatic crossover is induced by irradiation at progressively later stages in develop-

COMPARTMENTALIZATION IN IMAGINAL DISC DEVELOPMENT

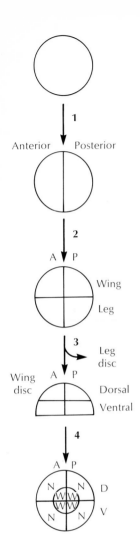

Fig. 22-11. Schematic representation of progressive compartmentalization of the presumptive mesothoracic region in *Drosophila*. *1*, Very early in development (prior to disc formation) anterior is separated from posterior. *2*, The next compartmentalization is between wing and leg formation. *3*, Physically separate wing and leg imaginal discs are formed, and the dorsal and ventral surfaces of the wing become compartmentalized. *4*, Still later the capacity to form the wing blades *(W)* is compartmentalized from the capacity to form the surrounding thoracic structure known as the notum *(N)*. Note that the distal structure (the wing blade) is compartmentalized at the center of the imaginal disc. (Modified from Morata, G., and P. A. Lawrence. 1977. Nature **265:**211.)

ment, the clones that are generated are restricted to progressively smaller compartments (Fig. 22-12). These successive compartmentalizations progressively restrict the developmental fate of cells in the wing disc. Not only is their geographical distribution restricted, but the cells in different compartments differ in characteristics such as mitotic rates, cell orientation, and cell affinities. Complex biochemical models have been proposed for the establishment of compartment boundaries by a dynamic balance in the rates of synthesis, diffusion, and destruction of specific regulatory molecules, but such models have not yet been verified experimentally.

Certain mutants observed in *Drosophila* may result from defects in compartment formation. The mutation engrailed *(en)* causes the posterior part of the wing to be transformed into a mirror image of the anterior region. It has been shown by somatic crossing-over experiments that the *en* allele produces the mutant phenotype only when it is present in the posterior compartment. When engrailed *(en/en)* clones are induced in heterozygous *(en/+)* wings, they have no effect unless they are located in the posterior compartment. Large mutant clones can be formed that fill most of the anterior compartment but do not cross the anterior-posterior boundary. These clones have no effect on wing development. However, mutant clones in the posterior compartment do show the mutant phenotype, and, in addition, they are able to cross the anterior-posterior boundary. The latter finding has been interpreted to indicate that the wild-type allele may control the ability of a cell to cross the anterior-posterior border and also control the expression of characteristics of posterior cells.

POSITIONAL INFORMATION IN IMAGINAL DISC DEVELOPMENT

Examination of the developmental potential of specific fragments of imaginal discs has demonstrated that cells at different locations within a disc differ from one another with regard to their developmental potential. If a mature imaginal disc is cut into fragments and these fragments are implanted into larvae which are then caused to meta-

Fig. 22-12. Restriction of cellular clones to compartments. **A,** Hypothetical imaginal disc showing a clone generated prior to compartmentalization spreading to both sides of a compartmental boundary. **B,** Same as **A** except that the clone was generated after the compartmental boundary was established. The marked cells grow to the edge of the compartment but do not cross the boundary. **C,** *Drosophila* wing, showing the boundary (dotted line) between the anterior and posterior compartments that is established at or before the cellular blastoderm stage. **D,** *Drosophila* wing with a large clone that covers most of the posterior surface but does not cross the anterior-posterior compartmental boundary. (From Crick, F. H. C., and P. A. Lawrence. 1975. Science **189:**340. Copyright 1975 by the American Association for the Advancement of Science.)

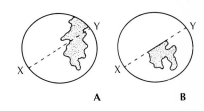

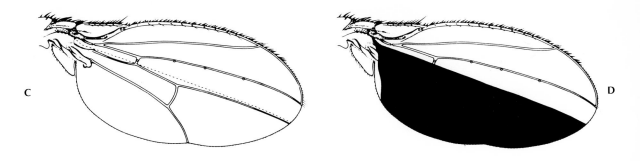

morphose, the fragments develop into specific parts of the adult structure. Fate maps of different imaginal discs can be drawn from the results of such experiments illustrating that specific disc regions contain information for the formation of particular parts of the adult structure that is derived from the disc. However, when imaginal discs are cut into fragments and these fragments are allowed to grow in an adult abdomen prior to their metamorphosis, a very different type of developmental potential is seen. For example, if a disc is bisected into two pieces and these are subjected to immediate metamorphosis, the complementary fragments give rise to complementary parts of the adult structure. However, if the fragments are allowed to grow before metamorphosis, one piece typically undergoes regeneration and produces a complete structure, whereas the complementary piece produces only a mirror-image duplicate of the part of the structure it would normally form. This pattern in which one fragment regenerates and the other duplicates is found when discs are cut in a wide variety of ways. Similar results are obtained when discs are bisected in situ; one cut surface regenerates and the other duplicates. It appears that the developmental capacities of the disc fragments are not determined in the strict sense, since daughter cells can produce parts of the adult structure other than those originally specified by the particular fragment. Although the commitment of a disc cell to form a particular kind of adult cell (e.g., leg versus wing) is determined and heritable, the commitment to form a particular part of the adult structure (e.g., anterior versus posterior) can be altered after cell proliferation. A number of workers have chosen to term the latter type of nonheritable commitment to form a specific part of an adult structure "specification" to distinguish it from determination that includes strict transmission of a commitment to progeny cells.

The mechanism whereby specification occurs may be the same in different discs. That is, the same signal may operate in different kinds of discs to convey positional information to cells. This is suggested by observations of certain developmental mutants. In the homeotic mutant antennapedia, in which various parts of the antenna are transformed into leg structures, it is found that any given antennal part is always replaced by a specific leg part. For example, the arista is always replaced by the fifth tarsal segment; the first antennal segment is always replaced by the coxa. This suggests that leg cells become specified by their position within the antennal disc. The mutant engrailed causes a mirror-image duplication to occur in both the wing and leg. Thus the defect that alters specification in these mutants appears to operate in both the wing and leg discs. The pattern of development of transdetermined cells within a disc also suggests that cells of different disc types can respond to the same positional signals. For example, when a transdetermined disc contains both leg and antennal structures, specific pairs of structures are always found to be contiguous, for example, tarsus and third antennal segment, implying that a common signal specifies tarsus in leg cells and third segment in antennal cells.

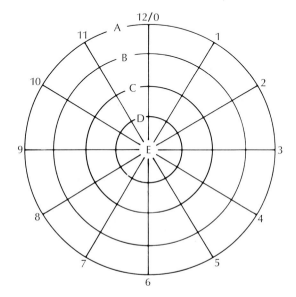

Fig. 22-13. Polar coordinate system of positional information. Each cell is postulated to have a positional value with respect to its location around the circle and with respect to its location on a radius. Positions *12* and *0* are identical so that the sequence is continuous. (From Bryant, S. V., and L. E. Iten. 1976. Dev. Biol. **50:**212.)

The findings of regeneration and duplication in complementary disc fragments that are cultured and metamorphosed has been explained in terms of a theory which postulates that the developmental capacity of disc cells is defined by their position with respect to a set of polar coordinates (Fig. 22-13). Each cell is postulated to have a positional value with respect to its location around the circle and with respect to its location on a radius. For discs that produce appendages, the center of the circle represents the distal part of the appendage and the periphery represents its proximal border. Thus the circle can be visualized as a cone, with the center of the circle as the apex of the cone. (This pattern coincides generally with the fate maps of the leg and wing discs, which are roughly concentric with proximal structures at the periphery and the distal tip of the appendage near the center of the map.)

The behavior of cells in the polar coordinate system is postulated to be governed by two rules. The first is the shortest intercalation rule and states that when cells with normally noncontiguous values are juxtaposed by cutting or grafting, growth occurs at the junction until cells with all the intervening positional values have been intercalated. In the case of the circular sequence of values, intercalation will always occur by the shortest route. For example, if cells with the values of two and five are juxtaposed, intercalation will occur by production of cells with values of three and four rather than in the other direction, going through the eight values from six to one. The second rule is the complete circle rule for distal regeneration. This states that cells at any radial level can give rise to cells with more central (distal) positional values as long as a complete circular sequence of values is present at the cut edge. For example, if all cells with values of C, D, and E are removed, they can be regenerated, provided that a complete circle of A and B cells remains.

The duplication and regeneration results obtained when wing discs are cut in three different ways and the resulting fragments are allowed to grow are shown in Fig. 22-14. When a disc is cut into 90- and 270-degree sectors (Fig. 22-14, *A*), the larger sector always undergoes regeneration, whereas the smaller sector duplicates, no matter which quadrant it consists of. This result has been explained by postulating that the cut edges come into contact and then undergo shortest-route intercalation. (It has been experimentally demonstrated that cut edges do fuse in wing disc fragments.) When a disc is cut into fragments with a single cut edge (Fig. 22-14, *B*), the duplication and regeneration observed have been explained to result from wound healing by closure of the cut edges followed by shortest-route intercalation. The final type of cut shown in the figure is one that separates the disc into a central fragment, and a peripheral fragment (Fig. 24-14, *C*). The central fragment duplicates, and the peripheral fragment regenerates. This can be explained by assuming that the cut surfaces of the fragments contract and then produce cells with distal positional values

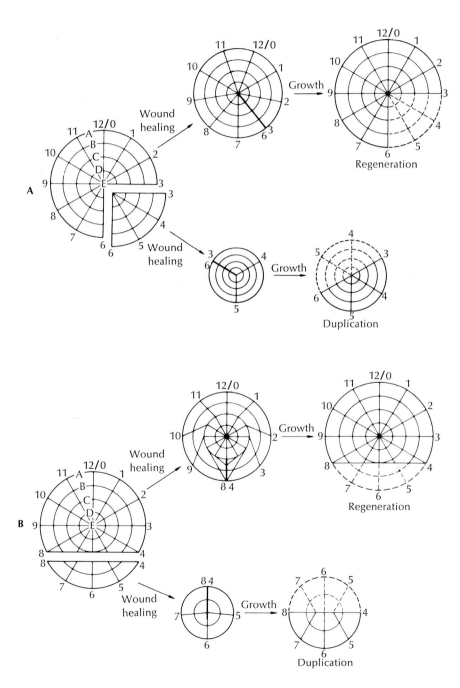

Fig. 22-14. Application of the polar coordinate model to regeneration and duplication results obtained after *Drosophila* wing discs are cut and the fragments are cultured.
A, Ninety- and 270-degree sectors of the wing disc undergo duplication and regeneration, respectively. These occur by means of shortest-route intercalation after wound healing juxtaposes positions *3* and *6*.
B, Fragments with a single cut edge undergo duplication and regeneration by shortest-route intercalation after wound healing juxtaposes positions *4* and *8*.
C, Central and peripheral fragments undergo duplication and regeneration, respectively. This occurs when the cut edges contract and give rise to cells with more distal (central) positional values. (From French, V., P. J. Bryant, and S. V. Bryant. 1976. Science **193:**969. Copyright 1976 by the American Association for the Advancement of Science.)

(i.e., undergo distal regeneration). This is possible because the cut surfaces contain the complete circular sequence of positional values, as required by the second rule.

The polar coordinate model of pattern regulation may apply to a number of systems in which removal of parts is followed by cell proliferation and addition of new pattern elements. The model has been used to explain the behavior of both newt and cockroach limbs after excision and grafting operations. Students wishing to pursue this topic further may consult the references at the end of the chapter.

OVERVIEW

Drosophila provides some excellent examples of the roles of genotype, hormonal action, cell determination, and cell position in development. The results of maternal-effect mutations emphasize the influence of specific gene products present in the egg on early embryonic development. The interactions of brain hormone, ecdysone, and juvenile hormone in larval development illustrate how chemical signals whose release is probably tied to various environmental factors can regulate the timing of maturational changes. Perhaps the most fascinating area of research in *Drosophila* development concerns various aspects of the development of the imaginal discs. Transdetermination of discs suggests that major changes in cell determination are possible and that such changes may be related to cell proliferation. The results of studies done with homeotic mutations suggest that cell determina-

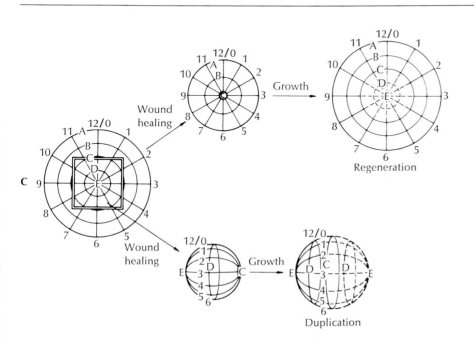

tion can be altered at relatively late stages of development and that the stability of cell determination may be regulated by specific gene products that control large banks of genes. Finally, evidence suggesting the existence of compartments and gradients of information in imaginal discs points up the potential importance of cell position on cell determination and pattern formation.

BIBLIOGRAPHY
Books and reviews

Beermann, W., ed. 1972. Results and problems in cell differentiation. Vol. 4. Developmental studies on giant chromosomes. Springer-Verlag, Inc., New York.

Bryant, P. J. 1974. Determination and pattern formation in the imaginal discs of *Drosophila*. Curr. Top. Dev. Biol. **8**:41.

Bryant, P. 1979. Pattern formation, growth control and cell interactions in *Drosophila* imaginal discs. Symp. Soc. Dev. Biol. **37**:295.

Counce, S. J., and C. H. Waddington, eds. 1973. Developmental systems: insects. Vol. 2. Academic Press, Inc., New York.

Crick, F. H. C., and P. A. Lawrence. 1975. Compartments and polyclones in insect development. Science **189**:340.

French, V., P. J. Bryant, and S. V. Bryant. 1976. Pattern regulation in epimorphic fields. Science **193**:969.

Fristrom, J. W., et al. 1970. *In vitro* evagination and RNA synthesis in imaginal discs of *Drosophila melanogaster*. *In* E. W. Hanly, ed. Problems in biology: RNA in development. University of Utah Press, Salt Lake City.

Garcia-Bellido, A. 1975. Genetic control of wing disc development in *Drosophila*. *In* Cell patterning. Ciba Foundation Symposium No. 29. Elsevier North-Holland, Inc., New York.

Garcia-Bellido, A., P. A. Lawrence, and G. Morata. 1979. Compartments in animal development. Sci. Am. **241**:102 (July).

Gehring, W. J. 1970. Problems of cell determination and differentiation in *Drosophila*. *In* E. W. Hanly, ed. Problems in biology: RNA in development. University of Utah Press, Salt Lake City.

Gehring, W. J. 1973. Genetic control of determination in the *Drosophila* embryo. Symp. Soc. Dev. Biol. **31**:103.

Hadorn, E. 1968. Transdetermination in cells. Sci. Am. **219**:110 (Nov.).

Kalthoff, K. 1979. Analysis of a morphogenetic determinant in an insect embryo (*Smittia Spec., Chironomidae, Diptera*). Symp. Soc. Dev. Biol. **37**:97.

Morata, G., and P. A. Lawrence. 1977. Homoeoic genes, compartments and cell determination in *Drosophila*. Nature **265**:211.

Morata, G., and P. A. Lawrence. 1978. Cell lineage and homeotic mutants in the development of imaginal discs of *Drosophila*. Symp. Soc. Dev. Biol. **36**:45.

Nüsslein-Volhard, C. 1979. Maternal effect mutations that alter spatial coordinates of the embryo of *Drosophila melanogaster*. Symp. Soc. Dev. Biol. **37**:185.

Postlethwait, J. H., and H. A. Schneiderman. 1973. Developmental genetics of *Drosophila* imaginal discs. Annu. Rev. Genet. **7**:381.

Slack, J. 1978. Chemical waves in *Drosophila*. Nature **271**:403.

Suzuki, D. T. 1970. Temperature-sensitive mutations in *Drosophila melanogaster*. Science **170**:695.

Ursprung, H., and R. Nothiger, eds. 1972. Results and problems in cell differentiation. Vol. 5. The biology of imaginal discs. Springer-Verlag, Inc., New York.

Selected original research articles

Bryant, P. J. 1975. Pattern formation in the imaginal wing disc of *Drosophila melanogaster*: fate map, regeneration and duplication. J. Exp. Zool. **193**:49.

Kalthoff, K., P. Hanel, and D. Zissler. 1977. A morphogenetic determinant in the anterior pole of an insect egg (*Smittia* spec., Chironomidae, Diptera). Localization by combined centrifugation and ultraviolet irradiation. Dev. Biol. **55**:285.

Kauffman, S. A., R. M. Shymko, and K. Trabert. 1978. Control of sequential compartment formation in *Drosophila*. Science **199**:259.

Rice, T. B., and A. Garen. 1975. Localized defects of blastoderm formation in maternal effect mutants of *Drosophila*. Dev. Biol. **43**:277.

CHAPTER 23

Development of the cellular slime mold *Dictyostelium discoideum*

☐ The developmental biology of the cellular slime mold *Dictyostelium discoideum* has received considerable attention in recent years. The popularity of this system for workers interested in both the molecular and cellular aspects of development stems in large part from its relative simplicity. Its advantages include easy cultivation, a small genome whose complexity is only 2% of that of mammalian cells and eleven times that of *E. coli,* and a simple developmental cycle during which cells of a single type differentiate without cell division, growth, or DNA synthesis into one of two alternate cell types.

Taxonomic classification of the cellular slime molds (the Acrasiales) has been confused because their evolutionary relationship to other organisms is unclear, but they are now commonly classified in the kingdom Protista along with bacteria, protozoa, fungi, and algae. In nature, they are found in the soil where bacteria are abundant, for example, in forest litter. In the laboratory they can be grown on bacteria or in a simple axenic medium (i.e., a medium lacking other living organisms).

The basic life cycle of *D. discoideum* (Fig. 23-1, center) occurs as follows. In the vegetative cycle, spores germinate to yield irregularly shaped, solitary amoeboid cells (sometimes called myxamoebae to distinguish them from true amoebae) that live on microorganisms and multiply by binary fission (Fig. 23-2, *A*). When their food supply becomes low, the amoebae become mutually attractive. Tens of thousands of them aggregate and travel in streams to a central point where they form a conical mound (Fig. 23-2, *B*). This mound of cells bends over on its side and produces a sluglike mass that is 2 to 4 mm long and is called a pseudoplasmodium, slug, or "grex" (Fig. 23-2, *C*). The slug is enclosed by an extracellular sheath that prevents loss or addition of cells. It shows polarity and has an anterior tip that contains up

LIFE CYCLE OF
DICTYOSTELIUM
DISCOIDEUM

679

to about 20% of the cells and is smaller in diameter than the rest of the slug.

Under dark, moist conditions the slug enters a migratory phase. It migrates with the tip leading toward the light and up a temperature gradient (in nature, toward the soil surface). As the slug moves, it travels through a sheath of slime that the cells deposit and which adheres to the substratum. After migrating (or immediately after aggregation if conditions are dry and light), the slug forms a fruiting body (Fig. 23-2, D and E). During the process of fruiting body formation, or culmination, the cells occupying the anterior 15% to 20% of the slug move downward and form a stalk while the posterior cells are lifted atop the stalk and differentiate into spore cells. A few cells remain at the base of the stalk and form the basal disc, whose cells are relatively undifferentiated. The stalk cells die, but the spores can remain viable for years and will germinate under the proper conditions, with each spore producing an amoeba. Cells are normally haploid throughout the entire cycle of vegetative growth and fruiting body formation, although diploid cells can also form fruiting bodies.

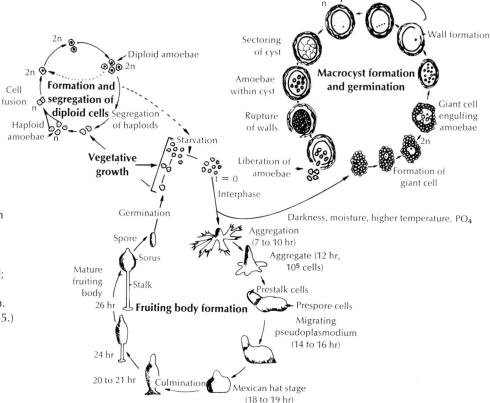

Fig. 23-1. Diagram showing stages in the life cycle of *D. discoideum*. Vegetative growth and fruiting body formation are depicted in center. Times are hours after beginning of starvation ($t = 0$). $n =$ Haploid; $2n =$ diploid. (Modified from Jacobson, A., and H. F. Lodish. 1975. Annu. Rev. Genet. **9:**145.)

Nongrowing amoebae have an alternate developmental fate (Fig. 23-1, right), a sexual stage, which occurs when they are grown in liquid instead of on a solid surface and under certain extreme environmental conditions. Under these conditions the cells do not undergo streaming when food becomes depleted, but instead they form loose aggregates. If the aggregates contain genetically identical cells, they usually die from lack of food, but if they contain cells of different mating types, they form a large macrocyst. A giant diploid cell forms in the center of the loose aggregate as a result of fusion of two cells, and this giant cell proceeds to engulf and digest the remaining cells of the aggregate. After all the cells have been engulfed, the giant cell forms a thick, multilayered cell wall and undergoes meiosis. Three of the four meiotic nuclei seem to be randomly destroyed, since an individual macrocyst produces only one class of offspring. Germination of the macrocyst occurs after a dormant period of 30 to 90 days. At this time the macrocyst becomes divided into several large proamoebae, which undergo mitosis and produce mature haploid amoebae, which are released from the macrocyst. The sexual stage could potentially be used for genetic studies of *D. discoideum,* but the inability to induce rapid germination of cysts has hampered efforts thus far.

D. discoideum also has a "parasexual" cycle (Fig. 23-1, left), which has been used for genetic studies. It results from the rare fusion of haploid cells to form binucleate cells, which may then undergo nuclear fusion to form diploid cells. These diploids often lose chromosomes and become aneuploid and eventually haploid again. The haploid cells may be of parental or recombinant genotype, depending on which chromosomes are lost (Fig. 23-3). The diploid and aneuploid cells also undergo mitotic crossing-over, in which parts of chromosomes are exchanged during somatic mitosis, as occurs normally during meiosis. Parasexual genetics has been used to map a number of mutations in *D. discoideum.* and it is hoped that genetic studies can eventually produce a fairly complete map of the *Dictyostelium* genome.

The events involved in the formation of fruiting bodies from vegetative myxamoebae of *D. discoideum* have been studied from a variety of different viewpoints. We will discuss three areas that have been well studied: (1) the aggregation of amoebae, (2) differentiation of amoebae into spore and stalk cells, and (3) structure of the genome of *D. discoideum* and its expression.

AGGREGATION OF AMOEBAE

Aggregation begins at about 6 to 7 hours after starvation of vegetative amoebae. The cells begin moving in a pulsatile manner toward a central point and then eventually form pulsating streams of aggregating cells. Such a coordinated type of behavior suggests that the cells

Fig. 23-2. Living preparations of *D. discoideum* showing stages in vegetative growth and fruiting body formation. **A,** Vegetative amoebae feeding on yeast. **B,** Amoebae undergoing aggregation. **C,** Migrating slug. **D,** Formation of fruiting body. **E,** Mature fruiting bodies. **A** is *Acrasis rosea;* **B** to **E** are *D. discoideum.* (From John Tyler Bonner, *The Cellular Slime Molds,* rev. and augmented 2nd edn. [copyright 1959 © 1967 by Princeton University Press]: Plates 6 and 7. Reprinted by permission of Princeton University Press.)

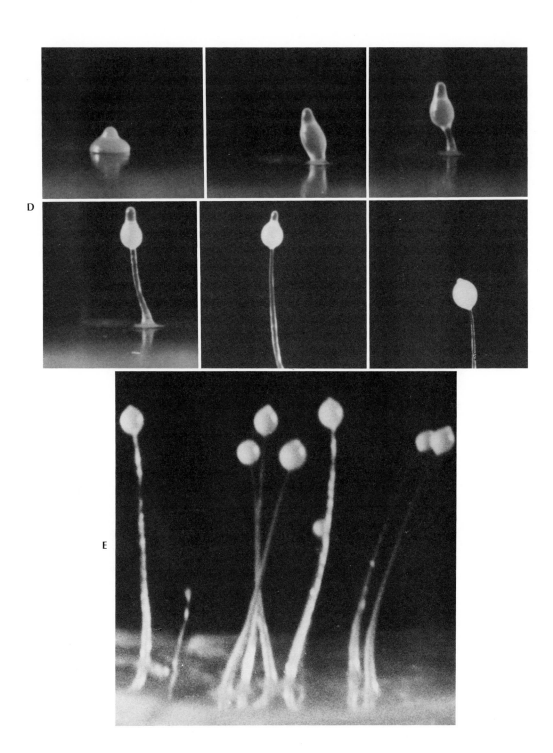

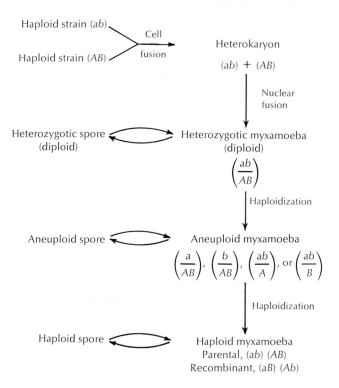

Fig. 23-3. Potential behavior of two unlinked genes, *Aa* and *Bb*, during the parasexual cycle of *D. discoideum*. The genotypes of the parental and daughter nuclei are given in parenthesis. (Modified from Sinha, U., and J. M. Ashworth. 1969. Proc. R. Soc. Lond. [Biol.] **173**:531.)

are responding to some factor in their environment. In one early experiment designed to test this possibility, cells were aggregated with a thin film of water over them. If the water was made to flow, amoebae upstream of the aggregating center moved about randomly while those downstream were directed toward the center. This result suggested that the center of the aggregate was producing a soluble attractant to which the starving cells were sensitive. This attractant was given the name "acrasin."

The existence of a chemical attractant was further indicated by the results of an experiment in which sensitive amoebae were placed on a thin film of agar overlying an aggregating group of cells. The amoebae on the agar aggregated in a pattern whose shape was identical to that of the lower group of cells. This finding implied that the aggregation center and the streams were producing a molecule that could affect the sensitive cells without direct cell contact.

Chemical identification of acrasin proved to be difficult, since the activity of acrasin-containing solutions was short-lived. This was because of the fact that aggregating cells also produce an enzyme that destroys acrasin. This enzyme was named acrasinase. After many years of experimentation, acrasin was finally identified as 3',5'-cyclic

AMP (cAMP) and acrasinase as cAMP phosphodiesterase, the enzyme responsible for converting cAMP to AMP.

The basic steps involved in aggregation have been shown to occur as follows. Aggregation is initiated by a small number of cells in the population that autonomously release periodic pulses of cAMP. Other cells in the population respond to the cAMP by moving toward the signal and also by releasing their own pulse of cAMP. After releasing a pulse of cAMP, amoebae are temporarily insensitive to it. During this refractory period, cAMP surrounding the cells is destroyed by a membrane-bound cAMP phosphodiesterase. This allows the cells to respond to the new pulse when it arrives. Each time a pulse of cAMP is detected, the cells undergo a surge of amoeboid movement in the direction of the side where their cAMP receptors are first filled.

Aggregating cells also form intercellular contacts and adhere in an end-to-end fashion. The cells acquire new cell surface antigens prior to aggregation. Univalent fragments of antibody to one of these antigens have been shown to block the formation of end-to-end aggregates.

Cellular properties that are required for aggregation, including membrane-bound cAMP receptors, membrane-bound phosphodiesterase, and membrane-bound aggregation sites, are acquired during the initial 6 to 10 hours after the beginning of starvation. This preaggregation phase is necessary for cells to become competent for aggregation. Cells can develop aggregation competence even when in an agitated suspension culture and will then undergo aggregation with little lag if transferred to a solid surface such as a coverslip.

DIFFERENTIATION OF AMOEBAE INTO STALK CELLS AND SPORE CELLS

As mentioned earlier in this chapter, construction of the fruiting body involves the differentiation of a single amoeboid cell type into two very different and highly differentiated cell types, the spore cell and the stalk cell (Fig. 23-4). Differentiation into one or other cell type is determined by a cell's position within the migrating slug. No matter what the overall size of a slug, the anterior 20% of cells become stalk and the posterior 80% become spore cells. This was originally demonstrated by experiments in which composite slugs were formed by combining parts of normal colorless slugs with parts of slugs that were red as a result of growing on a bacterium containing red pigment. In such slugs, anterior cells always formed the stalk, whereas posterior cells always formed spores.

Anterior prestalk and posterior prespore cells show some signs of predifferentiation prior to their actual differentiation into stalk and spore cells. Posterior cells have a characteristic structure, the prespore vesicle, which is absent from anterior cells, and posterior cells are

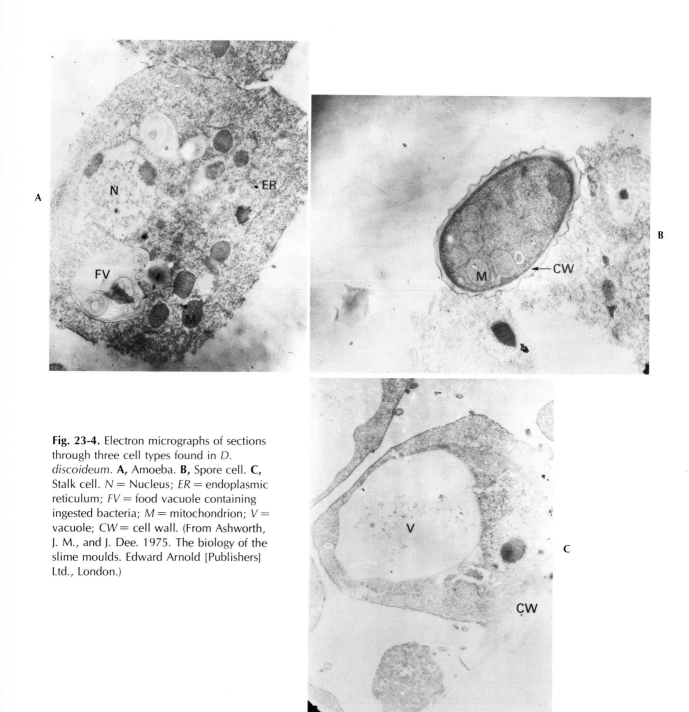

Fig. 23-4. Electron micrographs of sections through three cell types found in *D. discoideum*. **A,** Amoeba. **B,** Spore cell. **C,** Stalk cell. N = Nucleus; ER = endoplasmic reticulum; FV = food vacuole containing ingested bacteria; M = mitochondrion; V = vacuole; CW = cell wall. (From Ashworth, J. M., and J. Dee. 1975. The biology of the slime moulds. Edward Arnold [Publishers] Ltd., London.)

lighter in density than anterior cells. The posterior prespore cells also contain the enzyme uridine diphosphogalactose polysaccharide transferase, which prestalk cells do not. This enzyme is involved in synthesis of a mucopolysaccharide associated with spore cells.

The commitment of anterior and posterior cells to differentiate in specific directions is not irreversible, however. If a slug is cut into thirds and then allowed to migrate, each third will form a normal, though small, fruiting body. The anterior third, however, if not allowed to migrate, forms a fruiting body that is composed almost entirely of stalk cells (Fig. 23-5).

It has been suggested that the tip of the slug may act as a kind of organizing center and that after a slug is cut, the tip must undergo reorganization to direct the differentiation of a smaller group of cells. In the absence of migration, differentiation proceeds immediately, before new positional relationships can be established. The importance of the tip in organizing the slug has also been shown by experiments in which tips are grafted onto the sides of other slugs. This procedure results in the formation of several smaller slugs, each led by a grafted tip (Fig. 23-6). Removal of the tip at any time in development results in cessation of migration and morphogenesis until a new tip is formed from the remaining cells.

The factors responsible for differentiation of anterior cells into stalk cells and posterior cells into spore cells are unknown, but there are some properties of these two regions of the slug that may be rele-

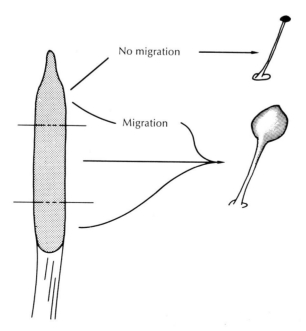

No migration

Migration

Fig. 23-5. Diagram showing fruiting body formation when *D. discoideum* is cut into thirds. After migration each third is able to form a normal fruiting body. If migration is not allowed, the anterior third is still able to differentiate, but in this case it forms a fruiting body composed primarily of stalk cells. (From Ashworth, J. M., and J. Dee. 1975. The biology of the slime moulds. Edward Arnold [Publishers] Ltd., London.)

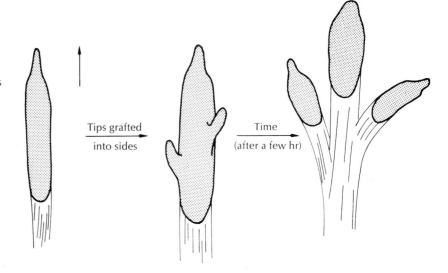

Fig. 23-6. Diagram showing the effect of grafting extra tips onto a migrating slug. Each tip organizes the formation of a small slug. (From Ashworth, J. M., and J. Dee. 1975. The biology of the slime moulds. Edward Arnold [Publishers] Ltd., London.)

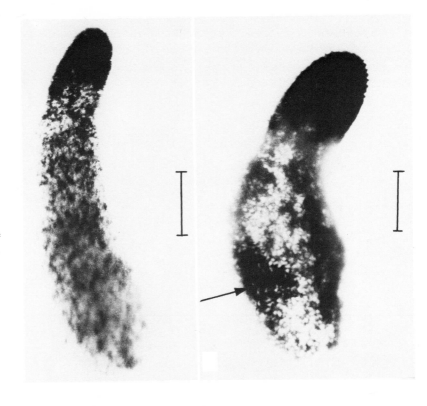

Fig. 23-7. *D. discoideum* slugs after histochemical staining for succinoxidase showing differential staining of anterior and posterior regions. The slug on right also contains a dark-staining region where a small hole (arrow) was made in a posterior region of the sheath, indicating that entry of the dye involved in the reaction is limited by the sheath. Scale bar = 0.1 mm. (From Farnsworth, P. A., and W. F. Loomis. 1974. Dev. Biol. **41**:77.)

vant to their differentiation. Slug cells continuously produce a cellulose-containing sheath and move forward relative to it. As a result, cells at the anterior end are surrounded by a thinner sheath than those located more posteriorly. This difference has been demonstrated by measurement of sheath thickness in electron micrographs of sections of slugs. The anterior 15% of the slug has been shown to stain more darkly than the posterior region in certain histochemical staining reactions. In the case of the reaction involving the enzyme succinoxidase, it has been demonstrated that the staining difference is not due to differential levels of succinoxidase in anterior and posterior cells but rather to the greater permeability of anterior cells to the dye used in the histochemical reaction (Fig. 23-7). This finding implies that substances released by slug cells may be present at a higher concentration in the posterior region than at the tip, where they would diffuse more readily through the sheath. It has been suggested that differences in the level of some such substance along the axis of the slug may provide a kind of positional information to which cells respond by differentiation into either stalk or spore cells. According to such a scheme, cells exposed to concentrations of the substance above a certain threshold value would respond by differentiating into spores.

Results of other experiments suggest that cAMP may be involved in the differentiation of stalk and spore cells. It has been shown by means of immunofluorescence assays that anterior cells have a greater concentration of bound cAMP than do posterior cells. Along the same line, it is known that the addition of high concentrations of cAMP to amoebae present at a density too low for aggregation causes the isolated cells to differentiate into stalklike cells. Specific diffusible promoters of stalk differentiation may also be involved.

Results obtained with a variety of experimental approaches have indicated that different genes are expressed at specific stages of *D. discoideum* development. For example, temperature-sensitive mutants have been obtained that are defective only in growth or only in development, as well as mutants that are defective in both. Hybridization studies done with purified single-copy *D. discoideum* DNA and RNA extracted from cells at different stages of the *D. discoideum* life cycle indicate that about 50% of the genome is expressed in the period including amoebal to midculmination stages (equivalent to about 16,000 genes 1000 nucleotides in length). About 19% of the genome is expressed during both vegetative and developmental stages, 11% to 12% of the genome is expressed only during negative growth, and 20% to 22% is expressed only during development.

Some specific examples of gene expression during fruiting body construction have been examined. Development of the fruiting body

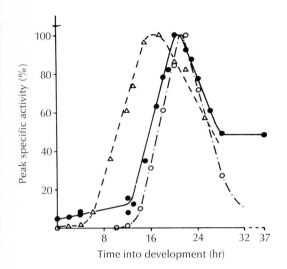

Fig. 23-8. Changes in the activities of three enzymes during the development of *D. discoideum.* △ = Trehalose-6-phosphate synthetase; ● = uridine diphosphoglucose pyrophosphorylose; ○ = uridine, diphosphogalactose polysaccharide transferase. (From Roth, R., J. M. Ashworth, and M. Sussman. 1968. Proc. Natl. Acad. Sci. U.S.A. **59:**1235.)

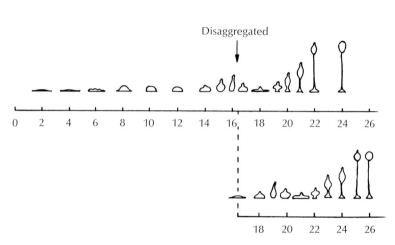

Fig. 23-9. Time course of morphogenesis of *D. discoideum* cells. Upper drawing shows stages of aggregate allowed to develop undisturbed. Lower drawing shows stages of cells disaggregated at 16½ hours of development and redeposited on fresh filters at their original cell density. (With permission from Newell, P. C., J. Franke, and M. Sussman. 1972. J. Mol. Biol. **63:**373. Copyright by Academic Press, Inc. [London] Ltd.)

requires the production of a number of polysaccharides not present in amoebae, and some of the enzymes in the metabolic pathways involved have been studied extensively. Changes in the activities of three of these enzymes during *D. discoideum* development are shown in Fig. 23-8. The activities of these enzymes are low in amoebae but rise dramatically a number of hours after development has been initiated by starvation. Increases seen in various enzyme activities appear to result from de novo protein synthesis and to require prior RNA synthesis, although this has not been directly demonstrated in all cases.

The activities of some developmentally regulated enzymes that accumulate at specific stages in *D. discoideum* development appear to be influenced by cellular aggregation and disaggregation and may thus be controlled by signals from the cell surface. When developing aggregates are shaken to produce a suspension of single cells and redeposited at too low a density for aggregation, enzyme activity does not increase beyond the level it had attained at the time of disaggregation. If cell aggregates are disaggregated as in the previous experiment but are then allowed to reaggregate, they do so rapidly, reaching within 2 to 3 hours the same morphogenetic stage that they may have

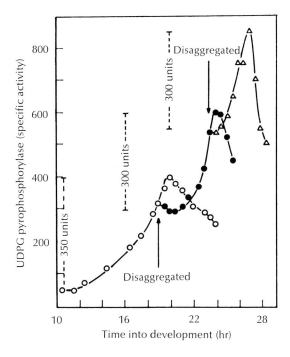

Fig. 23-10. Activity of uridine diphosphoglucose *(UDPG)* pyrophosphorylase in normally developing and disaggregated cells of *D. discoideum*. ○ = Undissociated control cells; ● = cells dissociated at 19 hours of development and redeposited on fresh filters; △ = cells dissociated at 19 hours and again at 23 hours and redeposited on fresh filters. (With permission from Newell, P. C., J. Franke, and M. Sussman. 1972. J. Mol. Biol. **63**:373. Copyright by Academic Press, Inc. [London] Ltd.)

taken as long as 18 hours to reach the first time (Fig. 23-9). In addition, enzyme activities in these reaggregated cells show a new increase, resulting in a final accumulation of about double the activity of normal fruiting bodies. Cells disaggregated and reaggregated a second time show an even further increase in enzyme activity (Fig. 23-10).

The relative simplicity of the *D. discoideum* genome has made its organization and mode of expression attractive objects of study, and in recent years considerable data have been accumulated regarding these subjects. The findings are especially interesting when viewed in the light of mechanisms that have been postulated for the regulation of gene expression in other eukaryotic cells, particularly the mechanisms envisioned in the Britten-Davidson model of gene expression (Chapter 7).

It has been determined from renaturation experiments that 70% of the DNA in *D. discoideum* is present in single copy and the remaining 30% is present in multiple copies, with the degree of repetition varying from 10^5 copies to as few as about 60 copies per genome. About 60% to 70% of the *D. discoideum* genome consists of single-copy DNA se-

STRUCTURE OF THE GENOME OF *DICTYOSTELIUM DISCOIDEUM* AND ITS EXPRESSION

quences with an average length of 1000 to 1200 nucleotide pairs interspersed with repetitive sequences with an average length of 250 to 400 nucleotide pairs. The DNA of *D. discoideum* also contains sequences of deoxythymidylic acid about 25 nucleotides long (poly[dT]$_{25}$). There are about 15,000 poly(dT)$_{25}$ sequences per genome (about one per gene), and they are interspersed with single-copy DNA sequences. As will be discussed, they are believed to be transcribed into poly(A)$_{25}$ sequences that are covalently linked to transcripts of single-copy DNA. The proposed arrangement of repeated sequences, single-copy sequences, and poly(dT) in the *Dictyostelium* genome is shown in Fig. 23-11. The arrangement of repeated and single-copy sequences is similar to that which has been found in some higher organisms and correlates with the arrangement of sequences predicted by the Britten-Davidson model of transcriptional control (Chapter 7).

The formation of mRNA in *D. discoideum* occurs by a process similar to, but somewhat simpler than, that found in mammalian cells (see Chapter 8). The nuclear precursor of mRNA in *D. discoideum* is only 20% longer than cytoplasmic mRNA, whereas hnRNA in mammalian cells is considered by many investigators to be at least three to four times larger than mRNA. The small precursor RNAs found in *D. discoideum* do not seem to be cleavage products derived from larger RNA molecules. Although the pre-mRNAs found in intact cells do not contain a 5′ triphosphate, as would be expected of primary transcripts, pre-mRNAs of the same small size are also made by isolated nuclei, and these do contain triphosphate residues at the 5′ terminus.

Most of the nuclear RNA in *D. discoideum* that is not a precursor to ribosomal RNA contains a sequence of poly(A) at its 3′ end, whereas only some of the hnRNA molecules in mammalian cells contain such a sequence. In contrast to the situation in mammalian cells, where

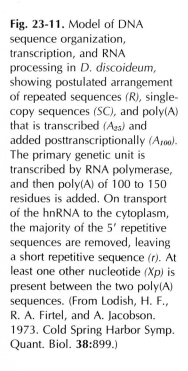

Fig. 23-11. Model of DNA sequence organization, transcription, and RNA processing in *D. discoideum,* showing postulated arrangement of repeated sequences *(R),* single-copy sequences *(SC),* and poly(A) that is transcribed *(A$_{25}$)* and added posttranscriptionally *(A$_{100}$).* The primary genetic unit is transcribed by RNA polymerase, and then poly(A) of 100 to 150 residues is added. On transport of the hnRNA to the cytoplasm, the majority of the 5′ repetitive sequences are removed, leaving a short repetitive sequence *(r).* At least one other nucleotide *(Xp)* is present between the two poly(A) sequences. (From Lodish, H. F., R. A. Firtel, and A. Jacobson. 1973. Cold Spring Harbor Symp. Quant. Biol. **38**:899.)

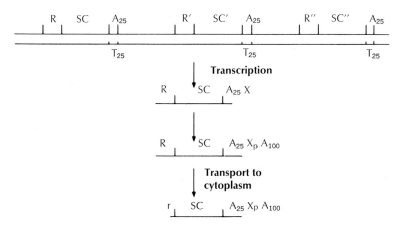

most of the hnRNA never reaches the cytoplasm, at least 70% of nuclear RNA containing poly(A) in *D. discoideum* is a precursor to cytoplasmic mRNA. This has been shown in two ways: (1) by experiments that determined the amount of pulse-labeled poly(A)-containing RNA transported from nucleus to cytoplasm and (2) by hybridizing nuclear poly(A)-containing RNA with cDNA made from mRNA with reverse transcriptase.

The nature of the RNA sequences that are not transported from nucleus to cytoplasm has been determined from experiments in which the types of sequences present in nuclear poly(A)-containing RNA are compared with those present in cytoplasmic poly(A)-containing RNA isolated from polysomes. More than 90% of the cytoplasmic mRNA hybridizes to *D. discoideum* DNA with kinetics indicating that it is complementary to single-copy DNA. In contrast, only about 75% of nuclear poly(A)-containing RNA is complementary to single-copy DNA. Since most nuclear poly(A)-containing RNA is transported to the cytoplasm, those sequences which are not transported must consist primarily of RNA transcribed from repetitive DNA. It has been shown that the repetitive sequences that do not reach the cytoplasm do not comprise a separate fraction of nuclear RNA but are located at the 5' ends of nuclear pre-mRNA and therefore must be removed from it prior to transport.

Two different poly(A) sequences are present in pre-mRNA and mRNA molecules of *D. discoideum*. When these RNAs are treated with ribonucleases that do not digest poly(A), two different sizes of poly(A) sequences are found—one about 100 nucleotides long and the other about 25 nucleotides long. These poly(A) sequences are both located near the 3' end of the RNA but are separated by at least one nucleotide that is not adenylic acid. The shorter tract of poly(A) is found in RNAs made by isolated nuclei, but the longer tract is not. The shorter sequence is believed to be transcribed from the $poly(dT)_{25}$ sequences present in the *D. discoideum* genome. The larger poly(A) tracts must be added posttranscriptionally, since there are no poly(dT) sequences in the genome large enough to code for them.

From these data a model has been put forth regarding mRNA production in *D. discoideum* (Fig. 23-11). The primary transcript is postulated to consist of the following sequences:

1. A sequence of about 300 nucleotides at or near the 5' end that is transcribed from repetitive DNA sequences
2. A sequence of about 1200 nucleotides that is transcribed from single-copy DNA
3. A sequence of $poly(A)_{25}$ near the 3' end that is transcribed from $poly(dT)_{25}$ tracts in the genome

This primary transcript is postulated to undergo two processing steps prior to its transport to the cytoplasm. Most of the sequences at the 5' end transcribed from repetitive DNA are removed, and a second poly(A) sequence about 100 residues long is added to the 3' end of the molecule. Transcription, processing, and transport take only about 4 minutes in *Dictyostelium*, compared to about 20 minutes in mammalian cells.

The function of the repetitive sequences at the 5' end and the poly(A)$_{25}$ at the 3' end of primary transcripts is unclear. The poly(dT)$_{25}$ from which the poly(A)$_{25}$ is transcribed could play a role in the termination of transcription. Since most of the repetitive sequence is not transported to the cytoplasm, its function must be related to events that occur in the nucleus. It is tempting to think that it may be a regulatory sequence of the type proposed in the Britten-Davidson model of gene expression. The degree of repetition suggests that there may be "families" of a few hundred genes each that share the same 5' repetitive sequence. Perhaps these genes undergo some kind of coordinate regulation. Alternately, the repeated 5' sequences could play a role in processing of the nuclear transcripts for transport to the cytoplasm.

As was stated at the start of this chapter, it is easy to understand the popularity of *D. discoideum* as a developmental system. Its beautiful and intriguing life cycle is basically simple and involves only a few different stages and cell types. The recent discovery of an alternate sexual cycle holds promise for genetic studies of *D. discoideum* development. Despite its relative simplicity, *D. discoideum* has many characteristics in common with higher eukaryotes. The organization of sequences in its genome is similar to that seen in a variety of multicellular eukaryotes, and the steps involved in gene expression are parallel to, although somewhat simpler than, those found in higher eukaryotes. *D. discoideum* development is regulated by some of the same mechanisms observed in the development of more complex organisms, such as hormonal signals (i.e., cAMP) and cellular position. The simplicity of the *D. discoideum* genome and the small number of cell types involved in its development will hopefully allow a precise analysis of the molecular events involved in the specific expression of genes that brings about cell differentiation.

BIBLIOGRAPHY
Books and reviews

Ashworth, J. M., and J. Dee. 1975. The biology of slime moulds. Edward Arnold (Publishers) Ltd., London.

Bonner, J. T. 1967. The cellular slime molds. Princeton University Press, Princeton, N.J.

Firtel, R. A., and A. Jacobson. 1977. Structural organization and transcription of the genome of *Dictyostelium discoideum*. Int. Rev. Biochem. 15:377.

Gerisch, G. 1968. Cell aggregation and differentiation in *Dictyostelium*. Curr. Top. Dev. Biol. 3:157.

Jacobson, A., and H. F. Lodish. 1975. Genetic control of development of the cellular slime mold *Dictyostelium discoideum*. Annu. Rev. Genet. 9:145.

Lodish, H. F., R. A. Firtel, and A. Jacobson. 1973. Transcription and structure of the genome of the cellular slime mold *Dictyostelium discoideum*. Cold Spring Harbor Symp. Quant. Biol. 38:899.

Loomis, W. F. 1975. *Dictyostelium discoideum:* a developmental system. Academic Press, Inc., New York.

Loomis, W. F. 1975. Polarity and pattern in *Dictyostelium. In* D. McMahon and C. F. Fox, eds. ICN-UCLA Symposium on Molecular and Cellular Biology. Vol. 2. Developmental biology. W. A. Benjamin, Inc., Menlo Park, Calif.

Selected original research articles

Beug, H., et al. 1973. Quantitation of membrane sites in aggregating *Dictyo-stelium* cells by use of tritiated univalent antibody. Proc. Natl. Acad. Sci. U.S.A. 70:3150.

Firtel, R. A. 1972. Changes in the expression of single copy DNA during development of the cellular slime mold *Dictyostelium discoideum*. J. Mol. Biol. 66:363.

Mackay, S. A. 1978. Computer simulation of aggregation in *Dictyostelium discoideum*. J. Cell Sci. 33:1.

Newell, P. C., J. Franke, and M. Sussman. 1972. Regulation of four functionally related enzymes during shifts in the developmental program of *Dictyostelium discoideum*. J. Mol. Biol. 63:373.

Pan, P., et al. 1974. Immunofluorescence evidence for distribution of cyclic AMP in cells and cell masses of the cellular slime molds. Proc. Natl. Acad. Sci. U.S.A. 71:1623.

CHAPTER 24

Aging

☐ The reader's first reaction may well be to ask why there is a chapter about aging in a book on development. Superficially, aging appears to be exactly the opposite of development—a wearing out and breaking down of everything that was formed and organized during development. Indeed, many researchers view aging as a degenerative process in which organization and function are gradually destroyed by stochastic (random chance) "hits" of one type or another, such as somatic mutations or accumulation of errors in protein synthesis. However, a substantial number of developmental biologists and gerontologists (those who study the biology of aging) are beginning to support a diametrically opposed view—that aging is an uninterrupted continuation of development, and that, far from being a random "wearing out" process, aging is brought about by specific regulatory mechanisms that are similar, at least in principle, to those responsible for other major events in the life cycle.

In this chapter aging is approached first in terms of descriptive biology and then in terms of proposals that have been offered to explain the observed phenomena. Relatively little is known concerning the actual mechanisms responsible for aging. However, this has not prevented the proliferation of an uncommonly large number of diverse theories of how aging works, each with its own highly vocal proponents. In fact, one reviewer recently referred somewhat sardonically to the "one investigator–one theory" approach to research in gerontology. That was an overstatement for effect, but there is, nevertheless, no generalized agreement among gerontologists concerning even the most fundamental aspects of aging.

One of the purposes of this chapter is to remind our readers that the field of developmental biology still contains many unanswered questions and areas of controversy. The subject of aging illustrates particularly well the diversity of different ways in which the same basic phenomenon can be interpreted and demonstrates the need for carefully planned experiments to distinguish among the many theoretical possibilities.

To those looking for firm answers and easily memorized principles, this chapter will be extremely frustrating. It is long and full of details, and, except for the basic descriptive biology of aging, virtually every point that is made is also argued against. The presence of this chapter in this book clearly indicates our bias toward interpreting aging in developmental and programmed terms, rather than in stochastic terms. However, current data are not strong enough to permit firm conclusions. Thus, to avoid misleading our readers, we have followed the description of each theory, including the ones we favor, with the arguments that are offered against it. We have tried for an honest portrayal of current thinking about a subject where most of the important discoveries appear not yet to have been made but which almost certainly will be made within the lifetimes of the students who use this book. It is our hope that our readers will be able to perceive in this chapter the sense of underlying excitement and challenge of future discovery that currently permeates research in aging, as well as in many other areas of developmental biology.

Although we all have a general concept of what aging consists of, it is surprisingly difficult to formulate a precise definition and even more difficult to find parameters that can be measured quantitatively as criteria of aging. Part of the problem is that aging is a slow, gradual process with few sharply defined landmark changes (such as menopause in the human female, which occurs too early in the total lifespan to be of maximum usefulness as a marker). A second source of difficulty is the variability that occurs from one individual to another in most of the traits that we commonly consider to be characteristic of aging, such as graying of hair, wrinkling of skin, and loss of physical strength. The pathology of aging is even more variable. Many pathological conditions such as cancer, cardiovascular diseases, and cataracts can be considered "age-related" in the sense that their incidence rises sharply with advanced age, but almost always they must be viewed in terms of increased vulnerability rather than as invariant symptoms of aging.

The lack of well-delineated markers makes it necessary to define aging either chronologically (the process of becoming old) or in terms of broad generalizations about the changes that are involved. Two basic concepts are usually included in the definition of aging: (1) a functional decline that leads to progressive loss of adaptation as time passes and (2) increased vulnerability to the forces of mortality, expressed statistically as a progressive increase in the probability of dying. Both these aspects of aging occur as smooth progressions that can be traced backward without interruption to relatively young ages. Although both changes are gradual and continuous, their mathemati-

DESCRIPTIVE BIOLOGY OF AGING

cal forms are different. Because of this, we will first consider each separately and then discuss the relationship between the two.

Loss of function

The aging process is characterized by a gradual loss of the organism's ability to cope with the situations that it encounters, both in everyday living and in times of special stress. In humans a wide variety of physiological functions have been shown to decline with age in an essentially linear fashion from about 30 years of age until senescence (Fig. 24-1). The extent of loss varies somewhat from one type of measurement to another, but on the average the amount of function remaining at age 80 is about half of that at age 30.

When such studies are pushed back to even younger ages, it appears that there is little, if any, stable adulthood between the completion of maturation and the beginning of age-related decline. Typical measurements of physical strength tend to increase with age until they reach a maximum at about the age of 25. In most cases a distinct decline can be detected by the age of 30. For mental functions such as the ability to learn new information the maximum appears to occur

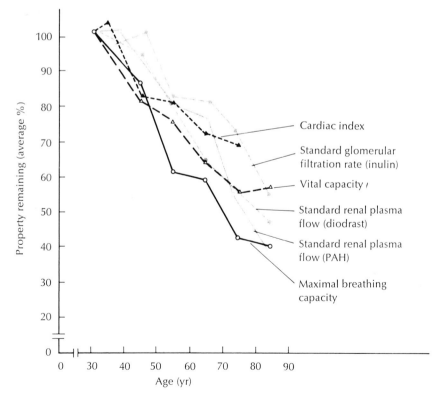

Fig. 24-1. Decline in efficiency of human physiological functions with age. Level at age 30 is arbitrarily assigned a value of 100%. (From Kohn, R. R. 1978. Principles of mammalian aging, 2nd ed. Prentice-Hall, Inc., Englewood Cliffs, N.J.)

Cardiac index

Standard glomerular filtration rate (inulin)

Vital capacity

Standard renal plasma flow (diodrast)

Standard renal plasma flow (PAH)

Maximal breathing capacity

even earlier. Thus typical students reading this book at about 20 years of age are at the peak learning capacity, if not already slightly "over the hill."

Although raw ability in most areas appears to reach its maximum and begin to decline soon after physical maturation, increased experience makes it possible to compensate effectively for the decline in absolute ability, at least for a while. How long this holds true depends on the type of activity. A professional football player may find it necessary to retire in his thirties, whereas a championship golfer may still be able to compete at 50, and the conductor of a symphony orchestra may still be able to do his job well at 80.

Closely related to adaptation and functional ability is the ability of an organism to remain alive. One of the basic characteristics of all species that undergo aging is a continually increasing probability of dying as they grow older. This can best be seen by plotting age-specific mortality (the probability of dying at a given age among those who have survived to that age) versus age. Because of the wide range of

Age-specific mortality and the Gompertz relationship

Fig. 24-2. Age-specific mortality in the United States, 1959 to 1961. **A,** Semilogarithmic plot of age-specific mortality (probability of not surviving an additional year among individuals who have arrived at a given age). The straight line exponential portion of the plot has been extrapolated to age zero to yield a hypothetical vulnerability to age-related death at the time of birth (M_o). Mortality doubles approximately every 8 years. The letters A to E are explained in the text. **B,** Linear plot of the same data. Mortality rates are plotted for every 5 years up to age 75 and every year above 75. Note the great increase in mortality at advanced ages, which tends to be deemphasized in the semilogarithmic plot. Also note that the region of deviation from exponential at advanced ages (labeled E in **A**) appears to be essentially linear in this plot. (Based on data from United States Life Tables 1959-1961, Vol. 1. No. 1. 1964. United States National Center for Health Statistics. Department of Health, Education, and Welfare, Public Health Service, Washington, D.C.)

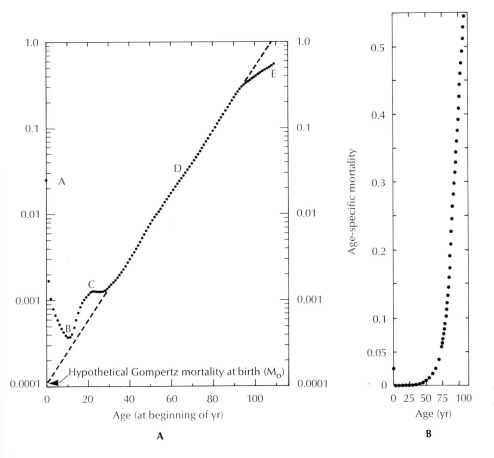

A

B

differences in age-specific mortality rate and the exponential increase in that rate with age, a semilogarithmic plot is normally used (Fig. 24-2, *A*), although a linear plot (Fig. 24-2, *B*) shows better the major impact that the exponential increase has at very advanced ages.

Five regions of particular interest (indicated by the letters *A* to *E*) can be identified in the plot of human age-specific mortality shown in Fig. 24-2, *A*. The very young are poorly adapted for survival and have a high mortality rate (*A*), primarily due to congenital defects and infectious diseases. For each year of life that is successfully completed, the probability of dying in the following year becomes less, up to the age of about 10, when the probability of dying is lower than at any other time in the entire life span (*B*). An increased mortality rate occurs among teenagers and young adults (*C*). This is primarily because of accidents, which are the major cause of death in that age range (Fig. 24-3). From about age 30 until advanced old age there is an almost perfect exponential increase in age-specific mortality (*D*). The name of Benjamin Gompertz, an insurance statistician, is often associated with this

Fig. 24-3. Mortality from accidents, cardiovascular diseases, and neoplasms in the United States (expressed as percentage of total mortality by age-group and sex). White area represents deaths from all other causes. (From Timiras, P. S. 1972. Developmental physiology and aging. Macmillan Publishing Co., New York.)

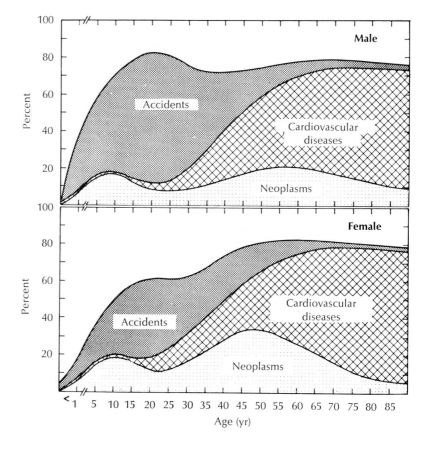

exponential relationship, which he first described in 1825. At ages above about 95, the rate of increase in mortality for the few people who survived that long appears to become less (*E*). This decrease could be due to existence of a small subpopulation of individuals who actually age at a slower rate, and thus dominate the small group of survivors left at very advanced ages. Alternately, the presence of data from a few individuals who have exaggerated their ages could also generate similar results. Accurate birth records are difficult to obtain for centenarians, and it is not unusual for ages to be advanced because of either faulty memory or the prestige associated with living to an unusually old age. There exist several areas in the world where claims are made for unusual longevity, but documentation is sparse, and many investigators have concluded that such "longevity" results from exaggeration under circumstances in which extreme old age is held in high esteem (such claims are reviewed in detail by Medvedev, 1974).

The exponential "Gompertz" curve for mortality, which extends without interruption from maturity to old age (Fig. 24-2, *A*) is generally considered to be an accurate representation of effects of aging. The straight line exponential portion of the age-specific mortality curve is commonly extrapolated to age zero to yield a hypothetical value for vulnerability to death due to "age-related" causes at birth (designated M_o in Fig. 24-2, *A*). The Gompertz relationship can then be expressed mathematically as $M_t = M_o e^{\alpha t}$, where M_t is age-specific mortality at age t, M_o is the extrapolated mortality rate at birth, and α is a proportionality constant.

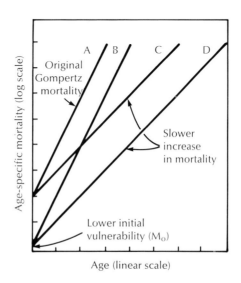

Fig. 24-4. Effects of M_o and α on mortality. *A*, Gompertz plot of mortality with relatively large M_o and α values. Mortality rises rapidly and results in a rather short life span. *B*, Effect of reducing M_o. Since α is unchanged, the curve rises as steeply as in *A*, but due to lower M_o the mortality is less at all ages, resulting in a modest increase in median and maximum life span. *C*, Effect of reducing α. Initial mortality rate is as high in *A*, but it increases more slowly, resulting in a significant extension in life span. *D*, Effect of reducing both M_o and α. Mortality starts lower and also increases more slowly, resulting in maximum increase in life span. Major differences in life spans of species, such as between mice and humans, reflect differences in both constants of the Gompertz equation (see Table 24-1).

Table 24-1. Comparison of initial vulnerability and doubling time of mortality rate for species with diverse life expectations*

Species	Life expectation (days)†	Initial vulnerability (per day)	Doubling time for mortality rate (days)
Blarina brevicauda (short-tailed shrew)	240	4.4×10^{-4}	87
Mus musculus (house mouse)	602	3.0×10^{-4}	220
Sigmodon hispidus (cotton rat)	514	2.2×10^{-4}	125
Oryzomys palustris (rice rat)	789	1.1×10^{-4}	197
Peromyscus leucopus (white-footed mouse)	1475	1.2×10^{-4}	447
Peromyscus californicus (California mouse)	1100	2.4×10^{-4}	441
Canis familiaris (beagle dog)	3617	2.7×10^{-5}	812
Equus caballus (thoroughbred mare)	6239	6.0×10^{-6}	1332
Homo sapiens (U.S. white female, 1969)	27,700	1.5×10^{-7}	3100

*From Sacher, G. A. 1978. In E. L. Schneider, ed. The genetics of aging. Plenum Press, Inc., New York.
†Life expectation is a calculated value, based only on Gompertz mortality, except for the shrew, for which the value is mean survival time. Deaths from birth to weaning are not included for the rodents.

As seen in Fig. 24-4, the two constants M_o and α both influence age-specific mortality. The rate of increase in mortality, determined by α, is considered to be an expression of the rate of aging. The absolute vulnerability to the forces of mortality is determined by M_o. A change in either of these values will affect life span. Reducing the value of M_o will reduce mortality at all ages and result in a longer average survival without affecting the rate at which age-specific mortality increases with age (Fig. 24-4, line *B*). Reducing the value of α slows the rate of increase in mortality and thus spreads aging over a greater period of time (Fig. 24-4, line *C*). Differences in life spans between species are likely to involve changes in both of the constants (Fig. 24-4, line *D*, and Table 24-1).

As will be discussed, a limited number of experimental manipulations have been found that alter life span. Typically, the primary effect of such treatments will be on only one of the constants of the Gompertz equation. For example, chronic administration of procaine hydrochloride to rats appears to reduce vulnerability to mortality (M_o) without altering the rate of aging (α), as shown in Fig. 24-5. Chronic caloric restriction, on the other hand, actually reduces the rate of aging (α), although there is also some apparent increase in vulnerability (M_o), presumably as an adverse side effect of the near starvation conditions needed to reduce the rate of aging (Fig. 24-6).

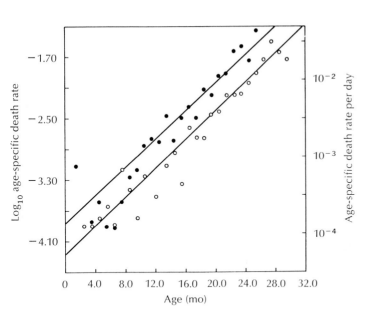

Fig. 24-5. Effect of procaine on mortality of rats. Long-term treatment of rats with procaine (○) reduces vulnerability to death at all ages as compared to controls (●), without altering the rate of increase in mortality. Thus the effect is on M_0 and not on α. (From Sacher, G. A. 1977. *In* Handbook of the biology of aging, edited by Caleb E. Finch and Leonard Hayflick, © 1977 by Litton Educational Publishing, Inc. Reprinted by permission of Van Nostrand Reinhold Co., New York.)

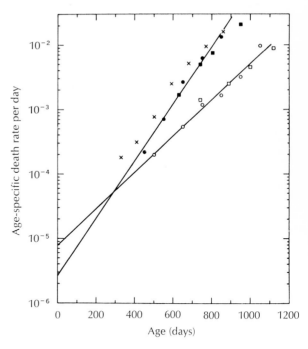

Fig. 24-6. Effect of caloric restriction on mortality of rats. Two restricted populations (○, □) are shown, together with their controls (●, ■) and a third "control" population (x). Caloric restriction significantly reduces the rate of increase of mortality (α) and also has a slight adverse effect on M_0. Note that the initial vulnerability in these experiments is significantly lower than that in Fig. 24-5. This could reflect either strain differences or differences in laboratory conditions. (From Sacher, G. A. 1977. *In* Handbook of the biology of aging, edited by Caleb E. Finch and Leonard Hayflick, © 1977 by Litton Educational Publishing, Inc. Reprinted by permission of Van Nostrand Reinhold Co., New York.)

For humans under the conditions in Fig. 24-2, the extrapolated probability of age-specific death during the first year of life is about 1 in 10,000 (compared to an actual mortality primarily due to non-age-related causes of about 1 in 40). The doubling time for the mortality rate (which is a function of α) is approximately 8 years. The very long life span of humans relative to most other animal species reflects both a lower M_0 value and a long doubling time for mortality rate (Table 24-1).

Relationship between function and mortality

The problem of how a linear loss of function that involves only a twofold change between ages 30 and 80 can be related to an exponential increase in mortality of 50- to 100-fold over the same age range has been the subject of speculation by many investigators. Several theories have been proposed suggesting that a relatively small reduction in vitality (function) can bring an organism below a "threshold level" needed for survival, particularly under conditions of acute stress. A detailed analysis of this relationship will not be undertaken here, but we will briefly summarize one theory that has been proposed by Bernard Strehler and Albert Mildvan.

This theory views survival as the process of overcoming a continual series of challenges, some mild and some severe, that are always being faced by the organism. Death occurs when a challenge is encountered that is too severe to be overcome. The amount of energy that an organism is able to expend to overcome a challenge is proportional to its functional capacity in other areas and declines linearly with age. Challenges of an intensity great enough to be lethal to a young, vigorous organism are assumed to be rare, and challenges that exceed a given intensity are assumed to increase exponentially* in frequency as the level of intensity decreases. This assumption predicts that a linear decline in the energy available to overcome challenges will result in an exponential increase in the probability of encountering a challenge that is too intense to be overcome. This relationship is diagrammed in Fig. 24-7.

---

*The actual assumption that is made is that the distribution of intensities of challenges is mathematically similar to the Maxwell-Boltemann distribution of energy levels among perfect gas molecules.

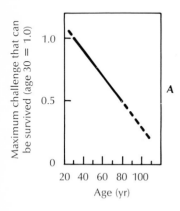

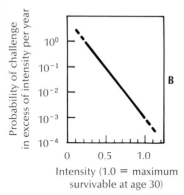

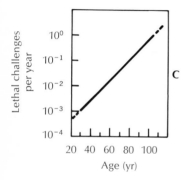

Fig. 24-7. Simplified representation of the Strehler-Mildvan theory of the relationship between linear decline in function and exponential increase in mortality. **A,** Linear decline in overall function (see Fig. 24-1) results in a linear decline in the intensity of a potentially fatal challenge that can be survived. It is assumed that survival is a measurement of the ability to achieve a level of energy output that exceeds the intensity of the challenge. The decline depicted here is 50% in the 50 years between ages 30 and 80, or a linear decline of 1% per year of the value at age 30. **B,** Probability of encountering a challenge that equals or exceeds a given intensity is assumed to increase exponentially as the intensity of the challenge becomes smaller. In these diagrams a 50% decrease in intensity is assumed to result in a 100-fold increase in probability of a challenge equaling or exceeding the intensity. **C,** Combining the probability distribution from **B** and the decline with age in ability to overcome challenges from **A** results in an exponential increase in the probability of encountering a lethal challenge (defined as one whose intensity exceeds the maximum effort that can be mustered against it). The frequency of lethal challenges (probability of dying) per year depicted here closely approximates the actual human mortality data in Fig. 24-2, *A*.

At present we know so little about the actual mechanisms of aging that proposals such as the Strehler-Mildvan theory must be regarded as strictly speculative. However, linear loss of function and exponential increase in mortality are firmly established experimental observations, and any alternative theories that are proposed must take them into account.

In any real population, some death occurs because of forces of mortality that do not change with age (e.g., deaths of small animals due to predators). Such deaths can be handled mathematically by adding a

Senescent and nonsenescent deaths

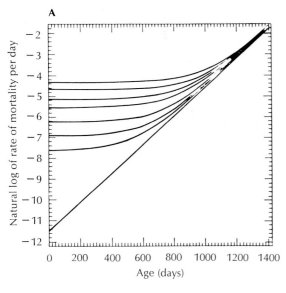 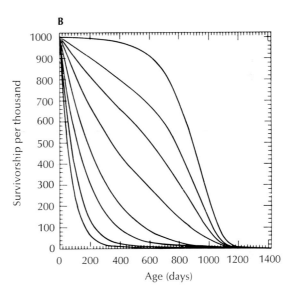

Fig. 24-8. Effect of nonsenescent deaths on the shape of mortality and survival curves. **A,** Age-specific mortality. Lowest line depicts "pure" Gompertzian mortality, increasing exponentially with age. Upper curves depict the effects of adding increasing levels of age-independent mortality. The result is a biphasic curve dominated by the constant mortality rate (β) early in life and by the exponentially increasing Gompertz mortality ($M_0 e^{\alpha t}$) later in life. **B,** Survival versus time for the populations whose mortality is depicted in **A.** Top curve depicts survival in a population subject only to senescent deaths. The death rate is initially low so that survival initially remains high. Mortality increases exponentially but does not have a significant effect on survival until about the middle of the life span, when survival begins to drop rather rapidly. The "rectangular" shape of this curve is characteristic of predominantly senescent death. As mortality that is independent of age increases (lower curves), the rectangular shape is lost and the curves approach the shape of a first-order decay curve. The lowest curve, for example, shows an essentially uniform decay of the population with a half-life of about 50 days for at least the first 600 days before Gompertz mortality becomes significant. (From Sacher, G. A. 1977. *In* Handbook of the biology of aging, edited by Caleb E. Finch and Leonard Hayflick, © 1977 by Litton Educational Publishing, Inc. Reprinted by permission of Van Nostrand Reinhold Co., New York.)

time-independent mortality factor to the Gompertz equation, as first proposed by William Makeham in 1867. In modern terms the Gompertz-Makeham equation can be expressed as $M_t = M_o e^{\alpha t} + \beta$, where β is the component of mortality that is independent of age. The effects of age-independent mortality on survival and mortality curves are illustrated in Fig. 24-8. In circumstances in which mortality rate is largely independent of age, the decline in number of survivors approaches the classical "first order" curve for the decay of a radioactive isotope with a precisely defined half-life.

Actual human survival curves under various medical and social conditions are shown in Fig. 24-9. Under optimum medical conditions, with nonsenescent deaths reduced to minimal levels (Fig. 24-9, top curve), survival approaches the theoretical "rectangular" form predicted by Gompertz kinetics (Fig. 24-8, *B*, top curve). At the other extreme, with minimal medical care, the survival curve begins to resemble that expected for an age-independent first-order decay (labeled "accidental" in Fig. 24-9). However, because of acquired immunity to infectious diseases, there are always some individuals who survive long enough to undergo senescent death. Table 24-2 demonstrates that modern medicine has greatly reduced premature deaths but not the rate of age-related increase in mortality.

Fig. 24-9. Survival curves for human populations under various medical and social conditions. Upper curve (Sweden, 1961 to 1965) has a minimum of "premature" deaths and approaches the maximum "rectangular" shape predicted from the Gompertz equation. At the other extreme the curve labeled "accidental" depicts age-independent death (first-order decay) with a half-life of about 13 years. Note that the lowest curves for actual populations (e.g., British India, 1921 to 1930) approach the "accidental" curve but differ from it in two areas: (1) higher than expected infant and child mortality and (2) a distinct component of senescent aging among the survivors of early death. (From Kohn, R. R. 1978. Principles of mammalian aging, 2nd ed. Prentice-Hall, Inc., Englewood Cliffs, N.J. Modified from Comfort, A. 1964. Ageing: the biology of senescence. Elsevier North Holland, Inc., New York.)

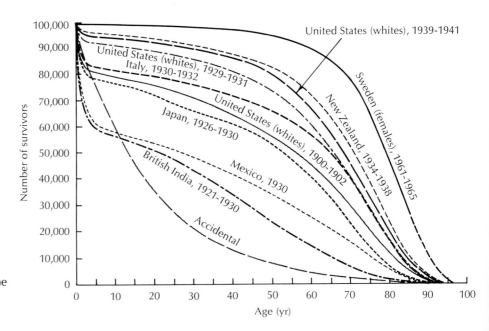

Table 24-2. Life expectancy at various ages in 1900-1902 and in 1974
(based on total population of United States)*

| Age (yr) | Life expectancy (average remaining years of life) | | Difference |
	1900-1902	1974	
At birth	48.2	71.9	23.7
1	55.2	72.2	17.0
5	55.0	68.4	13.4
15	46.8	58.6	11.8
25	39.1	49.3	10.2
35	31.9	39.9	8.0
45	24.8	30.9	6.1
55	17.9	22.7	4.8
65	11.9	15.6	3.7
75	7.1	9.8	2.7
85	4.0	5.7	1.7

*Based on data from Brotman, H. B. 1977. Gerontologist **17**:12.

Pathology of aging

It is sometimes difficult to distinguish between aging and the multitude of pathological changes whose incidence increases with age. The aging process is characterized by decreased ability to withstand challenges of any type and exponential increase in mortality due to virtually every possible cause (Fig. 24-10). Cardiovascular diseases and cancer are the two leading causes of death among the aged (Fig. 24-3), but if both could be completely eliminated, human life span would be extended by not more than 10 to 11 years because the frequency of death due to other causes also increases exponentially in old age. This is shown by the line *"All except atherosclerosis and neoplasms"* in Fig. 24-10. Thus curing specific diseases of aging reduces vulnerability (M_o) but has little or no effect on the rate of aging (α) of the Gompertz equation.

Degenerative diseases of the blood vessels are the most frequent cause of death in the extremely aged. Atherosclerosis is a lesion of the inner linings of arteries that involves the formation of enlarged "plaques" of cellular and fatty material, which in advanced cases may also become calcified. The term "arteriosclerosis" refers to a thickening and stiffening of the arteries (hardening of the arteries) that frequently accompanies atherosclerosis. The two terms are often confused and sometimes used interchangeably. However, it is atherosclerosis that is the major cause of death in old age. The atherosclerotic plaques severely restrict flow of blood, and sometimes, in combination with small blood clots, which may be carried through the circulation

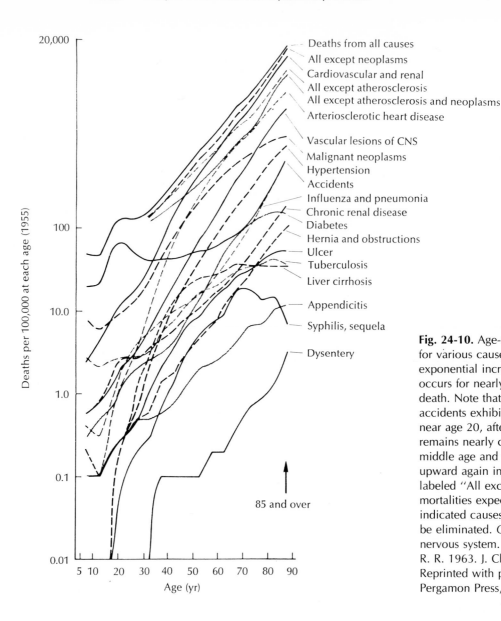

Fig. 24-10. Age-specific mortality for various causes of death. An exponential increase in mortality occurs for nearly all causes of death. Note that the curve for accidents exhibits a maximum near age 20, after which it remains nearly constant through middle age and turns sharply upward again in old age. Curves labeled "All except" are the mortalities expected if the indicated causes of death could be eliminated. *CNS* = Central nervous system. (From Kohn, R. R. 1963. J. Chronic Dis. **16**:5. Reprinted with permission from Pergamon Press, Ltd.)

from an injury elsewhere in the body, they completely block blood flow. When this occurs in the coronary artery, which supplies oxygenated blood to the heart muscle, the result is a heart attack, and when it happens in the brain, the result is a stroke.

Cancer is another disease whose incidence rises rapidly with advanced age, although not as rapidly as cardiovascular diseases. Thus while the rate of death due to cancer continues to rise (Fig. 24-10), the percentage of all deaths at a given age that are caused by cancer

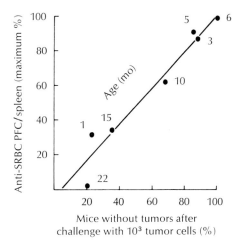

Fig. 24-11. Correlation between T cell-dependent humoral immunity and resistance to injected tumor cells in mice of various ages. The number of cells in the spleen that form antibodies against sheep red blood cells (anti-*SRBC* PFC) is plotted against the ability to reject tumors for mice of various ages. Very young mice (1 month) are deficient in both assays. A maximum is reached in both tests at 6 months, followed by a major decline with age. A strong linear correlation between the two responses is clearly evident. (From Makinodan, T. 1977. *In* Handbook of the biology of aging, edited by Caleb E. Finch and Leonard Hayflick, © 1977 by Litton Educational Publishing, Inc. Reprinted by permission of Van Nostrand Reinhold Co., New York.)

shows a decline (Fig. 24-3). Diminishing function of the immune system is apparently an important factor in the increased incidence of cancer in old age. Organ transplant patients whose cell-mediated immunity has been suppressed to prevent graft rejection have been found to have a much higher incidence of cancer than expected for their age. A process known as immune surveillance appears to recognize cellular changes and routinely destroy incipient cancer cells in normal, healthy young individuals. Experiments with mice have shown a close correlation between cell-mediated immunity and the ability to reject injected tumor cells at various ages (Fig. 24-11). Both these functions rise to a maximum in juveniles and young adults (3 to 6 months of age) and begin to decline significantly with advancing age (10 to 22 months).

Nonfatal age-related changes

There are a number of types of age-related pathological conditions that generally are not primary causes of death. Osteoporosis is a demineralization of bones that occurs in both sexes but tends to be a particularly severe problem in postmenopausal women, in whom bone loss averages about 7% per decade. The bones of very old individuals are sometimes reduced to thin shells (Fig. 24-12), and the risk of severe bone fractures resulting from seemingly minor accidents is a serious medical problem among the elderly. Arthritis and related degenerative diseases of the joints often severely limit mobility and ability to maintain an independent lifestyle among the elderly. Presbyopia, or inability to alter the focus of the eyes, makes it necessary for virtually everyone to use reading glasses or bifocals by the time they reach 50 years of age (Fig. 24-13). This condition is due to a stiffening of the lens of the eye so that the muscles are no longer able to change its shape enough to focus on close objects.

The term "senescence" is used to refer to the generalized decline in function that occurs during aging. The terms "senile" and "senility" are also sometimes used to refer to general characteristics of extreme old age, but more frequently they refer specifically to mental degeneration, particularly loss of ability to remember new information and impaired reasoning capacity. Extreme cases are often referred to as "senile dementia." Senility is frequently caused by restricted blood flow to the brain resulting from atherosclerosis or by permanent brain damage caused by anoxia during a stroke. However, there are also a variety of other organic brain syndromes that involve autoimmune phenomena (immune reactions directed against an individual's own

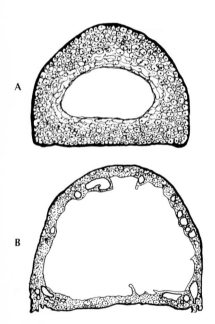

Fig. 24-12. Osteoporosis. **A,** Cross section of a finger bone of a 39-year-old woman. The bone is of normal thickness. **B,** Comparable cross section from a 90-year-old woman. The bone is reduced to a thin shell, and the medullary cavity is much enlarged. (From Andrew, W. 1971. The anatomy of aging in man and animals. Grune & Stratton, Inc., New York. By permission.)

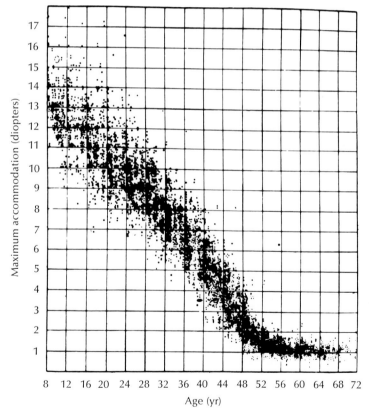

Fig. 24-13. Loss of ability to focus the eyes with age. Maximum focal accommodation (in diopters) is plotted against age for 4000 individuals. The loss with age is essentially linear. (From Strehler, B. L. 1977. Time, cells and aging, 2nd ed. Academic Press, Inc., New York.)

tissues) or deposits of proteinaceous or fatty material in the brain tissue. Many physically healthy old people are confined to nursing homes because of various forms of mental impairment.

Although all the conditions just discussed increase in incidence with age, none of them (with the possible exception of presbyopia) occur predictably enough to be considered markers for measuring the progress of aging. The same is also true for a variety of nonpathological changes such as graying of the hair and male pattern baldness. One physiological change that does seem to occur with predictable regularity is an increase in the degree of cross-linking of collagen. Chemical cross-links between the individual protein molecules in a collagen fiber continue to form and be converted to more stable forms as maturation and aging progress, gradually making the fiber tougher and tougher. The degree of cross-linking varies from one tissue to another, but for a given tissue and species of organism the degree of cross-linking has proved to be one of the more accurate physiological measurements of chronological age. A number of theories of aging have been constructed around the idea that increased cross-linking might restrict diffusion and increase stiffness to a point where function would be impaired. However, there is little firm evidence, other than increased cross-linking itself, to support such theories, and they have generally fallen into disfavor.

The patterns of aging that are observed among various mammals are similar, even for species with very different life spans, such as mice and humans. During the first quarter of the life span the animal grows and matures. Maximum levels of physiological function and reproductive capacity are achieved early in the second quarter of the life span and have begun to decline significantly by the time half the life span is completed. Female fertility is lost during the third quarter, and in both sexes there is a generalized functional decline, commonly characterized as "middle age." The last quarter of the life span is characterized by overt aging, including an overall decline in vitality and rapidly increasing mortality due to many different primary causes. The relative uniformity of the overall pattern of aging that is observed, irrespective of its absolute rate, has led a number of investigators to propose that there exists a specific "program" for aging whose basic pattern has been conserved during evolution, although the rate at which it proceeds has been altered quite freely. This concept will be discussed further later in this chapter.

Time sequence of aging

As mentioned briefly at the beginning of this chapter, the study of aging is characterized by an unusually large number of different theories and proposed mechanisms. In general, these can be categorized

PROPOSED MECHANISMS AND THEORIES OF AGING

as "stochastic" (due to random chance events) or programmed (happening according to a definite plan), although there are also some interesting combinations of both (e.g., programmed decline in repair of stochastically generated damage). One of the reasons for the existence of such a vast number of theories is the difficulty that has been experienced in attempts to sort out cause and effect. As we have already seen, aging can be characterized as a pleiotropic process – one that affects many different systems at the same time. In fact, virtually every aspect of normal physiological function undergoes at least some change during aging. Because of this, it is almost impossible to decide whether a particular change that is observed to occur during aging is simply a result of aging or whether it might play some kind of causative role in aging.

The true identity of the causative mechanism ultimately responsible for aging remains one of the major unsolved mysteries of modern biology, and investigators from many disciplines have sought the answer in their areas of specialty. Thus there exist theories of aging whose roots lie in the study of nutrition, metabolism, free radical chemistry, membrane biology, polymer science, physiological psychology, neurobiology, endocrinology, immunology, molecular biology, protein synthesis, genetics, somatic mutation, radiation biology, ultrastructure, cell biology, and mathematics. The diversity of the resultant theories makes it difficult even to classify all of them in a systematic manner. Rather than attempting to describe these many theories in detail, or even to list all of them, we will talk in terms of general types of theories and look in detail only at a limited number of theories and proposed mechanisms that are of particular interest.

In addition to the dichotomy between stochastic and programmed mechanisms of aging, it is also possible to make a distinction between "intrinsic" and "extrinsic" mechanisms. Intrinsic mechanisms are those which are built directly into the individual cells and tissues of an organism so that the aging of each component proceeds more or less independently, whereas extrinsic mechanisms are those imposed on the cells from the outside. Theories based on intrinsic mechanisms imply that each cell or tissue contains some kind of "clock" capable of causing aging in the absence of outside influences. Theories involving extrinsic mechanisms, on the other hand, are based on the assumption that aging is not an inherent property of most cells and tissues and that it is externally caused. Extrinsic mechanisms can be further subdivided into those in which aging is caused entirely by an external agent, such as radiation, and those in which a "pacemaker" organ (which contains an intrinsic aging clock) controls the aging of all other tissues in the body. In mammals, both the thymus gland and the neuroendocrine system have been suggested as possible pacemakers of aging.

Wearing out. Superficially there are enough similarities in the aging of men and machines that it seems natural to think of aging as a kind of "wearing out" at the biological level. An old person, like an old machine, tends to have many different parts that do not work well prior to the time that one particular part fails so completely that the entire unit can no longer function at all. Over the years there have been proposed many different theories based on wearing out, running down, and increased entropy.

"Entropy" is a term used by physical chemists to describe the state of randomness and equilibration that all systems tend to assume unless energy is put into them to maintain a more organized state. In one sense a living organism is a highly organized and highly improbable state of its environment. Energy must be expended continually to maintain that organization and prevent the components of the organism from reassuming their random positions in the environment. In this sense life can be considered as a constant battle against entropy, and aging can be viewed as the gradual victory of entropy.

However, a simple wearing out does not in itself explain aging. A vigorous young organism has the ability to use metabolic energy not only to maintain its state of organization but also to improve it. Alexander Leaf has expressed it this way: "A striking difference between our bodies and a high-quality mechanical device such as a bicycle or automobile is that we improve with usage, while the mechanical device can only deteriorate from wear and tear."* It is our everyday experience that physical strength and coordination are improved by vigorous training programs, manual dexterity is improved with practice, and knowledge and ability to solve problems are expanded by studying. Thus aging is not a simple wearing out process. Neither is it simply a loss of the body's ability to strengthen itself in response to increased levels of activity. Even in the elderly, conditioning programs involving regular exercise can usually improve significantly the physical fitness of previously inactive people. Nevertheless, the process of aging steadily diminishes both the level of fitness that can be achieved with maximum effort and the level that is maintained in the less active person.

Although simple "wearing out" theories are clearly not adequate in themselves to explain aging, a whole new class of "wearing out" mechanisms of aging has been proposed in recent years, organized

Stochastic mechanisms that are intrinsic in nature

---

*Leaf, A. 1975. Youth in old age. McGraw-Hill Book Co., New York.

around limitations in repair capacity. One of the currently most popular approaches is to attempt to link life span with ability to repair damaged DNA. Since the damaging agents are generally presumed to be of extrinsic origin (e.g., radiation), this theory will be described in the discussion of extrinsically caused aging.

Running down. A second group of classical theories is built around the general concept that each organism has a specific quota of living to accomplish and will age and die as that quota is used up. Most of these theories can be traced back to two types of metabolic observations made early in this century. The first is that cold-blooded animals grown at higher temperature show a decrease in life span that is roughly proportional to the increase in their metabolic rate so that total lifetime metabolism remains roughly the same at high or low temperature. The second is that among mammals, small species with high metabolic rates have short life spans, whereas larger species that metabolize less rapidly live longer. Although the correlation is inexact, particularly for humans, total metabolism per gram of tissue per lifetime is more or less constant for many different species with very different lifetimes.

Fig. 24-14. Relationship between body weight and life span for various species of mammals. (From Lamb, M. J. 1977. Biology of ageing. Blackie & Son Ltd., Glasgow.)

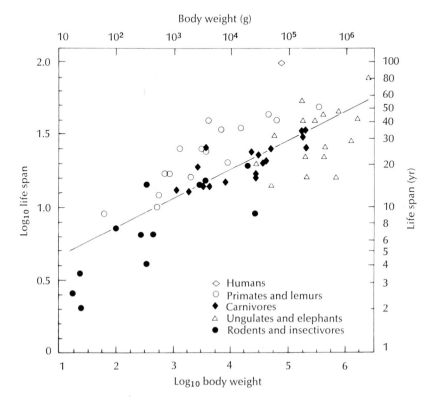

Since body weight is closely related to metabolic rate, some investigators have looked instead at the correlation between body weight and life span. Again, they have found a positive but imperfect correlation among the mammals (Fig. 24-14). Since humans live three times as long as they should on the basis of body weight, and since humans have disproportionately large brains, correlations of life span with brain size were also examined and were found to show less deviation than body weight (Fig. 24-15). A multivariate analysis of the effects of both brain and body weight on life span yields an equation which suggests a positive correlation between life span and brain size, with a penalty for having a body that is too big in proportion to the brain.

Without in any way saying what the mechanism is, such data suggest that a mammal's "quota" of life is somehow related to the size of its brain. However, all such data must be interpreted cautiously. For example, in terms of values established for mammals, birds live far too long for their brain size, body size, or metabolic rate. Likewise, mammals such as bats that enter true hibernation with lowered body temperature live longer than predicted by body or brain weight, suggesting that total lifetime energy metabolism may also be important.

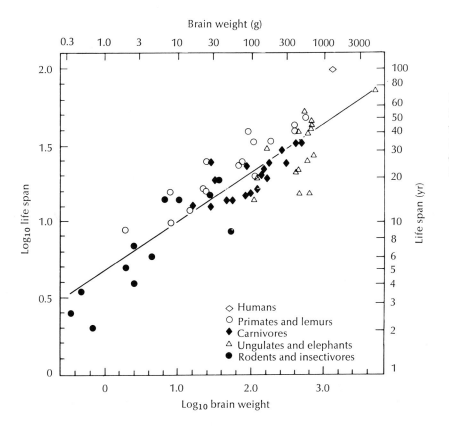

Fig. 24-15. Relationship between brain weight and life span for various species of mammals. (From Lamb, M. J. 1977. Biology of ageing. Blackie & Son Ltd., Glasgow.)

A possible role of lifetime energy dissipation is also suggested by the experiments mentioned earlier (Fig. 24-6) showing that chronic underfeeding significantly extends the total life span of rats (and mice) by slowing the rate of aging. In the original experimental design the animals are placed immediately after weaning on a diet that is nutritonally complete but with only enough calories for a very slow rate of growth. Such growth-restricted animals remain juvenile physiologically as well as in size for prolonged periods of time. When returned to a normal diet after as much as 1000 days on the restricted diet, most of the animals resume normal growth (although they often do not become as large as controls), mature sexually, and display normal fertility. Their postpubertal life spans are reduced somewhat, but their total life spans are significantly increased. Many diseases normally associated with aging are reduced in frequency, including cancer and respiratory diseases. However, the bones of the treated animals tend to become brittle, and they are prone to cataracts.

Calculations of total calories consumed per gram per lifetime show nearly equal values for restricted and control animals. However, data from the calorie-restriction experiments are also compatible with a number of other theories of aging. Cross-linking of collagen is delayed in the retarded animals. Maturation of endocrine function is delayed, suggesting a possible role of the hypothalamic-pituitary neuroendocrine axis (see discussion of pacemaker theories). Maturation of the immune system is also delayed, which is compatible with theories of autoimmunity and aging. The fact that the life span after sexual maturation is only slightly shortened can be interpreted to favor programmed aging, with the initiation of each new phase dependent on completion of the previous one. Finally, tryptophan deficiency has an effect similar to caloric deficiency both on sexual maturation and on total life span. It has been suggested that this is due not to the role of tryptophan as an essential amino acid but rather to its role as a precursor for the neurotransmitter, serotonin. This possibility can be fitted into various neuroendocrine theories of aging, which could also be related in some way to the apparent relationship between brain size and life span.

In at least one sense it is true that every organism has a "quota" of living. An aging "clock" of unspecified nature that increases age-specific mortality with a characteristic doubling time for each species is set into motion early in life. Thus far the only means that have been found to alter that clock are metabolic restrictions such as underfeeding and reducing body temperature (valid only in species in which it can be achieved without traumatic side effects). There are many other pharmacological treatments that appear to have minor effects on life span, but whenever these have been analyzed carefully, the effects have been found to be on vulnerability to mortality and not on the rate of aging. (This concept is analyzed in detail by Sacher, 1977.)

Free radical reactions, lipid peroxidation, membrane damage, lysosomal leakage, and lipofuscin accumulation. A complex set of theories has grown up around these topics, which are interrelated as follows. Free radicals are unstable chemical compounds with unpaired electrons that are generated during normal oxidative metabolism and can also be generated by radiation. They are capable of reacting with many types of biological molecules. Multiple double-bond systems such as those in polyunsaturated fatty acids are highly susceptible to free radical–induced peroxidation and polymerization. Lipid peroxidation, in turn, can disrupt the normal lipid bilayer structure of biological membranes. Lysosomes are intracellular vesicles that contain a variety of hydrolytic enzymes that are normally used for intracellular digestion within phagocytic and autophagic vesicles. Lysosomes have single membranes, which, according to the theory, are particularly susceptible to disruption by per-

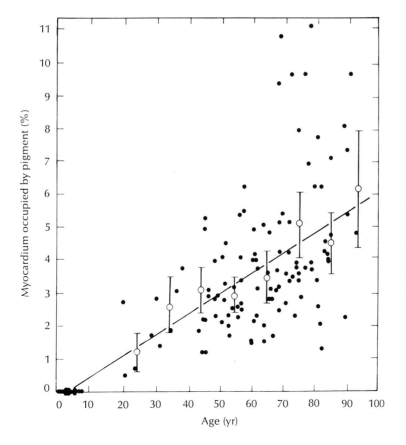

Fig. 24-16. Accumulation of lipofuscin with age in human heart muscle. ● = Precentage of the myocardium occupied by pigment in individual cases; ○ = means for 10-year periods; vertical bars = ±2 standard errors of the means. Although individual scatter is large, the means exhibit a definite increase with age that is nearly linear. (From Lamb, M. J. 1977. Biology of ageing. Blackie & Son Ltd., Glasgow.)

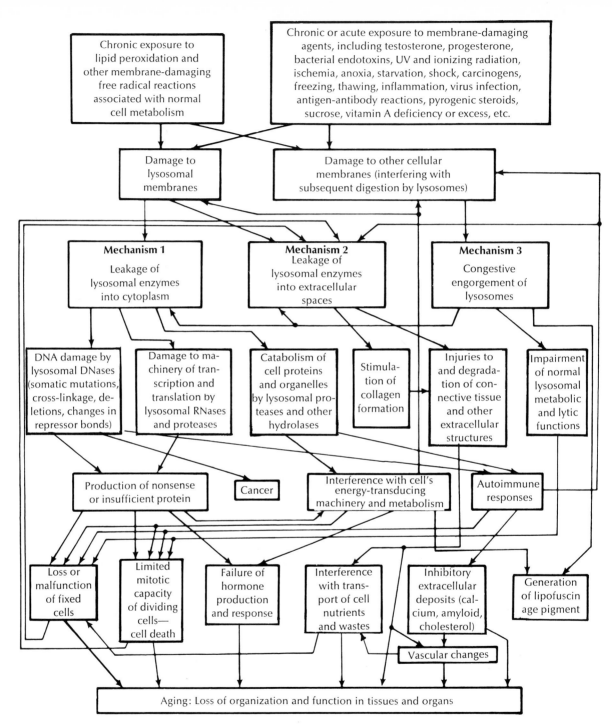

Fig. 24-17. Diagram summarizing possible interrelationships among free radicals, lipid peroxidation, membrane damage, lysosomal leakage, lipofuscin formation, and aging. (From Hochschild, R. 1971. Exp. Gerontol. **6:**153. Reprinted with permission from Pergamon Press, Ltd.)

oxidized lipids, with resultant leakage of lysosomal enzymes into the surrounding cytoplasm, as well as into extracellular spaces. Such leakage can presumably result in a multitude of kinds of intracellular and extracellular damage that ultimately lead to aging. This effect is further enhanced by accumulation in lysosomes of indigestible residues resulting from peroxidation and polymerization of membrane lipid structures. Such accumulations may cause further lysosomal leakage. Peroxidized lipids in combination with biological amines form a fluorescent pigment known as lipofuscin, which accumulates in substantial amounts in cells of aged individuals (Fig. 24-16). The complex diagram in Fig. 24-17 represents one investigator's attempt to show the relationship of all these changes to aging.

Lipofuscin accumulation itself appears to be relatively innocuous, since treatment with centrophenoxine, a drug that effectively dissipates lipofuscin from cells, has at best only relatively minor effects on aging as a whole. Likewise, treatment with a variety of antioxidants and free radical inhibitors, such as vitamin E and butylated hydroxytoluene (BHT), has only minor effect on aging or mortality. This may be due to the fact that normal cells already contain three different enzymatic mechanisms for dealing with intracellular peroxides and other free radical products—catalase, glutathione peroxidase, and superoxide dismutase. Theories of this type have a number of very active supporters, but thus far they have not been able to produce more than suggestive evidence for mechanisms of the type summarized in Fig. 24-17.

Cellular senescence. The phenomenon of cellular senescence or intrinsic loss of cellular capacity to reproduce has been widely studied both in cell culture systems and in experimental animals. Since both stochastic and programmed explanations have been proposed for cellular senescence, the basic phenomenon will be discussed here, with further elaboration of possible mechanisms in the section on programmed aging.

The limited reproductive capacity of cultured normal diploid cells has already been mentioned in Chapters 3 and 10. Although the phenomenon had been observed earlier, detailed studies by Leonard Hayflick and Paul Moorhead first called widespread attention to cellular aging in the early 1960s. They found that under precisely controlled conditions, the growth of cultures of human diploid cells can be divided into three distinct phases (Fig. 24-18). Phase I consists of the early establishment of a culture before it enters exponential growth; phase II is a period of exponential growth, which for fetal lung fibroblasts extends over approximately 50 population doublings; and phase III is characterized by a cessation of growth, with maintenance of

Fig. 24-18. Diagrammatic representation of origin, growth, and senescence of diploid cells in culture. Phase I is the period of slow growth while the primary culture (first growth in culture from original tissue) is getting started. Phase II refers to a period of rapid multiplication in which cell number increases exponentially with time. Phase III is a period when multiplication ceases and the culture eventually dies out. For human diploid fetal lung fibroblasts, phase III occurs after about 50 doublings. Recent reports indicate that with sufficient care nonmultiplying phase III cells can be maintained in culture for long periods of time without dying. (From Hayflick, L., and P. Moorhead. 1961. Exp. Cell Res. **25**:585.)

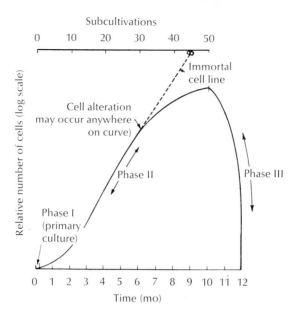

nonmultiplying cells for awhile, followed by degeneration. (Recent studies in other laboratories suggest that nonmultiplying phase III cells can be maintained for long periods with adequate culture media and conditions.)

The question of whether phase III is an expression of aging has remained controversial for many years. Early studies by Hayflick showed that for adult tissue, phase III occurred after fewer population doublings than for fetal tissue. Careful studies from a number of other laboratories have made it clear that doubling potential in culture is definitely lower for cells from older donors. In skin fibroblasts the loss of doubling potential is about 0.2 doublings per year of donor age, or about 16 doublings in an 80-year life span (Fig. 24-19). However, cells from very old individuals are still capable of a substantial number of doublings in culture, and forward projection of the regression analysis indicates that an individual would have to live to an age of over 200 for total loss of doubling potential of his skin fibroblasts. Thus, although cellular multiplication potential is clearly age-related, its exact relationship to aging is less clear. It is, nevertheless, widely studied as a potential model of somatic cell aging. In addition, it bears an interesting resemblance to the phenomenon of clonal aging observed in various unicellular organisms such as *Paramecium*, which must periodically undergo a sexual cycle to retain reproductive capacity. Asexual clones that are prevented from entering any kind of sexual cycle

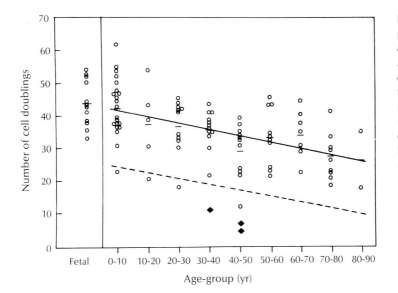

Fig. 24-19. Loss of doubling potential of cultured cells with age of the donor. All cultures (○) are of human skin fibroblasts derived from biopsy samples taken at a standardized location (middle of the inner side of the upper arm) to minimize variation due to the source of the cells. Short horizontal lines are the mean number of doublings in culture for each 10-year age-group. Note that individual variation is large and that cultures from some old individuals undergo more doublings before phase III than cultures from some young individuals. Solid line is a linear regression for the entire age range examined. If extended, it would reach zero cell doublings at about 220 years of age. Dotted line is the lower 95% confidence limit for the regression line. Three solid diamond-shaped symbols represent cells from patients with Werner's syndrome, which do not multiply well in culture. (From Norwood, T. H. 1978. *In* E. L. Schneider, ed. The genetics of aging. Plenum Publishing Corp., New York.)

gradually lose their ability to reproduce asexually after about 100 to 200 population doublings and ultimately die.

The mechanisms responsible for cellular senescence are not well understood. Experiments in which cells have been held in a nondividing state for long periods of time make it appear that the critical factor is how many doublings the cells have undergone, rather than the total amount of time that they have spent in a state of active metabolism. Environmental conditions modulate somewhat the number of doublings that a particular kind of cell will undergo in culture, but the basic counting mechanisms appear to be intrinsic. There is considerable variation in cellular life span in culture, both from species to species and among various tissues within the same species. There is not a good correlation between species life span and the number of doublings that cells from a particular species will undergo in culture. However, cells from rodents tend to undergo transformation into permanent lines rather easily, making it difficult to evaluate the phase III phenomenon when they are used. There are reports that diploid lines from rats have been maintained through large numbers of doublings in culture with only barely detectable changes in their karyotypes, but in theory even a point mutation might be enough for transformation into a permanent line.

Aging in culture is characterized by an increasing probability that cells will not divide again. Contrary to early reports, withdrawal from the cell cycle is not an abrupt synchronous event that occurs at the onset of phase III. Instead it occurs with increasing frequency as the overall cultures age. This can be seen both as a decline in the percentage of cells capable of incorporating tritiated thymidine into their nuclei and as a progressive decline in the percentage of cells capable of forming large colonies in clonal growth assays (Fig. 24-20).

Accumulation of lysosomes and residual bodies has been reported in phase III cells. However, experiments in which phase II cells are

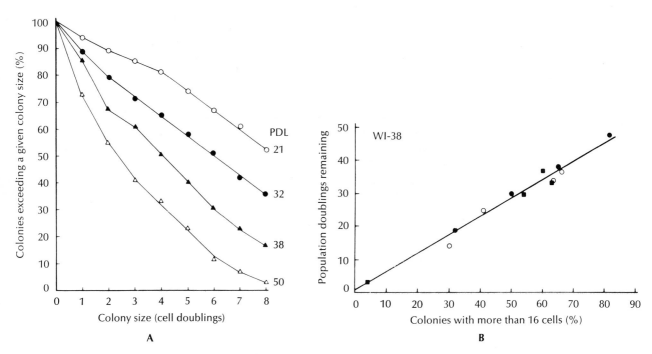

A

B

Fig. 24-20. Loss of colony-forming ability as cells approach phase III. A, Distribution of colony sizes (in number of doublings) at different population doubling levels (PDL). Widely separated single cells (WI-38) are inoculated into Petri dishes of culture medium and incubated until maximum colony size is obtained (2 to 3 weeks). All cells that attach are considered to be colonies of zero doublings. All colonies of two or more cells are counted as having completed one division, four or more cells, two divisions, etc. As the parental cultures age, an increasing fraction of their cells form only small, abortive colonies even after prolonged incubation. B, Decline in percentage of large colonies as phase III is approached. The number of colonies with 16 or more cells (four or more doublings) is plotted against population doublings remaining before phase III (determined by maintaining stock cultures until they reach phase III and counting backward to the cultures used as an inoculum for the clonal growth experiments). Different symbols indicate independent series of experiments. The fraction of colonies with more than 16 cells accurately predicts how many doublings the parental cultures will undergo before reaching phase III. (From Smith, J. R., O. Pereira-Smith, and P. I. Good. 1977. Mech. Ageing Dev. 6:283.)

maintained in a nondividing state by contact inhibition suggest that the accumulation results from the lack of division rather than causing it. When phase II cells with large accumulations of residual bodies are allowed to resume growth, there is rapid dilution of the residual bodies.

Only minor extension of life span can ordinarily be achieved by adjusting environmental conditions. Treatment of cultures with hydrocortisone throughout the entire culture period causes a statistically significant increase in the number of doublings achieved, but this seldom exceeds a 25% extension of the doubling potential of untreated controls. Although there was one report in the literature of a major extension of life span by treatment with vitamin E, neither the original investigators nor others have been able to repeat the experiment successfully.

Experiments in which phase II cells are fused to phase III cells with Sendai virus have shown that the senescent state is dominant (Table 24-3). In dikaryons formed by such fusions, neither the phase II nor the phase III nucleus will initiate new DNA synthesis, although phase II nuclei that are in the S period of the cell cycle at the time of fusion will finish that round of DNA synthesis. T98G is a human glial cell line that is immortal but does not exhibit transformed properties such as loss of growth arrest in crowded cultures. When T98G cells (or other similar lines) are fused to phase III human diploid fibroblasts and cultured at low density, the T98G nuclei behave in a manner similar to phase II diploid nuclei. However, if a phase III cell is fused with a fully transformed cell such as HeLa, the transformed phenotype

Table 24-3. DNA synthesis in heterokaryons produced by fusion of different types of cells*†

Nucleus A		Nucleus B	
From cell type	**DNA synthesis**	**From cell type**	**DNA synthesis**
Mortal phase II	+	Mortal phase II	+
Mortal phase III	−	Mortal phase II	− ‡
Mortal phase III	−	Immortal but subject to G_1 arrest	− ‡
Mortal phase III	+	Immortal transformed (no G_1 arrest)	+

*Based on data from Norwood, T. H., et al. 1974. Proc. Natl. Acad. Sci. U.S.A. **71:**2231; Norwood, T. H., et al. 1975. J. Cell Biol. **64:**551; and Stein, G. H., and R. M. Yanishevsky. 1979. Exp. Cell Res. **120:**155.

†"Mortal" refers to cells that enter phase III after a limited number of doublings in culture and "immortal" to those which have unlimited multiplication potential.

‡No new DNA synthesis is initiated, but synthesis that is in progress at the time of fusion is completed in the dikaryon.

is dominant, and both nuclei enter DNA synthesis. Thus there appear to be factors that can diffuse from one nucleus to the other and influence DNA synthesis either positively or negatively. The senescent phenotype is dominant, both over the normal phase II phenotype and over the T98G phenotype (which is normally immortal). Preliminary evidence suggests that the inhibitory signal from the phase III nucleus may be similar to regulatory signals that prevent entry of normal phase II cells into the cell cycle under conditions of crowding or growth factor deprivation (Chapter 17). This viewpoint is consistent with the "terminal differentiation" interpretation of phase III, which is discussed in the section on programmed aging later in this chapter. Fully malignant cells, such as HeLa, are able to override the inhibition of DNA synthesis, however, with a resultant reactivation of DNA synthesis in phase III nuclei in dikaryons with HeLa nuclei. A further analysis of the regulatory signals responsible for phase III and its reversal is clearly needed.

The possible role of loss of cellular reproductive capacity in aging of experimental animals has been investigated rather extensively. One approach has been to transplant rapidly reproducing tissues from an old animal to a younger one to determine whether cellular reproductive capacity continues beyond the end of the life span of the original donor animal. The answers that have been obtained thus far suggest that the ability of stem cells to reproduce extends far beyond the needs of the original donor animal but does not continue indefinitely. For example, red blood cell precursors can be transplanted into hosts whose own red blood cell precursors have been destroyed by irradiation. When the hosts grow old, further transplants can be made into young irradiated animals, and the cells can be kept in a state of active multiplication for prolonged periods. Similarly, skin grafts of contrasting color can be made into genetically compatible hosts and observed through several host generations. Also, mammary ductal epithelium can be transplanted into mammary fat pads that have had their own ductal tissue removed, and when the transplant expands to fill the entire fat pad, it can serve as the source for a new transplantation. In each case of this kind the transplants have eventually ceased to proliferate, but only after far more doublings than would have been needed to provide fully normal function throughout the entire life span of the original donor animal.

A particularly interesting situation has been observed with transplants of mammary ductal epithelium. The experiments are normally done by transplanting tissue into young females so that most of the ductal growth occurs during sexual maturation. When such growth ceases to occur after repeated transplants, the host animals containing the nongrowing tissue can be bred. Under the hormonal stimulus of pregnancy the ductal transplant undergoes differentiation into secretory alveoli (Chapter 14), and the cells in it once again enter a state of active proliferation. Thus in this particular case the ability to proliferate in response to one particular set of hormonal signals can be lost without affecting the ability of the same cells to proliferate in response to a different set of signals.

On the basis of available data it is difficult to forge a strong link between limited proliferative capacity and the overall process of aging. Direct limitation of proliferative capacity of stem cell lines does not appear to occur during normal aging, as discussed. In addition, age-related degenerative changes are observed in all types of cells in the body, including postmitotic cells, such as those of the central nervous system and skeletal muscles, that do not multiply at all in adult life. Thus, although the phase III phenomenon of cultured cells is interesting in itself, a convincing case has not yet been made for its proposed role as a primary mechanism of aging in the organism.

Error catastrophe and other molecular theories. In 1963 Leslie Orgel proposed a molecular mechanism for aging based on a positive feedback of errors in protein synthesis. This theory attracted the attention of many investigators, both in gerontology and in molecular biology, and has stimulated considerable experimental activity, much of which is still ongoing.

The basic concept of the error catastrophe theory, as it has been popularly named, is that certain proteins function catalytically in their own synthesis and that random errors in synthesis of those proteins may introduce new errors in the next round of protein synthesis (Fig. 24-21). This can create a positive feedback loop that continually increases the error frequency in synthesis of the catalytic proteins, which in turn leads to increased error frequency in the synthesis of all proteins. If the initial level of errors is sufficiently low and the feedback coupling is small (but not so small that the system stabilizes), the errors will accumulate exponentially like compound interest. Initially the accumulation will scarcely be perceptible, but as time passes and the exponential increase reaches its final stages, the errors appear to build up rapidly to catastrophic proportions (Fig. 24-22). In mathematical terms the accumulation will follow the equation $E_t = E_0 e^{\alpha t}$, where E_0 is the initial error frequency, E_t is the error frequency at time t, and α is a proportionality constant. This equation is identical in form to the Gompertz equation for increase in age-specific mortality with time.

Unfortunately, the theoretical predictions of the error catastrophe theory do not match observed changes in aging. The most reasonable expectation is that errors in protein synthesis should express them-

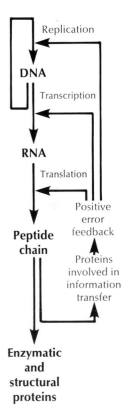

Fig. 24-21. Schematic diagram of the closed loops in biological information transfer that could potentially lead to a positive feedback of errors in protein synthesis.

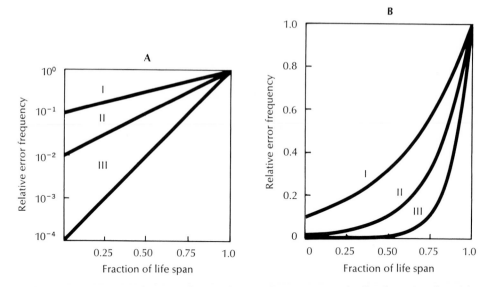

Fig. 24-22. Exponential accumulation of errors due to positive feedback. **A,** Semilogarithmic plot showing three possible rates of exponential error accumulation. In *I,* lifetime increase in errors is 10-fold; in *II,* 100-fold; and in *III,* 10,000-fold. **B,** Linear plots of error accumulation in the three models from **A.** None of these curves match the linear loss of function observed between ages 30 and 80 in humans, but curve I, which reflects the smallest lifetime increase in error frequency, comes closest. Curve III closely approximates human mortality data (Fig. 24-2) but predicts an exponential loss of function that is not observed.

selves as a decrease in function. However, as discussed at the beginning of this chapter, the loss of function with age is approximately linear and not exponential as predicted by the error catastrophe theory. Mortality does increase exponentially, but there does not appear to be any easy way to link mortality directly to errors in protein synthesis without also predicting an exponential loss of function.

Despite these theoretical objections, numerous experiments have been undertaken seeking direct molecular evidence for an error catastrophe. The usual approach has been to look for the accumulation of altered proteins during aging, either in experimental animals or in cultured cells approaching phase III. Parameters that are examined include the following:

1. The thermal stability of various enzymes is measured to detect amino acid substitutions that do not prevent enzyme function but make the protein easier to denature.
2. The specific activity or ratio of enzyme activity to enzyme protein is examined by comparing catalytic activity with the total amount of cross-reacting material (CRM) precipitated by anti-

body to the enzyme. (This assay is based on the assumption that most of the single amino acid substitutions that render enzymes catalytically inactive will not significantly alter their antigenic activity.)

3. Amino acid substitutions are looked for in purified proteins such as isoleucine in hemoglobin, which normally does not contain that amino acid. Negative results from such experiments must be interpreted cautiously, however, since except for very conservative substitutions (e.g., isoleucine in place of valine at the N-terminal end of the hemoglobin chain), amino acid substitutions may change the physical properties of proteins sufficiently so that they do not copurify with the unaltered proteins.

The results of such studies have been mixed. There is substantial evidence that altered proteins do accumulate in many systems during aging. However, such alterations may not be due to an error catastrophe, since they could also arise by other mechanisms such as somatic mutation or posttranslational modification. In addition, the error catastrophe theory predicts that newly synthesized proteins of all types will contain large numbers of errors as the loss of translational accuracy builds up to a crisis level. This does not appear to be the case, either in cultured cells or in experimental animals. Viruses that are dependent on host translational machinery replicate normally in phase III cells and in old animals. Transformation of phase III cells with small oncogenic viruses such as SV40, whose entire genome consists of only about six genes, results in rapid resumption of growth, which presumably could not happen if the phase III cell were suffering from an error catastrophe. Deliberate introduction of errors in protein synthesis by growth of cells in the presence of amino acid analogs that are incorporated into proteins generally does not hasten the occurrence of phase III or have any lasting effects after the analogs are removed. Although some enzymes lose activity and accumulate inactive crossreacting material during aging, many others do not. At least some proteins such as salivary α-amylase and hemoglobin appear to remain essentially unchanged during aging. Thus, although it may still be too soon for a final judgment, current data strongly suggest that a generalized error catastrophe does not occur during normal aging, either of cultured cells or of intact animals.

Numerous other molecular theories of aging have also been proposed, but none of these have attracted the wide attention given to the error catastrophe. Histone binding to DNA appears to become tighter during aging, possibly because of changes in the relative proportions of different histones that are present. It has been suggested that this may lead to a generalized decrease in gene expression, but studies of transcription have failed to produce any strong evidence for such a theory. DNA extracted from older animals tends to have more strand breaks, but it is not certain whether this reflects the state of the DNA in intact cells or the presence of higher levels of nucleases that cause the breaks during the extraction procedure. Interesting theories could be built around either alternative, but so far there is no strong evidence for an important role of either.

Thus, although our bias says that aging, like all other biological phenomena, must ultimately be explained at the molecular level, we are forced to the conclusion that no satisfactory molecular explanation has yet been offered. Such an explanation clearly remains one of the important challenges for future research on aging. However, before that is likely to happen, extensive research will be needed to gain a better understanding of how aging occurs at the cellular and organismic levels to provide an adequate basis for predictions at the molecular level that can then be subjected to experimental testing.

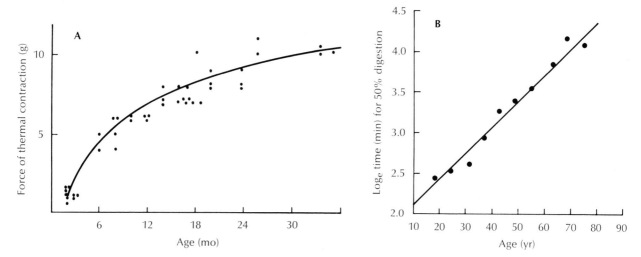

Fig. 24-23. Collagen cross-linking. **A.** Weight required to inhibit thermal contraction of rattail tendon as a function of age. Increased temperature disrupts hydrogen bonds between linear molecules in collagen fibers. As the molecules assume a random configuration, the tendon becomes shorter. Weight attached to the tendon causes the molecules to slide past one another and prevents the shortening. Covalent cross-links between the molecules prevent the sliding and permit small weights to be lifted during shortening. Thus the weight needed to prevent thermal contraction measures cross-linking. The biggest increase in cross-linking occurs early in life. **B,** Collagenase digestion of human tendon collagen as a function of age, expressed as the natural logarithm of the time for 50% digestion. Increased cross-linking slows the penetration of the enzyme into the collagen and thus increases digestion time. (From Kohn, R. R. 1978. Principles of mammalian aging, 2nd ed. Prentice-Hall, Inc., Englewood Cliffs, N.J.)

Collagen cross-linking. We mentioned earlier in this chapter that cross-linking of collagen increases with age and that some investigators have proposed that reduced diffusion of nutrients and waste products across basement membranes, which are composed largely of collagen, may be a causative factor in aging. Depending on viewpoint, this proposed mechanism can be considered to be intrinsic (it occurs entirely within the organism), extrinsic (it imposes effects on cells from the outside), or pacemaker (the rate of change of collagen cross-linking controls the rate of aging). In addition, proponents of neuroendocrine mechanisms of aging maintain that the cross-linking of collagen is ultimately controlled by the neuroendocrine pacemakers (e.g., the pituitary gland). Finally, the cross-linking of collagen can be viewed as part of the organism's developmental program, since it is essential to the development of normal adult strength and toughness. Proponents of this view suggest that aging may result from failure to turn off this program when it has proceeded far enough to satisfy adult needs.

Interestingly, the same data have also been used to argue that collagen cross-linking is probably not a major causative factor in aging. Examination of the relationship between cross-linking and age (Fig. 24-23) reveals that the greatest changes occur during maturation and that the changes between adulthood and old age are relatively small. Many investigators have suggested that the changes in collagen cross-linking that occur after maturation are not sufficient to account for the major loss of function that occurs during aging. In addition, there do not appear to be any firm data indicating that a significant reduction in diffusion actually occurs in old age. However, as mentioned earlier, quantitative measurement of collagen cross-linking provides an estimate of chronological age that is more accurate than almost any other physically quantifiable parameter. However, even this may be the result of an external pacemaker action, since when the synthesis of new collagen is artifically induced in an old animal, that collagen quickly acquires the degree of cross-linking expected for the age of the animal.

Radiation. Cumulative damage from natural background radiation has long been cited as a possible mechanism of aging. Exposure to graded doses of ionizing radiation, on either an acute (large single dose) or a chronic (small doses over an extended period of time) basis, leads to an increase in mortality that is additive with the natural forces of mortality. However, when causes of death are examined, it appears that the primary effect is due to enhancement of certain radiation-related diseases (such as cancer), rather than a general acceleration of the overall process of aging. Long-term studies of World

Stochastic mechanisms that involve extrinsic agents

War II atomic bomb survivors have failed to demonstrate more rapid progression of aging in general, despite increased mortality. For example, collagen cross-linking continues to correspond to chronological age. An interesting observation has come from a comparison of irradiated and normally aging mice. Pronounced graying of the hair is observed in black mice that are irradiated. However, graying of the hair is not a part of normal aging for that strain and apparently results from the killing of pigment stem cells by the irradiation, rather than from acceleration of aging.

Closely related to radiation theories are proposed mechanisms of aging based on accumulation of somatic cell mutations of unspecified origin. In support of such mechanisms, there does appear to be some increase in chromosomal abnormalities during aging. However, examination of the structures of specific proteins has failed to produce any evidence for the accumulation of somatic mutations during normal aging. For example, human hemoglobin contains no isoleucine other than trace amounts due to translational errors or somatic mutation. No significant increase in the isoleucine content can be detected as a function of age in control populations with a normal level of exposure to radiation. However, a significant increase was detected in Marshall Islanders who had been exposed to radiation from nuclear weapons testing 20 years earlier. Also, it has been found that the effect of radiation on the life span of the wasp *Habrobracon* is the same in haploid and diploid individuals, despite the greater sensitivity of the haploids to mutational effects.

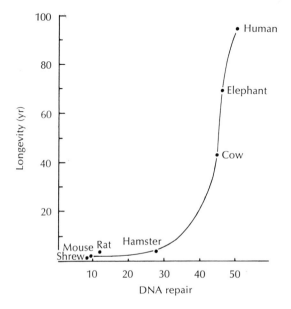

Fig. 24-24. Relationship between longevity (maximum life span) and ability to repair DNA by replacement of thymine dimers that are formed as a result of ultraviolet radiation. (From Sinex, F. M. 1977. *In* Handbook of the biology of aging, edited by Caleb E. Finch and Leonard Hayflick, © 1977 by Litton Educational Publishing, Inc. Reprinted by permission of Van Nostrand Reinhold Co., New York.)

A variety of mechanisms are known to exist for the repair of damage done to DNA by radiation and other causes, and it has been suggested that differences in repair capacity could alter the rate at which damage is accumulated and thus have an effect on life span. In one widely quoted study the rate of repair of damage caused to DNA by ultraviolet irradiation was found to correlate well with life span for a number of species with highly divergent life spans (Fig. 24-24). However, critics of this study point out that the short-life span animals that were studied were all rodents, which tend to be deficient in the particular kind of repair (excision of thymine dimers) that was studied, although they are not deficient in overall DNA repair. In addition, studies done on various species of primates with a 20-fold range of life spans appear to show no significant difference in repair of ultraviolet-induced damage to DNA.

Another possible involvement of DNA repair in aging might be a decline in repair activity with age. However, comparisons of DNA repair capacity in young and old animals have not revealed any significant change with age. There have been some reports of DNA repair deficiency in cultured cells from human patients with progeria, a disease that resembles accelerated aging in some (but not all) ways. However, further investigations appear to indicate that at least some patients with this rare disease have normal DNA repair. Xeroderma pigmentosum, a disease in which there is a definite loss of ability to repair ultraviolet-induced damage to DNA, does not seem to be associated with any accelerated aging, other than possibly in the skin, which is extremely susceptible to damage by sunlight. Cultured cells from xeroderma pigmentosum patients do not enter phase III prematurely. Thus, although there has been much speculation about the role of DNA repair in aging, there is at this time no firm evidence for its involvement.

Other deleterious agents. Over the years there have been numerous suggestions that various adverse environmental agents may contribute to the process of aging. Deuterium (heavy hydrogen, ^{2}H) is one that has been proposed and disproved by experimentation repeatedly, cropping up every few years in yet another theoretical paper. There appears to be no evidence for any acceleration of aging due to chronic exposure to heavy water or deuterium-containing compounds.

Toxic effects of metabolic products of intestinal bacteria were proposed as a cause of aging by Elie Metchnikoff soon after the turn of the century. No firm evidence for a direct effect on the rate of aging has ever been produced. However, it is interesting to note that cancer of the colon, whose incidence increases sharply with old age, has recently been shown to be correlated with lack of bulk in the diet and prolonged retention time for digestive wastes in the large intestine. Although current data are still primarily epidemiological, it appears that this age-related disease can be reduced greatly in incidence by eating a high-bulk diet that promotes rapid movement of wastes through the intestines.

Various other "age-related" diseases can also be linked at least partially to environmental causes. For example, hypertension has been associated with excess salt in the diet, kidney diseases with excess intake of cadmium, various kinds of neurological degeneration with chronic exposure to mercury, and increased risk of both cancer and cardiovascular diseases to marginal deficiency of the trace element selenium, just to mention a few. In all these cases we encounter the problem of attempting to distinguish between aging and age-related diseases. At present, there are no clearly identified extrinsic agents that accelerate the overall process of aging in experimental animals. Likewise, no therapeutic agents have been found that clearly slow the rate of aging (with the possible exception of metabolic effects such as reduced temperature in the case of poikilothermic animals and restricted feeding in the case of many [but not all] species, both poikilothermic and homeothermic).

Immortality of the germ line

One of the most difficult problems that must be dealt with in any type of stochastic theory of aging, whether it is based on intrinsic or extrinsic mechanisms, is the fact that the germ line is immortal as it is passed from one generation to the next. It is difficult to visualize any kind of stochastic "hit" that would not be equally damaging to both germ and somatic cells. Evolution seems to have achieved a stable compromise between mutation rates and selective survival so that enough mutational variation exists within a species (and its germ line) to permit evolution to progress and to permit the species to cope with changing environmental conditions while at the same time keeping the number of lethal or deleterious mutations at a manageable level. If the deleterious effects of age were accumulated in the germ line at the same rate as they are in somatic cells, the species would die out within a few generations, which is obviously not the case.

Many explanations have been offered for the immortality of the germ cells, ranging from more efficient repair mechanisms to a mysterious ability to rid themselves of all accumulated effects of aging at meiosis. Although the problem is often ignored in discussions of stochastic theories of aging, it is essential for any stochastic theory to include a means for achieving germ line immortality. Reduced susceptibility and/or enhanced repair are the mechanisms usually suggested. One investigator has recently suggested that the level of error correction required to maintain the germ line in a nearly error-free state is costly in terms of metabolic energy and that somatic aging might be the end result of an energy-saving evolutionary strategy that reduces somatic error correction to the minimum level that will keep the somatic cells functioning long enough for reproduction to occur. Although the end result of such a strategy is the accumulation of a lethal level of somatic defects, the energy that is saved may give the species a substantial evolutionary advantage over competing organisms with rigid error correction in all their cells.

The various pacemaker mechanisms of aging that have been proposed are best regarded as a combination of intrinsic and extrinsic mechanisms. The pacemaker organ, which controls the aging of all other tissues, is generally assumed to contain within itself some kind of intrinsic biological "clock" that is able to measure either metabolic or chronological time. The clock mechanism could either be programmed in nature or it could operate in response to stochastic events. Target tissues throughout the organism receive "extrinsic" signals from the pacemaker and undergo age-related changes in response to those signals. In theory, the extrinsic controls that are involved can be either positive (actively promoting aging) or negative (keeping the tissue "young" until the signals are withdrawn). Two major classes of pacemaker theories have been widely discussed in the gerontological literature—those in which the pacemaker is related to the brain, or more specifically to neuroendocrine control over the anterior pituitary, and those in which the pacemaker is related to the thymus gland or other aspects of the immune system.

Neuroendocrine pacemakers. Neuroendocrine theories of aging are derived primarily from three types of observations: (1) the apparent relationship between brain size and life span (discussed earlier), (2) neuroendocrine control of pituitary function (Chapter 14), and (3) the major role played by the anterior pituitary in key events of vertebrate life cycles such as growth and sexual maturation (Fig. 24-25). Much of the evidence offered in support of neuroendocrine theories of aging is indirect, but collectively it is substantial. Removal of the pituitary and maintenance of the animal with minimal hormonal supplements delays or prevents many processes that are normally age-related, including growth, collagen cross-linking, and sexual maturation. Underfeeding of rats and mice appears to depress function of the anterior pituitary, and it has been argued that the delay of maturation and the extension of total life span due to underfeeding are mediated by effects on the anterior pituitary.

Function of the anterior pituitary is regulated by neurosecretory

Pacemaker theories of aging

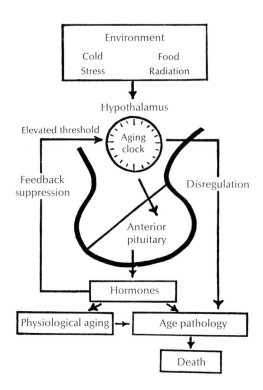

Fig. 24-25. Schematic representation of the hypothalamic-pituitary theory of aging. An intrinsic aging clock in the hypothalamus is proposed to serve as the pacemaker for aging of the rest of the body, acting primarily via the endocrine system. Altered responses to feedback from the body are also thought to be involved. Effects of the external environment on aging are assumed to be mediated by the hypothalamus. (From Everitt, A. V. 1973. Exp. Gerontol. **8**:265. Reprinted with permission from Pergamon Press, Ltd.)

releasing and inhibiting hormones produced by the hypothalamus, a region of the brain physically adjacent to the pituitary gland (Chapter 14). The neuroendocrine control system is in turn influenced by many factors, including complex feedback loops from the target organs for the pituitary hormones. Neurotransmitter substances, including catecholamines such as dopamine influence the function of the neuroendocrine cells that regulate the anterior pituitary. One of the changes that occurs during aging is an increase in the enzyme monoamine oxidase, which oxidizes dopamine to inactive products. It has been suggested that declining levels of dopamine or other specialized neurotransmitter substances may be the pacemaker responsible for many types of age-related changes. There is evidence that localized injection of a precursor of dopamine known as DOPA (dihydroxyphenylalanine) will temporarily reactivate the estrous cycle in aged female rats. Parkinson's disease, which is an age-related neurological disease of humans characterized by tremor and partial paralysis, responds favorably to treatment with DOPA. Side effects of the treatment that are observed in some but not all patients include a reactivation of sexual function and interest in male patients and episodes of cyclic uterine bleeding in postmenopausal female patients.

Although the data are not yet sufficient for a full evaluation, it is clear that the endocrine system does play an important role in many "age-related" changes. However, whether the role of the hypothalamic-pituitary axis is that of a true pacemaker, rather than simply one more component of a complex interrelated process of aging, remains to be proved. Obviously much more research is needed in this entire area.

Thymus/immune system. Major changes occur in the thymus gland and in the immune system during maturation and aging, and several investigators have suggested that such changes may play pacemaker roles in the process of aging. The immune system can be divided operationally into two parts. The first, which is responsible for circulating (humoral) antibodies and immune defense against free antigens and bacteria, involves cells derived from the bone marrow in mammals (or a special organ known as the bursa of Fabricius in birds) and is known as the B cell system. The second, which is responsible for cell-mediated immunity, including immune surveillance against incipient malignancy, destruction of altered and virus-infected host cells, and rejection of tissue grafts, involves cells derived from, or at least influenced by passage through, the thymus gland, and is known as the T cell system.

The T cell system undergoes a major decline during aging, as discussed earlier in this chapter in terms of cancer rejection (Fig. 24-11). In addition, there is a major increase in autoimmune reactions, in which T cell-mediated immunity becomes activated against cells within the body of the host. The thymus gland reaches maximum size relative to body weight very early in life, and by the age of puberty has begun to decline in absolute size and content of functional cells (Fig. 24-26). Because this decline occurs earlier than other degenerative aspects of aging, it has been suggested that the thymus could function as a pacemaker for other age-related changes and that its action could be mediated through the T cell immune system and manifested in such phenomena as decline of immune surveillance and increased autoimmunity.

Early removal of the thymus in mice leads to a degenerative condition known as "runting disease," which appears to be due to autoimmunity. The exact role of chronic autoimmune phenomena in human aging is not clear. There is a definite age-related increase in autoimmune diseases such as rheumatoid arthritis, and some investigators have suggested that most, if not all, the degenerative changes that occur during aging are actually the result of autoimmune phenomena. There are considerable data suggesting that T cells serve in some way to inhibit autoimmunity and that loss of full T cell competence can

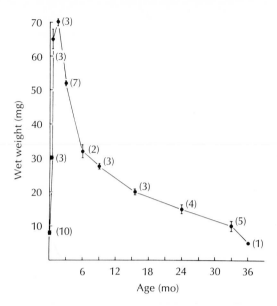

Fig. 24-26. Involution of the thymus gland. The wet weight of the thymus gland, freshly excised from mice, is plotted against age. Vertical bars represent one standard error and numbers in parentheses are numbers of mice tested at each age. (From Hirokawa, K. 1975. J. Immunol. **114:**1659.)

lead to autoimmunity. Recent data also suggest that chronic infection of tissues such as the brain with so-called slow viruses may lead to enhanced autoimmune damage to the infected tissues. One problem with the autoimmune theories of aging that has not yet been resolved, however, is the fact that acute autoimmune diseases occur more frequently in women than in men, despite the fact that women as a whole live longer than men.

There is no question that changes in the immune system occur during aging and influence at least some age-related diseases. However, the question of just how important immune phenomena will prove to be as direct or indirect causative agents for the pathology of aging remains to be seen. Although the possible pacemaker role of the immune system in aging remains largely conjectural, research on the relationship between immune function (or dysfunction) and aging is currently one of the most active areas of gerontological study and as such should be observed closely during the next few years.

Programmed aging

The fundamental concept of programmed aging is that the functional and morphological changes that accompany the aging process are a smooth continuation of the developmental program that generated the adult organism. However, beyond this basic concept the proposed means of achieving such programmed aging are surprisingly diverse and reflect some basic differences in philosophical approach. The most basic point of disagreement revolves around whether the primary evolutionary selection has been in favor of specific mechanisms of senescence or against the forces of mortality.

Evolution of senescence. Proponents of the view that evolution selects for senescence argue that to survive on a long-term basis, any species, whether unicellular or multicellular, must have highly refined error correction and repair mechanisms that will perpetually maintain its genetic integrity. Although selective disadvantage can eliminate defective cells (or individuals) at each round of replication and competition for environment, it is nevertheless essential that a reasonable fraction of genomes remain fully normal throughout each reproductive cycle. Such highly efficient repair mechanisms are evident in today's "primitive" organisms such as rapidly reproducing bacteria. The fact that one can obtain nearly exponential growth curves for such organisms argues that their repair mechanisms are indeed highly developed and that a high level of rejection of defective cells is not part of their normal life history.

New complications are encountered in the survival of multicellular organisms, such as growth control and the maintenance of some types of cells for long periods of time without replication of their DNA. However, it does not follow automatically that a limited life span is a necessary consequence of multicellularity, and one major line of thinking proposes that senescent death as we know it today exists only because it has a positive selective value in evolution. This view is usually traced back to August Weissman, who first argued in 1881 that even if a species of organisms possessed the ability to live forever, individuals would over a period of time suffer injuries that could not be fully repaired and that these would gradually render the species less fit and competitive. From that starting point, he argued that reproduction is necessary and that it is therefore necessary to eliminate the worn and damaged individuals so that they will not be competing for space and resources with more fully functional individuals. From this, it follows that senescent death that removes the parent from such competition is actually beneficial to the species and therefore has a positive selective value. Other proponents of the same general view have pointed out that even though fully mature progeny could presumably compete well with the damaged individuals, the presence of accumulated generations of defective adults might reduce the ability of immature offspring to reach maturity. Closely related to this view is the argument that eliminating the parents from further competition and reproduction after they have produced and reared their initial set of progeny maximizes the flow of new genetic combinations into the breeding populations and therefore permits evolution to progress more rapidly. Thus species that do not practice senescence are unlikely to remain competitive with species that do over a longer evolutionary period. This will be particularly true at times of environmental crisis, such as rapid changes in climate, when the ability to evolve rapidly may be essential for survival.

The þasic concept that senescent death has a positive evolutionary value and that it is thus actively selected for during evolution can be further subdivided into two viewpoints: (1) that specific deleterious "aging" genes are activated late in life to bring about the decline and eventual death of the organism or (2) that the repair and error correction mechanisms that operate efficiently earlier in life are turned off, allowing the forces of entropy to bring about the final demise of the organism. The question of whether aging has a positive evolutionary value is still being debated hotly in the most recent reviews and monographs on the subject.

One possible alternative to direct positive selection for senescence is the proposal that senescence may be a secondary consequence of positive selection for a gene that is valuable early in life but deleterious later. Excess cross-linking of collagen in old age might be an example of such an effect.

Evolution of longevity. Several investigators at present strongly favor the alternative point of view, which is that evolution is still in the process of lengthening life span and that repair and error correction mechanisms are still being perfected. Such arguments rely heavily on the assumption that relationships between brain and body size on the one hand and life span on the other have persisted unchanged through evolutionary time so that the life spans of ancestral forms can be estimated accurately from the fossil record of their brain and body sizes. If these estimates are valid, human life span has roughly doubled during the last few million years. In addition, somewhat earlier in evolution the life spans of primates relative to other types of mammals of comparable overall size also roughly doubled. It is argued that such an extension of life span is more likely to be due to positive evolutionary changes that enhance survival, rather than to reversal of a program for senescence that was acquired at an earlier time. However, it can also be argued that the life spans of most mammals are shorter for their body and brain sizes than those of other vertebrates such as birds and reptiles and that the evolution of longevity in primates and humans could be due to reversal of an evolutionary trend that had previously developed senescence to a maximum.

Several investigators have suggested that the evolution of a larger brain size relative to body size makes a longer life span mandatory. Development of a large brain requires a longer gestation time and a smaller size of litters. In addition, there may be selective advantages of a longer postnatal period of development in which a greater degree of learning can occur. If the considerations are valid, evolution of increased human longevity could be purely a side effect of positive selection for greater mental capacity. Whether or not this assumption is valid, we need to be cautious in interpreting cause-and-effect relationships among two seemingly very different traits whose evolution has been as closely linked as appears to be the case for cephalic development and life span in mammals.

Lack of evolutionary selection against aging. Another evolutionary interpretation that must be considered is the possibility that evolution has no effect, positive or negative, on events that occur in the later part of the life span, after progeny are produced and reared. Thus mutations that have deleterious effects late in life would not be selected against. This concept is sometimes described as "running out of program." Alex Comfort has used the following analogy: "In this we resemble a space probe that has been 'designed' by selection to pass Mars, but that has no further built-in instructions once it has done so, and no components specifically produced to last longer than that. It will travel on, but the failure rate in its guidance and control mechanisms will steadily increase—and this failure of homeostasis, or self righting, is exactly what we see in the aging organism."[*]

Still another closely related concept that has been proposed is an evolutionary strategy of dealing with a deleterious mutation by modifying its regulation so that it is not expressed until after the reproductive period is finished, as opposed to total elimination of that mutation. In cases in which satisfactory function could be maintained without the wild-type allele (e.g., in cases of duplicated genes), such a strategy would presumably have no selective disadvantage, and, if views expressed earlier about the advantages of removing parents from competition with their progeny are correct, the strategy might even offer a positive selective value.

Continuity of developmental program throughout life. One potential problem with all the evolutionary considerations discussed is that they focus primarily on length of life and on the events responsible for senescent death, without fully taking into account the remarkable uniformity of the sequence of events that take place at specific fractions of completion of the life span in mammals with life spans as diverse as those of mice and humans. Thus it can be argued that there must exist a sequential life program including aging that is conserved during evolution from one species of mammal to another, with differences only in the rate at which that program progresses.

---

[*]Comfort, A. Aug. 3, 1970. Time Magazine. p. 52.

Table 24-4. Human genes normally not expressed until adulthood*

Gene	Average age of expression
Peroneal atrophy (dominant)	24
Dominant muscular dystrophy	27
Dystrophia myotonica	39
Huntington's chorea	40
Hereditary glaucoma	42
Hairy rims of ears	40 to 50
Polycystic kidneys (adult onset)	40 to 50
Diabetes mellitus (adult type)	60

*Modified from Stern, C. 1973. Principles of human genetics, 3rd ed. W. H. Freeman & Co., Publishers, San Francisco. Data are approximate, since many of these traits are characterized by variable degrees of expression and apparent multiple forms of the genes. Onset typically occurs over a moderately wide but clearly narrower-than-random time period, as shown for Huntington's chorea in Fig. 24-27.

Such a program would include in sequence all the major events in the life span of an individual, beginning at fertilization and continuing through all the major steps of embryonic development, growth, sexual maturation, reproduction, senescent decline, and eventual death. In terms of mechanisms, it is theoretically possible either for all these events to be linked to a master clock, or pacemaker, that activates each in a predetermined order, or for each event to trigger the next one in the sequence. Experimental observations suggest that the latter may be at least partially the case. Embryonic induction, for example, is frequently dependent on prior stages of differentiation of both the inducing and the responding tissues and also requires morphogenetic movements to bring the two into close apposition. Similarly, sexual maturation and aging both seem to be significantly delayed when growth is retarded by underfeeding in rats. Over the years many investigators have proposed a linkage (generally vaguely defined) between cessation of growth in the adult and the beginnings of aging.

One interesting argument in favor of programmed aging is the increasing number of genes that are being found that are not activated until late in life (Table 24-4, Fig. 24-27). The fact that genes for specific traits such as hairy ears, Huntington's chorea, hereditary glaucoma, or adult-onset polycystic kidney disease do not reach full expression until many years after maturity has been achieved suggests that some type of coordinating "clock" mechanism continues to operate long after maturity has been reached. Thus, just as the events of embryonic development and maturation occur on a fairly precise timetable, so do the effects of genes whose overt expression is seen only relatively late in life. As discussed earlier, one theory of aging has even gone so far as to propose that aging is the result of delaying expression of deleterious

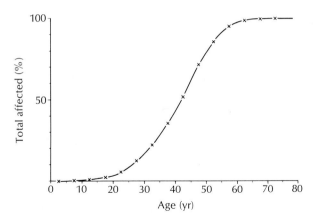

Fig. 24-27. Age at onset of symptoms for 762 patients with Huntington's chorea. A plot of those not yet showing symptoms would look much like a survival curve for senescent death. The median age for onset of symptoms is slightly over 40. (From Principles of human genetics, 3rd ed. by Curt Stern. W. H. Freeman & Co. Copyright © 1973.

genes to the postreproductive period when their selective value is no longer negative and might even be positive.

A number of investigators studying the phase III phenomenon in cultured cells are beginning to view it as a terminal differentiation phenomenon in which an actively multiplying cell differentiates into a stable nonproliferating form with characteristic morphological and physiological properties. It now appears that with proper care phase III cells can be kept metabolically active and "healthy" for long periods of time after they have withdrawn from the cell cycle. The fact that phase III cells can be returned to active multiplication by viral transformation appears to indicate that they do not suffer from any kind of intrinsic damage and that their nonmultiplication is due to regulatory rather than stochastic effects. As discussed earlier, studies of artificially produced binucleate cells suggest that a positive signal originating in the phase III cell is responsible for withdrawal from the cell cycle, but neither the nature of that signal nor its possible relationship to senescence in the intact animal is yet understood.

A detailed model in which senescence is the end result of a program of sequential activation and repression of genes has been proposed by M. S. Kanungo (Fig. 24-28). This model bears some resemblance to the Britten-Davidson model for control of transcription discussed in Chapter 7 (Fig. 7-11) except that many more sequential steps are involved, and a gradual buildup of products to critical threshold levels for activating the next step is assumed to be one of the timing mechanisms involved. Mutations that affect the rate of aging (e.g., species differences) are visualized as altering the rate of accumulation of such products. Obviously models of this type are still highly speculative, but they may be useful as starting points for the design of experiments seeking to test the validity of the programmed aging concept.

Genetics of aging. One of the major problems with programmed theories of this kind is the failure of geneticists to find mutations that alter the overall rate of aging, even when major efforts are made to identify them. In mice there are a number of mutant strains with shortened life spans, but in every case that has been examined carefully, it

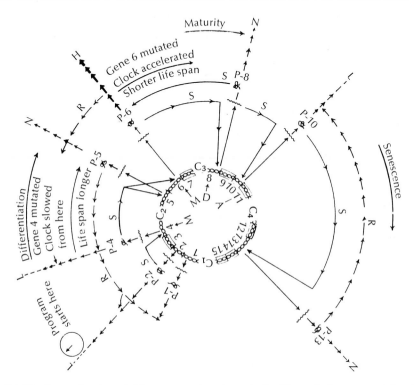

Fig. 24-28. Generalized model for progression through the life cycle by sequential activation and repression of specific genes. Four chromosomes (C_1, C_2, C_3, C_4) containing a total of 15 genes are shown. The genes in boxes (3, 7, 9, 11, 12, 14, 15) are not directly involved in the sequential program. The program starts with genes 1 and 2 active. P-2 stimulates gene 4, leading ultimately to repression of gene 2 and stimulation of genes 5 and 6. P-6 stimulates gene 8, whose product P-8 stimulates a further increase in P-6. P-8 also stimulates gene 10. P-10 in turn stimulates gene 13, whose product, P-13, represses gene 10 and brings about senescent decline. Mutation *(M)*, deletion *(D)*, and addition *(A)* of genes can be involved in the evolution of differences in life span. Although purely hypothetical, a scheme such as this illustrates how a complex set of feedback loops could lead to sequential changes, both qualitative and quantitative, in gene expression throughout the life span, including very gradual long-term changes. *P* = Protein (gene product) coded by a particular gene; *S* = stimulation; *R* = repression; *L* = decrease with time; *H* = increase with time; *N* = unchanged with time. (With permission from Kanungo, M. S. 1975. J. Theor. Biol. **53**:253. Copyright by Academic Press, Inc. (London) Ltd.)

has been found that the excess mortality is due to high incidence of a particular pathological condition, rather than acceleration of the overall aging program. A recent search using the small nematode *Caenorhabditis elegans*, whose total life span is only about 2 weeks, has been similarly unsuccessful. Several mutants with extended life spans were found, but in every case that was analyzed, they were found to involve defects in feeding, which extends life span in a manner similar to the underfeeding of rats and mice.

There are several so-called premature aging syndromes in humans, but in each case these appear to be caricatures of aging that incorporate only certain features of aging, rather than true acceleration of the overall aging program. The most spectacular is the Hutchinson-Gilford progeria syndrome, in which the average life span is about 11 years and all deaths occur before the age of 30. Major symptoms include growth retardation, loss of hair, stiff and enlarged joints, and general atherosclerosis. Victims of this disease have the appearance of miniature old people (Fig. 24-29). However, they do not accumulate lipofuscin age pigment, they do not get cataracts (which are common in true aging), and they have a low incidence of cancer. It is not certain whether this disease, which is rare, is an autosomal recessive or a dominant that arises as a new mutation each time it occurs.

Werner's syndrome is an autosomal recessive disease that causes its victims to die in their forties or fifties, looking very much older (Fig. 24-30). Afflicted individuals tend to have small stature and diminished sexual development, although some of both sexes have been fertile. Graying of the hair usually begins in the teens or early twenties, and by the thirties multiple symptoms are evident. Unlike progeria, Werner's syndrome is characterized by a high incidence of cataracts. Diabetes of an adult-onset type is also common, as are skin lesions, including chronic ulcerated sores at points of friction, particularly in the feet. The incidence of cancer is high, but the cancers are mostly sarcomas (cancers in tissues of mesodermal origin) rather than the carcinomas (cancers of epithelial cells of either ectodermal or endodermal origin) that are far more prevalent in natural aging. Thus, although both these diseases give their victims an appearance of premature

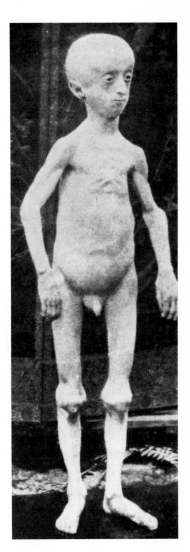

Fig. 24-29. A 17-year-old boy with the Hutchinson-Gilford progeria syndrome. In addition to the generally senile appearance, severe growth retardation accompanies this syndrome. Maximum height that is achieved seldom exceeds that of a normal 5-year-old. (From Herman, E. 1976. *In* A. V. Everitt and J. A. Burgess, eds. Hypothalamus, pituitary, and aging. Charles C Thomas, Publisher, Springfield, Ill. Originally from Gilford, H. 1904. Practitioner **73:**188.)

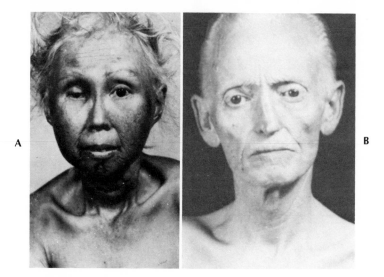

Fig. 24-30. Two patients with Werner's syndrome. Both appear to be much older than their actual ages. **A,** A 48-year-old Japanese-American woman. Cataracts were removed more than 10 years earlier, and the right eye was removed and replaced by a prosthesis after a severe attack of glaucoma. **B,** A 51-year-old white man. Note graying and loss of hair and aged appearance of the facial skin. Bilateral cataracts have been removed. (**A** from Epstein, C. J., et al. 1966. Medicine **45**:177; **B,** from Goodman, R. M., and R. J. Gorlin. 1977. Atlas of the face in genetic disorders, The C. V. Mosby Co., St. Louis.)

aging and share in common certain traits with normal aging, neither can be regarded as a simple acceleration of the aging program.

Although it has not been possible to identify single mutations that affect the rate of aging, there is evidence for polygenic effects on life span. Correlations in life span have been found both between parents and children and between identical twins. In addition, it is clear that each species has its own characteristic life span—presumably genetically determined.

FUTURE RESEARCH ON AGING

In the preceding sections we have examined a large number of different proposed mechanisms of aging and without exception have found something lacking in each of them. Stochastic and programmed theories have both been around for a long time, but in recent years there has been a massive proliferation of stochastic theories. Alex Comfort has suggested that this has been a consequence of the entry of many physicists into the study of aging after the discovery that radiation accelerates mortality in a fashion that is additive with natural aging. That observation, together with the superficial similarity between the human survival curve and multiple hit survival curves predicted by target theory, led to an exhaustive search both in theory and in practical experimentation for the nature of the hypothetical "hits" responsible for aging. In many cases that search was conducted naively, without a full understanding of the relationship of somatic aging to the rest of the life cycle or to the immortality of the germ line. None of the stochastic theories thus far generated have provided a convincing explanation for the phenomenon of senescence, although, as discussed, several such theories still must be regarded as viable possibilities.

Currently the pendulum is swinging in the other direction, with an increasing tendency to view senescence as some kind of programmed event or "terminal differentiation." Theoreticians who are attempting to view aging as a smooth continuation of the developmental program are at somewhat of a disadvantage. If, for example, we take the view that aging might be a special form of differentiation, we are left with the problem that we do not really know how differentiation works. We can say in vague general terms that there is selective control over gene expression and that microtubules and microfilaments are involved in the generation of particular morphological patterns. However, when it comes down to precise extrinsic and intrinsic regulatory and signal mechanisms that cause a cell to assume and maintain its particular set of properties, there are still vast gaps in our knowledge, and the theories that deal with these areas are still largely conjectural. Thus, if, as many investigators are currently suggesting, aging is in fact a smooth continuation of the developmental program, it may be necessary to learn far more about development in general to understand aging well. Conversely, however, we may also learn more about other aspects of development from the study of aging.

The usual course of events in research is progression from a large number of vague, generalized theories to a few better-substantiated theories and ultimately to the generalized acceptance of one particular theory that has stood the test of critical experimentation. Obviously this progression has not yet taken place in the study of aging, where we are still at the stage of proliferation of a multiplicity of poorly tested theories. At the present time much of the research effort appears to be devoted to finding evidence that favors particular specialized theories.

One of the critical needs for future research in this area is for analytically designed experiments that can begin to distinguish among some of the major classes of proposed mechanisms of aging. For example, does aging of a particular tissue proceed autonomously or is it under some kind of external pacemaker control? If pacemaker controls are involved, can the same effects be achieved in cell or organ cultures so that the pacemaker signals can be studied in detail? If tissue aging is autonomous under normal physiological conditions, can some kind of environmental manipulation such as addition of a protective agent or an inhibitor be found that will alter it, thus providing a handle for identifying the underlying mechanisms? It would be interesting, for example, to know whether intrinsic aging of tissues from warm-blooded animals would be slowed by a reduction in temperature.

A National Institute on Aging has recently been added to the National Institutes of Health of the United States Public Health Service, and increasing emphasis is now being given to basic research in gerontology. Hopefully such research will begin in the near future to provide answers for at least some of the most basic questions concerning aging. However, until that happens, it is necessary to leave unanswered most of the questions that have been raised in this chapter. It is hoped that these questions and similar unanswered questions from other chapters in this book will encourage some of our readers to enter research careers and ultimately to provide some of the answers that are needed, not only in the study of aging but also in many other areas of developmental biology.

BIBLIOGRAPHY
Books and reviews

Andrew, W. 1971. The anatomy of aging in man and animals. Grune & Stratton, Inc., New York.

Burnet, F. M. 1974. Intrinsic mutagenesis, a genetic approach to ageing. John Wiley & Sons, Inc., New York.

Comfort, A. 1964. Ageing: the biology of senescence. Holt, Rinehart & Winston, Inc., New York.

Cristofalo, V. J. 1972. Animal cell cultures as a model system for the study of aging. Adv. Gerontol. Res. 4:45.

Cristofalo, V. J., and E. Holeckova, eds. 1975. Advances in experimental medicine and biology. Vol. 53. Cell impairment in aging and development. Plenum Press, Inc., New York.

Cristofalo, V. J., J. Roberts, and R. C. Adelman. Advances in experimental medicine and biology. Vol. 61. Explorations in aging. Plenum Press, Inc., New York.

Daniel, C. W. 1977. Cell longevity: in vivo. In C. E. Finch and L. Hayflick, eds. Handbook of the biology of aging. Van Nostrand Reinhold Co., New York.

DeBusk, F. L. 1972. The Hutchinson-Gilford progeria syndrome. J. Pediatr. 80:697.

Epstein, C. J., et al. 1966. Werner's syndrome: a review of its symptomatology, natural history, pathologic features, genetics and relationship to the natural aging process. Medicine 45:177.

Everitt, A. V. 1973. The hypothalamic-pituitary control of ageing and age-related pathology. Exp. Gerontol. 8:265.

Everitt, A. V., and J. A. Burgess, eds. 1976. Hypothalamus, pituitary, and aging. Charles C Thomas, Publisher, Springfield, Ill.

Finch, C. E. 1976. The regulation of physiological changes during mammalian aging. Q. Rev. Biol. 51:49.

Finch, C. E., and L. Hayflick, 1977. Handbook of the biology of aging. Van Nostrand Reinhold Co., New York.

Gajdusek, D. C. 1977. Unconventional viruses and the origin and disappearance of kuru. Science 197:943.

Gershon, D., and H. Gershon. 1976. An evaluation of the 'error catastrophe' theory of ageing in the light of recent experimental results. Gerontology 22:212.

Goldstein, S. 1971. The biology of aging. N. Engl. J. Med. 285:1120.

Hochschild, R. 1971. Lysosomes, membranes and aging. Exp. Gerontol. 6:153.

Kanungo, M. S. 1975. A model for ageing. J. Theor. Biol. **53**:253.

Kirkwood, T. B. L. 1977. Evolution of ageing. Nature **270**:301.

Kohn, R. R. 1978. Principles of mammalian aging. 2nd ed. Prentice-Hall, Inc., Englewood Cliffs, N.J.

Lamb, M. J. 1977. Biology of ageing. Blackie & Sons, Ltd., Glasgow.

Leaf, A. 1975. Youth in old age. McGraw-Hill Book Co., New York.

Makinodan, T., and E. Yunis, eds. 1977. Immunology and aging. Vol. 1. Comprehensive immunology series. R. A. Good and S. B. Day, eds. Plenum Medical Book Co., New York.

Marx, J. 1974. Aging research. I. Cellular theories of senescence. Science **186**:1105.

Marx, J. 1974. Aging research. II. Pacemakers for aging? Science **186**:1196.

Medvedev, Z. A. 1974. Caucasus and Altay longevity: a biological or social problem? Gerontologist **14**:381.

Nichols, W. W., and D. G. Murphy, eds. 1977. Cellular senescence and somatic cell genetics. Vol. 2. Senescence: dominant or recessive in somatic cell crosses? Plenum Press, Inc., New York.

Rockstein, M., M. L. Sussman, and J. Chesky, eds. 1974. Theoretical aspects of aging. Academic Press, Inc., New York.

Ross, M. H. 1977. Dietary behavior and longevity. Nutr. Rev. **35**:257.

Rossman, I. 1971. Clinical geriatrics. J. B. Lippincott Co., Philadelphia.

Rothstein, M. 1979. The formation of altered enzymes in aging animals. Mech. Ageing Dev. **9**:197.

Sacher, G. A. 1977. Life table modification and life prolongation. In C. E. Finch and L. Hayflick, eds. Handbook of the biology of aging. Van Nostrand Reinhold Co., New York.

Sacher, G. A. 1978. Evolution of longevity and survival characteristics in mammals. In E. L. Schneider, ed. The genetics of aging. Plenum Press, Inc., New York.

Sacher, G. A. 1978. Longevity, aging, and death: an evolutionary perspective. Gerontologist **18**:112.

Schneider, E. L., ed. 1978. The genetics of aging. Plenum Press, Inc., New York.

Smith, K. C., ed. 1976. Aging, carcinogenesis, and radiation biology. Plenum Press, Inc., New York.

Strehler, B. L. 1977. Time, cells and aging, 2nd ed. Academic Press, Inc., New York.

Strehler, B. L., and A. S. Mildvan. 1960. General theory of mortality and aging. Science **132**:14.

Tice, R. R. 1978. Aging and DNA-repair capability. In E. L. Schneider, ed. The genetics of aging. Plenum Press, Inc., New York.

Timiras, P. S. 1972. Developmental physiology and aging. Macmillan Publishing Co., New York.

Walford, R. L. 1969. The immunologic theory of aging. Williams & Wilkins Co., Baltimore.

Wallace, D. C. 1975. A theory of the cause of aging. Med. J. Aust. **1**:829.

Selected original research articles

Cutler, R. G. 1975. Evolution of human longevity and the genetic complexity governing aging rate. Proc. Natl. Acad. Sci. U.S.A. **72**:4664.

Klass, M. 1977. Aging in the nematode *Caenorhabditis elegans*. Major biological and environmental factors influencing lifespan. Mech. Ageing Dev. **6**:413.

Martin, G. M., et al. 1970. Replicative lifespan of cultivated human cells. Lab. Invest. **23**:86.

Orgel, L. E. 1963. The maintenance of the accuracy of protein synthesis and its relevance to ageing. Proc. Natl. Acad. Sci. U.S.A. **49**:517.

Schneider, E. L., and Y. Mitsui. 1976. The relationship between *in vitro* cellular aging and *in vivo* human age. Proc. Natl. Acad. Sci. U.S.A. **73**:3584.

Sharma, H. K., and M. Rothstein. 1978. Age-related changes in the properties of enolase from *Turbatrix aceti*. Biochemistry **17**:2869.

Smith, J. R., et al. 1978. Colony size distributions as a measure of *in vivo* and *in vitro* aging. Proc. Natl. Acad. Sci. U.S.A. **75**:1353.

Spence, A. M., and M. N. Herman. 1973. Critical re-examination of the premature aging concept in progeria: a light and electron microscopic study. Mech. Ageing Dev. **2**:211.

Stein, G. H., and R. M. Yanishevsky. 1979. Entry into S phase is inhibited in two immortal cell lines fused to senescent human diploid cells. Exp. Cell Res. **120**:155.

SECTION SIX

Overview

25 What next?

☐ The previous 24 chapters are only a small sampling of the wealth of descriptive and mechanistic data that are available concerning development. In particular, we have elected to exclude from this book two areas of active mechanistic research—development and function of the immune system, and development and function of the nervous system. In each of these cases the system is highly specialized, and its development is extremely complex. In addition, current views regarding both systems are evolving so rapidly that it is difficult for those not working directly with them to keep up.

Rather than delving further into additional systems, we have chosen to stop at this point and attempt to summarize both what is known about development and what is yet to be learned. Cosmology is defined as the science of the total universe at all levels of organization, and what we are attempting to do in the chapter that follows can perhaps be called biocosmology, since we will be concerned with all levels of organization from subatomic particles to intact organisms, projected through time as each species moves through its life cycle, and viewed in parallel for many different species. We make no pretense of having all the answers or even of knowing all the questions that need to be answered in an area that broad. However, we will nevertheless attempt to analyze what we view as the major areas of knowledge and ignorance about development, together with a discussion of likely directions in future research, some of which we regard as frankly unpredictable.

CHAPTER 25

What next?

☐ In this chapter we seek to put into perspective what is currently known about the mechanisms of development, as well as to identify the critical questions that are in need of answers and to speculate on the future of research in this area. Since developmental biology is basically the study of how organisms progress through their life cycles, the extent to which we can understand the mechanisms of development is limited by our knowledge of the structure and function of the organisms themselves. Therefore, before we can properly ask what is known about development, we need first to ask what is known about the organism whose development is being studied.

Modern science has approached the problem of understanding living organisms from two very different starting points. On the one hand, beginning with the intact organism, it has moved steadily toward smaller and smaller structural and functional units (Fig. 25-1). A complex organism such as a human consists of a limited number of organs and organ systems, each of which in turn is composed of tissues and ultimately of cells. An adult human, for example, contains on the order of 10^{14} cells, of which there are more than a hundred easily distinguished types, and far more than that if subtle structural and functional differences are taken into account. Cells are often defined as the basic structural unit of life. However, each cell itself contains a hierarchy of yet smaller structural units, whose dimensions ultimately merge with those of molecules.

The second approach can be seen by working up from the bottom of Fig. 25-1. If we begin with modern concepts of chemistry and atomic physics, we are led to the conclusion that all living systems are composed of only three types of fundamental particles: electrons, protons, and neutrons. (We are not concerned here with the interconversion of matter and energy or with questions of even more fundamental particles.) The three subatomic particles are organized into atoms, of which not more than 25 to 30 kinds appear to be essential for life, although a few more may be found as low-level contaminants. As we progress to the level of molecules (excluding polymers for the moment), there are at least hundreds, and perhaps a thousand or more, of different compounds in a complex organism. At the level of biological polymers the number appears to be in the tens of thousands when we take into account all the genes that are expressed not only in terms of a protein product but also as hnRNA and mRNA, together with other nucleic acids, complex polysaccharides, etc.

One of the interesting properties of the scheme of organization

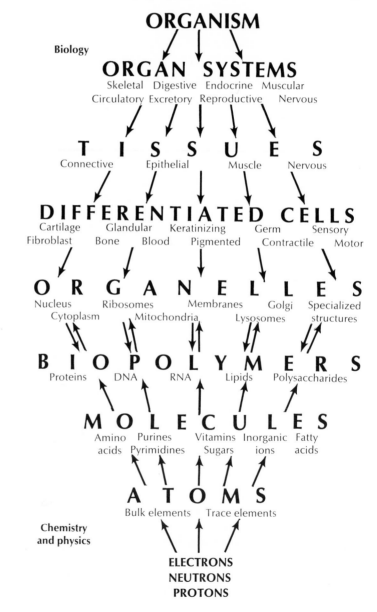

Fig. 25-1. Schematic diagram of the relationship of an organism to its structural components. Arrows pointing down indicate the progression of biological studies from the intact organism toward its component structures; arrows pointing up from the bottom indicate the progression of chemical and physical investigations from the simplest structure units toward increasing levels of complexity; double arrows indicate a region of overlap in which both chemical and biological investigations have been undertaken. Each level of organization includes illustrative examples in smaller type. These are not intended to be complete listings. Complexity is great in the center of the figure, and probably exceeds 10^4 different items in the case of biopolymers.

shown in Fig. 25-1 is that it is relatively simple at the top and bottom but extremely complicated in the middle. There is one organism at the top and a set of three fundamental particles at the bottom. However, in the center the number of components reaches the level of tens of thousands. In addition, these intermediate components interact not only vertically to generate the final structure but also horizontally among themselves to achieve the multitude of operational relationships required for metabolism, biosynthesis, and maintenance of complex structures and functions.

One of the major weaknesses of a scheme such as that in Fig. 25-1 is the fact that it does not emphasize adequately the dynamic aspects of life. As suggested in Chapter 24, a living organism can be viewed as a highly improbable state of its environment that is able to maintain its structural and functional integrity only by a continuous expenditure of energy. To do so requires intake from the environment of nutrients and oxygen (or light and photosynthetic precursors in the case of a plant), continual operation of a large number of biosynthetic, degradative, and energy-yielding metabolic pathways, and the discarding of waste materials into the environment. In addition, a large number of regulatory and homeostatic mechanisms are necessary to keep the entire system precisely balanced and coordinated. These, too, must be assessed as we seek to evaluate the current state of knowledge concerning structure and function of living organisms.

It is easy to speak glibly of seeking to understand biological processes in terms of mechanisms operating at the most fundamental levels of organization. Although valid in one sense, such statements must be viewed cautiously. While it is presumably true that all of life can be explained in terms of interactions among electrons, protons, and neutrons, the multiplicity of higher levels of organization that must be imposed on such an explanation makes it totally impractical to try to describe a living organism in those terms. Usually the best "mechanistic" understanding of a process is obtained by moving one or two levels of organization down the scheme in Fig. 25-1 from the level at which the basic phenomenon is observed.

Thus, for example, if we wish to describe the action of the heart in blood circulation, we would probably speak in terms of rhythmic contractions of muscle tissue and the opening and closing of valves. If our interest is specifically in contraction, we may speak of contractile fibers and perhaps of the sliding of thick and thin filaments past each other. If our interest is in the transduction of metabolic energy into motion, we may speak specifically of the molecular structures of actin and myosin, as well as of the roles of action potentials, calcium ions, and ATP in the process. If our interest is in fundamental chemical mechanisms, we may speak of electron orbitals, valence shells, and the formation of chemical bonds. Finally, if our interest is in atomic physics, we may be concerned with the fitting of protons and neutrons together into atomic nuclei and whether a particular combination will result in a stable nucleus. However, it is highly unlikely that we would attempt to speak of blood circulation directly in terms of atomic physics, chemical bonds, or actinomyosin fibers, even though all are ultimately part of the hierarchy of mechanisms that are involved in it.

We are currently at a point in the history of biology where investigations that originally began from the two extreme positions (chemical analysis and descriptive biology) have met, and in some cases even passed each other, at the center of Fig. 25-1. Thus "structures" such as ribosomes, membranes, and chromosomes are now being dealt with in chemical terms, and electron microscopists have found ways to "see" biologically significant molecules such as DNA, RNA, and proteins. Because of the extreme complexity of this middle region and the relative newness of studies into it, there are still some rather large gaps in our knowledge of the molecular and macromolecular basis for biological processes. This is particularly true for regulatory mechanisms, which often involve components that are present only in extremely small amounts. However, research in these areas is currently providing new information so rapidly that it is difficult to assimilate all the details and fit them into a broader perspective. Thus in the next few years we can expect many new links to be forged between well-defined molecules on the one hand and well-defined biological processes and structures on the other. However, at present, we cannot yet say that we understand all the connections.

When we turn to the study of development, we add yet another dimension of complexity—the progression of the organism through time and around its life cycle. Now, in addition to structure and function, we must also consider the net recruitment of matter from the environment and the generation of new organisms that are essentially identical to the parental organism. The special essence of development that distinguishes it from other aspects of biology consists primarily of the additional regulatory mechanisms that are involved. Development can be regarded as a combination of growth (both in size and cell number), cellular differentiation, and morphogenesis. None of these differ in principle from processes that go on continually during homeostatic maintenance of structure and function in a mature organism that is not overtly involved in development (other than aging and the maturation of gametes). Cell division and precisely controlled growth occur continuously in the replacement of tissues that undergo turnover. Differentiated products are continually made in highly specialized cells. Morphology is maintained despite substantial turnover of individual cells in some tissues and of macromolecules within cells that are not being replaced. The only real difference is that in development these processes lead not to maintenance of the status quo but rather to an orderly progression from one stage in the life cycle to the next.

It is clear from the material in the previous 24 chapters that quite a lot is known about the mechanical aspects of development and also that substantially less is known about the regulatory mechanisms. In the first section of this book we summarized phenomenological descriptions of embryonic development and cellular differentiation rather briefly. However, if we had chosen to dwell on details of those subjects, there is available a wealth of descriptive information that could easily have filled a book larger than this on each of those topics. In the next section we traced the flow of information from DNA to differentiated products. Although we were dealing with events occurring at the molecular level, once again most of the available information was purely descriptive. Very little is yet firmly known about the actual regulatory mechanisms that are involved. Current thinking suggests that a portion of the control over information flow probably occurs at the level of transcription. The involvement of NHC proteins in its regulation is suspected but not firmly proved. We lack even a vague idea of how the NHC proteins are in turn controlled. In addition, recent evidence suggests that events at other levels in the information flow pathway may also be of major importance, as described in Chapters 8 and 9.

As we moved to the third and fourth sections and focused our attention on cellular differentiation and morphogenesis, even the descriptive information became rather vague. For example, differentiated cells in culture retain a mysterious commitment to tissue specificity that persists even when they are not actively exhibiting a differentiated phenotype, but the most that we are yet able to do is to describe the phenomenon and attach to it the term "epigenotype"—we do not even understand the mechanisms that are involved, let alone the underlying control signals. Similarly, when we speak of morphogenesis, we know a little about the mechanical aspects such as the involvement of microtubules and microfilaments, selective adhesion, controlled growth, and selective cell death, but we lack even a fundamental basis for proposing specific regulatory mechanisms. This is reflected, for example, in references to vaguely defined morphogenetic "fields" or "gradients" that are almost metaphysical in nature.

Thus at this time the biggest void in our understanding of development is in the realm of regulatory signals and the mechanisms by which they achieve their effects on their target cells and tissues. The following are a few examples of the kinds of questions that we are not yet able to answer. What is the nature of the information in the gray crescent area that says "form the dorsal lip of the blastopore here"? What signal passes between the notochord and the neural ectoderm that says "form neural tissue"? What signal passes between the optic vesicle and the head ectoderm that says "form a lens and tell the ectoderm that grows over that lens to become a cornea"? What signal tells the mesoderm of the digits to condense and form cartilage and the mesoderm between the digits to die? What are the signals passed back and forth when salivary mesoderm is mixed with mammary epithelium that say "assume the shape of salivary tissue but function like mammary tissue"? What subtle signals tell the oviduct to respond to estrogen by making ovalbumin and the liver to respond by making vitellogenin? What signal attracts the germ cells to the gonads? What is the signal that increases the production of gonadotropins 11 to 14 years after birth? What is the signal that initiates the beginnings of senescent decline 25 to 30 years after birth? Also, how does each of these signals accomplish its effect and what in turn controls each of the signals?

Beyond these individual control circuits that deal with the separate components of development, there are also questions that must be asked about higher levels of organization. For an organism to succeed, all its parts must develop and function in harmony. This could be achieved either through evolutionary selection of many separate control systems that proceed independently at coordinated rates, or there could be a hierarchy of master coordinating controls, or the answer could lie somewhere in between, as suggested by currently incomplete data. It is clear that there exist central coordinating signals such as the progressive increase in thyroxine levels that triggers amphibian metamorphosis and the increase in gonadotropins that triggers puberty. On the other hand, there are also some aspects of development, such as limb formation, that appear to be nearly independent of what happens in the rest of the embryo once they have been initiated. However, until we know much more about the individual control systems that are involved in development, there is relatively little that we are going to be able to say about the mechanisms responsible for their synchronization.

In summary, development superimposes on the rest of biology an additional set of regulatory mechanisms that are concerned specifically with progression of the organism through the life cycle. To understand development, we must have a knowledge both of the general workings of living organisms and of the special control circuits that are exclusively concerned with movement through the life cycle. One further complication is the fact that not all organisms develop the same. What is true for human development may not be true for development of a mouse and is even less likely to be true for a chicken or a frog. For widely divergent species such as sea urchins, fruit flies, and humans, only a minimal number of basic principles are likely to be shared in common, and many of the details may be unrecognizably different.

Despite the rapid progress that is currently being made toward an understanding of basic mechanisms in many parts of biology, including some aspects of development, there are still substantial gaps in our knowledge, particularly in the area of regulatory interactions at the borderline between the chemistry of macromolecules and the structure and function of cellular organelles. With regard to development, we know almost nothing of the ultimate regulatory mechanisms of the life cycle, and in many cases we are still groping to understand the molecular mechanics that are involved in the phenomena that occur.

Does that then make this book a fraud? Have we wasted our readers' time for 24 chapters, only to tell them at the end that we do not have the answers? It is true that we do not know the answers. . . yet. However, the study of development is becoming one of the most exciting and dynamic fields in biology. In particular, a myriad of new techniques developed during the recent explosive advance of molecular biology and molecular genetics are now being applied to problems of development. It is reasonable to predict that the next quarter of a century will bring advances in our understanding of development that are every bit as spectacular as those which have happened in molecular genetics in the years since Watson and Crick first described the double helical structure for DNA in 1953. We have deliberately tried to bring our readers as close as possible to the frontiers of current research with its many unanswered questions so that they may share the excitement of discovery as the answers appear one by one in the research literature in the years to come.

At this point the temptation is strong to attempt to guess what the future will hold in the field of developmental biology. It is clear, for example, that regulatory mechanisms will receive more and more emphasis and that we will begin to understand the nature of the signals that trigger developmental processes. With that understanding will undoubtedly come increased ability to control at least some of the processes in question. We can probably even go so far as to say that many of the regulatory mechanisms will turn out to be closely related to currently known mechanisms for the control of gene expression. However, beyond that, it is probably best to wait and see what actually happens, since nature undoubtedly still has many surprises left in store for us.

The fact that modern biology still has enough left to be discovered so that it is unpredictable makes it an exciting field to follow, but at the same time, very frustrating for textbook writers. Each time that a new issue of any of the leading journals in the field is published, there is a distinct possibility that any statement in this book may become instantly obsolete. There have already been some major surprises while this book was being written. Probably the biggest was the discovery of noncoding sequence within the structural genes for well-characterized products of differentiation such as ovalbumin and β-globin. That discovery required modification of one of the most basic tenets of the "central dogma" of molecular biology—the colinear relationship between DNA nucleotide sequences and protein amino acid sequences.

Ranking a close second, although we have now had a little more time to become accustomed to it, was the discovery of overlapping gene sequences in small viruses, both of bacterial cells and of mammalian cells. The result of such overlaps is that two different proteins are coded from the same DNA sequence read in different frames.

Where or when the next surprise will occur cannot be predicted. Many large gaps exist in our knowledge of modern biology in general and of developmental mechanisms in particular. In addition, there are also many apparent conflicts within the existing data. As new data are gathered, the answers that are found may agree with our current biases, in which case we will label them as predictable and straightforward, or they may go contrary to our biases, in which case we will label them as surprises. Since our current knowledge of some aspects of development is still severely limited, there will undoubtedly be many cases in which our current thinking (or bias) will have to be revised as the new research data become available.

There is no question that much of the material covered in this book will ultimately be proved erroneous or incomplete. The only way we could have prevented that would have been to take a conservative approach that dealt only with firmly established facts. We have chosen instead to discuss many topics that lie at the threshold of new discovery to prepare our students to understand such discovery when it occurs. In our opinion the value of a book such as this is not primarily in what students can learn from it but rather in what it prepares students to learn in the future. Toward that goal, we strongly encourage our readers to make use of Appendix E and to learn to use the research literature to follow the latest advances in the mechanisms of development as they are reported. To the extent that we have achieved our goal, this point in the book should be labeled "the beginning" rather than "the end."

APPENDIX A

Renaturation reactions and cot curves

☐ This appendix provides a more detailed explanation of DNA renaturation reactions and their kinetics, which are described briefly in Chapter 5. Renaturation kinetics are used to determine the sequence complexity of DNA preparations and the amounts and kinds of repeated sequences that they contain. The basic experimental procedure, which is summarized in Fig. 5-10, consists of fragmenting and denaturing the DNA and then allowing it to renature under precisely controlled conditions.

MEASUREMENT OF RENATURATION RATES

Progress of the renaturation reaction can be monitored by a number of different methods. Single-stranded DNA absorbs ultraviolet light at 260 nm more strongly than double-stranded DNA, which exhibits the phenomenon of hypochromicity, or reduced absorbance, due to base pairing. Therefore, as renaturation proceeds, the more absorbent single strands are replaced by less absorbent double strands, and the amount of ultraviolet light that is absorbed decreases.

Renaturation reactions can also be followed by applying samples of the reaction mixture to hydroxyapatite columns at different times during the course of the reaction. At low salt concentrations, double-stranded DNA sticks to the hydroxyapatite while single-stranded DNA passes through the column. Thus the amount of DNA binding to the column indicates the extent of renaturation that has occurred. Hydroxyapatite is also useful for separating the reaction mixture into renatured and nonrenatured components. After the single-stranded DNA has all passed through the column, the bound double-stranded DNA can be eluted by washing the column with higher concentrations of salt. Alternately, nucleases that digest only single-stranded DNA can be used to distinguish hybridized from nonhybridized molecules.

DEFINITION OF TERMS

Renaturation occurs by random collisions of single-stranded fragments that contain complementary sequences. The rate is dependent on the effective concentration of each of the reacting species. That

759

effective concentration is determined by three variables: (1) the total concentration of the denatured DNA, (2) the complexity (sequence diversity) of the original DNA, and (3) the extent of sequence repetition in the original DNA. It is important to understand precisely the various terms and conventions that are used to describe renaturation reactions, since several of them are potential sources of confusion.

The term "concentration" is used to refer specifically to the total amount of DNA in solution, irrespective of its sequence diversity. Concentration is normally expressed in terms of moles per liter of nucleotides contained within the DNA in solution. This convention allows molar concentrations to be used without concern about the actual size of the fragmented DNA molecules. When it becomes necessary to refer to amounts of individual reacting species per unit volume, we will use the term "effective concentration." It is obvious that increasing the total concentration of DNA increases the effective concentrations of all the reacting species contained in the DNA preparation and therefore causes renaturation to proceed more rapidly.

Use of the term "complexity" can be particularly confusing. In renaturation studies, it refers specifically to the total number of nucleotide pairs needed to make one copy each of all the different nucleotide sequences that are contained within a genome or DNA preparation. This usage is based on the simple case in which a genome consists of one large DNA molecule and contains no repeated sequences. In this case the size of the genome as defined for renaturation studies is the same as its actual physical complexity in nucleotide pairs.

In DNA that contains only unique sequences, physical size is essentially a measure of complexity. A large genome contains more different kinds of sequences than a smaller one. At the same total DNA concentration, each sequence from a larger genome is at a lower effective concentration, and renaturation proceeds more slowly than for a smaller genome. In cases in which the genome is divided into several separate chromosomes, its complexity is defined as the total number of nucleotide pairs in all the different sequences in all the chromosomes. Operationally it does not matter whether the DNA starts out in one chromosome or in many, since in either case it is fragmented before the renaturation analysis is performed.

When a genome contains repeated sequences, the effective concentrations of those repeated sequences are higher than would be predicted from the total amount of DNA in the genome. Renaturation of the repeated sequences therefore proceeds more rapidly than that of the unique sequences. This further complicates the definition of genome complexity. The convention that has been adopted is to count each type of repeated sequence only once and to define complexity as the total number of base pairs of different sequences in the genome. This makes the complexity of the genome for purposes of analyzing renaturation significantly smaller than the total number of nucleotide pairs, or the physical size, of the genome.

Complexity defined in this way is not just a mathematical abstraction. As renaturation of the repeated sequences progresses, the effective concentrations of the remaining unreacted fragments decline steadily, until the final pair of fragments for any sequence (irrespective of its initial degree of repetition) is at the same concentration as the original unique fragments and therefore renatures at statistically the same rate as the unique sequences. This rate is determined by the total number of different sequences initially present and by the theoretical genome size that corresponds to one copy of each of these sequences. The only effect of the extra copies of the repeated sequences is to change the total concentration of DNA. They do not alter the effective concentrations of the species that ultimately determine the renaturation rate for the "unique" fraction (which also contains single copies of the repeated sequences).

As defined, the complexity of a genome in a renaturation reaction is determined not by the amount of DNA it contains but rather by the number of unique base sequences that are present in that DNA. Workers seeking to describe the complexity or diversity of a preparation of DNA find it convenient to do so in terms of the rate of renaturation of fragmented and denatured DNA. For this purpose they use a slight modification of some of the standard terminology of second-order reaction kinetics.

MATHEMATICAL DERIVATION OF COT CURVES

In particular, the product C_0t (initial concentration of reactants multiplied by time) is widely used, often written informally as "cot." Unfortunately, most authors either assume that the reader has a prior knowledge of the kinetics of a second-order reaction or else expect the reader to accept the direct relationship between cot values and diversity of base sequences without a detailed explanation.

The following discussion seeks to bridge that gap and explain the significance of the product C_0t without assuming a background in reaction kinetics. A certain amount of elementary calculus is needed for a rigorous mathematical derivation of the formulas involved. However, since calculus may not be a prerequisite for courses in which this book is used, a nonmathematical discussion of the concepts involved is also included after the formal derivation of the formulas.

A second-order reaction is one in which two molecules interact to form a product (or products). Both molecules are consumed by the reaction. Their concentrations are thus reduced as the reaction proceeds. The two molecules can be either identical or different. In the case of renaturation of DNA the reactants are complementary strands obtained by denaturing a DNA helix and are therefore different from one another. However, in most experiments their concentrations are equal, since the denaturation process yields comparable amounts of both strands. Equal concentrations of reactants are assumed in the calculations that follow. Minor modifications of the formulas are necessary in cases in which this is not true.

The rate of forward reaction is determined by the frequency of collision of a reactant molecule with its partner and by the probability that a collision will result in a reaction. Under conditions that are not close to equilibrium the reverse reaction can be ignored.

The following symbols are used for the derivations presented:

1. C_a and C_b are the concentrations of the two reactants, a and b.
2. $C = C_a = C_b$ is the concentration of each reactant under conditions where the concentrations of reactants are equal.
3. t is elapsed time since the reaction was initiated.
4. K is a second-order rate constant, which has a fixed value for any given pair of reactants and set of reaction conditions.

The rate of reaction, which in calculus notation is expressed as a differential, is equal to the product of the reaction constant and the concentrations of the two reactants (equation 1). Since concentration is declining, it has a negative sign.

$$\text{rate} = \frac{dC}{dt} = -KC_aC_b = -KC^2 \tag{1}$$

By rearranging and integrating the rate equation, we can obtain an equation for concentration (C) as a function of time (t) and initial concentration (C_0) as follows:

$$\int_{C_0}^{C} \frac{dC}{C^2} = -K\int_0^t dt \tag{2}$$

$$-\left[\frac{1}{C}\right]_{C_0}^{C} = -K\left[t\right]_0^t \tag{3}$$

$$\frac{1}{C} - \frac{1}{C_0} = Kt \tag{4}$$

$$\frac{1}{C} = \frac{1}{C_0} + Kt \tag{5}$$

$$\frac{1}{C} = \frac{1 + KC_0t}{C_0} \tag{6}$$

$$C = \frac{C_0}{1 + KC_0t} \tag{7}$$

It is often more convenient to refer to the fraction of the initial reactants remaining, rather than their absolute concentration. To accomplish this, equation 7 is rewritten as follows:

$$\frac{C}{C_0} = \frac{1}{1 + KC_0t} \tag{8}$$

It is difficult to arrive at an exact statement of this equation without using calculus or a long, roundabout process that in essence amounts to the derivation of the calculus formulas used to get from equation 2 to equation 3.

It is possible, however, to describe some of the general relationships in less mathematical terms, using equation 1 as a starting point:

1. The initial concentration of reactants determines how fast the reaction proceeds in its early phases (and throughout its course).
2. The rate of reaction declines fairly rapidly as the reaction progresses. When C is half of C_0, the reaction rate will be one fourth of the original rate, and by the time C becomes one tenth of C_0, the rate will be only one hundredth of the original.
3. The reaction constant, K, also determines how fast the reaction proceeds. Thus a pair of reactants with a high K value will react more rapidly than a pair with a low K value, if the initial concentration is the same in both cases.

Returning now to equation 8, we can verify that it has the properties needed to satisfy conditions 1, 2, and 3 above. The fraction remaining (C/C_0) is never greater than one and becomes smaller as K, C_0, or t is increased. The reaction rate declines rapidly with time. In a period of time of such length that the product KC_0t increases from zero to one, half of the original reactants are consumed. In a second time period of the same length, between $KC_0t = 1$ and $KC_0t = 2$, only an additional one sixth of the reactants are consumed (that is, C/C_0 changes from one half to one third).

It is also evident from equation 8 that the product C_0t determines the progress of any given reaction, since K is a constant for any pair of reactants under fixed reaction conditions. Thus neither absolute concentration nor time is critical. It is their product, C_0t, that determines how much of the reaction has been completed. The same relationship exists between the fraction remaining and the C_0t product whether there is a low initial concentration of reactants and a long reaction time or a high concentration and a short time.

If we now consider specifically the point at which the reaction is half completed, the following relationships emerge:

$$\frac{C}{C_0} = \frac{1}{1 + KC_0t_{1/2}} = \frac{1}{2} \tag{9}$$

$$1 + KC_0t_{1/2} = 2 \tag{10}$$

$$KC_0t_{1/2} = 1 \tag{11}$$

$$C_0t_{1/2} = \frac{1}{K} \tag{12}$$

Thus the C_0t product for half completion of the reaction is a simple function of the second-order rate constant for that reaction.

Most of the above discussion can be applied to any second-order reaction involving equal concentrations of two different reactants. When these formulas are applied to the renaturation of denatured DNA, the following conventions are normally observed:

1. Concentrations (both C and C_0) are expressed in terms of the total moles per liter of nucleotides present in the reaction mixture. The DNA is normally sheared to relatively small fragments (e.g., a single chain length of 400 nucleotides), but the exact size and number of reacting "molecules" is generally not determined precisely. Also, any diversity that may exist among the molecular species present is allowed to be reflected in the K term rather than the concentration term.
2. Time is expressed in seconds.
3. The K value is considered to be inversely proportional to the complexity of the genome being analyzed. The rationale for this assumption is that diversity increases the number of different reacting pairs present within a given total concentration of DNA and therefore reduces the available concentration of each. This results in a slower reaction rate and therefore a smaller K value. A reduced K value, in turn, means a larger cot value, according to the relationship in equation 12.

Fig. A-1. Time course of an ideal, second-order reaction, illustrating the features of the log cot plot. The equation represents the fraction of DNA that remains single stranded at any time after the initiation of the reaction. For this example, K is taken to be 1.0, and the fraction remaining single stranded is plotted against the product of initial concentration × time on a logarithmic scale. (From Britten, R. J., and D. E. Kohne. 1968. Science **161**:529. Copyright 1968 by the American Association for the Advancement of Science.)

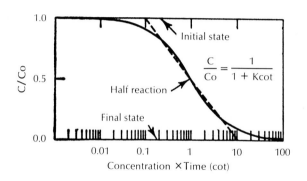

Experimental measurements of cot values, using the conventions outlined, have verified a direct proportionality between the complexity of the genome and cot values for one-half reassociation over an extremely wide range of complexities.

The relationship between cot values and the fraction of reactant remaining can best be seen by plotting the logarithm of cot against C/C_0 (Fig. A-1). The resulting curve is S-shaped, with its central section approximating a straight line (Fig. A-1, dotted line). Changes in the value of K result in displacement of the curve to the left or right but do not change its shape.

Fig. 5-11 shows experimentally determined cot values for genomes whose complexity varies from one nucleotide pair (polyadenylic acid plus polyuridylic acid) to 10^9 nucleotide pairs (the sum of all the unique sequences in calf thymus DNA). Excellent correspondence is seen between the experimentally determined cot values for one-half reaction and the complexity that is shown on a separate scale at the top of the figure. (Note that the ordinate in Fig. 5-11 is expressed in terms of the fraction reassociated [rather than reactants remaining] and therefore has 0 at the top and 1.0 at the bottom.)

Similar calculations can also be used to analyze DNA:RNA hybridization. There are two basic approaches, RNA excess and DNA excess. In RNA excess, the rate of hybridization of a small amount of radioactively labeled DNA is determined by the number of complementary RNA sequences in the reaction mixture. The product of initial RNA concentration × time, which is abbreviated "Rot" (or "rot," or sometimes "Crot," as in Fig. 7-10), is used in a manner similar to cot values to analyze the hybridization. Larger rot values are needed for hybridization of rare RNA sequences than for those in higher frequency classes. In DNA excess, tracer amounts of labeled RNA are added to a DNA renaturation mixture. The cot value at which the RNA becomes hybridized indicates the repetition class of the DNA from which the RNA was transcribed.

BIBLIOGRAPHY

Bostock, C. 1971. Repetitious DNA. Adv. Cell Biol. 2:153.

Britten, R. J., and D. E. Kohne. 1968. Repeated sequences in DNA. Science **161:** 529.

Britten, R. J., and D. E. Kohne. 1970. Repeated segments of DNA. Sci. Am. **222:**24 (April).

Hood, L. E., J. H. Wilson, and W. B. Wood. 1974. Molecular biology of eucaryotic cells, a problems approach, Vol. 1. W. A. Benjamin, Inc., Menlo Park, Calif.

APPENDIX B

Techniques for selection of cellular hybrids

☐ As outlined briefly in Chapter 11, selective media and/or culture conditions are widely used to isolate hybrid cells that contain genes contributed by two different parental cells from the background population of parental cells. The basic concept that is involved is simply to design a combination of medium and culture conditions that will support the multiplication of hybrids but will not support the multiplication of either parent. In theory, almost any two cellular characteristics that prevent multiplication by different mechanisms can be used. The only restrictions are (1) that each parental cell must possess a gene or characteristic capable of causing the other to grow and (2) that the gene or characteristic supporting growth must be dominant over the form that will not support growth when the two are contained in the same cell.

Perhaps the simplest selective system is one that employs two nutritional mutants that lack different enzymes in the same biosynthetic pathway. This has been done, for example, with auxotrophic mutants that are unable to synthesize the amino acid glycine and are therefore unable to multiply if glycine is not supplied in their culture medium. If a hybrid cell is constructed from two different auxotrophic mutant lines, each of which is unable to multiply without glycine because of a defect in a different gene, that hybrid will be able to multiply in a glycine-free medium that does not support growth of either parent. The reason is that each parent provides a functional gene that codes for the step that is blocked by mutation in the other parent. The hybrid thus contains functional genes for all the enzymes needed to synthesize glycine and does not require glycine in the culture medium for growth.

Although the use of complementary auxotrophic mutants is simple in theory, the procedures needed to obtain the necessary mutants (e.g., mutagenesis followed by selective killing of parental cells with BrdU and visible light) are complex, and this particular approach to the selection of hybrids is not widely used. Historically, the first, and still the most widely used, approach is the hypoxanthine-aminopterin-thymidine (HAT) medium described briefly in Chapter 11.

Use of the HAT medium takes advantage of the fact that mammalian cells have two alternative pathways for obtaining the nucleoside triphosphates that they need for nucleic acid synthesis. The first is de novo synthesis, in which the required purine and pyrimidine bases are synthesized within the cells from precursors that are biochemically very different. The second is the so-called salvage pathway, in which preformed bases or nucleosides from the diet are used after being converted to nucleotides by the addition of ribose phosphate to the free bases or by the addition of phosphate to the nucleosides.

Different sets of enzymes are involved in the two pathways so that it is possible to interfere with one without disturbing the other. De novo synthesis, both of purine nucleotides and of thymine nucleotides, requires the catalytic action of folic acid coenzymes and can be blocked by antimetabolites of folic acid such as aminopterin. Cells grown in the presence of aminopterin are unable to make their own purines or thymine and can survive only if supplied with a preformed purine ring structure (e.g., hypoxanthine, which is an effective precursor for both adenine and guanine) and with preformed thymine (which is supplied as the nucleoside, thymidine, so that it can be taken up effectively by the cells).

The salvage pathway for use of hypoxanthine involves the enzyme hypoxanthine-guanine phosphoribosyltransferase (HGPRT), which adds ribose and phosphate to hypoxanthine, converting it to inosinic acid, which in turn can be converted to adenosine monophosphate (AMP) and guanosine monophosphate (GMP). These in turn can be further converted to ATP, dATP, GTP, and dGTP for use both in energy metabolism and as precursors to RNA and DNA.

The salvage pathway for use of thymidine involves the enzyme thymidine kinase (TK), which adds a phosphate to thymidine, converting it to thymidine monophosphate (TMP), which in turn can be converted to TTP and used in DNA synthesis.

Mutant cells lacking the enzyme HGPRT can be selected for by growing cultures in the presence of a purine analog such as 8-aza-guanine under conditions where de novo purine synthesis is possible. Cells with functional HGPRT will incorporate lethal amounts of the analog and be killed. Cells lacking HGPRT will be unable to convert the purine analog to a nucleoside monophosphate and therefore will not have their nucleic acid synthesis disrupted by it. They will continue to grow relatively normally, relying entirely on de novo synthesis for their supply of purines. Clones of 8-azaguanine-resistant cells can be selected and analyzed, and populations of cells can be developed that lack the enzyme HGPRT and that have a low rate of back mutation.

Mutant cells lacking the enzyme TK can be selected by growing cultures in the presence of a thymidine analog such as 5-bromodeoxy-uridine under conditions that permit de novo synthesis of thymidine. Cells lacking TK do not convert the 5-bromodeoxyuridine to a nucleo-tide, and therefore their DNA metabolism is not disrupted by it. Stable mutants with low reversion rates can be selected by cloning and can be grown into large culture populations.

The HAT medium is a normal cell culture medium to which has been added hypoxanthine, aminopterin, and thymidine. Its most direct application is in the selection of hybrid cells that contain the combined genomes of a cell lacking HGPRT plus a cell lacking TK. Neither of the parental cell types is able to grow in HAT. All de novo synthesis, both of purines and of thymidine, is blocked by the aminopterin. Rela-tively few other cellular functions are disrupted by aminopterin, and good growth of most kinds of nonmutant cells can be obtained in the presence of aminopterin by supplying hypoxanthine, thymidine, and the amino acid glycine, whose synthesis is also blocked by aminop-terin.

The $HGPRT^-$ cell is unable to use the hypoxanthine in the HAT medium and thus has no purines available either for energy metabo-lism or nucleic acid synthesis. This cell, however, does carry a normal TK gene and is able to convert thymidine to TMP. The TK^- cell is una-ble to use the thymidine in the medium and therefore cannot make DNA. It does carry a normal HGPRT gene, however, and can use hy-poxanthine as a purine source.

If cellular fusion occurs, so that a hybrid cell is formed that con-tains both genomes, that cell will receive a good TK gene from the $HGPRT^-$ parent and a good HGPRT gene from the TK^- parent. Such a cell is therefore able to use exogenous sources of both hypoxanthine and thymidine and is fully capable of growth in the presence of ami-nopterin. These relationships are illustrated in Fig. B-1.

This selective growth advantage permits very rare hybridization events to be detected in the presence of large numbers of parental cells. The limiting factors are the rarity of reversions in the parental lines and the number of cultures the investigator is willing to screen in looking for a growing hybrid colony. The procedure is sensitive enough to detect cellular hybridization even without the use of inactivated Sendai virus, which greatly increases the frequency of hybridization.

One of the limitations in the use of HAT is that it requires develop-ment of a mutant strain of each of the parental lines that is defective either in HGPRT or TK. This largely limits its usefulness to established cell lines that can be grown for a sufficient number of cell generations to achieve the mutant selections. In the case of human diploid fibroblasts, however, this problem can be circumvented by use of cultures derived from individuals afflicted with the Lesch-Nyhan syndrome, an X chromosome–linked genetic disease in which there is a hereditary absence of HGPRT.

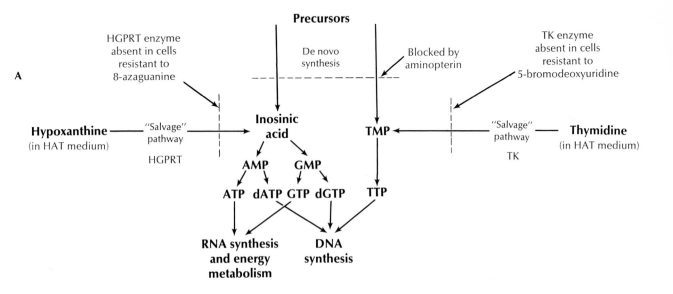

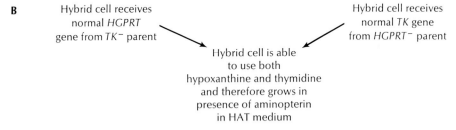

Fig. B-1. Biochemical basis for use of HAT medium to select for cellular hybrids. **A,** Inhibition of parental lines. De novo synthesis of purine and thymidine derivatives proceeds downward from top of diagram. Salvage pathways for purines and thymidine are shown at left and right, respectively. The *HGPRT⁻* parent cannot use exogenous hypoxanthine. The *TK⁻* parent cannot use exogenous thymidine. Aminopterin blocks both synthesis of purines and synthesis of thymidine. Thus neither parental line can multiply in HAT medium. **B,** Growth of hybrid cells. Each parent contributes a functional gene that complements the defect in the other parent. The hybrid cell can use both hypoxanthine and thymidine and thus can multiply in the presence of aminopterin.

Several hybridization techniques have been developed in which one parent is selected against by HAT while the other is selected against by other means. One is the use of lymphocytes (a type of white blood cell) as the second parent. Selection against the lymphocyte parent is achieved by eliminating all cells that do not attach to the culture surface. Selection against the other parent is with HAT. The cells that attach and are resistant to HAT are predominantly hybrids.

Another approach is the use of a contact-inhibited parent in combination with a HAT-sensitive parent that is not contact inhibited. Colonies that are resistant to HAT and not contact inhibited are likely to be hybrids. These can be detected as multilayered clusters of cells that continue to multiply on top of a monolayer of contact-inhibited cells.

A third approach is to use as the second parent a normal diploid cell that has ceased to multiply because of cellular senescence (Chapter 24). Cells that are capable of active multiplication in HAT medium are likely to be hybrids. This approach works well when the HAT-sensitive parent has transformed properties that are dominant over cellular senescence and are able to stimulate the senescent nucleus in the heterokaryon to reenter the cell cycle and form a synkaryon. However, it will not work if the HAT-sensitive parent is a normal diploid cell or even a "nontransformed" permanent line such as 3T3 that has retained density-dependent G_1 arrest (Chapter 17), since neither of these cell types is capable of reactivating the senescent nucleus in a heterokaryon (Chapter 24).

As a fourth approach, it is possible to use a small number of the HAT-resistant parental cells in combination with a large excess of the HAT-sensitive parental cells under conditions where high efficiency of hybridization can be achieved. In this case, relatively few of the HAT-resistant cells escape hybridization, and most of the HAT-resistant clones prove to be hybrids.

Finally, if all else fails, selection of a large number of clones and analysis of their properties can be used to detect hybrids. If there are differences in appearance between the parental cells, the hybrids will often have an intermediate morphology. The nuclei of hybrid cells contain more chromosomes than those of nonhybridized parents and thus tend to be larger and also to have more nucleoli (Fig. 11-11, E). However, so do hybrids formed by fusing two of the same parent.

Ultimately, suspected hybrids must be verified by karyotype analysis (Fig. 11-12) or biochemical analysis (Fig. 11-13), whether or not selective media have been employed. Even though selection is based on the use of highly stable mutants, there is always the possibility that a rare mutational event, perhaps only remotely related to the property used for selection, will permit a parental cell to multiply in the selective medium without hybridization. For example, a cell that is HGPRT negative might undergo a mutation that makes it resistant to the effects of aminopterin and therefore able to synthesize purines de novo despite the presence of aminopterin in the HAT medium.

BIBLIOGRAPHY

Littlefield, J. W. 1964. Selection of hybrids from matings of fibroblasts in vitro and their presumed recombinants. Science 145:709.

Ringertz, N. R., and R. E. Savage. 1976. Cell hybrids. Academic Press, Inc., New York.

APPENDIX C

Cell cycle and methods of analyzing it

☐ The cell cycle is the sequence of events that occurs during the time between the formation of a cell by the division of its parent and the cell's own division. The intracellular events that occur during the life of a cell are similar in parental and daughter cells and are therefore cyclic in nature. The most obvious stages in the life cycle of a cell are its beginning and its end, i.e., the division process. Formation of daughter cells from a parental cell begins with a division of nuclear materials (called mitosis, nuclear division, or karyokinesis) followed by cytoplasmic division (cytokinesis). (The term "mitosis" is sometimes applied to the entire division process of karyokinesis and cytokinesis.) The remainder of the cell cycle, which encompasses the period from division to division, is called interphase.

The interphase period can generally be subdivided into shorter periods on the basis of an additional event, the synthesis of DNA in the nucleus. Synthesis of nuclear DNA is restricted to a defined portion of the interphase period, which is called the S period (for synthesis). The S period can be identified by the increase in amount of nuclear DNA, which can be detected by means of spectrophotometric and/or fluorometric methods and also by the incorporation of radioactive DNA precursors such as ^{3}H-thymidine, which can be detected by means of scintillation spectrophotometry or autoradiography. In general, the S phase is preceded and succeeded by time periods called G_1 and G_2, respectively, during which other cellular events of interphase occur. The G stands for "gap," since at the time the cell cycle concept was being formulated, relatively little was known about the events that occurred during these periods. Even up to the present time, the events occurring during the G_1 and G_2 periods have proved to be refractory to experimental investigation.

The general cell cycle thus consists of the following periods: G_1, S, G_2 and division, or M (for mitosis). The cycle is frequently represented

772

in circular form, as shown in Fig. C-1. Portions of the general cell cycle are absent in some types of cells. For example, cells of early embryos often lack a G_1 period. As development proceeds, however, a G_1 period is established, and each type of differentiated cell assumes a characteristic cell cycle. Other types of cells such as amoebae and some established mammalian cell culture lines also lack a G_1 period. Presumably the cellular events that normally occur during G_1 and which prepare the cell for DNA synthesis are carried out in other portions of the cycle in such cells. One line of mammalian cells is known that has also eliminated the G_2 phase of the cell cycle, indicating that the cellular events occurring in G_2 may also take place at other stages of the cell cycle. This particular cell line thus appears to have a cell cycle consisting only of division and the S period.

The duration of the cell cycle is variable among different types of cells and among different sets of growth conditions for a specific type of cell. A typical set of values for mammalian cells in culture is a G_1 period of 8 to 10 hours, an S period of 6 to 8 hours, a G_2 period of 3 to 5 hours and mitosis of 1 hour. The total generation time for typical cells in culture is therefore about 18 to 24 hours. (By contrast, the cell line mentioned earlier that lacks both G_1 and G_2 periods has a generation time of only 10 to 12 hours.) Most differences in the duration of cell cycles are due to large variations in the length of the G_1 period, although some prolongation or contraction of other phases also has been observed. Some cells naturally withdraw from the cell cycle for variable periods of time (some rather permanently). Also, cultured cells may become "arrested" at some point in the cell cycle by conditions such as nutrient deprivation or density-dependent inhibition of growth. In general, such cells have not replicated their DNA and are therefore in a G_1-like state. The term G_0 is often used to identify the G_1-like condition of cells that are not traversing the cell cycle (Fig. C-1).

There are several methods of determining the duration of phases of the cell cycle. Unfortunately, most of the methods used involve manipulation or treatments of the cells that can alter one or more parameters of the phases being examined. Moreover, the precise status of the cell at the time of the experiment is often unknown. For example, analysis of the cell cycle in cultured mammalian cells could be affected by using cells that have just been subcultured, since such cells may exhibit more synchrony than a random population of cells. Or an unknown proportion of cells in a population may be in a G_0 state.

One commonly used method for analyzing the cell cycle is based on determining the percentage of mitotic cells that are radioactively labeled. This method relies on the fact that nuclear DNA synthesis is restricted to the S phase and that mitotic cells containing DNA that

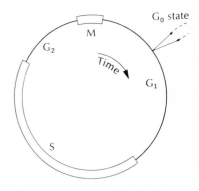

Fig. C-1. Diagram of the cell cycle represented in circular form. Dotted lines indicate length of time in the G_0 state. (Modified from Prescott, D. M. 1976. Reproduction of eukaryotic cells. Academic Press, Inc., New York.)

was labeled during the S phase can be recognized after autoradiography by the presence of condensed chromosomes covered by exposed silver grains. In this procedure a culture of cells is exposed to a radioactively labeled precursor of DNA (e.g., ³H-thymidine) for a brief period of time. (The culture is "pulsed" with radioactivity.) The radioactive precursor is then removed, and often a nonradioactive form of the same precursor is added to the culture. (The culture is "chased.") The radioactive precursor is incorporated into DNA only by those cells which are in the S phase of the cell cycle while the labeled precursor is in the culture medium. Cells in any portion of the S phase incorporate the labeled precursor, whereas cells in G_2, M, and G_1 do not. The experimental conditions used are important, since incorporation of too much ³H-thymidine or use of too high a concentration of an unlabeled thymidine for the chase can alter the duration of G_2 or S.

At various intervals after the pulse labeling of the cells, portions of the culture are fixed, attached to glass slides, and autoradiographed. The radioactive decay of the labeled precursor "exposes" the photographic emulsion. After chemical development of the emulsion to produce visible silver grains where it was exposed to radioactivity, the cells are stained and examined under a microscope. The mitotic cells present in samples taken soon after exposure to the isotope will be those cells which were already in M or G_2 at the time the radioactive precursor was present and will be unlabeled. Samples taken at later times will contain mitotic cells that were in some portion of S at the time of the pulse but then progressed to mitosis before undergoing fixation. Finally, samples taken at even later times will again fail to contain labeled mitoses, since mitotic cells in these samples were in G_1 at the time of the pulse. If the experiment is carried on for a sufficiently long time period, the entire sequence will then be repeated for the second cell cycle after the pulse. For each sample, the number of radioactive mitotic figures is expressed as a percentage of the total

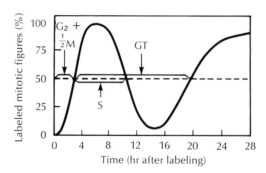

Fig. C-2. Hypothetical curve showing the percentage of labeled mitotic figures seen in successive samples following a pulse label with ³H-thymidine. GT = Total generation time. (Modified from Prescott, D. M. 1976. Reproduction of eukaryotic cells. Academic Press, Inc., New York.)

number of mitoses in the sample. An example of the type of results obtained in such an experiment is shown in Fig. C-2.

Several features of the data should be noted. First, the percent of labeled mitoses increases gradually rather than abruptly. Since the total length of time that cells spend in M is short relative to other segments of the cell cycle, the gradual increase presumably reflects primarily a heterogeneity in the duration of the G_2 phase among the different cells. The even more gradual slope of the downward side of the first peak represents the heterogeneity in length of G_2 plus S. The heterogeneity in length of subdivisions of the cell cycle also accounts for the failure of the curve to reach zero. The duration of the G_2 period is considered to be the time interval between the beginning of the pulse and the sample which has 50% labeled mitoses. (This interval is actually G_2 plus ½ M.) The duration of the S phase is considered to be the time interval between the samples containing 50% labeled mitoses on the ascending and descending sides of the first peak. The generation time (GT) is the interval between identical points on the first and second peaks. Since GT is equal to $G_1 + S + G_2 + M$ and the values for GT, S, and G_2 plus ½ M have been experimentally determined, simple subtraction gives the value for G_1 plus ½ M. The duration of M can be measured by direct observation of dividing cells or can be calculated from the percent of mitotic cells. (Duration of M equals $\frac{GT}{0.693}$ $\ln\left[\frac{M}{N} + 1\right]$ where M equals the number of mitotic cells and N equals the total number of cells counted. The complex form of this equation is due to the fact that each division generates two G_1 cells from one G_2 cell. Because of this, the numerical distribution of cells at various stages of the cycle is not a simple linear function of the lengths of the stages.)

A second method of analyzing the cell cycle relies on mathematical conversions of the percentage of cells in a given phase of the cell cycle to the duration of that phase. The overall generation time is measured by counting the cells at different intervals and determining the time required for population doubling. The percentage of cells in various phases of the cell cycle is determined by means of a flow microfluorometer.

In this procedure, cells in G_1, S, and G_2 plus M are differentiated from each other on the basis of the quantity of DNA they contain, and the cells in each category are counted. The cells to be analyzed are dispersed to form a suspension of single cells and fixed in that condition. The DNA is then stained with a fluorescent dye in a reaction that is quantitative (i.e., a constant quantity of dye is bound per unit of

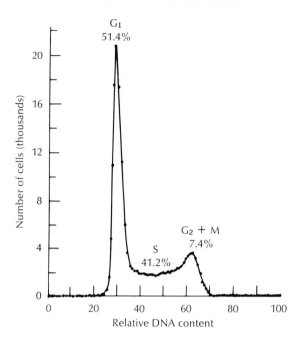

Fig. C-3. Experimental curve showing the distribution of DNA in a culture of CHO Chinese hamster ovary cells analyzed by flow microfluorometry. Percentages of cells in G_1, S, and G_2 were obtained by computer-fit analysis of areas under the DNA distribution curve. (Modified from Tobey, R. A., and H. A. Crissman. 1975. Exp. Cell Res. **93**:235.)

DNA). The stained, suspended cells are then forced through an orifice one at a time, in "single file." The fluorescent dye is excited by a laser, and the intensity of the resulting fluorescence of each individual cell is determined. More than 50,000 cells can be examined within a few minutes by an automated flow microfluorometer. The fluoresence characteristics of the cells are typically plotted in a manner similar to that shown in Fig. C-3. There are two major peaks of fluorescence that correspond to cells which have not yet doubled their DNA (G_1 cells) and to cells which have doubled their DNA (G_2 and M cells). The cells in the second peak have just two times the amount of fluorescence per cell as those in the first peak. The area between the two peaks reflects the cells that are synthesizing DNA (i.e., S cells). The percentage of cells in each phase of the cycle can be used (along with the GT) to calculate the duration of each phase.

The foregoing discussion is meant to be a general introduction to two methods of investigating the cell cycle. Those wishing a more detailed description of the problems involved in such analyses, a rigorous mathematical treatment of data, or alternative approaches to analyzing the phases of the cycle should consult the general reviews listed at the end of this appendix.

BIBLIOGRAPHY

Dean, P. N., and J. H. Jett. 1974. Mathematical analysis of DNA distributions derived from flow microfluorometry. J. Cell Biol. **60:**523.

Mitchison, J. M. 1971. The biology of the cell cycle. Cambridge University Press, Cambridge.

Nachtwey, D. S., and I. L. Cameron. 1968. Cell cycle analysis. Methods Cell Physiol. **3:**213.

Prescott, D. M. 1976. Reproduction of eukaryotic cells. Academic Press, Inc., New York.

Sisken, J. E. 1964. Methods for measuring the length of the mitotic cycle and the timing of DNA synthesis for mammalian cells in culture. Methods Cell Physiol. **1:**387.

APPENDIX D

Molecular self-assembly and catalyzed assembly

☐ Current evidence strongly suggests that molecular self-assembly is one of the fundamental mechanisms involved in the generation of organized biological structures and that it plays an important role in morphogenesis, particularly at the subcellular level. However, the model systems that have been studied in the greatest detail (e.g., assembly of virus particles) are relatively remote from the direct study of development. In addition, systems that involve self-assembly are dealt with in many different parts of this book without a specific focus on the self-assembly process as such. The purpose of this appendix is to draw together the diverse references to self-assembly and to examine in greater detail some of the model systems (mentioned briefly in the introduction to Section Four) that have been used to study the mechanisms of self-assembly. In addition, it examines the limitations of self-assembly and describes recent investigations of the mechanisms by which noncovalent association of macromolecules is promoted by specific accessory proteins that are not incorporated into the final structures.

DEFINITION AND MECHANISMS OF SELF-ASSEMBLY

It is desirable to begin by attempting to define self-assembly. The basic concept that underlies self-assembly is that the subunit molecules that are involved have properties such that formation of a highly structured complex from them is thermodynamically favored. The formation of covalent bonds is not involved in the aggregation processes (at least not initially). The forces that bring the individual component molecules together into an organized structure involve weaker forms of "bonding" such as hydrogen bonding, ionic bonding, and hydrophobic interactions. Collectively these molecular interactions must exert sufficient bonding force to overcome the dissipative effects of thermal agitation and make the fully assembled and organized structure a thermodynamically more favorable state than random dispersal of the component molecules.

Covalent bonds, which are formed by the sharing of an electron pair between two adjacent atoms, can be divided into two generalized types. The term "nonpolar" is used to describe those bonds in which the electron pair remains equally distributed around both atoms, whereas the term "polar" is used to describe the situation in which the electrons are preferentially attracted to one atom but remain associated with both to a sufficient extent to preserve the covalent bond. The term "ionization" is used to describe the situation when the electron pair becomes exclusively associated with one atom (or group of atoms) to form a negative ion, leaving the other atom (or group of atoms) as a positive ion, with no actual bond between the two other than electrical attraction.

Hydrogen has only one electron and thus can form only one true covalent bond. When that bond is to a more electronegative atom such as oxygen or nitrogen, it becomes polar, leaving the hydrogen atom with a weak positive charge. If there are other electronegative atoms nearby, the charges interact to form a "hydrogen bond," in which the hydrogen atom assumes a position almost midway between the atom it is covalently bound to and the second negative atom. Examples of hydrogen bonding can be seen in the DNA base pairs illustrated in Fig. 4-1, *C*. In all cases in that figure, the hydrogen atoms are covalently bonded to nitrogen atoms and interact either with oxygen atoms or nitrogen atoms in adjacent bases.

Hydrogen bonding also plays important roles in the orderly folding of polypeptide chains to form functional protein molecules. An example of this is the α-helix structure, in which a polypeptide chain is wound into a helical coil with about 3.6 amino acid residues per turn. The helical structure is stabilized by the formation of hydrogen bonds between adjacent turns of the coil. Specifically the bonding occurs between hydrogen atoms covalently bound to amide nitrogens in one turn of the helix and carbonyl oxygens in the next turn.

Hydrogen bonding also occurs between the hydrogen and oxygen atoms of adjacent water molecules, and between water molecules and adjacent biological molecules such as proteins or nucleic acids that have polar groups exposed on their surfaces. The presence or absence of such interactions is important in determining whether various biological polymers are water soluble. Proteins that have few polar groups exposed on their outer surfaces are generally not soluble and are referred to as "hydrophobic." Those with more exposed polar groups and a greater affinity for water are "hydrophilic."

Molecules that have exposed positive or negative ionized groups on their surfaces attract one another and form a weak bond known as an

Hydrogen bonds

Ionic bonds

ionic bond. An example of this is the interaction between histones and nucleic acids to form nucleosomes (Chapter 5). Under physiological conditions the phosphate groups in the DNA backbone ionize to give the molecule a net negative charge, and the basic amino acids in the histones ionize to give the histone molecules a net positive charge. The spatial distribution of these charges is such that the histones and DNA interact to yield the complex nucleosome structure described in Chapter 5 (formation of such a structure also requires cooperative interactions among the eight histone molecules that form the "core" of the nucleosome).

Hydrophobic bonding

The third type of bonding that is involved in the formation of biological structures is less precisely defined and is commonly referred to as "hydrophobic" bonding. Examples of this type of bonding can be seen in the formation of biological membranes (Figs. 16-1, 16-2, 16-4, and 16-5). Hydrophobic bonding occurs primarily among nonpolar molecules and among the nonpolar parts of molecules that have geographically separated polar and nonpolar regions, such as the phospholipids and proteins that are involved in membrane formation.

As commonly used, the term "hydrophobic bonding" actually involves a mixture of two distinct phenomena. The first is a relatively weak attractive interaction among the hydrophobic molecules or regions themselves. The second, and probably the more important in biological systems, is the tendency of water to form hydrogen bonds with itself and with other polar molecules and to exclude nonpolar substances into a separate insoluble phase (or into the "interior" of organized structures such as biological membranes, which are discussed in greater detail later).

Absence of specific energy source

Molecular interactions similar to those described provide the driving force for self-assembly. The definition of self-assembly usually states explicitly that no "external" source of energy is needed to accomplish the assembly. It is relatively easy to exclude gross energy input, such as hydrolysis of ATP during the assembly process, but a precise determination of the source of energy for "spontaneous" self-assembly is not always straightforward. Experimental study of self-assembly generally requires some kind of energy input (often in the form of altered environmental conditions) to dissociate the subunits of the structure under study so that its self-assembly can be studied. The presence of artificially separated subunits at higher than normal concentrations can in itself be viewed as a subtle form of energy input. However, self-assembly is generally said to occur when subunits that have been separated without undergoing gross chemical changes reassemble into a structure comparable to the original structure without the direct participation of any external energy-yielding molecular species.

The definition of self-assembly usually also requires that the subunits assemble themselves into the final structure without the participation of any catalytic molecules that are not components of the final structure. It is relatively easy to exclude assembly processes that require catalytic macromolecules. However, many difficult questions arise about requirements for specific inorganic ions and for specific physicochemical conditions such as ionic strength and pH. In addition, it is sometimes necessary to remove the denaturing agents gradually to achieve "self-assembly" of the dissociated subunits into a structure that resembles the original. In most cases the model systems that have been studied have at least some specific environmental requirements for the self-assembly process to occur. The critical question that actually needs to be evaluated in most cases is whether newly synthesized subunits in their natural physiological environment will interact "spontaneously" without the intervention of catalytic factors or energy sources to generate complex organized structures. In many cases, lack of detailed knowledge of localized microenvironments within the cells where the assembly processes are taking place makes such questions almost impossible to answer.

A final complication that needs to be dealt with is the problem of whether a template must be provided for the self-assembly process. A "simple" example can be seen in the self-assembly of crystals. In many cases a highly purified and carefully filtered solution will fail to form crystals when brought to a temperature and a concentration of solute that thermodynamically favors crystallization. In the absence of "seed crystals" the solute may remain in a thermodynamically unstable "supersaturated" condition. When a single seed crystal is added, such a solution will often crystallize rapidly. The extent to which "seed" structures are essential for molecular self-assembly is generally not clear, but since most self-assembly processes are somewhat similar to crystallization, it is likely that nucleation (providing the first bit of organized structure) may be important for many self-assembly processes.

A substantial number of biological self-assembly systems have been described in the literature. We will examine a number of these briefly and then consider in greater detail the assembly of bacteriophage T4, which has been particularly well studied both in molecular and in ultrastructural terms and which involves both self-assembly and catalyzed assembly.

DNA double helix

A DNA double helix can be viewed as a "structure" assembled from two subunits, a "positive" strand that contains coded genetic information and a "negative" strand that has a complementary base sequence and serves as a template for transcription (Chapter 4). Two types of bonding stabilize the DNA double helix and make it the thermodynamically favored form under physiological conditions. These are hydrogen bonding between the base pairs of complementary strands (Fig. 4-1) and hydrophobic interactions among the planar base pairs that are stacked tightly together in the center of the double helix (Fig. 4-2).

The DNA double helix is destabilized by high temperature, high salt concentrations, and extremes of pH. This results in separation of the two strands. Precise reassembly, including accurate rematching of hydrogen-bonded base pair sequences will occur "spontaneously," provided that the denaturing conditions are removed gradually and the single-stranded subunits are given sufficient time to locate precisely matching complementary subunits. Molecular hybridization experiments based on analysis of the rates and selectivity of DNA self-assembly have become major tools in the study of gene expression, as described in Chapters 5 and 7 and in Appendix A.

Secondary, tertiary, and quaternary structure of proteins

Biochemists use the term "primary structure" to refer to the amino acid sequence of a protein. "Secondary structure" refers to linearly repeated structures in one dimension, such as the α-helix described earlier. "Tertiary structure" refers to complex three-dimensional folding of a linear polypeptide chain to yield biologically active globular proteins. "Quaternary structure" refers to the aggregation together of more than one subunit to yield a final structure. Hydrogen bonding, ionic bonding, and hydrophobic bonding are all important in achieving the final structures of biologically active protein molecules. In certain cases, covalent bonds such as disulfide (S-S) cross-links are also involved in stabilizing final structures. Most of the coiling and folding to produce secondary and tertiary structures of proteins after their synthesis is probably "spontaneous," although precise data are generally not available. Strictly speaking, these phenomena are not "self-assembly," since only a single polypeptide chain is involved. However, the mechanisms that are involved appear to be identical to those involved in true self-assembly.

The acquisition of quaternary structure, on the other hand, does involve aggregation of two or more subunits, and it has been shown experimentally in a number of cases to involve spontaneous self-assembly. One of the best-studied examples is the enzyme lactate dehydrogenase, which is a tetramer. What makes lactate dehydrogenase particularly interesting is the fact that the tetramer may contain any numerical combination of two different types of monomers (A or B), depending on the tissue and developmental age of the organism. Five different isozymes, LDH-1 through LDH-5, have been isolated with the following compositions: LDH-1, B_4; LDH-2, AB_3; LDH-3, A_2B_2; LDH-4, A_3B; LDH-5, A_4. Pure preparations of A and B subunits can be obtained by gentle dissociation of LDH-5 and LDH-1, respectively. When such monomers are recombined under suitable conditions, spontaneous self-assembly yields all five of the possible LDH tetramers in amounts determined by the ratio of subunits in the input mixture.

Complex protein structures

Self-assembly of protein subunits appears to play an important role in the generation of numerous biological "structures" that are plainly visible with the electron microscope. The assembly of microtubules from tubulin dimers (Chapter 15) appears to be largely a self-assembly phenomenon. The story is complicated by the fact that binding and hydrolysis of GTP normally accompanies microtubule formation. However, the fact that microtubules containing a nonhydrolyzable analog of GTP (GMPPCP) will assemble spontaneously and can then be dissociated at low temperature and reassociated a second time at higher temperature suggests that self-assembly mechanisms play a major role in microtubule formation.

Collagen fibers can be dissociated to tropocollagen (which is itself a trimer) by treatment with high salt concentrations or acidic pH. The tropocollagen will reaggregate spontaneously under physiological conditions to yield fibrils with a 640 Å periodicity that are indistinguishable in the electron microscope from native collagen. The total picture is complicated, however, by the gradual formation of covalent cross-links in natural collagen. These cross-links, which involve modified lysine residues in adjacent chains, tend to make collagen fibers increasingly stable and resistant to degradation with advancing age (Chapter 24). Myosin from muscle is another example of a protein that can be isolated as a monomer and caused to reaggregate into structures that are rather similar to the naturally occurring aggregates—in this case the thick myosin filaments in native muscle tissue (Fig. 3-18).

Biological membranes

The basic structure of biological membranes is discussed at the beginning of Chapter 16. When placed into aqueous solutions, phospholipids, both natural and synthetic, readily form bilayer structures that are similar to natural membranes. These have all their hydrophobic fatty acid chains in an interior position, with the polar ends of the phospholipid molecules at the surface in contact with the aqueous phase. Synthetic membranelike structures that contain various mixtures of cholesterol and phospholipids comparable to normal membranes can also be caused to show self-assembly under laboratory conditions. When proteins that are normally found in association with membranes are added to such mixtures, they tend to assume their natural positions within the artificial membranes, with their hydrophobic nonpolar areas buried within the lipid bilayer and with their hydrophilic polar areas extending into the aqueous phase.

When suspensions of phospholipids are treated with ultrasonic vibration, the artificial membranes are transformed into hollow vesicles approximately 250 Å in diameter completely surrounded by a closed lipid bilayer membrane, and with a small amount of the aqueous phase trapped inside. These phospholipid vesicles, which are often referred to as "liposomes," have proved to be useful laboratory tools for transporting into cells highly polar substances that do not normally cross cell membranes well. The liposomes appear to fuse with the cellular membranes and "dump" the material trapped inside themselves into the cell. In addition, they are sometimes taken up by phagocytosis and broken down inside the cell. The use of liposomes has made it possible to inject many different substances into cells, including intact mRNA molecules, which have subsequently been shown to be translated by the protein-synthesis machinery of the recipient cells.

Bacterial ribosomes

As described in Chapter 4, bacterial ribosomes are highly complex structures. The small subunit (30S) contains a 16S RNA molecule and about 21 proteins, whereas the larger subunit (50S) contains two RNA molecules (28S and 5S) and about 33 different proteins. Although the complex architecture of these structures has not been fully worked out, techniques have been developed that permit each of the two major ribosomal subunits to be completely dissociated to a mixture of RNA and proteins and then reassembled to yield fully functional ribosomal subunits. Detailed studies suggest that there is a specific sequence in which the ribosomal subunits must be reassembled, with certain proteins binding directly to the isolated RNA, others binding to the initial RNA-protein complexes, and still others binding only after a complex core has been formed. A comparable degree of self-assembly has not yet been achieved with eukaryotic ribosomes.

Tobacco mosaic virus

The tobacco mosaic virus particle is a cylinder 3000 Å in length with an outer diameter of 180 Å and a hollow core 40 Å in diameter. It is composed of an RNA molecule approximately 6000 nucleotides in length plus more than 2000 identical protein subunits. The protein subunits are arranged in a helical stack with approximately 16 subunits per turn and a total of approximately 130 turns in an intact virus particle. The RNA molecule is wound helically in a groove in the stacked subunit structure at a distance of about 40 Å from the center of the cylinder. The virus particles can be dissociated into RNA and protein subunits, which readily self-assemble into infectious virus particles when the dissociating conditions are reversed. The protein subunits alone show a limited tendency to form helical stacks that are similar in nature to the intact virus particle, except that they are much shorter and less stable. Recent studies suggest that during assembly of tobacco mosaic virus particles, the RNA molecule is probably threaded through the central hole in the growing rod. This configuration is claimed to be essential for rapid elongation, presumably by allowing the RNA to be drawn through the rod to the assembly point as new protein subunits are added while keeping the remainder of the very long RNA molecule out of the way at the opposite (nongrowing) end.

The assembly of bacteriophage T4 is of great interest, both because of the many self-assembly steps that are involved and because studies of T4 assembly have begun to provide insight into molecular assembly steps that require the participation of accessory proteins that are not part of the final structure. Bacteriophage T4 is one of the most complex of the bacterial viruses. Its genome is large enough to code for approximately 160 to 170 "average-sized" genes (1000 nucleotide pairs each), and about 140 different genes have actually been identified. The mature virus particle has a highly complex morphology and contains about 40 different kinds of protein molecules. It is currently estimated that about 53 different gene products are essential for its morphogenesis, either as structural components or as accessory factors needed for the assembly process. In addition, there is an involvement of host (*E. coli*) proteins in the assembly process.

Extensive genetic, biochemical, and ultrastructural analysis over a period of many years has provided a richly detailed picture of the assembly of bacteriophage T4 from its component parts (Fig. D-1), although a number of key steps are still not fully understood. In particular, the use of conditionally lethal mutants (p. 144) that fail to make a functional gene product under nonpermissive conditions has made it possible to evaluate the roles played by individual gene products in the overall assembly process.

Three major assembly sequences operate independently to generate heads, tails, and tail fibers, which are then assembled into the final virus particles. Viral gene products that serve as structural components of the mature virus are shown in Fig. D-1 as squares or rectangles containing the gene designation (in most cases a number). The

SELF-ASSEMBLY AND CATALYZED ASSEMBLY IN BACTERIOPHAGE T4

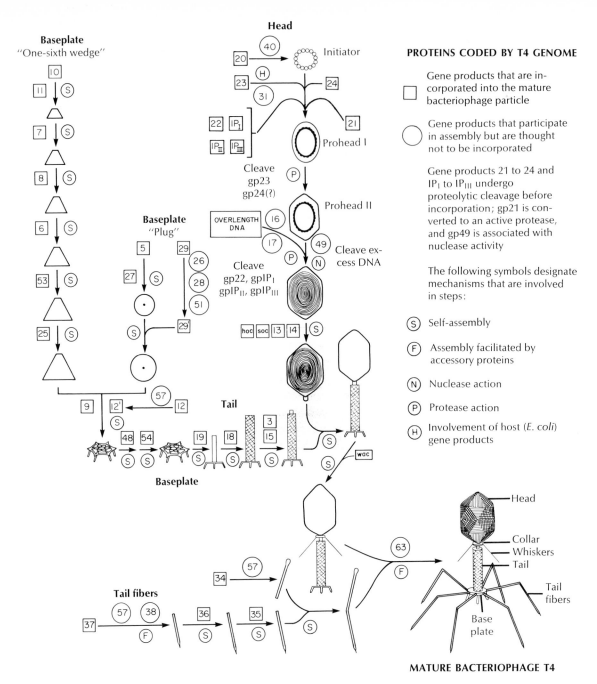

Fig. D-1. Assembly of bacteriophage T4. The major assembly steps as they are currently understood are outlined schematically. All the gene products shown in square or rectangular boxes are incorporated into the mature bacteriophage particle, although some are first partially degraded. Some of the indicated gene products, including IP$_I$, IP$_{II}$, IP$_{III}$, hoc, soc, and wac, are not strictly essential for the formation of infectious particles under laboratory conditions. (This assembly sequence is a composite of partial sequences described by Wood, W. B. 1979. The Harvey Lectures, series 73, 1977–1978, p. 203, and Murialdo, H., and A. Becker. 1978. Microbiol. Rev. **42**:529.)

normally used abbreviation "gp" (for gene product) has been omitted from the boxes for simplicity but is used elsewhere in the figure and in the text. Gene products that promote the assembly process but are not found in the mature virus particles are shown as circles containing the gene designation. These gene products are often spoken of as functioning catalytically, but it is not known whether true catalysis in a chemical sense is involved in any of the assembly steps. Small circles containing single letters indicate the type of process that is involved at each step, with the letter *S* designating self-assembly.

Sequential self-assembly

Self-assembly plays a major role in bacteriophage T4 morphogenesis, as can be seen from the large number of self-assembly steps that are identified in Fig. D-1. A particularly long self-assembly sequence occurs in tail formation, beginning with gp10 at the upper left and continuing through the joining of the head to the tail. In the entire tail assembly process, accessory proteins appear to be needed only in the modification of gp29 and the modification of gp12, prior to the self-assembly of these two modified products into the morphogenetic sequence. The assembly of the completed head onto the completed tail also proceeds by a self-assembly process, as does the addition of gp-wac (which forms the collar and whiskers). Several of the steps in the tail fiber assembly also occur by self-assembly.

One of the interesting features that has emerged from detailed studies of these self-assembly sequences is that with few exceptions each new addition can be made only after the previous one has been completed. The absence of one of the components early in an assembly sequence will generally block all subsequent additions. In the long self-assembly sequence involving the baseplate wedge and the tail, only gene products 9, 11, and 12 (modified by 57) can be added out of sequence. All others must be added one at a time in the proper order, or the assembly stops entirely. This phenomenon does not involve sequential availability of the building blocks, since nearly all of them are synthesized at the same time. Rather, it appears to involve conformational changes such that each time a subunit is added, the kinetic barrier to the next binding step is greatly decreased. The exact mechanisms responsible for this phenomenon of "heterocooperativity" have not been worked out, but the phenomenon appears to be an important aspect of the precisely organized sequential assembly procedure.

Roles of accessory proteins

In addition to providing insight into sequential self-assembly, studies of bacteriophage T4 morphogenesis have also provided information concerning three different types of accessory proteins that function in the assembly process: scaffolding proteins, hydrolytic enzymes, and promoters of noncovalent association. At least four different proteins play a scaffolding role in the formation of heads during bacteriophage T4 morphogenesis. The capsid (outer coat of the head) is an icosahedral structure composed primarily of gp23. Assembly is initiated at a vertex composed of gp20, and the remaining vertexes all contain gp24. In the mature virus particle the head is filled with DNA, but the initial step in assembly is formation of a "prohead" that contains no DNA. During this process the capsid is "supported" by internal structural elements, or "scaffolding," consisting of gp22 and the so-called internal proteins gpIP$_I$, gpIP$_{II}$, and gpIP$_{III}$.

Gene product 40 plays a catalytic role in forming the initiator vertex from gp20, and gp31 plays a catalytic role in the incorporation of the major component (gp23) into the capsid, but neither of these catalytic proteins is incorporated into the bacteriophage structure. A "host" protein, coded for by the *E. coli* genome and designated by the circled letter *H* in Fig. D-1, is also involved in the incorporation of gp23 into the capsid. After the capsid has been completed, a proteolytic enzyme (derived from gp21) removes fragments from gp23 and probably also from gp24. This allows structural rearrangement and enlargement of the capsid in preparation for DNA entry. At the time of entry of DNA into the head, gp22 and the internal proteins I, II, and III also undergo proteolysis, but substantial fragments are retained in the mature phage, as are degraded portions of gp21 itself.

The entry of DNA into the phage head is a complex and poorly understood process. The bacteriophage DNA is synthesized in linear units that are greater than one complete genome in length. Aided by gp16 and gp17 (which are thought not to have permanent structural roles), the DNA enters the head in a highly organized manner and appears to be packed into concentric coils. A nuclease activity (associated with gp49) clips the genome to the proper length. Head formation is then completed by adding gp13 and gp14, which are involved in attachment of the head to the tail. The overall process of head formation thus involves both scaffolding and the activity of proteases and nucleases. In addition, gp31 and gp40 play poorly defined roles in assembly of the prohead, and gp16 and gp17 are involved in the movement of DNA into the head.

The formation of tail fibers and their attachment to the rest of the virus particle provide some of the better understood examples of the involvement of accessory proteins in assembly processes. At the beginning of the tail fiber assembly sequence, gp37 is modified by interaction with gp57. Two of the modified gp37 molecules are then assembled together to form the major structural element of the distal half of

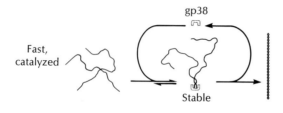

gp38

Fast,
catalyzed

Stable

A

Very slow,
uncatalyzed

gp37   Unstable
Intermediate
complex   C half fiber

Fig. D-2. Possible mechanisms for "catalyzed" assembly of components of bacteriophage T4. In each case it is known that accessory proteins promote the assembly steps that are depicted, but true catalysis has not been proved, and the representations of specific mechanisms in these drawings are purely speculative. **A,** Gene product 38 accelerates formation of the distal half of a tail fiber from two molecules of modified gp37. **B,** Whiskers (gp-wac) interact with tail fibers and help position the fibers for rapid addition to the rest of the bacteriophage. **C,** Gene product 63 functions catalytically to accelerate attachment of tail fibers to the baseplate. (From Wood, W. B. 1979. The Harvey Lectures, series 73, 1977-1978, p. 203.)

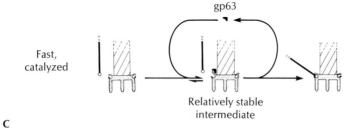

gp-wac   + Fast,
bimolecular   gp63
Fast,
unimolecular

B

+   Very slow, bimolecular

gp63

Fast,
catalyzed

Relatively stable
intermediate

C

Slow,
uncatalyzed

Relatively unstable
intermediate

a tail fiber. That assembly process occurs slowly in the absence of gp38 but is much more rapid when "catalyzed" by gp38. Fig. D-2, *A* is an artist's conception of how gp38 might be acting to bring the two modified gp37 molecules into a configuration that would permit them to interact more readily. (It must be emphasized that true catalysis has not yet been proved and that the actual mechanisms are not yet known for any of the facilitated assembly processes depicted in Fig. D-2.)

The attachment of completed tail fibers to the rest of the virus particle is facilitated by two viral gene products. One is a structural component of the virus, gp-wac, which forms the "collar" and the very thin "whiskers" attached to the mature bacteriophage at the base of the head. Mutant bacteriophages lacking the collar and whiskers are able to form infectious particles, but the addition of tail fibers is much slower and less efficient. The second assembly-promoting factor that is needed is gp63, in whose absence the addition of tail fibers is extremely slow. Unlike gp-wac, gp63 is not incorporated into the final virus particle. Speculative drawings showing how these two proteins may facilitate the attachment of tail fibers can be seen in Fig. D-2, *B* and *C*. The whiskers are known to interact with tail fibers, and it is thought that this interaction helps to bring the fiber into the right alignment for the gp63-catalyzed assembly step. The involvement of gp63 in that attachment is well documented, but the exact mechanisms that are involved remain conjectural, and the scheme shown in Fig. D-2, *C* must be labeled clearly as speculation.

The techniques for stepwise dissection of the bacteriophage T4 assembly process and the isolation and characterization of all the molecules that are involved in each individual step are now highly developed. It appears to be only a matter of time until speculative models of catalyzed assembly such as those in Fig. D-2 can be replaced with highly detailed descriptions of the exact molecular processes and interactions that are involved. Similar techniques are also beginning to be applied to the study of the assembly of many different types of supramolecular structures that are involved in cellular differentiation and other developmental phenomena in multicellular eukaryotic organisms. Model systems such as bacteriophage T4 are making major contributions to the development of the techniques that are needed for analysis of the more complex assembly processes that occur in higher organisms.

BIBLIOGRAPHY

Butler, P. J. G., et al. 1977. Configuration of tobacco mosaic virus RNA during virus assembly. Nature **265**:217.

Butler, P. J. G., and A. Klug. 1978. The assembly of a virus. Sci. Am. **239**:52 (Nov.).

Grant, P. 1978. Macromolecular assembly: the ontogeny of cell organelles. *In* Biology of developing systems. Holt, Rinehart & Winston, Inc., New York.

Miller, A. 1976. Self-assembly. *In* C. F. Graham and P. F. Wareing, eds. The developmental biology of plants and animals. W. B. Saunders,Co., Philadelphia.

Murialdo, H., and A. Becker. 1978. Head morphogenesis of complex double-stranded deoxyribonucleic acid bacteriophages. Microbiol. Rev. **42**:529.

Wood, W. B. 1979. Bacteriophage T4 assembly and the morphogenesis of subcellular structure. The Harvey Lectures. series 73, 1977-1978, p. 203.

Wood, W. B. and M. P. Conley. 1979. Attachment of tail fibers in bacteriophage T4 assembly: role of the phage whiskers. J. Mol. Biol. **127**:15.

Wood, W. B., and R. S. Edgar. 1967. Building a bacterial virus. Sci. Am. **217**:61 (July).

APPENDIX E

Introduction to the scientific literature

☐ By its basic nature an introductory textbook that includes a relatively broad range of topics is severely limited in the depth to which each of those topics can be explored. In addition, because of the inherent time lag involved in writing and publishing, the information in a textbook is already somewhat out-of-date when the book comes off the press, and it falls farther behind during the time interval between new editions.

If this book has been successful in stimulating reader interest in the rapidly progressing field of developmental biology, our readers will find it desirable to obtain additional and more current information from the scientific literature. The purpose of this appendix is to provide a brief introduction to the types of information that are available in the scientific literature at a typical university library and to provide instruction in locating the desired material among the overwhelming mass of information (and misinformation) that must be sifted through to find it. We will begin with a description of the various kinds of reference materials that are available and then proceed to a discussion of techniques and research aids for finding specific information.

CLASSIFICATION OF THE SCIENTIFIC LITERATURE

The following scheme somewhat arbitrarily divides the scientific literature into a number of broad categories that are not always as clearly distinct from one another as they are made to seem here. In essence, the classification follows the progression of a new discovery from the time when it is first reported to the scientific community until it becomes widely accepted general knowledge.

In most cases, research on a particular topic is done simultaneously in several different laboratories, with each making a series of small contributions to the total understanding of the problem under study. Dramatic breakthroughs are relatively rare, and major new concepts generally do not become evident until the results of many individual experiments are combined.

The individual findings are usually reported as they are discovered in highly technical and detailed original research reports. Broader concepts may emerge either in the discussion sections of original research reports written by investigators who believe that they have found key pieces to the puzzle, or else in review papers that seek to tie together research in many different laboratories. As such concepts become generally accepted by research specialists within a particular area, they are more broadly disseminated to the scientific community at large in the forms of monographs and reference books. At the same time, they begin to be incorporated into textbooks designed primarily for classroom teaching, and they may also make their way into popular publications intended for a broader audience. Each of these categories of scientific literature will be explored in greater detail.

Articles in professional journals. The accepted method of making a new finding available to other scientists is to publish a detailed article containing the experimental data and their interpretation in a scientific journal. Such articles are expected to contain all information needed for other investigators to repeat the experiments and to verify their validity. One of the limitations of original research reports is that each report generally represents only a small step forward, whose significance is difficult to understand without a general understanding of the field and the unanswered questions that existed at the time the report was published. A limited number of original research reports have been cited at the ends of individual chapters in this book.

Preliminary reports. New findings of special interest, particularly in highly competitive areas, are often announced as rapidly as possible in preliminary communications. These are generally short and contain less experimental detail than full-length journal articles. They are also published more quickly and sometimes not reviewed as carefully as longer articles. Some regular journals have special sections for preliminary reports, but most appear in special journals such as *Nature, Science, The Proceedings of the National Academy of Sciences U.S.A., Cell Biology International Reports*, and *Biochemical and Biophysical Research Communications*.

Abstracts. Short abstracts of papers presented at national meetings are published by several professional societies in journals such as *Journal of Cell Biology, Federation Proceedings*, and *In Vitro*. Abstracts can sometimes provide useful insight into current thinking in a particular investigator's laboratory, as well as the first printed announcement of new findings. However, their value is limited in that they contain almost no experimental detail and are often accepted for publication with little or no critical review. (Meeting abstracts should

Direct reports of original research

not be confused with abstracts of sjournal articles that are collected and published by several information services such as *Biological Abstracts*, *Chemical Abstracts*, and *Excerpta Medica*, as will be discussed later.)

Indirect reports of research

Review articles. Review articles bring together information from a number of original research papers and attempt to formulate overall concepts from that information. Frequently, when a series of individual discoveries results in the emergence of a major new concept, one of the investigators who has been involved in the research will prepare a review article that summarizes the development of the concept and the evidence for its validity. Such articles are often featured prominently at the beginning of issues of journals such as *Nature* or *Science* that have a broad general readership. Shorter reviews, focused primarily on the significance of single papers or small groups of papers, also appear regularly in the "News and Views" section of *Nature* and from time to time in other journals.

Review articles of a slightly different type that seek to summarize the current status of research on a particular topic are published regularly, particularly in collections with titles such as "Annual Review of . . . ," "Advances in . . . ," "Current Topics in . . . ," and "Recent Progress in . . ." Such articles usually focus both on the advances that have been made and on the unanswered questions that still remain in the particular area of study with which they deal. A substantial number of review articles of various types have been cited at the ends of individual chapters in this book.

Symposium volumes. Certain organizations such as the Society for Developmental Biology and the Cold Spring Harbor Laboratory for Quantitative Biology conduct regularly scheduled symposia on topics of current research interest and publish annual volumes of symposium proceedings. In addition, there are numerous symposia that are organized on a one-time basis around particular research topics. These also frequently result in the publication of a symposium volume. Symposium volumes typically contain a mixture of broad review articles and more detailed articles on new research that is being described for the first time.

Monographs. A monograph is similar to a symposium volume, except that no actual meeting has been held. An editor or an editorial panel asks a number of leading investigators in a particular field of study to prepare chapters on their areas of specialty, which are then assembled into a monograph, with varying amounts of editorial comment to link them together. The coverage of topics in monographs tends to be more thorough than in symposium volumes, and they can be excellent sources of advanced information. Depending on how well the editors have done their jobs, the coverage of the field may or may not be complete, and the articles may or may not fit together smoothly.

Collections of original research reports. Another approach that can sometimes be effective is to reprint in a single volume all the major research articles that have led to the development of a new concept. This approach works well only when the editor does a careful job of selecting the right articles and providing sufficient editorial discussion to emphasize the significance of each article and fill in the gaps between them. A particularly effective volume of this type, which was actually intended to be used as a textbook, is *Explorations in Developmental Biology*, edited by Chandler Fulton and Atilla O. Klein, Harvard University Press, Cambridge, 1976.

Reference books. As new information passes from the realm of recent discovery to generally accepted fact, it soon becomes incorporated into advanced treatises and reference texts, which are generally written by one or a few authors and which seek to deal with specific areas of knowledge in a comprehensive and uniform manner without the gaps and differences in style that characterize most monographs and symposium volumes. Such books are usually intended for use by specialists and advanced students within a narrowly defined field and are often detailed and difficult to follow without some prior knowledge of that field.

Textbooks. Books designed for use in classroom teaching are usually a good place to start learning a new field, and texts designed for more advanced courses can provide a good follow-up. However, textbooks in general tend to present conclusions with a minimum of evidence or data, and it is usually necessary to go to more advanced sources for the details after an understanding of the basics has been obtained from the textbook. A number of advanced textbooks and reference books have been cited at the ends of chapters in this volume.

Popularized reports. Articles written primarily for scientists in other fields and for science students, such as the articles in *Scientific American* and some of the lead articles in *Science*, often provide an excellent way to begin exploring new areas at a level that is comprehensible to the novice. A number of such articles have been cited in this book. Books and articles written for the general public may also be useful at times, but they are generally not critical in their approach and must be used cautiously. Important new findings often appear first in the *New York Times*, for example, but any report from a popular source must be viewed with skepticism until it has been verified in the regular scientific literature.

Other sources of information

INDEXING AND
ABSTRACTING
JOURNALS

The student who is using the research literature for the first time is likely to feel overwhelmed by its mass. Even within a highly restricted area of specialty, it is difficult to keep up with the new articles that are published and it is virtually impossible to do a retrospective survey of several years of publication without relying heavily on special library research aids. Fortunately, there are several types of indexing and abstracting journals that are designed for that purpose. However, each is designed for a particular field (e.g., biology, chemistry, medicine), and each is organized somewhat differently.

For purposes of illustration, we will show how several of these services can be used to locate articles on the role of the H-Y antigen in sex determination (Chapter 20). More specifically, we will examine ways of locating the following article:

Koo, G. C., et al. 1977. H-Y antigen: expression in human subjects with the testicular feminization syndrome. Science **196**:655.

In addition, as each indexing service is described, we will point out ways to find other articles related to the same topic and we will explore some of the difficulties that are likely to be encountered.

Biological Abstracts

Biological Abstracts publishes and indexes nearly 150,000 short abstracts of articles from all areas of biology each year. Issues are published twice monthly and are bound into semiannual volumes. The abstracts within each volume are numbered serially, and all indexing refers to the abstract number. Five different types of indexes are provided, in addition to a subject guide, both in individual issues and in the semiannual volumes.

Subject index. *Biological Abstracts* uses a computer to generate a subject index based on permutations of the article title plus added key words. In the case of the article by Koo et al. the words "white blood cells" have been added to indicate the type of cell that was examined for H-Y antigen. Following is the master entry in the computer that is used for indexing:

H-Y ANTIGEN EXPRESSION IN HUMAN SUBJECTS WITH THE TESTICULAR FEMINIZATION SYNDROME WHITE BLOOD CELLS/.

The computer selects keywords from the titles plus added words for all articles stored in its memory and generates an index that is alphabetized by key word. A segment of the title plus extra words that is approximately 55 letters plus spaces in length is printed, with the keyword near the middle and adjacent words to the left and right of it. The following nine permutations of the master entry appeared in the subject index of volume 64 of *Biological Abstracts:*

HITE BLOOD CELLS/ H-Y	ANTIGEN EXPRESSION IN HUMAN SUBJEC	50492
ATION SYNDROME WHITE	BLOOD CELLS/ H-Y ANTIGEN EXPRESSION	50492
SYNDROME WHITE BLOOD	CELLS/ H-Y ANTIGEN EXPRESSION IN HU	50492
OD CELLS/ H-Y ANTIGEN	EXPRESSION IN HUMAN SUBJECTS WITH	50492
TS WITH THE TESTICULAR	FEMINIZATION SYNDROME WHITE BLOOD	50492
ME WHITE BLOOD CELLS/	H-Y ANTIGEN EXPRESSION IN HUMAN SU	50492
ESTICULAR FEMINIZATION	SYNDROME WHITE BLOOD CELLS/ H-Y AN	50492
AN SUBJECTS WITH THE	TESTICULAR FEMINIZATION SYNDROME W	50492
EMINIZATION SYNDROME	WHITE BLOOD CELLS/ H-Y ANTIGEN EXPR	50492

Some of these permutations prove to be more useful than others. Because of the length of the master entry, none of the index entries clearly link H-Y antigen and testicular feminization. For a broad search of the literature the "H-Y" and the "testicular" entries are probably the most useful, since alphabetization proceeds to the right of the key word. If one looks under "antigen," entries with "H-Y" to the left could be anywhere in a rather long list of articles that have the word "antigen" in their titles.

The subject index provides the abstract number 50492. When that number is looked up in the abstracts section, the following is found.

50492. KOO, GLORIA C.*, STEPHEN S. WACHTEL*, PAUL SAENGER, MARIA I. NEW, HARVEY DOSIK, ANTHONY P. AMAROSE, ELIZABETH DORUS and VALERIO VENTRUTO. (Mem. Sloan-Kettering Cancer Cent., New York, N.Y. 10021, U.S.A.) **H-Y antigen: Expression in human subjects with the testicular feminization syndrome.** SCIENCE (WASH D C) 196(4290): 655-656. 1977.—Androgen-insensitive subjects with a 46,XY karyotype develop as phenotypic females despite the presence of testes. The white blood cells of these females type H-Y antigen–positive indicate that expression of the H-Y cell surface component is androgen-independent.

The abstract provides complete information on the authors, title, and journal reference for the article, as well as the address of the senior author and a brief summary of the information content of the article.

Several other abstracts are also indexed under the H-Y heading of the volume 64 subject index, including the following:

ES/ TESTIS DETERMINING   H-Y ANTIGEN IN XO MALES OF THE MOLE 26560

When the number 26560 is looked up, the following abstract is found:

26560. NAGAI, YUKIFUMI and SUSUMU OHNO. (Dep. Biol., City Hope Natl. Med. Cent., Duarte, Calif. 91010, USA.) **Testis-determining H-Y antigen in XO males of the mole-vole** (Ellobius lutescens). CELL 10(4):729-732. 1977.—The XO sex chromosome constitution was found in both sexes of the mole-vole (E. lutescens) belonging to the rodent family Microtinae. This enigmatic species has apparently been enduring a 50% zygotic lethality. The current serological study revealed the presence in XO males and the absence from XO females of H-Y (histocompatibility Y) antigen. In all the mammalian species studied the expression of H-Y antigen strictly coincided with the presence of testicular tissue and not necessarily with the presence of the Y chromosome. The testis-organizing function of the H-Y gene appears confirmed. In the mole-vole, X linkage of the testis-organizing H-Y gene is favored over its autosomal inheritance. Only X linkage of the H-Y gene creates a compelling evolutionary need to change the female sex chromosome constitution from XX to XO, and to abandon the dosage compensation by an X inactivation mechanism, so that the nonproductive $X^{H-Y}X$ zygote can be eliminated as an embryonic lethal. With regard to the electrophoretic mobilites of 3 X-linked marker enzymes, however, a genetic difference between the male-specific X^{H-Y} and the female-specific X was not detected. This might reflect a relatively recent speciation.

This abstract describes one of several unusual cases of sex determination that have helped strengthen the view that the H-Y antigen is a primary determinant of maleness.

One problem that is encountered with computer-generated indexes in general is that the computer often does strange things to alphabetization, particularly in cases in which hyphens or numbers are involved or when a term can be written either as one or two words. Hyphens are generally alphabetized before the letter *A*. Numbers may be either before or after letters, depending on the computer program. Thus "H-Y" and "H-2" are both generally found before "habit," but in *Biological Abstracts*, "H-Y" is before "H-2," whereas in *Index Medicus*, "H-2" comes before "H-Y."

Author index. The author index is especially useful for finding additional articles by investigators who are known to be working in a particular field. When the name Koo, G. C. is looked up in the author index of volume 64, two other abstracts in addition to 50492 are listed, and three more articles are listed under her name in the volume 63 author index. All these abstracts describe articles that are related in some manner to sex determination or H-Y antigen. When Wachtel, S. S. (second author of the paper in abstract 50492) is looked up in the author indexes, still other relevant abstracts are cited.

Biosystematic index. This index is organized first by taxonomic family and then by subject heading within that family. The categories tend to be too broad to be particularly useful, except for taxonomic studies. The two abstracts cited earlier were listed as follows:

```
PRIMATES   HOMINIDAE   IMMUN PATH    50492
RODENTIA   CRICETIDAE  GENET ANIMAL  26560
```

Generic index. Articles in which genus and species are included in the title are included in this index, subdivided by subject heading. Abstract 26560 was indexed as follows:

```
ELLOBIUS-LUTESCENS   GENET ANIMAL   26560
```

Cumulative concept index. This index lists all abstracts in an issue or a volume of *Biological Abstracts* that are related to various broadly defined subject areas. It is useful primarily to individuals who are striving to cover an area in depth, and then only if used regularly with each new issue. Abstract 50492 was one of several thousand that appeared under the concept "immunology-immunopathology" in the index to volume 64.

Subject guide. Abstracts are grouped together by subject in each individual issue of *Biological Abstracts*. A subject guide at the front of each issue tells where the abstracts on each subject start in that issue. Skimming through subject areas of new issues can be a useful way of keeping up with specific topics. However, one of the limitations is that each article is abstracted only once, and it is often difficult to predict in which of several subject areas it will be placed. For example, abstract 50492 appeared under "immunology - immunopathology," and abstract 26560 under "genetics and cytogenetics — animal."

Chemical Abstracts is similar in overall organization to *Biological Abstracts*. It is published weekly, with two complete volumes and over 400,000 abstracts per year. Odd-numbered issues contain abstracts on organic chemistry, biochemistry, and biological topics, and even-numbered issues are concerned with physical and inorganic chemistry. Thus there is an issue every 2 weeks concerned with topics related to the mechanisms of development. The organization of *Chemical Abstracts* is along chemical lines, and its coverage of biological topics is somewhat erratic, although many areas of molecular biology and biochemistry are covered well.

Chemical Abstracts

The paper by Koo et al. appeared as abstract 187252z in volume 86 of *Chemical Abstracts*. (The letter *z* is part of an error-correction scheme that helps the reader to know when the wrong abstract has been looked up.) The abstract is similar (but not identical) in form and content to the abstract in *Biological Abstracts*. It was included under the general subject heading of "mammalian pathological biochemistry," but that heading is far too broad to be useful in locating individual articles on special topics, even within a single issue. Several types of indexing are available.

Keyword index. Each individual issue of *Chemical Abstracts* contains a keyword index similar in principle to the index in *Biological Abstracts*, except that the computer is given only a few keywords or concepts to permute, rather than the entire article title. For abstract 187252z, there are three entries under "antigen," "HY," and "testicular feminization syndrome," respectively. In this case, "HY" is not hyphenated, and the entry "HY antigen testicular feminization syndrome" is found between "Huygens" and "hyacinth."

Subject index. Each complete volume (6 months) has a cumulative subject index with a rigorous hierarchy of arrangement so that one can be sure of finding all entries on a particular subject in one place. The entries for abstract 187252z are "antigens, H-Y, in testicular feminization" and "testis, disease or disorder, feminization, HY antigen expression in." In addition to semiannual indexes for each volume, *Chemical Abstracts* also publishes 5-year cumulative indexes.

Author index. The author index of *Chemical Abstracts* is somewhat more detailed than that of *Biological Abstracts* in that it gives the first name of the author and also the article title. The 5-year cumulative author indexes make it easier to look up all articles by an author who is known to be working on a particular topic of interest.

Chemical substance, chemical formula, ring structure, and patent indexes. These are specialized indexes that are useful primarily to chemists.

Index Medicus

Index Medicus is a computer-assembled listing of medically related references, arranged both by author and by subject. It is issued monthly and also on an accumulated basis once a year as *Cumulated Index Medicus*. It is convenient to use in that it provides a complete reference to each article in the subject and author listings, making it unnecessary to look up an abstract in a volume separate from the index, as must be done with *Biological Abstracts* and *Chemical Abstracts*. However, it does not abstract the articles, which makes it necessary to go to the original journals or to abstracting journals to gain more information about an article than is provided by its title.

The subject headings that are used are quite detailed; so the number of articles listed under a particular topic per year is generally manageable. Also each individual article is listed under as many different headings as are needed for full coverage of its subject material. A list of headings, including a guide to related headings that might contain additional references of interest, is published at the beginning of each year. New headings are added frequently as new research makes them appropriate. For example, beginning in January, 1979, "H-Y antigen" has its own heading. However, in 1977 it was necessary to look under "histocompatibility antigens—analysis" or under "testicular feminization—immunology" to find the article by Koo et al. Articles on the role of H-Y antigen in sex determination also were scattered under the headings "sex," "sex chromosomes," "sex determination," and "sex differentiation" in the 1977 *Cumulated Index Medicus,* although the Koo et al. article was not in any of them, probably because it did not have the right words in its title.

One of the limitations of *Index Medicus* is that it does not have any indexing beyond the rather extensive list of subject headings under which citations are classified. However, despite this problem and occasional difficulties in finding a category that includes the right set of articles, *Index Medicus* is often one of the most effective tools for surveying current publication in a particular area of specialty.

Science Citation Index consists of three distinct parts—a source index, a citation index, and a subject index, which are issued both quarterly and on a cumulative yearly basis. These three parts are used together in a number of different ways. In addition, there is a companion publication called *Index of Scientific Reviews*, whose organization is almost identical, except that it is limited to review articles.

Science Citation Index

Source Index. The source index is a listing by author of all articles published during the year (or quarter year). It can be used alone as an author index for articles published in a wide range of scientific fields. In addition, it can be used in conjunction with the citation index or the subject index to find articles on particular topics.

Citation Index. This index is entirely different from any we have discussed previously. It is organized around the premise that articles dealing with a particular topic can be identified by examining the references cited in their bibliographies. Thus, for example, we would assume that any article concerning the role of H-Y antigen in sex determination would be likely to cite "Wachtel, S. S., et al. 1975. Possible role for H-Y antigen in the primary determination of sex. Nature **257**: 235.," which was one of the first articles to propose that such a role existed. In the 1977 citation index, under the name "Wachtel, S. S." are listed a number of different articles that he has published, including "75 NATURE 257 235." Listed under that heading are abbreviated references to 32 different articles in the 1977 source index that contain Wachtel's 1975 *Nature* article in their bibliographies. The list includes KOO GC SCIENCE 196 655 77. Most of the other 31 articles also probably deal in some way with H-Y antigen or sex determination, since they have referenced the Wachtel paper. Complete bibliographic citations, including coauthors and article titles can be obtained from the source index.

Permuterm Subject Index. A subject index, similar in principle to the keyword indexes of *Biological Abstracts* and *Chemical Abstracts* has been generated for use with the source index by combining, in all possible combinations of two, the major terms in the titles of each of the articles in the source index. It is used in essentially the same manner as the other subject indexes. In theory, it avoids some of the problems encountered in the *Biological Abstracts* index, since it brings all the major words in the title together in groups of two. Its limitation, however, is that two words are not always enough to describe a concept or relationship that is being sought.

Current Contents

Current Contents is a weekly publication that is designed primarily to aid researchers or students to stay strictly up-to-date on the literature in their fields. The *Life Sciences* edition of *Current Contents* reproduces tables of contents from essentially all the major journals that publish articles related to the life sciences. It also includes quite a few books and monographs. Each issue has computer-generated subject and author indexes.

Other abstracts and indexes

In addition to those cited, there exist a wide variety of other abstracting and indexing services that serve various areas of specialty. Examples include *Genetics Abstracts, Psychological Abstracts,* and the extensive series of medically related abstracting journals published by *Excerpta Medica. Dissertation Abstracts* publishes abstracts of Ph.D. dissertations, often long before the information from the dissertation appears as a regular journal article. A number of scientific journals publish lists of articles related to their areas of specialty. For example, *Gerontology* and the *Journal of Gerontology* both publish bibliographies of recent articles on various aspects of aging. Thus, when the research aids already described do not prove satisfactory, the interested student should ask a librarian or a practicing scientist in the area of specialty that interests him what is the best way to locate current publications in that specialty.

Computerized searches

The information that is used to generate the various indexing and abstracting journals is all stored in computers, and most of it is available for customized computer searches, including the data banks for *Biological Abstracts, Chemical Abstracts, Index Medicus,* and *Science Citation Index* plus *Current Contents.* Most major university and medical school libraries have computer access to at least some of these data banks. By using appropriate combinations of subject headings and key words, it is often possible to generate a selected list of reference citations that have a high probability of being useful.

STRATEGY FOR LOCATING REFERENCES

There is no set strategy that will guarantee finding the desired references on a particular topic. For the student who is starting at the textbook level in a particular subject, the first step will probably be to progress to the advanced text and monograph level. Recently published review articles are an excellent way to get in touch with current thinking in a field. They can also provide valuable clues to finding the most current original research articles, including the names of investigators actively working in the field and the names of the journals in which new findings are most likely to be published. With that information, it is possible to use author indexes to locate new articles by those authors, and the tables of contents of the journals that are most likely to contain new information can be checked as each new issue appears. Also, through experience one learns which abstracting and indexing services are the most useful and which headings should be watched regularly.

Unfortunately, there is no fixed set of rules that can be followed in every case. Whenever a new topic is approached, it is necessary to start over with the development of a new scheme for efficient location of current research results. It is also necessary to revamp old strategies from time to time to accommodate changes that have occurred. For instance, in the example we have been following, it became important, beginning in January, 1979, to examine the subject heading "H-Y antigen" in *Index Medicus* each month.

The research aids described in this appendix have been invaluable in the preparation of this book, which would never have come into existence without them.

BIBLIOGRAPHY

Bottle, R. T., and H. V. Wyatt, eds. 1971. The use of biological literature, 2nd ed. Archon Books, Hamden, Conn.

Kirk, T. G., Jr. 1978. Library research guide to biology: illustrated search strategy and sources. Pierian Press, Ann Arbor, Mich.

INDEX

HOW TO USE THE INDEX

☐ This index is organized primarily in terms of nouns. For example, pages that describe the development of mammalian embryos are indexed under "Embryo, mammalian" rather than "Mammalian embryo." However, in cases where an adjective and a noun are commonly viewed as a single unit, the adjective is used for the primary listing (e.g., "Endoplasmic reticulum"). In cases of potential confusion, there is also a cross reference under the noun (e.g., "Reticulum, endoplasmic; *see* Endoplasmic reticulum").

Boldface numbers refer to major sections on specific topics. Italic numbers refer to information contained in tables, outlines, figures, or figure legends, separated from the main text material. The letter "n" after a page number refers to information contained in a footnote on that page.

All abbreviations used in this book are included in the index in normal alphabetical order, either with their meanings in parentheses or else with "see" references. The page on which a term is defined is indexed separately in cases where there are multiple references to that term. Numbers and Greek letters used as prefixes have been ignored in alphabetizing. Greek letters that are integral parts of names (e.g., Bacteriophage ϕX174) are alphabetized as if the name of the Greek letter were spelled out.

Intermedin; *see* Melanophore-stimulating hormone
Intersexuality, 610, **632-638**; *see also* Hermaphrodite; Pseudohermaphrodite
Interstitial cell–stimulating hormone (ICSH); *see* Luteinizing hormone
Intervening sequences, 115, 188-190, 188n, 252, 256-257, 271-272, 453-454, 758
  absence from histone genes, 190
  detection, 188-189
  globin genes, 188-189, 257, 271-272
  mRNA processing and, 189, 256-257, 271-272
  ovalbumin gene, 188-189, 453-454
    variation among individual chickens, 454
  splicing out posttranscriptionally, 189
  transcription, 189, 257, 271-272
Intestinal portals, *41*
Intestine, epithelium, 75
Intron, 188n
Ionic bond, 779-780
Ionophore, calcium
  acrosome reaction, 595
  egg activation, 601
Iris, transformation of cells to lens, 404
Iron deficiency, effect on translation of ferritin mRNA, 294
Isocitrate dehydrogenase, in myotubes of allophenic mice, 338-339
Isodesmosine, 310
Isolecithal egg, cleavage, 22, 23
Isoleucine-valine operon, 156, *156*
Isozymes, 224, 338, 369, 373, 782-783

J

Jelly coat, sea urchin egg, 595-596
Journals, scientific, 793-794
  indexing and abstracting, 796-802
Junctions, cellular, 72-75, 83-84, 425, 500-502
  in capillary endothelium, 83, *84*
  and induction, 425
  junctional complex, 72-74, *72, 73*
  in mammalian gastrulation, 500
  in *t⁹/t⁹* mutants, *500*, 500-502
  types of
    desmosome, *72, 73*, 74-75
    gap junction (nexus), *72, 74*, 75
    zonula adherens, *72, 73*, 74
    zonula occludens, *72, 72, 73*
"Junk" DNA, 115, 200-201
Juvenile hormone, 296-297, 663-665, *664*

K

Karyokinesis, 23, 347-349, 487; *see also* Cleavage, insect egg
  and differentiation, quantal cell cycle, 347, 349

Karyoplast; *see* Minicell
Karyotype, 105, 175, 221-228, 482, 616-619
  altered in established cell lines, 105
  analysis
    chromosome banding, 175
    colchicine used to arrest mitosis at metaphase, 482
  definition, *175*
  human
    normal, Giemsa banded, 175
    sex chromosome abnormalities, 221-228, 616-619
  of hybrid cells, 365, *366*
    selective chromosome loss, 365-367
  mouse, XO, 225
  and sex determination, 613-616, *614*
  X chromosome inactivation, 221-228
Kb (kilobase), 189n
Keratin
  fluorescent antibodies, 220
  multiple genes in chicken, 194
  synthesis by keratinocytes, 94
  unique-sequence gene spacers, 194
  used as marker of skin cell differentiation, 220
Keratinization, chalone control, 537, *538*
Keratinocytes, differentiated properties, 94
Kidney, 283, 416-417, 425-426, 537, 557-559, 625-626
  chalone action in, 537
  elimination of primitive forms, 283, 557-560; *see also* Pronephros; Mesonephros
  induction, 416-417, *417*, 425, *426*, 557-559, *558*
    cell contact needed, 425, *426*
    by foreign tissues, 416-417, *417*
    by uteric bud, 416-417, 559
  mesonephros, 416, 557-559, *558*, 625, 626
  metanephros, 416-417, 557-559, *558*, 625
  pronephros, 416, 557-559, *558*
Kilobase, definition, 189n
Kinase; *see also* Phosphorylation
  and growth control, 543-544
  in hormone action, *437, 438*
  initiation factor eIF-2 inactivated by, 290-291
  thymidine; *see* TK
Kinetics, second order, 761-765
Kinetin; *see* Cytokinins
Klinefelter's syndrome, 221, 226, 617-619, *618*
  relationship to X chromosome inactivation, 221, 226
  XXY karyotype, 226, 617
Kupffer cell, *61*

L

Labia majora, *627*, 628
Labia minora, 627, *627*
Labioscrotal swelling, 626-628, *627*
α-Lactalbumin, marker of mammary gland differentiation, 96, 448
Lactate dehydrogenase, assembly of subunits, 310, 782-783
Lactation; *see also* mammary gland, lactation
  suppression of ovulation during, 448
Lactogen, placental, *431*, 447, 451
  effect on mammary gland, 447, 451
Lactogenic hormone; *see* Prolactin
β-Lactoglobin, synthesis by mammary epithelium, 96
Lactose operon, 137, 147-150, *148, 149*, 155, 187, 260, 357-358
  BrdU effect on repressor binding, 357
  cAMP receptor protein, *146, 149*, 155
  catabolite repression, *148*, 155
  model system for differentiation, *149*, 149-150
  mRNA, leader sequences, 260
  negative induction, *146*, 147-149, *148*
  operator sequence, 187, 357, *358*
    as imperfect palindrome, 187, *358*
  repressor protein, 147-149, *148*, 187, 357
Lactose permease, effect on induction of lactose operon, *149*, 149-150
Lamellipodia, 493, *494, 495*
  cytochalasin B effect, 493, *495*
  microfilaments in, 493
Lampbrush chromosomes, 581, *582*
Larva, 11, 13, 15, 36, 88-89, 383, 387-390, 441-445, 501-502, 659-677; *see also* Metamorphosis
  amphibian, 11, 13, 441-445
  ascidian, 501-502
  Dentalium, 383, 383-384
  insect, 386-388, 662-677
  pluteus, sea urchin, 36, *36*, 389
  Tokophyra, 88, *89*
Lateral lip of blastopore, 30
Lateral plate; *see* Mesoderm, lateral
5′ Leader sequence
  eukaryotic mRNA, 260, 287
  prokaryotic mRNA, *151*, 260
Lectins
  cell agglutination, 541-542
  stimulation of cell division, 542-543
Lemming, Scandinavian wood, sex determination, *614*, *621*, 621-622
Lens, 403-404, 409-410, 488-489
  induction, 409, *410*
  iris cells transformed to, 404
  regeneration, *403*, 404
  vesicle, morphogenesis, *488*, 488-489
Lesch-Nyhan syndrome, 769
Levels of organizations, 3-4, 751-754, *752*

Mutation—cont'd
  rate—cont'd
    and size of eukaryotic genome, 166, 199-200
    steroid-resistant lymphoma cells, 440-441
    suppressor, 143-144
    T locus, 500, 500-502, *523*, 524-525
    talpid³, 525
    temperature sensitive, 662, 689
      Dictyostelium discoideum, 689
      Drosophila, 670-671
Mutational load, maximum number of genes in eukaryotic cells, 199-200
Myelin
  association with axons, 100-101, *100, 101*
  formation by Schwann cells, 55, 101, *101*
Myelocyte, structure, *62*
Myeloma cells
  antibody synthesis by, 106
  used to study differentiation, 106
Myoblast
  BrdU effect on transcription, 357, *358*
  changes in mRNA stability, 300
  differentiation in culture, 326-328, *328, 329*
  established line, differentiation in reconstituted cell with fibroblast cytoplasm, 373
  myotube formation from, 98, 326-328, *328, 329*, 336-338, *337, 338, 346,* 347-349
  quantal cell cycle, *346*, 347-349
    differentiation of cell line L₆E₉ without quantal cycle, 349
Myocardium, cell reaggregation, 515, *516*
Myofibrils, 80, *81*, 82, *97*, 98, *99*
Myoglobin
  marker for red skeletal muscle, 98
  possible message-specific translation factor, 294
Myokinase, 98
Myopus schisticolor, sex determination, *614, 621,* 621-622
Myosin
  association with microfilaments, 487
  filaments, self-assembly, 783
  mRNA
    accumulation prior to translation, 299
    altered stability during muscle differentiation, 300
    possible message-specific translation factor, 294
  muscle vs. nonmuscle, 76-77, 300, 487
  in myofibrils, 80, *81*
  quantal cell cycle, timing of synthesis, 348-349
  role in contraction, 82, *83*

Myosin—cont'd
  skeletal muscle, 80-82, *81, 83,* 98, 348-349
Myotome, *45, 47*
Myotubes, 79, *81, 97,* 98, 327-329, 336-338, 347-349
  formation from myoblasts, 98, 327-328, *328, 329,* 336-338, *337, 338,* 347-349
    in allophenic mice, 338, *339*
    quantal cell cycle, *346,* 347-349
Myxamoeba; *see* Amoeba, *Dictyostelium discoideum*

N

NAD (nicotinamide-adenine dinucleotide), inactivation of EF2 by diphtheria toxin and, 289
Nasobemia mutant, 668, *669*
Necrosis; *see* Death, cellular
Negative control
  transcription, prokaryotic cells, 145-153, *146*
  vs. positive control, eukaryotic transcription, 245-248
Negative staining, 476
Nerve growth factor (NGF), 531-536
  biological activities, 533-535, *534*
  chemical properties, 531-533
  discovery, 531
  entry into cytoplasm, 535
  homology with proinsulin, *532,* 532-533
  immunosympathectomy with antibodies to, 535
  receptor binding specificity, 535-536
  retrograde axonal transport, 535
  sources, 533
Nervous tissue, histological classification, 93
Neural cells, 100-102, 338-339, 491-493, 515, 521-524, 535-536
  reaggregation, 515, 521-524
  surface molecules, 522-524
Neural crest
  developmental fate, 44, *45*
  origin of melanocytes, 94
Neural fold, 37, *41,* 44, *47, 48*
Neural groove; *see* Neural fold
Neural plate, 37, *41,* 44, *48*
  cell reaggregation, 512, *513*
Neural tube, 44-45, *45, 46, 48, 488,* 488-489
  induction; see Induction, embryonic, primary
  morphogenesis, 488-489, *488*
    colchicine effect, 489
    cytochalasin B effect, 489
Neuralizing determinants, 409, 423
Neuroblast cells, 100, 491-493
  axon formation, 491-493, *492*
  colchicine effect, 491-493, *493*

Neuroblast cells—cont'd
  cytochalasin B effect, 491-493, *493*
Neuroblastoma cells, 101-102, 106, 338-339
  differentiation vs. division, 338-339
  neuronlike differentiation, 101-102
  use in study of differentiation, 106
Neurohypophysis; *see* Pituitary, posterior
Neurons, 55, 93, 100-103, 491-493, 515, 521-524, 533-536
  axon formation, 491-493
  differentiated properties, 100-103
  diverse morphologies, 101, *102, 103*
  embryologic origin, 100
  sensory, and nerve growth factor, 533, *534,* 535-536
  sympathetic and nerve growth factor, 533, 535-536
Neurosecretory cells, 433, *434*
Neurospora crassa, one gene—one enzyme relationship, 115
Neurotransmitters and aging, 734
Neurula, *16, 37,* 46; *see also* Neurulation
Neurulation, 44-45, *46, 48,* 488-489, *488*
Nexus; *see* Gap junction
N-formylmethionine; *see* Methionine, N-formyl
NGF; *see* Nerve growth factor
NHC proteins (nonhistone chromosomal proteins), 166-167, 171-172, 243-248, 253, 357, 440, 543-544, 536, 755
  binding to BrdU-substituted DNA, 357
  characterization, 171-172
  in chromatin, 166-167
  control of transcription, 243-248
    globin genes, 243, *244*
    histone genes, 243-246, *244*
    ovalbumin gene, *244, 245*
    role as positive activator, 246
  definition, 166
  effect on DNA-histone interaction, 247
  enzyme activities in, 171-172, *172*
  interaction with histones, 247
  phosphorylation, 173, 253, 543-544, 546
    and growth control, 543-544, 546
  progesterone-receptor complex binding to chromatin, 440
  regulatory role, 172-173, 243-248
  variability, 172
  what controls their synthesis? 246, 755
NMN (notochord-muscle-nerve complex), *501,* 502
Nonhistone chromosomal proteins; *see* NHC proteins
Nonsense mutation, 143-144
  suppression, 143-144
Nonsuppressible insulin-like activity (NSILA), 530
Norepinephrine, *431*
Normal diploid cells; *see* Diploid cells; Cultured cells, diploid